Lecture Notes in Computer Science 16169

Founding Editors

Gerhard Goos
Juris Hartmanis

The series Lecture Notes in Computer Science (LNCS), including its subseries Lecture Notes in Artificial Intelligence (LNAI) and Lecture Notes in Bioinformatics (LNBI), has established itself as a medium for the publication of new developments in computer science and information technology research, teaching, and education.

LNCS enjoys close cooperation with the computer science R & D community, the series counts many renowned academics among its volume editors and paper authors, and collaborates with prestigious societies. Its mission is to serve this international community by providing an invaluable service, mainly focused on the publication of conference and workshop proceedings and postproceedings. LNCS commenced publication in 1973.

Emanuele Rodolà · Fabio Galasso · Iacopo Masi
Editors

Image Analysis and Processing - ICIAP 2025 Workshops

23rd International Conference
Rome, Italy, September 15–19, 2025
Proceedings, Part I

 Springer

Editors
Emanuele Rodolà
Sapienza University of Rome
Rome, Italy

Fabio Galasso
Sapienza University of Rome
Rome, Italy

Iacopo Masi
Sapienza University of Rome
Rome, Italy

ISSN 0302-9743   ISSN 1611-3349 (electronic)
Lecture Notes in Computer Science
ISBN 978-3-032-11316-0   ISBN 978-3-032-11317-7 (eBook)
https://doi.org/10.1007/978-3-032-11317-7

Preface

This volume contains the papers accepted for presentation at the workshops and competitions hosted by the 23rd International Conference on Image Analysis and Processing (ICIAP 2025), held in Rome, Italy, from September 15 to 19, 2025. Organized alongside the main program by the Department of Computer Science of Sapienza University of Rome, these satellite events offered focused venues to explore emerging directions, prototype new ideas, and foster collaboration across subcommunities of computer vision, image analysis, pattern recognition, and machine learning. In this spirit, they complemented the main conference by providing flexible formats for concentrated discussion, community building, and rapid feedback on ongoing research. The initiative benefited from the strong backing of the central administration of the university and from the support of multiple institutional, academic, and industrial partners. The contributions of these partners helped shape and sustain the workshop program.

Across all workshops and competitions, 105 papers were accepted. This volume presents 54 of those 105 papers, drawn from the workshops listed below. Submissions underwent a double-blind evaluation: each paper received at least three independent reviews and was handled by workshop organizers and area chairs. The process included conflict-of-interest management, reviewer bidding and assignment, area chair oversight, and a meta-review with discussion, ensuring clarity, fairness, and scientific rigor. The accepted contributions collected in this volume capture both the breadth of topics addressed and the vitality of the involved communities, offering snapshots of current challenges, methods, and applications. Beyond disseminating results, these events encouraged benchmarking, dataset curation, reproducibility practices, and dialogue between theory and applications. We warmly thank all workshop organizers, program committee members, external reviewers, and the local team for their dedication, and we are grateful to all authors for their valuable contributions. We hope that the papers gathered here will serve as a useful reference and an inspiration for further research and collaboration across disciplines, institutions, and communities.

This first workshop volume includes papers from the following events:

- 4th Automatic Affect Analysis and Synthesis Workshop (3AS 2025)
- International Workshop on Advances in Drone Vision (ADV 2025)
- 3rd International Workshop on Artificial Intelligence and Radiomics in Computer-Aided Diagnosis (AIRCAD 2025)
- 2nd International Workshop on Computer Vision for Environment Monitoring and Preservation (CVEMP 2025)
- Workshop on Computer Vision and Generative Models for Medical Imaging (CVGMMI 2025)
- Human-Object Interaction: Integrating Egocentric and Exocentric Perspectives (HOI-EGO-EXO 2025)
- 4th International Workshop on Fine Art Pattern Extraction and Recognition (FAPER 2025)

- 2nd Workshop on Generation of Human Face and Body Behavior (GHB 2025)
- 1st International Workshop on Humans in the eXtended Artificial Intelligence Loop (HUARL 2025)

A companion volume will include the papers from the first workshops and competitions of ICIAP 2025.

September 2025

Emanuele Rodolà
Fabio Galasso
Iacopo Masi

Organization

General Chairs

Emanuele Rodolà	Sapienza University of Rome, Italy
Fabio Galasso	Sapienza University of Rome, Italy
Iacopo Masi	Sapienza University of Rome, Italy

Technical Program Chairs

Marco Cristani	University of Verona, Italy
Philippos Mordohai	Stevens Institute of Technology, USA
Paolo Soda	Università Campus Bio-Medico di Roma, Italy
Luisa Verdoliva	Università degli Studi di Napoli Federico II, Italy

Workshop Chairs

Giovanni Maria Farinella	University of Catania, Italy
Silvia Zuffi	IMATI-CNR, Italy

Tutorial Chairs

Alessio Del Bue	Italian Institute of Technology, Italy
Zorah Lähner	University of Siegen, Germany

Special Session Chair

Cosimo Distante	Institute of Applied Sciences and Intelligent Systems, CNR, Italy

Industrial Chairs

Ignas Budvytis	University of Cambridge, UK
Simone Scardapane	Sapienza University of Rome, Italy

Publicity and Social Media Chair

Antonino Furnari   University of Catania, Italy

Publication Chair

Danilo Avola   Sapienza University of Rome, Italy

Web Chairs

Antonio D'Orazio   Sapienza University of Rome, Italy
Daniele Pannone   Sapienza University of Rome, Italy

Local Arrangement Chairs

Danilo Avola   Sapienza University of Rome, Italy
Antonio D'Orazio   Sapienza University of Rome, Italy
Paolo Mandica   Sapienza University of Rome, Italy
Daniele Pannone   Sapienza University of Rome, Italy
Luca Scofano   Sapienza University of Rome, Italy
Daniele Trappolini   Sapienza University of Rome, Italy

CVPL Relation Chair

Luigi Cinque   Sapienza University of Rome, Italy

Steering Committee

Alberto Del Bimbo   University of Florence, Italy
Cosimo Distante   University of Salento, Italy
Nicu Sebe   University of Trento, Italy

Area Chairs

Video Analysis and Understanding

Mariella Di Miccoli   Polytechnic University of Catalonia, Spain
Niki Martinel   University of Udine, Italy

Pattern Recognition and Machine Learning

Simone Calderara   University of Modena and Reggio Emilia, Italy
Stéphane Lathuilière   Inria, Grenoble Alpes University, France
Riccardo Marin   TU Munich, Germany
Tatiana Tommasi   Politecnico di Torino, Italy

Deep Learning

Vishal Patel   Johns Hopkins University, USA
Indro Spinelli   Sapienza University of Rome, Italy

Multiview Geometry and 3D Computer Vision

Luca Cosmo   Ca' Foscari University of Venice, Italy
Federica Bogo   Meta AI, USA

Image Analysis, Detection, and Recognition

Tal Hassner   Meta AI, USA
Giuseppe Lisanti   University of Bologna, Italy

Multimedia

Lamberto Ballan   University of Padua, Italy

Biomedical and Assistive Technology

Simona Giunta   Sapienza University of Rome, Italy
Marco Leo   Institute of Applied Sciences and Intelligent Systems, CNR, Italy
Andrea Ravignani   Sapienza University of Rome, Italy

Digital Forensics and Biometrics

Irene Amerini   Sapienza University of Rome, Italy
Marilena De Marsico   Sapienza University of Rome, Italy

Image Processing for Cultural Heritage

Francesco Ragusa   University of Catania, Italy
Lorenzo Seidenari   University of Florence, Italy

Robot and Vision

Xavier Alameda-Pineda   Inria, Grenoble, France
Asako Kanezaki   TokyoTech, Japan

Brave New Ideas

Nicu Sebe   University of Trento, Italy
Fabrizio Silvestri   Sapienza University of Rome, Italy

AI and Science

Elena Agliari   Sapienza University of Rome, Italy
Elisa Tinti   Sapienza University of Rome, Italy

X-Realities

Andrea Giachetti   University of Verona, Italy
Marco Raoul Marini   Sapienza University of Rome, Italy

Embedded Vision

Luigi Capogrosso   University of Verona, Italy

Liasons

Latin America Liason

Olga Regina Pereira Bellon   Universidade Federal do Paraná, Brazil
Alexandre Xavier Falcão   University of Campinas, Brazil

Asia Liason

Luca Rossi   Hong Kong Polytechnic University, China

Endorsing Institutions

International Association for Pattern Recognition (IAPR)
Italian Association for Computer Vision, Pattern Recognition and Machine Learning
(CVPL)

Contents

International Workshop on Advances in Drone Vision (ADV 2025)

**3rd International Workshop on Artificial Intelligence and Radiomics
in Computer-Aided Diagnosis (AIRCAD 2025)**

**2nd International Workshop on Computer Vision for Environment
Monitoring and Preservation (CVEMP 2025)**

**Workshop on Computer Vision and Generative Models for Medical
Imaging (CVGMMI 2025)**

Human-Object Interaction: Integrating Egocentric and Exocentric Perspectives (HOI-EGO-EXO 2025)

1st International Workshop on Humans in the eXtended Artificial Intelligence Loop (HUARL 2025)

4th Automatic Affect Analysis and Synthesis Workshop (3AS 2025)

Workshop

4th Automatic Affect Analysis and Synthesis Workshop (3AS 2025)

In conjunction with ICIAP 2025—Rome, Italy, September 15–19, 2025

Workshop Organization

Organizers

Nadia Bianchi-Berthouze — University College of London, UK
Luigi Celona — University of Milano-Bicocca, Italy
Vittorio Cuculo — University of Modena and Reggio Emilia, Italy
Alessandro D'amelio — University of Milano, Italy
Javier Lorenzo Navarro — Universidad de Las Palmas de Gran Canaria, Spain

Program Committee

Francesco Agnelli — University of Milano, Italy
Leonardo Alchieri — Università della Svizzera Italiana, Switzerland
Hane Aung — Cornell University, USA
Giuseppe Boccignone — University of Milano, Italy
Jacopo Burger — University of Milano, Italy
Carlo Busso — University of Texas Dallas, USA
Pedro J. S. Cardoso — University of Algarve, Portugal
Gianluigi Ciocca — University of Milano-Bicocca, Italy
Lia Schmid — University of Heidelberg, Germany
Livia Del Gaudio — University of Modena and Reggio Emilia, Italy
Sabrina Patania — University of Milano-Bicocca, Italy
Michael Valstar — University of Nottingham, UK
Guoying Zhang — University of Oulu, Finland

Original Versus Zero-Shot Prompted AI Visuals: Examining Human Feedback to Corporate Sustainability Reporting

Alessandro Bruno[1]([⊠])[iD], Federica Ricceri[1][iD], and Chintan Bhatt[2][iD]

[1] Department of Business, Law, Economics, and Consumer Behaviour, IULM University, Milan, Italy
{alessandro.bruno,federica.ricceri}@iulm.it
[2] Department of Computer Science and Engineering, School of Technology, Pandit Deendayal Energy University, Gandhinagar 382007, GJ, India
chintan.bhatt@sot.pdpu.ac.in

Abstract. Aligned with the broad field of Affective Computing research, this work examines the capacity of Multimodal Large Language Models to elicit attentional patterns and cognitive states in viewers as they engage with visuals related to a specific domain: corporate sustainability reporting. Sustainability reports are documents officially released by companies and organisations to communicate actions to achieve sustainability goals. They are often lengthy documents, which pose challenges in capturing, orienting and retaining stakeholder attention. In this context, visual elements are commonly used not only to complement textual content but also to act as entry points that guide the reader's initial focus and/or to convey information. This study examines the eye movements of 41 observers during a webcam-based eye-tracking session as they view pictures from sustainability reports and their corresponding AI-generated versions with zero-shot prompting. Both visuals (original and AI-generated) are presented in the form of A/B tests. First, images are sourced from publicly available sustainability reports and captioned using Contrastive Language–Image Pretraining, a Vision-Language Model trained on image-text pairs, integrated within GPT-4o. Captions serve as zero-shot prompts provided to DALL-E3 for generating images from text. Perceptual features recorded include Time to First Fixation, Time to First Gaze, average gaze duration, number of fixations, total gaze count, and K-coefficient by looking into two time ranges: [0–3] and [0–10] seconds. Last, Wilcoxon signed-rank and paired t-tests are used to assess the statistical significance of similarity and divergence in attention and cognitive dynamics between the two conditions.

Keywords: Visual Perception · Eye-Tracking · Multimodal Large Language Models · Visuals · Sustainability Reporting · Corporate Social Responsibility

© The Author(s), under exclusive license to Springer Nature Switzerland AG 2026
E. Rodolà et al. (Eds.): ICIAP 2025 Workshops, LNCS 16169, pp. 5–16, 2026.
https://doi.org/10.1007/978-3-032-11317-7_1

1 Introduction

AC (Affective Computing) fosters the development of computing methods that allow machines to analyse and detect emotions, physiological signals, cognitive workload, and attention patterns from data. This work focuses on interpreting the cognitive signals and attentional patterns of individuals who observe visuals that convey sustainability messages. Eye movements reveal patterns accounting for visual attention mechanisms, including both bottom-up and top-down processes [3]. The former are stimulus-driven, while the latter are task-oriented, and indicate how a specific task can capture viewers' attention. HVS (Human Vision System) is a complex and sophisticated system that entails a path from the eyes to the Inferior Temporal (IT) and Posterior Parietal cortex [16]. It has been a long way since the introduction of computing methods addressing HVS mechanisms, as described by John K Tsotsos [28]. Duchowski and Krejtz employed the K-coefficient [11] for visualisation purposes, indicating the so-called ambient-focal dynamics in viewers observing a picture or a visual scene. The formulation of the K-coefficient is based on a parametric scale, with positive and negative values of K indicating focal or ambient viewing, respectively. Guo et al. [15] used coefficient K to analyse attention patterns in viewers assigned sub-tasks, supporting its use in cognitive workload analysis, key for affective computing interfaces. Biele et al. [5] claimed that emotions have a direct impact on visual attention, working as regulators and influencing the viewer's behaviour. They have highlighted that viewers' eye movements move from ambient to focal attention; at the same time, positive mood slows down that process, enhancing the exploratory process. Image Processing and Computer Vision addressed the analysis of cognitive and attentional paths using eye movements and the interpretation of both bottom-up and top-down visual attention processes from a predictive perspective [8,27]. "Corporate reports" refers to a broad range of documents, including accounting disclosures, press releases, and sustainability reports. Studies on the subject have shown that the presentation of visual materials in corporate reports represents a breakthrough in how companies convey outward-facing information [2]. Visuals have been employed in corporate reports to influence investors' evaluations or facilitate the interpretation of complex messages, due to their ability to grab viewers' attention and convey emotions [4] [17]. The same subject is studied by Fenk [12], who introduced a machine-supported approach to analyse photographs in sustainability reporting.

As stated by Majer et al. [19], visuals can evoke emotional responses that amplify the impact of sustainability messages. Pictures, infographics, and diagrams can drag viewers' attention and stimulate feelings in ways that words cannot [21]. Wedel and Pieter [29] fostered the discussion on visual elements, such as colour and layout, and the ways they influence how sustainability is perceived. Sustainability communication presents significant challenges, as organisations handle the pressure between maintaining transparency and achieving strategic communication objectives.

This paper presents experimental trials designed to investigate the differences and similarities between visuals from sustainability reports and their corresponding AI-generated versions. The contribution is threefold: 1) a new dataset including on-screen projection eye movement coordinates from 41 participants viewing pictures conveying sustainability messages; 2) new insights into Multimodal Large Language Models' performance in generating pictures communicating sustainability using zero-shot prompting supported by features, such as Time to First Fixation (TTFF), Time to First Gaze (TTFG), average gaze duration, number of fixations, total gaze count, and K-coefficient (spatial gaze entropy) by looking into two time ranges: [0–3] and [0–10] seconds; 3) a statistical validation of experimental results to assess convergence or divergence between AI-generated and original pictures from sustainability reports on the time ranges. The remainder of the paper addresses related works, materials and methods, experimental Results, and conclusions.

2   Related Works

This section aims to offer an overview of LLMs and MLLMs' employment in Affective Computing by looking at several disciplines and topics. Several methods will be surveyed in different research fields, with a noticeable lack of work regarding MLLMs and sustainability reporting from an Affective Computing perspective. Since the release of ChatGPT by OpenAI in November 2022, Large Language Models (LLMs) have brought a paradigm shift in Affective Computing research. On the one hand, some AC research focuses on text-based analysis and generation. On the other hand, a meaningful amount tackles the understanding and processing of perceptions and emotions through visuals. Li et al. [18] provided an in-depth analysis of emotional prompting by introducing EmotionPrompt, EmotionAttack, and EmotionDecode. They are textual and visual emotional stimuli working as additional prompts and grounded in psychological theories. The first is meant to enhance the AI model's performance. The second is used to impair the performance, while the third one explains the effect of emotional inputs. Paik et al. [26], studied the affective nature of generative AI by looking into the affective responses and perceptions of AI-generated news in visual journalism. They developed a codebook to map emotional response, journalistic ethics, and image features using a dataset consisting of 200 headlines and visuals. Paik et al. raised concerns about the journalistic integrity. Yu et al. [30] investigate the interplay between facial emotional expressions and user engagement, framed through the CASA (Computers Are Social Actors) theory. Grassini [14] worked on the emotional impact of AI-generated art, establishing a correlation between computational efforts and viewers' preferences. One hundred participants evaluated AI-generated artworks grouped into three quality tiers. The results showed that the subjects preferred those outputs generated with greater computational effort. LLMs can have a grasp of emotional intelligence. That is revealed by Li et al. [18]. with priming LLMs with emotional inputs leading to better answers. Automated metrics and trials conducted with 106 subjects confirmed it. Chen et al. [9] surveyed several studies on AI-driven emotional response detection in interactive installation art from a sustainable perspective, focusing on individual well-being. Zhang et al. [31] surveyed AC applications with LLMs, focusing on in-context learning, common sense reasoning, and advanced sequence generation as aspects fostering AU (affective understanding), AG (Affective Generation), and emotion modulation. Metha and Buntaion [20] studied the capacity of LLMs to detect emotion from images and assessed potential biases in generative models. They conducted trials using Google ViT and GPT4 Vision to detect emotions in pictures, revealing that generative AI models often produce images skewed toward negative emotional tones. As mentioned in the introductory section, sustainability reports are formal documents published periodically by companies that incorporate both text and visuals to describe environmental and social actions and impact [22]. LLMs have been used for the textual analysis of these reports, with the visual content within them also playing a significant role in conveying information. Zou et al. proposed [32] ESGReveal: An LLM-based approach for extracting structured data from ESG reports. Nakao et al. [25] thoroughly analysed image usage and trends in 1,000 sustainability reports using Amazon Rekognition [23] and deep learning models. Amazon Rekognition enabled the projection of images onto proxy variables related to sustainability. De Villiers et al. [10] presented a conceptual article where they draw a research agenda to examine AI's role in reshaping sustainability disclosures, and the criticality in the risks of AI facilitating greenwashing. Ahmad et al., [1] presented a theoretical discussion on the implications of AI in accounting practices and sustainability disclosures. Bronzini et al. [6] introduced a method to extract ESG-related information using LLMs. Large Language Models are, therefore, a means to parse text from reports. Moodaley and Telukdarie [24] delivered a systematic review on greenwashing, sustainability reporting, and AI. They specifically delved into deep learning methods to identify greenwashing from sustainability reports. Despite a growing number of scientific contributions on corporate sustainability reporting rely on MLLMs, there is a noticeable lack of experiments and trials examining the relationships between emotions, cognitive states, and sustainability reporting.

3   Materials and Methods

This section explores the methodology used to compare original visuals from sustainability reports with AI-generated visuals created through zero-shot prompting. The two following subsections provide insights into the new EVCSR (Eyeing Visuals from Corporate Sustainability Reports) dataset, which comprises eye movement spatial coordinates and perceptually relevant features, and describe the methodology pipeline spanning image captioning, zero-shot prompting, and statistical significance testing of the obtained results.

3.1   EVCSR Dataset

One of the contributions of this paper concerns the release of the EVCSR (Eyeing Visuals from Corporate Sustainability Reporting) Dataset [7]. It consists of on-screen projections of eye movements captured using a webcam-based eye-tracking

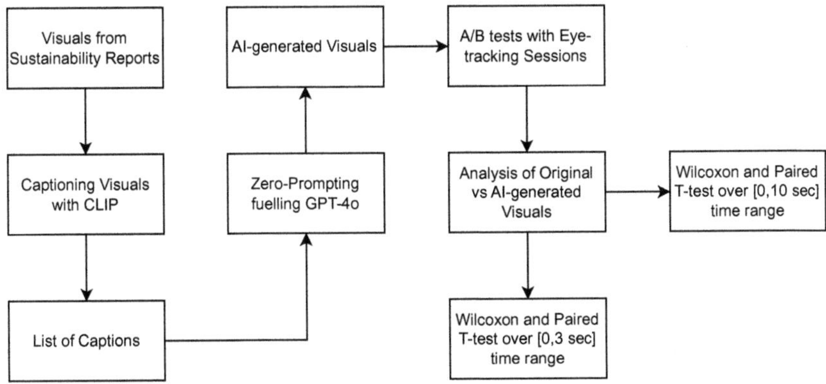

Fig. 1. The diagram above depicts the whole methodology's pipeline, starting with captioning visuals from sustainability reports to the analysis of eye-tracking session data.

platform, namely realeye.io, which allows for a sampling frequency 30 Hz. Eye-tracking sessions were allowed solely on laptops to ensure a higher accuracy rate during the calibration step. Forty-one subjects participated in the experimental session, with 17 males and 24 females. The participants' ages ranged from 19 to 59 years, with a mean value of approximately 32 years. The first quartile was 23 years, and the median age was 25 years. The majority of participants were young adults, with a smaller number of middle-aged individuals represented. The dataset of images consists of 20 pictures, equally split into original and AI-generated images. The original pictures were taken from three publicly available sustainability reports [13]. During the eye-tracking session, the 20 pictures were displayed as A/B tests. Each session lasted around 2.5 min, accounting for the calibration step, a 10-second view for each image pair, and a 1-second uniform grey picture to refresh each viewer's retina.

3.2 Methodology

As mentioned in the introductory section, this work aims to explore the capacities of MLLMs to generate visuals that resonate with original pictures from sustainability reports. In Fig. 1, the whole pipeline is provided. It encompasses image selection, figure captioning, zero-prompting, text-to-image generation, A/B testing with eye-tracking sessions, and statistical tests to assess the significance of results. A more detailed description follows:

- 10 pictures were taken from three publicly available sustainability reports [13]. The images aim to convey messages and claims that entail sustainability commitments in a specific application domain (flooring).
- The given pictures are processed by CLIP (Contrastive Language–Image Pre-training), a key component of GPT-4o that enables image understanding.

Fig. 2. AI-generated visual (on the left) vs Picture from the sustainability report (on the right)

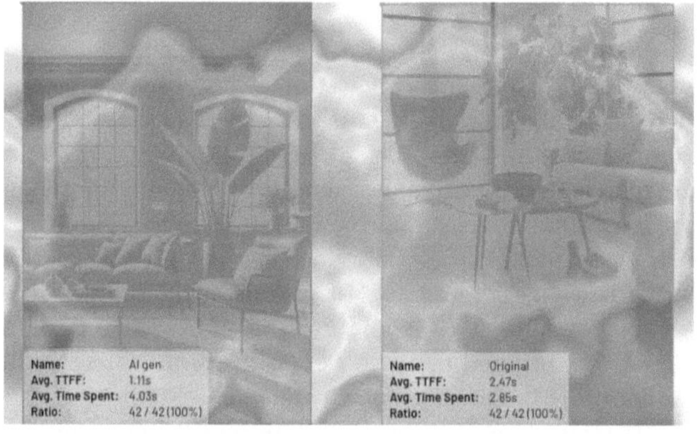

Fig. 3. Areas of Interest are represented by the green and orange bounding boxes framing the two visuals. (Color figure online)

CLIP is prompted with a static textual request: "Caption the given pictures", which guides it to generate relevant descriptions.

- The list of captions is then used as zero-shot prompts to fuel DALL·E3 in generating 10 visuals. Furthermore, to achieve visually faithful and unbiased image generation, certain words and adjectives were excluded that might influence the visual outcome. All captions were thoroughly analysed to avoid style-oriented words like "artistic," "abstract," "trendy," "futuristic"; camera and cinematic terms, such as "wide angle," "cinematic," "drone view,"; temporal or emotional framing like "sad," "happy," "from the 1950s".
- Each AI-generated visual is paired with the original image from the sustainability report and displayed side by side on a whiteboard, positioned symmetrically with equal distance from the centre. An example is shown in Fig. 2. The spatial arrangement was designed to facilitate A/B testing.

- AI-generated visuals and the original images alternated in their order of appearance, with each type occasionally presented first to control for order effects in visual attention.
- 10 panels are designed using the procedure described in the previous step, totalling 20 pictures: 10 original visuals from sustainability reports and 10 corresponding zero-shot prompted AI-generated versions.
- A/B tests are conducted with eye tracking. The webcam-based eye-tracking platform, realeye.io, is used to gather data from the participants. The experimental session enabled the collection of on-screen projections of eye-movement coordinates as participants viewed the visuals.
- Time to First Fixation (TTFF), Time to First Gaze (TTFG), average gaze duration, number of fixations, total gaze count, and K-coefficient (spatial gaze entropy) are gathered within two specific AOIs (Areas of Interest), corresponding to the two bounding boxes including the two visuals (Original and AI-generated) as visually depicted in Fig. 3.
- During the webcam eye-tracking session, A/B tests are shown for 10 s, with a uniform full grey frame displayed for 1 s to refresh viewers' retinas in between.
- Paired T-tests and Wilcoxon tests are run over the features listed above to determine meaningful divergence or convergence between AI-generated and original contents.

4   Experimental Results

This section explores perceptually and cognitively relevant factors that may indicate resonance between original and AI-generated content. As previously described, the realeye.io webcam-based eye-tracking tool was employed to gather data from 41 subjects (further details are in Sect. 3.1). The tool enables data recording frequencies of up 30 Hz, ensuring the collection of a certain number of gazes and fixation points during the trials. The following features were averaged over the 41 participants: TTFF (Time to First Fixation), TTFG (Time to First Gaze), Dwell Time, Gaze Number, and K-coefficient. Attentional patterns and cognitive states may be studied across different time scales to reveal insights into specific processes. Therefore, a multi-timescale resolution was adopted to examine attentional patterns within the following ranges: [0, 3 sec] and [0, 10 sec]. The former helps to explore bottom-up visual attention processes (stimulus-driven), perceptual encoding mechanisms, and immediate orienting responses. The latter pertains to sustained attention, mental effort regulation, and emotional modulation. The first 500 milliseconds were discarded from both ranges to avoid the so-called centre-bias, that is, the innate oculomotor behaviour that focuses on the centre of pictures even before significant processing of the content begins. As stated in Sect. 1, the primary objective of this work is to identify similarities and differences in attention and cognitive dynamics between original visuals from sustainability reports and their corresponding AI-generated versions using zero-shot prompting. Paired t-tests and Wilcoxon signed-rank tests were performed over the data as in Table 1 and in Table 2 to evaluate statistical significance and

Table 1. Eye-tracking metrics per image pair for AI-generated and original stimuli (3-sec exposure averaged over 41 participants).

Visual	Avg. TTFF	Avg. TTFG	Avg. Dwell Time	Fixations	Gaze	K-coeff.
Im1 AI Gen	0.46	0.64	1.48	116	1939	0.31
Im1 Original	0.87	1.28	1.06	88	1204	0.15
Im2 AI Gen	0.32	0.59	1.58	117	1943	0.11
Im2 Original	0.88	1.25	1.17	74	1091	-0.07
Im3 AI Gen	0.92	1.23	1.15	79	1172	0.27
Im3 Original	0.31	0.60	1.46	119	1893	-0.01
Im4 AI Gen	1.09	1.49	1.06	72	998	0.22
Im4 Original	0.33	0.59	1.57	122	1978	0.20
Im5 AI Gen	0.96	1.25	1.18	86	1228	0.31
Im5 Original	0.18	0.54	1.54	117	1855	0.18
Im6 AI Gen	0.93	1.11	0.98	71	1106	0.25
Im6 Original	0.33	0.59	1.47	124	1903	0.04
Im7 AI Gen	1.05	1.45	0.91	53	819	0.28
Im7 Original	2.17	2.42	3.45	269	4200	0.32
Im8 AI Gen	1.01	1.28	1.18	84	1348	0.13
Im8 Original	0.28	0.65	1.37	109	1658	0.00
Im9 AI Gen	0.97	1.18	1.05	79	1197	0.34
Im9 Original	0.34	0.68	1.38	110	1730	0.07
Im10 AI Gen	0.35	0.63	1.56	118	2031	0.16
Im10 Original	0.95	1.23	1.03	77	1104	0.04

to account for potential violations of the normality assumption. The two sets of data are analysed according to both tests in the following lines. Regarding the [0–3 sec] time range: The p-values of TTFF were 0.5515 (t-test) and 0.2754 (Wilcoxon). TTFG returned values of 0.6649 and 0.9219, while Dwell Time yielded 0.3247 and 0.6784, respectively. Concerning Fixations, the results were 0.2624 (t-test) and 0.3738 (Wilcoxon), and for Gaze, 0.3356 and 0.5940. Conversely, the K-coefficient showed a significant difference between conditions, with p-values of 0.0013 (t-test) and 0.0059 (Wilcoxon). These results imply that while most eye-tracking metrics did not differ significantly, the K-coefficient revealed a consistent and meaningful divergence between AI-generated and original visuals (see Fig. 4). Regarding the [0–10 sec] time range: Avg. TTFF reached the following p-values: 0.4649 (t-test) and 0.4922 (Wilcoxon), while for Avg. TTFG, they were 0.4662 and 0.6250. Both indicate no significant differences between original and AI-generated content. In contrast, Gaze Avg. Time Spent featured p-values of 0.0082 (t-test) and 0.0059 (Wilcoxon); Fixations showed 0.0064 and 0.0039; and Gaze produced p-values of 0.0151 and 0.0059. Significant differences are observed between conditions for the latter three metrics, particularly in sustained visual engagement (see Fig. 5).

Table 2. Eye-tracking metrics per image pair for AI-generated and original stimuli averaged over 41 participants (10-sec exposure).

Visual	Avg. TTFF	Avg. TTFG	Avg. Dwell Time	Fixations	Gaze	K-coeff.
Im1 AI Gen	0.56	0.80	4.83	406	6290	0.43
Im1 Original	1.78	2.08	3.43	300	4448	0.14
Im2 AI Gen	0.77	0.81	3.98	310	4969	0.37
Im2 Original	2.12	2.36	3.12	250	3883	0.17
Im3 AI Gen	1.59	1.91	4.60	383	5944	0.38
Im3 Original	0.34	0.62	3.82	317	4989	0.25
Im4 AI Gen	1.88	2.26	3.83	307	4822	0.30
Im4 Original	0.34	0.60	4.05	318	5104	0.30
Im5 AI Gen	1.74	1.84	4.23	374	5517	0.31
Im5 Original	0.48	0.86	3.84	328	4958	0.10
Im6 AI Gen	1.17	1.48	4.55	367	5729	0.38
Im6 Original	0.34	0.61	3.89	315	5040	0.35
Im7 AI Gen	0.43	0.97	4.57	372	5950	0.34
Im7 Original	0.38	0.65	1.65	127	2073	0.22
Im8 AI Gen	1.45	1.82	4.26	337	5407	0.52
Im8 Original	0.75	0.79	4.02	322	5073	0.16
Im9 AI Gen	1.41	1.57	4.28	351	5538	0.47
Im9 Original	0.64	0.90	3.52	302	4634	0.13
Im10 AI Gen	0.55	0.68	4.96	389	6505	0.23
Im10 Original	1.69	1.85	3.32	288	4264	0.14

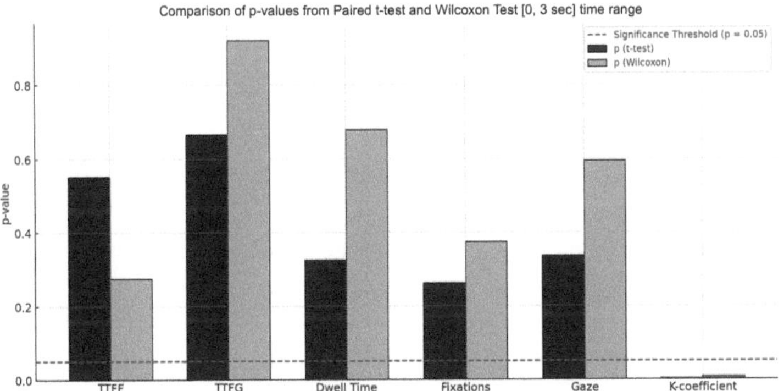

Fig. 4. P-values (y-axis) are shown for several features: TTFF, TTFG, Dwell Time, Fixations, Gaze, and K-coefficient recorded in the [0–3 sec] time range.

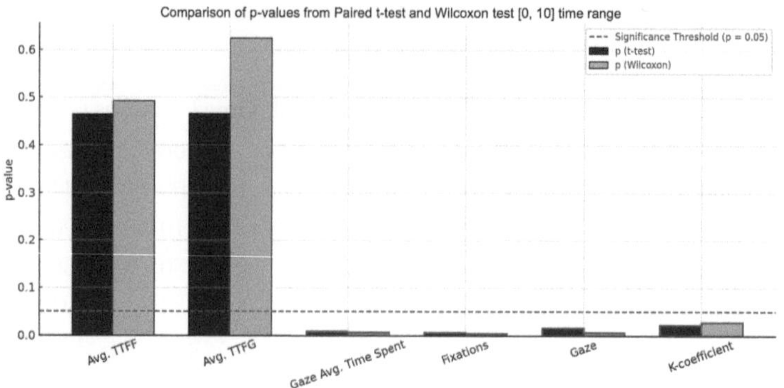

Fig. 5. P-values (y-axis) are shown for several features: TTFF, TTFG, Dwell Time, Fixations, Gaze, and K-coefficient recorded in the [0–3 sec] time range.

5   Conclusions

This paper aims to provide insights into attentional patterns and cognitive-related features, as represented by features such as gazes, fixations, dwell time, and k-coefficient, captured when observing original visuals from sustainability reports and the corresponding AI-generated visuals with zero-shot prompting. The primary goal is to explore MLLMs' capabilities in generating visual content that resonates with the original picture from sustainability reports. This goal is relevant to sustainability reporting research because it addresses a foundational, empirical question about how stakeholders visually attend to AI-generated versus human-designed content. The data gathered via realeye.io, an online webcam-based eye-tracking platform, revealed compelling aspects entailing two observation time ranges: [0–3 sec] and [0-10 sec]. For the shorter time range analysed, most features did not show any statistical significance between the two conditions, implying more similar feedback between the two visual categories. Only the K-coefficient, with a p-value lower than 0.05, revealed divergence between the original and the corresponding AI-generated version. Sticking to the K-coefficient meaning, it may suggest a different ambient–focal dynamic in viewers observing the two visuals conveying the same message. At the same time, fixations and gazes from the 41 participants do not reveal a specific trend, with some original pictures accounting for a higher number of gazes/fixations than those generated by DALL-e3, and vice versa. As the observation time progressed ([0-10 sec]), all features except TTFF and TTFG showed statistical significance between the AI-generated visual and the one from the sustainability reports. Noticeably, the number of fixations and gazes is higher for visuals generated by DALL-e3, suggesting either a more significant interest in the details of the image or a major cognitive effort in reading the visual content. The experimental results encourage further investigation of the matter from the perspectives of prompt engineering and affective computing, laying the groundwork for future studies into the communicative performance of AI-generated visuals in sustainability communication contexts. Factors such as seeds, randomness, and variety of MLLMs deserve attention as they can play a critical role in the outcome style, along with task-oriented prompts. The interplay between affective computing, attentional patterns, cognition and MLLMs reserves a lot of room for improvements.

References

1. Ahmad, V., Goyal, L., Arora, M., Kumar, R., Chythanya, K.R., Chaudhary, S.: The impact of ai on sustainability reporting in accounting. In: 2023 6th International Conference on Contemporary Computing and Informatics (IC3I), vol. 6, pp. 643–648. IEEE (2023)
2. Ang, L., Hellmann, A., Kanbaty, M., Sood, S.: Emotional and attentional influences of photographs on impression management and financial decision making. J. Behav. Exp. Financ. **27**, 100348 (2020)
3. Ardizzone, E., Bruno, A., Mazzola, G.: Visual saliency by keypoints distribution analysis. In: Maino, G., Foresti, G.L. (eds.) ICIAP 2011. LNCS, vol. 6978, pp. 691–699. Springer, Heidelberg (2011). https://doi.org/10.1007/978-3-642-24085-0_70
4. Beattie, V., Jones, M.: Corporate reporting using graphs: A review and synthesis. J. Account. Lit. **27**, 71–110 (2008)
5. Biele, C., Kopacz, A., Krejtz, K.: Shall we care about the user's feelings? influence of affect and engagement on visual attention. In: Proceedings of the International Conference on Multimedia, Interaction, Design and Innovation, pp. 1–8 (2013)
6. Bronzini, M., Nicolini, C., Lepri, B., Passerini, A., Staiano, J.: Glitter or gold? deriving structured insights from sustainability reports via large language models. EPJ Data Sci. **13**(1), 41 (2024)
7. Bruno, A.: EVCSR: Eye-Tracking & Visual Content Sustainability Reporting. https://github.com/alessandrobruno10/EVCSR/ (2025), gitHub repository, accessed 2025/11/27 15:14:31
8. Bruno, A., Gugliuzza, F., Ardizzone, E., Giunta, C.C., Pirrone, R.: Image content enhancement through salient regions segmentation for people with color vision deficiencies. i-Perception **10**(3), 2041669519841073 (2019)
9. Chen, X., Ibrahim, Z.: A comprehensive study of emotional responses in ai-enhanced interactive installation art. Sustainability **15**(22), 15830 (2023)
10. De Villiers, C., Dimes, R., Molinari, M.: How will ai text generation and processing impact sustainability reporting? critical analysis, a conceptual framework and avenues for future research. Sustainability Accounting, Manag. Policy J. **15**(1), 96–118 (2024)
11. Duchowski, A.T., Krejtz, K.: Visualizing dynamic ambient/focal attention with coefficient K. In: Burch, M., Chuang, L., Fisher, B., Schmidt, A., Weiskopf, D. (eds.) ETVIS 2015. MV, pp. 217–233. Springer, Cham (2017). https://doi.org/10.1007/978-3-319-47024-5_13
12. Fenk, L.: Environmental photograph use in corporate sustainability reporting: a machine-supported visual content analysis. Bus. Strateg. Environ. **34**(1), 1097–1112 (2025)
13. Forbo Flooring Systems: Sustainability report 2023. Tech. rep., Forbo Flooring Systems (2023), https://view.publitas.com/forboflooring-com/sustainability-report-2023/page/1, Accessed 3 June 2025

14. Grassini, S.: Computational power and subjective quality of ai-generated outputs: the case of aesthetic judgement and positive emotions in ai-generated art. Inter. J. Human–Comput. Interact., 1–10 (2024)
15. Guo, Y., Pannasch, S., Helmert, R.J.: Eye movements in extended tasks: analyses of ambient/focal attention with coefficient k. In: 2022 Symposium on Eye Tracking Research and Applications, pp. 1–7 (2022)
16. Iwata, M.: Visual association pathways in human brain. Tohoku J. Exper. Med. **161**(Supplement), 61–78 (1990)
17. Kanbaty, M., Hellmann, A., He, L.: Infographics in corporate sustainability reports: providing useful information or used for impression management? J. Behav. Exp. Financ. **26**, 100309 (2020)
18. Li, C., et al.: The good, the bad, and why: unveiling emotions in generative ai. arXiv preprint arXiv:2312.11111 (2023)
19. Majer, J.M., Henscher, H.A., Reuber, P., Fischer-Kreer, D., Fischer, D.: The effects of visual sustainability labels on consumer perception and behavior: a systematic review of the empirical literature. Sustainable Product. Consumpt. **33**, 1–14 (2022)
20. Mehta, M., Buntain, C.: Emotional images: assessing emotions in images and potential biases in generative models. arXiv preprint arXiv:2411.05985 (2024)
21. Metag, J., Schäfer, M.S., Füchslin, T., Barsuhn, T., Kleinen-von Königslöw, K.: Perceptions of climate change imagery: Evoked salience and self-efficacy in Germany, Switzerland, and Austria. Sci. Commun. **38**(2), 197–227 (2016)
22. Michelon, G., Pilonato, S., Ricceri, F.: Csr reporting practices and the quality of disclosure: an empirical analysis. Crit. Perspect. Account. **33**, 59–78 (2015)
23. Mishra, A.: Machine learning in the AWS cloud: Add intelligence to applications with Amazon Sagemaker and Amazon Rekognition. John Wiley & Sons (2019)
24. Moodaley, W., Telukdarie, A.: Greenwashing, sustainability reporting, and artificial intelligence: a systematic literature review. Sustainability **15**(2), 1481 (2023)
25. Nakao, Y., Ishino, A., Kokubu, K., Okada, H.: Exploring visual communication in corporate sustainability reporting: using image recognition with deep learning. Corp. Soc. Responsib. Environ. Manag. **31**(4), 3210–3234 (2024)
26. Paik, S., et al.: The affective nature of ai-generated news images: Impact on visual journalism. In: 2023 11th International Conference on Affective Computing and Intelligent Interaction (ACII), pp. 1–8. IEEE (2023)
27. Tliba, M., et al.: Satsal: a multi-level self-attention based architecture for visual saliency prediction. IEEE Access **10**, 20701–20713 (2022)
28. Tsotsos, J.K.: A computational perspective on visual attention. MIT Press (2021)
29. Wedel, M., Pieters, R.: A review of eye-tracking research in marketing. Rev. Marketing Res. **4**, 123–147 (2007). https://doi.org/10.1108/S1548-6435(2007)0000004009
30. Yu, J., Dickinger, A., So, K.K.F., Egger, R.: Artificial intelligence-generated virtual influencer: examining the effects of emotional display on user engagement. J. Retail. Consum. Serv. **76**, 103560 (2024)
31. Zhang, Y., et al.: Affective computing in the era of large language models: A survey from the nlp perspective. arXiv e-prints pp. arXiv–2408 (2024)
32. Zou, Y., et al.: Esgreveal: an llm-based approach for extracting structured data from esg reports. J. Clean. Prod. **489**, 144572 (2025)

EEG-Based Mental Stress Detection: A Comparative and Explainable Study Across Tasks and Subjects

Francesco Agnelli, Giorgio Blandano, Jacopo Burger[✉],
Alessandro D'Amelio, Giuseppe Facchi, Omar Ghezzi,
Raffaella Lanzarotti, and Lia Schmid

PHuSe Lab, Department of Computer Science, University of Milan, Milan, Italy
{agnelli.francesco,blandano.giorgio,burger.jacopo,damelio.alessandro,
facchi.giuseppe,ghezzi.omar,lanzarotti.raffaella,schmid.lia}@unimi.it

Abstract. In this work, we investigate the problem of recognizing mental stress using electroencephalography (EEG) and machine learning, with the explicit objective of developing an interpretable and effective model across different task conditions. To this end, we implement a supervised learning pipeline on two publicly available EEG datasets, SAM40 and WAUC, the latter of which involves controlled tasks with varying cognitive and physical demands. Following standard preprocessing, we extract a broad range of well-established EEG-derived features -including spectral power, entropy-based complexity measures, and hemispheric asymmetries- and train an XGBoost classifier to distinguish between stress levels. To interpret the model's decision process, we employ SHAP (SHapley Additive exPlanations), which quantifies the contribution of each feature to the model's predictions at both individual and population levels. Our results reveal that a restricted subset of features, particularly gamma-band asymmetries and entropy measures, consistently dominates the decision process across subjects. Moreover, the set of important features varies systematically with changes in physical load, suggesting the existence of condition-specific EEG patterns associated with stress.

Keywords: EEG · Interpretability · SHAP · Stress · Physiological Sensing

1 Introduction

Stress is a psychophysiological response to real or perceived threats to homeostasis, triggered by internal or external stressors [18]. While acute stress can be adaptive, enhancing alertness and mobilizing resources, chronic stress is associated with an increased risk of both mental and physical disorders. Elevated stress levels lead to physiological dysregulation, measurable through sensitive biomarkers that can be used to quantify an individual's stress. As a result, there has been

E. Rodolà et al. (Eds.): ICIAP 2025 Workshops, LNCS 16169, pp. 17–28, 2026.
https://doi.org/10.1007/978-3-032-11317-7_2

growing interest in the automatic assessment of stress levels using non-invasive physiological measurements [4,8,30].

At the neurobiological level, the stress response involves the amygdala and anterior cingulate cortex (ACC), which assess threats and activate the hypothalamic-pituitary-adrenal (HPA) axis and autonomic nervous system (ANS), leading to the release of glucocorticoids and catecholamines. Understanding stress as a dynamic interaction between physiology, cognition, and prediction lays the groundwork for objective monitoring. In particular, electroencephalography (EEG) offers a non-invasive way to assess stress in real time and has become an essential component in the design of human-centered systems, particularly in high-stakes domains such as aviation, driving, and operational safety [10].

While common approaches emphasize predictive performance, there is growing recognition that model interpretability is equally critical for deployment in sensitive applications. Despite this increasing interest, few studies systematically evaluate the alignment between model-driven explanations and known neurocognitive correlates of workload [21]. Understanding which features drive classification decisions can improve both system transparency and neuroscientific insight. Existing EEG-based classifiers often struggle with generalization due to substantial inter-subject variability [22], and tend to focus on isolated cognitive paradigms rather than complex, multimodal demands. To address these limitations, we investigate EEG-based stress assessment in more ecologically valid settings using the SAM40 [9] and the WAUC dataset [1]; the latter specifically includes tasks performed under varying cognitive and physical load conditions.

We extract a broad set of EEG-derived features, including spectral power, entropy-based complexity, and hemispheric asymmetry measures, and train an XGBoost classifier [7] to distinguish stress levels. To analyze the decision process, we apply SHAP (SHapley Additive exPlanations) [17], which quantifies the contribution of each feature to model predictions at both individual and population levels.

Our study systematically evaluates model performance and feature relevance across intra-subject, mixed-subject, and leave-one-subject-out (LOSO) configurations. We show that a restricted subset of features -particularly gamma-band asymmetries and wavelet-based entropy- dominates the decision process across participants. Interestingly, the explainability analysis reveals that the relative importance of the selected features changes significantly depending on the condition -specifically, the level of physical load.

Overall, we present a principled, model-agnostic framework for interpretable EEG-based stress classification, contributing to the broader field of explainable physiological computing. Our approach integrates interpretable machine learning with an in-depth analysis of EEG features, capturing both inter- and intra-subject dynamics under realistic dual-task conditions while identifying robust neurophysiological correlates of stress.

2   Related Work

The use of ML in EEG-based stress detection faces challenges such as small dataset sizes and variability among subjects. As reviewed in [13], classical approaches like Support Vector Machines (SVMs) are still widely used for their robustness and ability to generalize in high-dimensional, low-sample settings.

Tree-based ensemble models have also shown promising results. For instance, [23] reported over 98% accuracy on the SAM40 dataset [9] using a stacking ensemble of Random Forest and k-Nearest Neighbors (k-NN), demonstrating the effectiveness of ensemble learning in EEG applications.

Deep learning (DL) methods are increasingly outperforming traditional ML approaches in EEG analysis, thanks to their capacity to automatically learn relevant features from raw data [11,19]. Applications of DL in EEG embrace a variety of diverse domains: emotion recognition [28], motor imagery [24], seizure detection [29], sleep staging [27], brain disorders detection [5] and stress assessment [25]. In the context of stress, Saadati et al. [20] integrated functional Near Infrared Spectroscopy (fNIRS) with EEG signals and adapted a convolutional neural networks (CNN) to handle multimodal input, achieving promising results with a classification accuracy of 89.5%. Similarly, the study in [26] introduced a hybrid architecture combining deep recurrent networks (RNN) and 3D CNNs to jointly process raw and spectral EEG data for binary mental workload classification, reporting an average cross-task accuracy of 88.9%. In a related approach, Kwak et al. [15] proposed a temporal attention mechanism based on Long Short-Term Memory (LSTM) networks to capture both local and global temporal structures in EEG signals, achieving 90.8% accuracy on their proprietary dataset. Chakladar et al. [6] employed a grey wolf optimizer in combination with a deep BLSTM-LSTM architecture to estimate multiple levels of stress, reaching 86.33% and 82.57% accuracy in "no task" and "multitasking" scenarios, respectively.

Despite their strong performance, deep learning models remain largely opaque and lack interpretability, an essential requirement in neurophysiological research and applied cognitive monitoring. For this reason, our work focused on interpretable ML approaches, enabling transparent analysis of stress-related EEG patterns and individual variability.

3   Proposed Method

This section outlines the methodological pipeline (adopted dataset, EEG preprocessing, feature extraction, model training, and interpretability analysis) used to determine stress levels from EEG signals via standard machine leaning algorithms, with a focus on interpretability.

3.1   Datasets

SAM40. The SAM40 dataset [9] contains EEG data from 40 healthy subjects (mean age: 21.5). Participants underwent a series of tasks designed to elicit cognitive stress:

- Stroop Test: Conflict between ink color and word meaning.
- Mirror Image Recognition: Visual symmetry task under varying difficulty.
- Arithmetic Problem Solving: Mental math under time pressure.
- Relaxation Phase: Resting baseline for comparison.

Each session included three trials, after which participants self-reported stress on a 1–10 scale. EEG was recorded with a 32-channel Emotiv Epoc Flex 128 Hz. The dataset includes preprocessed EEG signals, so no additional artifact removal was necessary. The original preprocessing pipeline involved bandpass filtering (0.5–45 Hz), smoothing via Savitzky-Golay filtering, and wavelet thresholding.

WAUC. The WAUC dataset [1] includes approximately 60 min of continuous EEG data recordings from 48 participants (mean age: 27.4), split into two groups performing physical load tasks either on a treadmill or stationary bike. Participants completed the MATB-II multitasking protocol under no physical activity (baseline), moderate activity (treadmill at 3 km/h or bike at 50 rpm), or high activity (treadmill at 5 km/h or bike at 70 rpm).

Tasks included system monitoring, tracking, and resource management. Each session lasted 10 min, followed by a 5-minute break. Participants provided NASA-TLX and Borg scale self-assessments. EEG was recorded using an 8-channel Enobio system 500 Hz, with electrodes placed over frontal, temporal, and parietal regions. Two labeling strategies were employed: task-based labels from the original cognitive workload protocol and self-reported scores from post-task NASA-TLX questionnaires.

3.2   Preprocessing and Feature Extraction

Preprocessing. All EEG signals underwent a standardized preprocessing pipeline consisting of band-pass filtering to eliminate unwanted frequency components, downsampling 250 Hz to reduce computational load while preserving signal fidelity, and channel-wise normalization to correct for amplitude differences across electrodes and ensure consistent signal scaling.

To further enhance signal quality, two artifact removal methods were applied. The first, wICA [12], improves upon traditional ICA by applying wavelet thresholding to individual components. This approach is particularly effective in low-channel configurations like WAUC (8 channels), as it attenuates artifacts without discarding entire components.

The second method, ASR (Artifact Subspace Reconstruction) [14], dynamically detects and suppresses artifacts by comparing signal variance to a subject-specific baseline, derived from segments involving only physical activity. This enables the isolation of cognitive activity from movement-induced interference. A high threshold ($k = 30$) was used to preserve genuine neural activity while effectively mitigating noise.

These complementary techniques were selected to ensure thorough artifact removal while maintaining the integrity of neural signals, thereby supporting robust downstream analysis.

EEG Features Extraction. EEG features were extracted from each preprocessed segment using a unified pipeline applied to both datasets (SAM40 and WAUC), with adjustments to their specific recording setups. Segmentation was performed using 5-second windows with 50% overlap for SAM40 and 5-second non-overlapping windows for WAUC. From each segment and channel, eight types of features were computed, encompassing spectral, connectivity, and nonlinear descriptors.

Spectral power was estimated using Welch's method with Hamming windows (50% overlap), computed for canonical frequency bands: theta (4–8 Hz), alpha (8–12 Hz), beta (13–30 Hz), and gamma (30–50 Hz). To account for intersubject variability in absolute power, spectral values were normalized by the total broadband power (1–50 Hz) within each trial.

Hemispheric asymmetry was derived from lateralized homologous electrode pairs (e.g., F3/F4, P3/P4), computed as the signed difference in normalized spectral power between the left and right hemisphere. This measure reflects lateralized cortical activation patterns often associated with stress and affective processing.

Inter-channel functional connectivity was assessed via magnitude-squared coherence, estimated pairwise between selected electrodes for each frequency band. Coherence quantifies the phase-consistent coupling of EEG signals, which has been linked to cognitive load and task synchronization.

To capture temporal complexity beyond second-order statistics, we extracted a set of nonlinear features from the raw EEG time series. Hjorth parameters - activity, mobility, and complexity- were computed to quantify signal variance, dominant frequency, and the degree of frequency modulation, respectively.

The Higuchi fractal dimension, a measure of self-similarity and signal irregularity across scales, was estimated using Higuchi's algorithm with a maximum scaling factor of $k_{\max} = 10$.

In addition, three entropy-based metrics were employed to characterize different aspects of signal complexity: Shannon entropy, calculated from a 10-bin amplitude histogram, capturing distributional unpredictability; wavelet entropy, derived from the energy distribution of discrete wavelet transform coefficients (Daubechies-4, 4 decomposition levels), quantifying multi-resolution signal disorder; and approximate entropy, computed with embedding dimension $m = 2$ and similarity threshold $r = 0.2 \times \text{std}$, measuring temporal regularity and pattern predictability.

To enable interpretability and ablation analysis, features were structured hierarchically into macro categories (e.g., PSD, asymmetry, coherence), further subdivided by frequency band or metric type, and organized by electrode location. Final feature dimensionality depended on the number of EEG channels: 380 features per segment for SAM40 (32 channels) and 120 for WAUC (8 channels). All features were standardized using z-score normalization on the training set within each cross-validation fold.

3.3   Experimental Setup

SAM40. Preliminary experiments were conducted on the SAM40 dataset to evaluate the feasibility of EEG-based stress classification. Two binary classification scenarios were considered:

- high vs. no stress
- low vs. high stress.

Concerning the second scenario, the model was also tested in a task-specific fashion, both within and across cognitive tasks (e.g., training on Stroop, testing on Mirror).

WAUC. Experiments were subsequently extended to the more comprehensive WAUC dataset, which offers a larger sample size and longer recordings, allowing for more detailed modeling and explainability analyses.

In particular, three experimental pipelines were evaluated on the WAUC dataset to investigate the effects of various preprocessing techniques on classification performance: wICA with spectral features only ($m1$), wICA with full feature set ($m2$), and ASR with full feature set ($m3$).

Each of these configurations was tested under three validation strategies.

- *Intra-subject* validation was performed using 5-fold cross-validation within each subject to assess individual-level performance.
- *Mixed-subject* validation involved repeated 5-fold cross-validation on the pooled dataset to evaluate general performance across participants.
- *LOSO strategy* was used to assess generalization to entirely unseen individuals.

Classification performance was assessed using accuracy and ROC-AUC as primary metrics across all configurations. Hyperparameters were optimized via grid search.

4   Experiments and Results

4.1   Stress Detection Results

Results on SAM40. In Table 1, we report the results of the experiments on the SAM40 dataset using a standard 80–20 training-test split. In the first binary classification setting, discriminating between high stress and no stress conditions, a test accuracy of 72% and a ROC-AUC of 80% were achieved.

More interesting insights emerge in the second binary setting, where the goal is to distinguish between low and high stress levels. Here, we explored both intra-task and inter-task generalization between the Stroop and Mirror tasks. When training the model on samples from the Stroop task and testing it on the same task (intra-task), we obtained an accuracy of 74% and a ROC-AUC of 82%. However, when evaluating the same model on the Mirror task (inter-task), the accuracy remained stable at 74%, but the ROC-AUC dropped to 69%, indicating reduced generalization in terms of discriminative confidence.

Conversely, when training on the Mirror task and testing on it (intra-task), we observed even better performance with 85% accuracy and a ROC-AUC of 87%. Yet, in the inter-task setting, testing on the Stroop task, the performance degraded considerably, with accuracy dropping to 64% and ROC-AUC to 61%.

Results on WAUC. Table 2 summarizes the classification results on the WAUC dataset across three validation protocols, highlighting the impact of feature richness and artifact suppression (ASR).

In the intra-subject setting, where models are trained and tested within individuals, combining all features with ASR yields near-ceiling performance (ROC-AUC = 0.986, accuracy = 0.952), indicating excellent subject-specific discriminability. In the mixed-subject scenario, with cross-validation across pooled data from all participants, the ASR-enhanced pipeline maintains strong generalization (ROC-AUC = 0.868, accuracy = 0.783), clearly outperforming other configurations. Conversely, in the leave-one-subject-out (LOSO) setting, which tests generalization to unseen individuals, performance drops sharply across all models (ROC-AUC close to 0.5), underscoring the difficulty of capturing inter-subject invariants in EEG-based stress detection.

Overall, the findings emphasize the importance of ASR and rich feature sets for boosting classification performance, particularly in settings where intra- or mixed-subject generalization is sufficient. Nevertheless, achieving robust generalization across individuals remains a key challenge in EEG-based stress detection.

Table 1. Classification performance on the SAM40 dataset. *High vs No Stress*: standard 80/20 train/test split. *High vs Low Stress*: intra- and inter-task generalization across Stroop and Mirror tasks.

Configuration	TrainROC-AUC	TestAcc	TestROC-AUC
High vs No Stress	95%	72%	80%
High vs Low Stress			
Intra-task: Stroop/Stroop	97%	74%	82%
Inter-task: Stroop/Mirror	–	74%	69%
Intra-task: Mirror/Mirror	89%	85%	87%
Inter-task: Mirror/Stroop	–	64%	61%

4.2   Explainability Analysis

Building on the best-performing configuration -ASR preprocessing combined with the full EEG feature set, we leverage the XGB's native compatibility with SHAP (SHapley Additive exPlanations), which facilitates both global and local interpretability of model decisions.

Table 2. ROC-AUC and accuracy for the three adopted protocols (Intra-subject, Mixed-subject and Leave-One-Subject-Out) and feature extraction methods (m1: wICA+spectral features, m2: wICA+full features set, m3: ASR+full feature set) on the WAUC dataset. Mean and standard deviation are reported. Best results in **bold**.

Protocol	Method	ROC-AUC	Accuracy
Intra-subject	$m1$	0.773 ± 0.0289	0.704 ± 0.0269
	$m2$	0.896 ± 0.0240	0.828 ± 0.0236
	$m3$	**0.986 ± 0.0068**	**0.952 ± 0.0125**
Mixed	$m1$	0.562 ± 0.0003	0.542 ± 0.0003
	$m2$	0.594 ± 0.0029	0.563 ± 0.0020
	$m3$	**0.868 ± 0.0005**	**0.783 ± 0.0007**
LOSO	$m1$	**0.520 ± 0.0317**	**0.520 ± 0.0317**
	$m2$	0.489 ± 0.0220	0.489 ± 0.0220
	$m3$	0.506 ± 0.0374	0.509 ± 0.0267

To gain insights into how EEG features contribute to stress classification, SHAP values were computed using TreeSHAP [16] under two complementary perspectives on the WAUC dataset:

- across the three levels of physical workload to examine condition-specific patterns
- averaged across all subjects to capture population-level relevance.

To address the high dimensionality of the feature space, variables were grouped into interpretable domains (e.g., spectral, connectivity), and their SHAP values were aggregated and visualized using bar and box plots.

The results, summarized in Figs. 1 and 2, highlight both condition-dependent modulation of feature relevance and generalizable EEG biomarkers of mental stress. Figure 1 illustrates SHAP-based feature importance across three physical load conditions: All Physical Load, No Physical Load, and High Physical Load. In the absence of physical activity, entropy-based features (e.g., wavelet entropy) dominate, suggesting that cleaner EEG signals allow complexity metrics to more effectively capture cognitive stress. Under high physical load, the model shifts its reliance toward spectral asymmetries, coherence, and Hjorth parameters, likely reflecting an adaptation to muscle-induced artifacts by prioritizing more robust signal descriptors. When aggregating across all physical loads, a hybrid profile emerges, with a balanced contribution from both entropy and connectivity-related features. These patterns underscore the importance of contextual factors, such as physical effort, in shaping the discriminative power and interpretability of EEG-based stress models.

Complementing this analysis, Fig. 2 presents the mean SHAP values of grouped EEG features across all subjects. Asymmetry in the gamma band emerges as the most influential feature, followed by wavelet entropy and beta-band asymmetry, emphasizing the joint importance of spectral imbalance and signal complexity. Gamma coherence and asymmetries in the alpha and beta bands also show substantial contributions, highlighting the role of inter-regional communication and hemispheric dynamics. Nonlinear features such as approximate entropy and Higuchi's fractal dimension (HFD) rank in the mid-range, while conventional spectral power descriptors (e.g., PSD in alpha, beta, and theta) are the least informative. Together, these findings indicate that the model favors relational and complexity-based features over traditional power measures when identifying markers of mental stress from EEG data.

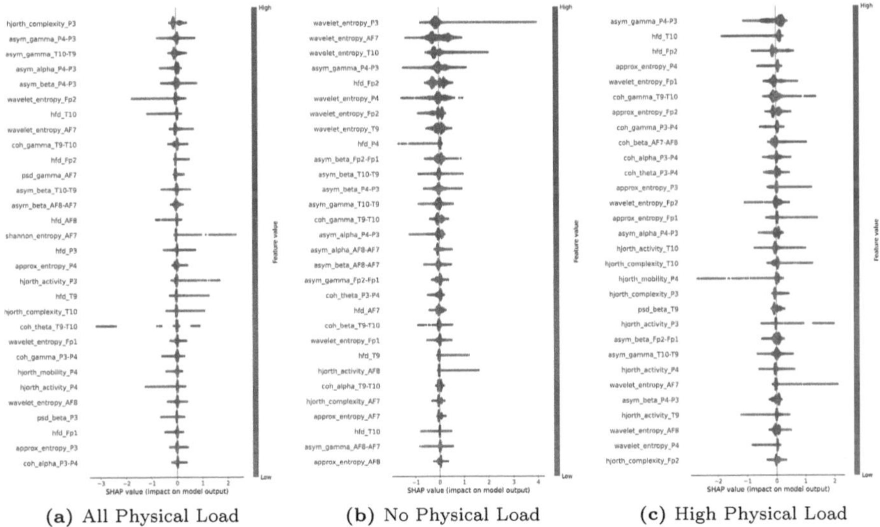

(a) All Physical Load   (b) No Physical Load   (c) High Physical Load

Fig. 1. SHAP-based feature importance across three physical load conditions on the WAUC dataset: All Physical Load, No Physical Load, and High Physical Load.

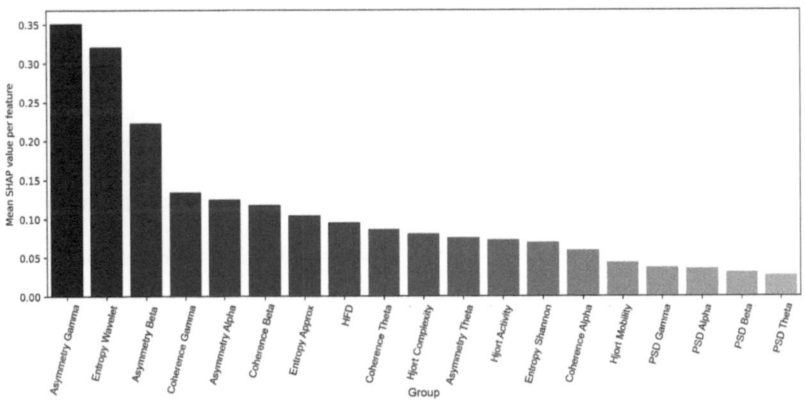

Fig. 2. Mean feature importance aggregated across all subjects

5 Conclusion

This study combines well-established machine learning techniques with interpretable methods to develop an explainable EEG-based stress classification approach. The aim is to evaluate the impact of different variables -namely, the performed task or condition and the presence of physical load- on the generalization capabilities of a machine learning model. We assessed the performance of XGBoost classifiers trained on a diverse set of EEG-derived features, including spectral power, entropy-based complexity measures, and hemispheric asymmetries, using the publicly available SAM40 and WAUC datasets. Model interpretability was achieved through SHAP, which enabled us to quantify feature contributions to classification decisions at both individual and population levels.

Our findings indicate that a restricted subset of features, particularly gamma-band asymmetries and wavelet-based entropy, dominated the decision process if results are aggregated across subjects and load conditions. This suggests the presence of precise neurophysiological markers of stress, even in the face of inter-subject variability. Moreover, our interpretability framework facilitated a systematic characterization of both subject-specific and global patterns of feature relevance, offering insights that would remain inaccessible in black-box models. Interestingly, results on the WAUC dataset show that when stratifying by specific conditions (e.g., the amount of physical load), the importance of certain features varies significantly. This highlights the need for developing models specifically tailored to the condition at hand. These findings align with recent research emphasizing the importance of addressing the data distribution shifts in EEG datasets caused by changes in condition (e.g., subject shift, environment shift, device shift) [2,3].

Future research directions include extending the framework to regression-based modeling of continuous subjective responses and incorporating domain adaptation techniques to improve generalization across distribution shifts.

Acknowledgments. This work was carried out at the Department of Computer Science "Giovanni degli Antoni", University of Milan – Via Celoria 18, Milan. It was partially funded by the National Plan for NRRP Complementary Investments (PNC, established with the decree-law 6 May 2021, n. 59, converted by law n. 101 of 2021) in the call for the funding of research initiatives for technologies and innovative trajectories in the health and care sectors (Directorial Decree n. 931 of 06-06-2022) - project n. PNC0000003 - AdvaNced Technologies for Human-centrEd Medicine (project acronym: ANTHEM). This work reflects only the authors' views and opinions, neither the Ministry for University and Research nor the European Commission can be considered responsible for them.

References

1. Albuquerque, I., et al.: Wauc: a multi-modal database for mental workload assessment under physical activity. Front. Neurosci. **14**, 549524 (2020)
2. Barbera, T., et al.: On using ai for eeg-based bci applications: problems, current challenges and future trends. arXiv preprint arXiv:2506.16168 (2025)
3. Barmpas, K., Panagakis, Y., Zoumpourlis, G., Adamos, D.A., Laskaris, N., Zafeiriou, S.: A causal perspective on brainwave modeling for brain-computer interfaces. J. Neural Eng. **21**(3), 036001 (2024)
4. Boccignone, G., et al.: pyVHR: a python framework for remote photoplethysmography. PeerJ Comput. Sci. **8**, e929 (2022)
5. Burger, J., Cuculo, V., D'Amelio, A., Grossi, G., Lanzarotti, R.: ECoGNet: an eeg-based effective connectivity graph neural network for brain disorder detection. In: 2025 International Joint Conference on Neural Networks (IJCNN). IEEE (2025)
6. Chakladar, D.D., Dey, S., Roy, P.P., Dogra, D.P.: Eeg-based mental workload estimation using deep blstm-lstm network and evolutionary algorithm. Biomed. Signal Process. Control **60**, 101989 (2020)
7. Chen, T., et al.: Xgboost: extreme gradient boosting. R package version 0.4-2 **1**(4), 1–4 (2015)
8. Ghezzi, O., Boccignone, G., Grossi, G., Lanzarotti, R., D'Amelio, A.: CliffPhys: Camera-based respiratory measurement using clifford neural networks. In: European Conference on Computer Vision. pp. 221–238. Springer (2024). https://doi.org/10.1007/978-3-031-73013-9_13
9. Ghosh, R., et al.: Sam 40: dataset of 40 subject eeg recordings to monitor the induced-stress while performing stroop color-word test, arithmetic task, and mirror image recognition task. Data Brief **40**, 107772 (2022)
10. Hernández-Sabaté, A., Yauri, J., Folch, P., Piera, M.À., Gil, D.: Recognition of the mental workloads of pilots in the cockpit using eeg signals. Appl. Sci. **12**(5), 2298 (2022)
11. Ismail Fawaz, H., Forestier, G., Weber, J., Idoumghar, L., Muller, P.-A.: Deep learning for time series classification: a review. Data Min. Knowl. Disc. **33**(4), 917–963 (2019). https://doi.org/10.1007/s10618-019-00619-1
12. Jung, T.P., et al.: Removing electroencephalographic artifacts by blind source separation. Psychophysiology **37**(2), 163–178 (2000)
13. Katmah, R., Al-Shargie, F., Tariq, U., Babiloni, F., Al-Mughairbi, F., Al-Nashash, H.: A review on mental stress assessment methods using eeg signals. Sensors **21**(15) (2021)
14. Kothe, C.A.E., Jung, T.P.: Artifact removal techniques with signal reconstruction,28 Apr (2016), uS Patent App. 14/895,440
15. Kwak, Y., Kong, K., Song, W.J., Min, B.K., Kim, S.E.: Multilevel feature fusion with 3d convolutional neural network for eeg-based workload estimation. IEEE Access **8**, 16009–16021 (2020)
16. Lundberg, S.M., et al.: From local explanations to global understanding with explainable ai for trees. Nat. Mach. Intell. **2**(1), 56–67 (2020)
17. Lundberg, S.M., Lee, S.I.: A unified approach to interpreting model predictions. In: Guyon, I., Luxburg, U.V., Bengio, S., Wallach, H., Fergus, R., Vishwanathan, S., Garnett, R. (eds.) Advances in Neural Information Processing Systems 30, pp. 4765–4774. Curran Associates, Inc. (2017)
18. McEwen, B.S.: Protective and damaging effects of stress mediators. N. Engl. J. Med. **338**(3), 171–179 (1998)

19. Rim, B., Sung, N.J., Min, S., Hong, M.: Deep learning in physiological signal data: a survey. Sensors **20**(4), 969 (2020)
20. Saadati, M., Nelson, J., Ayaz, H.: Convolutional neural network for hybrid fNIRS-EEG mental workload classification. In: Ayaz, H. (ed.) AHFE 2019. AISC, vol. 953, pp. 221–232. Springer, Cham (2020). https://doi.org/10.1007/978-3-030-20473-0_22
21. Sabbagh, D., Ablin, P., Varoquaux, G., Gramfort, A., Engemann, D.A.: Predictive regression modeling with meg/eeg: from source power to signals and cognitive states. Neuroimage **222**, 116893 (2020)
22. Saha, S., Baumert, M.: Intra-and inter-subject variability in eeg-based sensorimotor brain computer interface: a review. Front. Comput. Neurosci. **13**, 87 (2020)
23. Shikha, S., Sethia, D., Sreedevi, I.: Ensemble Classifier for EEG-Based Stress Classification: An Empirical Study on Stacking Classifiers, pp. 389–400 (Dec 2024)
24. Venkatachalam, K., Devipriya, A., Maniraj, J., Sivaram, M., Ambikapathy, A., et al.: A novel method of motor imagery classification using eeg signal. Artif. Intell. Med. **103**, 101787 (2020)
25. Yang, S., Yin, Z., Wang, Y., Zhang, W., Wang, Y., Zhang, J.: Assessing cognitive mental workload via eeg signals and an ensemble deep learning classifier based on denoising autoencoders. Comput. Biol. Med. **109**, 159–170 (2019)
26. Zhang, P., Wang, X., Zhang, W., Chen, J.: Learning spatial-spectral-temporal eeg features with recurrent 3d convolutional neural networks for cross-task mental workload assessment. IEEE Trans. Neural Syst. Rehabil. Eng. **27**(1), 31–42 (2018)
27. Zhang, X., et al.: Automated multi-model deep neural network for sleep stage scoring with unfiltered clinical data. Sleep and Breathing **24**(2), 581–590 (2020). https://doi.org/10.1007/s11325-019-02008-w
28. Zhang, Y., et al.: An investigation of deep learning models for eeg-based emotion recognition. Front. Neurosci. **14**, 622759 (2020)
29. Zhao, W., Wang, W.: Seizurenet: a model for robust detection of epileptic seizures based on convolutional neural network. Cognitive Comput. Syst. **2**(3), 119–124 (2020)
30. Ziaratnia, S., Laohakangvalvit, T., Sugaya, M., Sripian, P.: Multimodal deep learning for remote stress estimation using cct-lstm. In: Proceedings of the IEEE/CVF Winter Conference on Applications of Computer Vision, pp. 8336–8344 (2024)

Seeing Beyond: Unlocking Image Emotion with Contextual Depths

Federico Cozzi[1]([envelope]) [iD], Andrea D'Eusanio[1] [iD], and Giuseppe Boccignone[2] [iD]

[1] Emotiva, Milano, Italy
{federico.cozzi,andrea.deusanio}@emotiva.it
[2] PHuSe Laboratory—Dipartimento di Informatica, Università degli Studi di Milano,
Via Celoria 18, Milano 20133, Italy
giuseppe.boccignone@unimi.it
http://www.emotiva.it

Abstract. Can a machine truly grasp the myriad emotions an image evokes? We dive into this fascinating challenge, exploring how contextual information can dramatically improve the recognition of emotions in images. Our journey begins with EmoSet, a rich dataset already annotated for the feelings images stir within us. We then expand its horizons, carefully crafting both concise and expansive textual descriptions for each visual. These added narratives act as crucial guides, unearthing semantic and emotional nuances that may remain hidden in the image alone. When we fused these new text embeddings with visual features to train a baseline model, the results were compelling: a nearly 5% boost in accuracy on the manually annotated subset of our dataset. This significant improvement, achieved even with relatively straightforward contextual additions, underscores a vital insight. It highlights how even simple forms of contextual enrichment can meaningfully contribute to emotion classification and, more broadly, emphasizes the profound importance of multimodal inputs for truly understanding the affective content of images.

Keywords: Evoked Emotion Recognition · Affective Computing · Multimodal Learning

1   Introduction

While most prior research has focused on recognizing the emotional state of people *depicted* in images, primarily through facial expressions or body posture, our focus lies in predicting *evoked emotions*: the affective response that a viewer experiences when observing an image, regardless of whether any person is portrayed.

State-of-the-art models for evoked emotion recognition often rely solely on raw visual content, which can lead to misinterpretation when key contextual cues, such as irony, cultural symbols, or narrative elements, are ambiguous or absent. These shortcomings are especially pronounced in domains such as advertising, where affective meaning is frequently implied rather than explicitly shown.

It is important to note that, from the standpoint of affect theory, the majority of current models is by and large underpinned by assumptions grounding in

© The Author(s), under exclusive license to Springer Nature Switzerland AG 2026
E. Rodolà et al. (Eds.): ICIAP 2025 Workshops, LNCS 16169, pp. 29–40, 2026.
https://doi.org/10.1007/978-3-032-11317-7_3

Basic Emotion Theory (BET [9]). However, recent advancements in emotion sciences [2] give evidence of a more nuanced and complex picture. Differently from BET, psychological construction theories (PCT) of emotion [2] give evidence that while each emotional event is unique, a common set of fundamental psychological operations underlies the processing of every emotional event. These operations include *core affect* (reflecting valence and arousal) and, crucially, *conceptualization* (generating meaning by integrating external with internal signals via associations with past experiences). Germane to the work presented here, PCT offers the hypothesis that language functions as a context in emotion perception [1]. Even in the "simple" case of portrayed emotion recognition, empirical studies have shown that emotions are not events that broadcast precise information on the face, and facial behaviors, viewed in isolation will be ambiguous as to their emotional meaning. Structural information from the face is necessary, but not sufficient, for emotion perception.

Interestingly enough, similar evidence has been provided by affective computing models (cf. Section 2).

We argue that if context is crucial for interpreting an emotion that is visibly present on a face, it is even more critical for predicting an emotion that is merely evoked by a scene. In spite of such compelling evidence, most large-scale evoked emotion datasets lack rich, multimodal annotations needed to model this context.

To bridge this gap, and grounding in the language-as-context hypothesis [1], we address the research question (RQ1) of whether textual descriptions offer a lightweight and scalable method to encode affective context that may not be visually explicit or easily understandable (Sect. 3). A complementary research question (RQ2) is whether more general, off-the-shelf multimodal models can provide valuable contextual enrichment without task-specific fine-tuning. More specifically, our contribution can be summarised as follows.

- We extend EmoSet [23], a dataset of images annotated with evoked emotions, by enriching each image with contextual embeddings (short and long textual descriptions) generated by a multimodal large language model (MLLM) with vision capabilities.
- These textual descriptions are then processed to extract embedded textual features and integrated with visual features; we show (Sect. 4) that incorporating these text embeddings into a baseline visual classifier improves classification accuracy by an average of $+5\%$ points on the manually annotated subset (RQ1).
- A detailed ablation study is performed to compare the effectiveness of both visual and textual embedding branches (RQ2).

Figure 1 provides at a glance a high-level overview of our approach.

To sum up and to the best of our knowledge, this work uniquely leverages a general MLLM to generate both short and long textual descriptions, which are then used to augment the dataset. Our ablation studies further clarify how different single modalities contribute to enhancing model performance in a scalable, plug-and-play manner.

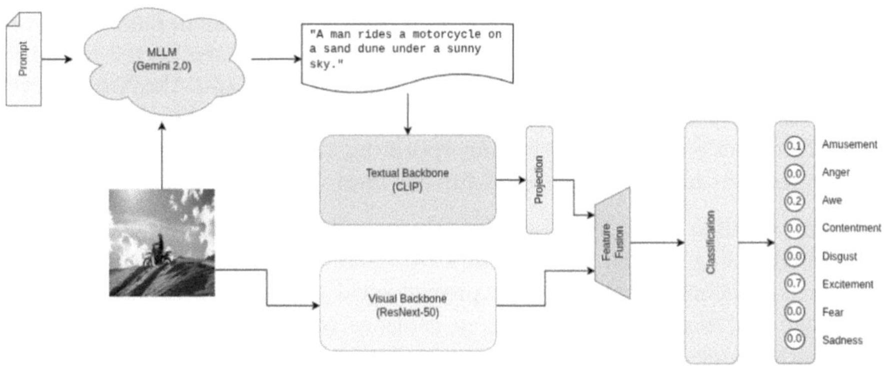

Fig. 1. The proposed approach: each image is processed with external MLLM to produce accurate caption and long description. Then, the image is fed into a CNN backbone and the text context representation is fed into a text encoder. The two obtained feature vectors are concatenated before classification.

2   Related Works

Evoked Emotion Recognition from Visual Stimuli. Evoked emotion recognition, predicting a viewer's emotional response to an image, is a growing area in affective computing [4]. Early datasets like *ArtPhoto* and *Abstract Paintings* [17] established the concept by linking images with viewer-reported reactions across eight basic emotions. Subsequent efforts introduced large-scale datasets such as the Visual Sentiment Ontology (VSO) [3] and *ImageEmotion* [25], which expanded to eight discrete emotion categories. While early methods using global image descriptors like GIST and SIFT achieved moderate success (50–60% accuracy [17]), Convolutional Neural Networks (CNNs) significantly boosted accuracy, often exceeding 70% [25].

These studies confirm the viability of predicting emotion from global image features, while also highlighting the limitations of relying solely on visual information [4].

Context in Evoked Emotion. In evoked emotion tasks, context plays a central role in shaping emotional perception. Unlike depicted emotion recognition, which focuses on facial expressions or body language, evoked emotions are often driven by scene semantics, composition, and cultural or situational context. As such, subtle signals like irony, contrast, or implied narratives can be crucial. Studies on datasets such as EMOTIC and CAER have shown that models incorporating both local (e.g., facial features) and global (e.g., background) context significantly outperform their visual-only counterparts [14,15]. For instance, CAER-Net improved F1-score from 32% to 40% (+25%) on depicted-emotion tasks by integrating scene and face features [15], while context-enhanced pipelines on EMOTIC achieved 20–44% relative mAP improvements over face-only baselines [14,16]. These findings emphasize that scene context helps resolve ambiguities that might otherwise lead to incorrect inferences. EmoSet [23] is a unique dataset because, unlike EMOTIC or AffectNet, it doesn't require human faces or expressions in every image, reflecting modern media's diversity (see Sect. 3.1). Initial tests with standard architectures like ResNet50, VGG-16, and AlexNet achieved around 75% top-1 classification accuracy on the manually-labeled subset. The original EmoSet paper showed that incorporating manually annotated attributes, such as scene or object labels, could further improve performance.

Textual Context in Multimodal Emotion Modeling. Researchers are increasingly using textual input to improve visual models' ability to understand emotions, especially when audio or physiological data isn't available. Combining image and text embeddings (from captions or tags) has shown consistent 3–7% performance gains in tasks like sentiment analysis and emotional retrieval [18,20]. These multimodal models typically fuse image and text features before classification. Recent advancements utilize Large Language Models (LLMs) to generate descriptive captions that capture both semantic and emotional nuances, leading to richer emotion representation. However, there's limited research on how different forms of textual context, such as short versus long descriptions, impact evoked emotion classification. The SOTA in this area is EmoVIT [21], which uses visual instruction tuning to fine-tune a Vision Transformer to generate detailed, emotion-centric explanations for images. While powerful, this approach requires a specialized model specifically tuned for emotion explanation.

3   Data and Method

3.1   The EmoSet Dataset

EmoSet [23] is a recent, publicly available benchmark designed specifically for evoked emotion recognition from visual stimuli.[1] Unlike many emotion datasets that focus on portrayed emotion, inferring the emotion of a person within the image, EmoSet targets the affective response elicited in a viewer when observing an image, making it highly suitable for applications in marketing, content analysis, and affective computing.

The dataset includes over 3M natural images sourced from platforms like Flickr and Unsplash, annotated with one of eight discrete emotion labels: *amusement, anger, awe, contentment, disgust, excitement, fear*, and *sadness*. A subset of 118K images has been carefully labeled by human annotators who have passed an "empathy screening". The images in this split have been labeled by 10 annotators each, and only the labels with a 70% agreement rate between the annotators have been kept. Other than the target evoked emotions, the dataset includes also labels for some low and high level attributes that can be helpful for the classification or the understanding of the image, like the brightness and colorfulness of the image or the annotation of scene, objects, facial expressions and human actions (when applicable).

---

[1] https://github.com/JingyuanYY/EmoSet.

3.2   Method

The proposed method consists of three key components (cf., Fig. 1): (1) extraction of visual features via a CNN, (2) generation and embedding of contextual textual descriptions using a multimodal large language model (MLLM) and pre-trained encoders, and (3) a fusion strategy to combine the two modalities for final emotion prediction.

Visual Feature Extraction. For the visual stream, we employ a **ResNeXt-50_32x4d** backbone [22], pre-trained on ImageNet [7]. This network is trained to extract high-level representations that capture spatial and semantic patterns relevant to emotion elicitation. The final classification layer of the pre-trained network is replaced to match our emotion label space. The output of the penultimate layer is used as the 2048-dimensional visual feature vector, denoted $\mathbf{v}_v \in \mathbb{R}^{2048}$.

Textual Context Generation and Embedding. To enrich the dataset with contextual information, we generated both short (caption) and long descriptions for each image using Google's Gemini Flash 2.0 model [11]. These descriptions are designed to capture latent semantic and emotional cues that may not be evident in the raw pixel data. It's also important to note that these description has been obtained in a zero-shot fashion, without prompting the MLLM with examples or reasoning techniques.

We then convert these textual image descriptions into vector representation using a text embedding model, specifically CLIP encoder from the `ViT-B/32` pre-trained model [19]. The final hidden state is used as sentence embedding, which is denoted as $\mathbf{v}_t \in \mathbb{R}^{D_t}$, where D_t is the dimensionality of the text encoder output (e.g., 512 for CLIP). Additional tests with different backbones are reported in Sect. 4.

Multimodal Fusion and Classification. A critical step in our multimodal pipeline is aligning the feature spaces of the visual and textual modalities. The raw embeddings from the text encoder $\mathbf{v}_t$ are optimized for general language understanding, not specifically for our emotion task, and have a different dimensionality than the visual features $\mathbf{v}_v$. To address this issue, we introduce a trainable **projection layer**. This small neural network acts as an adapter, mapping the text embedding into a new space which is both dimensionally compatible with the visual features and more semantically aligned with the emotion recognition task. The projection $\mathbf{v}_t \mapsto \mathbf{v}'_t$ is defined as:

$$\mathbf{v}'_t = \mathrm{LayerNorm}(\mathrm{ReLU}(\mathbf{W}_p\mathbf{v}_t + \mathbf{b}_p)) \tag{1}$$

where $\mathbf{W}_p$ and $\mathbf{b}_p$ are the learnable weight matrix and bias of a linear layer, respectively, and LayerNorm stabilizes the training by normalizing the activations. This projection transforms $\mathbf{v}_t \in \mathbb{R}^{D_t}$ into a projected vector $\mathbf{v}'_t \in \mathbb{R}^{512}$.

The final multimodal feature vector, $\mathbf{v}_f$, is created by concatenating the visual features and the projected textual features:

$$\mathbf{v}_f = \mathrm{concat}(\mathbf{v}_v, \mathbf{v}'_t) \tag{2}$$

This combined vector is then passed to a fully connected classifier, composed of a two-layer MLP (e.g., 1024 -> 512) with ReLU activation, followed by a softmax output layer for the final emotion prediction.

Training Details. As our work addresses a multi-class classification problem, we train the model by minimizing a standard Cross-Entropy loss function. For optimization, we utilized the Adam optimizer [13] with a weight decay of 1×10^{-5} and a batch size of 32. The overall training procedure follows a two-phase fine-tuning protocol designed to stably transfer knowledge from the pre-trained backbones.

Phase 1: Training the Classification Head. We freeze all layers of both the visual (ResNeXt-50) and textual backbones. In this phase, only the randomly initialized classification and projection layers are trained. This crucial step prevents the large, chaotic gradients from these new layers from causing "catastrophic forgetting" and destabilizing the powerful, pre-trained feature extractors. We train this head with a learning rate of 1×10^{-4} until the validation loss plateaus.

Phase 2: Fine-tuning the Backbones. Once the classifier head has converged, we then unfreeze the top-most layers of the backbones. Based on preliminary experiments, we determined that unfreezing the final $M = 4$ residual blocks of the ResNeXt-50 and the top $N = 4$ layers of the text encoder provided an effective trade-off between performance and training efficiency. To gently adapt these newly-trainable layers, the learning rate for this phase is reduced by an order of magnitude to 1×10^{-5}. Training is governed by an early stopping mechanism with a patience of 5 epochs, monitored on the validation loss. We also employ a learning rate scheduler, specifically a 'ReduceLROnPlateau' scheduler with a factor of 0.2 and a patience of 2 epochs, which adjusts the rate based on the same validation metric.

To train our models, we used the manually annotated EmoSet split, retaining its original training, validation, and test sets. This approach allowed for a more precise comparison with other published models, ensuring replicability of results and effectively benchmarking performance. Our primary model is trained using automatically generated short descriptions as text inputs.

Data Augmentation. To prevent both networks from overfitting on the training data, we applied data augmentation techniques to both the visual and textual modalities. For image augmentation, we utilized a pipeline consisting of resizing, random cropping, and AutoAugment [5]. For textual augmentation, we combined two distinct approaches:

- Full-description based paraphrasing: This sentence-based approach leverages a Large Language Model (LLM) to generate paraphrased iterations from the original, comprehensive long caption. This method enhances and diversifies vocabulary and offers the additional benefit of incorporating or highlighting details previously omitted in the initial caption [12].
- Synonym word substitution: In this word-based technique, certain vocabulary items are replaced with synonyms using the WordNet database.

4   Experimental Analysis and Results

In the following, to evaluate our hypothesis that textual context enhances evoked emotion recognition, we focus on top-1 accuracy on the manually labeled EmoSet split, for direct comparison with prior work.

4.1   Ablation Studies and Experiments

To determine the optimal method for integrating visual and textual modalities, we performed a detailed ablation study. We explored unimodal baselines, as well as several fine-tuning approaches. The results are summarized in Table 1.

Table 1. Ablation study on different model backbones and fusion impact.

Model Configuration	Top-1 Accuracy (%)
Unimodal	
Visual-Only (ResNet-50)	74.0
Visual-Only (ResNeXt-50)	76.6
Visual-Only (RegNetY-16)	76.7
Text-Only (CLIP)	77.1
Text-Only (BERT)	76.7
Multimodal	
Fine-tuned RexNeXt-50 + frozen CLIP	80.2
Fine-tuned RexNeXt-50 and CLIP	**81.5**

Choice of Visual and Textual Backbones. The selection of appropriate feature extraction backbones is critical to the performance of our multi-modal framework. We conducted an ablation study to determine the optimal visual and textual backbones, evaluating each modality's performance independently. The results of this analysis are summarized in Table 1.

For the visual branch, we evaluated three prominent CNN architectures: ResNet-50, ResNeXt-50, and RegNetY-16. We included ResNet-50 as it serves as the backbone in the original EmoSet baseline method [23], providing a direct point of comparison. Our findings indicate that the performance of these backbones on our task correlates with their established efficacy on large-scale classification benchmarks. Specifically, RegNetY-16 achieved the highest accuracy, but at the cost of significantly increased training time. Consequently, we selected ResNeXt-50 for our final model, as it presented the most favorable trade-off between performance and computational efficiency.

For the textual branch, we compared two widely-used transformer-based encoders: BERT (specifically, bert-base-uncased [8]) and the text encoder from CLIP (ViT-B/32) [19]. As shown in Table 1, the CLIP text encoder yielded superior performance. Notably, we also observed that it led to faster convergence during the initial training phase. We attribute this advantage to CLIP's contrastive pre-training on 400 million image-text pairs. This process inherently learns text representations that are aligned with visual concepts, providing a more effective initialization for our vision-and-language task than BERT's general-purpose text pre-training. Therefore, we selected the CLIP text encoder as the textual backbone for all subsequent experiments.

Backbones Finetuning Strategies. To determine the optimal depth for fine-tuning our model's backbones, we performed an ablation study during the second training phase. We evaluated how the number of unfrozen layers in both the text and visual encoders affects downstream performance. The results of this analysis are summarized in Table 2. For the CLIP text encoder, we evaluated configurations with 0 (frozen), 1, 2, 4, 8, and all 12 layers unfrozen. As reported in Table 2, performance steadily improves as more layers are fine-tuned, peaking at 8 layers. Fine-tuning the entire 12-layer encoder led to a decrease in performance, a phenomenon we attribute to overfitting given the limited size of our training data. While fine-tuning 8 layers yielded the highest performance, the gain over using 4 layers was marginal. Therefore, to achieve a favorable balance between model performance and training efficiency, we selected the 4-layer configuration for all subsequent experiments. For the visual backbone (ResNeXt), we experimented with unfreezing the top 1, 2, 3, and 4 residual blocks. Our results demonstrate a clear positive correlation between the number of fine-tuned blocks and overall performance, with the 4-block configuration yielding the best results. Consequently, we adopted this strategy for the visual backbone in our final model architecture.

Table 2. Backbones finetuning strategies

Backbone	0	1	2	3	4	8	12
CLIP	74.0	76.3	76.5	–	76.7	76.8	76.3
ResNeXt-50	–	75.1	76.1	76.1	76.6	–	–

Note: Column headers indicate the number of encoding layers (CLIP) or ResNeXt blocks (ResNeXt-50) fine-tuned for the respective models. A dash (–) signifies that the specific configuration or fine-tuning level is not applicable or data are not available for that model.

4.2   Discussion: Overall Performance and SOTA Comparison

The unimodal results reveal two key findings. First, within the visual modality, ResNeXt-50 (76.6%) offers a strong performance baseline without the computational overhead of RegNetY-16 (76.7%). Second, the text-only model using the CLIP encoder (77.1%) surpasses all visual-only models, indicating that the textual descriptions are highly informative and serve as a powerful predictive signal on their own. Crucially, our proposed multimodal approach significantly outperforms every unimodal baseline. Fusing a fine-tuned ResNeXt-50 with a frozen CLIP text encoder boosts performance to 80.2%. By fine-tuning both backbones, our final model achieves a top-1 accuracy of 81.5%. This represents a substantial gain of 4.4% points over the best-performing single modality (Text-Only CLIP), clearly demonstrating the synergistic benefit of integrating visual and textual information.

Table 3. Performance comparison against state-of-the-art models on the EmoSet manual test set (top-1 accuracy on 8 classes).

Method	Core Pipeline/Model	Top-1 Accuracy (%)
Our Work	**ResNeXt-50 + Text Context Embedder**	**81.5**
EmoVIT [21]	InstructBLIP + Emotion Instruction Tuning	83.4
A4Net [24]	ConvNeXt-V2 + Attribute-Aware Fusion	79.5
Fine-tuned CLIP [10]	Fine-tuned ViT-B/16	78.4
EmoSet Baseline [23]	*ResNet-50*	*74.0*

As detailed in Table 3, our proposed multimodal model achieves a top-1 accuracy of 81.5% on the EmoSet test set. This result positions our method as highly competitive among state-of-the-art approaches, surpassed only by EmoVIT [21] (83.4%). Notably, EmoVIT's performance relies on complex LLM reasoning and the use of external low- and high-level attributes (provided by EmoSet, see Sect. 3.1). In contrast, our model's competitive performance, achieved without these explicit attributes, suggests this semantic information can be learned by the vision backbone during training (low level attributes) or captured by MLLM during description extraction.

Error Analysis and Class Confusion. A detailed error analysis reveals a distinct pattern: the majority of misclassifications occur between semantically related classes, such as amusement and awe, or fear and disgust (see Fig. 2a). This pattern points towards a challenge with fine-grained separation rather than a fundamental failure in representation learning. To investigate this further, we evaluated the top-2 accuracy, which reached 94.0%. The significant 12.5% point improvement over top-1 accuracy demonstrates that in many "incorrect" cases, the ground-truth label is the model's second choice. This phenomenon is characteristic of tasks with high inter-class similarity. Indeed, a qualitative review of the EmoSet training data (Fig. 3) reveals numerous instances where visually similar images are assigned different, though semantically proximal, labels by human annotators. We therefore argue that the model's performance should not be interpreted as a failure but as a reflection of the inherent ambiguity in the task. For instance (Fig. 2b), its internal representations correctly capture (as in [23]) a binary categorization of the core-affect dimension of valence (pleasant with 98.0% accuracy, unpleasant 95%). Indeed, the model's behavior on the fine-grained task mirrors the nuanced—and sometimes subjective—distinctions made by the human annotators.

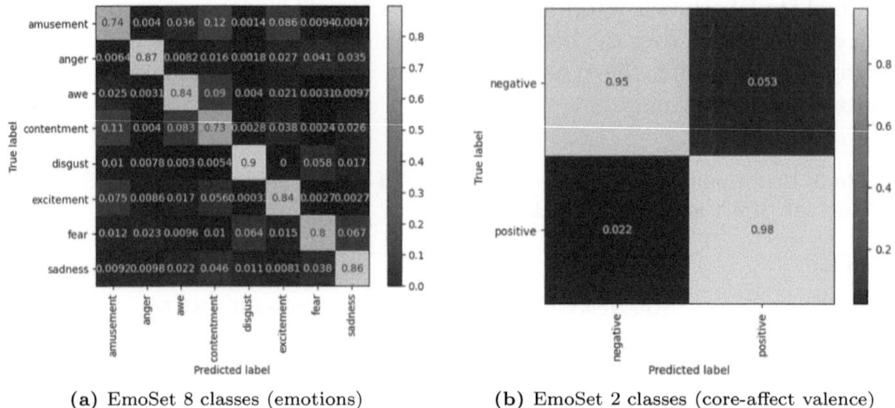

(a) EmoSet 8 classes (emotions)   (b) EmoSet 2 classes (core-affect valence)

Fig. 2. Classification confusion matrix of the model on 8 emotions and 2 classes of the discretised core-affect valence dimension.

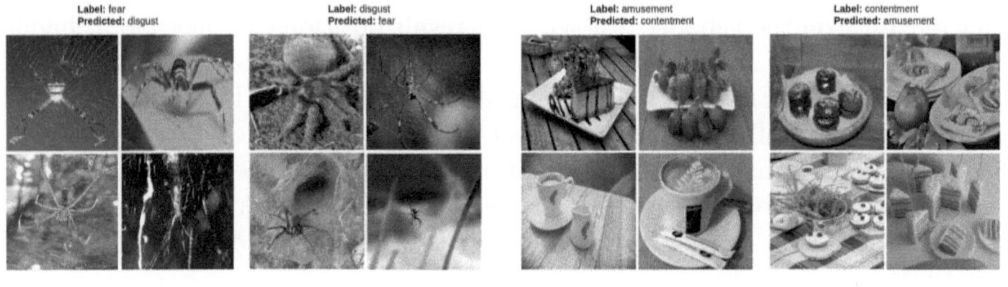

(a) Fear-Disgust confusion.   (b) Amusement-Contentment confusion.

Fig. 3. Confusion examples between similar classes.

5   Conclusion

Our primary contribution lies in empirically demonstrating that even a relatively straightforward combination of visual and textual features can lead to significant performance gains (81.5% top-1 accuracy on the EmoSet manual test set, nearly a 5% point improvement over a strong visual-only baseline) clearly showcasing the power of integrating diverse forms of information. These promising results underscore a critical insight: for complex tasks like evoked emotion recognition, relying solely on visual information is often insufficient (RQ1). The context provided by text, even when machine-generated, offers crucial semantic and narrative cues that help resolve ambiguities in scenes that might otherwise appear visually similar but evoke distinct emotional responses. Our detailed ablation studies further revealed that fine-tuning both the visual and textual components of the model is essential for creating a cohesive and effective joint representation (RQ2). This has significant practical implications, suggesting that future affective computing systems, especially in fields like marketing and human-computer interaction, should be designed with multimodal capabilities from the outset. Further, our error analysis provided a more nuanced view of the model's performance indicating that many of the model's "errors" aren't random misclassifications. Instead, they represent semantically meaningful confusions between closely related emotions, such as disgust and fear. This suggests that our model has learned a feature space that closely mirrors human emotional perception, while also highlighting the inherent ambiguity and subjectivity that are often present in the task of evoked emotion recognition itself.

Despite these promising results, we acknowledge some limitations. The contextual information we used was generated by an LLM; as such, it may not perfectly capture the intricate nuances of human interpretation. An interesting avenue might be the integration of observer's goal-directed attention [6]. Additionally, our study focused solely on visual and textual modalities. Future work should explore the integration of other valuable contextual sources, such as audio, cultural information, or even a viewer's personal background, to build even more robust and personalized emotion recognition systems. Eventually, applying this multimodal framework to real-world datasets from advertising campaigns and exploring multi-label classification to better account for the complex reality of mixed emotional responses.

References

1. Barrett, L.F., Lindquist, K.A., Gendron, M.: Language as context for the perception of emotion. Trends Cogn. Sci. **11**(8), 327–332 (2007)
2. Barrett, L.F., Satpute, A.B.: Historical pitfalls and new directions in the neuroscience of emotion. Neurosci. Lett. **693**, 9–18 (2019)
3. Borth, D., Ji, R., Chen, T., Breuel, T., Chang, S.F.: Large-scale visual sentiment ontology and detectors using adjective noun pairs. In: Proceedings of the 21st ACM international conference on Multimedia, pp. 223–232 (2013)
4. Campos, V., Jou, B., Giro-i Nieto, X.: From pixels to affect: a survey on visual sentiment analysis. IEEE Trans. Aff. Comp. **12**(3), 617–637 (2017)
5. Cubuk, E.D., Zoph, B., Mane, D., Vasudevan, V., Le, Q.V.: Autoaugment: learning augmentation strategies from data. In: IEEE Conf. Comput. Vis. Pattern Recog, pp. 113–123 (2019)
6. D'Amelio, A., Lucchi, M., Boccignone, G.: Scanddm: generalised zero-shot neurodynamical modelling of goal-directed attention. In: Del Bue, A., Canton, C., Pont-Tuset, J., Tommasi, T. (eds.) ECCV 2024 Workshops, pp. 234–244. Springer Nature Switzerland, Cham (2025). https://doi.org/10.1007/978-3-031-91578-9_17
7. Deng, J., Dong, W., Socher, R., Li, L.J., Li, K., Fei-Fei, L.: Imagenet: a large-scale hierarchical image database. In: IEEE Conference on Computer Vision and Pattern Recognition, pp. 248–255 (2009)

8. Devlin, J., Chang, M.W., Lee, K., Toutanova, K.: Bert: Pre-training of deep bidirectional transformers for language understanding. arXiv preprint arXiv:1810.04805 (2018)

9. Ekman, P., Cordaro, D.: What is meant by calling emotions basic. Emot. Rev. **3**(4), 364–370 (2011)

10. Gao, G., Li, C., Wang, Y., Chen, S., Li, Y., Tu, G., Zhang, S.: CLIP-emotion: a zero-shot-based facial and scene emotion recognition. In: Proc. ICASSP, pp. 1–5 (2023)

11. Gemini Team, G.: Gemini: A family of highly capable multimodal models. arXiv preprint arXiv:2312.11805 (2023)

12. Kim, H., et al.: Fine-tuning clip text encoders with two-step paraphrasing. arXiv preprint arXiv:2402.15120 (2024)

13. Kingma, D.P., Ba, J.: Adam: a method for stochastic optimization. arXiv preprint arXiv:1412.6980 (2014)

14. Kosti, R., Alvarez, J.M., Recasens, A., Lapedriza, A.: Context based emotion recognition using emotic dataset. IEEE Trans. Pattern Anal. Mach. Intell. **42**(11), 2755–2766 (2020)

15. Lee, J., Kim, S., Kim, S., Park, J., Sohn, K.: Context-aware emotion recognition networks (2019). https://arxiv.org/abs/1908.05913

16. Limami, F., Hdioud, B., Oulad Haj Thami, R.: Contextual emotion detection in images using deep learning. Front. Art. Intell. **7**(2024) (2024)

17. Machajdik, J., Hanbury, A.: Affective image classification using features inspired by psychology and art theory. In: Proc. 18th ACM International Conference on Multimedia, pp. 83–92 (2010)

18. Niu, T., Zhu, S.J., Pang, L., El-Saddik, A.: Mvsa: a multi-view sentiment analysis dataset. In: Proc. 2016 ACM on Multimedia Companion, pp. 313–318 (2016)

19. Radford, A., et al.: Learning transferable visual models from natural language supervision. In: International Conference on Machine Learning, pp. 8748–8763 (2021)

20. Wang, Y., Wang, W., Huang, C.C.: Multimodal sentiment analysis using hierarchical fusion with context modeling. In: Proc. 2019 International Conference on Multimodal Interaction, pp. 239–243 (2019)

21. Xie, H., Peng, C.J., Tseng, Y.W., Chen, H.J., Hsu, C.F., Shuai, H.H., Cheng, W.H.: Emovit: Revolutionizing emotion insights with visual instruction tuning. In: IEEE Conf. Comput. Vis. Pattern Recog. pp. 26596–26605 (2024)

22. Xie, S., Girshick, R., Dollár, P., Tu, Z., He, K.: Aggregated residual transformations for deep neural networks. In: IEEE Conference on Computer Vision and Pattern Recognition, pp. 1492–1500 (2017)

23. Yang, J., Huang, Q., Ding, T., Lischinski, D., Cohen-Or, D., Huang, H.: Emoset: a large-scale visual emotion dataset with rich attributes. In: International Conference on Computer Vision, pp. 20383–20394 (2023)

24. Yang, S., Wang, Y., Li, Y., Wang, L., Shan, S.: A4Net: attribute-aware affective network for visual emotion recognition. IEEE Trans. Circuit Syst, Video Technol. (2024)

25. You, Q., Luo, J., Jin, H., Yang, J.: Building a large-scale dataset for image emotion recognition: the fine-grained and the abstract. In: AAAI, vol. 30 (2016)

Zero-Shot Evaluation of Commercial Software and State-of-the-Art FER Models on Standardized Datasets

José Salas-Cáceres[1]([envelope]) [iD], Javier Lorenzo-Navarro[1] [iD],
Modesto Castrillón-Santana[1] [iD], Patricia Picazo-Peral[2] [iD],
and Sergio Moreno-Gil[2] [iD]

[1] Universidad de Las Palmas de Gran Canaria, Instituto Universitario SIANI,
Las Palmas de Gran Canaria, Spain
`jose.salas@ulpgc.es`
[2] Universidad de las Palmas de Gran Canaria, Instituto Universitario de Turismo y
Desarrollo Sostenible TIDES, Las Palmas de Gran Canaria, Spain

Abstract. Commercial facial expression recognition tools, such as FaceReader 9©, are often used as off-the-shelf solutions in applied research and industry. However, their real-world generalization capacity, especially in dynamic and unconstrained environments, is rarely scrutinized. This study evaluates the zero-shot performance of FaceReader 9 on two standardized dynamic datasets, RAVDESS and CREMA-D, and compares its results with several publicly available state-of-the-art FER models. The results reveal that FaceReader 9 is significantly outperformed across all metrics, with accuracy levels close to random chance on the more challenging dataset. In contrast, even static models trained on general-purpose datasets perform markedly better, and a dynamic model specifically trained on the evaluation datasets achieves a substantial performance gain. These findings emphasize the limitations of commercial FER systems in dynamic contexts and highlight the value of task-specific training and temporal modeling for robust emotion recognition.

Keywords: Facial Expression Recognition · Biometry · Validation · FaceReader

1 Introduction

Facial Expression Recognition (FER) is a key task in various disciplines, including psychology, marketing, and human-computer interaction [22]. There are multiple methods for collecting information related to human affect. For instance, researchers in fields such as tourism and hospitality often rely on questionnaires or customer reviews to assess users' experiences and emotional responses. However, these approaches are inherently limited, as participants are aware of the data collection process, potentially introducing self-reporting bias.

In contrast, automatic FER systems can capture emotional reactions unobtrusively and in real time [4]. Nonetheless, they introduce significant technical challenges related to data acquisition, model generalization, and processing

E. Rodolà et al. (Eds.): ICIAP 2025 Workshops, LNCS 16169, pp. 41–52, 2026.
https://doi.org/10.1007/978-3-032-11317-7_4

pipelines. To address these complexities, researchers opt to employ commercial integrated software solutions that claim to provide robust emotion analysis capabilities out of the box.

Several commercial platforms offer such solutions, including EmoVU by Eyeris Technology [2], AFFDEX 2.0 by Affectiva [1], and FaceReader by Noldus [24]. These tools typically leverage visual data to estimate attributes such as emotional expression, age group, and gender, among others. They are designed to operate in a zero-shot setting, meaning they do not require retraining on the target dataset prior to deployment. Many of these solutions claimed to achieve an almost perfect accuracy in detecting emotions [24] .However, these commercial systems are often trained on proprietary datasets and employ closed-source architectures, making independent validation and benchmarking difficult. Consequently, external evaluations must be performed using licensed software and standard public datasets, a task not always possible as some of the software are limited to not function in recorded data [13].

In the machine learning community, FER remains an open research problem. Numerous state-of-the-art models have been proposed and benchmarked on publicly available datasets such as RAF-DB [16] and AffectNet [23]. These datasets enable reproducible comparisons between models under standardized evaluation protocols.

In this work, FaceReader 9© (FR) by Noldus is evaluated and compared to several state-of-the-art FER models using a zero-shot setting. The goal is to assess the generalization capabilities of FaceReader and determine its relative performance on dynamic datasets when benchmarked against modern deep learning-based approaches.

2   Related Work

In the field of FER, a common distinction is made between *static* and *dynamic* approaches [17]. This differentiation is based on the temporal scope of the data processed by the model or presented on the dataset. *Static models* operate on individual frames, predicting the emotional state from a single, temporally isolated image. In contrast, *dynamic models* are designed to analyze sequences of frames, generating a single prediction by exploiting temporal information across an entire video segment. The same goes with databases, which portray single frames of an emotion in the static case or a whole sequence of the expression. Although the boundary between these two categories can be fluid, since dynamic datasets can be treated as static by processing each frame independently, it remains an important distinction, particularly when characterizing datasets. Dynamic datasets typically provide richer contextual information, often including synchronized audio, speech transcripts, or motion cues, which are generally absent in static datasets. These additional modalities can be crucial for accurate emotion recognition, especially in realistic scenarios where multimodal approaches can lead to better performance due to the extra information they provide.

Another common categorization applied to FER datasets concerns the recording conditions under which the data were collected. In this regard, datasets are typically classified into two groups [27]: *laboratory-controlled* datasets, such as RAVDESS [18], and *in-the-wild* datasets, such as MELD [29]. The former are captured in highly controlled environments, where lighting, pose, and background are carefully managed to ensure optimal visibility and consistency across samples. These conditions facilitate the extraction of clear and unambiguous expressions. In contrast, in-the-wild datasets are collected under more naturalistic and unconstrained conditions, where variations in recordings are common. While these datasets better reflect the variability of real-world scenarios and often contain more authentic emotional expressions, they also introduce significant challenges for automated recognition systems due to the degraded or incomplete visual information.

As previously mentioned, in this paper we compare the performance of the commercial software FaceReader 9, which, for internal validation, was evaluated on two publicly available datasets: the Amsterdam Dynamic Facial Expression Set (ADFES) [32] and the Warsaw Set of Emotional Facial Expression Pictures (WSEFEP) [26]. According to the developers, the software achieved accuracies of 99.3% and 95.7% on ADFES and WSEFEP, respectively. These high performance levels on controlled datasets have been partially corroborated by external evaluations. For instance, in [14], three commercial FER systems were tested on both highly standardized and in-the-wild datasets. The standardized set was constructed by aggregating four datasets: Karolinska Directed Emotional Faces [21], ADFES, WSEFEP, and the Radboud Faces Database [15], while the in-the-wild set consisted of the Static Facial Expressions in the Wild (SFEW) dataset [7]. The obtained results highlighted a clear performance gap between controlled and real-world conditions. FaceReader 8, for instance, achieved up to 97% accuracy on the standardized dataset but only 31% on the SFEW dataset. Moreover, the study reported several issues related to face detection by FaceReader when applied to non-standardized or uncontrolled data. A broader evaluation of commercial FER tools was presented in [8], where eight systems were tested on a dataset comprising 937 videos from two sources: one capturing spontaneous expressions (UT-Dallas) and another with posed expressions (BU-4DFE). In this evaluation, accuracy scores ranged from 48% to 62%, with FaceReader 7 reaching a True Positive Rate (TPR) of 57.31%. Finally, in [13], a comparative analysis between GPT-4o, a general-purpose large language model with multimodal capabilities, and FaceReader 7 was conducted using the WSEFEP dataset. The results indicated that GPT-4o outperformed the commercial software in the task of facial emotion recognition, even within this highly controlled and standardized environment. These studies suggest that, while commercial FER software such as FaceReader demonstrates high accuracy under ideal, controlled conditions, its generalization to more complex, real-world environments remains limited. This highlights the need for more extensive evaluations of such tools using dynamic and standard datasets.

3   Datasets

This section presents a comprehensive overview of the datasets utilized in this study, detailing their origin, composition, and key characteristics relevant to the experimental design and analysis.

3.1   RAVDESS

The Ryerson Audio-Visual Database of Emotional Speech and Song (RAVDESS) is an audiovisual dataset first introduced in [18]. It comprises recordings of 24 professional actors (12 male and 12 female) performing eight emotional states: neutral, calm, happy, sad, angry, fearful, disgusted, and surprised. The dataset contains a total of 7,356 recordings, with each actor contributing 104 unique clips. Each emotion was portrayed twice, once with normal intensity and once with heightened intensity, except for the neutral emotion, which was recorded only once per actor. The dataset is well-balanced in terms of gender representation, however, there is a predominance of Caucasian individuals, with limited representation of other ethnic groups. RAVDESS is divided into two subsets: one in which actors sing, and another in which they speak. For the purposes of this study, only the speech subset was used, as it more closely resembles realistic interactive scenarios.

As mentioned previously, each emotion was expressed at two different intensity levels. In this study, no distinction was made between the two meaning both were treated as equivalent instances of the same emotional category. All recordings were captured in a professional studio environment, under controlled lighting conditions, using high-quality cameras and microphones, lasting between 3 and 5.5 s. Representative frames illustrating the overall visual quality of the dataset are shown in Fig. 1. RAVDESS is widely used and recognized within the affective computing community, as highlighted in [20].

3.2   CREMA-D

The Crowd-sourced Emotional Multimodal Actors Dataset (CREMA-D) is a comprehensive audiovisual corpus comprising 7,442 clips performed by 91 actors, 48 male and 43 female, as described in [12]. The dataset demonstrates an intentional effort to balance actor demographics in terms of gender, age, and ethnicity.

Each actor was instructed to articulate a predefined set of 12 sentences designed to elicit a diverse range of emotional expressions. The dataset includes six emotion categories, each portrayed at four distinct intensity levels. While CREMA-D provides annotations for emotional intensity, the present study focuses solely on emotion classification. All intensity levels were treated equally, given their relatively uniform distribution across samples.

Although the recordings were conducted under controlled conditions, the overall visual quality of the videos is somewhat limited by the camera's native resolution of 960×760 pixels. Representative frames illustrating this quality are presented in Fig. 2.

Fig. 1. Example of frames seen in RAVDESS. Extracted from [18].

4   Methodology

4.1   Data Preparation

Although FaceReader operates on still images of individual faces, the software is claimed to be capable of analyzing the dynamic progression of emotions in video sequences. For convenience, all video clips corresponding to the same identity were concatenated, resulting in a single continuous video per subject. These aggregated videos were then imported into the FaceReader 9 software. Finally, the output was exported in `.csv` format for further analysis.

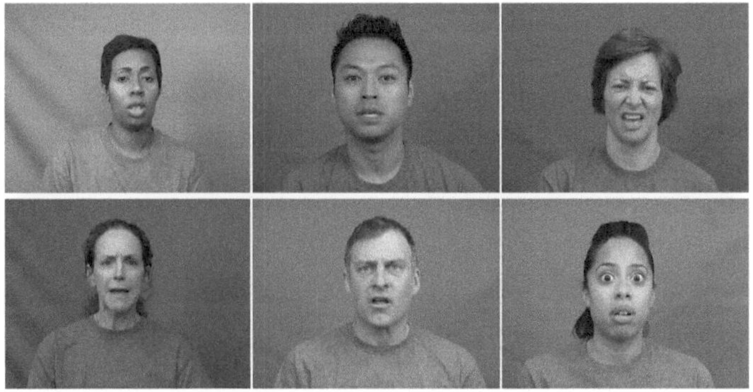

Fig. 2. Example of frames seen in CREMA-D.

For the other tested models, frames were extracted one by one from the concatenated videos. Then, as will be described in the following sections, facial regions were extracted from each image to be processed by the evaluated models.

4.2   Data Pipeline

We follow a processing structure commonly adopted in biometric and affective computing systems, which is also shared by the commercial software under consideration (Fig. 3). This pipeline consists of three main stages:

1. **Face Detection and Alignment**: The initial stage involves detecting and aligning the face within each frame. For the experimental setup presented in this study, face detection was performed using RetinaFace [6] on both datasets to ensure consistent input across models. In contrast, FaceReader employs its proprietary face detection module [19,25,36]. This module has limitations and tends to fail in certain situations where lighting conditions and image quality are suboptimal [3,14]. Because of the impossibility of bypassing this module when using FaceReader, RetinaFace was not applied when testing on it. However, due to the controlled nature of the environments in which the datasets were recorded, this is not expected to pose a significant issue.
2. **Face Modeling**: In the second stage, a feature representation of each face is generated. The approach to feature extraction varies across models; some models compute embeddings using other architectures like DeepFace [33] or VGG face [5], while others operate directly on raw pixel data. FaceReader utilizes a deep neural network described in [11] to generate a synthetic representation of the face, extracting 468 facial landmarks. Subsequently, Principal Component Analysis (PCA) is applied to reduce the dimensionality of this representation [19,25].
3. **Facial Expression Classification**: In the final stage, emotion classification is performed using either the extracted embeddings or the original pixel-based input, depending on the model. Although FaceReader applies PCA to the set of facial landmarks, it is reported that classification is ultimately performed on the facial image itself, following the methodology outlined in [11]. In all models evaluated in this study, the classification task involves recognizing the six basic emotions defined by Ekman in [9], in addition to the neutral expression.

4.3   Evaluation

Since the datasets do not exhibit significant class imbalance, accuracy was selected as the evaluation metric. This metric reflects the overall percentage of correct predictions made by each model.

As previously described, all video clips corresponding to the same identity were concatenated prior to processing by the models and the FaceReader software. This decision introduced a minor challenge: due to the absence of explicit

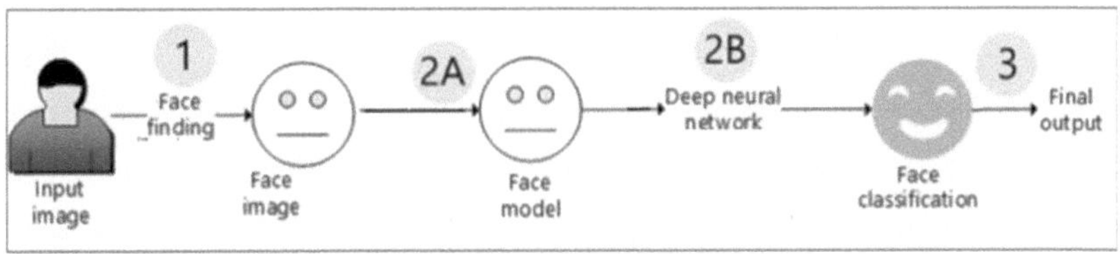

Fig. 3. Facereader9© pipeline to achieve emotion classification. Extracted from the User manual [25].

boundaries between original video segments, some model predictions spanned multiple source videos. This method does not influence the predictions themselves, as none of the evaluated models rely on previous outputs to generate new predictions. Moreover, any potential misalignment can be easily resolved by truncating the predicted emotion intervals to fit within the corresponding video segment. An additional issue, not caused by the concatenation itself but by the static nature of the models, which are designed to process individual frames independently, was that multiple predictions were sometimes generated for a single video segment. To address this issue, the evaluation method illustrated in Fig. 4 was employed. The figure shows three horizontal timelines representing: the ground-truth labels associated with each video segment, the predictions produced by the model, and the final, selected predictions used for evaluation.

The procedure followed to extract the selected predictions is detailed below:

1. First, the temporal boundaries of the current sub-video segment under evaluation are retrieved. For the i-th video segment v_i, this includes the start time t_{si}, end time t_{ei}, and its associated emotion label l_i.
2. Next, all model predictions p_j that satisfy any of the following three conditions are selected. If necessary, their temporal bounds are truncated to ensure they lie within the range $[t_{si}, t_{ei}]$. A prediction p_j, with start and end times t_{sj} and t_{ej} respectively, is retained if:
   - $t_{sj} \geq t_{si}$ and $t_{ej} \leq t_{ei}$: the prediction lies entirely within the sub-video.
   - $t_{sj} \geq t_{si}$ and $t_{ej} \geq t_{ei}$: the prediction starts within the sub-video but extends beyond its end.
   - $t_{sj} \leq t_{si}$ and $t_{ej} \leq t_{ei}$: the prediction starts before the sub-video but ends within its bounds.
3. Finally, among the selected predictions, the one with the longest duration within the interval $[t_{si}, t_{ei}]$ is chosen and treated as the unique prediction for that video segment. Then it is compared to the true label l_i.

5   Results and Discussion

Table 1 presents the accuracy achieved by each model, following the evaluation methodology described in the previous section. In addition, the performance of

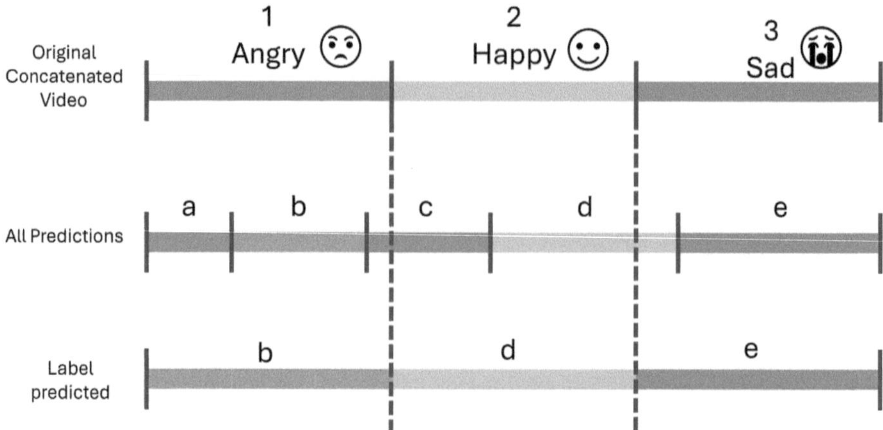

Fig. 4. Graphical explanation of the method used to assign the label to each video.

Table 1. Accuracy achieved by tested models on the RAVDESS and CREMA-D datasets.

Model	RAVDESS (%)	CREMA-D (%)	Trained Dataset
DAN [34]	49.76	39.21	RAF-DB [16]
DAN [34]	60.34	49.07	Affectnet [23]
SwinFace [30]	49.44	42.55	RAF-DB
ResMasknet [28]	53.04	40.88	FER2013 [10]
AP-ViT [35]	44.31	39.17	RAF-DB
FaceReader 9	22.20	17.88	Private database
Embracenet* [31]	88.10	80.27	RAVDESS and CREMA-D

a multimodal dynamic model trained directly on the RAVDESS and CREMA-D datasets is included as a reference.

The results clearly indicate that the commercial software FaceReader underperforms in comparison to all other evaluated models. Specifically, FaceReader achieved an accuracy of 22.20% on RAVDESS and 17.88% on CREMA-D, more than 20% points below the second-lowest performing model in both datasets. The result on CREMA-D is particularly noteworthy, as it is only marginally higher than the expected accuracy of a random classifier (16.66%), suggesting limited predictive capability under these conditions.

Among the tested static models, the best performance was obtained by DAN trained on AffectNet, which achieved 60.34% accuracy on RAVDESS and 49.07% on CREMA-D. However, as expected, the dynamic multimodal model specifically trained on these datasets substantially outperformed the static models, increasing the performance in almost 30% in both cases.

Finally, it is worth noting that, across all models, the performance on RAVDESS consistently exceeded that on CREMA-D. This discrepancy is likely due to the greater complexity and higher intra-class variance present in CREMA-D, which features a larger number of subjects with more diverse ethnicities and backgrounds. This factors contribute to making it a more challenging dataset.

6   Conclusion

This study has presented a comparative evaluation of several state-of-the-art models for FER against a commercial software, FaceReader 9 by Noldus. The evaluation was conducted under zero-shot conditions using two dynamic and publicly available datasets: RAVDESS and CREMA-D. The results demonstrate a clear performance gap between FaceReader 9 and the tested state-of-the-art models, with the commercial software consistently underperforming in both datasets. Specifically, FaceReader 9 achieved an accuracy of only 22.20% and 17.88% in RAVDESS and CREMA-D respectively, performing close to chance level in the latter case and trailing more than 20% points behind the next-lowest scoring model.

These findings suggest that the ability of generalization to dynamic, diverse, and more ecologically valid datasets of FaceReader 9 is substantially limited. In contrast, publicly available SOTA models trained on large-scale datasets such as AffectNet, demonstrated considerably higher accuracy in the same zero-shot setting. Moreover, a dynamic model specifically trained on the RAVDESS and CREMA-D datasets achieved significantly superior results, nearly doubling the performance of the best static models. This highlights the importance of domain adaptation and the value of training on task-specific data when addressing the complexities of FER in real-world scenarios.

Overall, the findings underscore the need for critical assessment of commercial emotion analysis tools when deployed in dynamic and heterogeneous settings. They also reaffirm the efficacy of current research models, which, despite being open-source and freely available, often outperform proprietary solutions in terms of generalization and robustness.

Acknowledgments. This publication is part of the project PID2021-122402OB-C22, funded by MCIN/ AEI/10.13039/501100011033/FEDER, EU, the ACIISI-Gobierno de Canarias and FEDER under project ULPGC Facilities Net and Grant EIS 2021 04, and by the Consejería de Universidades, Ciencia e Innovación y Cultura (Gobierno de Canarias) and the European Social Fund Plus (FSE+) under the funding framework for doctoral research.

Data Availability Statement. The data supporting this research is available in various ways: **CREMA-D**: Data deposited in an open repository (https://github.com/CheyneyComputerScience/CREMA-D). **RAVDESS**: Data deposited in an open repository (https://zenodo.org/records/1188976#.YFZuJ0j7SL8).

References

1. Affdex 2.0: A real-time facial expression analysis toolkit - imotions. https://imotions.com/support/document-library/affdex-2-0-a-real-time-facial-expression-analysis-toolkit/, Accessed 18 June 2025
2. Technology — eyeris. https://www.eyeris.ai/technology, Accessed 18 June 2025
3. Burgess, R., Culpin, I., Costantini, I., Bould, H., Nabney, I., Pearson, R.M.: Quantifying the efficacy of an automated facial coding software using videos of parents. Front. Psychol. **14** (2023). https://doi.org/10.3389/fpsyg.2023.1223806
4. Canal, F.Z., Müller, T.R., Matias, J.C., Scotton, G.G., de Sa Junior, A.R., Pozzebon, E., Sobieranski, A.C.: A survey on facial emotion recognition techniques: a state-of-the-art literature review. Inf. Sci. **582**, 593–617 (2022)
5. Cheng, S., Zhou, G.: Facial expression recognition method based on improved vgg convolutional neural network. Int. J. Pattern Recognit Artif Intell. **34**(07), 2056003 (2020). https://doi.org/10.1142/S0218001420560030
6. Deng, J., Guo, J., Zhou, Y., Yu, J., Kotsia, I., Zafeiriou, S.: Retinaface: single-stage dense face localisation in the wild. CoRR abs/ arXiv: 1905.00641 (2019)
7. Dhall, A., Kaur, A., Goecke, R., Gedeon, T.: Emotiw 2018: audio-video, student engagement and group-level affect prediction. In: Proceedings of the 20th ACM International Conference on Multimodal Interaction, ICMI 2018, pp. 653–656. Association for Computing Machinery, New York (2018). https://doi.org/10.1145/3242969.3264993
8. Dupré, D., Krumhuber, E.G., Küster, D., McKeown, G.J.: A performance comparison of eight commercially available automatic classifiers for facial affect recognition. PLoS ONE **15**(4), e0231968 (2020)
9. Ekman, P., et al.: Basic emotions. Handbook Cognition Emotion **98**(45–60), 16 (1999)
10. Goodfellow, I.J., et al.: Challenges in representation learning: a report on three machine learning contests. Neural Netw. **64**, 59–63 (2015)
11. Gudi, A., Tasli, H.E., den Uyl, T.M., Maroulis, A.: Deep learning based facs action unit occurrence and intensity estimation. In: 2015 11th IEEE International Conference and Workshops on Automatic Face and Gesture Recognition (FG), vol. 06, pp. 1–5 (2015). https://doi.org/10.1109/FG.2015.7284873
12. Keutmann, M.K., Moore, S.L., Savitt, A., Gur, R.C.: Generating an item pool for translational social cognition research: methodology and initial validation. Behav. Res. Methods **47**(1), 228–234 (2015)
13. Kramer, R.S.S.: Identifying basic emotions and action units from facial photographs with ChatGPT. J. Nonverbal Behav. **49**(2), 289–306 (2025)
14. Küntzler, T., Höfling, T.T.A., Alpers, G.W.: Automatic facial expression recognition in standardized and non-standardized emotional expressions. Front. Psychol. **12**, 627561 (2021)
15. Langner, O., Dotsch, R., Bijlstra, G., Wigboldus, D.H.J., Hawk, S.T., van Knippenberg, A.: Presentation and validation of the radboud faces database. Cognition Emotion **24**(8), 1377–1388 (2010). https://doi.org/10.1080/02699930903485076
16. Li, S., Deng, W.: Reliable crowdsourcing and deep locality-preserving learning for unconstrained facial expression recognition. IEEE Trans. Image Process. **28**(1), 356–370 (2019)
17. Li, S., Deng, W.: Deep facial expression recognition: a survey. IEEE Trans. Affect. Comput. **13**(3), 1195–1215 (2022). https://doi.org/10.1109/TAFFC.2020.2981446

18. Livingstone, S.R., Russo, F.A.: The ryerson audio-visual database of emotional speech and song (ravdess): a dynamic, multimodal set of facial and vocal expressions in north american english. PLOS ONE **13**(5), 1–35 (2018). https://doi.org/10.1371/journal.pone.0196391
19. Loijens, L., Krips, O.: Facereader methodology note. White paper, Noldus Information Technology, Wageningen, The Netherlands (2018). https://info.noldus.com/free-white-paper-on-facereader-methodology
20. Luna-Jiménez, C., Kleinlein, R., Griol, D., Callejas, Z., Montero, J.M., Fernández-Martínez, F.: A proposal for multimodal emotion recognition using aural transformers and action units on ravdess dataset. Appli. Sci. **12**(1) (2022). https://doi.org/10.3390/app12010327
21. Lundqvist, D., Flykt, A., Öhman, A.: Karolinska Directed Emotional Faces (KDEF) [Database record. APA PsycTests (1998)
22. Manosso, F.C., Domareski Ruiz, T.C.: Using sentiment analysis in tourism research: a systematic, bibliometric, and integrative review. J. Tourism, Heritage Serv. Marketing **7**(2), 16–27 (2021). https://doi.org/10.5281/zenodo.5548426
23. Mollahosseini, A., Hasani, B., Mahoor, M.H.: Affectnet: a database for facial expression, valence, and arousal computing in the wild. IEEE Trans. Affect. Comput. **10**(1), 18–31 (2019). https://doi.org/10.1109/TAFFC.2017.2740923
24. Noldus: Facereader — facial expression analysis — noldus. https://noldus.com/facereader, Accessed 18 June 2025
25. Noldus Information Technology: FaceReader Reference Manual, Version 9. Noldus Information Technology, Wageningen, The Netherlands (2021). https://www.noldus.com/facereader, version 9; deep learning-based algorithms for face finding, facial modeling, expression and action unit analysis
26. Olszanowski, M., Pochwatko, G., Kuklinski, K., Scibor-Rylski, M., Lewinski, P., Ohme, R.K.: Warsaw set of emotional facial expression pictures: a validation study of facial display photographs. Front. Psychol. **5**(2014) (2015). https://doi.org/10.3389/fpsyg.2014.01516
27. Pan, B., Hirota, K., Jia, Z., Dai, Y.: A review of multimodal emotion recognition from datasets, preprocessing, features, and fusion methods. Neurocomputing **561**, 126866 (2023). https://doi.org/10.1016/j.neucom.2023.126866
28. Pham, L., Vu, T.H., Tran, T.A.: Facial expression recognition using residual masking network. In: 2020 25th International Conference on Pattern Recognition (ICPR), pp. 4513–4519 (2021). https://doi.org/10.1109/ICPR48806.2021.9411919
29. Poria, S., Hazarika, D., Majumder, N., Naik, G., Cambria, E., Mihalcea, R.: MELD: a multimodal multi-party dataset for emotion recognition in conversations. In: Korhonen, A., Traum, D., Màrquez, L. (eds.) Proceedings of the 57th Annual Meeting of the Association for Computational Linguistics, pp. 527–536. Association for Computational Linguistics, Florence, Italy (Jul 2019). https://doi.org/10.18653/v1/P19-1050, https://aclanthology.org/P19-1050/
30. Qin, L., et al.: Swinface: a multi-task transformer for face recognition, expression recognition, age estimation and attribute estimation. IEEE Trans. Circuits Syst. Video Technol. **34**(4), 2223–2234 (2024). https://doi.org/10.1109/TCSVT.2023.3304724
31. Salas-Cáceres, J., Lorenzo-Navarro, J., Freire-Obregón, D., Castrillón-Santana, M.: Multimodal emotion recognition based on a fusion of audiovisual information with temporal dynamics. Multimedia Tools and Applications (2024). https://doi.org/10.1007/s11042-024-20227-6

32. van der Schalk, J., Hawk, S.T., Fischer, A.H., Doosje, B.: Moving faces, looking places: validation of the Amsterdam dynamic facial expression set (ADFES). Emotion **11**(4), 907–920 (2011)
33. Serengil, S., Ozpinar, A.: A benchmark of facial recognition pipelines and co-usability performances of modules. J. Inform. Technol. **17**(2), 95–107 (2024). https://doi.org/10.17671/gazibtd.1399077, https://dergipark.org.tr/en/pub/gazibtd/issue/84331/1399077
34. Wen, Z., Lin, W., Wang, T., Xu, G.: Distract your attention: multi-head cross attention network for facial expression recognition. Biomimetics **8**(2) (2023). https://doi.org/10.3390/biomimetics8020199, https://www.mdpi.com/2313-7673/8/2/199
35. Xue, F., Wang, Q., Tan, Z., Ma, Z., Guo, G.: Vision transformer with attentive pooling for robust facial expression recognition. IEEE Trans. Affect. Comput. **14**(4), 3244–3256 (2023). https://doi.org/10.1109/TAFFC.2022.3226473
36. Zafeiriou, S., Zhang, C., Zhang, Z.: A survey on face detection in the wild: Past, present and future. Comput. Vis. Image Underst. **138**, 1–24 (2015)

Privacy Through Data-Efficiency: Sparse Temporal Difference Videos for Emotion Classification Utilizing Vision Transformers

Fabian Krause[1]([✉]) [iD], Robert Brunstein[2], and Sebastian Stober[2] [iD]

[1] Unmanned Aircraft Systems - Institute of Flightsystems - German Aerospace Center (DLR e.V.), Brunswick 38108, Germany
fabian.krause@dlr.de

[2] AILab - Otto-von-Guericke University Magdeburg, Magdeburg 39106, Germany

Abstract. Video emotion classification often utilizes fully RGB data, which can raise privacy and bias concerns and has an impact on autonomous systems that rely on data-efficient processing. As a remedy, we propose multiple approaches to use temporal difference videos as a sparse alternative representation. We aim to evaluate whether RGB data is needed to achieve computer vision goals. To this end, we first create an RGB baseline on Video Emotion Classification benchmarks with a Vision Transformer for videos. We evaluate the impact of underlying motion versus appearance features and the influence of the temporal position on classification accuracy in RGB material. Second, we introduce and evaluate our different approaches of temporal difference videos as input. Third, we propose an extension with a SSL reconstruction task, which we evaluate in transfer learning. In summary, we demonstrate new possibilities for utilizing temporal difference videos as a sparse alternative to fully colored RGB video data, thereby promoting privacy and data efficiency.

Keywords: temporal difference videos · data efficiency · motion versus appearance

1 Introduction

Emotion Classification is mostly based on combined multimodal features from audio and video data. If video is used as exclusive information source it is used as fully colored RGB data which can raise privacy concerns and practical limitations in data-efficiency. However, video also offers motion, a dynamic, temporal aspect between frames. In contrast to extracting static content from video data [21], we explore the possibility of leveraging such sparse temporal information, i.e. motion data with simple means of difference images. Hereby, we evaluate whether fully colored RGB data is needed to achieve Computer Vision goals in emotion classification. Additionally, we examine whether the motion itself or appearance characteristics are contributing to the classifiers performance. These results should provide an indication towards the possibility to generalize results in different domains or other lightweight data reduction methods.

E. Rodolà et al. (Eds.): ICIAP 2025 Workshops, LNCS 16169, pp. 53–64, 2026.
https://doi.org/10.1007/978-3-032-11317-7_5

We suspect RGB data to be especially unfavorable regarding real-world applications. It is data expensive that can pose a challenge to limited hardware. For example, a companion computer in a drone or a robot is limited by space and its power consumption. Performing an emotion classification to gain information if a person is in distress in a Search and Rescue scenario or to differ it from an emotionless mannequin or scare crow can pose a challenge. Current architectures rely on intermediate compression bottlenecks or modalities other than video to provide solutions [21]. However, other modalities are not necessarily available in a practical setting and architectural bottlenecks could omit necessary details. Besides architectural aspects, RGB data is also unfavorable in the area of distributed and communicative systems. Conventional video streaming uses a high bandwidth that affects the ability of a robot to communicate other information. Furthermore, RGB data can be considered privacy and secrecy invading. A clear facial image of a person as depicted in Fig. 1 'A' allows for an easy recognition. Additionally, a drone might unintentionally pick up information that is supposed to be secret, such as critical infrastructure or transport of dangerous goods. Following the data minimization principle of the General Data Protection Regulation (GDPR), we must consider whether fully colored RGB material is needed to achieve computer vision goals in practical applications.

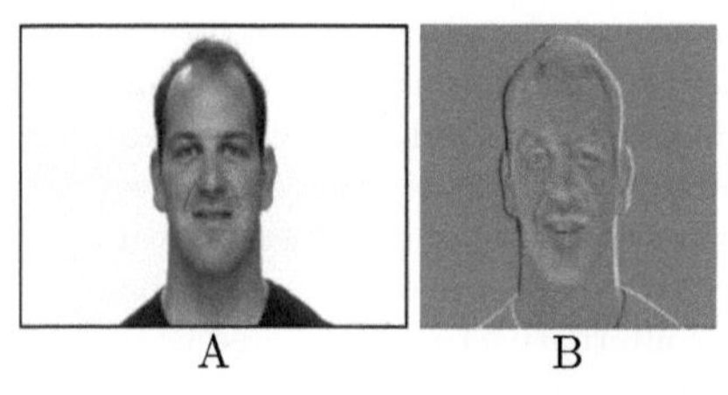

Fig. 1. Samples at different timestamps from a RAVDESS [17] video. A is leaning towards 'neutral' although it has a ground truth of 'happiness'. B displays an exemplary difference image that reduces the characteristic features to shape.

Furthermore, motion can be beneficial for the task itself. Figure 1 'A' highlights an example in Emotion Classification we believe to be improved by including motion. Although an emotion can be captured in a static image, the emotion presents itself as an observable change in behavior, a motion. A single image leaves ambiguity for humans to interpret [2]. By viewing image 'A' in Fig. 1, we could assume a neutral, happy, or sad emotion. However, its ground truth is 'happiness'. Similar observations can be true for object detection or action recognition in a context of drones. Especially in urban environment, current algorithms struggle with appearance based clutter [11]. We could utilize motion as an effective filter in a noisy environment.

We propose temporal difference videos as an alternative to fully colored RGB data. Depending on the motion and temporal difference between two frames, a sparse image is often created by significantly reducing temporal redundancy. Figure 1''B' highlights an example. As a result, it enables potential to reduce needed memory and to highlight specific motion cues. More importantly, a model might be less likely to focus on a bias introduced by appearance or static background. Privacy protection might be strengthened by reducing the exposed data to shape. Although some characteristic features are still present, it could be more difficult for untrained people to identify a person based on shape compared to more recognizable features like hair color, freckles or clothes.

We present various kinds of difference video calculation approaches targeting local and global motion information. As we want to focus primarily on the input, we limit ourselves to a plain architecture and classification task. We explore the potential benefits of our approach initially in video emotion classification as it provides a controlled environment, an established relation to motion and foreign baselines.

2   Related Work

Although a need to differentiate between appearance and temporal information was identified in recent literature [7,12], approaches focus on architectures instead of input representations. To the best of our knowledge, difference images are seldom used beyond the difference of two consecutive frames as a basis for Transformer based deep learning approaches. Instead, sampling and self-supervised approaches are dominantly used for video based tasks.

Approaches utilizing video as a source are often faced with memory size problems. As mitigation, they combine temporal and spatial dimensions by sampling or introducing intermediate compression bottlenecks [21]. Slicing only the temporal domain is one effective way to reduce needed memory [14,15]. However, it focuses solely on temporal resolution and keeps temporal redundancy. Slicing in both domains randomly can be an alternative, as demonstrated with shards by Akbari et al. [1]. By sampling spatial patches first and padding them with temporal close neighbors, one can create cubicles or tublets [3,16]. Both approaches reduce temporal redundancy and reduce exposed data to cutouts. However, they introduce an assumption that either the position of interest is vaguely known in the video or the whole video describes the motion. This assumption could hold in emotion classification, but it fails for drone detection. Slicing and embedding depend on each other. Smaller slices allow for a more minimal embedding network, while a larger slice needs a more complex encoding network to remain computable. Most often, approaches use pretrained models to compress input images in a feature sequence. However, the quality depends on the bottleneck. Contrarily, minimal embedding leaves the interpretation of patches to the backbone architecture, which can determine importance. Empirical studies show both embedding approaches on a similar level [4,13]. However, those studies could suffer from a bias on appearance as discussed by Kumar et al. [14].

A reduction to motion could provide a more privacy and data-efficient approach. However, motion is often targeted as a learning task in Self-supervised learning (SSL) [10,20] and seldom chosen as exclusive input. Optical flow maps are often created or used as an additional representation [7,18]. For example, Lorre et al. [18] evaluates RGB images against optical flow and difference images. They are calculated based on differences between consecutive gray-scaled images. Their experiments show that optical flow performs superior to their difference images and both exceed RGB images. Contrarily, Carreira et al. [6] experiment with different architectures and report optical flow to be significantly worse than RGB. However, their combination as an inflated 3D Convolutional Network architecture exceeds RGB results. As a remedy to optical flow computation assumptions and its computational costs, Tian et al. [22] proposes to learn motion maps directly as a SSL reconstruction task. They argue, that high-level semantics can influence motion cues beyond assumptions like brightness constancy. Nevertheless is the calculation of optical flow an additional and potentially costly step.

3   Approach

We present an overview of our approaches for video classification in the following. The experimental setup for each approach is based on a vanilla Vision Transformer for video and further described in Sect. 4.

3.1   Appearance Based Classification

We examine the impact of fully colored RGB data alone on a transformer architectures performance. Hereby, we establish a baseline to compare our temporal difference image based approaches against. Furthermore, we hypothesize that classification accuracy strongly depends on the appearance features and thus the position of a sampled frame in a video.

We use a prepended class token to predict a class distribution with a single linear layer followed by a softmax activation function. Categorical Cross Entropy serves as our loss function. We establish two baselines. First, we slice a single frame that allows us to examine the influence of appearance in more detail. Second, we slice eight frames from different positions to provide a directly comparable baseline to difference videos. A single frame omits the temporal aspect, whereas eight frames potentially allow the model to learn some degree of motion information from fully colored input. Besides evaluating the overall performance of a trained classifier, we also examine the model's performance on each frame along the temporal axis.

3.2   Difference Images Based Classification

We examine whether we can utilize temporal difference images alone to highlight motion information. Our training approach is similar to Sect. 3.1. However, we explore several options to calculate difference images. Every approach defines a single fixed or relative anchor A. The anchor A is used as a reference point to subtract sampled frames I_k ($I_k^{diff} = A - I_k$). We aim to evaluate the influence of different approaches to calculate difference videos based on a single fixed anchor or temporally close relative anchors. We expect the latter to provide a rather local view on motion features and a fixed anchor to provide a global change. By selecting a fixed anchor, we aim to provide the model with a static reference point to which it can relate. Furthermore, we examine whether the anchor position influences model performance.

First Anchor Inspired by linear predictive coding, this position represents difference images as a constant forward distance with different time steps.

Middle Anchor We select a frame in the middle of the video showing likely characteristic features of an emotion.

Random Anchor We randomly uniform sample an anchor from the whole video. Compared to first anchor or middle anchor, we reduce our prior assumptions in this method. The influence an anchor position can have on final model performance is minimal. As an advantage, this method provides more variety in terms of training samples.

Predecessor We follow existing approaches and utilize frame-to-frame differences. In our terminology, we do not choose a single anchor for the whole video but rather choose for every sampled frame its predecessor frame as anchor.

Aggregated In addition to predecessor, we aggregate single difference images between sampled timestamps and we divide them by the number of aggregated frames. Hereby, we include a lossy compressed version of the whole video in our difference-based model. The magnitude of the compression is determined by the number of frames sampled from the video.

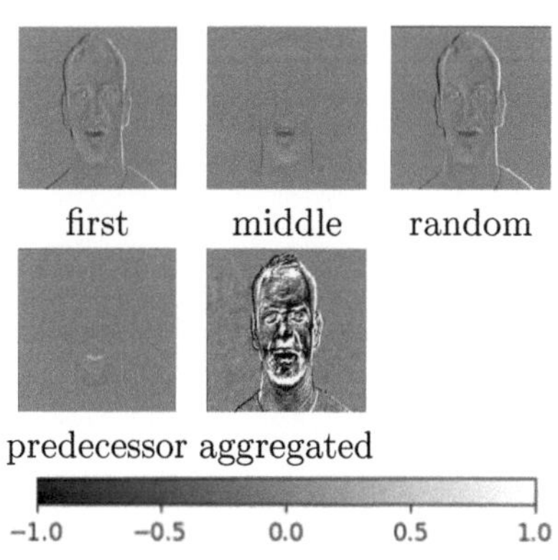

Fig. 2. Exemplary difference images at frame 50

Examples of the calculated difference images from the same timestamp can be seen in Fig. 2. We can generate difference images that show facial features and the shape of a head. In particular, we potentially isolate characteristic movement such as a moving mouth. Depending on the temporal difference towards the anchor, changes might be stronger or weaker. Middle anchor shows less intense facial features. However, this observation is likely caused by the temporal proximity of the sampled frame towards the anchor and not a property of the anchor selection. In contrast, predecessor constantly has minimal but sometimes prominent features. It has similar disadvantages as appearance-based approaches, which may depend heavily on sampling for their performance. In aggregated it is visible that features are more intense compared to our other methods. However, they are not necessarily more characteristic. Although the head itself can be clearly recognized, an eventually specific movement of the mouth, as observable in all previous methods, is more difficult to spot.

3.3   Self-supervised Learning

Additionally to our classification-based loss, we aim to evaluate a SSL based task. As established by differential coding, we can dissect a video into a single RGB image and its subsequent differences images. Inspired by this idea, we use our difference images to reconstruct the anchor image as our SSL task. Given a sequence of difference images, the models is tasked to predict the RGB Anchor. We aim to create a sufficiently challenging but not impossible task. Choosing any arbitrary image from the video as a reconstruction target would likely pose a hard challenge. Our model would miss a prior and could default to a mean image. Staying in the data distribution of difference images could be beneficial. However, such a reconstruction is ill-fitted as difference images are sparse.

Our training approach follows our general approach but we swap the linear layer for a decoder. The class token should provide a valuable and compressed representation. If the subsequent decoder is to create a valid image, the bottleneck needs to provide all information of the targeted image instance. We use a single dense layer prior to a total of five transpose convolution layers. Depending on the targeted resolution, we use combinations of stride two and kernel sizes between three and seven in the first four layers. The final transposed convolution layer has a stride of one and a kernel size of three to create a refined final resolution. We use mean absolute error (MAE) as a loss function in all reconstruction tasks. We found it creates more contoured images than mean squared error (MSE) in initial experiments.

4   Experimental Setup

Our architecture follows the original Vision Transformer [8], but we transfer it to videos similar to Bertasius et al. [4]. Based on our datasets the process includes slicing a specific number of frames from the video, preprocessing them, calculating the difference images, dividing them into non-overlapping patches, and finally embedding them.

We chose CREMA-D [5] (7k videos) and RAVDESS [17] (3k videos) as benchmark datasets. Both datasets allow us to focus on emotion specific motion. They were recorded in a studio scenario with emotions performed by actors. Every emotion is recorded frontal, showing only the actor's face and neck in front of a neutral background. Although the datasets are small compared to other approaches, we argue that it is still suitable for our approach on difference images. Our preprocessing should provide for a sufficient amount of input variations in training. We split our datasets based on the videos into 80% train and 20% test set. As we have only a small amount of data available, we omitted a validation set and repeated our runs multiple times.

We extract n frames out of total N frames of each video with a restricted uniform random sampling along the temporal axis. Our restriction ensures an inclusion of timestamps across the whole video. We first divide the video into $s = \lfloor \frac{N}{n} \rfloor$ equally long segments and then sample within these segments. Our slicing method combines the common approaches of unrestricted uniform and equidistant sampling [21]. It allows for a utilization of most video data while simultaneously ensuring the model receives input along the whole video.

Motivated by our datasets, our preprocessing starts by cutting the video into a quadratic shape as the majority of relevant facial information is in the center. Second, the frames are resized to be 282 pixels in width and height, from which a 256 pixel cutout is randomly extracted during training. At test time, we swap the random crop to a center crop.

We extract all tokens as 2D patches starting from the top left corner in the first frame and continuing to the bottom right corner in the last frame. The tokens are embedded with a minimal linear embedding. Our positional encoding is either learnable or a sine-cosine positional embedding. Following Bertasius et al. [4], we prepend a class token to the sequence, which serves as a compressed representation of the video. It is later used for classification tasks and as a feature vector for our SSL restoration task. We train from scratch as difference images stem from a significantly different input distribution than most models have been pretrained on.

We use a hyperparameter grid search with educated guesses to iteratively train each variation on a NVIDIA A40. We use a training schedule with a linear decay to 1×10^{-6}. An initial warmup is performed with about 5% of the total number of steps. Our default patch size is 16. Our full hyperparameters after the grid search are a 200 Epochs linear learning rate decay from 1×10^{-5} with a warmup of 1000 steps, L2 regularization with a factor of 0.01 and a dropout of 0.02. Our Transformer uses 8-layers and a sine-cosine embedding for RGB input and a learnable embedding for difference images.

5   Evaluation and Discussion

Table 1 provides an overview of our best selected results and available, comparable work. We confirm a hyperparameter configuration and our results to be reliable with multiple runs and display its accuracy mean and standard deviation. Our own baseline is only 1% superior in Crema-D compared to our temporal difference video approaches. In comparison to other approaches, we achieve results that indicate a comparable performance with transformers [9] but lack if the whole video is included [19]. Our approach, which utilizes self-supervised learning with a random anchor, shows successful transfer learning to RAVDESS. In summary, temporal difference video approaches achieve a slightly worse, similar accuracy than colored approaches, while a SSL approach seems beneficial. Furthermore, multiple images perform superior to a single image. Specific aspects are discussed further in the following.

5.1   Single Versus Multiple RGB Images

We suspect the higher test accuracy of multiple compare to single RGB images to be caused by characteristic features showing dependent of the frame position, as hypothesized in Sect. 3.1. To examine our hypothesis, we use our baseline

Table 1. Our final results on difference images with their respective standard deviation. Used modalities (mod) are indicated by audio (a), video (v) and difference images (d). Best results per dataset excluding multimodal approaches are bold.

method	mod	accuracy in %	
		CREMA-D	RAVDESS
single RGB	v	57.7 ± 0.9	–
multiple RGB	v	63.7 ± 0.5	–
first anchor class	d	60.3 ± 0.5	–
random anchor class	d	57.0 ± 0.3	–
random anchor SSL	d	62.1 ± 0.6	$\mathbf{83.6 \pm 4.9}$
Embracement [19]	a+v	80.3	88.1
VGG-face + LSTM [19]	v	**74.0**	82.2
MATER [9]	v	51.4	58.2

models to calculate the predicted class for every available frame position on our test set. Instead of sampling, we calculated classification results on every position ($\approx 105\text{k}$ images) on our test set. Figure 3 shows the resulting mean accuracy at a normalized frame position of 100 frames. Accuracy is clearly dependent of the frame position. Early frames in the first third are significantly worse in their accuracy. We suspect, either the emotion itself has not started yet or it is starting in a state showing neutral, uncharacteristic features and transferring to an aroused state.

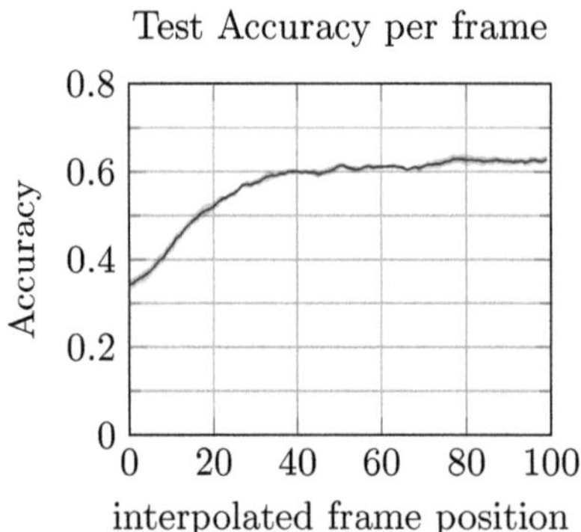

Fig. 3. Mean test accuracy per normalized frame position

Position dependent accuracy provides an explanation for our prior observations of eight frames exceeding a single frame and small deviations in our training behavior. By uniform randomly sampling a single frame from each video in the test set and repeating this process three times, we approximate our test accuracy with about 4% (4500 samples) of our whole population of 105k possible positions. It can occur that one epoch receives qualitatively worse samples than a following epoch. Especially, as we know earlier frames to offer less emotion specific information. By using eight frames, we elevate the amount of input for our model and our random test size. Most importantly, we use a restricted uniform random sampling with multiple frames, as described in Sect. 3.1. Thereby, we ensure an equally distributed coverage of the whole video length. As a result, we can be confident that approximately two third of frame positions are sampled in the last two thirds of the video. These positions show emotion specific features, as displayed in Fig. 3. We conclude that our baseline using eight frames can rely on a more qualitative input. Thereby, eight frames are superior to a single frame because of their sampling and video emotion detection relies in our case on specific characteristic features instead of connecting frames to a motion.

5.2 Difference Images Based Classification

In Table 2 we present our final classification results with the difference between local and global motion information, the anchor positions and our SSL approach. It demonstrates the variant of difference images has an impact on classification accuracy. We are showing the mean and standard deviation of three repeated runs per configuration. We believe our chosen results to be reliable, as they show only a minor standard deviation.

The performance gap of predecessor can be explained by a lower informational content of input images. To illustrate this, we assume samples from a video with 30 fps. By utilizing predecessor computation, we obtain difference images that depict a timestep of approximately 33 milliseconds. While this timespan is great for spotting small changes, it is ill-suited to capture a complex motion that spans over multiple seconds. Eight short changes are not characteristic enough to sufficiently describe a single emotion in our model configuration. As suspected, it suffers similar sampling problems as colored material. We expect an improvement if the number of included images is increased. Aggregated is one option to do so. If we include small changes and build an average over up to 500 milliseconds, we can include significantly more information in a single difference image. Hence, learnable characteristics are more likely present which is supported by aggregated performing significantly better than predecessor. Additionally, we noticed aggregated to smooth outliers. A fixed anchor is beneficial in difference image calculation as it allows for input across the whole video with different temporal steps. Thus, it provides a global view compared to the local view of predecessor. As a result, it can capture the emotion in its whole change of arousal more easily. The results obtained by our variant of a fixed middle anchor strengthens this observation. However, the global view is independent of the anchor position.

Our findings indicate that the model does not learn specifics of the motion itself. It rather focuses on characteristic features and shapes highlighted by the difference images. This indication is supported by first anchor performing superior to all other approaches. Most likely the difference from a rather neutral state in the first frame towards later positions with aroused states emphasizes characteristic features.

5.3 Self-supervised Learning

We show the quality of our SSL approach with a downstream task on first and random anchor in two variants in Table 2. First, we directly test the quality of our feature space with linear probing by swapping the decoder for a single linear layer as classification head. Our model is frozen during training and only

Table 2. Single best results of each variant and their respective standard deviation.

method	accuracy in %	
	CREMA-D	RAVDESS
baseline single RGB	57.7 ± 0.9	
baseline multiple RGB	63.7 ± 0.5	
predecessor	46.6 ± 0.4	
aggregated	53.8 ± 0.8	
middle anchor	54.5 ± 0.9	
random anchor	57.0 ± 0.3	
first anchor	60.3 ± 0.5	
first anchor SSL probing	31.3 ± 0.2	33.3 ± 0.1
random anchor SSL probing	46.6 ± 0.1	42.6 ± 0.2
first anchor SSL tuning	60.1 ± 12.3	75.0 ± 6.9
random anchor SSL tuning	62.1 ± 0.6	83.6 ± 4.9

the linear layer can be adapted. Second, we allow the whole model to adapt its weights (fine-tune). We repeat a training run multiple times, similar to our classification approach in Sect. 5.2. However, we only have to consider different learning rates as hyperparameter in our downstream task. In addition to our downstream evaluation on CREMA-D, we also test our method in transfer learning on RAVDESS. In both datasets fine tuning shows superior results compared to linear probing. We assume our SSL approach is able to provide a suitable initialization of weights, but not necessarily a qualitative feature space. A shift from focusing on appearance in the rgb reconstruction towards utilizing information for classification is needed. Random anchor outperforms first anchor in all cases. We suspect this fact to be caused by the larger variations random anchor was trained with and the reconstruction target. While first anchor is well suited for direct classification, it is in a disadvantage in reconstruction. The first frame of a video shows most often neutral features. As a result, the reconstruction task focuses on neutral features and the general appearance which is not suitable for classification. Random anchor does not have this disadvantage as its reconstruction targets stem from the whole video. Our findings support this statement as in linear probing first anchor is significantly worse than random anchor.

6   Conclusion and Future Work

Temporal difference videos can be utilized for Vision Transformer-based deep learning in emotion classification. By solely using temporal difference videos, we achieved an accuracy close to our baseline on colored RGB material. Therefore, we showed a possibility of enabling a privacy enhancing and data efficient format with a small performance loss on a nonspecialist architecture.

We evaluated temporal difference videos to reduce needed input data in emotion classification for privacy and data efficiency. We used a Vision Transformer for videos to create a baseline in emotion classification as a first step towards utilizing temporal difference videos. Additionally, we showed that accuracy is driven by characteristic features instead of the underlying motion. We evaluated different approaches to calculate difference videos as an alternative representation. Our approach with a fixed anchor in the first position performs sufficiently well. Furthermore, we demonstrated a SSL reconstruction task based on difference videos. Our approach using a random anchor shows positive tendencies towards valuable learned weights. We were able to fine-tune on CREMA-D and RAVDESS rapidly. Additionally, we could successfully transfer learned features to unseen classes in RAVDESS.

In future work, we aim to further evaluate data-efficient video formats as a valuable representation, especially in object detection and action recognition from the point of view of a drone. We have observed that difference videos are often spatially sparse and highlight object shapes. This property can be used to design more efficient and targeted architectures. Furthermore, it could counter appearance clutter and enable training from artificial data with a lower computational effort. Lastly, we want to investigate the relation and potentially transfer learning to event cameras.

References

1. Akbari, H., et al.: Vatt: transformers for multimodal self-supervised learning from raw video, audio and text. In: Advances in Neural Information Processing Systems, vol. 34, pp. 24206–24221. Curran Associates, Inc. (2021)
2. Ambadar, Z., Schooler, J.W., Cohn, J.F.: Deciphering the enigmatic face: the importance of facial dynamics in interpreting subtle facial expressions. Psychol. Sci. **16**(5), 403–410 (2005). http://www.jstor.org/stable/40064238
3. Arnab, A., Dehghani, M., Heigold, G., Sun, C., Lučić, M., Schmid, C.: Vivit: a video vision transformer. In: Proceedings of the IEEE/CVF International Conference on Computer Vision (ICCV), pp. 6836–6846 (October 2021)
4. Bertasius, G., Wang, H., Torresani, L.: Is space-time attention all you need for video understanding? In: Proceedings of the 38th International Conference on Machine Learning. Proceedings of Machine Learning Research, 18–24 Jul, vol. 139, pp. 813–824. PMLR (2021). https://proceedings.mlr.press/v139/bertasius21a.html
5. Cao, H., Cooper, D.G., Keutmann, M.K., Gur, R.C., Nenkova, A., Verma, R.: Crema-d: crowd-sourced emotional multimodal actors dataset. IEEE Trans. Affect. Comput. **5**(4), 377–390 (2014). https://doi.org/10.1109/TAFFC.2014.2336244
6. Carreira, J., Zisserman, A.: Quo vadis, action recognition? a new model and the kinetics dataset. In: Proceedings of the IEEE Conference on Computer Vision and Pattern Recognition (CVPR) (July 2017)
7. Cho, S., et al.: Dual prototype attention for unsupervised video object segmentation. In: Proceedings of the IEEE/CVF Conference on Computer Vision and Pattern Recognition (CVPR), pp. 19238–19247 (June 2024)
8. Dosovitskiy, A., et al.: An image is worth 16x16 words: Transformers for image recognition at scale. ArXiv abs/ arXiv: 2010.11929 (2020)

9. Ghaleb, E., Niehues, J., Asteriadis, S.: Joint modelling of audio-visual cues using attention mechanisms for emotion recognition. Multimedia Tools Appli. **82**, 11239–11264 (2023). https://doi.org/10.1007/s11042-022-13557-w

10. Goncalves, L., Leem, S.G., Lin, W.C., Sisman, B., Busso, C.: Versatile audio-visual learning for emotion recognition. IEEE Trans. Affective Comput., 1–14 (2024). https://doi.org/10.1109/TAFFC.2024.3433386

11. Hartmann, N., Rüter, J., Jünger, F.: Ensuring safety of deep learning components using improved image-level property selection for monitoring. In: AIAA SCITECH 2025 Forum. American Institute of Aeronautics and Astronautics, Inc. (2025). https://doi.org/10.2514/6.2025-2512

12. He, S., Ding, H.: Decoupling static and hierarchical motion perception for referring video segmentation. In: Proceedings of the IEEE/CVF Conference on Computer Vision and Pattern Recognition (CVPR), pp. 13332–13341 (June 2024)

13. Jaegle, A., Gimeno, F., Brock, A., Vinyals, O., Zisserman, A., Carreira, J.: Perceiver: general perception with iterative attention. In: Meila, M., Zhang, T. (eds.) Proceedings of the 38th International Conference on Machine Learning. Proceedings of Machine Learning Research, 18–24 Jul, vol. 139, pp. 4651–4664. PMLR (2021). https://proceedings.mlr.press/v139/jaegle21a.html

14. Kumar, A., Kumar, A., Vineet, V., Singh Rawat, Y.: A Large-Scale Analysis on Self-Supervised Video Representation Learning. arXiv e-prints arXiv:2306.06010 (Jun 2023). https://doi.org/10.48550/arXiv.2306.06010

15. Lei, J., Berg, T., Bansal, M.: Revealing single frame bias for video-and-language learning. In: Proceedings of the 61st Annual Meeting of the Association for Computational Linguistics. pp. 487–507. Association for Computational Linguistics, Toronto, Canada (Jul 2023). https://doi.org/10.18653/v1/2023.acl-long.29

16. Liu, Z., et al.: Video swin transformer. 2022 IEEE/CVF Conference on Computer Vision and Pattern Recognition (CVPR), pp. 3192–3201 (2021)

17. Livingstone, S.R., Russo, F.A.: The ryerson audio-visual database of emotional speech and song (ravdess) (2018). https://doi.org/10.5281/zenodo.1188976

18. Lorre, G., Rabarisoa, J., Orcesi, A., Ainouz, S., Canu, S.: Temporal contrastive pretraining for video action recognition. In: Proceedings of the IEEE/CVF Winter Conference on Applications of Computer Vision (WACV) (Mar 2020)

19. Salas-Cáceres, J., Lorenzo-Navarro, J., Freire-Obregón, D., Castrillón-Santana, M.: Multimodal emotion recognition based on a fusion of audiovisual information with temporal dynamics. Multimedia Tools Appli. (2024). https://doi.org/10.1007/s11042-024-20227-6

20. Schiappa, M.C., Rawat, Y.S., Shah, M.: Self-supervised learning for videos: a survey. ACM Comput. Surv. **55**(13s) (2023). https://doi.org/10.1145/3577925

21. Selva, J., Johansen, A.S., Escalera, S., Nasrollahi, K., Moeslund, T.B., Clapés, A.: Video transformers: a survey. IEEE Trans. Pattern Anal. Mach. Intell. **45**(11), 12922–12943 (2023). https://doi.org/10.1109/TPAMI.2023.3243465

22. Tian, Y., Che, Z., Bao, W., Zhai, G., Gao, Z.: Self-supervised motion representation via scattering local motion cues. In: Vedaldi, A., Bischof, H., Brox, T., Frahm, J.-M. (eds.) ECCV 2020. LNCS, vol. 12359, pp. 71–89. Springer, Cham (2020). https://doi.org/10.1007/978-3-030-58568-6_5

Decoding Facial Expressions in Video: A Multiple Instance Learning Perspective on Action Units

Livia Del Gaudio(✉) [iD], Vittorio Cuculo [iD], and Rita Cucchiara [iD]

University of Modena, Reggio Emilia, Italy
{liviadel.gaudio,vittorio.cuculo,rita.cucchiara}@unimore.it

Abstract. Facial expression recognition (FER) in video sequences is a longstanding challenge in affective computing and computer vision, particularly due to the temporal complexity and subtlety of emotional expressions. In this paper, we propose a novel pipeline that leverages facial Action Units (AUs) as structured time series descriptors of facial muscle activity, enabling emotion classification in videos through a Multiple Instance Learning (MIL) framework. Our approach models each video as a bag of AU-based instances, capturing localized temporal patterns, and allows for robust learning even when only coarse video-level emotion labels are available. Crucially, the approach incorporates interpretability mechanisms that highlight the temporal segments most influential to the final prediction, providing informed decision-making and facilitating downstream analysis. Experimental results on benchmark FER video datasets demonstrate that our method achieves competitive performance using only visual data, without requiring multimodal signals or frame-level supervision. This highlights its potential as an interpretable and efficient solution for weakly supervised emotion recognition in real-world scenarios.

Keywords: Affective Computing · Facial Expression Recognition · Multiple Instance Learning

1 Introduction

Automatic facial expression recognition (FER) has gained increasing attention in the field of computer vision, enabling machines to understand human emotions and interact more naturally with people [25]. Facial expressions are among the most intuitive and natural means by which humans convey intentions and internal states in everyday life [28]. Consequently, systems for automatic FER have seen success across diverse real-world applications, such as human–computer interaction, student engagement detection, and healthcare services [17,31].

The majority of these achievements have been driven by deep neural networks trained on static images, which excel at learning discriminative representations [32,33]. However, facial expressions are inherently dynamic phenomena: they arise from complex muscle activations across different facial regions, and their natural evolution is continuous rather than instantaneous [2,21]. Thus,

E. Rodolà et al. (Eds.): ICIAP 2025 Workshops, LNCS 16169, pp. 65–76, 2026.
https://doi.org/10.1007/978-3-032-11317-7_6

capturing the dynamics of facial expressions over time is essential for a reliable understanding, particularly in video-based FER, which must handle temporal dependencies, pose changes, and occlusions encountered in real-world settings.

In emotion science, facial expressions are canonically described using Action Units (AUs), standardized units defined by the Facial Action Coding System (FACS) [12]. Each AU represents a specific muscle contraction or relaxation, offering a psychologically grounded representation of facial behavior. Working with AU activations enables structured, interpretable analysis of facial dynamics and supports realistic expression synthesis with fine-grained control [6,8].

Despite these benefits, many FER video datasets provide only coarse, video-level emotion labels. Such weak supervision limits the effectiveness of methods that require frame-level annotations. To address this, we propose the adoption of a Multiple Instance Learning (MIL) approach [22], where each video is represented as a bag of temporal segments drawn from AU time series, allowing learning to proceed with only bag-level emotion labels.

An additional feature of this approach is its built-in interpretability in the temporal domain. By operating directly on AU time series and leveraging attention-based aggregation within the Multiple Instance Learning (MIL) framework, the model identifies which temporal segments of a video are most relevant for emotion recognition. This allows for coarse-grained localization of emotionally salient moments without requiring frame-level annotations. Such temporal transparency is especially valuable in domains where understanding the *when* behind a prediction is critical. For example, in mental health settings, temporal interpretability could support clinicians in monitoring emotional variability across time in patients with mood or developmental disorders. In fields like marketing or user experience research, it can help isolate moments in the video where users exhibit strong affective responses to stimuli. Similarly, in education and social robotics, knowing when emotional shifts occur enables systems to respond more appropriately and adaptively to human behavior.

Our contributions are fourfold. First, we introduce a pipeline that models facial dynamics as multivariate time series of Action Units, capturing expressive patterns in video. Second, we adopt a weakly supervised MIL formulation that enables training from coarse video-level emotion labels. Third, we incorporate an interpretable attention mechanism that enables temporal localization of emotion-relevant events within a sequence. Finally, we validate our approach on benchmark video FER datasets, demonstrating both competitive performance and the ability to support qualitative analysis of emotional dynamics in an interpretable manner.

2   Related Work

Facial Expression Recognition in Videos. Facial Expression Recognition (FER) in videos aims to capture emotional dynamics by leveraging both spatial and temporal features. Traditional approaches relied on hand-crafted descriptors such as LBP, HOG, and SIFT, coupled with classifiers like SVMs or Random

Forests [7,34]. However, these struggled with capturing nuanced temporal variations. The rise of deep learning has enabled more robust modeling: CNNs extract spatial features, while RNNs, particularly LSTMs, model temporal dependencies in expression sequences [17]. More recent works explore spatiotemporal architectures such as 3D-CNNs [27,37], and attention-based transformers that learn to focus on salient frames and facial regions. Benchmark datasets like AFEW and the EmotiW challenge [9] have driven progress by offering standardized video-based FER evaluations. Despite these advances, FER in videos remains challenging due to factors like occlusion, head pose variation, and the need for real-time inference [15].

Action Units in FER. Facial Action Units (AUs), derived from the Facial Action Coding System (FACS) [12], provide a comprehensive set of facial muscle movements that can be combined to describe a wide range of expressions. Integrating AUs into FER has enhanced the interpretability and granularity of predictions. Recent works adopt multi-task learning approaches that jointly model expression categories and AU activations [16,19,35], improving the recognition of subtle or compound expressions. By associating neural activations with specific AUs, these models also enhance the transparency of FER decisions. Publicly available datasets like DISFA [23] and BP4D [36] provide AU annotations at the frame level, enabling supervised learning of AU-aware FER systems and facilitating robust benchmarking.

Multiple Instance Learning in Vision. Multiple Instance Learning (MIL) addresses learning scenarios where only weak labels are available at the bag level (collections of instances) rather than at the instance level [11]. MIL is particularly valuable in vision tasks where fine-grained annotation is costly or ambiguous, such as in weakly supervised object detection [30] or in computational pathology for whole slide image (WSI) classification [3,20]. In WSI, each slide is treated as a bag of image patches, and MIL frameworks allow learning from global slide-level diagnoses without requiring pixel-level annotations. In the context of FER, few attempts have been made to adopt MIL approaches, particularly in video or physiological settings where emotional labels are often available only at the clip level [29]. Recent deep MIL approaches use attention or pooling mechanisms to weight the contribution of each instance (*e.g.*, frame) in a bag [13], enabling models to focus on the most informative segments while being trained end-to-end. This approach is well-suited to emotion recognition in unconstrained videos, where expressions are sparse, transitions are gradual, and labeling every frame is impractical.

Interpretability in Affective Computing and FER. Interpretability is a key concern in affective computing, especially for applications in healthcare, education, and human-computer interaction. Traditional black-box models hinder user trust and limit usability in sensitive domains. To mitigate this, many works incorporate attention mechanisms that generate saliency maps or focus on semantically meaningful facial regions [1,24]. AU-based explanations further enhance interpretability by linking predictions to known facial movements, grounding model outputs in human-understandable terms. More recent efforts employ advanced strategies such as ROI-based hierarchical attention [26] and disentangled representations [18], aiming to separate semantic attributes from latent encodings. These methods improve transparency, but balancing accuracy, generalizability, and interpretability remains an open challenge.

3 Proposed Method

3.1 Problem Formulation

Let V denote a video sequence of T frames depicting a human face. Our goal is to classify the video into one of K discrete emotion categories (e.g., happy, angry, neutral). Rather than relying on raw visual features, we use a functional representation based on facial Action Units (AUs), which capture the intensity of muscle activations at each frame. Using an AU extractor, we represent each frame t as a d-dimensional vector $x_t \in \mathbb{R}^d$, where each dimension corresponds to the intensity of a specific AU. The entire video is thus encoded as a multivariate time series:

$$\mathbf{X} = \{x_1, x_2, \ldots, x_T\} \in \mathbb{R}^{T \times d}.$$

Given a dataset $\mathcal{D} = \{(\mathbf{X}^{(i)}, y^{(i)})\}_{i=1}^{N}$, where each video $\mathbf{X}^{(i)}$ is labeled with a single video-level emotion class $y^{(i)} \in \{1, \ldots, K\}$, our objective is to learn a function f_θ that maps each AU time series to an emotion prediction:

$$\hat{y}^{(i)} = f_\theta(\mathbf{X}^{(i)}).$$

The challenge arises from the fact that emotion-relevant cues may only appear in brief temporal segments of the video, while only global labels $y^{(i)}$ are provided. We therefore cast this task as a weakly supervised multivariate time series classification problem under a Multiple Instance Learning (MIL) formulation as recently proposed in TimeMIL [4].

In this framework, each video $\mathbf{X}$ is treated as a *bag* composed of T instances (the AU vectors at each time step), and the label is assigned at the bag level. We assume that for each class $c \in \{1, \ldots, K\}$, the bag is positive for class c if and only if at least one instance within it contains discriminative evidence for that class. Under this multi-class, one-vs-rest MIL setting, the class-wise MIL condition is:

$$y_{i,c} = \begin{cases} 1, & \text{if } \sum_{t=1}^{T} y_{i,t}^{(c)} > 0, \\ 0, & \text{otherwise,} \end{cases}$$

where $y_{i,t}^{(c)} \in \{0, 1\}$ indicates whether the instance at time t provides evidence for class c. These instance-level labels are latent and not provided during training.

The model architecture consists of three components: (i) an instance-level feature extractor f that embeds each AU vector x_t into a latent space, (ii) a

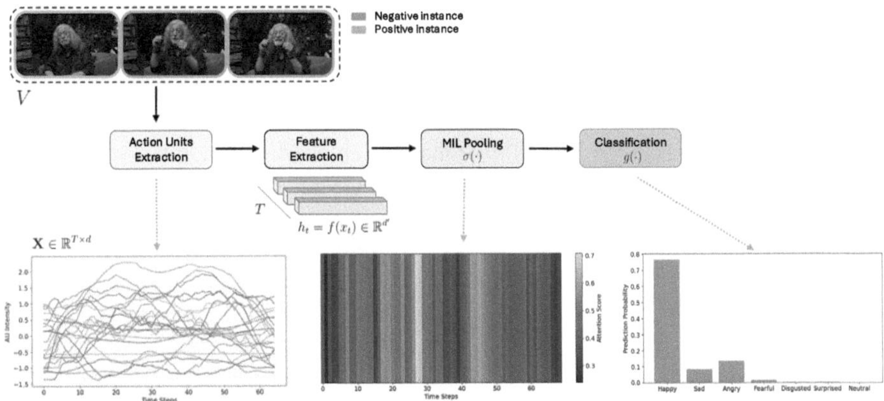

Fig. 1. Overview of the proposed FER pipeline. A video consisting of multiple frames (some negative, some positive instances) undergoes automatic Action Unit (AU) detection to yield a time-series feature matrix. Each temporal slice is mapped via a feature extractor and the embeddings are aggregated via a multiple-instance learning (MIL) pooling operator producing a bag-level representation that is finally classified into emotion categories.

pooling mechanism σ that aggregates the instance embeddings into a bag-level representation $z = \sigma(\{f(x_t)\})$, and (iii) a classifier g that maps z to a probability distribution over classes. The final prediction is:

$$\hat{y} = \arg\max_{c \in \{1,\ldots,K\}} g_c(\sigma(\{f(x_t)\}_{t=1}^{T})).$$

A crucial aspect of this formulation is its capacity for *inherent interpretability*. Because predictions are derived from aggregating instance-level representations, the model can naturally identify which temporal segments (*i.e.*, which subsets of AU activations) are most relevant to a given emotion prediction. When the aggregation function is attention-based, the model produces a weighted importance score for each time step, enabling the localization of emotionally salient moments within the video. These scores can be interpreted as soft evidence of where, in the temporal progression of facial expressions, the model finds discriminative cues. This property is particularly valuable in affective computing applications such as healthcare and behavioral analysis, where both the outcome and the reasoning behind the prediction are important.

3.2   Pipeline Overview

The proposed method performs emotion recognition in videos by modeling facial dynamics as multivariate time series of Action Units (AUs), and classifying each sequence using a weakly supervised, time-aware learning framework. The pipeline is composed of four main stages, as shown in Fig. 1: AU extraction, instance construction, temporal encoding and pooling, and final video-level classification.

1. AU Extraction. Given a video sequence of T frames depicting a human face, we apply a pre-trained facial behavior analysis tool (*e.g.* Py-Feat [5]) to extract frame-wise intensities of a fixed set of $d = 20$ Action Units (AUs), such as those defined in the Facial Action Coding System (FACS). This produces a multivariate time series $\mathbf{X} \in \mathbb{R}^{T \times d}$, where each vector x_t captures the facial muscle configuration at time t.

2. Instance Construction. Rather than processing the entire sequence holistically, we treat each AU vector (or optionally, short overlapping temporal windows) as an instance in a multiple instance learning (MIL) setup. This produces a bag $\mathcal{B} = \{x_1, x_2, \ldots, x_T\}$, where the instances are temporally ordered but only weakly labeled at the bag (*i.e.*, video) level. This setup supports learning in scenarios where only coarse video-level emotion labels are available.

To extract discriminative local temporal features from this AU time series, we employ a convolutional encoder based on an Inception-style architecture. This encoder is designed to operate directly on the multivariate AU sequence and applies multiple one-dimensional convolutions with varying kernel sizes in parallel. This configuration allows the model to capture temporal patterns of differing durations and frequencies, which is particularly important given that emotional expressions can vary in both their temporal dynamics and AU combinations. The output of the Inception block is a sequence of embedded instance vectors $h_t \in \mathbb{R}^{d'}$, one per time step, encoding the localized context and interaction of AUs within short temporal windows.

3. Temporal Encoding and Pooling. The resulting sequence of instance embeddings is then passed to a time-aware MIL pooling module inspired by the TimeMIL [4] framework. In this stage, the sequence is processed by a multi-head self-attention transformer that integrates contextual information across time and enriches each instance representation based on its relationship to other points in the sequence. To preserve temporal ordering, a crucial aspect of emotion expression, the transformer incorporates learnable Wavelet Positional Encodings (WPE), which provide a compact and expressive encoding of relative positions and temporal frequency components. A special learnable class token is appended to the sequence and functions as a global aggregator: it attends to the full set of instance embeddings and accumulates information relevant for the classification.

4. Classification and Interpretability. The final output of the pooling transformer is a fixed-dimensional bag-level representation derived from the class token, which is then passed through a classification head to produce emotion predictions over a fixed set of seven categories: *Angry, Disgust, Fear, Happy, Neutral, Sad,* and *Surprise.* This weakly supervised formulation allows the model to learn discriminative temporal patterns without requiring frame-level or segment-level emotion annotations.

In addition to producing accurate emotion predictions, the model offers inherent interpretability in the temporal domain. Specifically, the self-attention weights associated with the class token, used during pooling, serve as a natural

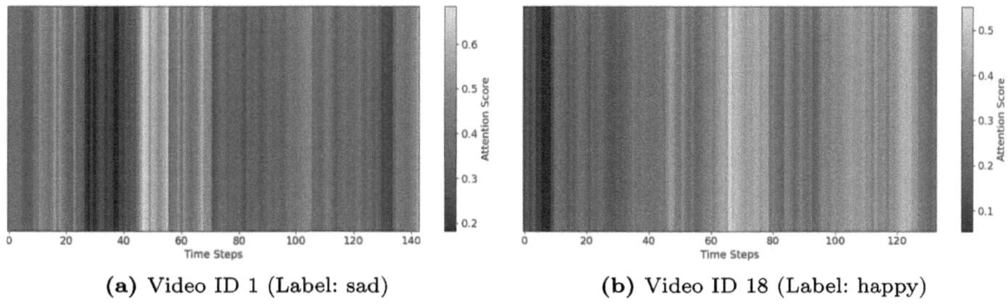

(a) Video ID 1 (Label: sad)   **(b)** Video ID 18 (Label: happy)

Fig. 2. Saliency timelines visualizing the self-attention weights associated with the class token, which localize the emotionally relevant time steps in the facial expression sequences of two videos from the DFEW dataset [14].

mechanism for temporal localization, indicating which time steps in the video most strongly influenced the final classification. These attention weights can be visualized as a saliency timeline, as shown in Fig. 2, highlighting the emotionally relevant segments of the facial expression sequence. While the model does not explicitly disentangle the contribution of individual Action Units (AUs) to the final prediction, its ability to localize key moments over time provides valuable insight into the dynamics of facial expressions, and offers a level of interpretability that aligns with the temporal unfolding of emotions.

To extract meaningful temporal segments from the self-attention weights, we apply a post-hoc analysis to the attention map produced during inference. Specifically, we consider the attention weights assigned by the class token to each time step in the input sequence. Formally, for a given input sequence $X_i = [x_i^1, x_i^2, \ldots, x_i^T]$, the attention mechanism computes:

$$A_i = \mathrm{softmax}\left(\frac{x_i^{\mathrm{cls}} W_h^Q (X_i W_h^K)^\top}{\sqrt{d_k}} \right) \in \mathbb{R}^T$$

where x_i^{cls} is the class token representation, W_h^Q and W_h^K are learned projection matrices for the query and key, and the softmax is applied over the temporal dimension. The resulting attention vector A_i encodes the relative importance of each time step $t \in [1, T]$ in contributing to the final prediction.

To robustly identify emotionally salient intervals, we process A_i through a smoothing and thresholding pipeline. First, the attention vector is normalized and convolved with a moving average followed by a Gaussian filter to reduce noise and emphasize consistent patterns of high attention over time. A dynamic threshold is then computed as the p-th percentile (e.g., 90th) of the smoothed attention values, and time points exceeding this threshold are marked as salient. This procedure yields a set of time intervals during which the model attends most strongly to the input, effectively localizing emotionally informative moments.

To better understand the behavior of the model, we visualize the attention weights over time for a set of test videos. These attention maps, shown in Fig. 3, reveal that the model learns to assign higher importance to segments con-

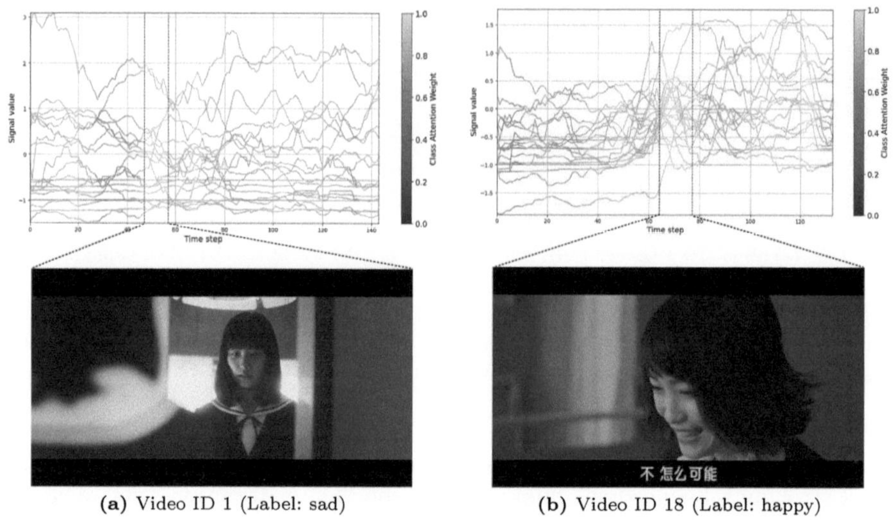

(**a**) Video ID 1 (Label: sad)   (**b**) Video ID 18 (Label: happy)

Fig. 3. Qualitative examples of Action Units timeseries showing time intervals during which the model attends most strongly to the input video frames. Highlighted regions correspond to the most emotionally informative moments localized by the model's attention mechanism.

taining pronounced or prototypical facial expressions, often aligning with peak expression moments in the video. Such attention-based localization is especially valuable in longer sequences with neutral or transitional expressions.

4   Experiments

We evaluate the proposed method on two benchmark datasets for video-based facial emotion recognition and conduct quantitative and qualitative analyses to validate the effectiveness of the AU-based TimeMIL pipeline. The experimental protocol includes ablation studies on pooling and positional encoding mechanisms, and interpretability analyses based on attention mechanism.

4.1   Datasets and Preprocessing

We conduct experiments on the AFEW [10] and DFEW [14] datasets, two widely used benchmarks for dynamic facial expression recognition in the wild. The AFEW (Acted Facial Expressions in the Wild) dataset consists of 773 training and 383 validation clips collected from movies and TV series, labeled with seven emotion categories: *Angry, Disgust, Fear, Happy, Neutral, Sad,* and *Surprise.* Videos in AFEW vary in length, facial pose, lighting, and occlusion, posing challenges typical of real-world emotion recognition. The DFEW (Dynamic Facial Expressions in the Wild) dataset extends this challenge with a larger and more diverse collection of videos sourced from movies and television, offering a richer distribution of emotional content. It contains 16,372 in-the-wild video clips sourced from over 1,500 films, and each clip is labeled with one of seven basic expressions. Clips exhibit challenging variations in illumination, occlusion, and head pose; the recommended five-fold cross-validation protocol is adopted for performance evaluation.

For both datasets, we use Py-Feat [5] to extract frame-level Action Unit (AU) intensities, resulting in a multivariate time series of 20 AU features for each video. To accommodate variable-length videos during training, we adopt dynamic padding based on the batch, where shorter sequences are zero-padded to match the longest sequence in the batch. Corresponding binary masks are generated to exclude the padded regions from self-attention computations. No additional facial landmark features or pixel-level data are used.

4.2   Evaluation Metrics

To assess classification performance across potentially imbalanced emotion categories, we adopt both unweighted average recall (UAR) and weighted average recall (WAR) as evaluation metrics. UAR corresponds to the average of per-class recall values, providing a class-balanced assessment, whereas WAR accounts for class distribution and reflects overall accuracy weighted by class frequencies.

Formally, let C be the number of classes, R_c the recall for class c, and N_c the number of samples in class c. Then:

$$\text{UAR} = \frac{1}{C} \sum_{c=1}^{C} R_c \quad \text{and} \quad \text{WAR} = \frac{1}{N} \sum_{c=1}^{C} N_c \cdot R_c,$$

where $N = \sum_{c=1}^{C} N_c$ is the total number of samples.

4.3   Results and Analysis

To assess the contribution of each component of the proposed architecture, we perform studies on two key design choices: the MIL pooling strategy and the positional encoding mechanism.

First, we compare different MIL pooling strategies: mean pooling, max pooling, attention-based pooling, conjunctive pooling, and the proposed time-aware attention pooling. Results, summarized in Table 1, indicate that pooling choice significantly impacts performance. On AFEW, max pooling achieves the best overall performance (37.86% WAR and 34.34% UAR), while attention-based and time-aware pooling yield competitive results with slightly lower UAR. On DFEW, the proposed time-aware attention pooling outperforms all other methods in terms of WAR (52.01%), while conjunctive pooling achieves the highest UAR (42.05%). These findings suggest that while simpler pooling methods may be sufficient on smaller datasets like AFEW, time-aware strategies offer more consistent benefits on larger and more varied datasets like DFEW, likely due to their ability to model temporal relevance more explicitly.

Table 1. Comparison of different MIL pooling strategies on AFEW and DFEW datasets. Results are reported as Weighted Average Recall (WAR) and Unweighted Average Recall (UAR), in percentage.

Pooling Strategy	AFEW		DFEW	
	WAR (%)	UAR (%)	WAR (%)	UAR (%)
Mean Pooling	35.51	31.30	51.94	40.87
Max Pooling	**37.86**	**34.34**	51.82	40.41
Attention-based Pooling	36.81	33.27	51.38	40.93
Conjunctive Pooling	35.77	33.44	51.91	**42.05**
Time-aware Attn Pooling	36.29	33.86	**52.01**	41.11

Second, we evaluate three types of positional encodings: none (no positional embedding), sinusoidal encodings (as in the original Transformer), and wavelet positional encodings (WPE). As shown in Table 2, the model with WPE yields the best performance on the AFEW dataset (39.16% WAR and 36.47% UAR), suggesting that the multi-resolution properties of wavelet-based encoding better capture the temporal dynamics of short, expressive sequences. Conversely, on the larger and more diverse DFEW dataset, the sinusoidal encoding performs best (52.53% WAR and 42.23% UAR), indicating its robustness and generalizability. These results highlight the importance of selecting a positional encoding scheme that aligns with the temporal complexity and scale of the input data.

Table 2. Performance comparison of different positional encoding strategies on AFEW and DFEW datasets.

Positional Encoding	AFEW		DFEW	
	WAR (%)	UAR (%)	WAR (%)	UAR (%)
None (no PE)	37.86	34.34	52.01	41.11
Sinusoidal PE	36.03	32.84	**52.53**	**42.23**
Wavelet PE (WPE)	**39.16**	**36.47**	51.03	40.58

5 Conclusion

In this work, we introduced a novel pipeline for emotion recognition in videos that models the temporal evolution of facial dynamics using Action Units (AUs) as time series features. Our approach leverages a Multiple Instance Learning (MIL) framework which integrates contextual information across time. By construction, the output is inherently interpretable in the time domain, as it enables accurate localization of the segments most relevant to the classification decision. Experimental results on two in-the-wild FER datasets demonstrate competitive performance and support the effectiveness of modeling facial dynamics as time series for emotion recognition. Future work will focus on incorporating techniques that can also disentangle the contributions of individual AUs to the final predictions and integrate multiple modalities.

Acknowledgments. We acknowledge ISCRA for awarding this project access to the LEONARDO supercomputer, owned by the EuroHPC Joint Undertaking, hosted by CINECA.

References

1. Albanie, S., Nagrani, A., Vedaldi, A., Zisserman, A.: Emotion recognition in speech using cross-modal transfer in the wild. In: ACM MM (2018)
2. Boccignone, G., Bodini, M., Cuculo, V., Grossi, G.: Predictive sampling of facial expression dynamics driven by a latent action space. In: SITIS 2018 (2018)
3. Campanella, G., et al.: Clinical-grade computational pathology using weakly supervised deep learning on whole slide images. Nat. Med. **25**(8), 1301–1309 (2019)
4. Chen, X., et al.: Timemil: advancing multivariate time series classification via a time-aware multiple instance learning. In: ICML (2024)
5. Cheong, J.H., Jolly, E., Xie, T., Byrne, S., Kenney, M., Chang, L.J.: Py-feat: Python facial expression analysis toolbox. Affect. Sci. **4**(4), 781–796 (2023)
6. Cuculo, V., D'Amelio, A.: OpenFACS: an open source FACS-based 3D face animation system. In: Zhao, Y., Barnes, N., Chen, B., Westermann, R., Kong, X., Lin, C. (eds.) ICIG 2019. LNCS, vol. 11902, pp. 232–242. Springer, Cham (2019). https://doi.org/10.1007/978-3-030-34110-7_20
7. Cuculo, V., Lanzarotti, R., Boccignone, G.: Using sparse coding for landmark localization in facial expressions. In: EUVIP 2014, pp. 1–6 (2014)
8. Cuculo, V., Lanzarotti, R., Boccignone, G.: The color of smiling: computational Synaesthesia of facial expressions. In: Murino, V., Puppo, E. (eds.) ICIAP 2015. LNCS, vol. 9279, pp. 203–214. Springer, Cham (2015). https://doi.org/10.1007/978-3-319-23231-7_19
9. Dhall, A., Goecke, R., Ghosh, S., Joshi, J., Sikka, K.: Emotiw 2019: video and group emotion recognition challenges. In: Proceedings of ACM ICMI (2019)
10. Dhall, A., Goecke, R., Joshi, J., Sikka, K., Gedeon, T.: Emotion recognition in the wild challenge 2014: baseline, data and protocol. In: Proceedings of ACM ICMI, pp. 461–466 (2014)
11. Dietterich, T.G., Lathrop, R.H., Lozano-Pérez, T.: Solving the multiple instance problem with axis-parallel rectangles. Artif. Intell. **89**(1–2), 31–71 (1997)
12. Ekman, P., Friesen, W.V.: Facial Action Coding System: A Technique for the Measurement of Facial Movement. Consulting Psychologists Press (1978)
13. Ilse, M., Tomczak, J., Welling, M.: Attention-based deep multiple instance learning. In: ICML (2018)
14. Jiang, X., et al.: Dfew: a large-scale database for recognizing dynamic facial expressions in the wild. In: ACM MM (2020)
15. Kahou, S.E., Michalski, V., Konda, K., Memisevic, R., Pal, C.: Recurrent neural networks for emotion recognition in video. In: Proceedings of ACM ICMI (2015)
16. Kollias, D., Zafeiriou, S., et al.: Deep affect prediction in-the-wild: aff-wild database and challenge, deep architectures, and beyond. Int. J. Comput. Vision **127**(6–7), 803–830 (2019)
17. Li, S., Deng, W.: Deep learning for video-based facial expression recognition: a survey. IEEE Trans. Affect. Comput. (2020)
18. Lin, J., Pan, S., Lee, C.S., Oviatt, S.: An explainable deep fusion network for affect recognition using physiological signals. In: Proceedings of the 28th ACM International Conference on Information and Knowledge Management (2019)

19. Liu, Y., Zhang, X., Zhou, J., Fu, L.: SG-DSN: a semantic graph-based dual-stream network for facial expression recognition. Neurocomputing **462**, 320–330 (2021)
20. Lu, M.Y., Williamson, D.F., Chen, T.Y., Chen, R.J., Barbieri, M., Mahmood, F.: Data-efficient and weakly supervised computational pathology on whole-slide images. Nat. Biomed. Eng. **5**(6), 555–570 (2021)
21. Mai, X., et al.: All rivers run into the sea: unified modality brain-inspired emotional central mechanism. In: ACM MM (2024)
22. Maron, O., Lozano-Pérez, T.: A framework for multiple-instance learning. In: Advances in Neural Information Processing Systems, vol. 10 (1997)
23. Mavadati, S.M., Mahoor, M.H., Bartlett, K., Trinh, P., Cohn, J.F.: Disfa: a spontaneous facial action intensity database. IEEE Trans. Affect. Comput. **4**(2), 151–160 (2013)
24. Mollahosseini, A., Hasani, B., Mahoor, M.H.: Affectnet: a database for facial expression, valence, and arousal computing in the wild. IEEE Trans. Affect. Comput. **10**(1), 18–31 (2017)
25. Picard, R.W.: Affective Computing. MIT Press (2000)
26. Sun, X., Xia, P., Zhang, L., Shao, L.: A ROI-guided deep architecture for robust facial expressions recognition. Inf. Sci. **522**, 35–48 (2020)
27. Thuseethan, S., Rajasegarar, S., Yearwood, J.: DEEP3DCANN: a deep 3DCNN-ANN framework for spontaneous micro-expression recognition. Inf. Sci. **630**, 341–355 (2023)
28. Vinciarelli, A., et al.: Bridging the gap between social animal and unsocial machine: a survey of social signal processing. IEEE Trans. Affect. Comput. **3**(1), 69–87 (2011)
29. Wang, H., et al.: Rethinking the learning paradigm for dynamic facial expression recognition. In: CVPR (2023)
30. Wang, X., Yan, Y., Tang, P., Bai, S., Li, X.: Revisiting multiple instance neural networks. In: AAAI (2018)
31. Wang, Y., et al.: A survey on facial expression recognition of static and dynamic emotions. arXiv preprint arXiv:2408.15777 (2024)
32. Wang, Y., et al.: MGR 3 net: multigranularity region relation representation network for facial expression recognition in affective robots. IEEE Trans. Industr. Inf. **20**(5), 7216–7226 (2024)
33. Yang, W., Yu, J., Chen, T., Liu, Z., Wang, X., Shen, J.: Multi-threshold deep metric learning for facial expression recognition. Pattern Recogn. **156**, 110711 (2024)
34. Zeng, Z., Pantic, M., Roisman, G.I., Huang, T.S.: A survey of affective recognition methods: audio, visual, and spontaneous expressions. IEEE TPAMI **31**(1), 39–58 (2009)
35. Zhang, J., Che, W., Bai, Y., Liu, Q.: MFDNET: multi-feature fusion and dual loss network for facial expression recognition. Inf. Sci. **515**, 191–202 (2020)
36. Zhang, X., et al.: BP4D-spontaneous: a high-resolution spontaneous 3d dynamic facial expression database. Image Vis. Comput. **32**(10), 692–706 (2014)
37. Zhao, S., et al.: A two-stage 3D CNN based learning method for spontaneous micro-expression recognition. Neurocomputing **448**, 276–289 (2021)

A Pilot Study Exploring the Alignment of Humans and CNN During Perception of Social Interactions

Guido Vallarino[1], Lucia Schiatti[2,3,6](✉), Matteo Moro[1], Yen-Ling Kuo[4], Mengmi Zhang[5], Monica Gori[2], Boris Katz[6], Andrei Barbu[6], and Alessio Del Bue[3]

[1] DIBRIS & MaLGa Center, University of Genoa, Genova, Italy
[2] UVIP, Istituto Italiano di Tecnologia, Genova, Italy
[3] PAVIS, Istituto Italiano di Tecnologia, Genova, Italy
`lucia.schiatti@iit.it`
[4] Computer Science Department, University of Virginia, Charlottesville, VA, USA
[5] Deep NeuroCognition Lab, Nanyang Technological University, Singapore, Singapore
[6] CSAIL & CBMM, Massachusetts Institute of Technology, Cambridge, MA, USA

Abstract. As Artificial Neural Networks (ANNs) reach human and super-human performance, the convergence between the visual strategies learned by ANNs and human perception is a fundamental step to allow their deployment as models of biological vision and their integration into assistive and rehabilitation technologies. In this work, we investigate the alignment between visual strategies learned by an ANN, specifically the Temporal Shift Module (TSM), and human gaze-based visual attention, during classification of social interactions from videos. We present preliminary results comparing the similarity between saliency maps generated by the ANN model after fine-tuning on the social interaction classification task and ground-truth fixation maps of human gaze data.

Keywords: Visual Attention · Social Interactions · ANNs

1 Introduction

The success of Artificial Neural Networks (ANNs) in solving a range of visual tasks at a human-like level raised interest in exploring whether visual strategies learned by deep networks are aligned with those learned by humans [9]. In fact, such an alignment is a desirable property to exploit ANNs as interpretable working models in the context of cognitive science [17], as well as to facilitate their integration into assistive applications [20]. In this regard, given that some ANNs are known to resemble, to a good approximation, the human visual cortex [21], the capability of correlating models' outcomes to humans' neural and behavioral data holds the potential to exploit ANNs as in-silico testbeds to model the effect of sensory and neuro-developmental disorders (*e.g.*, visual issues and autism) on vision-related competencies, in the developing child. Such an application is particularly intriguing when considering complex visuo-cognitive abilities, interlinked with the development of social cognition and social interaction

© The Author(s), under exclusive license to Springer Nature Switzerland AG 2026
E. Rodolà et al. (Eds.): ICIAP 2025 Workshops, LNCS 16169, pp. 77–88, 2026.
https://doi.org/10.1007/978-3-032-11317-7_7

skills. The latter are crucial for human overall development and are known to be affected by several atypical developmental conditions.

In this work, our goal is to explore the potential of ANNs, and, in particular, Convolutional Neural Network (CNN) models, to be used in the context of assessment and rehabilitation of social skills, by measuring their alignment with human visual social attention, *i.e.*, visual attention during perception of social interactions from videos. The long-term goal is to investigate which ANN architectures best support the emergence of cognitive abilities, with a specific focus on social cognition. Indeed, such knowledge is required to effectively adopt ANNs in the context of assistive and rehabilitation technologies, *e.g.*, to simulate the effect of sensory and cognitive impairments on visual attention and visuo-cognitive skills, as well as to predict the outcomes of suitable training strategies. To compare ANN models' alignment to humans' visual strategy when perceiving social interactions, we designed a cognitive test and collected Eye-Tracking (ET) data from 9 human participants, who were asked to watch videos and classify the type of observed social interaction, among "help", "hinder", and "no interaction". We provide preliminary results about the similarity between a CNN-based architecture and humans during perception of social interactions from goal-oriented actions. In particular, we compare humans' and model's classification performance and visual strategy (*i.e.*, saliency maps). To this aim, we employ a CNN model, the Temporal Shift Module (TSM) [23], selected for its ability to model temporal information relevant to social interactions, and we extract visual explanations from the ANN via Gradient-based Localization (Grad-CAM) [22].

Our work's main contribution is experimental. Indeed, while we use existing computational architectures and evaluation metrics, we: i) introduce a novel visual task, *i.e.*, classification of goal-oriented social interactions, to investigate model-human alignment, and ii) we illustrate a methodological approach to design experiments that can be used to test visuo-cognitive abilities of humans and ANNs, in parallel. To the best of our knowledge, although the problem of ANN-human alignment has been previously tackled in visual tasks such as object recognition [8], this is the first study focusing on visual perception of social interactions. A further contribution is that we focus on videos rather than images as model's input, given that complex dynamic stimuli are more likely to be suitable within the assessment of humans' social interaction skills, considering, for instance, applications in the context of visual and cognitive rehabilitation.

2   Related Work

2.1   ANNs for Assistive and Rehabilitation Technologies

Current ANNs are trained to solve vision-related tasks which are also relevant in the assessment and rehabilitation of sensory issues (*e.g.*, visual impairments) and cognitive disorders (*e.g.*, autism). Furthermore, it has been shown that the internal representation of ANNs resemble, to a first approximation, primate neural recordings [21]. Previous work highlighted how the development of computer vision models that are also neurally aligned, and that exploit human-like visual strategies to solve visual tasks, could be crucial not only for their interpretability, but to facilitate their effective integration into clinical applications [20]. Several assistive applications of computer vision currently adopt a so-called sensory substitution paradigm, replacing humans (*e.g.*, visually impaired individuals) in performing specific tasks, rather than improving their ability to access and process visual information [2]. Conversely, a human-model similarity is desirable to exploit ANNs' potential as clinical rehabilitation rather than substitution tools. Indeed, neurally and human-aligned models could predict the effect of perceptive and cognitive issues on a range of vision-related functional tasks, including higher-level competencies, such as social cognition and social skills, and to plan and monitor rehabilitation interventions, accordingly. Towards this goal, previous works investigated, for instance, the correspondence between the hierarchical information flow in an ANN and that of the ventral visual system, observed with magnetic resonance imaging and electrocorticography [29]. Similarly, different ANN-based architectures applied in the computer vision domain, *i.e.*, CNN and Transformer, could be investigated to test their alignment to neural and behavioral observations, in solving relevant visual and visuo-cognitive tasks.

2.2   Alignment of ANN and Human Strategies During Visual Tasks

Recently, several works started to investigate the similarity between visual strategies learned by models and those adopted by humans [9]. In [28], a systematic evaluation of deep features across different architectures and tasks is presented. The authors found that deep features are more effective than classic metrics in modeling low-level perceptual similarity, thus suggesting that the latter is an emergent property shared across deep visual representations. In [27], the role of context information in visual recognition was investigated, benchmarking experiments with ANNs against psychophysical data collected from humans. In [26], a biologically inspired ANN was presented to approximate the human mechanisms integrating bottom-up (stimuli-driven) and top-down (task-dependent) signals during visual search. Performance was evaluated against human behavior assessed via eye-tracking. To cope with the lack of shared strategies for a quantitative assessment of the model's human-like behavior, [15] introduced a test benchmark enriched with human annotations to measure the AI-human visual alignment (i.e., abstaining from making incorrect decisions and wavering between correct prediction and abstention) and the model's reliability, for image classification tasks.

In addition to the evaluation of emerging human-like strategies against ground-truth data from human experiments, a different branch of literature investigates the use of human data in the model training loop, to directly guide a human-like behavior [7,19]. For instance, [8] evaluated 84 state-of-the-art ANNs with respect to human strategies for object recognition and found that they are less aligned with humans as the categorization accuracy improves. To cope with this, they introduced a training routine (called neural harmonization) to automatically align the model's and humans' visual strategies. In the present work, we focus on the first approach, using human gaze data exclusively as a ground-truth to assess the similarity of the model's "visual attention" emerging during the fine-tuning to learn to classify categories of social interactions.

2.3 Modeling Social Interactions

Although recognizing social interactions is a capability often attributed to higher-level social cognition, such as theory of mind, recent work in cognitive science suggests that the core of social interaction is visual, i.e., the recognition of sophisticated relationships, such as social interactions, occurs through automatic visual processing [10]. This evidence is supported by the fact that recognition of social interactions can be well modeled by bottom-up computational algorithms, and corroborated by neuroscientific findings that underpin the representation of social interactions in the visual cortex [16]. The hypothesis, introduced by [16], that the human visual system represents even higher-level core components of social interaction, such as goal compatibility among agents (help vs. hinder) paves the way to the exploration of ANN-human alignment beyond classic visual tasks (*e.g.*, object recognition), as in the case for the task presented in this paper, i.e., classification of social interactions from videos. Based on the assumption that core components of social interactions are represented in the visual system, and therefore they could be automatically perceived from simple visual cues rather than requiring complex social inference and visual understanding processes, we question whether such representations could be learned automatically by ANNs.

The literature related to computational models of social understanding relies on a very broad definition of "social", typically focusing on tasks such as group activity recognition or gaze direction prediction. Anyhow, previous attempts to model complex social interactions among agents were made, for instance, using a Bayesian inference approach to detect help vs. hinder [25], or exploiting Markov Decision Processes to incorporate social concepts into agents' goals [24]. In [18], the first dataset of physically-grounded abstract social events, validated with human experiments, was introduced as a benchmark supporting the development of models able to recognize complex interactions. The design of such a dataset relies on the evidence that rich interactions can be evoked in human perception even by simple geometric shapes that move independently. Following this line of research, here we investigate the visual processing strategies emerging from a CNN model trained to perceive different types of social interactions, *i.e.*, help and hinder, and their alignment with humans.

3 Methods

3.1 Dataset

The dataset used in this work comprises: i) a set of videos depicting social interactions between two virtual agents performing goal-oriented actions (see Fig. 1), and ii) gaze data collected from an experiment with human participants, who were asked to watch videos and classify the type of social interaction.

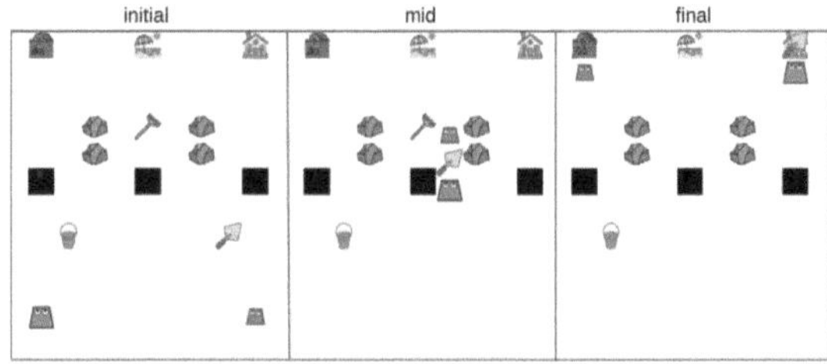

Fig. 1. The initial, middle, and final frame of an "help" video. The mid frame displays the green agent helping the blue one to carry the trowel to the final location. (Color figure online)

Social Interaction Videos. The dataset includes 108 videos depicting 11×11 2D maps, with a resolution of 700×700 pixels. Each map includes three objects and three locations, and two agents performing a distinct goal-oriented action (i.e., carrying one object to one location) out of three possible object/location pairs (see Fig. 1). One agent (the green one) can be also assigned with a social goal, corresponding to a social behavior toward the blue agent out of three options: help, hinder, or no interaction. Baseline physical trajectories were generated automatically using breadth-first search (BFS) with preferred order to break the tie. These were manually modified to generate help and hinder trajectories. A total of 108 videos (36 for each social interaction class) was obtained by different combinations of maps, and agents' physical and social goals.

Gaze Data. To assess the human-model similarity during perception of social interactions, we collected gaze data from 9 healthy adults (average age: 30.6 ± 9.3) during an eye-tracking experiment. The experiment was implemented using Tobii Pro Lab software and gaze data was recorded 60 Hz from the Tobii Nano screen-based eye tracker, mounted at the bottom of a 24-in. Dell U2413f monitor (1920 × 1200 resolution). During the test, subjects were asked to watch the video stimuli described above, and then select the observed interaction among help, hinder and no interaction. To assess the social perception understanding over time, three conditions were implemented, displaying videos at three different temporal cutoffs, *i.e.*, 33%, 66%, and 100% of the original video length, resulting in a total of 108 trials. Videos were randomly ordered and ranged from a few seconds to approximately 30 s in duration. The experimental protocol was approved by the Institutional Review Board, and written informed consent was obtained from all participants prior to their participation in the study.

3.2 Action Recognition Model Architectures

A Convolutional Neural Network (CNN)-based architectures for action classification was employed in this study, *i.e.*, the Temporal Shift Module (TSM) [23]. In order to assess the feasibility of the proposed approach and the performance of TSM, which is a model suitable to best capture temporal dynamics, on the considered task, we initially compared it to the Inflated 3D ConvNet (I3D) [6], which is a widely adopted architecture for video-based action recognition. The latter extends the 2D convolutional filters and pooling kernels of the Inception Module [23] into the temporal dimension, enabling the model to learn spatio-temporal features from raw video inputs.

On the other hand, the TSM model chosen for this work introduces an efficient mechanism to model temporal information within standard 2D CNNs. TSM divides each input video into N clips and it samples one frame from each. A CNN backbone (ResNet-50 [11]) extracts feature maps for each sampled frame. A portion of the feature channels is then shifted forward or backward along the temporal dimension using a bi-directional temporal shift strategy. This operation allows temporal context to be integrated without introducing additional computational cost. The resulting tensor is used to classify the entire video sequence.

Training Procedure. Both I3D and TSM models were initialized with weights pre-trained on the Kinetics-400 dataset [14] and fine-tuned on the custom social interaction video dataset. I3D was implemented using the publicly available PyTorch version, while TSM was trained using the MMAction2 toolbox [1].

An initial set of experiments was carried out to assess the baseline performance of both I3D and TSM in classifying social interactions. The train/test split was designed to be balanced with respect to social interactions and video scenarios, resulting in 90 train and 18 test videos. Based on these preliminary experiments, the best-performing architecture (TSM) was selected for subsequent analysis. Then, the dataset was re-split to be consistent with the human experiment, ensuring that the videos used for testing the model matched those presented to human participants during the classification task. At this level, two strategies were implemented: (A) training only on full-length videos and testing on 33%, 66%, and 100% length videos; and (B) conducting separate trainings for each of the three video length versions (33%, 66%, and 100%), resulting in three distinct models, each tested on videos of the corresponding length.

For all fine-tuning procedures, a 4-fold cross-validation was applied to the training set to ensure robustness and generalization. For both architectures (I3D and TSM), each video was sub-sampled to 16 equally spaced frames to balance memory efficiency with information preservation. To address the limited size of the dataset, data augmentation was performed by rotating each video of three different angles (*i.e.*, 90, 180 and 270). This resulted in a final training set of 360 videos for the preliminary experiments and 288 videos for the second training phase. Augmented versions of the test set were excluded during evaluation.

3.3   Comparison of Humans and Model Attention Maps

Grad-CAM: To compare human visual attention with activations resulting from the TSM during classification of social interactions, we leveraged the Gradient-weighted Class Activation Mapping (Grad-CAM) technique [22]. Grad-CAM generates class-specific heatmaps by computing the gradient of the target

Table 1. Model's accuracy in the social interaction classification task. Accuracy of each fold is reported together with the mean (μ) $\pm$ standard deviation (σ) value across the different folds (last row) for both I3D and TSM.

Fold	I3D			TSM		
	Train	Val	Test	Train	Val	Test
1st	0.98	0.67	0.67	0.92	0.98	1.00
2nd	1.00	0.46	0.56	1.00	0.87	0.89
3rd	1.00	0.78	0.56	0.94	0.93	0.94
4th	0.99	0.62	0.67	0.94	0.91	1.00
$\mu \pm \sigma$	0.99 ± 0.01	0.63 ± 0.13	0.62 ± 0.06	0.95 ± 0.03	0.92 ± 0.05	0.96 ± 0.05

class score with respect to the activations of the last convolution layer. The resulting activation maps highlight the most influential regions of the input frames for the model's classification decision. In our case, Grad-CAM was applied to each sampled frame of the input video, resulting in a sequence of heatmaps indicating frame-level relevance toward the final prediction.

Saliency Maps: Humans' and model attention maps were compared by computing: i) an overall spatio-temporal map for each video (disregarding the dynamic temporal dimension) and ii) by analyzing the similarity dynamic trend over time. Specifically, to perform the static analysis, for each video the Grad-CAM maps were computed over the sampled frames, summed and normalized to produce a single spatio-temporal saliency map representing model attention. Human attention maps were derived from gaze data collected during the eye-tracking experiment. Fixations coordinates were extracted using the Tobii Pro Lab built-in filter, and aggregated across participants for each video to generate a fixation density map, estimating the distribution of human visual attention.

Temporal Analysis: In addition to global similarity analysis, we investigated the temporal dynamics of attention throughout each video. Indeed, previous studies suggest that human and model attentions often align more strongly at specific moments of interaction [3]. To enable temporal comparison, we segmented each video into 16 equal-length temporal clips, consistent with the temporal partitioning used by the ANN models, and we extracted ground truth human fixation maps for each corresponding segment of the videos. We performed a qualitative comparison of human-model attention similarity taking into account differences within and outside the social Times of Interest (TOIs), i.e. the portions of the videos when a socially relevant behavior is displayed.

Saliency Maps' Similarity Evaluation Metrics: To quantify differences between human and model overall and dynamic saliency maps, we adopted a subset of metrics widely used to assess similarity and dissimilarity among both human-derived and model-derived saliency maps [4]. These metrics collectively provide a comprehensive evaluation of both spatial overlap and distributional similarity between human and model attention. Specifically, we chose two location-based metrics, *i.e.*, AUC-Judd [13] and NSS [5], and two distribution-

Table 2. Comparison of model vs. human social interaction classification accuracy (mean ± std) on videos cut at 33%, 66%, and 100% of their original length. Model's results are reported for both training strategies A and B (as explained in Sect. 3.2).

	33%	67%	100%
TSM (training strategy A)	0.40 ± 0.08	0.75 ± 0.03	0.90 ± 0.06
TSM (training strategy B)	0.50 ± 0.16	0.80 ± 0.14	
humans		0.59 ± 0.11	0.79 ± 0.10	0.91 ± 0.10

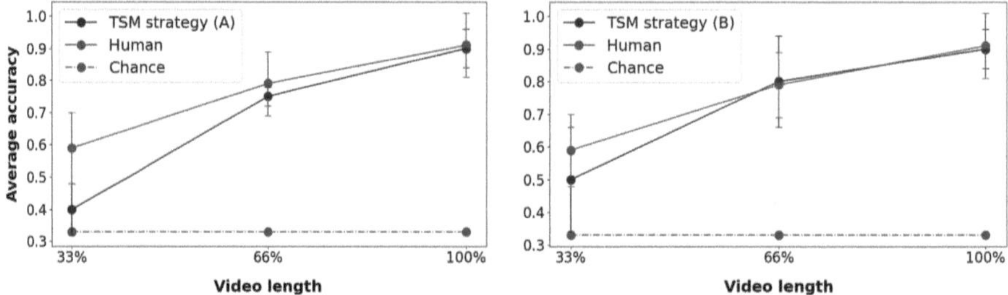

Fig. 2. Visualization of model vs. human social interaction classification accuracy. As larger portion of the video are available, the model reaches human-like performance with both fine-tuning strategies A and B.

based metrics *i.e.*, CC [12] and KL-Divergence (KLD) [5]. Among the considered evaluation metrics, only the KLD is a dissimilarity metric.

4 Results and Discussion

4.1 Model Selection

As described in the previous section, we initially trained two CNN-based architectures, I3D and TSM, to classify social interactions from videos. By looking at results shown on Table 1, we observe a higher average test accuracy for TSM (0.96 ± 0.05) compared to I3D (0.62 ± 0.06). As I3D results suggest the presence of overfitting, we selected TSM for subsequent experiments.

4.2 Human vs. Model Social Interaction Classification Accuracy

After modifying the training/test split with the procedure described in Sect. 3.2, we conducted experiments based on the two training strategies (A and B). Table 2 reports the model's test accuracies for the social interaction classification task using different input video portions (33%, 67%, and 100%), expressed as mean ± standard deviation across the four cross-validation folds. As can be observed, for both strategies, the performance improves when larger portions of video are available, and it aligns with humans' classification accuracy for full-length videos (Fig. 2). Strategy B generally outperforms strategy A in leading

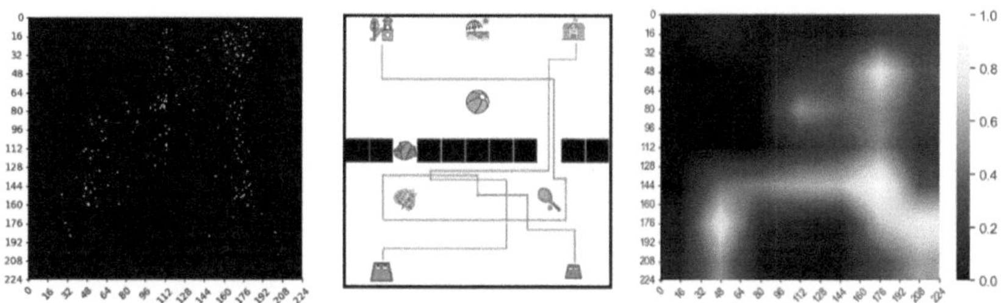

Fig. 3. From left to right: humans' fixations map for a video; initial video frame with the agents' trajectories highlighted; TSM saliency map computed via GradCAM.

Table 3. Evaluation metrics between humans and model's overall saliency maps. Values are averaged across all videos for the three classes: help, hinder and no interaction.

Classes	AUC Judd ↑	NSS ↑	CC ↑	KL Divergence ↓
Help	0.61 ± 0.10	0.40 ± 0.33	0.16 ± 0.14	1.72 ± 0.41
Hinder	0.69 ± 0.06	0.76 ± 0.31	0.34 ± 0.14	1.28 ± 0.31
No interaction	0.64 ± 0.07	0.47 ± 0.28	0.18 ± 0.10	1.53 ± 0.31
TOTAL	0.65 ± 0.08	0.53 ± 0.34	0.22 ± 0.15	1.53 ± 0.35

to a model's correct identification of social interactions from partial videos (33% and 67%), closing the gap with human performance. This is likely due to the better alignment between training and test distributions.

4.3 Similarity Between Human and Model Attention Maps

Table 3 reports the values of similarity metrics between humans and model's overall saliency maps, in terms of average values among all videos belonging to the same social interaction class (*i.e.*, help, hinder, and no interaction). Specifically, each metric is computed by comparing the model's attention map extracted via Grad-CAM on a full-length video with the ground-truth saliency map of humans' fixations. Figure 3 shows an example of the discrete map of humans' fixations (left) and corresponding model's attention map (right) for a single video.

By looking at the overall saliency maps comparison, AUC and NSS show the highest resemblance between model-generated and human attention maps, with highest similarity for the "hinder" class. NSS results, however, exhibit higher variance. This could be explained by the fact that NSS is sensitive to false positives (*i.e.*, Grad-CAM activations on pixels with no corresponding human fixation points) and relative differences in saliency. Indeed, Grad-CAM activations appear to be generally more distributed over time and on single frames than human fixations. Considering the distribution-based metrics, both CC and KLD values confirm the trend observed by using AUC and NSS (highest/lowest similarity/dissimilarity for the hinder class), even though the absolute values for

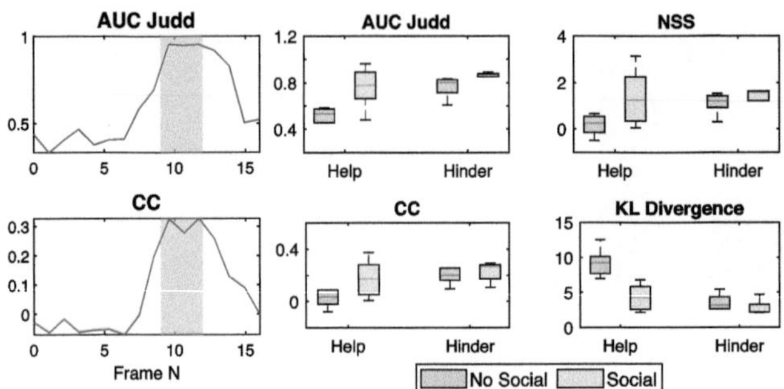

Fig. 4. Evaluation metrics over time: (a) example of the dynamic trend of the similarity metrics across frames for one video, with key times of video social interaction (social TOIs) highlighted in gray; (b) box plot comparison between average similarity metrics' values computed within (Social) and outside (No Social) social TOIs, respectively. (Color figure online)

CC are significantly lower. As KLD is penalized by false negatives (*i.e.*, Grad-CAM missed activations on pixels where human fixations are present), while CC is equally affected by false positives and false negatives, CC lower absolute values could be explained by the fact that the weight of false positives is higher in this setting: the saliency maps derived via Grad-CAM exhibit broader and more diffuse areas of attention than human gaze-based ones. Figure 4 shows results related to the saliency evaluation metrics over time. Specifically, we report the similarity values between human and model saliency maps over time and their qualitative correlation with social TOIs, *i.e.*, key times of social interactions during videos. Figure 4a shows an example of the dynamic trends of the human-model saliency maps' similarity over time for a single video of "help" category. As highlighted on the plot, the human-model similarity, quantified by AUC, NSS and CC, is maximal (and dissimilarity, quantified by KLD is minimal) during frames corresponding to social TOIs, displayed as grey shaded areas. Such a trend is confirmed when looking at all videos (Fig. 4b). Indeed, the distribution of similarity metrics values is on average higher (lower for dissimilarity KLD) within social TOIs (green), compared to the rest of frames (blue), where no social interaction is occurring.

5   Conclusion

In this work we investigated the similarity between visual strategies learned by a CNN-based model and humans, during classification of social interactions from videos. Our preliminary results suggest that there could be an emerging similarity between the model and human visual social attention maps, especially when looking at model's activation over time. In future work, we plan to extend the analysis to different ANNs architectures, *e.g.*, by comparing Transformers vs. CNNs. It is worth noting that, in our work, human gaze data is not used within the model fine-tuning. Conversely, the model takes as input the video stimuli alone, together with the corresponding label of social interaction class. Indeed, rather than forcing the model's alignment to humans, our overreaching goal is to investigate ANNs' capability of supporting the emergence of human-like cognitive abilities, with a focus on visual social attention and social cognition.

Aknowledgment. This research has received funding from the European Union's Horizon 2020 research and innovation programme under the Marie Sklodowska-Curie grant agreement No 896415 (TIRESIA), and it was supported by the Center for Brains, Minds and Machines, NSF STC award CCF-1231216 and by the EU program NextGenerationEU and the Ministry of University and Research, National Recovery and Resilience Plan, Mission 4, Component 2, Investment 1.5, project RAISE (ECS00000035).

References

1. https://github.com/open-mmlab/mmaction2
2. Barbu, A., Banda, D., Katz, B.: Deep video-to-video transformations for accessibility with an application to photosensitivity. Pattern Recogn. Lett. **137**, 99–107 (2020)
3. Boccignone, G., Cuculo, V., D'Amelio, A., Grossi, G., Lanzarotti, R.: Give ear to my face: modelling multimodal attention to social interactions. In: Leal-Taixé, L., Roth, S. (eds.) ECCV 2018. LNCS, vol. 11130, pp. 331–345. Springer, Cham (2019). https://doi.org/10.1007/978-3-030-11012-3_27
4. Bylinskii, Z., et al.: MIT saliency benchmark (2015)
5. Bylinskii, Z., Judd, T., Oliva, A., Torralba, A., Durand, F.: What do different evaluation metrics tell us about saliency models? IEEE Trans. Pattern Anal. Mach. Intell. **41**(3), 740–757 (2018)
6. Carreira, J., Zisserman, A.: Quo vadis, action recognition? A new model and the kinetics dataset. In: Proceedings of the IEEE Conference on Computer Vision and Pattern Recognition (CVPR) (2017)
7. Cartella, G., et al.: Trends, applications, and challenges in human attention modelling. In: IJCAI (2024)
8. Fel, T., Felipe, I., Linsley, D., Serre, T.: Harmonizing the object recognition strategies of deep neural networks with humans. Adv. Neural. Inf. Process. Syst. **35**, 9432 (2022)
9. Geirhos, R., Meding, K., Wichmann, F.A.: Beyond accuracy: quantifying trial-by-trial behaviour of CNNs and humans by measuring error consistency. Adv. Neural. Inf. Process. Syst. **33**, 13890–13902 (2020)
10. Hafri, A., Firestone, C.: The perception of relations. Trends Cogn. Sci. **25**(6), 475–492 (2021)
11. He, K., Zhang, X., Ren, S., Sun, J.: Deep residual learning for image recognition. In: Proceedings of the IEEE Conference on Computer Vision and Pattern Recognition, pp. 770–778 (2016)
12. Jost, T., Ouerhani, N., Von Wartburg, R., Müri, R., Hügli, H.: Assessing the contribution of color in visual attention. Comput. Vis. Image Underst. **100**(1–2), 107–123 (2005)

13. Judd, T., Durand, F., Torralba, A.: A benchmark of computational models of saliency to predict human fixations. In: MIT Technical report (2012)
14. Kay, W., et al.: The kinetics human action video dataset. arXiv preprint arXiv:1705.06950 (2017)
15. Lee, J., et al.: VisAlign: dataset for measuring the alignment between AI and humans in visual perception. Adv. Neural. Inf. Process. Syst. **36**, 77119–77148 (2023)
16. McMahon, E., Isik, L.: Seeing social interactions. Trends Cogn. Sci. **27**(12), 1165–1179 (2023)
17. Mertens, L.P.: Modeling social cognition and its neurologic deficits with artificial neural networks. In: Proceedings of the 25th International Conference on Multimodal Interaction, pp. 726–730 (2023)
18. Netanyahu, A., Shu, T., Katz, B., Barbu, A., Tenenbaum, J.B.: Phase: physically-grounded abstract social events for machine social perception. In: Proceedings of the AAAI Conference on Artificial Intelligence, vol. 35, pp. 845–853 (2021)
19. Peterson, J.C., Abbott, J.T., Griffiths, T.L.: Evaluating (and improving) the correspondence between deep neural networks and human representations. Cogn. Sci. **42**(8), 2648–2669 (2018)
20. Schiatti, L., et al.: Modeling visual impairments with artificial neural networks: a review. In: 2023 IEEE/CVF International Conference on Computer Vision Workshops (ICCVW), pp. 1979–1991 (2023)
21. Schrimpf, M., Kubilius, J., Lee, M.J., Murty, N.A.R., Ajemian, R., DiCarlo, J.J.: Integrative benchmarking to advance neurally mechanistic models of human intelligence. Neuron **108**(3), 413–423 (2020)
22. Selvaraju, R.R., Cogswell, M., Das, A., Vedantam, R., Parikh, D., Batra, D.: Grad-CAM: visual explanations from deep networks via gradient-based localization. Int. J. Comput. Vision **128**, 336–359 (2020)
23. Szegedy, C., et al.: Going deeper with convolutions. In: Proceedings of the IEEE Conference on Computer Vision and Pattern Recognition (CVPR) (2015)
24. Tejwani, R., et al.: Incorporating rich social interactions into MDPs. In: 2022 International Conference on Robotics and Automation (ICRA), pp. 7395–7401. IEEE (2022)
25. Ullman, T., Baker, C., Macindoe, O., Evans, O., Goodman, N., Tenenbaum, J.: Help or hinder: Bayesian models of social goal inference. Adv. Neural Inf. Process. Syst. **22** (2009)
26. Zhang, M., Feng, J., Ma, K.T., Lim, J.H., Zhao, Q., Kreiman, G.: Finding any waldo with zero-shot invariant and efficient visual search. Nat. Commun. **9**(1), 3730 (2018)
27. Zhang, M., Tseng, C., Kreiman, G.: Putting visual object recognition in context. In: Proceedings of the IEEE/CVF Conference on Computer Vision and Pattern Recognition, pp. 12985–12994 (2020)
28. Zhang, R., Isola, P., Efros, A.A., Shechtman, E., Wang, O.: The unreasonable effectiveness of deep features as a perceptual metric. In: Proceedings of the IEEE Conference on Computer Vision and Pattern Recognition, pp. 586–595 (2018)
29. Zhou, Q., Du, C., He, H.: Exploring the brain-like properties of deep neural networks: a neural encoding perspective. Mach. Intell. Res. **19**(5), 439–455 (2022)

Decoding Affective States from fMRI Using Automatically Labeled Multi-modal Movie Stimuli

Andrea Corsico, Giorgia Rigamonti[(✉)] , Simone Zini , Luigi Celona ,
and Paolo Napoletano

University of Milano-Bicocca, viale Sarca 336, 69121 Milan, Italy
{andrea.corsico,giorgia.rigamonti,simone.zini,luigi.celona,
paolo.napoletano}@unimib.it

Abstract. Emotions, feeling states arising from physiological changes, and sentiments, subjective evaluations as positive, negative, or neutral, are central to human behavior, decision-making, and mental health. These affective states are shaped by naturalistic stimuli, such as movies, often outside conscious awareness. Understanding how external content evokes emotions and is reflected in brain dynamics is key to advancing affective neuroscience and identifying neural markers of conditions like stress, depression, and ADHD. Unlike task-based or resting-state paradigms, naturalistic approaches such as film viewing offer greater ecological validity by engaging distributed networks involved in perception, emotion, and social cognition. In this study, we analyze existing fMRI data from the Algonauts 2025 challenge, derived from the CNeuroMod project, where participants watched episodes of Friends while undergoing brain imaging during naturalistic stimulus presentations. We investigate how dialogue audio and subtitles contribute to valence-based emotional processing. Sentiment labels (positive, neutral, negative) were automatically generated: audio was analyzed using a Wav2Vec2.0 speech emotion model estimating arousal, valence, and dominance, while subtitle sentiment was derived using VADER and Flair. These features were aligned with fMRI acquisition to examine how auditory and linguistic components map onto brain networks involved in emotion and social cognition, including the amygdala, insula, orbitofrontal cortex, and superior temporal sulcus. Our findings show that acoustic and combined sentiment models better track cortical dynamics in salience and default mode networks, while subtitle-based models yield weaker, localized effects. This highlights the value of multimodal sentiment features for studying affective brain responses during naturalistic viewing.

Keywords: fMRI · Affective Decoding · Naturalistic Stimuli · Speech Emotion Recognition · Subtitle Sentiment Analysis · Audio Sentiment Analysis

1 Introduction

Emotions—integrated feeling states driven by physiological changes—and sentiments—subjective evaluations of experiences as positive, negative, or

E. Rodolà et al. (Eds.): ICIAP 2025 Workshops, LNCS 16169, pp. 89–100, 2026.
https://doi.org/10.1007/978-3-032-11317-7_8

neutral—play central roles in shaping human behavior, decision-making, and social interactions [3]. These affective responses are continuously modulated by naturalistic stimuli, such as films, music, and language. While individuals consciously select much of their media consumption, they may remain unaware of how these choices shape their emotional and sentiment states. Moreover, exposure to unsolicited content, particularly on social media, can impact emotional well-being, with evidence linking excessive or dysregulated consumption to adverse mental health outcomes, especially among adolescents [9,14,20].

Understanding the relationship between external stimuli, evoked emotional responses, and the corresponding neural dynamics is essential for identifying biomarkers associated with mental health conditions such as stress, depression, and Attention Deficit Hyperactivity Disorder (ADHD) [12]. Naturalistic stimuli such as movies, speech, and music are increasingly used in affective neuroscience for their ecological validity and ability to engage complex neural processes [19]. Unlike task-based or resting-state paradigms, which either oversimplify or suffer from poor subject engagement, naturalistic paradigms preserve the richness of real-world perception and cognition [22]. Such paradigms offer improved test-retest reliability and reduced inter-subject variability [7], making them especially valuable in affective neuroscience applications [17]. In response, several open-access neuroimaging datasets with emotional content have been curated to support research in this domain [10].

Basic emotion theories typically classify six universal emotions—happiness, sadness, anger, fear, surprise, and disgust—each associated with characteristic neural signatures [6]. Sentiments, although related, reflect more enduring, cognitively mediated evaluations, often expressed through language. Recent developments in sentiment analysis have expanded from text-based sources like social media posts and reviews [26] to include the linguistic analysis of movie subtitles [2], enabling new opportunities to link stimulus content with viewer affective responses.

The human brain operates dynamically, with ongoing neural activity modulated both spontaneously and in response to external stimuli [16]. For example, as illustrated in Fig. 1, watching a movie simultaneously activates regions in the visual cortex and emotion-processing areas.

Functional Magnetic Resonance Imaging (fMRI) is a noninvasive neuroimaging technique that enables the investigation of such brain dynamics by detecting hemodynamic changes associated with neural activity. Thanks to its millimeter-scale spatial resolution, fMRI is particularly effective in mapping distributed brain networks involved in emotional processing [13,25].

Controlled studies using brief video or audio clips have demonstrated that emotional states are encoded in brain activity patterns [11]. However, naturalistic stimuli offer greater ecological validity and more reliable detection of these patterns across repeated exposures [21]. Emotions are transient, multifaceted responses elicited by both internal and external events and are known to involve automatic changes across physiological, behavioral, and experiential domains [3]. While individuals exhibit personalized responses to the same stimuli – due to

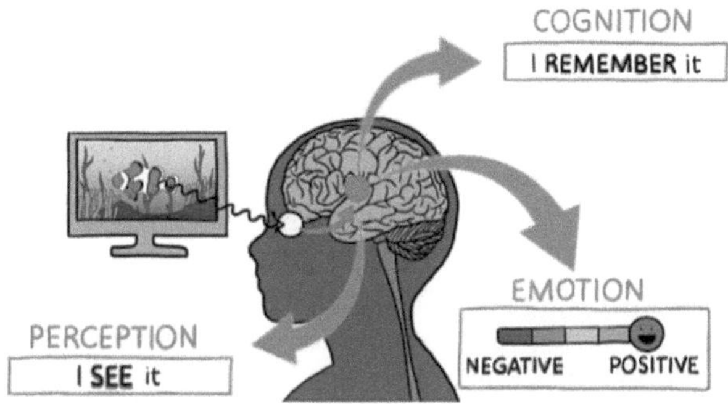

Fig. 1. Watching a movie engages multiple brain functions simultaneously, activating regions involved in perception, emotion processing, and cognition.

differences in context, physiology, and personality – there is a shared component in how emotionally evocative content influences neural activity, especially when examined at the group level [5].

This study explores how emotional valence—derived from audio and linguistic features of audiovisual narratives—is represented in cortical brain activity. Participants view naturalistic video content (Friends episodes) while fMRI data are acquired. Although participants experience the full audiovisual stimulus, our analysis isolates the influence of the audio signal and linguistic features (derived from subtitles) on emotional valence processing. We use the Schaefer 1000 atlas to parcellate cortical activity, enabling correlation of emotional valence time series with functional network dynamics. Given that the Schaefer 1000 atlas excludes subcortical regions critical for automatic affective responses (e.g., amygdala, hippocampus), our analysis emphasizes conscious, cognitive, and regulated components of emotional processing—captured through cortical network activity patterns.

The key contributions of this work are:

- We perform audio sentiment analysis using a transformer model to estimate continuous emotion dimensions, and text sentiment analysis using complementary lexicon-based (VADER) and machine learning–based (Flair) methods.
- We investigate how audio and subtitle modalities shape cortical emotional responses during naturalistic movie viewing.
- We identify functional networks involved in valence tracking, highlighting the salience, default mode, control, and language-temporal networks, using the Schaefer 1000 cortical parcellation.

The remainder of this manuscript is structured as follows: Sect. 1 contains Introduction and Literature Review. Section 2 explains Methodology. Section 3 illustrates the experimental rationale, analysis design, and experimental results. Section 4 concludes the paper.

2   Methodology

We hypothesize that the emotional content of movies evolves over time, dynamically influencing the emotional and corresponding brain states of individuals. Since movies are composed of multiple elements, such as audio, visuals, and dialogues (subtitles), each of these components can modulate a viewer's emotional state, in addition to their own pre-existing affective baseline.

To investigate whether brain activity in regions commonly associated with sentiment processing correlates with the emotional trajectory of a movie, we first define the analysis pipeline (Fig. 2) and then utilize open-access fMRI datasets. However, these datasets typically do not provide time-resolved sentiment labels. To address this, we generate time-based sentiment annotations using features extracted from the movie itself, specifically from the audio and subtitle tracks. We employ three different methods to estimate valence from both modalities. The resulting valence scores are averaged to produce a unified sentiment score for each time segment. These scores are then temporally aligned with the corresponding fMRI data, and correlation analyses are performed. Finally, the correlation results are mapped onto functional brain networks—based on Yeo's 7-network parcellation—to enable a network-level interpretation of the brain's response to dynamic emotional content.

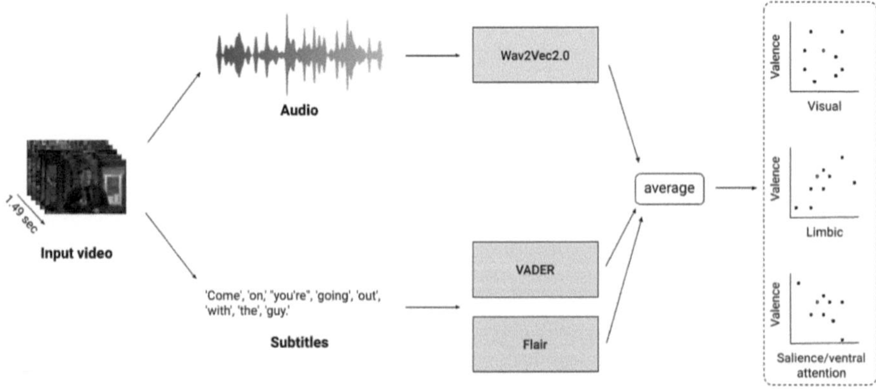

Fig. 2. Frame chunks of 1.49 s are iteratively sampled from a video. For each video chunk, audio and subtitles are extracted. The audio is processed using Wav2Vec2.0 to estimate valence, while the subtitles are analyzed with VADER and Flair to obtain text-based valence scores. The resulting valence scores are averaged to obtain a final sentiment score per chunk. These scores are then aligned with corresponding fMRI data, which are grouped into functional brain networks for network-based interpretation of brain activity.

2.1 Dataset

In our experiments, we use a subset of the data from the Algonauts 2025 project.[1] The challenge dataset is derived from the CNeuroMod project,[2] which offers extensive single-subject fMRI recordings collected during both controlled experiments and naturalistic stimuli presentations. Due to its breadth and the variety of multimodal inputs, CNeuroMod serves as an excellent resource for developing and evaluating fMRI encoding models capable of generalizing across conditions. The subset employed in this work consists of approximately 80 h of multimodal cinematic stimuli paired with corresponding fMRI responses.

In particular, we focus on the CNeuroMod Friends dataset, where participants viewed all episodes from the first six seasons of the sitcom Friends. The stimuli include temporally aligned visual frames, audio tracks, and textual transcripts (as illustrated in Fig. 3a). Neural responses are captured as whole-brain fMRI time series from four participants (sub-01, sub-02, sub-03, sub-05), projected onto the Montreal Neurological Institute (MNI) template [4] and summarized into signals from 1,000 functionally defined brain parcels [18] (Fig. 3b). The fMRI data were recorded with a repetition time (TR) of 1.49 s, resulting in one fMRI sample for every 1.49 s of stimulus presentation.

This pilot study investigates the feasibility of classifying affective content in fMRI data using sentiment labels extracted from subtitles and audio. We hypothesize that if sentiment classification proves effective in this context—characterized by diverse and dynamic audiovisual content—it is likely to succeed in scenarios involving more structured narratives or dialogue-driven films. In future work, we plan to extend the study to include additional movies and leverage the full dataset for more comprehensive analysis.

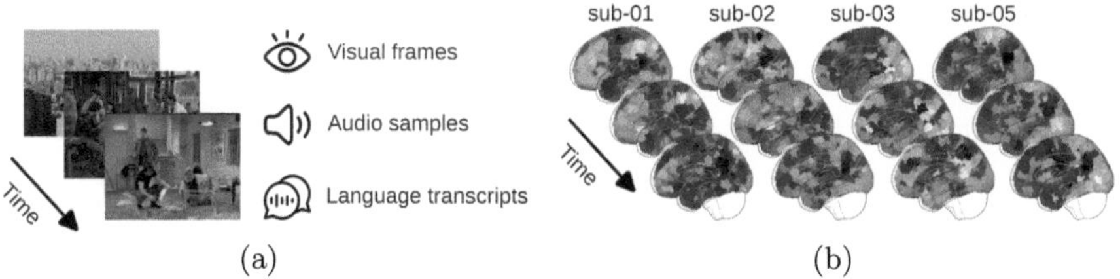

Fig. 3. Challenge data. (a) Multimodal movie stimuli including visual frames, audio tracks, and time-stamped language transcripts. (b) Whole-brain fMRI time series recorded from four subjects in response to the movie stimuli.

---

[1] https://algonautsproject.com/.
[2] https://www.cneuromod.ca/.

2.2 Subtitle and Audio Processing

Audio samples were extracted from the videos and segmented into consecutive 1.49-second utterances. The segments were saved as .wav files, with a sampling rate of 22,050 Hz. No further processing was performed.

Movie dialogues (subtitles) were provided as .tsv files containing timestamped transcripts of the spoken content in the movie stimuli. Each .tsv file segmented the dialogue into consecutive 1.49-s chunks, matching the audio sample segmentation, with each row corresponding to one such interval. This segmentation aligns with the fMRI acquisition protocol, where one volume was recorded every 1.49 s (i.e., the repetition time, TR, was 1.49 s). If no speech occurred during a given interval, the corresponding row in the .tsv file was left empty.

2.3 Sentiment Analysis

After obtaining the audio utterances and corresponding text sentences, the next step is to perform sentiment analysis on this data.

For sentiment analysis of the audio utterances, we employ a Wav2Vec2.0 transformer model fine-tuned on the MSP-Podcast corpus to estimate the continuous emotional dimensions of arousal, dominance, and valence. These models have achieved state-of-the-art performance in speech emotion recognition, outperforming convolutional baselines (e.g., CNN14) and closing the valence gap (concordance correlation coefficient, CCC $\approx$ 0.638 on MSP-Podcast) [24]. Notably, transformer-based architectures demonstrated greater robustness to noise and cross-domain variability, exhibited reduced gender bias, and implicitly captured linguistic content—particularly improving valence estimation—despite being trained solely on raw audio [24]. The outputs of this audio model are raw valence scores in the range [0,1] for each dimension, reflecting emotional polarity per utterance.

For sentence-level sentiment analysis of the subtitles, we adopt both lexicon-based and machine learning–based approaches from the field of Natural Language Processing (NLP). Lexicon-based methods are among the simplest and most commonly used techniques for analyzing sentiment in textual data [23]. These methods rely on a predefined sentiment lexicon (dictionary), where words are annotated with their context-independent semantic orientation. Since these methods do not require any training, they are fast and interpretable, but are limited by the vocabulary covered in the lexicon.

In contrast, machine learning–based sentiment analyzers train models capable of generalizing to previously unseen words and contexts. In our study, we use two widely adopted sentiment analyzers: the Valence Aware Dictionary for Sentiment Reasoning (VADER) [8] and Flair [1]. VADER is a lexicon-based sentiment analyzer that computes the polarity of a sentence based on the weighted sentiment scores of its constituent words [8]. It also incorporates heuristic rules to account for intensity modifiers such as punctuation, capitalization, and emojis. The overall sentiment score is calculated by summing the intensity of each word in the sentence [8].

Flair, on the other hand, is a machine learning–based sentiment analyzer built on a character-level LSTM neural network. It uses contextual word embeddings and considers both word and character sequences to predict sentiment labels. Flair relies on pre-trained models to perform sentiment classification and leverages vector representations of words to infer sentiment based on their contextual sequence [1].

3   Experiments and Results

The objective of this study is to examine how emotional valence, as shaped by audiovisual narratives, relates to patterns of cortical brain activity. Specifically, we investigate how audio and linguistic features from multimedia stimuli contribute to conscious emotional processing within the cerebral cortex. Participants view episodes of Friends, a socially rich, emotionally expressive television series. Although participants experience the full audiovisual material, our analysis isolates the contributions of the audio signal and the linguistic content (captured via subtitle text) to the temporal dynamics of emotional valence. Emotional responses are quantified along the valence dimension (positive, neutral, or negative) over time.

This work builds on prior evidence showing that emotionally salient narratives engage distributed neural circuits implicated in affective and social processing [15]. While traditional models emphasize the role of limbic structures such as the amygdala, hippocampus, and hypothalamus in emotional experience, our approach focuses on cortical networks, given the constraints of the Schaefer 1000 atlas, which excludes subcortical regions. We therefore interpret correlations between emotional valence and cortical activity as markers of conscious, cognitive, and regulated aspects of emotional processing, reflecting the integration of cortical outputs with subcortical contributions that are not directly measured. Using fMRI, we correlate time series of computationally estimated emotional valence (derived from audio and subtitle-based linguistic features) with BOLD activity across 1000 cortical parcels.

Our analysis emphasizes functional networks associated with cortical emotional processing: the salience network (detection of emotionally relevant content), the default mode network (self-referential and internal emotional appraisal), the control network (regulation and modulation of affect), and the language-temporal network (semantic-emotional processing of audio-linguistic input). We assess how these networks track the temporal dynamics of emotional valence, with particular attention to expected hemispheric asymmetries and temporal co-activation patterns.

This approach provides insight into how auditory and linguistic information drive conscious emotional experiences, highlighting the cortical mechanisms of valence tracking while acknowledging the methodological limitation of not directly capturing subcortical, automatic emotional processes.

3.1   Results

In this section, we present the results of the Pearson correlation analysis between the estimated valence and fMRI signals for each video chunk. This analysis aims to assess whether the valence dynamics of the stimuli are reflected in brain activity within specific cortical regions. A qualitative overview of the results is provided in the accompanying Fig. 4, which shows cortical maps of correlation values averaged across stimuli and subjects. Correlations were averaged across different stimuli and subjects. We report both combined correlations—aggregated across all valence estimation methods—and correlations obtained from each individual method.

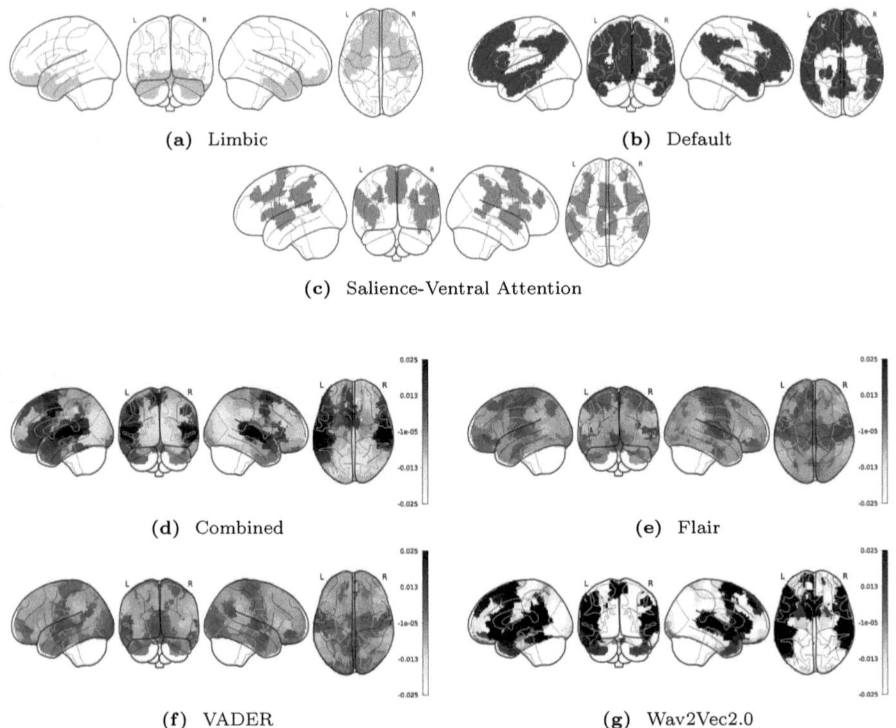

Fig. 4. Cortical activation maps associated with emotional valence during viewing of audiovisual narratives. The top panels illustrate canonical functional networks from the Yeo parcellation—(a) Limbic, (b) Default mode, and (c) Salience-Ventral Attention—which are reference networks considered most strongly associated with emotional processing. The bottom panels show correlations between estimated valence (from audio and linguistic features) and BOLD activity across 1000 cortical parcels for (d) Combined model, (e) Flair, (f) VADER, and (g) Wav2Vec2.0. Cool colors (black-red) indicate positive valence associations, while warm colors (yellow-white) indicate negative valence associations. (Color figure online)

To aid interpretation, Fig. 4 (panels a–c) also depicts reference maps of three canonical functional networks—Limbic, Default mode, and Salience-Ventral Attention—which are considered most strongly associated with emotional processing. These network maps provide anatomical context for understanding the spatial distribution of valence-related correlations in the subsequent panels (d–g).

The cortical maps reveal distinct patterns across models. The Combined and Wav2Vec2.0 models display stronger and more spatially specific correlations compared to VADER and Flair. The Combined model shows widespread bilateral engagement across frontal, temporal, and parietal cortices, while Wav2Vec2.0 highlights more localized regions with stronger associations.

In particular, Wav2Vec2.0 exhibits focal areas of high correlation consistent with regions implicated in salience and attentional processing, although precise anatomical localization is inferred from functional network associations rather than directly labeled in the figure. By contrast, VADER and Flair exhibit weaker and more uniformly distributed correlations, with no pronounced regional hotspots. This pattern suggests that models relying primarily on linguistic features, such as VADER and Flair, may be less effective at capturing dynamic valence signals that align with brain activity during audiovisual stimulation.

To further interpret these results, we summarized the 1000 cortical parcels into seven canonical functional networks using the Yeo parcellation [25]: Cont (control network), Default (default mode network), DorsAttn (dorsal attention network), Limbic (limbic network), SalVentAttn (salience/ventral attention network), SomMot (somato-motor network), and Vis (visual network).

Table 1 presents the rankings of valence associations across these networks for each model, where 1 indicates the strongest positive valence association and 7 the strongest negative. The rankings indicate that both Wav2Vec2.0 and the Combined model associate positive valence primarily with the salience/ventral attention and default mode networks, and negative valence with the visual and

Table 1. Ranking of functional network associations with valence-related cortical activation for each model. For each network and model, **1** indicates the strongest positive association with valence (more positive), and **7** indicates the strongest negative association with valence (more negative).

Method	Wav2Vec 2.0	VADER	Flair	Combined
Cont	6	5	4	7
Default	2	4	3	3
DorsAttn	3	6	6	6
Limbic	4	3	2	4
SalVentAttn	1	7	7	2
SomMot	5	1	1	1
Vis	7	2	5	5

control networks. These findings support the interpretation that acoustic features (in Wav2Vec2.0) or multimodal features (in the Combined model) better capture valence signals linked to emotional salience detection and self-referential processing. In contrast, VADER and Flair show their strongest positive valence associations in the somato-motor network and exhibit a more diffuse distribution of valence associations across other networks. These network-level patterns are consistent with the spatial distributions observed in the cortical maps and underscore the added value of integrating acoustic or multimodal features to capture affective brain responses during naturalistic audiovisual experiences.

4   Conclusion

This study demonstrates the feasibility of linking sentiment information derived from audio and linguistic features of naturalistic audiovisual narratives to cortical brain activity. By combining audio-based sentiment analysis (via Wav2Vec2.0) with subtitle-based sentiment analysis (via VADER and Flair), we identified time-varying emotional valence signals and mapped their associations with cortical network dynamics during movie viewing. Using the Schaefer 1000 atlas, our findings reveal that distinct functional networks—including the salience, default mode, control, and somato-motor networks—track the temporal evolution of emotional valence, reflecting key components of affective processing.

Moreover, our results highlight that models incorporating acoustic-prosodic features, such as Wav2Vec2.0 and the combined approach, capture more extensive and bilateral engagement of cortical networks, particularly within salience and default mode regions. This suggests that prosodic and multimodal cues provide richer, temporally aligned information about dynamic emotional content compared to purely text-based sentiment measures. In contrast, VADER and Flair exhibited weaker and more uniformly distributed associations, underscoring the limitations of relying solely on subtitle-derived sentiment for modeling affective brain responses during complex audiovisual experiences.

Together, these findings emphasize the added value of integrating multiple modalities of emotional information when investigating the neural correlates of affective experience in ecologically valid settings. Future work should investigate how individual differences in emotional perception, attention, and regulation influence these neural representations and how these insights can inform affective computing and mental health applications.

Acknowledgments. Financial support from ICSC – Centro Nazionale di Ricerca in High Performance Computing, Big Data and Quantum Computing, funded by European Union – NextGenerationEU.

This work was partially funded by the National Plan for NRRP Complementary Investments (PNC, established with the decree-law 6 May 2021, n. 59, converted by law n. 101 of 2021) in the call for the funding of research initiatives for technologies and innovative trajectories in the health and care sectors (Directorial Decree n. 931 of 06-06-2022) - project n. PNC0000003 - AdvaNced Technologies for Human-centrEd Medicine (project acronym: ANTHEM) (https://fondazioneanthem.it/). This work reflects only the authors' views and opinions, neither the Ministry for University and Research nor the European Commission can be considered responsible for them.

References

1. Akbik, A., Bergmann, D., Blythe, T., Rasul, K., Schweter, S., Vollgraf, R.: Flair: an easy-to-use framework for state-of-the-art NLP. In: Proceedings of NAACL-HLT 2019 (2019)
2. Barbieri, F., Camacho-Collados, J., Neves, L., Espinosa-Anke, L., et al.: Tweeteval: unified benchmark and comparative evaluation for tweet classification. In: Proceedings of the 2020 International Conference on Language Resources and Evaluation (LREC) (2020)
3. Barrett, L.F.: The theory of constructed emotion: an active inference account of interoception and categorization. Philos. Trans. Roy. Soc. B: Biol. Sci. **372**(1718) (2017)
4. Brett, M., Johnsrude, I.S., Owen, A.M.: The problem of functional localization in the human brain. Nat. Rev. Neurosci. **3**(3), 243–249 (2002)
5. Chang, L., Jolly, E., Cheong, J., Burnett, C., et al.: Emotion as an emergent property of neuromodulation networks. Neuroimage **226**, 118755 (2021)
6. Cowen, A.S., Keltner, D.: Emotion semantics show both cultural variation and universal structure. Proc. Natl. Acad. Sci. **118**(37), e2003167118 (2021)
7. Finn, E.S., Bandettini, P.A.: Can brain state be manipulated to emphasize individual differences in functional connectivity? Nat. Neurosci. **23**, 885–893 (2020)
8. Hutto, C., Gilbert, E.: Vader: a parsimonious rule-based model for sentiment analysis of social media text. In: Proceedings of the International AAAI Conference on Web and Social Media, vol. 8, pp. 216–225 (2014)
9. Keles, B., McCrae, N., Grealish, A.: A systematic review: the influence of social media on depression, anxiety and psychological distress in adolescents. Int. J. Adolesc. Youth **25**(1), 79–93 (2020)
10. Khosla, M., Rane, H., et al.: The emotion dataset: brain responses to emotional movies. Sci. Data **8**(1), 1–11 (2021)
11. Kragel, P.A., LaBar, K.S.: Decoding the nature of emotion in the brain. Trends Cogn. Sci. **20**(6), 444–455 (2016)
12. Li, L., Sun, M., Qi, M., Li, Y., Li, D.: Neural correlates of emotional working memory predict depression and anxiety. Front. Neurosci. **19**, 1574901 (2025)
13. Mahrukh, R., Shakil, S., Malik, A.S.: Sentiments analysis of FMRI using automatically generated stimuli labels under naturalistic paradigm. Sci. Rep. **13**(1), 7267 (2023)
14. Marengo, D., Sindermann, C., Elhai, J.D., Montag, C.: Exploring the association between problematic social media use and mental health outcomes: a meta-analysis. Comput. Hum. Behav. **129**, 107396 (2022)
15. Nummenmaa, L., Glerean, E., Viinikainen, M., Jääskeläinen, I.P., Hari, R., Sams, M.: Emotions promote social interaction by synchronizing brain activity across individuals. Proc. Natl. Acad. Sci. **109**(24), 9599–9604 (2012)
16. Raichle, M.E.: The brain's default mode network. Annu. Rev. Neurosci. **38**, 433–447 (2015)
17. Saarimäki, H.: Emotions are more than feelings: brain activity underlying the perception of emotion in naturalistic settings. Curr. Opin. Psychol. **41**, 115–121 (2021)

18. Schaefer, A., et al.: Local-global parcellation of the human cerebral cortex from intrinsic functional connectivity MRI. Cereb. Cortex **28**(9), 3095–3114 (2017)
19. Sonkusare, S., Breakspear, M., Guo, C.: Naturalistic stimuli in neuroscience: critically acclaimed. Trends Cogn. Sci. **23**(8), 699–714 (2019)
20. Twenge, J.M., Spitzberg, B.H., Campbell, W.K.: Mental health, self-injury, and suicidality among us adolescents after the advent of smartphone technology: trends and possible mechanisms. Child Adolescent Psychiatric Clin. **29**(4), 589–603 (2020)
21. Vanderwal, T., Eilbott, J., Castellanos, F.X.: Individual differences in functional connectivity during naturalistic viewing conditions. Neuroimage **157**, 521–530 (2017)
22. Vanderwal, T., Eilbott, J., Castellanos, F.X.: Movies in the magnet: naturalistic paradigms in developmental functional neuroimaging. Dev. Cogn. Neurosci. **47**, 100893 (2021)
23. Verma, B., Thakur, R.S.: Sentiment analysis using lexicon and machine learning-based approaches: a survey. In: Tiwari, B., Tiwari, V., Das, K.C., Mishra, D.K., Bansal, J.C. (eds.) Proceedings of International Conference on Recent Advancement on Computer and Communication. LNNS, vol. 34, pp. 441–447. Springer, Singapore (2018). https://doi.org/10.1007/978-981-10-8198-9_46
24. Wagner, J., et al.: Dawn of the transformer era in speech emotion recognition: closing the valence gap. IEEE Trans. Pattern Anal. Mach. Intell. 1–13 (2023)
25. Yeo, B.T., Krienen, F.M., Sepulcre, J., et al.: The organization of the human cerebral cortex estimated by intrinsic functional connectivity. J. Neurophysiol. **106**(3), 1125–1165 (2011)
26. Zhang, L., Wang, S., Liu, B.: Deep learning for sentiment analysis: a survey. IEEE Trans. Knowl. Data Eng. **30**(10), 2022–2043 (2018)

International Workshop on Advances in Drone Vision (ADV 2025)

Workshop

International Workshop on Advances in Drone Vision (ADV 2025)

In conjunction with ICIAP 2025—Rome, Italy, September 15–19, 2025

Workshop Organization

Organizers

Gennaro Vessio   University of Bari Aldo Moro, Italy
Giovanna Castellano   University of Bari Aldo Moro, Italy
Pasquale De Marinis   University of Bari Aldo Moro, Italy
Mario Vento   University of Salerno, Italy

Program Committee

Corrado Mencar   University of Bari Aldo Moro, Italy
Ciro Castiello   University of Bari Aldo Moro, Italy
Davide Moroni   Institute of Information Science and Technologies, National Research Council, Italy
Vincenzo Polizzi   STARS Laboratory, University of Toronto, Canada
Alessia Saggese   University of Salerno, Italy
Davide Boscaini   Fondazione Bruno Kessler, Italy
Wencheng Zhu   Tianjin University, China

Application of Object-Based Image Analysis to Drone Data for the Monitoring of Methane Emissions in Landfills

Maurizio De Molfetta[1] , Giovanni Dimauro[2] , Donatello Fosco[3]([✉]) ,
Bruno Notarnicola[1] , and Pietro Alexander Renzulli[1]

[1] Ionian Department of Law, Economics and Environment, University of Bari Aldo Moro, 70121 Bari, BA, Italy
[2] Department of Informatica, University of Bari Aldo Moro, 70125 Bari, BA, Italy
[3] Department for the Promotion of Human Science and Quality of Life, San Raffaele Roma University, Via di Val Cannuta, 247, 00166 Rome, Italy
donatello.fosco@uniba.it

Abstract. Reducing anthropogenic methane (CH_4) emissions is recognised as one of the most effective strategies for mitigating global warming in the near term. Among anthropogenic sources, landfills represent a major and persistent contributor to CH_4 emissions. Current monitoring techniques—such as flux chambers in Europe and Surface Emission Monitoring (SEM) in the United States—suffer from critical limitations, including low spatial resolution, limited coverage, and inaccessibility to key emission zones such as steep slopes and infrastructure components. To address these limitations, this study presents an operational workflow based on high-resolution unmanned aerial vehicle (UAV) imaging and Object-Based Image Analysis (OBIA). The method integrates RGB and multispectral data acquired via UAVs within the QGIS environment to enable semi-automatic mapping of biogas extraction wells and the identification of areas with a higher probability of methane emissions. This approach enhances spatial and temporal coverage, enabling improved detection of anomalies such as vegetation stress, surface fissures, and depressions. By overcoming the limitations of traditional ground-based monitoring, the proposed UAV-OBIA framework provides a robust, non-invasive, and scalable tool for modern landfill management. The integration of remote sensing and image analysis supports more targeted methane mitigation strategies, contributes to improved operational safety, and aligns with broader climate objectives related to greenhouse gas emission reduction.

Keywords: Methane emission · landfill monitoring · high-resolution UAV imaging · Object-Based Image Analysis · UAV based sensing

1 Introduction

The reduction of anthropogenic methane (CH_4) emissions represents one of the most effective strategies for mitigating global warming in the short term. Methane is a potent greenhouse gas with a global warming potential (GWP) significantly greater than that of

E. Rodolà et al. (Eds.): ICIAP 2025 Workshops, LNCS 16169, pp. 105–115, 2026.
https://doi.org/10.1007/978-3-032-11317-7_9

carbon dioxide (CO_2), exerting a warming effect 86 times stronger over a 20-year time horizon and 30 times stronger over 100 years [1].

Anthropogenic CH_4 emissions are responsible for approximately 60% of the total atmospheric methane concentration and have contributed to nearly half of the observed global temperature increase, which has risen by 1.2 °C above pre-industrial levels [2].

Among the main anthropogenic sources of methane, landfills represent a substantial and persistent contributor [3].

Currently, the most widely adopted techniques for monitoring surface methane emissions from landfills include the flux chamber method, predominantly used in Europe, and Surface Emission Monitoring (SEM), commonly employed in the United States.

In Europe, the flux chamber method involves measuring the change in CH_4 concentration within small, sealed chambers placed directly on the landfill surface. Initially developed in the 1960s [4] and later formalised by UK technical guidance in 2010 [5], this technique is valued for its operational simplicity and relatively low cost [6]. Nonetheless, several studies [7–10] have demonstrated that the flux chamber method tends to capture only a small portion of the hotspots on the landfill body. The method in fact is not suitable for detecting emissions from plant infrastructure (e.g. gas extraction wells) or from fissures in inaccessible areas, such as steep lateral slopes, which often constitute significant emission sources [11].

In the United States, Surface Emission Monitoring (SEM) is employed as an alternative for assessing diffuse methane emissions across landfill surfaces. This method entails the use of portable instruments to perform point-based measurements of CH_4 concentration along predefined grid or pathway patterns. Compared to flux chambers, it shares several limitations, including sparse sampling and a low probability of detecting highly localised emission sources. Furthermore, SEM effectiveness can be compromised by meteorological variability during data collection, which alters the site's emission profile [12–15].

Both techniques, therefore, suffer from substantial limitations in terms of spatial and temporal coverage, often failing to identify critical high-emission areas (hotspots). Low sampling density and the inability to access or monitor certain landfill regions significantly constrain the reliability and completeness of these traditional approaches.

Given the inherent complexity of landfill environments, more comprehensive and integrated monitoring methodologies are needed. Advances in remote sensing technologies, particularly those involving unmanned aerial vehicles (UAVs), offer promising alternatives. High-resolution aerial imaging, when integrated with data from environmental sensors, can substantially enhance both spatial and temporal resolution, thereby improving the detection and quantification of methane emissions. UAV-based approaches also reduce human exposure to hazardous conditions, increasing safety and efficiency.

In this context, the use of drones equipped with optical or multispectral sensors, integrated with Object-Based Image Analysis (OBIA) approaches, is emerging as an effective methodology for non-invasive monitoring with high spatial resolution of landfill surfaces [16, 17].

The OBIA approach allows the segmentation of images into homogeneous objects in terms of shape, colour, and context, overcoming the limitations of pixel-based analyses, particularly in cases where it is necessary to distinguish between artificial elements (wells) and natural anomalies (vegetation stress, water stagnation, etc.) [18].

Numerous recent studies demonstrate the potential to identify emission zones through the analysis of vegetation anomalies using specific indices, such as the Normalized Difference Vegetation Index (NDVI) and the Soil-Adjusted Vegetation Index (SAVI), as well as thermal alterations and topographic variations derived from digital models [19–24].

In summary, the integration of high-resolution drone imaging with multi-sensor data fusion presents a powerful, cost-effective, and scalable solution for modern landfill monitoring.

The present study introduces an operational OBIA workflow implemented within the QGIS environment. It is applied to RGB and multispectral data acquired via drones, with the aim of semi-automatically mapping biogas extraction wells and identifying areas with a higher probability of methane emissions. These areas can subsequently be subjected to further investigation through the integration of additional data acquired using drone-mounted tuneable diode laser absorption spectroscopy (TDLAS) sensors.

This approach goes beyond the limitations of traditional ground-based techniques, providing a more detailed and multidimensional view of both methane emissions and operational risks. Such innovations support more sustainable waste management practices and contribute to global efforts to reduce greenhouse gas emissions in line with international climate targets. By enabling the identification of high-probability emission areas and the mapping of gas extraction infrastructure, the proposed OBIA workflow contributes to improved site management and enhanced landfill gas capture efficiency.

2 Materials and Method

2.1 Flight Parameters and Characteristics of the RGB Ortho-Mosaic

The survey was carried out over a selected portion of the "Malagrotta" landfill, which has been the subject of previous environmental studies and analyses. Photogrammetric acquisitions were carried out using DJI Matrice 4E, a multirotor drone designed for surveying and mapping applications. A high-resolution RGB camera, model M4E, with a maximum resolution of 5280 × 3956 pixels, was used for image acquisition. The camera is equipped with a fixed focal length lens of 12.29 mm.

Image stability during flight is ensured by a three-axis mechanical gimbal (tilt, roll, pan). The drone is also equipped with a Real-Time Kinematic (RTK) positioning system, which enables high-precision geolocation of the imagery.

Flight missions were carefully planned to ensure homogeneous coverage of the area of interest, while maintaining a high density of image overlap along and between flight paths (overlap and sidelap). This configuration is essential to optimise the outcomes of photogrammetric processing based on Structure-from-Motion (SfM) techniques, enhancing the accuracy of 3D reconstruction and the generation of high-resolution Digital Surface Models (DSM) and ortho-mosaics. Table 1 summarises the main information relating to flight.

The RGB ortho-mosaic was generated using Agisoft Metashape Professional (v. 2.1.1.17641), based on the aligned images and high-quality depth maps. A mild smoothing filter was applied during processing to preserve fine surface details while avoiding

Table 1. Summary of key flight parameters.

Parameter	value
Average flight altitude (AGL)	27.9 m
Total number of images acquired	592
Camera model	M4E (5280 × 3956 px)
Ground Sampling Distance (GSD)	8.52 mm/pixel
Image overlap (longitudinal/lateral)	>80%
Surveyed area	0.0505 km^2

excessive generalisation. In Table 2, the main descriptive parameters of the ortho-mosaic are summarised.

Table 2. Main characteristics of the orthomosaic.

Parameter	value
Final dimensions	37,168 × 33,741 pixels
Radiometric resolution	3 bands (RGB, 8-bit per band)
Geographic coordinate system	WGS 84 (EPSG:4326)
Surface model used	Mesh derived from depth maps
Interpolation	Enabled
Blending	Mosaic mode with automatic gap filling
Ortho-mosaic processing time	~29 min
Final file size	18.71 GB

These specifications make the ortho-mosaic suitable for object-based spatial analysis (OBIA), where the radiometric and geometric integrity of the images is a fundamental requirement for the accurate identification of ground objects, such as biogas extraction wells or potentially emissive anomalous areas.

2.2 Application of the OBIA Protocol

The OBIA operational workflow was implemented within the QGIS environment using the Processing, SCP, and Orfeo Toolbox (OTB) plugins. The initial phase—also the most time-consuming—was the segmentation process, performed using the "Segmentation" tool provided by the OTB suite. Among the available algorithms, the MeanShift method was selected following comprehensive comparative tests with alternative techniques, including Large MeanShift and Watershed. MeanShift offered the best compromise between segmentation quality, background noise reduction, processing time, and system resource requirements.

This non-parametric algorithm, suitable for both clustering and segmentation tasks, is particularly well-adapted for processing medium- to high-resolution ortho-mosaics. It simultaneously considers the spatial coordinates and spectral values of the pixels, which in this study were acquired across multiple bands (RGB, NIR, and Red Edge).

The segmentation was carried out directly on the RGB ortho-mosaic using a set of parameters, which are summarised in Table 3.

Table 3. Segmentation parameters used.

Parameter	Description	value
Spatial radius	Maximum spatial distance between pixels (px)	10
Range radius	Maximum spectral difference (Digital Number)	30
Minimum region size	Minimum segment size (pixels)	300

The output of the segmentation was a vector shapefile (.shp), which served as the foundation for the subsequent classification stage. Following the segmentation, a supervised classification was carried out using the "Train Images Classifier" and "Image Classifier" modules. A training dataset was manually created as a shapefile layer, populated with Regions of Interest (ROIs) delineated on the orthomosaic. These ROIs corresponded to the visible wellhead valves of the biogas extraction wells. The selection of wellhead valves as classification targets was informed by their distinctive spectral and geometric characteristics, the absence of other similar metallic objects in the surveyed area, and their clear recognisability in nadir-view RGB imagery. A total of 17 ROIs were used for training.

Due to the high spectral homogeneity of the target class, this limited sample size was sufficient to achieve effective training. The classification was performed using the K-Nearest Neighbours (KNN) algorithm, with $k = 5$, a value selected following preliminary tuning over a range from 3 to 9. Although the NDVI index was initially considered as a pre-filter to exclude vegetated regions before segmentation, this step was ultimately deemed unnecessary due to the satisfactory performance and efficiency of the segmentation process.

A 5-fold cross-validation approach was adopted to assess the model's robustness and to mitigate overfitting. Class imbalance was addressed through weighted sampling. The final classification yielded a binary thematic map distinguishing between well and non-well segments, which was then used for evaluation and mapping purposes. The methodological flowchart is summarised in Fig. 1.

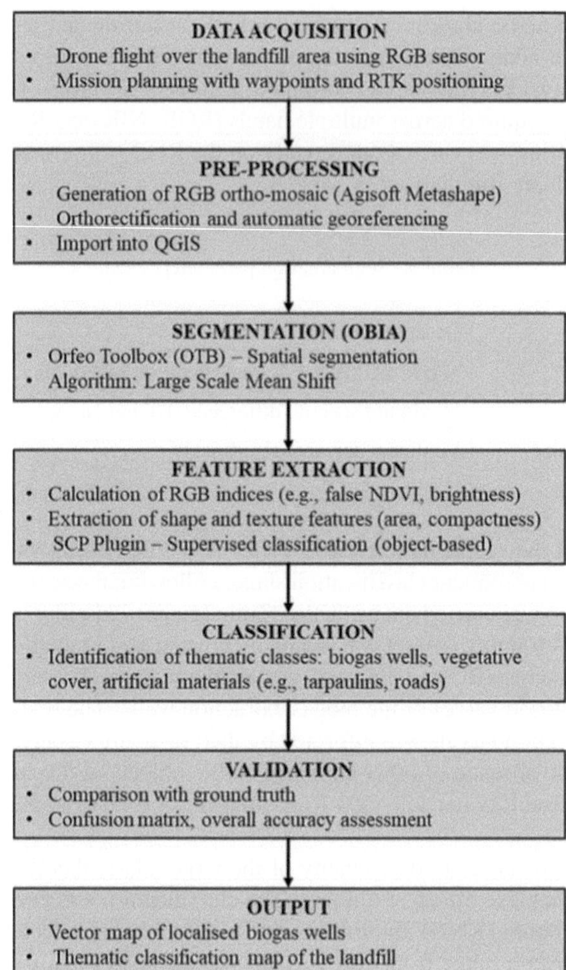

Fig. 1. Flowchart of the OBIA Methodology for RGB Drone Data Processing.

2.3 Performance Metrics

In this study, model performance was assessed using three fundamental classification metrics: precision, recall, and the F1-score. These metrics provide a comprehensive understanding of the algorithm's ability to correctly identify relevant instances while managing false detections.

Precision quantifies the proportion of correctly identified positive cases among all instances labelled as positive by the model. In other words, it reflects the accuracy of positive predictions, indicating how many of the detected targets are true positives rather than false alarms.

Recall, also referred to as sensitivity, measures the model's effectiveness in retrieving all actual positive instances within the dataset. It is defined as the ratio of true positive detections to the total number of actual positives, highlighting the completeness of the classification with respect to the positive class.

The F1-score serves as a harmonic mean of precision and recall, offering a balanced single metric that considers both false positives and false negatives. This measure is particularly useful when the dataset is imbalanced or when it is important to find a trade-off between precision and recall. Mathematically, these metrics are defined as:

$$Precision = \frac{TP}{TP + FP} \tag{1}$$

$$Recall = \frac{TP}{TP + FN} \tag{2}$$

$$F1 - score = 2\frac{Precision * Recall}{Precision + Recall} \tag{3}$$

where TP represents true positives, FP false positives, and FN false negatives. By focusing on these three metrics, the evaluation highlights both the reliability of positive predictions and the model's capacity to capture relevant targets, providing a clear and effective framework to compare algorithmic performance in detection tasks.

3 Results and Discussion

3.1 Geometric and Radiometric Quality of the Photogrammetric Processing

The high number of tie points (601,513) and image projections (over 2.3 million) enabled a robust spatial reconstruction, with a mean reprojection error of 0.806 pixels. This level of accuracy is consistent with the precision standards required for detailed environmental and engineering surveys.

Camera positions were estimated with an average horizontal (X-Y) accuracy of 8.14 cm. However, the vertical (Z-axis) error was significantly higher, reaching 84.54 cm. This discrepancy is likely due to the absence of ground control points (GCPs) and the use of an automated flight configuration, which limits georeferencing accuracy along the vertical axis. This suggests that, for applications requiring high absolute accuracy, the integration of GCPs is strongly recommended.

The ortho-mosaic exhibits good radiometric uniformity, with accurate colour representation and clear distinction between different surface types. This is essential for the application of Object-Based Image Analysis (OBIA) techniques, which rely on the segmentation of images into objects that are homogeneous in terms of colour, shape, and texture. In particular, the spatial resolution allows for the identification of fine details, while radiometric consistency facilitates surface classification. However, the limited absolute accuracy may affect the precise geolocation of identified objects, highlighting the importance of using Ground Control Points (GCPs) to improve the overall accuracy of the model. In this case, GCPs were not employed due to the UAV's integrated RTK system, which was deemed sufficient given the minimal morphological and altimetric variation of the survey area, which was found to be completely flat.

3.2 Evaluation OBIA Analysis

The geometric and radiometric quality of the ortho-mosaic makes the acquired data suitable for the application of OBIA methodologies. In particular, the spatial resolution allows for the identification of fine details, while radiometric consistency facilitates surface classification. However, the limited absolute accuracy may affect the precise geolocation of the identified objects, emphasising the importance of using Ground Control Points (GCPs) to enhance the overall accuracy of the model. In this specific case, GCPs were not employed due to the presence of an RTK system integrated into the UAV, which was considered sufficient given the minimal morphological and altimetric variation of the surveyed area, which was found to be completely flat.

The results demonstrated high accuracy in discriminating the target objects, validating the effectiveness of the OBIA methodology for automated identification of biogas wells in complex environments. Cross-validation with independent test samples allowed estimation of classification performance metrics, yielding a precision of 0.87, recall of 0.82, and an overall F1-score of 0.84. These values indicate a favourable balance between minimising false positives and maximising target detection completeness.

The algorithm successfully detected all biogas extraction wells within the study area, demonstrating robust object recognition capabilities even under conditions of partial occlusion or surface interference (e.g., vegetation or debris). Cross-validation results confirmed a high degree of classification consistency, with an F1-score standard deviation below 3%, indicating the model's reliability across different subsets. These findings highlight the value of OBIA in leveraging both radiometric and geometric cues for accurate object identification, particularly in complex environments where pixel-based methods often struggle due to noise or over-segmentation.

The described approach is summarised in Fig. 2, which illustrates the location of the biogas extraction wells (a) and the identification of relevant features (b).

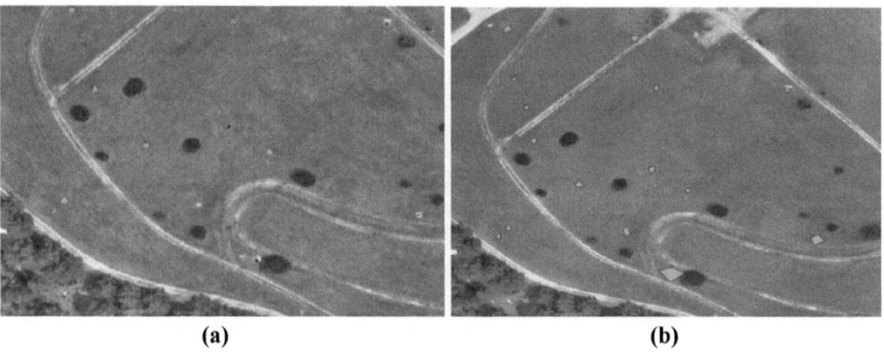

(a)   (b)

Fig. 2. Location of biogas extraction wells within the study area (a) and identification of relevant features using the Object-Based Image Analysis (OBIA) approach (b).

4   Conclusion

This study demonstrated the effectiveness of an operational workflow based on Object-Based Image Analysis (OBIA) techniques applied to RGB data acquired via UAVs, for non-invasive landfill monitoring and the automated identification of biogas extraction wells and areas with a higher probability of methane emissions.

The photogrammetric processing produced a robust three-dimensional reconstruction, supported by a high number of tie points (601,513) and over 2.3 million image projections, with a mean reprojection error of 0.806 pixels. This level of precision meets the standards required for high-resolution environmental and engineering surveys. However, absolute vertical accuracy was significantly lower (average error of 84.54 cm) compared to horizontal accuracy (8.14 cm), highlighting the importance of integrating Ground Control Points (GCPs) in scenarios requiring precise georeferencing, despite the use of an onboard RTK system.

The resulting ortho-mosaic exhibited good radiometric uniformity and clear differentiation between surface types—both critical for the effective application of the OBIA approach. Classification of target objects yielded high accuracy, with an average precision of 0.87, recall of 0.82, and a total F1-score of 0.84, indicating a favourable balance between reducing false positives and ensuring comprehensive target detection.

Overall, the results confirm that the integration of high-resolution drone data with object-based analysis methods provides a robust, scalable, and low-impact solution for modern landfill monitoring.

This study focuses on the operational use of OBIA with drone-acquired data but does not include direct comparisons with pixel-based or deep learning models such as CNNs or U-Nets. OBIA was chosen for its interpretability, easy integration with open-source GIS platforms like QGIS, and strong performance in high-resolution settings. Future work will explore benchmarking against alternative models to validate and potentially enhance accuracy through hybrid or ensemble methods.

The workflow relies on indirect indicators to identify methane emission hotspots. While suitable for initial detection, it does not provide quantitative CH_4 flux estimates. Integration with direct sensing technologies, such as UAV-mounted TDLAS systems, is required for precise measurement and mitigation planning.

A further limitation is the lack of comparison with methods like semantic segmentation via deep learning. Future developments will incorporate and test advanced machine learning models on annotated datasets to assess performance.

Finally, although the model performed well without Ground Control Points (GCPs), precise geolocation would benefit from their inclusion to reduce vertical error.

This approach overcomes the limitations of traditional ground-based techniques by enhancing spatial and temporal coverage, reducing operator exposure to hazardous conditions, and supporting more targeted methane mitigation strategies. The adoption of such innovative tools contributes to more sustainable landfill management and aligns with global climate targets for the reduction of greenhouse gas emissions.

5 Disclosure of Interests

The authors have no competing interests to declare that are relevant to the content of this article.

Acknowledgements. This research was co-funded by the Complementary National Plan PNC1.1 "Research initiatives for innovative technologies and pathways in the health and welfare sector" D.D. 931 of 06/06/2022, DARE – DigitAl Lifelong pRevEntion initiative, code PNC0000002, CUP B53C22006420001.

References

1. Johnson, D., Heltzel, R.: Methane emissions measurements of natural gas components using a utility terrain vehicle and portable methane quantification system. Atmos. Environ. **144**, 1–7 (2016)
2. International Energy Agency (IEA). World Energy Outlook (2023)
3. European Environment Agency (EEA). Annual European Union greenhouse gas inventory 1990–2020 and inventory report 2022. Submission to the UNFCCC (2022)
4. Pearson, J.E., Jones, G.E.: Emanation of radon 222 from soils and its use as a tracer. J. Geophys. Res. **70**(20), 5279–5290 (1965)
5. Environment Agency (EA). Monitoring landfill gas surface emissions: LFTGN 07. Bristol, UK: Environment Agency (2010)
6. Lucernoni, F., Rizzotto, M., Tapparo, F., Capelli, L., Sironi, S., Busini, V.: Use of CFD for static sampling hood design: an example for methane flux assessment on landfill surfaces. Chemosphere **163**, 259–269 (2016)
7. Börjesson, G., Danielsson, Å., Svensson, B.H.: Methane fluxes from a Swedish landfill determined by geostatistical treatment of static chamber measurements. Environ. Sci. Technol. **34**(18), 4044–4050 (2000)
8. Environmental Protection Agency (EPA). EPA Handbook. Optical and Remote Sensing for Measurement and Monitoring of Emissions Flux of Gases and Particulate Matter (2018)
9. Wong, C.L.Y.: Analysis of the number of flux chamber samples and study area size on the accuracy of emission rate measurements. J. Air Waste Manag. Assoc. **68**(10), 1103–1117 (2018)
10. Czepiel, P.M., et al.: The influence of atmospheric pressure on landfill methane emissions. Waste Manag. **23**(7), 593–598 (2003)
11. Gonzalez-Valencia, R., Magana-Rodriguez, F., Cristóbal, J., Thalasso, F.: Hotspot detection and spatial distribution of methane emissions from landfills by a surface probe method. Waste Manag. **55**, 299–305 (2016)
12. Omidi, A., et al.: Most Landfill Methane Emissions Escape Detection in EPA21 Surface Emission Monitoring Surveys. Preprint (2025)
13. Abichou, T., Bel Hadj Ali, N., Amankwah, S., Green, R., Howarth, E.S.: Using ground- and drone-based Surface Emission Monitoring (SEM) data to locate and infer landfill methane emissions. Methane **2**(4), 440–451 (2023)
14. Bel Hadj Ali, N., Abichou, T., Green, R.: Comparing estimates of fugitive landfill methane emissions using inverse plume modeling obtained with Surface Emission Monitoring (SEM), Drone Emission Monitoring (DEM), and Downwind Plume Emission Monitoring (DWPEM). J. Air Waste Manag. Assoc. **70**(4), 410–424 (2020)

15. Kormi, T., Mhadhebi, S., Ali, N.B.H., Abichou, T., Green, R.: Estimation of fugitive landfill methane emissions using surface emission monitoring and Genetic Algorithms optimization. Waste Manag. **72**, 313–328 (2018)
16. Brovelli, M.A., Oxoli, D., Ballabeni, A.: UAV-based landfill monitoring for environmental and safety assessment. Remote Sens. **12**(3), 385 (2020)
17. Maimaitijiang, M., Ghulam, A., Sidike, P., et al.: UAV-based remote sensing for precision agriculture: a comprehensive review. IEEE Geosci. Remote Sens. Mag. **8**(3), 47–70 (2020)
18. Blaschke, T.: Object based image analysis for remote sensing. ISPRS J. Photogramm. Remote Sens. **65**(1), 2–16 (2010)
19. Sedano-Cibrián, J., de Luis-Ruiz, J.M., Pérez-Álvarez, R., Pereda-García, R., Tapia-Espinoza, J.D.: 4D Models generated with UAV photogrammetry for landfill monitoring thermal control of Municipal Solid Waste (MSW) landfills. Appl. Sci. **13**(24), 13164 (2023)
20. Lewis, A.W., Yuen, S.T.S., Smith, A.J.R.: Detection of gas leakage from landfills using infrared thermography - applicability and limitations. Waste Manag. Res. **21**(5), 436–447 (2003)
21. Mønster, J., Kjeldsen, P., Scheutz, C.: Methodologies for measuring fugitive methane emissions from landfills – a review. Waste Manag. **87**, 835–859 (2019)
22. Scheutz, C., Samuelsson, J., Fredenslund, A.M., Kjeldsen, P.: Quantification of multiple methane emission sources at landfills using a double tracer technique. Waste Manag. **31**(5), 1009–1017 (2011)
23. Kissas, K., Ibrom, A., Kjeldsen, P., Scheutz, C.: Methane emission dynamics from a Danish landfill: the effect of changes in barometric pressure. Waste Manag. **138**, 234–242 (2019)
24. Fosco, D., De Molfetta, M., Renzulli, P., Notarnicola, B.: Progress in monitoring methane emissions from landfills using drones: an overview of the last ten years. Sci. Total Environ. **945**, 173981 (2024)

UKANWeed: An Application of Kolmogorov-Arnold Networks to Weed Mapping

Pasquale De Marinis[(✉)], Elena Tavoletti, Gennaro Vessio, and Giovanna Castellano

Department of Computer Science, University of Bari Aldo Moro, Bari, Italy
{pasqualede.marinis,elena.tavoletti,gennaro.vessio,
giovanna.castellano}@uniba.it

Abstract. Effective weed monitoring is a critical component of precision agriculture, directly impacting crop health and yield. This study presents UKANWeed, an application-specific adaptation of the recently proposed UKAN architecture, designed for image segmentation in agricultural settings. UKANWeed integrates Kolmogorov-Arnold Networks (KAN) into its layers, resulting in a lightweight yet highly efficient model that can distinguish weeds from crops. Evaluated on the WeedMap dataset, UKANWeed achieves an F1-score of 86.3, outperforming the widely used UNet architecture while requiring significantly fewer parameters. Additionally, we investigated the behavior of UKANWeed and UNet under extreme model compression, revealing a lower bound in the representational capacity below which task performance degrades sharply. The compactness and accuracy of UKANWeed make it suitable for deployment on edge devices such as drones, enabling real-time, in-field weed detection and crop monitoring—an essential step toward scalable, autonomous precision agriculture. The code is available at https:// github.com/pasqualedem/UKANWeed.

Keywords: Precision Agriculture · Weed Mapping · Unmanned Aerial Vehicles · Semantic Segmentation · Kolmogorov-Arnold Networks

1 Introduction

Agriculture is fundamental to human sustenance, and advancements in agricultural machinery and techniques have significantly improved crop yields. Among these, weed management is crucial for removing unwanted plants that compete with cultivated crops. Effective weed control enhances field productivity and promotes sustainable agricultural practices [21].

Drones, also known as unmanned aerial vehicles (UAVs), have become invaluable tools across various fields, including crowd analysis [3,8], wildfire detection [9], and precision agriculture [6,20], offering both versatility and cost-effectiveness. These devices can capture high-resolution imagery and data from agricultural fields, enabling farmers to monitor crop development, detect diseases and pests, and optimize irrigation strategies. By providing timely and accurate

E. Rodolà et al. (Eds.): ICIAP 2025 Workshops, LNCS 16169, pp. 116–127, 2026.
https://doi.org/10.1007/978-3-032-11317-7_10

information, drones help reduce operational costs, increase crop yield, and minimize the use of inputs such as water, fertilizers, and pesticides. Traditional methods, including manual field inspections or satellite-based remote sensing, cannot match the level of detail and immediacy offered by drone-based data collection.

UAVs are capable of conducting real-time aerial surveys of crops, allowing farmers to make rapid, informed decisions. Furthermore, drones can efficiently cover large areas within hours—a task that would typically require days or even weeks using conventional approaches. This time-saving capability enables farmers to conserve resources and respond more quickly to field conditions.

Recent research has suggested that deep learning models can be applied to semantic segmentation and weed identification in aerial images acquired by drones [2,18,19]. However, despite these notable advances, automatic weed detection remains a complex challenge. Deep learning techniques are not yet widely adopted in agriculture, primarily due to the extensive manual annotation required to generate large volumes of training data. This issue is especially pronounced in agricultural datasets, where labeling plants in field images is a labor-intensive process. Additionally, these models often demand substantial computational resources, presenting obstacles to deployment on drones, which are constrained by limited processing capabilities and energy autonomy. The demand for real-time performance, combined with these limitations, highlights the need for lightweight solutions specifically designed for weed mapping applications.

In this paper, we propose UKANWeed, a lightweight deep learning architecture for semantic segmentation in weed mapping from drone-acquired imagery. UKANWeed is based on UKAN, the encoder–decoder architecture recently introduced by Li et al. [13] for medical image segmentation, and adapts it to the agricultural domain. Specifically, UKANWeed inherits the symmetric UNet-style design [16] and replaces selected convolutional blocks with Kolmogorov–Arnold Network (KAN) modules to improve functional expressiveness while maintaining a compact parameter count—an essential property for edge deployment on UAVs. KAN layers are based on the Kolmogorov-Arnold representation theorem [10], which states that any multivariate continuous function can be represented as a finite composition of univariate continuous functions. Unlike traditional layers that use linear combinations, KAN layers parameterize these univariate functions using splines, allowing the network to approximate complex functions more accurately and efficiently.

This architecture addresses the dual challenge of achieving high predictive accuracy and enabling deployment on resource-constrained platforms. Evaluation is performed on the WeedMap dataset [17], which features multispectral drone imagery of sugar beet fields, offering a realistic and demanding benchmark for weed detection.

The rest of this paper is organized as follows: Sect. 2 reviews related work in weed mapping and KAN layers; Sect. 3 describes the dataset used for evaluation; Sect. 4 details the UKANWeed architecture and its components; Sect. 5 presents the experimental setup and results; and Sect. 6 concludes the paper with a summary of findings and future work.

2   Related Work

Advancements in computer vision and remote sensing have revolutionized precision agriculture by addressing tasks such as disease and pest identification, assessing abiotic stress, monitoring crop growth, predicting yields, and mapping weeds.

Weed mapping, a semantic segmentation task, involves assigning each pixel in an image to either the weed or crop class. Deep learning algorithms have demonstrated superior performance compared to traditional techniques in this domain. Early work by Dos Santos et al. [19] highlighted the effectiveness of Convolutional Neural Networks (CNNs), such as AlexNet, over Support Vector Machines (SVMs) and Random Forests. Lottes et al. [14] further advanced the field by employing a CNN with dual decoders for stem detection and plant segmentation, showing promising results on the BoniRob and UAV datasets.

The integration of multispectral imagery, which captures detailed information about plant health and species, enhances the accuracy of deep learning models compared to those relying solely on RGB imagery. For example, UNet has been successfully employed to separate weeds from crops and soil [5]. WeedNet, based on the SegNet architecture and trained on the WeedMap dataset, is another successful application in this domain [18]. The WeedMap dataset, comprising multispectral images of sugar beet fields in Germany and Switzerland, has become a benchmark in weed mapping studies [17]. We also use this dataset due to its benchmark status, although we restrict our analysis to the RGB-relevant channels.

On the other hand, KAN layers have emerged as a promising approach to enhance the expressive power of neural networks [13]. These layers are based on the Kolmogorov-Arnold representation theorem, which states that any continuous multivariate function can be expressed as a finite combination of continuous univariate functions [10]. KAN has been proven to be effective in Computer Vision tasks, such as image classification [4], and has been successfully applied to medical imaging [11].

A technically effective solution for semantic segmentation is the integration of KAN layers into the core of a UNet, addressing the difficulty that encoder-decoder architectures face in extracting comprehensive information from high-dimensional feature representations. Notably, Ma et al. [15] demonstrated that the use of KAN for remote sensing is a successful strategy, achieving accuracy levels surpassing the current state of the art. Given the similarities between remote sensing and drone vision tasks, it is worthwhile to investigate whether the application of KANs is also advantageous in precision agriculture. In particular, we employ the UKAN implementation provided by Li et al. [11].

3   Materials

The experiments were conducted on the WeedMap dataset [17], specifically using RGB images generated from data collected by the RedEdge camera. For each

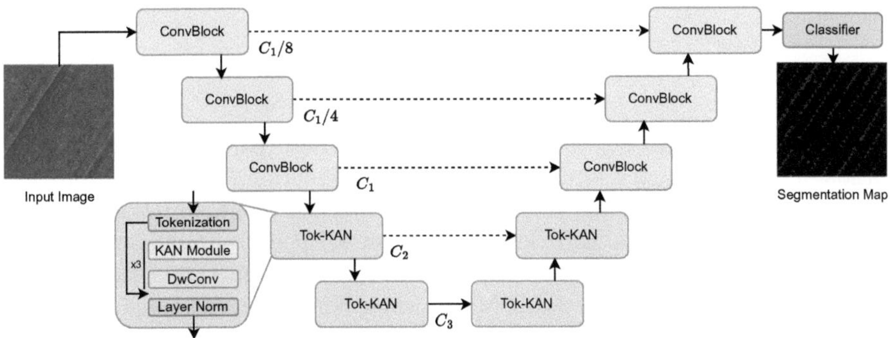

Fig. 1. UKANWeed architecture. The encoder consists of three convolutional blocks, followed by two Tokenized-KAN blocks, and the decoder mirrors the encoder with skip connections. A ConvBlock is a sequence of two convolutional layers, each followed by batch normalization and a ReLU activation function. The Classifier is a 1×1 convolutional layer that outputs the final segmentation map.

field, the dataset curators constructed a composite RGB image, referred to as an ortho-mosaic, by stitching together multiple photographs captured by drones. This ortho-mosaic was subsequently divided into tiles of size 480×360 pixels.

Images from fields 000, 001, 002, and 004 were used for model training, while images from field 003 were reserved for testing. An additional split was created for validation, with 20% of the training images set aside for this purpose.

It was observed that many training images contained a disproportionately high number of background pixels, leading to a class imbalance. Consequently, training images consisting solely of background were excluded from the dataset to mitigate this issue.

4   Methods

In this work, we base our UKANWeed framework on UKAN, the architecture proposed by Li et al. [11], initially tested on medical imaging datasets. In that context, UKAN achieved superior performance compared to the conventional UNet architecture, even with reduced computational cost.

This performance improvement is due to the architectural innovation at the core of UKAN: the replacement of traditional convolutional or MLP-based modules with KAN (Kolmogorov-Arnold Network) blocks. Unlike MLPs, which learn weights for linear combinations followed by activation functions, KANs learn parameters for more complex functions. This enhances the expressive power of the network, a property theoretically supported by the Kolmogorov-Arnold representation theorem. This theorem states that any continuous multivariate function can be expressed as a finite combination of continuous univariate functions, formally:

$$f(x) = \sum_q \phi_q \left(\sum_p \psi_{pq}(x_p) \right)$$

The functions ϕ_q and ψ_{pq} are spline functions whose parameters are learned, although in practice, ϕ_q is often set to the identity function I. This result provides a stronger theoretical foundation than the Universal Approximation Theorem typically associated with MLP layers: while an MLP can approximate a target function, a KAN layer can, in principle, learn the exact function. Therefore, one may expect KANs to outperform MLPs when both have the same number of parameters.

4.1   UKANWeed

The UKANWeed architecture—depicted in Fig. 1—follows a fully symmetric encoder-decoder design with skip connections between corresponding encoder and decoder levels. Each encoder layer reduces the image height and width by half while the number of output channels is treated as a hyperparameter. For the first three convolutional layers, only the number of channels C_1 in the third layer is explicitly defined; the previous two layers use one-quarter and one-eighth of that number, respectively.

Each encoder block consists of a ConvBlock, which is a sequence of two convolutional layers, each followed by batch normalization and a ReLU activation function. The ConvBlock is followed by a max pooling operation that reduces the spatial dimensions by half and a ReLU activation function:

$$X^{(i+1)} = \text{MaxPool}\left(\text{ReLU}\left(\text{ConvBlock}(X^{(i)}) \right) \right)$$

where $X^{(i)}$ is the input tensor to the i-th encoder block, and MaxPool denotes the max pooling operation. The final two encoder levels are implemented as Tokenized-KAN blocks, each consisting of a Tokenization Module, a KAN Module, and a Post-KAN Module.

The Tokenization Module is responsible for transforming spatial feature maps into token representations suitable for downstream processing [7]. Let the input batch be denoted as $I \in \mathbb{R}^{B \times C \times H \times W}$, where B is the batch size, C the number of channels, and $H \times W$ the spatial dimensions. A convolutional layer with a kernel size of 3×3—aligned with the token size—is applied, producing output feature maps $I' \in \mathbb{R}^{B \times C' \times H' \times W'}$, where C' is the number of output channels, and H', W' are determined by the convolution stride and padding. Each feature map is then reshaped into a matrix $T \in \mathbb{R}^{(H'W') \times C'}$, where each row corresponds to a token and each column to a feature dimension. Concatenating these matrices for all images in the batch yields a global token matrix $T_{\text{batch}} \in \mathbb{R}^{(B \cdot H' \cdot W') \times C'}$. The dimensions B, H', W' are stored to enable later de-tokenization.

The KAN Module applies a series of nonlinear transformations to the token matrix while preserving its shape. Let the input to this module be $T^{(0)} = T_{\text{batch}} \in \mathbb{R}^{N \times D}$, where $N = B \cdot H' \cdot W'$ and $D = C'$. The matrix passes through three

KAN layers, each applying a learnable structured transformation $\boldsymbol{\Phi}_i$ followed by a depth-wise convolution ($DWConv$) [1]:

$$T^{(i+1)} = DWConv(\boldsymbol{\Phi}_i(T^{(i)})) \quad \text{for } i = 0, 1, 2$$

Each $\boldsymbol{\Phi}_i$ computes a sequence of K structured nonlinear transformations Φ_k based on functional decomposition. Each transformation, in matrix form, can be expressed as:

$$\Phi_k = \begin{pmatrix} \phi_{k,1,1}(\cdot) & \phi_{k,1,2}(\cdot) & \cdots & \phi_{k,1,n_k}(\cdot) \\ \phi_{k,2,1}(\cdot) & \phi_{k,2,2}(\cdot) & \cdots & \phi_{k,2,n_k}(\cdot) \\ \vdots & \vdots & \ddots & \vdots \\ \phi_{k,n_k+1,1}(\cdot) & \phi_{k,n_k+1,2}(\cdot) & \cdots & \phi_{k,n_k+1,n_k}(\cdot) \end{pmatrix}$$

where n_k is the hidden size in the k-th transformation, and each $\phi_{k,i,j}(\cdot)$ is an activation function parameterized by learned weights. Given that each function operates on a single feature dimension, they can be parameterized as splines, which are piecewise polynomial functions that can approximate complex nonlinear relationships. The output of the KAN module is a transformed token matrix $T^{(i+1)} \in \mathbb{R}^{N \times D}$, where $N = B \cdot H' \cdot W'$ and $D = C'$. This transformation does not alter the dimensionality of the input matrix, enabling efficient processing of the tokenized data.

The Post-KAN Modules revert the token matrix to its spatial form and apply a skip connection. The matrix $T^{(3)} \in \mathbb{R}^{(B \cdot H' \cdot W') \times C'}$ is reshaped back into the tensor $X \in \mathbb{R}^{B \times C' \times H' \times W'}$. A skip connection is then established with the tensor before the KAN module:

$$\text{Output} = LN(T^{(0)} + KAN(T^{(0)}))$$

where $KAN(T^{(0)})$ is the output of the KAN module and LN denotes layer normalization. These operations preserve the spatial resolution and feature dimensionality, allowing the processed data to integrate seamlessly with the rest of the encoder-decoder architecture. The two Tokenized-KAN blocks operate respectively at the C_2 and C_3 channels.

The decoder has a structure that mirrors the encoder, with the first two layers consisting of Tokenized-KAN blocks. These blocks contain the same number of channels as the corresponding encoder layers. The decoder layers progressively upsample the feature maps, doubling their spatial dimensions at each step while halving the number of channels. The upsampling is achieved through interpolation, followed by a ConvBlock that reduces the number of channels. The skip connections from the encoder layers are summed with the corresponding decoder layers, allowing the model to leverage high-resolution features from the encoder:

$$X^{(E+i+1)} = \text{ReLU}\left(\text{ConvBlock}\left(\text{Upsample}(X^{(E+i)}) + X^{(i)}\right)\right)$$

where E is the number of encoder layers, i is the index of the decoder layer, and $X^{(i)}$ is the output of the corresponding encoder layer.

The final segmentation map is produced by a 1×1 convolutional layer, referred to as the Classifier, which reduces the number of channels to the number of classes in the segmentation task.

4.2 Architectural Variants

The comparison was carried out using multiple hyperparameter configurations for both UKANWeed and UNet, specifically by varying the number of channels $[C_1, C_2, C_3]$ in the three stages of the encoder and decoder. We denote the models as UNet-C_3 and UKANWeed-C_3, where C_3 refers to the number of channels in the deepest layer of the encoder and decoder. In the original UKAN paper, the hidden sizes were set to $[128, 160, 256]$. To better accommodate resource-constrained scenarios, we introduced five additional, lighter configurations: $[64, 80, 128]$, $[32, 40, 64]$, $[16, 20, 32]$, $[8, 10, 16]$ and $[8, 8, 8]$. This approach enables us to examine the performance of UKANWeed on the weed mapping task with a minimal number of parameters.

For UNet, we not only replicated these smaller configurations but also added two larger ones—$[192, 240, 384]$ and $[256, 320, 512]$—to evaluate whether increasing the model capacity could enable UNet to match or exceed the performance of UKANWeed.

Given the exclusive multiclass nature of the task and the presence of significant class imbalance, with respect to UKAN, we replaced the standard cross-entropy loss with the Focal Loss [12] as the loss function, which is defined as:

$$\mathcal{L}_{FL} = -\alpha_t \cdot (1 - p_t)^\gamma \cdot \log(p_t)$$

where α_t is a balancing factor, p_t is the predicted probability of the actual class, and γ is a focusing parameter that adjusts the rate at which easy examples are down-weighted.

5   Experiments

The experiments were conducted to evaluate the performance of UKANWeed in comparison to UNet on the WeedMap dataset. The primary objectives were to assess the segmentation accuracy of both architectures and to analyze their computational complexity in terms of model size and number of operations. Additionally, we investigated the capacity lower bounds of both architectures by testing multiple configurations with varying numbers of channels. This analysis aimed to identify the minimum model complexity required to achieve satisfactory performance on the weed mapping task.

5.1   Setup

The training parameters were aligned with those used by Li et al. [11]. Specifically, the Adam optimizer was employed with $\beta_1 = 0.9$, $\beta_2 = 0.999$, and a weight

Table 1. Segmentation performance per class (F1 Score (%) ↑) and model complexity (Params ↓, GFlops ↓). BG stands for Background. The number next to each model name indicates the number of channels in the deepest encoder/decoder layer (i.e., C_3).

Model	BG	Crop	Weed	Mean	Params (M)	GFlops
UNet 8	98.51	77.26	16.57	64.11	0.0075	0.0708
UKANWeed 8	98.49	78.22	26.79	67.83	0.0143	0.0905
UNet 16	98.59	75.33	20.69	64.87	0.0129	0.0749
UKANWeed 16	98.64	73.82	5.15	59.20	0.0268	0.1004
UNet 32	97.95	68.72	35.84	67.50	0.0471	0.2619
UKANWeed 32	98.51	76.71	15.10	63.44	0.1030	0.3641
UNet 64	98.51	82.54	66.97	82.67	0.1795	0.9722
UKANWeed 64	98.64	84.64	71.41	84.90	0.4036	1.3812
UNet 128	98.61	83.87	71.42	84.63	0.7001	3.7383
UKANWeed 128	**98.82**	**86.95**	73.22	**86.33**	1.5976	5.3741
UNet 256	98.60	84.78	70.79	84.72	2.7646	14.6520
UKANWeed 256	98.74	85.30	70.02	84.69	6.3567	21.1951
UNet 384	98.69	85.39	73.04	85.71	6.1934	32.7411
UNet 512	98.74	85.51	**73.90**	86.05	10.9866	58.0057

decay coefficient $\lambda = 1 \times 10^{-4}$. The learning rate was adjusted via scheduling reduced on plateau, which is a technique that reduces the learning rate when the validation loss plateaus. This approach is beneficial for fine-tuning the model and preventing overfitting. We performed a grid search to determine the optimal learning rate for each model configuration. The learning rate was set to vary between 1×10^{-2} and 1×10^{-4} for non-KAN layers, and between 1×10^{-2} and 1×10^{-3} for KAN layers. The minimum learning rate was set to 1×10^{-6}. The model was trained for 500 epochs with a batch size of 2 images, stopping after 20 epochs without improvement on the validation set. We fixed the γ parameter of the Focal Loss function at 2.0. For the α parameter, we evaluated two configurations: (1) setting α to be the inverse of the class frequencies computed from the current training image and (2) assigning a uniform value of 1.0. All experiments were conducted using an NVIDIA RTX 4090 GPU.

5.2 Results

As an accuracy metric, the F1-score was selected, consistent with previous studies on the same topic. This choice is motivated by its robustness to class imbalance, a typical challenge in this task. Model complexity was evaluated using two indicators: the number of parameters and the number of operations measured in GFlops. The evaluation was conducted on the images from field 003, with an inference batch size of 8 images.

Table 1 presents the per-class and mean F1-scores for all evaluated models, along with their computational complexity, quantified in terms of parameter count (in millions) and inference cost (in GFlops) on a 512×512 input image.

When comparing UKANWeed and UNet architectures with equivalent hidden dimensionality, UKANWeed consistently exhibits greater parameter counts and higher computational costs. This is an expected outcome, as UKANWeed replaces the MLP blocks of UNet with KAN blocks, which are inherently more complex and parameter-intensive.

However, when controlling for model complexity (i.e., comparing configurations with similar parameter counts and GFlops), UKANWeed demonstrates superior performance. Notably, UKANWeed-128 achieves the highest mean F1-score of 86.33 with only 1.60M parameters and 5.37 GFlops, outperforming UNet-256, which reaches 84.72 with 2.76M parameters and 14.65 GFlops. Moreover, UKANWeed performance saturates at 128 channels, while UNet requires 512 channels to reach its peak, indicating that UKANWeed attains comparable or better accuracy with significantly smaller models.

UKANWeed-128 slightly outperforms UKANWeed-256 in the mean F1-score despite having fewer parameters. This may be due to overfitting in the larger model, which could have memorized training data rather than generalizing well. UKANWeed-128 balances expressiveness and regularization more effectively, resulting in improved generalization performance.

Both UNet and UKANWeed show performance degradation below 64 channels. Specifically, UKANWeed-32 achieves a mean F1-score of 63.44, compared to 67.50 for UNet-32. The performance drop in UKANWeed-32 is primarily attributed to a drastic reduction in F1-score for the weed class (15.10%), in contrast to 35.84% for UNet-32. This suggests that UKANWeed-32 fails to learn meaningful features for weed detection. Nonetheless, UKANWeed-32 outperforms UNet-32 on the crop class (76.71% vs. 68.72%), highlighting class-specific disparities.

Interestingly, UKANWeed-8 performs better than both UKANWeed-16 and UKANWeed-32, suggesting convergence difficulties in low-capacity UKANWeed variants, possibly due to the architectural complexity introduced by the KAN layers. In contrast, UNet exhibits more stable performance as model capacity increases, with a clear positive trend except for UNet-8 and UNet-16, which yield comparable results, thereby defining an empirical lower performance bound for this architecture.

5.3 Qualitative Evaluation

Following the quantitative evaluation, Fig. 2 presents a qualitative assessment of the segmentation results on selected WeedMap images, which encompass varying levels of weed and crop presence. This visual comparison includes the two top-performing configurations—UKANWeed-128 and UNet-512—as well as two low-capacity variants, UKANWeed-8 and UNet-16. Consistent with the metrics reported in Table 1, UKANWeed-128 and UNet-512 demonstrate robust performance, effectively distinguishing between crops and weeds under diverse conditions. In contrast, the low-capacity models demonstrate limited discriminative capability, frequently misclassifying crop edges and small plants as weeds, which

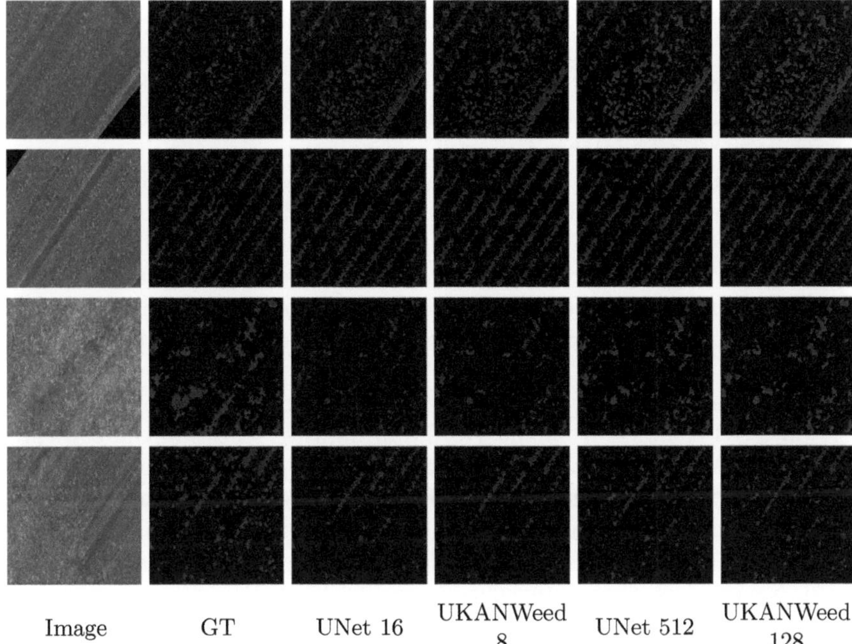

<table>
<tr><td>Image</td><td>GT</td><td>UNet 16</td><td>UKANWeed 8</td><td>UNet 512</td><td>UKANWeed 128</td></tr>
</table>

Fig. 2. Comparison between the original image, ground truth (GT), and predictions from different models: UNet-16, UNet-512, UKANWeed-8, and UKANWeed-128 on different scenarios.

highlights their inability to go beyond generic plant segmentation. These models likely underfit due to insufficient representational power.

6   Conclusion

This study investigated the integration of Kolmogorov-Arnold Networks into a UNet architecture, referred to as UKANWeed, for the task of weed segmentation in RGB drone imagery, using the WeedMap dataset. The aim was to assess whether the theoretically superior expressive power of KANs translates into empirical advantages in a resource-constrained agricultural setting. Furthermore, a study on the boundaries of model capacity was conducted by evaluating multiple configurations of UKANWeed and UNet, ranging from models with very low capacity (8 channels) to those with high capacity (512 channels).

The results demonstrate that UKANWeed consistently outperformed standard UNet architectures in terms of segmentation accuracy (mean F1-score), while requiring comparable or even lower computational resources. In particular, UKANWeed-128 achieved higher accuracy than UNet-256, despite having significantly fewer parameters and a lower computational cost. Notably, it also outperformed the larger UKANWeed-256 model, suggesting that increasing capacity beyond a certain point may lead to diminishing returns or even overfitting, rather than improved generalization. Conversely, both low-capacity variants of UKAN-Weed and UNet struggled to learn meaningful features—especially for the weed class—highlighting the existence of a minimum capacity threshold necessary for effective segmentation in this domain.

These findings provide empirical support for the utility of KAN layers in semantic segmentation tasks with limited data and hardware constraints. By more effectively capturing the underlying functional structure of image features, KAN-enhanced networks can achieve higher accuracy without a proportional increase in computational cost. This makes them especially well-suited to drone-based agricultural applications, where both inference efficiency and model fidelity are critical.

Despite the promising results, some limitations remain. This study focused solely on RGB channels from the WeedMap dataset, reflecting the constraints of low-cost UAVs but omitting potentially valuable multispectral information. Future work should assess the impact of multispectral integration on segmentation performance. Additionally, while model complexity was analyzed in terms of parameters and GFlops, no real-world deployment on edge devices was performed. Nonetheless, the lightweight nature of UKANWeed-128 suggests strong potential for deployment on platforms like NVIDIA Jetson or ARM-based boards with neural accelerators. Future research should include deployment and profiling to evaluate runtime, energy efficiency, and integration in practical scenarios.

Acknowledgments. The research of P. De Marinis is supported by a Ph.D. fellowship funded under the Italian "D.M. n. 352, April 9, 2022" – NRRP, Mission 4, Component 2, Investment 3.3 – co-supported by Exprivia S.p.A. (CUP H91I22000410007).

Disclosure of Interests. The authors declare that they have no competing interests relevant to the content of this article.

References

1. Cao, J., et al.: DO-Conv: depthwise over-parameterized convolutional layer. IEEE Trans. Image Process. **31**, 3726–3736 (2022)
2. Castellano, G., De Marinis, P., Vessio, G.: Weed mapping in multispectral drone imagery using lightweight vision transformers. Neurocomputing **562**, 126914 (2023)
3. Castellano, G., Cotardo, E., Mencar, C., Vessio, G.: Density-based clustering with fully-convolutional networks for crowd flow detection from drones. Neurocomputing **526**(C), 169–179 (2023)
4. Cheon, M.: Demonstrating the Efficacy of Kolmogorov-Arnold Networks in Vision Tasks (2024). arXiv:2406.14916 [cs]
5. Chicchón Apaza, M.Á., Monzón, H.M.B., Alcarria, R.: Semantic segmentation of weeds and crops in multispectral images by using a convolutional neural networks based on U-net. In: Botto-Tobar, M., Zambrano Vizuete, M., Torres-Carrión, P., Montes León, S., Pizarro Vásquez, G., Durakovic, B. (eds.) ICAT 2019. CCIS, vol. 1194, pp. 473–485. Springer, Cham (2020). https://doi.org/10.1007/978-3-030-42520-3_38

6. Daponte, P., et al.: A review on the use of drones for precision agriculture. IOP Conf. Ser. Earth Environ. Sci. **275**(1), 012022 (2019). Publisher: IOP Publishing

7. Dosovitskiy, A., et al.: An image is worth 16x16 words: transformers for image recognition at scale. arXiv preprint arXiv:2010.11929 (2020)

8. Khan, M.A., Menouar, H., Hamila, R.: Revisiting crowd counting: state-of-the-art, trends, and future perspectives. Image Vis. Comput. **129**, 104597 (2023)

9. Kinaneva, D., Hristov, G., Raychev, J., Zahariev, P.: Early forest fire detection using drones and artificial intelligence. In: 2019 42nd International Convention on Information and Communication Technology, Electronics and Microelectronics (MIPRO), pp. 1060–1065. IEEE (2019)

10. Kolmogorov, A.N.: On the representation of continuous functions of several variables by superpositions of continuous functions of a smaller number of variables. Am. Math. Soc. (1961)

11. Li, C., et al.: U-KAN makes strong backbone for medical image segmentation and generation. arXiv preprint arXiv:2406.02918 (2024)

12. Lin, T.Y., Goyal, P., Girshick, R., He, K., Dollár, P.: Focal loss for dense object detection. In: Proceedings of the IEEE International Conference on Computer Vision, pp. 2980–2988 (2017)

13. Liu, Z., et al.: KAN: Kolmogorov–Arnold Networks (2024)

14. Lottes, P., Behley, J., Chebrolu, N., Milioto, A., Stachniss, C.: Joint stem detection and crop-weed classification for plant-specific treatment in precision farming. In: 2018 IEEE/RSJ International Conference on Intelligent Robots and Systems (IROS), pp. 8233–8238. IEEE (2018)

15. Ma, X., Wang, Z., Hu, Y., Zhang, X.: Kolmogorov-Arnold network for remote sensing image semantic segmentation. arXiv preprint arXiv:2501.07390 (2025)

16. Ronneberger, O., Fischer, P., Brox, T.: U-net: convolutional networks for biomedical image segmentation. In: Navab, N., Hornegger, J., Wells, W.M., Frangi, A.F. (eds.) MICCAI 2015. LNCS, vol. 9351, pp. 234–241. Springer, Cham (2015). https://doi.org/10.1007/978-3-319-24574-4_28

17. Sa, I., et al.: A large-scale semantic weed mapping framework using aerial multispectral imaging and deep neural network for precision farming. Remote Sens. **10**(9), 1423 (2018)

18. Sa, I., et al.: weedNet: dense semantic weed classification using multispectral images and MAV for smart farming. IEEE Robot. Autom. Lett. **3**(1), 588–595 (2018). Conference Name: IEEE Robotics and Automation Letters

19. dos Santos Ferreira, A., Freitas, D.M., da Silva, G.G., Pistori, H., Folhes, M.T.: Weed detection in soybean crops using ConvNets. Comput. Electron. Agric. **143**, 314–324 (2017). Publisher: Elsevier

20. Vougioukas, S.: Agricultural robotics. Annu. Rev. Control Robot. Autonom. Syst. **2**, 365–392 (2019)

21. Wiles, L.J.: Beyond patch spraying: site-specific weed management with several herbicides. Precision Agric. **10**(3), 277–290 (2009)

A Weakly-Supervised Learning Approach for RGB Crop Detection Using UAV Imagery

Pasquale De Marinis$^{(\boxtimes)}$ (iD), Federico Canistro, Gennaro Vessio (iD), and Giovanna Castellano (iD)

Department of Computer Science, University of Bari Aldo Moro, Bari, Italy
{pasqualede.marinis,federico.canistro,gennaro.vessio,
giovanna.castellano}@uniba.it

Abstract. Precision agriculture increasingly relies on accurate crop monitoring to optimize yields and resource use, yet faces significant challenges due to the scarcity of labeled data and the high cost of advanced imaging technologies. To address these limitations, we propose a novel weakly-supervised learning approach for crop detection in aerial RGB imagery, in which pseudo-labels are automatically generated through zero-shot segmentation—thus minimizing the need for manual annotation. Our pipeline combines the Segment Anything Model for zero-shot segmentation, DBSCAN for clustering and label inference, and Faster R-CNN with a ResNet-101 backbone for object detection. This strategy enables effective crop detection even in scenarios with little or no human supervision, offering a scalable and cost-efficient solution for diverse agricultural contexts. We evaluate our approach using a newly collected dataset of drone-based RGB images, which comprises vineyards, orchards, olive groves, and wheat fields. Experimental results demonstrate high precision, recall, and F1 scores across crop types, validating the robustness and applicability of the proposed method in real-world agricultural environments. Our findings highlight the potential of integrating modern self-supervision techniques with object detection frameworks to enhance sustainable and data-efficient precision farming.

Keywords: Precision Agriculture · Weakly-Supervised Learning · Crop Detection · Segment Anything · Drones

1 Introduction

Precision agriculture represents a transformative approach to farming, leveraging advanced technologies to optimize crop production while promoting environmental sustainability [6,26]. By accounting for the spatial and temporal variability of agricultural fields, precision agriculture integrates GPS, remote sensing, and data analytics to enhance resource management and maximize yield potential. This innovation is crucial to meeting the increasing global demand for food with minimal ecological impact.

In recent years, advances in Artificial Intelligence (AI) have further revolutionized precision agriculture by enabling data-driven decision-making [7].

E. Rodolà et al. (Eds.): ICIAP 2025 Workshops, LNCS 16169, pp. 128–139, 2026.
https://doi.org/10.1007/978-3-032-11317-7_11

AI techniques facilitate the analysis of large-scale agricultural data to support real-time monitoring and predictive analytics through Internet of Things devices. However, despite these technological breakthroughs, widespread adoption remains limited due to the high cost of specialized equipment and the need for expert knowledge. Moreover, many existing approaches depend on multispectral or hyperspectral imaging, which—while effective—can be prohibitively expensive and inaccessible for small to medium-sized farms. The limited availability of labeled data for model training further constrains the deployment of AI-driven solutions in diverse agricultural settings.

Unmanned Aerial Vehicles (UAVs), commonly referred to as drones, have emerged as a valuable tool in precision agriculture, offering a cost-effective and flexible platform for collecting high-resolution imagery of agricultural fields [23,28]. Equipped with advanced sensors, UAVs enable detailed assessments of crop health, soil conditions, and environmental factors. Their ability to provide timely and localized data empowers farmers to make informed decisions about irrigation, pest control, and resource allocation, ultimately improving both productivity and sustainability. However, fully exploiting UAV data still poses challenges, particularly in terms of data processing and analysis, which often require complex algorithms and significant computational resources.

To address these limitations, this paper proposes a novel *weakly-supervised learning* approach for crop detection using UAV-captured RGB imagery. Weakly-supervised learning refers to learning paradigms in which supervision signals are noisy, incomplete, or automatically generated [33]. In our case, pseudo-labels are generated through zero-shot segmentation and clustering, allowing for model training without the need for manual annotation. Specifically, we integrate the Segment Anything Model (SAM) [17] for zero-shot segmentation, DBSCAN [10] for feature clustering and label inference, and Faster R-CNN [24] with a ResNet-101 backbone [14] for object detection. A key strength of our approach lies in the use of standard RGB images, which are more accessible and cost-effective than multispectral alternatives.

Unlike conventional pipelines, our method does not rely on manually labeled data for initial training. Instead, segmentation masks generated by SAM are clustered using DBSCAN to create pseudo-labels that guide the training of the detector. This design enhances scalability and adaptability, making the approach suitable for real-world agricultural scenarios where labeled data is often unavailable or costly to obtain.

We evaluate the proposed pipeline on a newly curated dataset composed of drone-captured RGB images from various agricultural contexts, including vineyards, orchards, olive groves, and wheat fields. Our experiments demonstrate that the method effectively detects and classifies different crop types with high precision and recall, even in the presence of variability in field conditions.

The remainder of this paper is structured as follows. Section 2 reviews related work in precision agriculture, with a focus on UAV-based sensing and deep learning. Section 3 introduces our proposed approach in detail. Section 4 describes the dataset, experimental setup, and evaluation results. Finally, Sect. 5 summarizes our contributions and outlines directions for future research.

2   Related Work

Recent advancements in computer vision and remote sensing have significantly influenced the development of precision agriculture. Among these, UAVs have emerged as indispensable tools, enabling detailed crop health assessment and precise agricultural land mapping [18]. The increasing adoption of UAVs is driven by their ability to enhance yield quality while reducing health risks associated with manual pesticide application. Equipped with various sensors and imaging systems, UAVs support tasks such as crop monitoring, height estimation, pesticide spraying, and soil analysis [22]. UAV platforms typically comprise key components such as an airframe, propulsion and control systems, power supply, navigation modules, and payloads (e.g., cameras and multispectral sensors), which work in unison to support diverse agricultural missions [1,3,8].

In parallel, deep learning techniques have played a transformative role in agricultural data analysis. Convolutional Neural Networks (CNNs) have proven highly effective in processing UAV-captured imagery for tasks such as plant disease detection, pest identification, and nutrient deficiency assessment [2,25]. By automatically learning hierarchical features, these models outperform traditional machine learning approaches like Support Vector Machines and Decision Trees in both scalability and representational power [19,29,31]. The synergy between UAV-based data collection and deep learning-based analysis has laid the foundation for automated, large-scale crop monitoring systems.

Compared to satellite imagery, UAVs offer higher spatial and temporal resolution, making them ideal for detecting subtle variations in vegetation. Their integration with CNN-based models enables fine-grained crop segmentation and classification, which supports targeted and timely agricultural interventions [4,5,12,13]. Object detection architectures such as Faster R-CNN [24] and YOLO [16] have been widely adopted in agricultural contexts, balancing detection accuracy and inference speed for real-time applications.

Several recent studies have also explored the use of UAVs and deep learning for unsupervised or weakly-supervised analysis. Ivošević et al. [15] provide an overview of UAV deployments in real agricultural settings, emphasizing their practical benefits. Osco et al. [20] investigate the potential of the Segment Anything Model for remote sensing applications, showcasing its ability to perform zero-shot segmentation on aerial imagery. Furthermore, research by Petti and Li [21] and Zhou et al. [32] highlights the growing interest in weakly-supervised learning for crop segmentation, where limited or noisy supervision is leveraged to train reliable models.

Unlike previous studies that rely on human-annotated data or predefined class-level supervision, our work adopts a *pseudo-supervised* paradigm, wherein segmentation masks are generated automatically through SAM and refined using

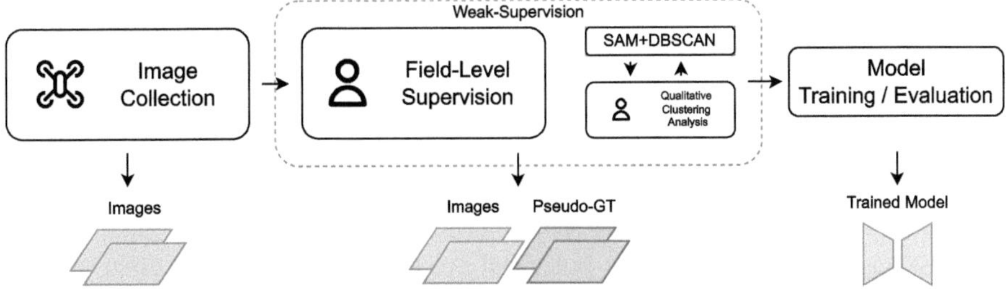

Fig. 1. Illustration of the comprehensive pipeline employed in this study, highlighting the sequential integration of three steps: drone imagery acquisition, weak supervision and object detection model training. The weak supervision is characterized by the field-level human labeling of crop types, Segment Anything for crop segmentation, and DBSCAN for clustering. A qualitative cluster analysis is performed to identify the correct number of clusters, given that SAM may segment additional elements such as weeds or rocks. The final object detection model is trained using the pseudo-labels generated by the previous steps.

clustering techniques. This setup corresponds to the *inaccurate supervision* category of weak supervision, in which models are trained on noisy or auto-generated labels instead of ground-truth annotations.

Our contribution lies in combining recent advances—SAM for zero-shot segmentation, DBSCAN for cluster-based label generation, and Faster R-CNN for robust detection—into a unified pipeline designed for practical deployment. Unlike multispectral approaches, which require costly sensors and calibration, our method operates solely on RGB imagery, making it scalable and accessible to non-specialist users. The ability to generate pseudo-labels without expert intervention enables our model to generalize to diverse agricultural contexts where annotated datasets are unavailable, thereby addressing a critical bottleneck in real-world precision agriculture.

3   Method

Figure 1 presents an overview of the proposed pseudo-supervised pipeline for crop detection in UAV-based RGB imagery. The method addresses the challenges posed by limited annotated data in agricultural domains by combining zero-shot segmentation, unsupervised clustering, and supervised object detection based on automatically generated pseudo-labels.

The pipeline begins with the Segment Anything Model [17], a vision transformer based architecture [9] capable of performing segmentation in a zero-shot fashion. SAM is particularly well suited for agricultural scenarios, where the availability of labeled data is often scarce. Its zero-shot capabilities enable it to identify potential crop regions directly from RGB images, making it adaptable to previously unseen crop types and diverse environmental conditions.

After the initial segmentation stage, the detected regions are grouped based on their visual and geometric features. These features include the size, color, and texture of the bounding boxes derived from SAM's output. For this task, we adopt the DBSCAN clustering algorithm [11], enhanced with Principal Component Analysis (PCA) [27] to reduce dimensionality and focus on the most informative features. DBSCAN is particularly effective in agricultural contexts due to its ability to identify clusters of varying densities and handle complex, irregular patterns without the need to predefine the number of clusters.

In this context, SAM acts as the generator of pseudo-ground truth by performing segmentation to isolate potential crop regions. DBSCAN then clusters these segments into coherent groups, to which class labels are assigned at the cluster level. This process results in a set of labeled regions that serve as the basis for training the object detection model. Notably, SAM may segment additional elements such as weeds or rocks; as a result, the number of identified clusters—and hence the number of assigned labels—can exceed the initial set of target crop types. In this phase, we perform a qualitative cluster analysis to determine the correct number of clusters, ensuring that the final labels accurately reflect the crop types present in the imagery. This step is crucial for maintaining the integrity of the pseudo-labels, as it prevents misclassification of non-crop elements. This pseudo-supervised approach eliminates the need for manually annotated data while still approximating realistic crop boundaries and class distributions.

The final phase of the pipeline involves object detection using Faster R-CNN [24], with a ResNet-101 backbone [14]. This architecture is widely recognized for its high accuracy in object detection tasks, particularly in cluttered or complex environments such as agricultural fields. Its Region Proposal Network efficiently identifies candidate regions, which are then refined and classified by the downstream network. The ResNet-101 backbone enables deep feature extraction, capturing subtle differences between crop types, even when visual cues are minimal or ambiguous.

The primary role of Faster R-CNN is not simply to replicate the pseudo-ground truth but to enhance detection accuracy beyond the SAM+DBSCAN results. While SAM+DBSCAN is effective for segmentation and initial clustering, it does not provide the precise object detection or classification necessary for practical applications in agricultural monitoring. The object detection model improves the detection step by refining bounding box predictions and class labels through deep feature extraction, which SAM+DBSCAN does not inherently provide. By integrating Faster R-CNN, we significantly enhance detection accuracy, filter out noise from the pseudo labels, and reduce misclassification errors that may occur when using SAM+DBSCAN alone. In other words, it acts as an additional processing layer.

The integration of SAM, DBSCAN, and Faster R-CNN yields a pipeline that is both effective and scalable for large-scale crop detection. SAM provides a fast and flexible method for initial segmentation, DBSCAN organizes the output into semantically coherent groups, and Faster R-CNN delivers accurate, instance-level detection. This synergy allows the system to operate efficiently in real-world scenarios, offering a practical solution for crop monitoring in data-scarce agricultural environments.

Fig. 2. Example RGB tiles representing the four crop types in our proposed dataset.

4   Experiments

4.1   Dataset and Experimental Setting

The dataset employed in this study comprises RGB images acquired by a drone over several agricultural fields located in Foggia, Italy. It includes four primary crop types: vineyards, orchards, olive groves, and wheat fields. In total, the dataset contains 40,000 images, each with a resolution of 512×512 pixels.[1] The data were collected using a DJI AIR 2S drone flying at an altitude of 40 m between 9:00 a.m. and 1:00 p.m., at an average speed of 5 m/s. Figure 2 illustrates sample images representing the different crop types.

To ensure rigorous model evaluation, the dataset was split *field-wise* into training (70%), validation (15%), and test (15%) sets. This strategy guarantees that image tiles originating from the same physical field are not shared across splits, thus preventing data leakage and ensuring robust generalization. Ground truth annotations were generated via zero-shot segmentation using the Segment Anything Model, followed by DBSCAN clustering to infer crop-type labels, as described in Sect. 3. In addition to the four main crop categories, the clustering process also identified two auxiliary classes: *rocks* and *weeds*.

The ortho-mosaic generated from drone footage was partitioned into 512×512 pixel tiles to balance computational efficiency with spatial detail. Non-agricultural elements, such as roads, were manually removed using the LabelMe

---

[1] The dataset is publicly available at http://doi.org/10.5281/zenodo.12607112.

annotation tool [30], ensuring that the dataset remained focused on relevant vegetation features. All pixel values were normalized using z-score normalization based on the global mean and standard deviation computed across the entire dataset.

SAM was applied in evaluation mode without prompts, leveraging its pre-trained generalization capabilities to identify candidate regions without the need for domain-specific fine-tuning. DBSCAN, combined with PCA, was then used to cluster the resulting segments based on visual and geometric properties. Specifically, we retained the top 20 principal components, which accounted for approximately 85% of the total variance. This dimensionality was selected based on the explained variance curve, which exhibited a clear elbow point. DBSCAN hyperparameters were set to $\varepsilon = 0.5$ and a minimum sample size of 5, empirically chosen to balance sensitivity and noise robustness.

The object detection model was implemented using Faster R-CNN with a ResNet-101 backbone [14,24], pre-trained on the COCO dataset. The model was optimized using stochastic gradient descent with a learning rate of 0.005, momentum of 0.9, and weight decay of 0.0005. The learning rate was dynamically adjusted using `ReduceLROnPlateau` based on the validation loss, and early stopping was employed to halt training after three epochs without improvement. These settings were determined through preliminary experimentation on the validation set to ensure both efficiency and generalization.

All experiments were conducted on a workstation equipped with an Intel Core i7 processor, 32 GB RAM, and an NVIDIA GeForce RTX 3080 Ti GPU. The software stack was based on PyTorch, utilizing the `torchvision` library extensively for model components and data handling.

4.2   Results

To evaluate the performance of our crop detection models, we report precision, recall, and F1 scores alongside the Intersection over Union (IoU) to assess localization accuracy.

Our proposed approach, which combines SAM for zero-shot segmentation, DBSCAN for clustering, and Faster R-CNN with a ResNet-101 backbone, demonstrates strong performance across various crop types, as shown in Table 1. These results highlight the pipeline's effectiveness in managing the complexities of UAV-based agricultural imagery. For comparison, a baseline configuration using DBSCAN directly on pixel values followed by Faster R-CNN yielded the lowest scores, providing a reference point for evaluating the contributions of SAM and the ResNet architecture.

To further validate the contribution of each component, we conducted ablation studies exploring alternative configurations. Replacing ResNet-101 with ResNet-50 led to a moderate decrease in performance, suggesting that deeper feature extraction is beneficial for handling visually similar crop types. Additionally, substituting DBSCAN with K-means clustering reduced the detection scores, confirming that DBSCAN's ability to model arbitrarily shaped and vari-

Table 1. Detection performance metrics (Precision, Recall, and F1 score) across different model configurations and crop types at an IoU threshold of 0.5.

Method	Class	Precision	Recall	F1 score
SAM+DBSCAN+ResNet-101	Olive grove	0.83	0.76	0.67
	Orchard	0.80	1.00	0.91
	Vineyard	0.84	0.64	0.73
	Wheat	0.73	0.68	0.75
	Weeds	0.83	0.68	0.75
	Rocks	0.80	0.56	0.66
SAM+DBSCAN+ResNet-50	Olive grove	0.75	0.69	0.61
	Orchard	0.73	0.97	0.83
	Vineyard	0.76	0.58	0.66
	Wheat	0.66	0.62	0.68
	Weeds	0.75	0.62	0.68
	Rocks	0.73	0.51	0.60
SAM+K-means+ResNet-50	Olive grove	0.64	0.59	0.52
	Orchard	0.62	0.83	0.71
	Vineyard	0.65	0.49	0.56
	Wheat	0.56	0.53	0.58
	Weeds	0.64	0.53	0.58
	Rocks	0.62	0.43	0.51
DBSCAN+ResNet-101	Olive grove	0.53	0.46	0.50
	Orchard	0.45	0.55	0.50
	Vineyard	0.52	0.45	0.48
	Wheat	0.47	0.52	0.49
	Weeds	0.50	0.47	0.48
	Rocks	0.45	0.39	0.41

ably dense clusters is better suited to the inherent variability of agricultural imagery.

The combined results from the ablation studies and the baseline model provide a comprehensive understanding of how individual architectural choices affect performance. Among all configurations tested, the SAM+DBSCAN+ResNet-101 model consistently achieved the highest precision, recall, and F1 scores, demonstrating its robustness and effectiveness in real-world scenarios.

Performance variations across classes can be attributed to several factors. First, the class imbalance played a role: overrepresented classes, such as *orchards*, benefited from a larger number of training examples, while underrepresented classes, like *wheat*, performed slightly worse. Second, intra-class variability— resulting from changes in lighting, camera angle, or growth stage—made it more challenging for the model to generalize within certain crop types. Finally, some classes, such as *vineyards* and *orchards*, exhibit similar visual characteristics, which can lead to occasional misclassifications.

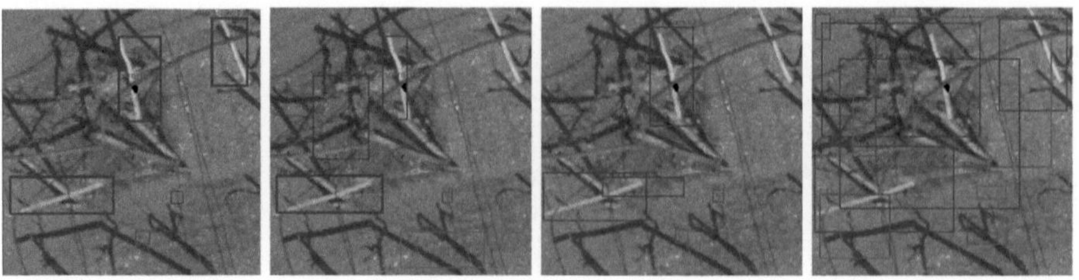

Fig. 3. Qualitative comparison between model predictions on the same tile. From left to right: SAM+DBSCAN+ResNet-101, SAM+DBSCAN+ResNet-50, SAM+K-means+ResNet-50, and DBSCAN+ResNet-101. Blue regions represent *orchards*, purple regions *rocks*, and red regions *vineyards*. (Color figure online)

Figure 3 provides a qualitative comparison of predictions from the four main configurations on the same tile. The SAM+DBSCAN+ResNet-101 model yields more consistent and well-localized predictions, particularly in distinguishing between visually similar classes, thereby reaffirming the quantitative results.

Overall, the integration of SAM, DBSCAN, and Faster R-CNN with ResNet-101 results in a robust and generalizable detection framework. SAM's contextual segmentation capabilities, DBSCAN's flexible clustering, and ResNet-101's deep feature representation collectively contribute to high detection accuracy across diverse crop types. These findings confirm the pipeline's potential for supporting scalable, label-efficient crop monitoring in precision agriculture.

5   Conclusion

This study presented a robust weakly-supervised learning framework that integrates the Segment Anything Model for segmentation, DBSCAN for clustering, and Faster R-CNN with a ResNet-101 backbone for object detection. The proposed approach achieved high precision, recall, and F1 scores across multiple crop types, demonstrating its effectiveness in real-world agricultural scenarios. By combining these advanced machine learning techniques, the framework enables accurate and efficient crop detection and classification, offering practical benefits for agricultural productivity and resource management.

Despite these promising results, the study has some limitations. The model's performance was evaluated using a dataset specifically collected with drones over agricultural fields in Southern Italy. As such, its generalizability to different environmental conditions, geographic regions, or unrepresented crop types remains to be assessed. Nonetheless, the approach represents a significant advancement in the application of unsupervised learning techniques to precision agriculture, particularly by addressing the challenge of limited labeled data.

Future work may explore several directions to build upon these findings. Expanding the dataset to include a wider variety of crops and environmental scenarios would allow for more comprehensive validation. The inclusion of additional contextual variables, such as soil moisture, temperature, and vegetation indices, could further enhance prediction accuracy and support informed decision-making. Moreover, evaluating alternative clustering strategies or detection architectures could lead to performance enhancements, especially in more heterogeneous or visually complex environments.

In conclusion, this research underscores the potential of integrating advanced machine learning techniques in precision agriculture, paving the way for more efficient and scalable crop monitoring solutions. The application of drones for data collection has proven to be highly effective, enabling the capture of high-resolution imagery essential for accurate analysis. By addressing the challenges of labeled data scarcity and complex agricultural environments, the study contributes to the ongoing efforts to enhance agricultural practices through technology driven solutions.

Acknowledgments. The research of P. De Marinis is supported by a Ph.D. fellowship funded under the Italian "D.M. n. 352, April 9, 2022" – NRRP, Mission 4, Component 2, Investment 3.3 – for the Ph.D. project "Computer Vision techniques for sustainable AI applications using drones", co-supported by Exprivia S.p.A. (CUP H91I22000410007).

Disclosure of Interests. The authors declare that they have no competing interests relevant to the content of this article.

References

1. Aslan, M.F., Durdu, A., Sabanci, K., Ropelewska, E., Gültekin, S.S.: A comprehensive survey of the recent studies with UAV for precision agriculture in open fields and greenhouses. Appl. Sci. **12**(3) (2022)
2. Bouguettaya, A., Zarzour, H., Kechida, A., Taberkit, A.M.: Deep learning techniques to classify agricultural crops through UAV imagery: a review. Neural Comput. Appl. **34**(12), 9511–9536 (2022)
3. Castellano, G., Cotardo, E., Mencar, C., Vessio, G.: Density-based clustering with fully-convolutional networks for crowd flow detection from drones. Neurocomputing **526**, 169–179 (2023)
4. Castellano, G., De Marinis, P., Vessio, G.: Applying knowledge distillation to improve weed mapping with drones. In: 2023 18th Conference on Computer Science and Intelligence Systems (FedCSIS), pp. 393–400 (2023)
5. Castellano, G., De Marinis, P., Vessio, G.: Weed mapping in multispectral drone imagery using lightweight vision transformers. Neurocomputing **562**, 126914 (2023)
6. Cisternas, I., Velásquez, I., Caro, A., Rodríguez, A.: Systematic literature review of implementations of precision agriculture. Comput. Electron. Agric. **176**, 105626 (2020)
7. Coulibaly, S., Kamsu-Foguem, B., Kamissoko, D., Traore, D.: Deep learning for precision agriculture: a bibliometric analysis. Intelli. Syst. Appl. **16**, 200102 (2022)
8. Delavarpour, N., Koparan, C., Nowatzki, J., Bajwa, S., Sun, X.: A technical study on UAV characteristics for precision agriculture applications and associated practical challenges. Remote Sens. **13**(6), 1204 (2021)
9. Dosovitskiy, A., et al.: An image is worth 16x16 words: transformers for image recognition at scale. arXiv preprint arXiv:2010.11929 (2020)

10. Ester, M., Kriegel, H.P., Sander, J., Xu, X., et al.: A density-based algorithm for discovering clusters in large spatial databases with noise. In: KDD, vol. 96, pp. 226–231 (1996)
11. Hahsler, M., Piekenbrock, M., Doran, D.: DBScan: fast density-based clustering with R. J. Stat. Softw. **91**(1), 1–30 (2019)
12. Hall, O., Dahlin, S., Marstorp, H., et al.: Classification of maize in complex small-holder farming systems using UAV imagery. Drones **2**(3) (2018)
13. Hasan, M., Tanawala, B., Patel, K.: Deep learning precision farming: tomato leaf disease detection by transfer learning. In: Proceedings of 2nd International Conference on Advanced Computing and Software Engineering (ICACSE) (2019)
14. He, K., Zhang, X., Ren, S., Sun, J.: Deep residual learning for image recognition. In: Proceedings of the IEEE Conference on Computer Vision and Pattern Recognition, pp. 770–778 (2016)
15. Ivošević, B., Kostić, M., Ljubičić, N., Grbović, Ž., Panić, M.: Application of unmanned aerial systems to address real-world issues in precision agriculture. In: Unmanned Aerial Systems in Agriculture, pp. 51–69. Elsevier (2023)
16. Jiang, P., Ergu, D., Liu, F., Cai, Y., Ma, B.: A review of YOLO algorithm developments. Procedia Comput. Sci. **199**, 1066–1073 (2022)
17. Kirillov, A., et al.: Segment anything. In: Proceedings of the IEEE/CVF International Conference on Computer Vision, pp. 4015–4026 (2023)
18. Mogili, U.R., Deepak, B.B.V.L.: Review on application of drone systems in precision agriculture. Procedia Comput. Sci. **133**, 502–509 (2018)
19. Olaniyi, E., Chen, D., Lu, Y., Huang, Y.: Generative adversarial networks for image augmentation in agriculture: a systematic review. J. Agric. Sci. Technol. **47**(3), 123–138 (2023)
20. Osco, L.P., et al.: The segment anything model (SAM) for remote sensing applications: from zero to one shot. Int. J. Appl. Earth Obs. Geoinf. **124**, 103540 (2023)
21. Petti, D., Li, C.: Weakly-supervised learning to automatically count cotton flowers from aerial imagery. Comput. Electron. Agric. **194**, 106734 (2022)
22. Rahman, M.F.F., Fan, S., Zhang, Y., Chen, L.: A comparative study on application of unmanned aerial vehicle systems in agriculture. Agriculture **11**(1) (2021)
23. Rejeb, A., Abdollahi, A., Rejeb, K., Treiblmaier, H.: Drones in agriculture: a review and bibliometric analysis. Comput. Electron. Agric. **198**, 107017 (2022)
24. Ren, S., He, K., Girshick, R., Sun, J.: Faster R-CNN: towards real-time object detection with region proposal networks. In: Advances in Neural Information Processing Systems, vol. 28 (2015)
25. Saranya, T., Deisy, C., Sridevi, S., Anbananthen, K.S.M.: A comparative study of deep learning and Internet of Things for precision agriculture. Eng. Appl. Artif. Intell. **122**, 106034 (2023)
26. Sharma, A., Jain, A., Gupta, P., Chowdary, V.: Machine learning applications for precision agriculture: a comprehensive review. IEEE Access **9**, 4843–4873 (2020)
27. Shlens, J.: A tutorial on principal component analysis. arXiv preprint arXiv:1404.1100 (2014)
28. Sishodia, R.P., Ray, R.L., Singh, S.K.: Applications of remote sensing in precision agriculture: a review. Remote Sens. **12**(19), 3136 (2020)
29. Tugrul, B., Elfatimi, E., Eryigit, R.: Convolutional neural networks in detection of plant leaf diseases: a review. J. Agric. Sci. Technol. **47**(3), 123–138 (2023)
30. Wada, K.: LabelMe: Image Polygonal Annotation with Python. GitHub repository (2016)
31. Wu, H.T.: Developing an intelligent agricultural system based on long short-term memory. Mob. Netw. Appl. **26**(3), 1397–1406 (2021)

32. Zhou, R., et al.: Weakly supervised semantic segmentation in aerial imagery via cross-image semantic mining. Remote Sens. **15**(4), 986 (2023)
33. Zhou, Z.H.: A brief introduction to weakly supervised learning. Natl. Sci. Rev. **5**(1), 44–53 (2018)

ConvNeXt Based Architecture for Crowd Counting in Highly Crowd Environments

Gennaro Percannella, Alessia Saggese$^{(\boxtimes)}$, and Mario Vento

University of Salerno, Fisciano, Italy
{pergen,asaggese,mvento}@unisa.it

Abstract. People analytics in crowded environments is important for public safety, event management, and urban planning. Anyway, traditional methods typically fail in high density crowds, due to occlusions and low-resolution people to be counted. This is much more true in case the videos are acquired by a camera mounted on board of a drone. Within this context, we propose a novel approach combining point-based formulations with ConvNeXt architecture and an enhanced Fine-Grained Feature Pyramid (eFGFP) for multi-scale management, including an additional layer at 1/4 resolution, improving the handling of small objects in complex scenarios. ConvNeXt enhances feature extraction and representation, while eFGFP preserves fine-grained information across different scales. Evaluated on 19 datasets with over 43,000 images, our architecture demonstrates impressive performance compared to state-of-the-art methods, offering a robust solution for crowd counting in highly crowded environments.

Keywords: Crowd Counting · ConvNeXt · Fine-Grained Feature Pyramid

1 Introduction

People analytics in highly crowded environments is a critical task across various domains, including public safety, event management, and urban planning. However, traditional methods based on people or head detectors, as demonstrated in [10,17], tend to fail as crowd density increases. This is primarily due to severe occlusions, the dynamic and unpredictable nature of crowds, and the extremely low resolution of individuals in images—often reduced to just a few pixels per person. These challenges are further amplified in aerial surveillance scenarios, where drone-mounted cameras offer a wide field of view but at the cost of reduced spatial resolution and increased perspective distortion. Despite these limitations, drone vision presents a unique opportunity for scalable and flexible crowd monitoring, especially in large-scale outdoor events or emergency situations. To fully leverage this potential, robust crowd counting models must be designed to operate effectively under the constraints of aerial imagery, including motion blur, varying altitudes, and rapidly changing viewpoints.

E. Rodolà et al. (Eds.): ICIAP 2025 Workshops, LNCS 16169, pp. 140–151, 2026.
https://doi.org/10.1007/978-3-032-11317-7_12

Additionally, given the time required for manually producing the ground truth in this kind of scenarios, it typically consists of single point per each person (typically on the head), necessitating the automatic generation of bounding boxes through heuristics. A widely adopted approach to face the above mentioned issues is to formulate the problem in terms of density map estimation [8]: the idea is to establish a relationship between crowd density and the visual features of the image. Starting from the raw image, a density map is computed using a Convolutional Neural Network (CNN), and the estimated count is obtained by summing the predicted density map.

One of the pioneering works in this domain is [30], where the authors proposed a single-column CNN with six convolution layers. Single-column methods, however, struggle with scale variations in object sizes, leading to the development of multi-column models. The first multi-column CNN (MCNN), proposed in [32], featured a three-layer architecture with varying receptive fields in each column. Each column generates a density map that matches the ground truth map's dimensions, and these outputs are concatenated to form the final density map. More recently, PCCNet [9] was proposed as a three-column network to simultaneously learn the density map, density class, and segmentation map in crowd images, sharing encoded features at different stages. Despite their effectiveness, multi-column models often face computational overhead. To address this, the Switch CNN (SCNN) [1] employs a 3-column network with different kernel sizes and a classifier network that selects one column at inference time based on crowd density. With the advent of self-attention mechanisms, the Vision Transformer (ViT) has been successfully applied to crowd counting. TransCrowd [15] and CounTr [2] employ a pure transformer model, converting each image into fixed-size patches, which are flattened and used as input for a ViT encoder. The output then feeds a regression head to predict the total count within the image. Despite these advances, most methods rely on the automatic generation of density maps, typically created by placing a Gaussian kernel at each annotated point and summing the kernels. However, this approach can generate density maps that differ significantly from the ground truth. To address these issues, DM-Count [25] avoids Gaussian smoothing and instead uses Optimal Transport to measure the similarity between the normalized predicted density map and the normalized ground truth density map. Even without Gaussian kernels, ground truth density maps may still contain artifacts, and the generated density map must be converted to head points. To address this challenge and leverage the advantages of object detection methods, the authors of [24] formulated the problem in terms of point detection. Their network, namely Point to Point (P2P), directly predicts a set of point proposals to represent heads inside an image, aligning with human annotation results. Using VGG as the backbone and a Feature Pyramid Network (FPN) for multi-scale management, P2P demonstrates promising results, even compared to state-of-the-art technologies. A similar approach has been proposed in [19], where the problem is formulated in terms of head detector, and a VMamba based architecture is proposed (instead than VGG) as encoder, combined with an advanced FPN, namely a High-level Semantic Supervised Feature

Pyramid Network for gradually integrating information from different scales and partially preserving high-level semantic features.

From the analysis of the literature, it is evident that the point detection-based formulation proposed in P2P is highly promising. Currently, more effective representation backbones, such as VMamba [19] and transformers [2], could potentially offer better performance in terms of contextual information extraction and semantic understanding. VMamba, for instance, leverages Visual State-Space (VSS) blocks with the 2D Selective Scan (SS2D) module, which facilitates the gathering of contextual information from various sources and perspectives. Similarly, transformers like TransCrowd [15] utilize self-attention mechanisms to effectively extract semantic crowd information. However, these choices would significantly increase computational burden, potentially hindering the real-time application of such networks.

Other than the choice of the backbone, another critical issue pertains the neck for multi-scale management. Indeed, Feature Pyramid Network (FPN)-based methods, widely adopted in crowd counting architecture [24], suffer from intrinsic flaws related to channel reduction, which leads to loss of semantic information and can cause aliasing effects.

In order to face with the above mentioned open challenges, in this paper we propose to inherit the advantages of a point-based formulation while combining a ConvNeXt architecture as the backbone with a Fine-Grained Feature Pyramid (FGFP) for multi-scale management. ConvNeXt has been chosen since it incorporates elements from transformer architectures, such as self-attention mechanisms, while retaining the simplicity and efficiency of convolutional networks. This hybrid design allows ConvNeXt to leverage the strengths of both CNNs and transformers, resulting in improved feature extraction and representation. Furthermore, the use of depth-wise convolution in ConvNeXt significantly reduces computational complexity, making the model more efficient and scalable. FGFP has been chosen instead since it is designed to enhance the extraction of fine-grained information within feature maps by maintaining richer semantic information across different scales. Furthermore, differently from the original version of the FGFP, in order to also deal with more complex scenarios in which people are very small, we introduce a further layer in the FGFP, so that the model can leverage features also at 1/4 of the original input image resolution, and not only at 1/8, 1/16 and 1/32. We will thus refer to this layer as enhanced FGFP, namely eFGFP.

The performance of the proposed architecture has been computed over 19 different publicly available datasets, with a test set composed of more than 43,000 images. The achieved results, even compared with the state of the art, confirm the effectiveness of the proposed approach.

2   Proposed Method

The overall architecture of the proposed solution is illustrated in Fig. 1. The raw input image is first processed by the ConvNeXt backbone to extract initial feature maps. These feature maps are then fed into a neck, specifically an enhanced version of a Fine-Grained Feature Pyramid (eFGFP) module, which performs multi-scale feature fusion. The enhanced feature maps are subsequently used to generate point proposals representing heads within the image, aligning with human annotation results by means of the Hungarian Algorithm.

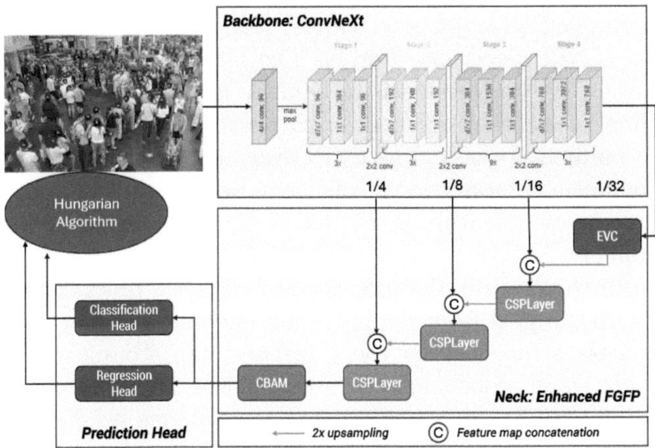

Fig. 1. Overview of the proposed architecture: ConvNeXt is used as backbone and combined with an enhanced version of FGFP. The resulting features maps are used to generate point proposals representing people heads within the image, aligning with human annotation results by means of the Hungarian Algorithm.

2.1 Backbone

The original version of P2PNet employs a VGG network. However, transformers have gained significant popularity in recent years due to their impressive performance across various tasks. Anyway, despite their advantages, transformers come with a substantial computational burden. To balance the benefits of transformer architectures with computational efficiency, we propose using ConvNeXt [18] as the backbone. ConvNeXt is a modernization of the convolutional ResNet architecture: in fact, it incorporates elements from transformer architectures, such as self-attention mechanisms and image *patchification*, while retaining the simplicity and efficiency of convolutional networks. This hybrid approach allows ConvNeXt to capture complex patterns and relationships within the data, enhancing the overall performance of the model without paying in terms of computational costs.

2.2 Neck

In contrast to P2PNet, which employs a traditional Feature Pyramid Network (FPN) as its neck, we propose to use an enhanced Fine-Grained Feature Pyramid (eFGFP) [20]. The FGFP is designed to enhance the extraction of fine-grained information within feature maps by maintaining richer semantic information across different scales. The main idea is that the deepest features typically contain the most abstract feature representations, which are instead scarce in the shallow features. Thus, FGFP allows the deepest features to be used to regulate all frontal shallow features simultaneously. This is achieved through the combination of the Explicit Visual Center (EVC) [22] block and a set of Cross Stage Partial Layers (CSPLayers) (see Fig. 1). The EVC is a generalized intra-layer feature regulation method designed to capture global long-range dependencies while preserving local corner region information of the input image, which is crucial for dense prediction tasks. The CSPLayer, on the other hand, is a module designed to enhance computational efficiency and feature learning. This layer addresses redundancy in feature extraction while maintaining a strong representational capability.

The $\frac{1}{32}$ feature map is initially processed by the EVC block. Subsequently, the information is up-sampled from the input module and fused with the $\frac{1}{16}$ feature map. This process is repeated for the $\frac{1}{8}$ feature map. To address the challenge of detecting very small objects in complex scenarios, we introduce an additional CSPLayer in the FGFP at $\frac{1}{4}$ of the original input image resolution. This allows the model to leverage features at $\frac{1}{4}$, $\frac{1}{8}$, $\frac{1}{16}$, and $\frac{1}{32}$ resolutions, improving the overall feature representation and making the model more robust in diverse and challenging environments. This is why we will refer to the considered decoder as eFGFP, namely enhanced FGFP.

Finally, the neck includes a spatial-channel attention mechanism based on the Convolutional Block Attention Module (CBAM), which strengthens the capability of the proposed architecture to adaptively highlight the most relevant spatial regions and feature channels. The channel attention (CA) module emphasizes features that are most relevant for recognizing people and ignores irrelevant ones, adaptively focusing on the most informative channels. Furthermore, the spatial attention (SA) mechanism, which identifies the most relevant regions in the image, is in charge of guiding the model to focus on regions where people are likely located. The fusion of spatial and channel attention modules allows the model to adaptively highlight the most relevant spatial regions and feature channels, thereby boosting detection in complex, very crowded scenarios and in presence of variegate backgrounds and thus scenarios.

2.3   Prediction Head

The enhanced feature maps are used to generate the set $\hat{P} = \{\hat{p}_j | j \in \{1, .., M\}\}$ of point proposals $\hat{p}_j$, each one representing a people head detected by the network within the image. For each point, the network also predicts a confidence score $\hat{c}_j$. Thus, an additional prediction head of the model also outputs $\hat{C} = \{\hat{c}_j | j \in \{1, .., M\}\}$. Let also be $p_i = (x_i, y_i)$ the head point of the $i-th$ individual in the scene. Thus, the ground truth collection of points can be referred as: $P = \{p_i | i \in \{1, .., N\}\}$. At training time, the matching between $\hat{P}$ and P is performed using the Hungarian Algorithm. Anyway, instead of only considering the distance in pixels between the ground truth p_i and the detected point $\hat{p}_j$, as in P2PNet [24] we also consider the confidence score, so as to give priority to those detected points having a higher confidence score. In a more formal way, the distance $d(p_i, \hat{p}_j)$ is computed as:

$$d(p_i, \hat{p}_j) = \tau ||p_i - \hat{p}_j||_2 - \hat{c}_j \qquad (1)$$

where $||.||_2$ is the l_2 distance and τ is used to balance the importance of the pixel distance with respect to the confidence score.

Finally, once the association has been performed, we combine the Euclidean loss $\mathcal{L}_{loc}$ to supervise the point regression, and the Cross Entropy loss $\mathcal{L}_{cls}$ to train the proposal classification: $\mathcal{L} = \mathcal{L}_{loc} + \lambda \mathcal{L}_{cls}$, where λ is a weight to balance the effect of the regression loss.

3 Experimental Results

In order to validate the proposed approach, we used the 19 datasets summarized in Table 1. For each dataset, the table provides the image resolution (height $\times$ width), the total number of images, the total count (TC) of individuals across all images (so as to have an idea about the complexity of the dataset), the minimum (MinC) and maximum count (MaxC) of individuals in a single image, and the average count (AC). Additionally, the number of images allocated to the test set (TS) is specified. Where available, the original partition into training and test sets has been preserved. For datasets such as GCC, MICC People Counting, NWPU-Crowd, SmartCity, and UCF-CC-50, which did not include predefined partitions, we randomly selected 20% of the samples for the test set. In total, the training set is composed by $119,967$ images (approximately 73% of the total dataset), while the remaining $42,902$ images were used for testing. The datasets cover both sparse and dense conditions (ranging from 0 to 25,791 individuals in a single image), indoor and outdoor settings, and diverse environments including cities, buildings, malls, laboratories, stadiums, and train stations.

The performance of the proposed approach was evaluated using two standard metrics: Mean Absolute Error (MAE) and Mean Squared Error (MSE). These metrics are defined as follows:

$$MAE = \frac{1}{N} \cdot \sum_{i=1}^{N} \left| C_i^{pred} - C_i^{gt} \right| \S; \quad MSE = \sqrt{\frac{1}{N} \cdot \sum_{i=1}^{N} \left| C_i^{pred} - C_i^{gt} \right|^2} \qquad (2)$$

where N represents the number of images and C_i^{pred} and C_i^{gt} denote the predicted and ground truth counts of individuals in the i-th image, respectively.

In order to also evaluate the performance of the proposed approach without any bias due to the number of people in each image, we also consider a normalized version of MAE and MSE, namely the nMAE (Normalized Mean Absolute Error) and nMSE (Normalized Mean Squared Error). These metrics differ with respect

Table 1. Datasets used in our experimentation. For each dataset, we report the year of publication (Y), the image resolution (expressed in terms of Height x Width), the number of images, the total count (TC) of persons, the minimum (MinC) and the maximum count (MaxC) in a single image, the average count (AC) and the number of images used for the test set (TS).

Dataset	Y	Res.	Images	TC	MinC	MaxC	AC	TS
Beijing-BRT [6]	2019	360 × 640	1,280	16,795	1	64	13	560
CityUHK-X	2017	384 × 512	3191	106,783	2	134	33	688
Crowd Surveillance [29]	2019	Variable	13,945	386,513	2	1,420	35	3,605
DroneCrowd [28]	2021	1080 × 1920	33,960	4,864,280	25	455	145	9,000
FDST [7]	2019	Variable	15,003	394,081	9	57	27	6,000
GCC [27]	2019	1080 × 1920	15,210	7,625,843	0	3,995	501	3,042
JHU-CROWD++ [23]	2019	910 × 1430	4,372	1,515,005	0	25,791	346	1,600
Mall [5]	2012	480 × 640	2,000	62,316	13	53	31	1,200
MICC [3]	2014	480 × 640	3,358	17,630	0	28	5	672
NWPU [26]	2020	Variable	3,609	2,133,375	0	20,033	418	722
ShanghaiTech Part A [32]	2016	Variable	482	241,677	33	3,139	501	182
ShanghaiTech Part B [32]	2016	768 × 1024	716	88,488	9	578	123	316
ShanghaiTech-RGBD [14]	2019	1080 × 1920	2,193	144,512	6	234	66	1,000
SmartCity [31]	2018	1080 × 1920	50	369	1	14	7	10
UCF-CC50 [11]	2013	Variable	50	63,974	94	4,543	1,279	10
UCF-QNRF [12]	2018	Variable	1,535	1,251,642	49	12,865	815	334
UCSD [21]	2008	158 × 238	2,000	49,885	11	46	25	1,200
Venice [16]	2019	720 × 1280	167	35,902	86	421	215	87
VSCrowd [13]	2022	1080 × 1920	62,938	2,344,276	1	296	37	13,902

to the counterparts since the average absolute errors are normalized by the number of people C_i^{gt} in the i-th image under analysis:

$$nMAE = \frac{1}{N} \sum_{i=1}^{N} \frac{|C_i^{pred} - C_i^{gt}|}{C_i^{gt}}; \quad nMAE = \sqrt{\frac{1}{N} \sum_{i=1}^{N} \frac{|C_i^{pred} - C_i^{gt}|^2}{(C_i^{gt})^2}} \quad (3)$$

Finally, in order to also evaluate a measure of the localization and not only of the counting, we consider the nAP metric introduced in [24], which is calculated based on the Average Precision, which is the area under the Precision-Recall (PR) curve.

Table 2 provides a comprehensive summary of the performance obtained for the proposed architecture, denoted as ConvNeXt + eFGFP. It is important to note that for the metrics MAE, MSE, nMAE, and nMSE, lower values indicate better performance. Conversely, for the nAP metric, higher values are better. Additionally, the table presents the results obtained by the state-of-the-art baseline model, P2PNet [24]. The performance of P2PNet has been assessed using the original model as trained and provided by the authors. In the same table, we also present the performance achieved by employing various state-of-the-art backbones, including CNN architectures such as MobileNet, EfficientNet, and the proposed ConvNeXt, as well as VMamba and Swin-T. For these evaluations, the original FPN neck has been utilized. It is observed that no single architec-

Table 2. Overall performance in terms of MAE, MSE, nMAE, nMSE and nAP of the proposed Architecture (ConvNeXt + eFGFP) compared with state of the art baseline considered in this paper (namely P2PNet), other state of the art methods on the same test sets and trained in the same condition, and the different considered variants, in terms of backbone and neck.

Method	MAE ↓	MSE ↓	nMAE ↓	nMSE ↓	nAP ↑
P2PNet [24]	61.85	182.15	48.3	25.5	35.3
MobileNet	28.2	150.3	17.4	22.1	53.8
EfficientNet	18.7	128.2	12.6	14.6	61.6
Swin-T	18.5	92.2	6.5	14.4	63.3
VMamba	17.0	91.6	6.4	13.2	68.1
ConvNeXt	16.9	103.9	8.3	13.4	65.7
VMamba + IFI	16.9	91.5	6.4	13.2	63.6
VMamba + eFGFP	16.1	92.0	6.5	12.5	67.9
ConvNeXt + IFI	15.9	92.0	6.5	12.4	61.7
Proposed (ConvNeXt + eFGFP)	**13.7**	**79.5**	**4.8**	**10.7**	**68.2**

ture consistently outperforms all others across all indices. However, ConvNeXt and VMamba demonstrate better performance compared to the other architectures. Consequently, we have chosen to evaluate both of them with different necks, including the proposed eFGFP and the recently introduced Implicit Feature Interpolation (IFI) [4], which represents an enhanced version of the original FPN.

We can observe that for the VMamba architecture, the usage of different decoders does not significantly affect performance, as there is only a negligible or even no improvement across the different indices. Conversely, both IFI and eFGFP have a positive impact on the performance of ConvNeXt. Specifically, the chosen eFGFP decoder substantially enhances all the considered indices compared to the counterpart FPN: +3.2 for MAE, +24.4 for MSE, +3.2 for nMAE, +2.7 for nMSE, and +2.5 for nAP.

In general, our method achieves the lowest MAE and MSE values, demonstrating impressive accuracy compared to all other considered methods. The normalized metrics (nMAE and nMSE) further confirm the robustness of our approach across various scenarios.

The results obtained over each single dataset by the proposed approach are reported in Tables 3 and 4, together with a comparison with ConvNeXt and the original FPN, and the same considered backbone combined with IFI. In bold we can see the best index over each dataset. From Table 3 we can see that the proposed approach obtains the best results in terms of MAE, MSE and nAP on 10 out of 19 datasets, while in 3 datasets two of the three indices are the best ones, and in 4 datasets one index is the best one. Note that there is only a single dataset, namely Crowd surveillance, characterized by sparse and slightly crowded environments (from 2 to 1,420 persons in a single image), in which the original FPN outperforms the other necks over the three indices. Note that the impressive performance achieved by the proposed approach is also confirmed by the results reported in Table 3, in terms of nMAE and nMSE.

Table 3. Performance obtained by ConvNeXt over the different datasets, combined with different decoders, namely the original FPN, IFI, and the considered eFGFP in terms of MAE, MSE, and nAP. The best result for each dataset is in bold.

Test Set	ConvNeXt + FPN			ConvNeXt + IFI			ConvNeXt + eFGFP		
	MAE↓	MSE↓	nAP↑	MAE↓	MSE↓	nAP↑	MAE↓	MSE↓	nAP↑
Beijing-BRT	1.93	3.27	76.0	**1.89**	**2.88**	74.2	1.91	3.21	**77.2**
Crowd surveillance	**5.39**	**15.65**	**76.0**	8.22	27.50	69.2	6.68	18.00	74.3
City-UHK-X	4.57	7.01	61.2	4.13	6.23	57.5	**3.94**	**6.00**	**64.7**
DroneCrowd	21.03	26.54	42.7	18.90	24.90	38.0	**16.90**	**21.10**	**44.3**
FDST	**1.54**	**2.12**	81.9	2.47	3.53	78.7	2.11	3.46	**82.6**
GCC	65.83	165.45	66.1	56.50	150.90	57.1	**44.29**	**128.80**	**66.4**
JHU-CROWD++	74.57	304.93	44.1	81.50	309.50	40.0	**72.70**	**285.80**	**47.0**
Mall	3.09	3.78	67.4	**2.44**	3.18	63.0	2.53	**3.13**	**73.5**
MICC	0.24	0.52	75.4	0.46	0.76	75.5	**0.10**	**0.32**	**78.2**
NWPU	92.36	566.05	48.2	89.60	413.90	43.6	**80.70**	**347.12**	**50.0**
ShanghaiTech-A	**63.47**	**118.63**	54.4	68.50	124.14	50.0	68.00	124.00	**59.5**
ShanghaiTech-B	17.95	28.26	72.8	18.90	30.67	70.0	**14.90**	**26.50**	**76.0**
ShanghaiTechRGBD	6.98	**9.45**	75.4	8.70	20.00	69.2	**5.90**	11.70	**76.3**
SmartCity	**5.50**	**8.70**	50.5	14.20	16.72	28.0	9.50	14.20	42.3
UCF-CC50	242.60	349.03	33.2	241.30	433.60	32.4	**233.40**	**275.10**	**38.8**
UCF-QNRF	123.43	205.60	45.9	116.02	205.40	40.9	**103.80**	**177.42**	**49.1**
UCSD	2.09	2.73	59.1	**2.07**	**2.64**	57.4	2.12	2.82	**71.2**
Venice	112.87	131.23	52.5	116.80	133.80	51.3	**90.60**	**104.40**	**60.3**
VSCrowd	7.23	14.57	**79.5**	6.76	10.63	74.0	**6.27**	**10.52**	78.2
Overall	16.90	103.90	65.7	15.97	92.00	61.7	**13.70**	**79.50**	**68.2**

Table 4. Performance obtained by ConvNeXt over the different datasets, combined with different decoders, namely the original FPN, IFI, and the considered eFGFP in terms of nMAE and nMSE. The best result for each dataset is in bold.

Test Set	ConvNeXt + FPN		ConvNeXt + IFI		ConvNeXt + eFGFP	
	nMAE↓	nMSE↓	nMAE↓	nMSE↓	nMAE↓	nMSE↓
Beijing-BRT	3.9	14.7	**3.1**	**14.4**	3.8	14.6
Crowd surveillance	**6.4**	**13.2**	20.0	20.0	8.5	16.4
City-UHK-X	2.8	13.2	2.2	12.0	**2.1**	**11.4**
DroneCrowd	3.0	15.3	2.7	46.5	**1.9**	**12.3**
FDST	**0.5**	**5.7**	1.5	9.1	1.7	7.8
GCC	3.6	13.2	3.0	11.4	**2.1**	**8.9**
JHU-CROWD++	14.7	22.8	15.2	24.9	**12.9**	**22.2**
Mall	1.3	9.5	**0.9**	**7.5**	0.9	7.8
MICC	0.7	4.5	1.4	8.4	**0.2**	**1.9**
NWPU	21.2	22.5	11.3	21.8	**7.9**	**19.6**
ShanghaiTech-A	**4.5**	**14.6**	4.9	15.8	4.8	15.6
ShanghaiTech-B	3.2	14.5	3.8	15.3	**2.8**	**12.0**
ShanghaiTechRGBD	**1.5**	10.0	7.0	13.4	2.4	**9.1**
SmartCity	100.0	76.3	100.0	100.0	100.0	100.0
UCF-CC50	5.4	18.5	8.4	18.4	**3.4**	**17.8**
UCF-QNRF	3.92	17.1	3.9	11.3	**2.9**	14.4
UCSD	0.7	6.5	0.7	6.4	**0.6**	**6.1**
Venice	25.7	44.9	26.0	46.5	**16.3**	**36.1**
VSCrowd	3.36	11.6	1.8	10.9	**1.7**	**10.1**
Overall	8.3	13.4	6.5	12.4	**4.8**	**10.7**

A few examples of the proposed approach are reported in Fig. 2. In order to evaluate the possibility to run the proposed architecture in real time, we also characterize the size of the model and the frame rate. The model is characterized by 97.67 M parameters and over a server equipped with AMD EPYC 7282 16-Core Processor and NVIDIA GPU A100 it is able to achieve about 11 Frames per seconds, thus confirming the feasibility to run it in real time.

Fig. 2. Examples of our method in action. In the first row, the ground truth points are reported, while in the second row we show the output of the proposed approach.

4   Conclusions

In this paper, we have presented a novel approach to people analytics in highly crowded environments by leveraging the strengths of the ConvNeXt architecture combined with an enhanced Fine-Grained Feature Pyramid (eFGFP) for multi-scale management. Infact, the proposed architecture effectively integrates the advantages of both convolutional networks and transformer-based models, resulting in improved feature extraction and representation. The use of depthwise convolution in ConvNeXt significantly reduces computational complexity, making the model suitable for deployment in real time with limited computational resources. Additionally, the eFGFP design ensures the retention of rich semantic information across different scales, further improving the model's performance in complex scenarios. These design choices collectively enable robust crowd counting in dynamic and challenging scenarios, such as aerial monitoring from moving drones. This makes our approach particularly suitable for real-world applications in public safety, event surveillance, and urban planning, where scalability, accuracy, and efficiency are essential.

References

1. Babu Sam, D., Surya, S., Venkatesh Babu, R.: Switching convolutional neural network for crowd counting. In: CVPR 2017
2. Bai, H., He, H., Peng, Z., Dai, T., Chan, S.H.G.: Countr: an end-to-end transformer approach for crowd counting and density estimation. In: ECCV Workshops (2023)
3. Bondi, E., Seidenari, L., Bagdanov, A.D., Del Bimbo, A.: Real-time people counting from depth imagery of crowded environments. In: AVSS (2014)
4. Chen, I., Chen, W.T., Liu, Y.W., Yang, M.H., Kuo, S.Y.: Improving point-based crowd counting and localization based on auxiliary point guidance. In: ECCV (2025)
5. Chen, K., Loy, C.C., Gong, S., Xiang, T.: Feature mining for localised crowd counting. In: BMVC, vol. 1, p. 3 (2012)
6. Ding, X., Lin, Z., He, F., Wang, Y., Huang, Y.: A deeply-recursive convolutional network for crowd counting. In: ICASSP 2018
7. Fang, Y., Zhan, B., Cai, W., Gao, S., Hu, B.: Locality-constrained spatial transformer network for video crowd counting. In: ICME 2019
8. Gao, G., Gao, J., Liu, Q., Wang, Q., Wang, Y.: CNN-based density estimation and crowd counting: a survey. arXiv abs/2003.12783 (2020)
9. Gao, J., Wang, Q., Li, X.: PCC net: perspective crowd counting via spatial convolutional network. In: TCSVT 2019
10. Greco, A., Saggese, A., Vento, B.: A robust and efficient overhead people counting system for retail applications. In: ICIAP, pp. 139–150 (2022)
11. Idrees, H., Saleemi, I., Seibert, C., Shah, M.: Multi-source multi-scale counting in extremely dense crowd images. In: CVPR 2013
12. Idrees, H., et al.: Composition loss for counting, density map estimation and localization in dense crowds. In: ECCV 2018
13. Li, H., et al.: Video crowd localization with multifocus gaussian neighborhood attention and a large-scale benchmark. IEEE Trans. Image Process. **31**, 6032–6047 (2022)
14. Lian, D., Li, J., Zheng, J., Luo, W., Gao, S.: Density map regression guided detection network for RGB-D crowd counting and localization. In: CCVPR (2019)
15. Liang, D., Chen, X., Xu, W., Zhou, Y., Bai, X.: Transcrowd: weakly-supervised crowd counting with transformers. Sci. China Inf. Sci. **65**(6), 160104 (2022)
16. Liu, W., Salzmann, M., Fua, P.: Context-aware crowd counting. In: CVPR 2019
17. Liu, Y., Shi, M., Zhao, Q., Wang, X.: Point in, box out: beyond counting persons in crowds. In: Computer Vision and Pattern Recognition, pp. 6469–6478 (2019)
18. Liu, Z., Mao, H., Wu, C.Y., Feichtenhofer, C., Darrell, T., Xie, S.: A convnet for the 2020s. arXiv preprint arXiv:2201.03545 (2022)
19. Ma, H.Y., Zhang, L., Shi, S.: VMAMBACC: a visual state space model for crowd counting. arXiv abs/2405.03978 (2024)
20. Ma, H.Y., Zhang, L., Wei, X.Y.: FGENet: fine-grained extraction network for congested crowd counting. In: MMM 2024. LNCS, vol. 14556, pp. 43–56. Springer-Verlag, Heidelberg (2024). https://doi.org/10.1007/978-3-031-53311-2_4
21. Mahadevan, V., Li, W., Bhalodia, V., Vasconcelos, N.: Anomaly detection in crowded scenes. In: Conference on Computer Vision and Pattern Recognition (2010)
22. Quan, Y., Zhang, D., Zhang, L., Tang, J.: Centralized feature pyramid for object detection. IEEE Trans. Image Process. **32**, 4341–4354 (2023)

23. Sindagi, V.A., Yasarla, R., Patel, V.M.: JHU-crowd++: large-scale crowd counting dataset and a benchmark method. In: PAMI 2020
24. Song, Q., et al.: Rethinking counting and localization in crowds: a purely point-based framework. In: ICCV 2021
25. Wang, B., Liu, H., Samaras, D., Nguyen, M.H.: Distribution matching for crowd counting. Adv. Neural. Inf. Process. Syst. **33**, 1595–1607 (2020)
26. Wang, Q., Gao, J., Lin, W., Li, X.: NWPU-crowd: a large-scale benchmark for crowd counting and localization. In: PAMI 2020
27. Wang, Q., Gao, J., Lin, W., Yuan, Y.: Learning from synthetic data for crowd counting in the wild. In: CVPR 2019
28. Wen, L., et al.: Detection, tracking, and counting meets drones in crowds: a benchmark. In: CVPR 2021
29. Yan, Z., et al.: Perspective-guided convolution networks for crowd counting. In: ICCV 2019
30. Zhang, C., Li, H., Wang, X., Yang, X.: Cross-scene crowd counting via deep convolutional neural networks. In: CVPR 2015
31. Zhang, L., Shi, M., Chen, Q.: Crowd counting via scale-adaptive convolutional neural network. In: WACV 2018
32. Zhang, Y., Zhou, D., Chen, S., Gao, S., Ma, Y.: Single-image crowd counting via multi-column convolutional neural network. In: CVPR 2016

3rd International Workshop on Artificial Intelligence and Radiomics in Computer-Aided Diagnosis (AIRCAD 2025)

Workshop

3rd International Workshop on Artificial Intelligence and Radiomics in Computer-Aided Diagnosis (AIRCAD 2025)

In conjunction with ICIAP 2025—Rome, Italy, September 15–19, 2025

Workshop Organization

Chairs and Organizers

Albert Comelli	Fondazione Ri.MED, Italy
Cecilia Di Ruberto	University of Cagliari, Italy
Andrea Loddo	University of Cagliari, Italy
Lorenzo Putzu	University of Cagliari, Italy
Alessandro Stefano	CNR of Cefalù, Italy
Luca Zedda	University of Cagliari, Italy

Program Committee

Seyed-Ahmad Ahmadi	NVIDIA, Germany
Viviana Benfante	University of Palermo, A.R.N.A.S.
xxx	Civico Di Cristina Benfratelli Hospitals, Italy
Monica Bianchini	University of Siena, Italy
Roberto Cannella	University of Palermo, Italy
Renato Cuocolo	University of Salerno, Italy
Giuseppe Cutaia	University of Palermo, Italy
Navdeep Dahiya	Georgia Institute of Technology, USA
Mario D'Acunto	National Research Council, Italy
Angelo Genovese	University of Milano, Italy
Marco Grangetto	University of Torino, Italy
Jon Ander Gómez Adrián	Universitat Politècnica de València, Spain
Riccardo Laudicella	University of Messina, Italy
Salvatore Livatino	University of Hertfordshire, UK
Mario Molinara	University of Cassino and Southern Lazio, Italy
Davide Moroni	National Research Council, Italy
Paolo Napoletano	University of Milan, Bicocca, Italy
Antonio Parziale	University of Salerno, Italy
Giovanni Pasini	Sapienza University of Rome, Italy
Luca Pireddu	CRS4, Italy
Mubashara Rehman	University of Udine, Italy
Giorgio Russo	IBSBC-CNR, Italy
Giuseppe Salvaggio	University of Palermo, Italy
Mattia Savardi	University of Brescia, Italy
Alberto Signoroni	University of Brescia, Italy
Arnaldo Stanzione	University of Naples Federico II, Italy
Nicola Strisciuglio	University of Twente, Netherlands
Enzo Tartaglione	Télécom Paris, Institut Polytechnique de Paris, France

Francesco Tortorella   University of Salerno, Italy
Lorenzo Ugga   University of Naples Federico II, Italy
Federica Vernuccio   University of Palermo, Italy
Pierpaolo Alongi   University of Palermo, Italy
Antonio Lo Casto   University of Palermo, Italy
Rosario Corso   University of Palermo, Italy

A Fully Automatic Patch-Based Deep Learning Pipeline for Mass and Microcalcification Detection in Mammography: A Preliminary Study

Giovanni Pasini[1,2]($\boxtimes$) iD, Nicolò Lauciello[2,3] iD, Alessia Finti[1] iD, Giorgio Russo[2] iD, Franco Marinozzi[1] iD, Fabiano Bini[1] iD, and Alessandro Stefano[2] iD

[1] Department of Mechanical and Aerospace Engineering, Sapienza University of Rome, Eudossiana 18, 00184 Rome, Italy
giovanni.pasini@uniroma1.it

[2] Institute of Bioimaging and Complex Biological Systems, National Research Council, 90015 Cefalù, Italy

[3] Department of Earth and Marine Sciences, University of Palermo, Via Archirafi 22, 90123 Palermo, Italy

Abstract. This study proposes a fully automatic, patch-based pipeline for the classification and localization of masses and microcalcifications in mammographic images. Leveraging the EfficientNetB6 convolutional neural network, the method classifies image patches into one of three categories, healthy tissue, mass, or calcification, based on their lesion content. The CBIS-DDSM dataset was used, containing over 3500 annotated images. Preprocessing included removing irrelevant content, enhancing image quality through normalization, filtering, and contrast enhancement. Patches (528×528) were extracted with 50% overlap and labeled if they contained at least 20% of a lesion. The network was trained using transfer learning followed by fine-tuning, with data augmentation applied to improve generalizability. During inference, patches were extracted with 85% overlap, classified, and reassembled into full-size probability maps, which were smoothed to aid visualization.

The model achieved robust performance, with an average accuracy of 81%, F1-score of 81%, and area under the ROC curve (AUC) of 94% across all classes. These results highlight the effectiveness of EfficientNetB6 in extracting meaningful features from mammograms, even with limited lesion visibility. The generated probability maps serve as visual aids to support clinical decision-making and enable integration with techniques like Grad-CAM for further interpretability. The approach enables efficient processing (~3 min per image), making it viable for real-time applications. This framework represents a step towards an automated, interpretable, and efficient mammographic reporting system.

Keywords: Mammography · Deep Learning · Patch-Based Classification

E. Rodolà et al. (Eds.): ICIAP 2025 Workshops, LNCS 16169, pp. 157–165, 2026.
https://doi.org/10.1007/978-3-032-11317-7_13

1 Introduction

Breast cancer is the most diagnosed cancer among women worldwide, second only to lung cancer in total cases. It affects about 10 to 12% of women and causes around 500,000 deaths each year [1]. Women aged 40 to 49 experience the highest rates, accounting for roughly 41%, while only about 21% of cases are in women over 71. Often, early-stage breast cancer doesn't show obvious symptoms, but warning signs can include a firm lump, changes in breast shape, nipple irregularities, or unexplained weight loss. Thanks to advances in treatment and screening efforts, death rates have been gradually decreasing. To improve understanding and classification of breast lesions, the World Health Organization updated its system in 2003, providing clearer categories for different types like intraductal proliferative lesions, papillary formations, mesenchymal and fibroepithelial tumors, and rarer types like lymphomas and secondary cancers [2, 3]. Also, physical signs like palpable masses and microcalcifications, though not true tissue types, are important clues seen in imaging tests when assessing breast cancer risk [4]. Diagnosing breast cancer usually involves reviewing the patient's medical history, performing a physical exam, and conducting imaging scans. Mammography remains the primary tool for early detection, especially when no symptoms are present; however, dense breast tissue can sometimes hide tumors, leading to missed or mistaken diagnoses [5, 6] Using digital breast tomosynthesis, which creates close to three-dimensional images of the breast, helps reduce the problem of tissue overlap and improves the visualization of suspicious areas [7, 8]. Furthermore, advances in CAD (Computer Aided Systems) have helped in refining mass and microcalcifications early detection, improving diagnostic precision [9]. Within this scope, Deep Learning (DL) radiomics has shown remarkable success in extracting hierarchical features from raw imaging data, surpassing traditional machine learning radiomics systems [10], and automating the diagnostic procedures. Indeed, traditional radiomics [11, 12], even if more explainable, is affected by several challenges [13, 14] and still needs the intervention of humans in the workflow [15]. Commonly used DL architectures involve Convolutional Neural Networks (CNN), such as U-Net [16], ResNet [17] used either to detect and segment masses and microcalcifications in mammographic images or classify the nature of the identified lesions between benign and malignant. Following this line, the aim of this study is to develop a fully automatic patch-based pipeline to detect masses and microcalcifications from mammographic images using EfficientNetB6 [18] CNN, and generate probability maps, whose aim is to guide the clinician to focus on which part of the image is more likely to find a lesion and distinguish it from healthy tissue.

2 Materials and Methods

2.1 Dataset

The CBIS-DDSM dataset, available from The Cancer Imaging Archive (TCIA), contains digitized mammograms categorized into microcalcifications (1872 images) and masses (1696 images) from 753 and 891 patients, respectively. Of the 3568 total images, 2111 are benign and 1457 malignant. Each patient has multiple IDs linked to different views (e.g., CC, MLO) and lesions, with at least two image sets per case: a full mammogram and a lesion-focused view, including a segmented Region of Interest (ROI) and a cropped image. ROIs were delineated through manual and semiautomatic methods (modified Chan-Vase region detection). Some patients show both lesion types. Images are in Digital Imaging and Communication in Medicine (DICOM) format with tomosynthesis extensions. Accompanying CSV metadata includes breast density (BI-RADS), side, view, lesion type and ID, shape, margins, distribution, pathology, BI-RADS score, and subtlety rating (1–5), offering a rich resource for breast cancer imaging studies [19].

2.2 Image Pre-processing

Each mammographic scan is pre-processed before passing through the CNN model according to the following steps: i) image cleaning, and ii) image enhancement.

Image Cleaning

Images are cleaned of any non-breast relevant content such as mammogram supports and view indications. Firstly, the procedure involves the identification of the foreground pixels, including breast, mammogram supports and view indications, through thresholding of pixels with values greater than 0. Then, the generated masks are rescaled to smaller size by a factor of 0.1 to improve computational speed of subsequent morphological operations. Binary erosion with a disk structure (radius $= 10$), small objects removal and bounding box computation are adopted to identify the largest breast tissue component, then binary closing and binary dilation with disk structures (radius $= 10$) are used to refine the final masks. Finally, masks are rescaled back to their original size and applied to the original image, then both images and masks are left oriented.

Image Enhancement

Images are normalized through truncation normalization, firstly clipping the gray levels between the 5th and 99th percentiles, then normalizing them with min-max normalization [20]. Successively, a median filter is used to reduce noise in the images with a square window of size 7, and Contrast Limited Adaptive Histogram Equalization (CLAHE) is used to improve contrast.

2.3 Training Procedure

A new patch-based dataset is created starting with pre-processed images. Firstly, the size of images is made divisible by the chosen patch size (528×528) through padding if needed, then patches are extracted with 50% overlap and labeled based on the content. Patches that contained at least the 20% of the total mass or calcification are labeled "mass" or "calcification", accordingly. Patches that do not contain any lesion are labeled as healthy and completely black patches are discarded. The final dataset contained a total amount of 7006 training patches (2304 masses, 2351 calcifications, 2351 healthy) and 2008 testing patches (706 masses, 651 calcifications, 651 healthy). An EfficientNetB6 model pretrained on ImageNet [21] dataset is chosen as baseline, and transfer learning followed by fine tuning were used. For transfer learning, the final classification block is replaced with a GlobalAveragePooling2D, a Dense layer of 1024 neurons, a Relu activation function, a Dropout layer with 30% percentage and a final Dense layer with 3 neurons (equal to the number of classes) and SoftMax activation function. A training schedule comprising of an initial learning rate of 1×10^{-3} a final learning rate of 1×10^{-8} a Polynomial decay scheduler with power of 0.9, batch size equal to 16, Adam Optimizer and Categorical Cross Entropy loss function, is chosen. For fine tuning, an initial learning rate of 5×10^{-5} a final learning rate of 1×10^{-8} a Polynomial decay scheduler with power of 1, batch size equal to 8, Adam Optimizer and Categorical Cross Entropy loss function, is chosen. For both transfer and fine tuning, total epochs are set to 50. Moreover, data augmentation was applied in the form of left and right flipping, upside down flipping, brightness change, rotation (range -20°, + 20°), gaussian noise addition and contrast change. Each augmentation type is triggered with a probability of 0.33 [22].

2.4 Classification Pipeline

The full classification pipeline involves, i) image importing, ii) image cleaning, iii) image enhancement, iv) patch extraction with 85% overlap and black patches removal, v) patch classification, vi) probability maps generation for lesion detection. Each patch composing the original image is classified either healthy, mass or calcification, and a classification probability is associated with that patch and to all the pixels it contains. Then, probability patches are reassembled to form a probability image with size equal to the original image, and overlapping regions are averaged. Finally, the probability image is smoothed with gaussian blurring to improve visualization. The pipeline is shown in Fig. 1.

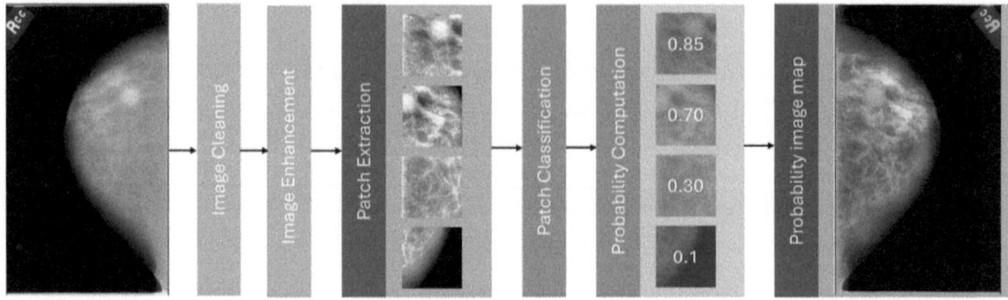

Fig. 1. Classification pipeline

3 Results

Overall, performance of the model to classify patches are reported both averaged and considering a One vs All approach. Precision, recall, f1-score, and Area Under Curve (AUC) to distinguish healthy tissue from mass and calcification reached 81%, 83%, 82% and 95%, accordingly. Precision, recall, f1-score, and AUC to distinguish masses from healthy tissue and calcification reached 83%, 81%, 82% and 94%, accordingly. Precision, recall, f1-score, and AUC to distinguish calcifications from healthy tissue and masses reached 80%, 81%, 80% and 94%, accordingly. On average, an AUC of

94%, and an accuracy of 81% were reached. Metrics are reported in Table 1. Moreover, Receiver Operating Characteristic (ROC) curve and Confusion Matrix are reported in Fig. 2. Furthermore, probability maps for lesion detection are shown in Fig. 3.

Table 1. Performance metrics for patch classification, One vs All and average.

Classification	Precision	Recall	F1-score	AUC	Accuracy
Healthy vs mass and calcification	81%	83%	82%	95%	-
Mass vs healthy and calcification	83%	81%	82%	94%	-
Calcification vs healthy and mass	80%	81%	80%	94%	-
Average	81%	81%	81%	94%	81%

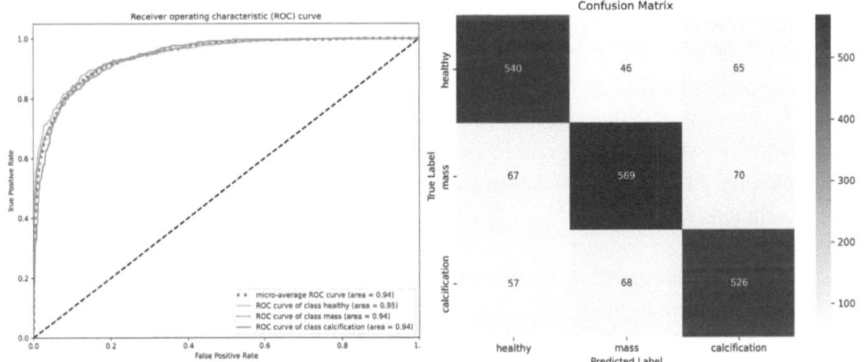

Fig. 2. Roc curve on the left and confusion matrix on the right

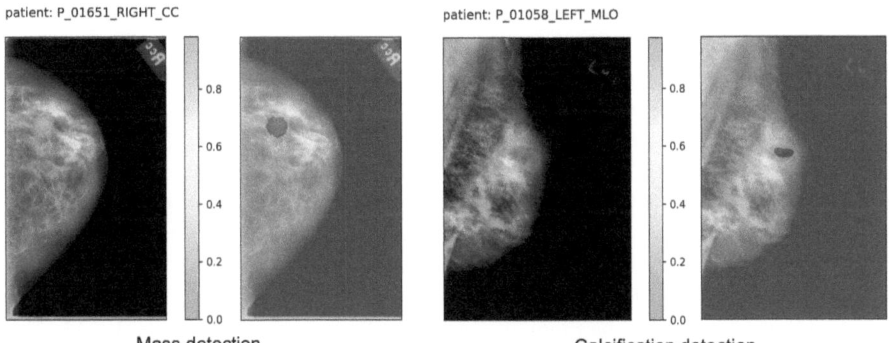

Fig. 3. On the left, mass detection probability map and segmentation overlayed on probability map for patient P_01651_RIGHT_CC. On the right, calcification detection and segmentation overlayed on probability map for patient P_01058_LEFT_MLO.

4 Discussion

This study aimed to develop a completely automatic patch-based lesion detection pipeline for mammographic images to localize masses and microcalcification through probability maps. Images needed preprocessing to improve classification performance; therefore, image cleaning and image enhancement were adopted. The former was used to remove any potential source of error during model training and model inference. Indeed, mammography supports and view indications, which do not carry any breast information, can lead to a misclassification. The latter was used to normalize images via truncation normalization, reduce noise and improve image contrast, making masses and microcalcifications much more visible [23]. Patch extraction with 50% overlap and a 20% content threshold for labeling as mass or calcification were used to help the model adapt to varied views, without overflowing with redundant and too similar patches (overlap greater than 50%), and perform reliably during inference. Moreover, online data augmentation with probability of 0.33 guaranteed training samples variability and improved model robustness. EfficientNetB6 was chosen for its ability to extract robust and meaningful features, as evidenced by its strong performance on the test set. Indeed, high performance metrics were reached, such as 81% accuracy, 81% precision, 81% recall, 81% f1-score and 94% AUC. Moreover, EfficientNetB6 has unique compound scaling capability, which optimizes the network's depth, width, and resolution simultaneously. This joint scaling strategy enables EfficientNet to achieve an optimal trade-off between performance and resource efficiency, reducing the number of parameters and computational operations (measured as FLOPS) compared to traditional architectures.

Our results are comparable to those of Shen et al. [24], who achieved average 89% accuracy (88% for malignant mass, 83% for benign mass, 73% for malignant calcification, 81% for benign calcification and 96% for background) using a five-class classifier (background, benign/malignant calcification, benign/malignant mass) on curated CBIS-DDSM patches. However, their ROI sampling strategy, extracting 10 patches with $\geq$ 90% overlap around each centered lesion, differs from our 20% lesion-content threshold, which may pose a greater classification challenge. Indeed, patch sampling strategy and choosing the correct patch size are an open issue as reported in [25]. However, our choice is justified by the need to detect even small lesions. The potential overclassification of background as lesion, resulting from the 20% cut-off, was intentional, aiming to favor sensitivity, and will be further refined in future works. Furthermore, an 85% overlap was used during inference to ensure smooth transitions between patch probabilities, aiding reconstruction of the final probability map, which was then refined with gaussian smoothing. To accelerate the whole process, fully black patches are excluded, enabling classification in ~ 90ms per patch and ~ 3 min per image with 16GB gpu, fast enough for clinical use. Finally, probability maps can help clinicians identify the key image regions influencing the prediction and could be combined with Grad-CAM [26] in future developments. In addition, a fully automated approach ensures reproducibility and reduces clinicians' working load [27]. Moreover, probability maps can also be thresholded to generate segmentations and bounding boxes for further classification of masses and microcalcifications (benign vs. malignant) [28]. Furthermore, a system of this kind can be implemented in matRadiomics [29] to enhance its functionalities. The main limitations are the lack of an external test set and the single-center design, whereas multicenter studies are recommended [30].

5 Conclusion

In conclusion, this study aims to develop a completely automatic patch-based detection system for differentiating and detecting masses, microcalcifications and healthy tissue on mammographic images with high performance metrics (81% accuracy, 94% AUC, 81% f1-score and 81% precision), and it is a step forward to develop a complete fully automatic reporting system for mammography exams.

References

1. Benson, J.R., Jatoi, I., Keisch, M., Esteva, F.J., Makris, A., Jordan, V.C.: Early breast cancer. Lancet **373**, 1463–1479 (2009). https://doi.org/10.1016/S0140-6736(09)60316-0
2. Tan, P.H., et al.: The 2019 world health organization classification of tumours of the breast. Histopathology **77**, 181–185 (2020). https://doi.org/10.1111/HIS.14091
3. Muller, K., Jorns, J.M., Tozbikian, G.: What's new in breast pathology 2022: WHO 5th edition and biomarker updates. J Pathol Transl Med. 56, 170–171 (2022). https://doi.org/10.4132/JPTM.2022.04.25
4. Kim, S., Tran, T.X.M., Song, H., Park, B.: Microcalcifications, mammographic breast density, and risk of breast cancer: a cohort study. Breast Cancer Res. **24**, 1–11 (2022). https://doi.org/10.1186/S13058-022-01594-0
5. Sternlicht, M.D.: Key stages in mammary gland development: the cues that regulate ductal branching morphogenesis. Breast Cancer Res. **8**, 1–11 (2005). https://doi.org/10.1186/BCR 1368
6. Biswas, S.K.; et al.: The Mammary Gland: Basic Structure and Molecular Signaling during Development. International Journal of Molecular Sciences 2022, Vol. 23, Page 3883. 23, 3883 (2022). https://doi.org/10.3390/IJMS23073883
7. Gao, Y., Moy, L., Heller, S.L.: Digital breast Tomosynthesis: update on technology, evidence, and clinical practice. Radiographics **41**, 321–337 (2021). https://doi.org/10.1148/RG.202120 0101
8. Richman, I.B., et al.: Comparative effectiveness of digital breast tomosynthesis for breast cancer screening among women 40–64 years old. JNCI: J. Nat. Cancer Inst. **113**, 1515–1522 (2021). https://doi.org/10.1093/JNCI/DJAB063
9. Esposito, D., Paternò, G., Ricciardi, R., Sarno, A., Russo, P., Mettivier, G.: A pre-processing tool to increase performance of deep learning-based CAD in digital breast Tomosynthesis. Health Technol (BERL). **14**, 81–91 (2024). https://doi.org/10.1007/S12553-023-00804-9/TABLES/3
10. Stefano, A., et al.: Comparative evaluation of machine learning-based radiomics and deep learning for breast lesion classification in mammography. Diagnostics 2025 **15**, 15, 953 (2025). https://doi.org/10.3390/DIAGNOSTICS15080953
11. Scavuzzo, A., et al.: Radiomics analyses to predict histopathology in patients with metastatic testicular germ cell Tumors before post-chemotherapy retroperitoneal lymph node dissection. J. Imaging 2023 **9**, 213. 9, 213 (2023). https://doi.org/10.3390/JIMAGING9100213

12. Nepi, V., Pasini, G., Bini, F., Marinozzi, F., Russo, G., Stefano, A.: MRI-based radiomics analysis for identification of features correlated with the expanded disability status scale of multiple sclerosis patients. In: Lecture Notes in Computer Science (including subseries Lecture Notes in Artificial Intelligence and Lecture Notes in Bioinformatics), vol. 13373 LNCS, pp. 362–373 (2022). https://doi.org/10.1007/978-3-031-13321-3_32

13. Pasini, G.: Assessing the robustness and reproducibility of CT radiomics features in non-small-cell lung carcinoma. Lecture Notes Comput. Sci. (Include. Subser. Lect. Notes Artif. Intell. Lect. Notes Bioinfo). **14366**, 39–48 (2024). https://doi.org/10.1007/978-3-031-51026-7_4

14. Stefano, A.: Challenges and limitations in applying radiomics to PET imaging: possible opportunities and avenues for research. Comput. Biol. Med. **179**, 108827 (2024). https://doi.org/10.1016/J.COMPBIOMED.2024.108827

15. Torrisi, S.E., et al.: Assessment of survival in patients with idiopathic pulmonary fibrosis using quantitative HRCT indexes. Multidiscip. Respir. Med. **13** (2018). https://doi.org/10.1186/S40248-018-0155-2,

16. Gerbasi, A., et al.: DeepMiCa: Automatic segmentation and classification of breast MIcroCAlcifications from mammograms. Comput. Methods Programs Biomed. **235**, 107483 (2023). https://doi.org/10.1016/J.CMPB.2023.107483

17. Donnelly, J., Moffett, L., Barnett, A.J., Trivedi, H., Schwartz, F., Lo, J., Rudin, C.: AsymMirai: interpretable mammography-based deep learning model for 1–5-year breast cancer risk prediction. Radiology **310** (2024). https://doi.org/10.1148/RADIOL.232780

18. Tan, M., Le, Q. V.: EfficientNet: rethinking model scaling for convolutional neural networks. In: 36th International Conference on Machine Learning, ICML 2019. 2019-June, 10691–10700 (2019)

19. Sawyer-Lee, R., Gimenez, F., Hoogi, A., Rubin, D.: CBIS-DDSM - the cancer imaging Archive (TCIA). https://www.cancerimagingarchive.net/collection/cbis-ddsm/. Accessed 27 May 2025. https://doi.org/10.7937/K9/TCIA.2016.7O02S9CY

20. Schwarzhans, F., et al.: Image normalization techniques and their effect on the robustness and predictive power of breast MRI radiomics. Eur. J. Radiol. **187**, 112086 (2025). https://doi.org/10.1016/J.EJRAD.2025.112086

21. Deng, J., Dong, W., Socher, R., Li, L.J., Li, K., Fei-Fei, L.: ImageNet: a large-scale hierarchical image database. In: 2009 IEEE Conference on Computer Vision and Pattern Recognition, CVPR 2009, pp. 248–255 (2009). https://doi.org/10.1109/CVPR.2009.5206848

22. Isensee, F., Jaeger, P.F., Kohl, S.A.A., Petersen, J., Maier-Hein, K.H.: NnU-Net: a self-configuring method for deep learning-based biomedical image segmentation. Nat. Methods **18**, 203–211 (2021). https://doi.org/10.1038/S41592-020-01008-Z

23. Mat Radzi, S.F., et al.: Impact of image contrast enhancement on stability of radiomics feature quantification on a 2D mammogram radiograph. IEEE Access. **8**, 127720–127731 (2020). https://doi.org/10.1109/ACCESS.2020.3008927

24. Shen, L., Margolies, L.R., Rothstein, J.H., Fluder, E., McBride, R., Sieh, W.: Deep learning to improve breast cancer detection on screening mammography. Sci. Rep. 2019 9:1. 9, 1–12 (2019). https://doi.org/10.1038/s41598-019-48995-4

25. Quintana, G.I., Li, Z., Vancamberg, L., Mougeot, M., Desolneux, A., Muller, S.: Exploiting patch sizes and resolutions for multi-scale deep learning in mammogram image classification. Bioengineering 2023 **10**, 534. 10, 534 (2023). https://doi.org/10.3390/BIOENGINEERING10050534

26. Selvaraju, R.R., Cogswell, M., Das, A., Vedantam, R., Parikh, D., Batra, D.: Grad-CAM: visual explanations from deep networks via gradient-based localization. Int. J. Comput. Vis. **128**, 336–359 (2020). https://doi.org/10.1007/S11263-019-01228-7

27. Stefano, A., Vitabile, S., Russo, G., Ippolito, M., Farletta, M., et al.: A fully automatic method for biological target volume segmentation of brain metastases. Int. J. Imaging Syst. Technol., **26**, 29–37. https://doi.org/10.1002/ima.22154

28. Comelli, A., Stefano, A., Benfante, V., Russo, G.: Normal and abnormal tissue classification in positron emission tomography oncological studies. Pattern Recognit Image Anal. **28**, 106–113 (2018). https://doi.org/10.1134/S1054661818010054/
29. Pasini, G., Bini, F., Russo, G., Comelli, A., Marinozzi, F., Stefano, A.: matRadiomics: a novel and complete radiomics framework, from image visualization to predictive model. J. Imag. 2022 **8**, 221. 8, 221 (2022). https://doi.org/10.3390/JIMAGING8080221
30. Pasini, G., Stefano, A., Russo, G., Comelli, A., Marinozzi, F., Bini, F.: Phenotyping the histopathological subtypes of non-small-cell lung carcinoma: how beneficial is radiomics? Diagnostics. **13**, 1167 (2023). https://doi.org/10.3390/DIAGNOSTICS13061167

Preliminary Radiomics-Based Classification of Tumor Nuclei in Melanoma Histopathology

Alessia Finti[1]([envelope]) [ORCID], Alessandro Stefano[2] [ORCID], Giovanni Pasini[1] [ORCID], Giorgio Russo[2] [ORCID], Franco Marinozzi[1] [ORCID], and Fabiano Bini[1] [ORCID]

[1] Department of Mechanical and Aerospace Engineering, Sapienza University of Rome, Rome, Italy
alessia.finti@uniroma1.it

[2] Institute of Bioimaging and Complex Biological Systems, National Research Council (IBSBC-CNR) Cefalù, Palermo, Italy

Abstract. Melanoma is, nowadays, the most aggressive kind of skin tumor. Even if advanced therapies and diagnosis techniques have been developed, metastatic cases still have a poor prognosis. Diagnosis of melanoma is complicated and confounded by the histopathological heterogeneity in samples and the absence of affordable biomarkers. The advent of innovative analytical techniques in medical imaging, such as radiomics and pathomics, poses hope in providing the needed assistance in diagnostic accuracy. In this paper, a preliminary radiomic workflow has been proposed for a subset of 100 annotated Regions of Interest (ROIs) from the PUMA dataset, comprising primary and metastatic melanoma samples. The goal was to differentiate tumor nuclei from non-tumor nuclei using radiomic features extracted from Whole Slide Images (WSIs). Feature extraction was performed using PyRadiomics, and LASSO regression was applied to carry out feature selection and reduce the features count from 102 to 57. The selected features were used to train a Random Forest classifier, whose output was assessed in terms of accuracy, precision, recall, specificity, F1-score, and area under the ROC curve (AUC). The model exhibited 74.56% accuracy, 73% precision, 77% recall, 72% specificity, 75% F1-score and an AUC of 0.82, which indicates a good discriminatory power. The results validated the capability of radiomics in classifying nuclei on melanoma WSIs. Even if limited by using a subset of the PUMA dataset and pre-annotated nuclei, this pipeline is significant in laying the foundation for further integration of radiomics and Artificial Intelligence (AI) in digital pathology.

Keywords: Radiomics · Pathomics · Melanoma

1 Introduction

Skin cancers represent the most common group of cancers worldwide [1]. Even if several progresses have been made in diagnosis and therapies, metastatic melanoma keeps recording poor diagnosis [2].

Melanoma diagnosis remains challenging due to the histopathological and clinical heterogeneity of melanoma samples among patients, as well as the lack of biomarker

E. Rodolà et al. (Eds.): ICIAP 2025 Workshops, LNCS 16169, pp. 166–174, 2026.
https://doi.org/10.1007/978-3-032-11317-7_14

reliability [3]. In melanoma, early detection and monitoring are considered crucial for individuals with atypical nevi [4]. Similar challenges have also been identified in other medical fields, as well as the importance of assessing the early progression of a disease [5, 6].

Current diagnostic guidelines recommend initial non-invasive dermoscopy clinical examination and, together with the examination of patient's clinical history and conditions, microscopic examination of melanoma diagnostic features and histopathological characteristics is performed through Whole Slide Images (WSI) analysis of the biopsy [3, 7].

To date, several innovative technologies have been proposed to help melanoma diagnosis. Feature extraction approaches such as radiomics and pathomics must be mentioned. Radiomics has already been used for the extraction of high-dimensional quantitative features in melanoma Positron Emission Tomography/Computerized Tomography (PET/CT) images [8]. In wider terms, PET and PET/CT radiomics offers unique insights into tumor biology and treatment response [9, 10]. Furthermore, the combination of Deep Learning (DL) techniques and pathomics analysis based on cell nuclei features from WSIs has been proposed [11].

Nowadays, the application of radiomics techniques to WSIs, with potential in image segmentations of texture and nuclei in various cancers is emerging [12, 13].

In this context, the current study describes a preliminary implementation of a radiomic workflow, for feature identification and nuclei classification, applied to a reduced subset of PUMA Dataset Melanoma WSIs. The proposed workflow aims to classify and distinguish tumor nuclei from other nuclei classes in both primary and metastatic WSIs.

2 Materials and Methods

2.1 Dataset

Panoptic segmentation of nUclei and tissue in advanced MelanomA (PUMA) dataset was selected for the current study [14]. As described in [15], PUMA dataset contains 155 primary and 155 metastatic melanoma H and E-stained Regions of Interest (ROI) in.tif format with nuclei and tissue annotations, in.GeoJSON format, from a single melanoma referral institution. Schuveling et al. [15] describes PUMA dataset as the first melanoma specific dataset, laying the groundwork for developing melanoma-specific nuclei and tissue segmentation models to be used for assessing diagnostic and predictive biomarker development.

Only a portion of this dataset has been released for public use: the available data is composed of 97429 nuclei in 103 primary and 103 metastatic melanoma H&E-stained ROIs.

From the available subset, a reduced sample composed of 50 primary ROIs and 50 metastatic ROIs was selected to conduct a preliminary classification analysis. This reduction was made to execute a first assessment of the proposed workflow. Figure 1 contains a visual example of the employed histological primary and metastatic ROIs, together with the overlayed annotations visualized through opensource pathology viewer QuPath [12].

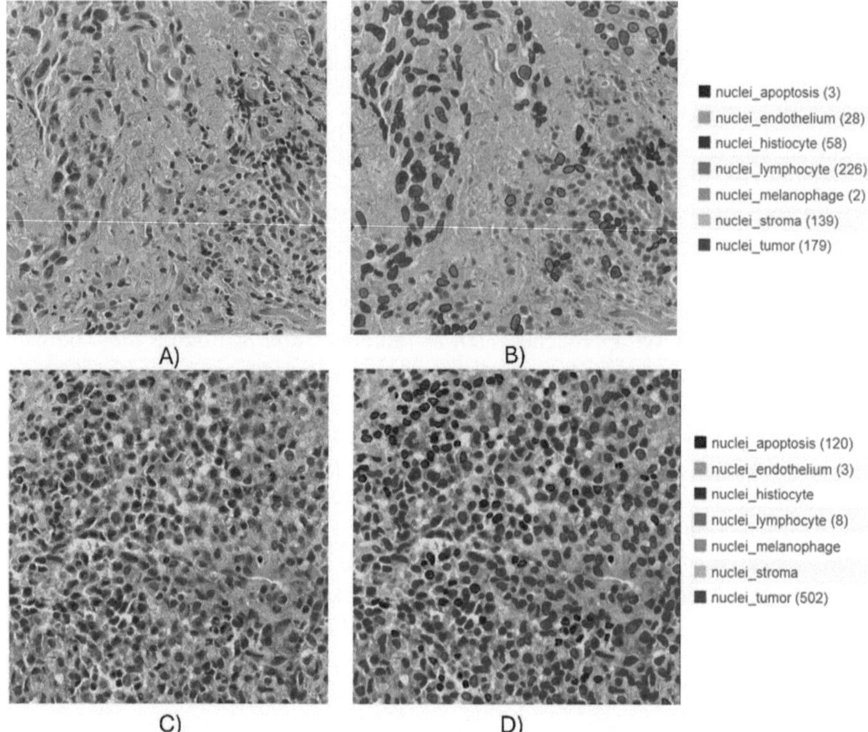

Fig. 1. Example of analyzed tissue ROIs [11]. (A) Example of primary non-metastatic tissue ROI. (B) Same primary ROI with annotated nuclei visualized in QuPath [12]. (C) Example of metastatic tissue ROI. (D) Same metastatic ROI with annotated nuclei visualized in QuPath.

More specifically, as shown in Fig. 1, nuclei annotations of both primary and metastatic ROIs were jointly considered in the developed workflow. Nuclei annotations divide nuclei into ten distinct classes, namely: *"nuclei_tumor"*, *"nuclei_lymphocyte"*, *"nuclei_plasma_cell"*, *"nuclei_hystiocyte"*, *"nuclei_melanophage"*,*"nuclei_neutrophil"*, *"nuclei_stroma"*,*"nuclei_endothelium"*, *"nuclei_epithelium"* and*"nuclei_apoptosis"*.

To develop a classifier able to distinguish tumor nuclei from other types, all the non-tumor annotated nuclei were grouped as a unique class of "non tumor" nuclei.

2.2   Radiomic Feature Extraction and Selection

To ensure the reproducibility and consistency, PyRadiomics was used for radiomic feature extraction. Pyradiomics is a popular open-source Python package for large-scale radiomic analyses [16]. It incorporates the standards proposed by the Imaging Biomarker Standardization Initiative (IBSI)[17], aiming at standardizing feature definitions and enhancing interoperability across different datasets and imaging studies.

Given the large amount of extracted radiomic features, feature selection was performed to exclude non-informative predictors for the subsequent classifier training.

Dimensionality reduction allowed to reduce the risk of overfitting and to simplify the classifier training. Feature selection was performed using LASSO regression with cross-validation through MATLAB's *lassoglm* function. For feature selection, both the LambdaMinMSE criterion, Lambda value with the minimum MSE, and Lambda1SE, Largest Lambda value such that MSE is within one standard error of the minimum MSE, were assessed.

2.3 Classifier Training

After feature selection with LASSO method and Lambda1SE criterion, a supervised classification model based on Random Forest algorithm was trained. Once the majority class (tumor nuclei) was under-sampled to avoid class imbalance, the dataset containing extracted and selected features was split into training and test set according to a 70% training and 30% testing partition. The selected features were normalized in terms of the standard score (subtraction of the mean and division by the standard deviation, both determined on the training set only) to ensure consistent scaling within the training and test set.

Random Forest classifier was chosen after empirical comparisons with alternative classifiers such as AdaBoost and Gradient Boosting, whose performance was slightly worse. The classifier was implemented in MATLAB through *the TreeBagger* function with 100 decision trees and out-of-bag estimation for internal model validation. The performance of the model was evaluated on the test set, setting a binary classification with labels 0 (non-tumor nuclei) and 1 (tumor nuclei). From confusion matrix different performance indices were then extracted, such as accuracy, precision, recall (or sensitivity), specificity, F1-score and area under the Receiver Operating Characteristic (ROC) curve (AUC). Finally, ROC curve was plotted to visually judge the discriminatory ability of the model.

3 Results

Using the Lambda1SE criterion, LASSO regression feature selection resulted in a total radiomic feature count of 57 from an original set of 102. This criterion was meant to help us achieve the right balance between a simple model, with fewer features, but good predictive performance. The cross-validation curve shown in Fig. 2 illustrates the MSE trend as a function of Lambda values, with vertical line indicating the locations of minimum MSE (LambdaMinMSE) and location of 1-SE cutoff (Lambda1SE) used as the final feature subset.

As stated in the Material and Methods section, selected features were used to train a Random Forest classifier after splitting dataset in training and test set.

Confusion Matrix is reported in Fig. 3, enhancing a good balance among classes, with the prevalence of correct predictions for both tumor nuclei class (Positive class) and non-tumor nuclei (Negative class). The distribution of False Negatives and False Positives is comparable.

The Random Forest performance metrics are reported in Table 1. More specifically, the model has shown good performance on the test set with a 74.56% accuracy. Precision, Sensibility, Specificity and F1-Score were also computed. These indicators also

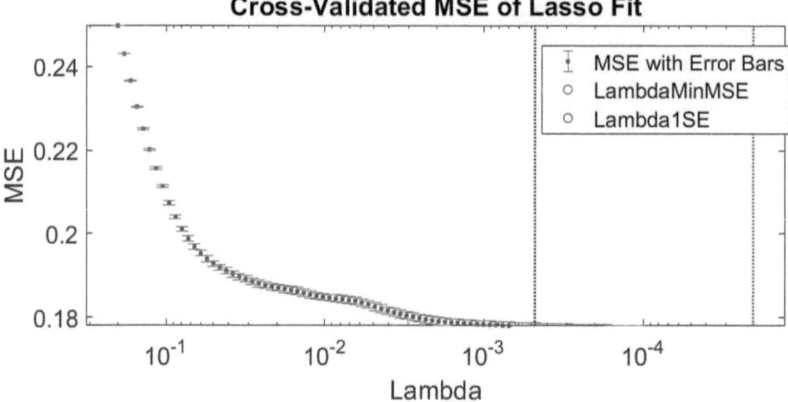

Fig. 2. Cross-validation results for the LASSO model. The mean squared error (MSE) is plotted against the regularization parameter Lambda. The dashed green line marks the Lambda value that minimizes the MSE (LambdaMinMSE) and the dashed blue line marks a more conservative choice based on the one-standard-error rule (Lambda1SE), which favors a more compact and stable set of features.

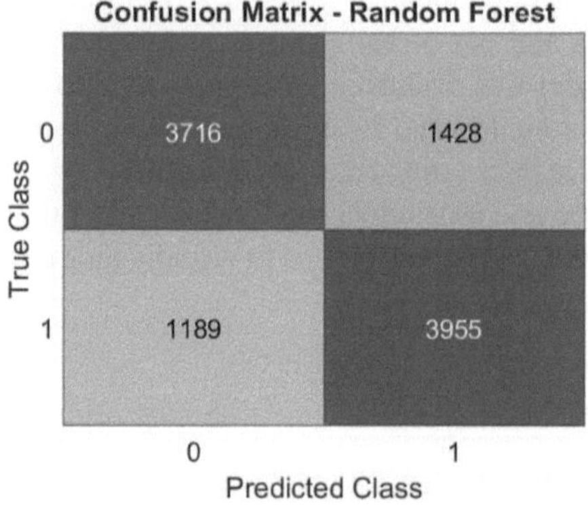

Fig. 3. Confusion matrix of the Random Forest classifier on the test set. This matrix shows how many nuclei were correctly or incorrectly classified into the two categories (tumor and non-tumor). The balance between true positives and false positives gives insight into the classifier's performance.

confirmed classifier reliability with a 73% value for precision, 77% in sensibility, 72% in specificity and a 75% F1-score.

In addition, the AUC was assessed to be 0.82, indicating an excellent discriminating ability of the model between tumor and non-tumor nuclei. The ROC curve illustrated in Fig. 4 shows a clear separation from the random line (AUC = 0.5), indicating that the probabilities predicted by the classifier are well calibrated to truth.

Table 1. Performance metrics of the Random Forest classifier. Metrics computed on the test set include Accuracy, Precision, Recall (Sensitivity), Specificity, F1-score, and the Area Under the ROC Curve (AUC). The values confirm the good discriminative power of the model.

Metric	Value
Accuracy	74.56%
Precision	73%
Sensibility (Recall)	77%
Specificity	72%
F1-score	75%

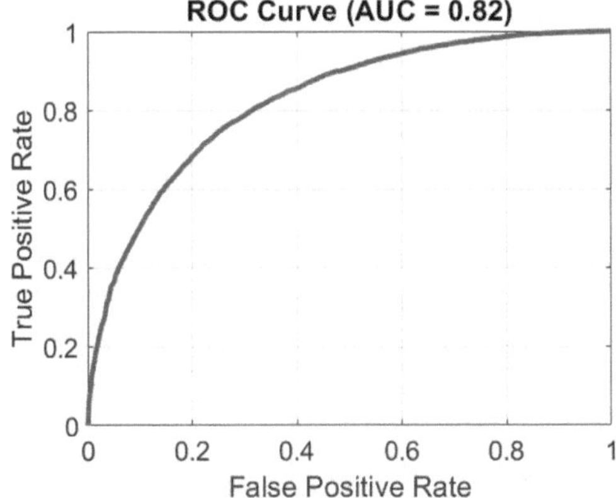

Fig. 4. ROC curve of the classifier. The curve demonstrates the trade-off between true positive rate and false positive rate across different classification thresholds. The AUC value of 0.82 indicates strong discriminative performance of the model in distinguishing tumor nuclei from non-tumor nuclei.

4 Discussions

The experimental results validate the usefulness of the proposed radiomic workflow for classifying tumor and non-tumor nuclei from histopathological images of melanoma. Even if the findings of this study refer to a preliminary analysis, they support the potential of radiomics as a tool for nuclei classification in WSIs.

LASSO regression for feature selection was useful to reduce the dimensionality of the problem while keeping the key features, with the model retaining only 57 relevant features from a total of 102. Although this reduction may cause loss of information, the new features are advantageous for more interpretability and generalizability of the model in clinical applications.

The Random Forest classifier achieved consistent performance across all main evaluation metrics, with a total accuracy was 74.56%. The model also reached an F1-score of 75%, indicating the model's ability to correctly classify both classes into the two categories of nuclei. In precision (73%) and recall (77%), a good balance was achieved between limiting false positives and maximizing the number of detected tumor nuclei.

The confusion matrix indicates a reasonable balance of false positives and false negatives, demonstrating a lack of excessive bias towards one of the two classes, which is important for clinical application, where both parameters are important.

The AUC was 0.82, indicating very good discriminative ability. This measure shows that the classifier was effective at classifying instances across these two classes, with good, predicted probabilities. The visual inspection of the ROC curve also confirms that there is a clear separation from the random baseline (AUC $= 0.5$), enhancing the model's robustness.

However, the study has few limitations. First, it relies on a subset of the PUMA dataset, which is detailed but represents only a portion of the available dataset. Moreover, even if PUMA dataset is very well annotated and curated, all the data comes from a single institution. This could lead to bias and the impact of the generalizability of a model on data from other clinical centers or from different scanning protocols. The study also uses pre-annotated nuclei; in real word applications this step requires accurate nuclei segmentation, and the accuracy of the segmentation step will influence the classification performance. Thus, the integration of a deep learning workflow for nuclei segmentation could further extend the proposed approach's usability.

In the future, the approach will be validated on the entire PUMA dataset and on external datasets from other institutions to test and validate its generalizability, which could be improved by testing the model on external datasets. Furthermore, data will be investigated through other hybrid models that combine radiomics and deep learning techniques to enhance classification reliability and clinical applicability. Eventually, once validated, the proposed approach could be adopted as a decision-supporting tool to assist clinicians in the identification of tumor nuclei and to aid the existing diagnostic protocols. As stated by [3] about the several difficulties in diagnosing Melanoma, the proposed framework could reduce the subjectivity of visual interpretation of WSIs and improve the diagnostic workflow.

5 Conclusions

The proposed radiomic workflow has offered encouraging preliminary results for the classification of tumors versus non-tumor nuclei from histopathological melanoma image sets. From a subset sample of the publicly available PUMA dataset, radiomic feature extraction was combined with feature selection through LASSO and classification by Random Forest, achieving good performance metrics with an AUC of 0.82. The results thus point out the value of radiomics, also when applied to WSIs, to assist in interpreting complex tissue structures.

The pipeline showed a certain degree of robustness and general practicality, despite relying on pre-annotated nuclei and data from a single institution. Yet, important limitations are posed by the relatively small size of the dataset, the omission of a segmentation step, and the single center nature of the data set. Future developments would focus on validating the workflow on the entire PUMA dataset and on external cohorts, incorporating automatic nuclei segmentation and assessing hybrid methods that combine radiomics with deep learning to further improve classification performance and clinical translation.

Acknowledgments. This research has received no external funding.

Disclosure of Interests. The authors have no competing interests to declare that are relevant to the content of this article.

References

1. International Agency for Research on Cancer (IARC): Skin Cancer. https://www.iarc.who.int/cancer-type/skin-cancer. Accessed 27 May 2025
2. Sundararajan, S., Thida, A.M., Yadlapati, S., Mukkamalla, S.K.R., Koya, S.: Metastatic Melanoma. In: StatPearls. StatPearls Publishing, Treasure Island (FL) (2025)
3. de Paula Alves Coelho, K.M., de Macedo, M.P., Lellis, R.F., de Pinheiro-Junior, N.F., Rocha, R.F., Xavier-Junior, J.C.C.: Dermatopathology Committee of the Brazilian Society of Pathology, S.P., Brazil: Guidelines for diagnosis and pathological report of melanocytic skin lesions — recommendations from the Brazilian Society of Pathology. Surgical Exp. Pathol. **8**, 3 (2025). https://doi.org/10.1186/s42047-025-00178-4
4. Karp, P., Karp, K., Kądziela, M., Zajdel, R., Żebrowska, A.: The importance of early detection and prevention of atypical skin lesions and other melanoma risk factors in a younger population. Cancers **16**, 4264 (2024). https://doi.org/10.3390/cancers16244264
5. Case, A.H., et al: Defining the pathway to timely diagnosis and treatment of interstitial lung disease: a US Delphi survey. BMJ Open Resp. Res. **10** (2023). https://doi.org/10.1136/bmjresp-2022-001594
6. Banna, G.L., et al.: Predictive and prognostic value of early disease progression by PET evaluation in advanced non-small cell lung cancer. Oncology **92**, 39–47 (2016). https://doi.org/10.1159/000448005
7. Pavlidis, E.T., Pavlidis, T.E.: Diagnostic biopsy of cutaneous melanoma, sentinel lymph node biopsy and indications for lymphadenectomy. World J Clin Oncol. **13**, 861–865 (2022). https://doi.org/10.5306/wjco.v13.i10.861
8. Amrane, K., et al.: Review on radiomic analysis in 18F-fluorodeoxyglucose positron emission tomography for prediction of melanoma outcomes. Cancer Imaging **24**, 87 (2024). https://doi.org/10.1186/s40644-024-00732-5
9. Stefano, A.: Challenges and limitations in applying radiomics to PET imaging: Possible opportunities and avenues for research. Comput. Biol. Med. **179**, 108827 (2024). https://doi.org/10.1016/j.compbiomed.2024.108827
10. Bauckneht, M., et al.: [18F]PSMA-1007 PET/CT-based radiomics may help enhance the interpretation of bone focal uptakes in hormone-sensitive prostate cancer patients. Eur. J. Nucl. Med. Mol. Imaging **52**, 2076–2086 (2025). https://doi.org/10.1007/s00259-025-07085-6
11. Kim, R.H., et al.: Deep learning and pathomics analyses reveal cell nuclei as important features for mutation prediction of BRAF-mutated melanomas. J. Invest. Dermatol. **142**, 1650-1658.e6 (2022). https://doi.org/10.1016/j.jid.2021.09.034
12. Sheikh, T.S., Cho, M.: Segmentation of variants of nuclei on whole slide images by using radiomic features. Bioengineering **11**, 252 (2024). https://doi.org/10.3390/bioengineering11030252
13. Radulović, M., et al.: Bridging histopathology and radiomics toward prognosis of metastasis in early breast cancer. Microsc. Microanal. **30**, 751–758 (2024). https://doi.org/10.1093/mam/ozae057
14. Schuiveling, M.: Melanoma Histopathology Dataset with Tissue and Nuclei Annotations (2025)

15. Schuiveling, M., et al.: A novel dataset for nuclei and tissue segmentation in melanoma with baseline nuclei
16. Van Griethuysen, J.J.M., et al.: Computational radiomics system to decode the radiographic phenotype. Can. Res. **77**, e104–e107 (2017). https://doi.org/10.1158/0008-5472.CAN-17-2104
17. Zwanenburg, A., et al.: The image biomarker standardization initiative: standardized quantitative radiomics for high-throughput image-based phenotyping. Radiology **295**, 328–338 (2020). https://doi.org/10.1148/radiol.2020200712

Mask-Aware Transformers Enable Robust Learning from Incomplete Volumetric Medical Imaging

Camillo Maria Caruso[iD], Riccardo Bruni, and Valerio Guarrasi[✉][iD]

Unit of Artificial Intelligence and Computer Systems, Department of Engineering,
Università Campus Bio-Medico di Roma, Rome, Italy
`valerio.guarrasi@unicampus.it`

Abstract. Incomplete or sparsely sampled volumetric scans are pervasive across medical imaging modalities: patient motion, shortened acquisition protocols, and hardware constraints frequently result in missing slices that undermine downstream analysis. Conventional deep learning pipelines either discard these studies or rely on voxel-wise interpolation, potentially introducing artefactual signals. We present a Mask-Aware Vision Transformer (MAViT), a modality-agnostic architecture that learns directly from incomplete 3D volumes without synthetic reconstruction. MAViT leverages a binary slice-availability mask to identify corrupted patches and selectively suppress their contribution within each self-attention block, effectively guiding feature aggregation while mitigating the impact of missing data. To benchmark robustness, we synthetically corrupt brain MRI volumes from the Alzheimer's Disease Neuroimaging Initiative with different slice-drop rates. Despite being trained on heterogeneous missing-slice patterns, MAViT achieves state-of-the-art performance on Alzheimer's disease classification, surpassing 3D interpolation-based and 2D slice-wise baselines. These findings indicate that mask-aware modelling offers a valuable approach to learning from incomplete volumetric data, readily extending beyond brain MRI to other imaging modalities.

Keywords: Vision Transformers · Slice corruption · Masked attention · Alzheimer's · Medical Imaging · Incomplete data modeling

1 Introduction

Three-dimensional imaging modalities, e.g., Magnetic Resonance Imaging (MRI), Computed Tomography (CT), Positron Emission Tomography (PET), have become indispensable to contemporary clinical workflows [18,28]. Each produces stacks of 2D slices that collectively form a 3D representation of anatomy or physiology. In practice, however, these acquisitions are often structurally incomplete: patient motion, breath-hold failures, shortened protocols aimed at

E. Rodolà et al. (Eds.): ICIAP 2025 Workshops, LNCS 16169, pp. 175–186, 2026.
https://doi.org/10.1007/978-3-032-11317-7_15

throughput or dose reduction, hardware malfunctions, and outright study termination can all result in *missing slices* or heavily corrupted sections within the volume [5]. The prevalence of such gaps is particularly acute in cohorts with limited compliance, e.g., pediatric, geriatric, critically ill, or cognitively impaired patients, where scan interruptions are common and re-acquisition is impractical [19,21].

Deep learning (DL) pipelines for medical image analysis have yielded impressive gains in tasks ranging from disease classification to organ segmentation and radiomics-style risk scoring [26]. Yet virtually all state-of-the-art architectures assume a fully sampled input grid [5,17]. When confronted with incomplete scans, these models either fail or rely on a preprocessing stage that attempts to *reconstruct* the missing data, most commonly through voxel-wise interpolation (nearest-neighbour, linear, B-spline) or learning-based inpainting. These synthetic fill-in strategies can hallucinate structures, blur fine details, and propagate domain shifts, ultimately compromising diagnostic fidelity and model robustness [15,25].

We address this problem at the model rather than the preprocessing level and introduce Mask-Aware Vision Transformer (MAViT), a modality-agnostic architecture that learns directly from incomplete volumes. MAViT ingests a binary slice-availability mask alongside the raw voxels and injects mask-conditioned tokens into every self-attention block [2–4], enabling the network to focus on anatomically valid regions while ignoring corrupted slices. This design eliminates the need for synthetic reconstruction and is inherently applicable to any stacked-slice modality. We demonstrate MAViT on T1-weighted brain MRI from the Alzheimer's Disease Neuroimaging Initiative (ADNI) [13], a benchmark rich in neurodegenerative variation, but the core methodology readily transfers to other volumetric imaging modalities, e.g., CT and PET.

The remainder of this manuscript is organized as follows: Sect. 2 reviews existing strategies for handling incomplete volumetric data; Sect. 3 details the MAViT model and the experiments conducted; Sect. 4 presents the data on which MAViT is tested on; Sect. 5 presents the obtained results; Sect. 6 concludes with clinical implications, limitations, and avenues for future work.

2   Background

Robust analysis of volumetric medical images in the presence of missing slices has historically been pursued through pre-reconstruction: the raw scan is first "repaired" to a dense grid and only then forwarded to downstream tasks such as classification or segmentation. This section reviews the dominant classes of such methods, highlights their limitations, and motivates our mask-aware alternative.

Classical pipelines depend on voxel-wise interpolation, i.e., *nearest-neighbour*, *linear*, and *B-spline*, because these filters are differentiable, readily available, and preserve the tensor shape expected by conventional CNNs [14,20]. Figure 1 illustrates the effects of voxel-wise interpolation methods on a brain MRI volume with missing slices. In both panels, slices from index 116 to 134 (7% of the total

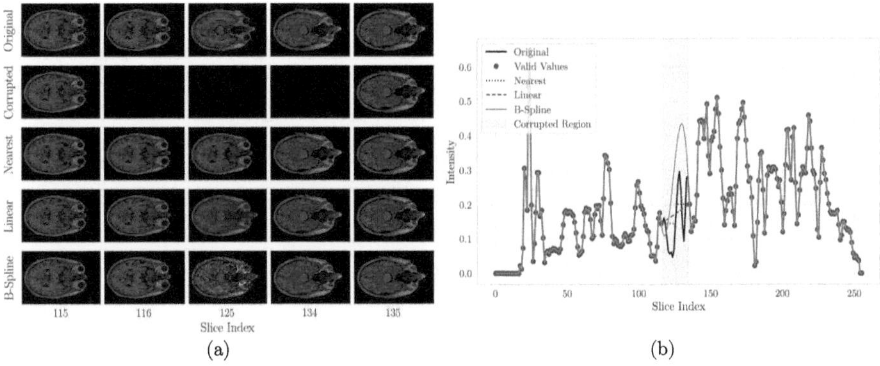

Fig. 1. Panel (a): Comparison of interpolation methods applied to structurally incomplete MRI volumes. From top to bottom: original scan, corrupted scan with missing slices, nearest-neighbour interpolation, linear interpolation and B-spline interpolation. Panel (b): Mean intensity profile along the z-axis for a representative voxel column. The shaded area indicates the region of missing slices. Different interpolation trajectories are shown for linear, nearest-neighbor, and B-spline methods.

volume) have been artificially removed to simulate slice corruption. Panel (a) shows axial views of selected slices for the original, corrupted, and reconstructed volumes using three common interpolation methods: nearest-neighbour, linear, and B-spline. While all methods restore the missing region, they introduce visible artefacts that can distort anatomical fidelity. Panel (b) quantifies these effects by plotting voxel intensity values along a fixed spatial location across the full slice stack. Compared to the original signal, the interpolated curves exhibit distinct failure modes: (i) stepwise discontinuities with nearest-neighbour interpolation, (ii) excessive smoothing with linear interpolation that may erase subtle features, and (iii) oscillatory artefacts with B-spline interpolation that can hallucinate non-existent structures. These distortions become more pronounced with increasing missingness and may bias radiomic features or degrade the performance of downstream deep learning models.

Vision Transformers (ViTs) [7] and their 3D variants (UNETR [12], Swin-UNETR [24], TransBTS [27]) have achieved SOTA performance in multiple volumetric tasks owing to their capacity for global attention. Nevertheless, they inherit the assumption of fully sampled inputs; corrupted scans are typically padded via interpolation before training and inference. To our knowledge, no prior ViT3D incorporates the slice-availability mask *during inference* to modulate attention weights.

The shortcomings of both handcrafted and learned reconstruction motivate a **model-level** alternative. By injecting a binary mask into every self-attention layer, our MAViT can discard corrupted slices without synthesising new voxels and preserve global anatomical context through long-range attention.

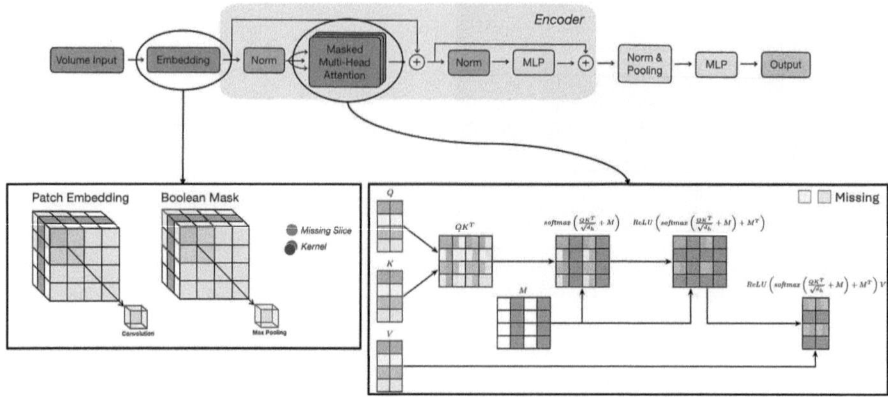

Fig. 2. Schematic representation of the MAViT architecture. The top panel shows the overall structure, which follows a ViT-like design with a patch embedding module, a transformer encoder featuring masked multi-head self-attention, and an MLP classification head. The bottom left panel illustrates the embedding process: input volumes are divided into non-overlapping 3D patches, and a parallel masking branch detects missing slices and generates a binary mask aligned with the patch grid. The bottom right panel details the masked attention mechanism, where attention weights are modulated by the binary mask to suppress contributions from missing or unreliable regions.

3   Methods

This section describes the design and implementation of our MAViT model, along with baseline comparisons using 2D slice-wise analysis and 3D interpolation-based variants. We also outline the missingness simulation protocols and the experimental setup used to evaluate performance.

3.1   MAViT

The proposed MAViT model, illustrated in Fig. 2, is an evolution of the ViT [7] tailored for handling volumetric medical data with missing slices. Its architecture follows the standard ViT backbone, consisting of an embedding block that partitions the input volume into non-overlapping patches and encodes them, a transformer encoder that incorporates a masked multi-head self-attention mechanism, and a multilayer perceptron (MLP) head for classification.

To enable robust inference in the presence of corrupted or missing regions, MAViT modifies the self-attention computation to account for patch-level missingness. The attention weights are modulated using a binary slice-availability mask M, where each entry indicates whether a given patch is valid. The modified attention operation is defined as:

$$\text{Attn}(Q, K, V) = \text{ReLU}\left(\text{softmax}\left(\frac{QK^T}{\sqrt{d_h}} + M\right) + M^T\right) \cdot V$$

where Q, K, V are the query, key, and value matrices respectively, and d_h is the dimensionality of each attention head.

This formulation effectively suppresses contributions from missing or unreliable regions, allowing the model to focus on informative inputs only. To support this mechanism, the embedding block is paralleled by a masking pathway that transforms an initial slice-level missingness map into a patch-level binary mask. As depicted in the left part of Fig. 2, the information about missing slices is downsampled via a max-pooling operation that mirrors the patch extraction kernel, ensuring alignment with the embedded patch sequence. The resulting mask is then incorporated into the attention computation, as illustrated on the right side of the figure.

3.2 2D Slice-Wise Baseline (ViT2D)

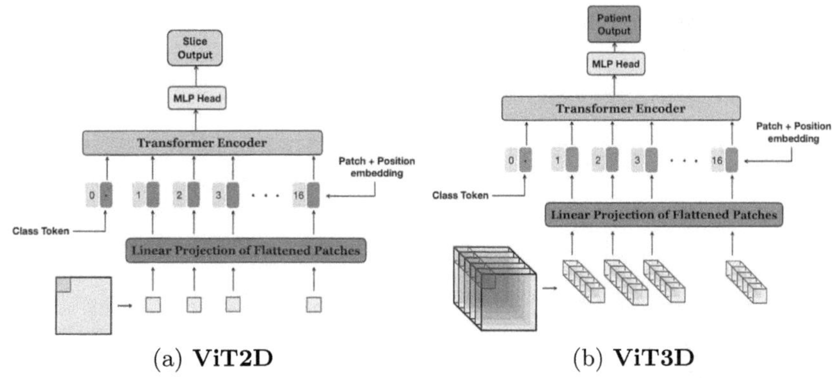

Fig. 3. Comparison between ViT2D and the ViT3D architecture. ViT2D processes individual 2D slices independently and produces slice-level predictions, while ViT3D operates directly on 3D volumes and performs patient-level classification. The proposed MAViT model is a ViT3D model integrating masked self-attention.

For baseline comparison, we implemented a 2D Vision Transformer (ViT2D) [7]. Here, each single slice in the volumetric scan is analyzed independently, resulting in slice-level prediction, as shown on the left side of Fig. 3. To derive a patient-level prediction, these slice-wise outputs are aggregated using majority voting, following standard practice in slice-based approaches [18]. In this framework, missing slices can be entirely excluded during both training and inference, offering an alternative to interpolation-based strategies. In such cases, the final patient-level output is determined solely by the majority vote among the available slices. Nonetheless, the 2D approach suffers from the limitation of not incorporating any global volumetric context, potentially overlooking inter-slice dependencies and cross-sectional patterns that could be informative for the final prediction.

3.3   Interpolation-Based (ViT3D)

To highlight the limitations of synthetic reconstruction and benchmark MAViT's advantage, we also tested the ViT3D [12] relying on the interpolation of missing slices prior to analysis, with classical techniques: Nearest-Neighbour, Linear, and B-spline interpolation [15,18]. These pipelines used identical architectures to MAViT, but without masked attention. Missing slices were interpolated during preprocessing, enabling the model to operate on a fully dense input but at the cost of potentially introducing artefactual signals.

3.4   Missingness Simulation

To emulate scenarios of incomplete scans, synthetic slice corruption was introduced by removing contiguous axial slices along the z-axis, both in the training and the test sets. Four levels of missingness were simulated (1%, 5%, 7%, and 12%), reflecting challenges common in clinical practice such as motion artefacts, scan interruptions, or acquisition constraints [18,23].

3.5   Experimental Setup

All experiments were conducted using stratified five-fold cross-validation at the subject level, ensuring balanced class distributions across folds and preventing data leakage between training and testing sets. Within each fold, 20% of the training subjects were further set aside as a validation set to guide early stopping and model selection.

All models were initialized with weights pre-trained on ImageNet, and were trained using the Adam optimizer with an initial learning rate of 1×10^{-3}. Early stopping was applied based on the validation AUC, with a patience threshold to prevent overfitting. Training was carried out using PyTorch on NVIDIA A100 GPUs. Due to the computational demands of volumetric processing, a batch size of 256 was used for the 2D slice-based models, while the 3D models were trained with a batch size of 4.

Slice-based models (ViT2D) were trained using individual 2D axial slices resized to 224×224 pixels, whereas volumetric models (ViT3D, MAViT, and interpolation-based variants) operated directly on whole MRI volumes resized to $128 \times 128 \times 128$ voxels. The same data splits and corruption maps were consistently used across all experiments to ensure comparability.

Performance was evaluated at the subject level, using five metrics: accuracy, precision, recall, F1-score, and area under the receiver operating characteristic curve (AUC). AUC was selected as the primary performance metric due to its robustness to class imbalance and independence from classification thresholds [11]. All reported results are expressed as mean $\pm$ standard deviation, averaged over the five cross-validation folds.

4   Materials

This study utilized data from the ADNI [13], a large-scale longitudinal initiative aimed at developing clinical, imaging, genetic, and biochemical biomarkers for early detection and monitoring of Alzheimer's disease. T1-weighted structural MRI scans were collected across four successive ADNI phases (ADNI1, ADNI-GO, ADNI2, and ADNI3), involving scanners from multiple vendors, including GE Healthcare, Philips, and Siemens. To maintain consistency and prevent bias from repeated measurements, only one scan per subject was selected, either from the baseline or screening visit. The final dataset comprised 687 unique subjects, divided into two diagnostic categories: 364 cognitively normal (CN) controls and 323 Alzheimer's disease (AD) patients, as classified by ADNI protocols [13].

Preprocessing involved several steps to standardize input data and reduce inter-scan variability. Volumes were initially reoriented to a standardized anatomical reference frame, followed by intensity-based thresholding and bounding-box cropping to remove non-informative regions. Voxel intensities within each volume were normalized using min-max normalization, scaling values to the $[0, 1]$ range to mitigate scanner-specific signal variations and enhance training stability.

In the 3D pipelines, to achieve spatial uniformity, the volumes were resampled to a resolution of $128 \times 128 \times 128$ voxels. Resizing was conducted through interpolation exclusively along the in-plane dimensions, i.e., x and y, preserving anatomical integrity, while zero-padding was applied along the axial dimension, i.e., z-axis, to maintain continuity. In the comparative 2D pipeline, slices were resized to 224×224 pixels to align with input specifications of the pre-trained ViT2D model.

Data augmentation was applied to both 3D and 2D pipelines during training, including random rotations ($\pm 15°$), flipping, and up to $90°$ rotations to simulate different head positions while preserving anatomical plausibility.

5   Results

Table 1 reports the classification performance of all evaluated models across increasing levels of structural incompleteness. Results are averaged over five cross-validation folds and presented as standard classification metrics: AUC, accuracy, recall, precision, and F1-score.

In structurally complete volumes (0% missing slices), the ViT3D model outperformed the slice-based ViT2D baseline in every metric, achieving an AUC of 93.58% and an accuracy of 88.72%. This demonstrates the advantage of maintaining full volumetric context over slice-wise analysis. Importantly, under this condition of no missing data, the ViT3D architecture becomes functionally equivalent to both MAViT and interpolation-based methods, as no masking or interpolation is required. Thus, all volumetric models operate on the same complete input and converge to comparable performance, isolating the effect of missingness as the critical factor in subsequent comparisons.

Table 1. Classification performance at various missing rates. Metrics are reported as mean ± standard deviation, averaged across five cross-validation folds and expressed as percentages. Best values are highlighted in bold black, second-best in bold blue.

Missing Rate	Approach	AUC	Accuracy	Recall	Precision	F1-Score
0%	ViT2D	91.21 ± 1.55	85.32 ± 2.10	83.72 ± 2.44	84.92 ± 2.33	84.36 ± 2.11
	ViT3D	**93.58 ± 1.20**	**88.72 ± 1.83**	**87.41 ± 2.01**	**88.13 ± 1.72**	**87.75 ± 1.87**
1%	ViT2D	88.77 ± 1.15	80.59 ± 1.65	79.34 ± 1.71	80.20 ± 1.84	79.68 ± 1.78
	ViT3D + Nearest-Neighbour	**92.31 ± 3.19**	82.10 ± 2.63	81.82 ± 2.71	82.63 ± 2.55	81.85 ± 2.66
	ViT3D + Linear	**92.39 ± 3.01**	**83.61 ± 3.25**	**83.48 ± 3.12**	**83.60 ± 3.28**	**83.52 ± 3.19**
	ViT3D + B-Spline	90.45 ± 2.57	**82.62 ± 3.42**	**82.66 ± 3.30**	82.64 ± 3.41	**82.57 ± 3.34**
	MAViT	91.62 ± 2.08	82.23 ± 3.35	81.50 ± 3.40	**82.70 ± 3.30**	82.00 ± 3.35
5%	ViT2D	88.24 ± 1.25	79.93 ± 1.50	78.50 ± 1.60	79.10 ± 1.70	78.80 ± 1.65
	ViT3D + Nearest-Neighbour	90.97 ± 2.55	**83.15 ± 3.37**	**82.83 ± 3.45**	**83.64 ± 3.30**	**82.93 ± 3.40**
	ViT3D + Linear	**92.11 ± 2.84**	81.76 ± 3.10	81.85 ± 3.15	82.29 ± 3.18	81.65 ± 3.13
	ViT3D + B-Spline	**92.05 ± 0.99**	**82.54 ± 2.27**	**82.55 ± 2.20**	**82.63 ± 2.30**	**82.46 ± 2.25**
	MAViT	90.22 ± 3.09	81.15 ± 3.92	81.18 ± 3.85	81.48 ± 3.95	81.02 ± 3.90
7%	ViT2D	88.45 ± 3.24	77.52 ± 2.11	76.97 ± 2.24	77.20 ± 2.32	77.00 ± 2.26
	ViT3D + Nearest-Neighbour	90.29 ± 1.89	81.61 ± 4.82	**82.83 ± 3.45**	**83.64 ± 3.30**	**82.93 ± 3.40**
	ViT3D + Linear	**90.83 ± 2.65**	**82.23 ± 3.49**	82.30 ± 3.42	**82.93 ± 3.40**	**82.05 ± 3.44**
	ViT3D + B-Spline	90.73 ± 1.21	81.08 ± 2.24	80.92 ± 2.18	81.83 ± 2.27	80.85 ± 2.22
	MAViT	**90.94 ± 2.82**	**83.31 ± 2.63**	81.43 ± 2.71	81.41 ± 2.65	81.26 ± 2.68
12%	ViT2D	85.82 ± 0.87	76.42 ± 2.30	76.13 ± 2.45	76.20 ± 2.49	76.00 ± 2.32
	ViT3D + Nearest-Neighbour	88.60 ± 2.76	80.51 ± 4.63	80.34 ± 4.72	81.22 ± 4.55	80.27 ± 4.68
	ViT3D + Linear	**90.08 ± 1.00**	80.98 ± 2.70	**82.16 ± 2.78**	82.51 ± 2.65	**82.22 ± 2.72**
	ViT3D + B-Spline	89.23 ± 2.80	**82.23 ± 4.08**	81.67 ± 4.15	**83.44 ± 4.00**	81.75 ± 4.05
	MAViT	**90.87 ± 0.91**	**82.69 ± 2.75**	**82.43 ± 2.80**	**82.81 ± 2.70**	**82.53 ± 2.72**

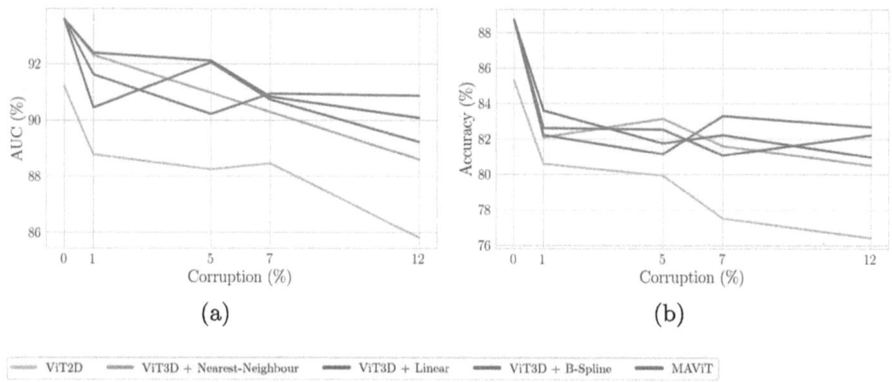

Fig. 4. (a) and (b) report the trends of AUC and Accuracy, respectively, across increasing levels of structural incompleteness.

As missing slices were introduced, performance decreased across all models, but the degree of degradation varied with the handling strategy. At low levels of missingness (1%), the ViT3D variants using interpolation (linear or nearest-neighbour) preserved high AUC scores (above 92%), slightly outperforming the MAViT model, which achieved an AUC of 91.62%. Notably, the B-spline interpolation variant yielded marginally lower performance, consistent with its known tendency to introduce oscillatory artefacts in structurally irregular regions.

At moderate missingness (5% and 7%), the performance of interpolation-based methods and the MAViT model remained similar, reflecting their comparable ability to mitigate mild corruption. However, differences became more pronounced under severe corruption (12% missing slices). In this high-missingness regime, the MAViT model exhibited superior robustness, maintaining an AUC of 90.87% compared to 90.08% for linear interpolation, 89.23% for B-spline, and 88.60% for nearest-neighbour interpolation. The ViT2D baseline, which discards corrupted slices, consistently underperformed relative to the volumetric approaches across all levels of incompleteness, highlighting the importance of preserving global volumetric context to ensure stable and reliable predictions under challenging conditions.

Figure 4 visualizes the trends in AUC and accuracy across different missingness levels. All models exhibited declining performance as anatomical integrity degraded, but the MAViT showed a notably more stable trajectory, particularly in high-missingness regimes. This stability underscores the model's ability to preserve discriminative capacity even when a substantial portion of the volume is missing.

These results highlight the advantage of integrating structural validity directly into the model's architecture. While interpolation strategies are effective for mild to moderate corruption, they become unreliable as anatomical gaps widen. In contrast, the Mask-Aware attention mechanism allows the model to selectively focus on valid, uncorrupted regions, avoiding artefacts introduced by synthetic interpolation and enhancing robustness under severe incompleteness.

6 Conclusions

This work addresses the critical challenge of structural incompleteness in volumetric medical imaging, a common and clinically relevant artifact that compromises the reliability of automated diagnostic systems. We proposed the MAViT model capable of directly learning from corrupted anatomical volumes without relying on synthetic reconstruction. By integrating validity maps into the self-attention mechanism, the model selectively attends to intact regions, preserving anatomical integrity and avoiding distortive reconstruction biases.

Our results, validated on brain MRI data from the ADNI dataset, demonstrate that while traditional interpolation-based methods offer acceptable performance at low missingness levels, they deteriorate substantially as the extent of missing data increases. In contrast, MAViT consistently maintains competitive or superior classification accuracy, even in high-missingness regimes (up to 12% of contiguous slices removed), underscoring its robustness and clinical applicability.

The comparative evaluation with 2D slice-based transformers and multiple interpolation-based strategies further highlights the advantages of model-level adaptation over conventional preprocessing. By embedding structural validity directly into the model's architecture, our approach establishes a principled framework for robust learning from incomplete volumetric data, a pervasive yet often overlooked challenge in real-world clinical workflows.

Limitations include evaluation on a single disease cohort, therefore, future work will focus on evaluating the generalizability of the Mask-Aware attention mechanism across different imaging modalities, e.g., CT, PET, and anatomical regions, and exploring its integration with generative models for uncertainty-aware reconstruction, also considering multimodal scenarios [1,6,8–10,16,22]. Additionally, we aim to extend this framework to tasks such as segmentation and multi-class classification, ultimately contributing to the development of clinically faithful AI systems that embrace, rather than ignore, the imperfections of real-world medical data. Another current limitation is the reliance on a predefined binary slice-availability mask, which in future iterations could be estimated automatically from the input data using learned or heuristic-based detection modules.

Acknowledgments. Camillo Maria Caruso is a Ph.D. student enrolled in the National Ph.D. in Artificial Intelligence, XXXVII cycle, course on Health and life sciences, organized by Università Campus Bio-Medico di Roma.

This work was partially founded by: i) Università Campus Bio-Medico di Roma under the program "University Strategic Projects" within the project "AI-powered Digital Twin for next-generation lung cancEr cAre (IDEA)"; ii) PRIN 2022 MUR 20228MZFAA-AIDA (CUP C53D23003620008); iii) PRIN PNRR 2022 MUR P2022P3CXJ-PICTURE (CUP C53D23009280001); iv) PNRR MUR project PE0000013-FAIR; v) PNRR M6/C2 project PNRR-MCNT2-2023-12377755.

Resources are provided by the National Academic Infrastructure for Supercomputing in Sweden (NAISS) and the Swedish National Infrastructure for Computing (SNIC) at Alvis @ C3SE, partially funded by the Swedish Research Council through grant agreements no. 2022-06725 and no. 2018-05973.

Disclosure of Interests. The authors have no competing interests to declare that are relevant to the content of this article.

References

1. Caragliano, A.N., et al.: Doctor-in-the-loop: an explainable, multi-view deep learning framework for predicting pathological response in non-small cell lung cancer. Image Vision Comput. 105630 (2025)
2. Caruso, C.M., et al.: Not another imputation method: a transformer-based model for missing values in tabular datasets. arXiv preprint arXiv:2407.11540 (2024)
3. Caruso, C.M., Guarrasi, V., Ramella, S., Soda, P.: A deep learning approach for overall survival prediction in lung cancer with missing values. Comput. Methods Programs Biomed. **254**, 108308 (2024)
4. Caruso, C.M., Soda, P., Guarrasi, V.: MARIA: a multimodal transformer model for incomplete healthcare data. Comput. Biol. Med. **196**, 110843 (2025)
5. Chao, Z., Kim, H.J.: Slice interpolation of medical images using enhanced fuzzy radial basis function neural networks. Comput. Biol. Med. **110**, 66–78 (2019)

6. Di Teodoro, G., et al.: A graph neural network-based model with out-of-distribution robustness for enhancing antiretroviral therapy outcome prediction for HIV-1. Comput. Med. Imaging Graph. **120**, 102484 (2025)

7. Dosovitskiy, A., et al.: An image is worth 16x16 words: transformers for image recognition at scale. arXiv preprint arXiv:2010.11929 (2020)

8. Francesconi, A., et al.: Class balancing diversity multimodal ensemble for Alzheimer's disease diagnosis and early detection. Comput. Med. Imaging Graph. **123**, 102529 (2025)

9. Guarrasi, V., et al.: A systematic review of intermediate fusion in multimodal deep learning for biomedical applications. Image Vision Comput. 105509 (2025)

10. Guarrasi, V., et al.: Multimodal explainability via latent shift applied to COVID-19 stratification. Pattern Recogn. **156**, 110825 (2024)

11. Hanley, J.A., McNeil, B.J.: The meaning and use of the area under a receiver operating characteristic (ROC) curve. Radiology **143**(1), 29–36 (1982)

12. Hatamizadeh, A., et al.: UNETR: transformers for 3D Medical image segmentation. In: Proceedings of the IEEE/CVF Winter Conference on Applications of Computer Vision, pp. 574–584 (2022)

13. Jack, C.R., Jr., et al.: The Alzheimer's Disease Neuroimaging Initiative (ADNI): MRI methods. J. Magn. Resonan. Imaging: Off. J. Int. Soc. Magn. Resonan. Med. **27**(4), 685–691 (2008)

14. Jones, S.E., et al.: Three-dimensional mapping of cortical thickness using Laplace's equation. Hum. Brain Mapp. **11**(1), 12–32 (2000)

15. Lehmann, T.M., Gonner, C., Spitzer, K.: Survey: interpolation methods in medical image processing. IEEE Trans. Med. Imaging **18**(11), 1049–1075 (1999)

16. Mogensen, K., et al.: An optimized ensemble search approach for classification of higher-level gait disorder using brain magnetic resonance images. Comput. Biol. Med. **184**, 109457 (2025)

17. Moraes, T., Amorim, P., Da Silva, J.V., Pedrini, H.: Medical image interpolation based on 3D Lanczos filtering. Comput. Methods Biomech. Biomed. Eng. Imaging Visual. **8**(3), 294–300 (2020)

18. Moshe, Y.H., et al.: Handling missing MRI data in brain tumors classification tasks: usage of synthetic images vs. duplicate images and empty images. J. Magn. Resonan. Imaging **60**(2), 561–573 (2024)

19. Pardoe, H.R., et al.: Motion and morphometry in clinical and nonclinical populations. Neuroimage **135**, 177–185 (2016)

20. Pieper, S., et al.: 3D slicer. In: 2004 2nd IEEE International Symposium on Biomedical Imaging: Nano to Macro (IEEE Cat No. 04EX821), pp. 632–635. IEEE (2004)

21. Power, J.D., et al.: Spurious but systematic correlations in functional connectivity MRI networks arise from subject motion. Neuroimage **59**(3), 2142–2154 (2012)

22. Ruffini, F., et al.: Multi-dataset multi-task learning for COVID-19 prognosis. In: Linguraru, M.G., et al. (eds.) MICCAI 2024. LNCS, vol. 15012, pp. 251–261. Springer, Cham (2024). https://doi.org/10.1007/978-3-031-72390-2_24

23. Sinha, H., Raamana, P.R.: Solving the pervasive problem of protocol non-compliance in MRI using an open-source tool mrQA. Neuroinformatics **22**(3), 297–315 (2024)

24. Tang, Y., et al.: Self-supervised pre-training of swin transformers for 3D medical image analysis. In: Proceedings of the IEEE/CVF Conference on Computer Vision and Pattern Recognition, pp. 20730–20740 (2022)

25. Thévenaz, P., et al.: Interpolation revisited [medical images application]. IEEE Trans. Med. Imaging **19**(7), 739–758 (2002)

26. Wang, J., Wang, S., Zhang, Y.: Deep learning on medical image analysis. CAAI Trans. Intell. Technol. **10**(1), 1–35 (2025)
27. Wang, W., Chen, C., Ding, M., Yu, H., Zha, S., Li, J.: TransBTS: multimodal brain tumor segmentation using transformer. In: de Bruijne, M., et al. (eds.) MICCAI 2021. LNCS, vol. 12901, pp. 109–119. Springer, Cham (2021). https://doi.org/10. 1007/978-3-030-87193-2_11
28. Zou, Q., Ahmed, A.H., Nagpal, P., Priya, S., Schulte, R.F., Jacob, M.: Variational manifold learning from incomplete data: application to multislice dynamic MRI. IEEE Trans. Med. Imaging **41**(12), 3552–3561 (2022)

A Generalized Gray Level Co-occurrence Matrix for Rotation-Invariant Texture Detection in Radiomics

Haoyue Chen[1]([envelope]) [iD], Rosario Corso[2] [iD], Albert Comelli[3] [iD],
and Anthony Yezzi[1] [iD]

[1] School of Electrical and Computer Engineering, Georgia Institute of Technology,
Atlanta, GA, USA
{hchen765,anthony.yezzi}@gatech.edu
[2] Dipartimento di Matematica e Informatica, Università degli Studi di Palermo,
Palermo, Italy
rosario.corso02@unipa.it
[3] Ri.MED Foundation, Via Bandiera 11, Palermo 90133, Italy
acomelli@fondazionerimed.com

Abstract. This paper presents a generalized Gray Level Co-occurrence Matrix (GLCM) framework that extends traditional formulations to a continuous, differentiable spatial domain. By replacing discrete pixel displacements with continuous vectors of constant L2-norm, the method achieves rotational invariance and improved robustness across varying resolutions. A radius-based neighborhood criterion ensures consistent spatial coverage, while flexible directional sampling—uniform or random—enhances texture representation. The differentiable formulation supports integration with gradient-based optimization and learning frameworks. Experimental results demonstrate that the proposed GLCM generalization offers greater stability and accuracy, with broad applicability in medical imaging, materials analysis, and texture-driven segmentation.

Keywords: Generalized GLCM · Texture Analysis · Rotational Invariance

1 Introduction

Texture analysis plays a pivotal role in medical imaging by enabling quantitative assessment of tissue heterogeneity that often remains invisible to the human eye [3]. By transforming complex image patterns into measurable descriptors, texture analysis enhances diagnostic accuracy, disease characterization, and treatment planning. With the rise of radiomics, medical imaging has evolved into a data-rich domain, where high-dimensional image features are extracted for predictive modeling and decision support.

At the core of many radiomics pipelines [1] lies the Gray Level Co-occurrence Matrix (GLCM), which quantifies spatial relationships between pixel intensities and has proven effective across various imaging modalities [4]. Despite its utility, traditional GLCM is limited by its reliance on fixed, discrete spatial displacements, lowering its robustness to changes in orientation, scale, and resolution. Addressing these limitations requires a more flexible and continuous formulation capable of integrating seamlessly with modern computational frameworks.

In this paper, we introduce a generalized, continuous GLCM framework that enhances rotational invariance, spatial flexibility, and even differentiability. This approach supports integration with machine learning algorithms and gradient-based optimization techniques, bridging critical gaps in current texture analysis and radiomics methodologies.

2   Related Work

Texture analysis encompasses a broad class of computational techniques aimed at quantifying variations in image intensity patterns—capturing both prominent and subtle textures. In neuroimaging and other clinical domains, texture-based radiomics features have demonstrated strong utility in lesion segmentation, disease grading, and longitudinal monitoring of conditions such as brain tumors and epilepsy [6]. Tamura et al. [13] identified perceptual attributes—coarseness, contrast, directionality, and roughness—that align well with human vision and form the basis for many modern texture descriptors. Despite the divergence between digital and perceptual analysis, many features correlate with subjective assessments and support accurate classification across diverse imaging tasks.

Among statistical texture measures, gray-level spatial dependence metrics—especially those based on the GLCM—are foundational. Introduced by Haralick et al. [5], GLCM features have demonstrated strong discriminative power across domains including histopathology, satellite imagery, and medical imaging. Their effectiveness is further supported by integration with statistical classification methods such as Kullback–Leibler divergence-based similarity [9].

Radiomics represents a data-driven framework for extracting image-derived biomarkers in a high-throughput, reproducible manner [8]. By treating clinical images as quantitative data sources, radiomics facilitates prognostic and predictive modeling. Standard pipelines—including acquisition, segmentation, feature extraction, and modeling—face challenges such as inter-protocol variability, segmentation bias, and feature redundancy [7]. Among second-order texture features, GLCM-based metrics remain highly effective in encoding spatial patterns, especially when fused with clinical and molecular data for personalized treatment strategies. Nonetheless, large-scale validation and standardized reporting remain critical [14].

Classical GLCM, however, is limited by discrete pixel offsets, which constrain its adaptability across varying orientations and resolutions. While foundational work by Haralick et al. [4] defined widely used GLCM features, these are inherently rigid. Subsequent studies have proposed extensions to higher-dimensional data [12] and domain-specific adaptations for osteoarthritis [10]. The Image Biomarker Standardization Initiative (IBSI) [14] has also highlighted the importance of methodological reproducibility in radiomics.

To address these limitations, our work introduces a generalized GLCM framework based on continuous, differentiable spatial displacements. By computing cooccurrence statistics over spatially weighted circular neighborhoods, our method enhances rotational invariance and local texture fidelity while supporting seamless integration with modern learning architectures.

3   Motivation

The primary motivation for this research on generalizing GLCM entries lies in creating a more flexible and versatile framework that operates effectively within continuous space. Traditional GLCM methods are constrained by discrete pixel orientations and fixed spatial relationships, which can limit adaptability in contexts requiring finer, smoother gradations.

Enhanced Flexibility in Continuous Space. By redefining GLCM entries to work in a continuous domain, we enable a flexible structure that captures texture features across a broader array of orientations and spatial resolutions. This flexibility is crucial for applications in fields like medical imaging or materials science, where textures are not strictly aligned with predefined orientations or uniform distances.

Differentiable Region of GLCM Values. A continuous, differentiable region of values for GLCM entries provides smoother transitions and allows gradient-based methods to be applied effectively. This is especially beneficial for machine learning models that rely on differentiable inputs for feature extraction and training. A generalized, continuous GLCM model thus facilitates integration into advanced analytical frameworks, supporting more complex applications that require both flexibility and smooth computational consistency.

Utilization of Constant L2 Norm Displacements and Minimum Ball Size for Cartesian Grids

The flexibility to use displacements defined by a constant L2 norm rather than the traditional L-infinity norm allows for a more robust and rotationally invariant assessment of texture features, enhancing the robustness of GLCM in variable spatial contexts.

4   Generalization of GLCM

Traditional GLCM

The GLCM was first introduced by Haralick et al. [5] as a statistical method to describe texture in digital images. It captures the spatial relationship of gray levels by counting how often pairs of pixel intensities occur at a specified distance and orientation. This method became foundational in texture analysis, particularly for tasks involving pattern recognition, segmentation, and radiomics feature extraction.

Quantization of the Grayscale Image. Let the image domain be denoted by $\Omega \subset \mathbb{Z}^2$, and let the grayscale image be represented by a function $I : \Omega \to [0, 255]$, which maps each pixel coordinate $(i, j) \in \Omega$ to its intensity value.

To reduce the number of gray levels, we quantize I into N discrete levels. Let the quantized image be denoted by $I_q : \Omega \to \{0, 1, \ldots, N-1\}$, defined as:

$$I_q(i, j) = \left\lfloor \frac{I(i, j)}{256/N} \right\rfloor (256/N)$$

where $\lfloor \cdot \rfloor$ denotes the integer part of its argument (truncation).

GLCM Construction. Select a region of interest (ROI) within the image domain, denoted by $\Omega_{\mathrm{ROI}} \subseteq \Omega$. This is the set of pixel coordinates used for GLCM computation:

$$\Omega_{\mathrm{ROI}} = \{(i, j) \in \Omega \mid \text{pixel lies within the ROI}\}$$

For each pair of pixels (i_1, j_1) and (i_2, j_2) within Ω_{ROI}, where the pair is separated by a displacement vector $\boldsymbol{\delta} = (\delta x, \delta y)$, representing the spatial relationship between pixels:

$$(i_2, j_2) = (i_1 \pm \delta x, j_1 \pm \delta y),$$

the co-occurrence matrix is calculated as:

$$P(g_1, g_2 \mid \boldsymbol{\delta}) = \sum_{(i_1,j_1),(i_2,j_2)} \begin{cases} 1, & \text{if } I_q(i_1, j_1) = g_1 \text{ and } I_q(i_2, j_2) = g_2, \\ 0, & \text{otherwise.} \end{cases}$$

where:

- g_1 and g_2 are the quantized gray levels of the pixel pairs.
- $\theta = \arctan\left(\frac{\delta y}{\delta x}\right)$ measures the angle of the displacement.

Generalized Method

For a given pixel (i, j) in an image, the following steps outline a generalized approach to computing GLCM by integrating pixels within a defined displacement (Fig. 1).

Displaced Neighborhood Definition. For the reference pixel (i, j), it can be transformed into spatial coordinates (x_i, y_j) by mapping the discrete pixel indices to continuous spatial coordinates. We first define a local neighbor-

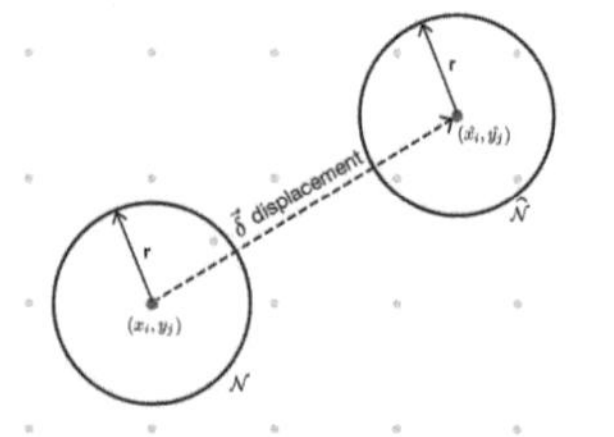

Fig. 1. Generalization of GLCM

hood $\mathcal{N}$ around the point (x_i, y_j) using a ball of radius r. Next, we apply a displacement vector $\boldsymbol{\delta} = (\delta x, \delta y)$ to shift the entire neighborhood by this vector, yielding a displaced neighborhood $\hat{\mathcal{N}}$, which is centered at:

$$(\hat{x}_i, \hat{y}_j) = (x_i + \delta x, y_j + \delta y)$$

Note that $(\hat{x}_i, \hat{y}_j)$ will be another pixel location only when δx and δy are both integers.

Weight Assignment. We now have the displaced neighborhood $\hat{\mathcal{N}}$ as a circular region centered at the displaced center $(\hat{x}_i, \hat{y}_j)$ with radius r. For each point $(\hat{x}, \hat{y})$ within $\hat{\mathcal{N}}$ and the region of interest, we assign a spatially dependent weight w, where $w = 1$ at the displaced center $(\hat{x}_i, \hat{y}_j)$ and gradually decreases to $w = 0$ as we move toward the displaced neighborhood boundary. For example, in the following experiment, we employ a quadratic weight function defined as:

$$w(\hat{d}) = \begin{cases} 1 - \dfrac{\hat{d}^2}{r^2}, & \text{if } \hat{d} \leq r \\ 0, & \text{if } \hat{d} > r \end{cases} \tag{1}$$

where $\hat{d} = \sqrt{(\hat{x}_i - \hat{x})^2 + (\hat{y}_j - \hat{y})^2}$, indicates the distance between the points in the displaced neighborhood and the displaced center in Euclidean distance. This formulation ensures that pixels near the center retain significantly higher importance, while those farther from the center experience a more rapid attenuation in influence, and those outside the displaced neighborhood have no influence at all.

Radius r Threshold. To ensure that there is **at least one pixel** within the displaced neighborhood, the key is to define r in a way that accommodates the pixel grid's resolution and ensures non-zero neighborhood coverage. To determine the minimum radius r, we selected a 2×2 reference grid cell region with spatial resolutions Δx defining the cell dimensions (in general cases, we may choose $\Delta x = 1$). Here are three cases:

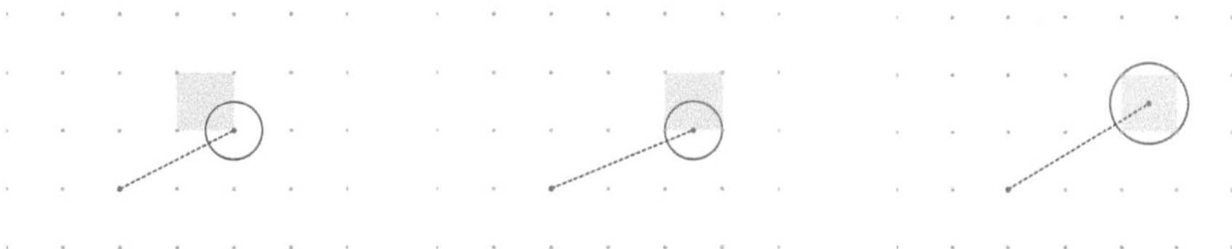

Fig. 2. Centered at a Grid Point

Fig. 3. Centered Between Two Grid Points

Fig. 4. Centered Within the Interior of a Grid Cell

Case 1: Displaced Neighborhood Centered at a Grid Point

In the simplest case, where the neighborhood is centered at a grid point (Fig. 2), we observe that, with one pixel already contained within the region, the radius may be assigned any arbitrary value.

Case 2: Displaced Neighborhood Centered Directly Between Two Grid Points

If the displaced neighborhood is centered directly between two grid points, such as at the midpoint (Fig. 3), the minimum radius r required to ensure that $\hat{\mathcal{N}}$ encompasses at least one pixel is given by

$$r = \frac{1}{2}\Delta x$$

Case 3: Displaced Neighborhood Centered in the Interior of Four Vertices

In most cases, however, the region will be located within the grid cell region (Fig. 4). Consider the extreme case, we assume $\hat{\mathcal{N}}$ is centered at the cell center, with the closest pixels located at a distance of:

$$\sqrt{(\frac{1}{2}\Delta x)^2 + (\frac{1}{2}\Delta x)^2} = \frac{\sqrt{2}}{2}\Delta x$$

Therefore, the minimum radius r required for this configuration is

$$r = \frac{\sqrt{2}}{2}\Delta x$$

Summary. To ensure that the displaced neighborhood includes at least one pixel within the displaced neighborhood:

- **Case 1:** $r > 0$ if the center is at a grid point.
- **Case 2:** $r \geq \frac{1}{2}\Delta x$ if the center is directly between two grid points.
- **Case 3:** $r \geq \frac{\sqrt{2}}{2}\Delta x$ if the center is inside the reference grid cell.

From above statement, if the chosen r satisfies Case 3, then Case 1 or Case 2 is already satisfied, which indicates that the minimum radius for this neighborhood definition is given by

$$r_{\min} = \frac{\sqrt{2}}{2}\Delta x \tag{2}$$

GLCM Calculation. To calculate the GLCM, we consider each of the K pixels locations $(\hat{x}_k, \hat{y}_k)$, with corresponding pixel indices $(\hat{i}_k, \hat{j}_k)$, that are contained within the displaced neighborhood $\hat{\mathcal{N}}$. By choosing $r \geq \frac{\sqrt{2}}{2}\Delta x$, we ensure that $K \geq 1$, as described above.

Let $I_q(i,j)$ represent the quantized gray level of pixel (i,j). We pair the reference pixel's gray level $I_q(i,j)$ with the quantized gray levels $I_q(\hat{i}_k, \hat{j}_k)$. For each pair $(I_q(i,j), I_q(\hat{i}_k, \hat{j}_k))$, we increment the corresponding GLCM entry:

$$\text{GLCM}(I_q(i,j), I_q(\hat{i}_k, \hat{j}_k)) + = \frac{w(\hat{d}_k(x,y))}{\sum_{l=1}^{K} w(\hat{d}_l(x,y))} \tag{3}$$

where $\hat{d}_k(x,y) = \sqrt{(x - \hat{x}_k)^2 + (y - \hat{y}_k)^2}$. This formulation ensures that each displaced pixel contributes proportionally to the GLCM based on its spatial alignment with the intended displacement magnitude. Rather than relying on a single offset direction, this approach aggregates contributions from multiple directions and locations captured within the displaced neighborhood, enabling a net co-occurrence measurement that is both smoother and more robust. The normalization factor in the denominator ensures that the total contribution from the displaced neighborhood sums to one (just as would the contribution from a single offset in traditional GLCM), preserving the probabilistic interpretation.

5   Properties of the Generalized GLCM Framework

Flexible Sampling of Directions

Compared to the traditional GLCM, where the displaced point is restricted to coincide with exact grid locations, our generalized GLCM framework enables flexible sampling by considering all pixels within the displaced neighborhood. As a result, the center of the displaced neighborhood no longer needs to align with grid points, allowing for more continuous and directionally adaptive texture analysis.

Continuous and Differentiable Matrix Entries

In the classic GLCM, the matrix entries represent counts of pixel pairs with quantized gray levels. In the generalized version, matrix entries are defined using weighted contributions within displaced neighborhoods. To ensure differentiability, we must choose a differentiable weighting function such as Eq. (1). However, we must also keep the normalization factor for the GLCM increment in Eq. (3) in mind as well. If we consider the case of two pixels in the displaced neighborhood with weights w_1 and w_2, and denote the GLCM increment from the first pixel by $f(w_1, w_2) = \frac{w_1}{w_1 + w_2}$, then its partial derivatives are:

$$\frac{\partial f}{\partial w_1} = \frac{w_2}{(w_1 + w_2)^2}, \quad \frac{\partial f}{\partial w_2} = -\frac{w_1}{(w_1 + w_2)^2}$$

These derivatives remain finite and well-defined as long as $w_1 + w_2 > 0$, which implies that the neighborhood must include at least one point with a strictly positive weight. This condition can be satisfied by selecting a radius r such that the displaced neighborhood always contains at least one point strictly within its interior (i.e., not allowing cases where all contributing points lie along the boundary). Thus, in cases were a contributing point lies along the boundary of the band (and therefore is attributed a weight of zero), we can be sure that at least one other contributing piont lies strictly within the interior with a non-zero weight in order to guarantee a non-zero noralization factor.

This property can be guranteed by choosing the neighborhood radius r to be strictly greater than the minimum radius $r_{\min}$ derived in Eq. (2). Such a choice ensures that the generalized GLCM is smooth and differentiable, making it compatible with gradient-based optimization methods commonly used in machine learning frameworks.

Approach GLCM as $r \to 0$

As the radius r becomes smaller and approaches 0, the GLCM converges to a more local structure, similar to the traditional GLCM. As $r \to 0$, we consider only the nearest neighbors, so:

$$r \to 0 \implies \text{local structure similar to traditional GLCM}$$

Thus, the generalized GLCM behaves like the traditional GLCM for very small r, while retaining its differentiable and continuous properties.

Rotational Invariance as $\frac{\delta}{r} \to 0$

When the displacement ratio $\frac{\delta}{r} \to 0$, the GLCM becomes rotationally invariant. This happens because:

– At small displacements (compared to the radius r), the contribution of displaced neighborhood becomes independent of the exact angle.
– This leads to a texture descriptor that captures the overall texture without being sensitive to the specific direction or rotation of the image.

This kind of rotational invariance is particularly useful in applications where textures might appear at different orientations.

Constant Displacement Using L2 Norm

The traditional GLCM is based on pixel pairs with fixed spatial displacement in specific directions (e.g., horizontal, vertical, diagonal). In new generalization, we now can use a constant Euclidean distance ($L2\ norm$) between the central points and points within its displaced neighborhood. For each central point $(\hat{x}_i, \hat{y}_j)$, we consider all points in $(\hat{x}, \hat{y})$ to build our weight function, such that $d = \sqrt{(\hat{x}_i - \hat{x})^2 + (\hat{y}_j - \hat{y})^2}$. This means that they are evaluated within a circular neighborhood of radius r, rather than specific directional displacements.

6   GLCM-Based Radiomics Features

We denote g_1 and g_2 as gray levels of reference and displaced pixels. The normalized co-occurrence matrix is defined as:

$$glcm(g_1, g_2) = \frac{GLCM(g_1, g_2)}{\sum GLCM(g_1, g_2)}.$$

The following definitions apply [5,11]:

– ϵ: Small constant for stability ($\approx 2.2 \times 10^{-16}$).
– N_g: Number of quantized gray levels.
– $glcm_x(g_1) = \sum_{g_2} glcm(g_1, g_2), \quad glcm_y(g_2) = \sum_{g_1} glcm(g_1, g_2)$
– $\mu_x = \sum g_1 \cdot glcm_x(g_1), \quad \mu_y = \sum g_2 \cdot glcm_y(g_2)$

- σ_x, σ_y: The standard deviations of $glcm_x$, $glcm_y$, respectively.
- $glcm_{x+y}(k) = \sum_{g_1+g_2=k} glcm(g_1, g_2)$, $glcm_{x-y}(k) = \sum_{|g_1-g_2|=k} glcm(g_1, g_2)$
- $H_X = -\sum glcm_x \log_2(glcm_x + \epsilon)$, $\quad H_Y = -\sum glcm_y \log_2(glcm_y + \epsilon)$
- $H_{XY} = -\sum glcm \log_2(glcm + \epsilon)$
- $H_{XY1} = -\sum glcm \log_2(glcm_x \cdot glcm_y + \epsilon)$
- $H_{XY2} = -\sum glcm_x \cdot glcm_y \log_2(glcm_x \cdot glcm_y + \epsilon)$

The GLCM-based radiomics features used in this study are summarized as below:

Table 1. GLCM-Based Radiomics Features

Feature	Formula	Feature	Formula
Auto-correlation	$\sum glcm(g_1, g_2) \cdot g_1 g_2$	JointAverage	$\sum glcm(g_1, g_2) \cdot g_1$
ClusterProminence	$\sum glcm(g_1, g_2)(g_1 + g_2 - \mu_x - \mu_y)^4$	Dis-similarity	$\sum glcm(g_1, g_2) \cdot |g_1 - g_2|$
ClusterTendency	$\sum glcm(g_1, g_2)(g_1 + g_2 - \mu_x - \mu_y)^2$	Contrast	$\sum glcm(g_1, g_2)(g_1 - g_2)^2$
Correlation	$\frac{\sum glcm(g_1,g_2)(g_1-\mu_x)(g_2-\mu_y)}{\sigma_x \cdot \sigma_y}$	DifferenceAverage	$\sum k \cdot glcm_{x-y}(k)$
DifferenceEntropy	$-\sum glcm_{x-y}(k) \log(glcm_{x-y}(k) + \epsilon)$	DifferenceVariance	$\sum (k - DA)^2 \cdot glcm_{x-y}(k)$
ClusterShade	$\sum glcm(g_1, g_2)(g_1 + g_2 - \mu_x - \mu_y)^3$	JointEnergy	$\sum glcm(g_1, g_2)^2$
JointEntropy	$-\sum glcm(g_1, g_2) \log(glcm(g_1, g_2) + \epsilon)$	Homo-geneity	$\sum \frac{glcm(g_1,g_2)}{1+(g_1-g_2)^2}$

7 Results

This section presents the outcomes of GLCM calculations and analyses performed on a pediatric CT brain image [2], focusing on a region of interest spanning from $[200, 200]$ to $[300, 300]$ by choosing num value to be 32 (Table 1). The analyses were conducted for both the original image and a 45-degree rotated version, which demonstrated improved precision and robustness in texture representation. The displacement will also rotate 45 degrees.

Rotational Invariance Assessment

To evaluate rotational invariance, the Frobenius Norm and Correlation Coefficient between the GLCM matrices of the original and rotated images were computed. The results indicate:

- Frobenius Norm: The proposed method achieved lower Frobenius Norm values as the chosen radius of displaced region goes larger, signifying enhanced rotational invariance in texture representation (Fig. 5).
- Correlation Coefficient: The proposed method yielded higher correlation coefficients as radius goes bigger compared to traditional approaches, reflecting greater consistency across rotations (Fig. 6).

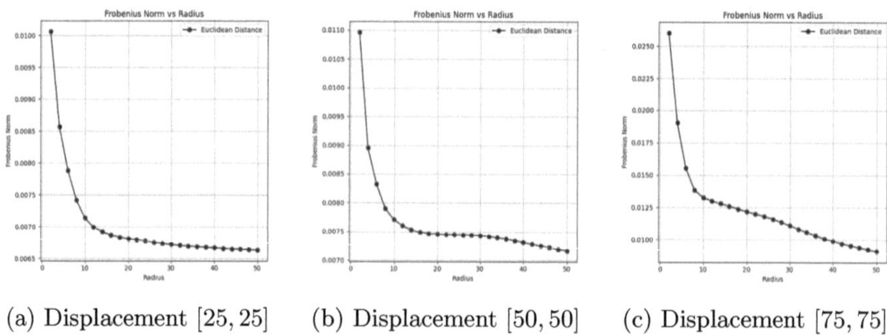

(a) Displacement $[25, 25]$ (b) Displacement $[50, 50]$ (c) Displacement $[75, 75]$

Fig. 5. Frobenius Norm vs Radius at Various Displacements

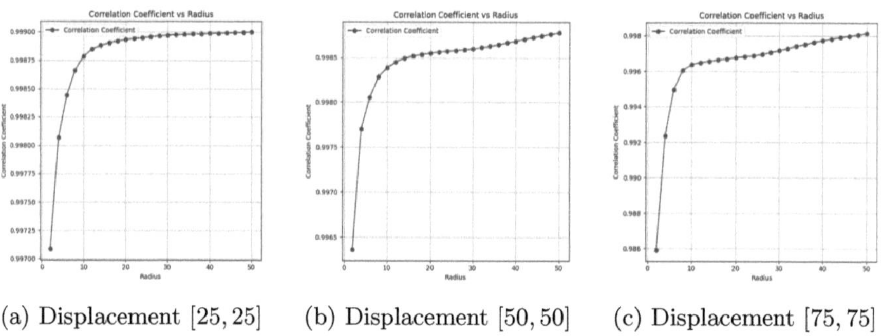

(a) Displacement $[25, 25]$ (b) Displacement $[50, 50]$ (c) Displacement $[75, 75]$

Fig. 6. Correlation Coefficient vs Radius at Various Displacements

Co-occurrence-Based Texture Features

Texture features were calculated for both the original and rotated images. The proposed method consistently provided stable and reliable feature values, further emphasizing its robustness under rotational transformations. These findings highlight the effectiveness of the proposed method in accurately representing texture features while ensuring rotational consistency, making it well-suited for advanced radiomics analyses.

To better represent the relationship between the features, we also compute the relative errors using $RE = \frac{|A-B|}{|A|}$. Since correlation is already normalized, the relative error for correlation is computed as $RE_{\text{correlation}} = |A - B|$.

Figure 7 shows the relative error for displacement $[50, 50]$, representative of the general trend across displacements. Results for other displacements were similar and are omitted for clarity.

As shown in Fig. 7, relative errors remain consistently low, confirming that the proposed method preserves feature values under rotation and achieves rotational invariance.

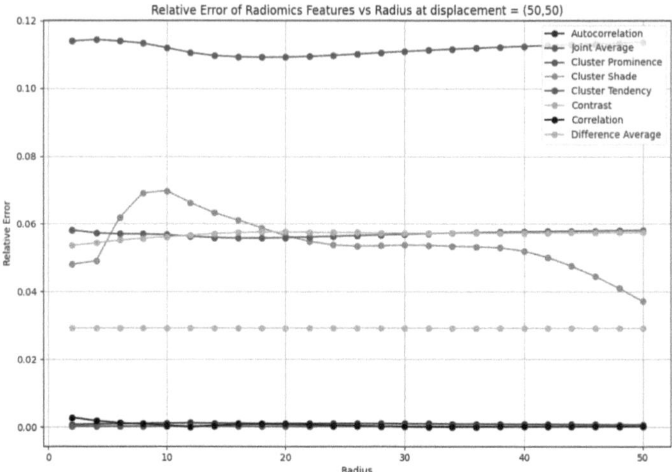

Fig. 7. Relative error between original and rotated images for displacement $[50, 50]$.

8 Conclusion and Future Work

We presented a generalized GLCM framework that extends traditional discrete formulations into a continuous, differentiable, and rotationally invariant model. By defining spatial neighborhoods with displacement vectors and radial weighting functions, our method enables co-occurrence computation over circular regions in 2D images. This formulation improves spatial coverage and supports integration with gradient-based machine learning pipelines. Experiments in 2D showed that our method maintains stable Frobenius norms and correlation coefficients across increasing neighborhood radii, confirming robustness to scale and rotation.

Future directions include:

- **Aggregated Averaged GLCM:**
  Single-direction GLCMs often yield unstable feature values—particularly for higher-order metrics like *Cluster Shade*. To improve rotational invariance and feature stability, we propose an aggregated GLCM variant that averages co-occurrence matrices across directions.
- **3D Extension and Radiomics Evaluation:**
  Generalize the method to 3D using spherical neighborhoods and compare radiomics features against traditional GLCM for clinical prediction tasks.
- **Active Contour Integration And Deep Learning Applications:**
  Embed the differentiable GLCM into region-based active contour models to guide texture-driven segmentation and incorporate GLCM priors into neural networks for improved classification and segmentation, particularly under limited data and in texture-rich domains.

Acknowledgment. R.C. is a member of the "Gruppo Nazionale per l'Analisi Matematica, la Probabilità e le loro Applicazioni" (INdAM). Anthony Yezzi and Haoyue Chen were supported by Army Research Office Grant W911NF-22-1-0267.

References

1. Ali, M., et al.: Prostate cancer detection: performance of radiomics analysis in multiparametric MRI. In: Image Analysis and Processing – ICIAP 2023 Workshops, Proceedings, Part II, vol. 14224, pp. 83–92 (2023). https://doi.org/10.1007/978-3-031-51026-7_8

2. Bickle, I.: Normal CT brain - pediatric. Radiopaedia.org (2017). https://doi.org/10.53347/rID-53337, https://radiopaedia.org/cases/53337, case study, Radiopaedia.org. (Accessed 21 Jan 2025)

3. Gillies, R.J., Kinahan, P.E., Hricak, H.: Radiomics: images are more than pictures, they are data. Radiology **278**(2), 563–577 (2016). https://doi.org/10.1148/radiol.2015151169

4. Haralick, R.M., Shanmugam, K., Dinstein, I.: Textural features for image classification. IEEE Trans. Syst. Man Cybern. **SMC-3**(6), 610–621 (1973). https://doi.org/10.1109/TSMC.1973.4309314

5. Haralick, R.M., Shanmugam, K., Dinstein, I.: Textural features for image classification. IEEE Trans. Syst. Man Cybern. **SMC-3**(6), 610–621 (1973). https://doi.org/10.1109/TSMC.1973.4309314

6. Kassner, A., Thornhill, R.E.: Texture analysis: a review of neurologic MR imaging applications. AJNR Am. J. Neuroradiol. **31**(5), 809–816 (2010). https://doi.org/10.3174/ajnr.A2061

7. Kumar, V., Gu, Y., Basu, S., et al.: Radiomics: the process and the challenges. Magn. Reson. Imaging **30**(9), 1234–1248 (2012). https://doi.org/10.1016/j.mri.2012.06.010

8. Lambin, P., et al.: Radiomics: extracting more information from medical images using advanced feature analysis. Eur. J. Cancer **48**(4), 441–446 (2012). https://doi.org/10.1016/j.ejca.2011.11.036

9. Ojala, T., Pietikäinen, M., Harwood, D.: A comparative study of texture measures with classification based on featured distributions. Pattern Recogn. **29**(1), 51–59 (1996)

10. Peuna, A., Hekkala, J., Haapea, M., et al.: Variable angle GLCM analysis of T2 maps reveals degenerative changes in knee osteoarthritis. J. Magn. Reson. Imaging **47**(5), 1316–1327 (2018). https://doi.org/10.1002/jmri.25881

11. Radiomics: Gray level co-occurrence matrix (GLCM) features (nd). https://pyradiomics.readthedocs.io/en/latest/features.html#module-radiomics.glcm. Accessed 3 Dec 2024

12. Sebastian, B.V., Unnikrishnan, A., Balakrishnan, K.: Gray level co-occurrence matrices: generalisation and some new features. arXiv preprint (2012). https://doi.org/10.48550/arXiv.1205.4831

13. Tamura, H., Mori, S., Yamawaki, T.: Textural features corresponding to visual perception. IEEE Trans. Syst. Man Cybern. **8**(6), 460–473 (1978). https://doi.org/10.1109/TSMC.1978.4309999

14. Zwanenburg, A., Leger, S., Vallières, M., Löck, S.: The image biomarker standardisation initiative: standardised quantitative radiomics for high-throughput image-based phenotyping. Radiology (2020). https://doi.org/10.1148/radiol.2020191145

Machine Learning in Mammography Classification: A Radiomics-Based Evaluation and Developments

Nicolò Lauciello[1,2]([✉]) , Giovanni Pasini[2,3] , Giorgio Russo[2] ,
Franco Marinozzi[3] , Fabiano Bini[3] , and Alessandro Stefano[2]

[1] Department of Earth and Marine Sciences, University of Palermo, Via Archirafi 22, 90123 Palermo, Italy
nicolo.lauciello@unipa.it
[2] Institute of Bioimaging and Complex Biological Systems, National Research Council (IBSBC-CNR), Contrada Pietrapollastra-Pisciotto, 90015 Cefalù, Italy
[3] Department of Mechanical and Aerospace Engineering, Sapienza University of Rome, Eudossiana 18, 00184 Rome, Italy

Abstract. Breast cancer remains the second leading cause of cancer-related mortality among women worldwide. Early detection and accurate classification of breast lesions, including masses and microcalcifications, are therefore essential to improving diagnostic precision and patient outcomes.

This study investigates the performance of three supervised machine learning (ML) classifiers—Linear Discriminant Analysis (LDA), Support Vector Machine (SVM), and K-Nearest Neighbours (KNN)—in the context of a radiomics-based approach to mammographic image classification. The analysis was conducted using the matRadiomics toolbox on 1,219 mammograms obtained from the publicly available CBIS-DDSM dataset.

All three classifiers—LDA, SVM, and KNN—were tested on the same mammographic subsets of masses and microcalcifications. LDA achieved the highest overall performance in terms of AUC (69.7% for microcalcifications; 62.9% for masses) and F1-score (0.741 and 0.638, respectively). While KNN showed slightly higher accuracy on masses (60%), its F1-score (0.576) remained lower than LDA. For microcalcifications, both SVM and KNN reached comparable F1-scores (0.741 and 0.742), though LDA retained the best AUC and overall balance. These findings highlight both the potential and the current limitations of traditional ML-based radiomics techniques in breast cancer classification. They also underscore the importance of exploring more advanced artificial intelligence approaches, such as deep learning and ensemble methods, to enhance diagnostic accuracy and clinical applicability in mammographic screening.

Keywords: Machine Learning · Breast Cancer · Classification

© The Author(s), under exclusive license to Springer Nature Switzerland AG 2026
E. Rodolà et al. (Eds.): ICIAP 2025 Workshops, LNCS 16169, pp. 199–209, 2026.
https://doi.org/10.1007/978-3-032-11317-7_17

1 Introduction

Breast cancer is the most common malignancy among women worldwide, affecting approximately 10–12% of the female population and accounting for nearly 500,000 deaths each year [1]. Although recent advances in screening modalities—such as tomosynthesis, ultrasound, and magnetic resonance imaging (MRI)—have improved diagnostic precision [2, 3], significant limitations remain. The complex anatomical architecture of the breast can hinder accurate image interpretation, often leading to false positives and false negatives that may adversely affect clinical decision-making and patient outcomes. In this context, radiomics has emerged as a promising approach for extracting quantitative features from medical images—features that go beyond what is visible to the human eye—and correlating them with clinical and prognostic data [4–7]. By capturing aspects such as shape, texture, and intensity, radiomics supports the development of predictive models for lesion classification and outcome prediction [8–13]. When integrated with machine learning (ML), radiomics offers a non-invasive strategy to enhance diagnostic accuracy and reduce inter-observer variability in radiological assessments. This study explores a radiomics-based workflow using the matRadiomics toolbox to classify breast lesions [14]; 1,219 mammograms from the publicly available CBIS-DDSM dataset [15] were analyzed, three ML classifiers were evaluated to determine their effectiveness in distinguishing between benign and malignant lesions. By assessing the predictive performance of these models, the study aims to contribute to the development of robust, reproducible, and automated decision-support systems for breast cancer diagnosis.

2 Materials and Methods

2.1 Dataset

The CBIS-DDSM database [15], hosted by The Cancer Imaging Archive (TCIA), is a curated and standardized update of the original DDSM dataset. It contains over 10,000 digitized mammographic images from 1,566 patients, systematically organized into two primary lesion categories: microcalcifications and masses. For each case, a segmented region of interest (ROI) highlighting the lesion and a cropped image focused on the lesion are provided.

In total, the dataset includes 1,872 images with microcalcifications and 1,696 with masses, corresponding to 753 and 891 patients, respectively. Of the 3,568 annotated images, 2,111 are labeled benign and 1,457 malignant. Each case is accompanied by detailed clinical annotations in CSV format, including BI-RADS scores, breast density (scale 1–4), lesion characteristics (shape, margins, and distribution), scan metadata (laterality and projection view), pathology labels and subtlety ratings.

2.2 Toolbox for Radiomics Analysis: MatRadiomics

MatRadiomics [14, 16] is an IBSI (Image Biomarker Standardization Initiative) - compliant platform designed to streamline the radiomics workflow. Built on MATLAB and

Python, it integrates PyRadiomics for feature extraction, and traditional ML algorithms for predictive modeling, specifically: Linear Discriminant Analysis (LDA), Support Vector Machine (SVM) and K-Nearest Neighbors (KNN). Key features include DICOM image visualization, target segmentation [17], and image preprocessing e.g., CLAHE (Contrast Limited Adaptive Histogram Equalization) and Median Filter. With its intuitive GUI (Graphical User Interface), MatRadiomics offers a comprehensive solution for medical image analysis and predictive modeling, thanks to a highly reproducible pipeline that enables a complete radiomics workflow.

2.3 Preprocessing

This study employed two image enhancement techniques—Contrast Limited Adaptive Histogram Equalization (CLAHE) and Median Filtering—to improve the visual quality of mammograms and support more accurate lesion classification. CLAHE is a contrast enhancement method that operates locally on small regions (or tiles) of the image, rather than globally. By redistributing the pixel intensity values within each tile, CLAHE enhances local contrast while preventing excessive noise. To further refine image quality, Median Filtering is used as a noise reduction technique, replacing each pixel's value with the median of its neighboring pixels within a defined sliding window. Unlike linear filters, this non-linear approach effectively suppresses outlier noise while preserving important image features such as lesion edges and boundaries. Combined, these preprocessing methods enhance the visibility of anatomical structures and pathological features, reducing variability and improving the reliability of subsequent radiomics analysis and classification models [18].

2.4 Features Extraction and Selection

To enhance radiomics analysis, three image transformations were applied: Original, Wavelet, and Laplacian of Gaussian (LoG). While original images retain raw structural data, wavelet transforms decompose them into frequency components to capture fine textures, and LoG filtering improves edge detection and highlights subtle intensity variations [19].

Feature selection was then carried out using a sequential approach combining correlation analysis and logistic regression [20]. Features were first ranked based on point biserial correlation with the binary outcome (benign vs. malignant), then refined through stepwise logistic regression to retain the most predictive and non-redundant variables.

2.5 Machine Learning Predictive Models

2.5.1 LDA

Linear Discriminant Analysis (LDA) is a supervised machine learning technique used for classification tasks, its main objective is to project data onto a lower-dimensional space that maximizes class separability by minimizing intra-class variance and maximizing inter-class variance. In practice, LDA computes a linear boundary based on class distributions and applies it to new data to predict class membership. Beyond improving classification performance, it also reduces dimensionality and enhances model interpretability, making it well-suited for medical imaging applications where feature sets are often large and complex [21].

2.5.2 SVM

Support Vector Machine (SVM) is a supervised learning algorithm commonly used for binary classification. Its core objective is to identify the optimal hyperplane that best separates the two classes by maximizing the margin—i.e., the minimum distance between the hyperplane and the nearest data points from each class [22]. Regarding the settings of the SVM model, through the matRadiomics tool, a "Radial Basis Function" (RBF) kernel is applied; while, regarding hyperparameters, that is the rule applied:

$\gamma = \frac{1}{Numberoffeatures}$, where "$\gamma$" stands for gamma and the parameter "c" is defaulted to 0.1.

2.5.3 KNN

K-Nearest Neighbors (KNN) is a non-parametric, instance-based classification algorithm that assigns class labels based on the majority class among the k closest samples in the feature space. It operates on the principle that similar data points tend to appear near each other. For each test instance, the algorithm identifies its k nearest neighbors from the training data—typically using Euclidean distance—and classifies it according to the most frequent class among those neighbors [23]. However, its performance is sensitive to the choice of k and can become computationally expensive with large datasets [24]. In this study, different k values (5, 10, 15, 25, 50, 75, 100, 150) were tested to evaluate their impact on classification performance. The value k = 75 was selected for reporting results, as it yielded the best trade-off in terms of accuracy, AUC, and F1 score across the models. This choice is also supported by the empirical rule suggesting k $\approx \sqrt{N}$, where N is the number of samples in the dataset.

2.6 Pipeline

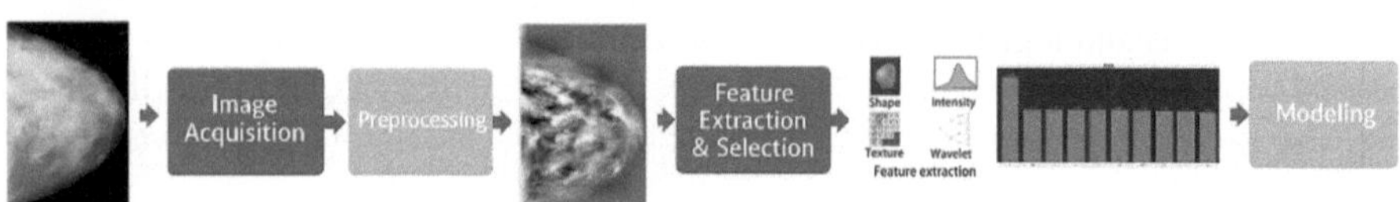

Fig. 1. Schematic representation of the radiomics pipeline adopted for mammographic lesion classification.

The pipeline, shown in Fig. 1, includes preprocessing to enhance image quality, extraction of radiomics features, feature selection to reduce dimensionality, then used to develop and evaluate classification models.

3   Results

Model performance was assessed using 10-fold cross-validation to ensure robustness and generalizability. Evaluation metrics included accuracy, precision, recall, F1-score, and the area under the Receiver Operating Characteristic (ROC) curve (AUC). These were reported in a comparative table to provide a comprehensive overview of classification performance. Confusion matrices were also used to detail the distribution of true positives, false positives, true negatives, and false negatives for each classifier.

In addition, ROC curves were plotted to visualize the discriminatory capability of each model. Performance comparisons were conducted across the three classifiers (SVM, LDA, and KNN, tested across multiple K values), highlighting the most effective configuration based on both threshold-independent and threshold-dependent metrics.

3.1   Classification Performances: Metrics Overview

The performance evaluation of the classifiers—LDA, SVM, and KNN—was conducted separately for the two lesion types (masses and microcalcifications), using Precision, Recall, F1-score, AUC, and Accuracy as primary metrics, shown in Table 1.

Table 1. Performance of classifiers for both masses and calcifications.

Masses	Precision	Recall	F1-score	AUC	Accuracy
LDA	56.8%	72.8%	63.8%	62.9%	59.1%
SVM	54.5%	59.2%	56.7%	61.9%	55.3%
KNN	58.9%	56.3%	57.6%	63.3%%	60%

Calcifications	Precision	Recall	F1-score	AUC	Accuracy
LDA	64.5%	87.3%	74.1%	69.7%	66.3%
SVM	64.2%	87.3%	73.9%	67%	66.2%
KNN	64.6%	87.6%	74.2%	69%	65.7%

For mass classification, LDA achieved the best balance between recall (72.8%) and F1-score (63.8%), outperforming the other models in both sensitivity and overall classification quality. KNN recorded the highest accuracy (60%) and AUC (63.3%), suggesting stronger discriminative capability, albeit with slightly lower recall. SVM, while slightly trailing in all metrics, remained competitive and consistent. In the case of microcalcifications, all three models performed considerably better. KNN achieved the best overall F1-score (74.2%) and recall (87.6%), closely followed by LDA and SVM. Accuracy and AUC values were also higher across all models, with LDA and KNN slightly outperforming SVM in most metrics. Overall, the results indicate that LDA and KNN exhibit superior performance, especially in handling microcalcifications, while SVM showed moderate yet stable performance across both tasks.

3.2 ROC Curve Analysis

To further assess the discriminatory ability of each classifier, ROC curves were plotted (see Figs. 2, 3 and 4). These curves illustrate the trade-off between true positive rate (sensitivity) and false positive rate across different classification thresholds.

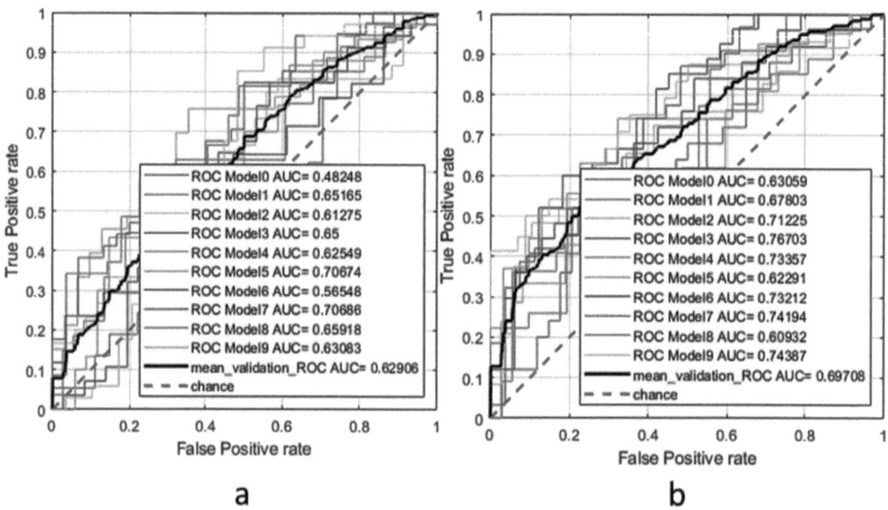

a   b

Fig. 2. Roc Curve LDA Model; a) masses, b) calcifications.

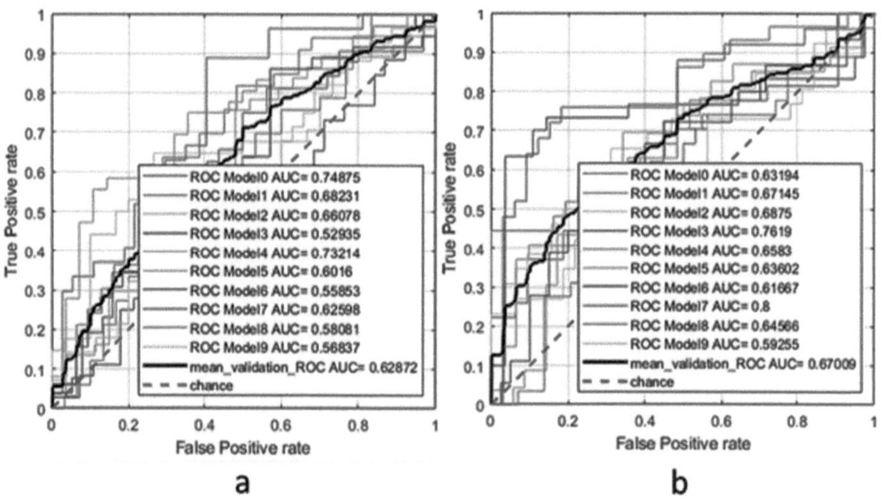

a   b

Fig. 3. Roc Curve SVM Model; a) masses, b) calcifications.

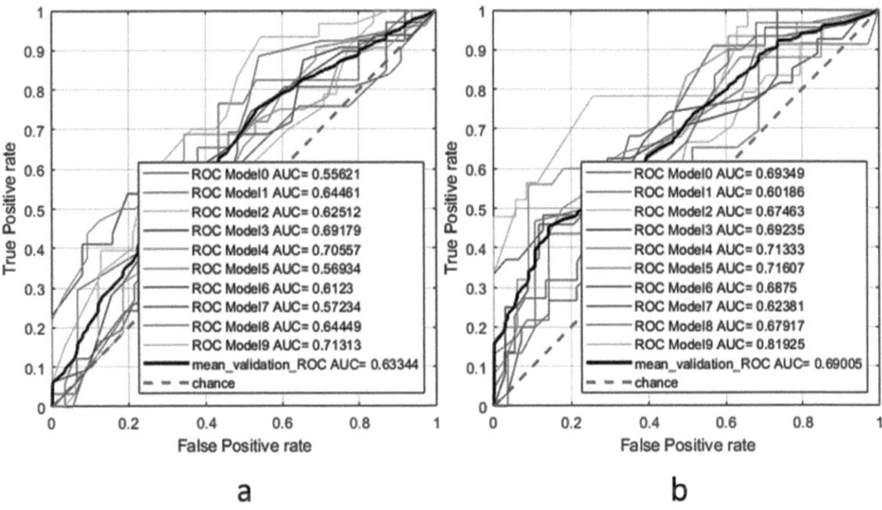

Fig. 4. Roc Curve KNN Model; a) masses, b) calcifications.

3.3 Confusion Matrix

To provide a detailed view of classification outcomes, confusion matrices were computed for each model. Figures 5, 6 and 7 illustrate the distribution of true positives, false positives, true negatives, and false negatives, offering deeper insight into model performance beyond aggregated metrics.

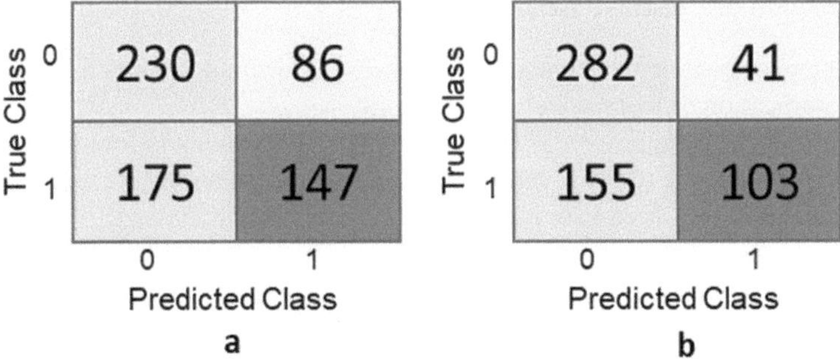

Fig. 5. Confusion Matrix for LDA model: a) masses, b) calcifications.

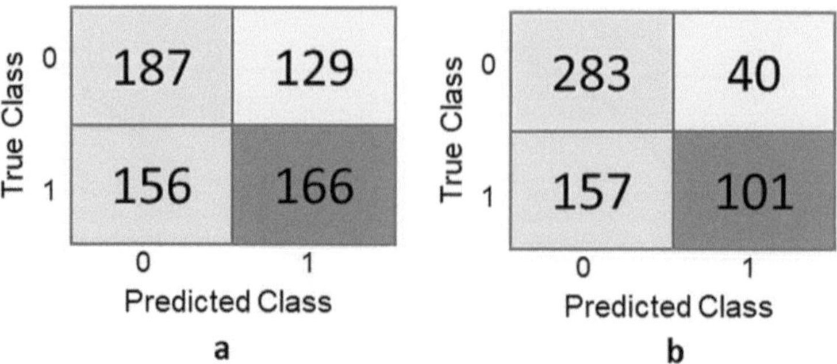

Fig. 6. Confusion Matrix for SVM model: a) masses, b) calcifications.

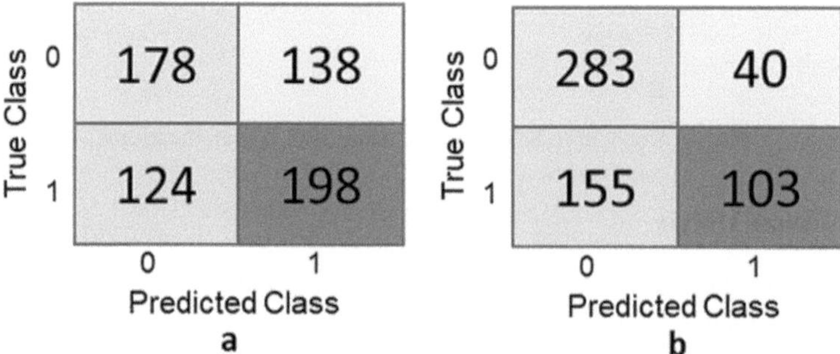

Fig. 7. Confusion Matrix for KNN model: a) masses, b) calcifications.

3.4 Discussion of Classifier Behavior

The performance comparison across classifiers reveals distinct patterns in their behavior for the classification of masses and calcifications.

LDA achieved the highest recall in both lesion types, particularly for masses (72.8%), suggesting its tendency to favor sensitivity and minimize false negatives. This is supported by its relatively high true positive rates, especially in the case of calcifications (TP = 282), although this came at the cost of a higher number of false positives (FP = 155).

SVM, while slightly more balanced in its confusion matrix structure, showed lower overall performance in terms of F1-score and accuracy, particularly in masses (F1 = 56.7%). Its relatively high number of false negatives (FN = 129) in this category indicates a tendency to under-detect true lesions, impacting its clinical reliability.

KNN with k = 75, selected based on empirical testing, showed the most balanced performance overall for masses, with the highest accuracy (60%) and a favorable trade-off between precision and recall. For calcifications, all three classifiers performed similarly in terms of F1-score (~74%), with minor variations in true/false positives and negatives, suggesting that this task is generally less challenging for all models when compared to mass detection.

Notably, across classifiers and lesion types, calcification detection consistently achieved higher recall and F1 scores, indicating that the extracted radiomics features are more discriminative for this class. This may reflect the more distinct textural or morphological patterns of calcifications compared to masses.

4 Discussions

This study confirms the feasibility of applying classical machine learning algorithms to radiomics-based classification of mammographic lesions, focusing on the CBIS-DDSM dataset. Through comparative evaluation of three supervised classifiers—LDA, SVM, and KNN—across the tasks of mass and calcification classification, we observed significant variability in model behavior and performance.

LDA demonstrated the most favorable recall values, particularly for masses, indicating its strength in correctly identifying malignant lesions. However, its higher false-positive rates suggest a trade-off between sensitivity and specificity.

KNN (with k = 75) yielded the best overall accuracy in mass classification (60%), indicating a more balanced decision boundary.

Conversely, SVM showed comparatively lower performance, especially in masses, possibly due to the limited discriminative capacity of the feature space for this lesion type. All classifiers achieved substantially higher performance in detecting calcifications than masses, with F1-scores around 74% and AUC values close to 70%. This discrepancy suggests that calcifications present more distinctive radiomics patterns than masses, which are typically more heterogeneous in appearance. Such heterogeneity likely contributes to reduced classifier performance on mass classification—a finding consistent with prior work [25].

Despite the rigorous 10-fold cross-validation approach adopted to ensure generalizability, the overall classification results, particularly for masses, remain modest (AUC ~62% for LDA). These findings are below those reported in prior literature on other datasets, where LDA reached AUCs up to 88% [26]. However, comparisons should be made cautiously, as dataset differences—especially in segmentation quality, lesion types, and annotation protocols—significantly influence performance outcomes [27].

In addition, as shown in similar research, it will be necessary to test other ML classifiers to compare and validate or not the results of the proposed models, and define future optimal directions [28].

These challenges emphasize the need for more advanced models capable of learning hierarchical features directly from the imaging data. In light of these findings, future work will focus on deep learning (DL) methods, which have shown promising results in similar contexts. Recent studies have reported AUCs exceeding 89% using DL pipelines on the CBIS-DDSM dataset [28], and even higher when combining CNNs with classical classifiers like SVM (up to 98.46% AUC on curated subsets) [29, 30]. Such hybrid architectures may overcome the limitations of handcrafted features and better capture the complex visual patterns characteristic of mammographic lesions.

5 Conclusions

In conclusion, while traditional machine learning classifiers offer an interpretable and computationally efficient baseline, the integration of DL models emerges as a necessary progression to achieve clinically meaningful performance in breast lesion classification.

This shift reflects the broader trend toward AI-driven decision support systems in medical imaging. The combination of DL architectures with classical ML approaches holds promise for enhancing diagnostic accuracy while reducing dependence on handcrafted feature engineering. Such advancements could significantly improve clinical workflows and decision-making in breast cancer diagnostics.

Disclosure of Interests. The authors have no competing interests to declare that are relevant to the content of this article.

References

1. Benson, J.R., Jatoi, I., Keisch, M., Esteva, F.J., Makris, A., Jordan, V.C.: Early breast cancer. Lancet **373**, 1463–1479 (2009). https://doi.org/10.1016/S0140-6736(09)60316-0
2. Lee, S.E., Yoon, J.H., Son, N.-H., Han, K., Moon, H.J.: Screening in patients with dense breasts: comparison of mammography, artificial intelligence, and supplementary ultrasound. Am. J. Roentgenol. **222** (2024). https://doi.org/10.2214/AJR.23.29655
3. Bancroft, A., Santa Cruz, J., Levett, K., Nguyen, Q.D.: Incidental breast hemangioma on breast MRI: a case report. Cureus (2024). https://doi.org/10.7759/cureus.57903
4. Rizzo, S., et al.: Radiomics: the facts and the challenges of image analysis. Eur. Radiol. Exp. **2** (2018). https://doi.org/10.1186/s41747-018-0068-z
5. Kumar, V., et al.: Radiomics: the process and the challenges. Magn. Reson. Imaging **30**, 1234–1248 (2012). https://doi.org/10.1016/j.mri.2012.06.010
6. Russo, G., et al.: Feasibility on the use of radiomics features of 11[C]-MET PET/CT in central nervous system tumours: preliminary results on potential grading discrimination using a machine learning model. Curr. Oncol. **28**, 5318–5331 (2021). https://doi.org/10.3390/curroncol28060444
7. Gillies, R.J., Kinahan, P.E., Hricak, H.: Radiomics: images are more than pictures. They are data. Radiology **278**, 563–577 (2016). https://doi.org/10.1148/radiol.2015151169
8. Stefano, A.: Challenges and limitations in applying radiomics to PET imaging: possible opportunities and avenues for research. Comput. Biol. Med. **179**, 108827 (2024). https://doi.org/10.1016/j.compbiomed.2024.108827
9. Banna, G.L., et al.: Predictive and prognostic value of early disease progression by PET evaluation in advanced non-small cell lung cancer. Oncology (Switzerland) **92**, 39–47 (2017). https://doi.org/10.1159/000448005
10. Vernuccio, F., et al.: Diagnostic performance of qualitative and radiomics approach to parotid gland tumors: which is the added benefit of texture analysis? Brit. J. Radiol. **94** (2021). https://doi.org/10.1259/bjr.20210340
11. Torrisi, S.E., et al.: Assessment of survival in patients with idiopathic pulmonary fibrosis using quantitative HRCT indexes. Multidiscip. Respir. Med. **13** (2018). https://doi.org/10.1186/s40248-018-0155-2
12. Laudicella, R., et al.: [68Ga]DOTATOC PET/CT radiomics to predict the response in GEP-NETs undergoing [177Lu]DOTATOC PRRT: the theragnomics concept. Cancers (Basel) **14**, 984 (2022). https://doi.org/10.3390/cancers14040984
13. Comelli, A., Stefano, A., Benfante, V., Russo, G.: Normal and abnormal tissue classification in positron emission tomography oncological studies. Pattern Recogn. Image Anal. **28**, 106–113 (2018). https://doi.org/10.1134/S1054661818010054
14. Pasini, G., Bini, F., Russo, G., Comelli, A., Marinozzi, F., Stefano, A.: MatRadiomics: a novel and complete radiomics framework, from image visualization to predictive model. J. Imaging **8** (2022). https://doi.org/10.3390/jimaging8080221

15. Sawyer-Lee, R., Gimenez, F., Hoogi, A., Rubin, D.: Curated breast imaging subset of digital database for screening mammography (CBIS-DDSM). DataCite Commons (2016)
16. Bini, F., et al.: Preclinical implementation of MatRadiomics: a case study for early malformation prediction in Zebrafish model. J. Imaging **10**, 290 (2024). https://doi.org/10.3390/jimaging10110290
17. Stefano, A., et al.: A fully automatic method for biological target volume segmentation of brain metastases. Int. J. Imaging Syst. Technol. **26**, 29–37 (2016). https://doi.org/10.1002/ima.22154
18. Sukassini, M.P., Velmurugan, T.: Noise removal using morphology and median filter methods in mammogram images. In: Proceedings of the 3rd International Conference on Small and Medium Business (2016)
19. Vial, A., et al.: The role of deep learning and radiomic feature extraction in cancer-specific predictive modelling: a review. Transl. Cancer Res. **7**, 803–816 (2018). https://doi.org/10.21037/tcr.2018.05.02
20. Barone, S., et al.: Hybrid descriptive-inferential method for key feature selection in prostate cancer radiomics. Appl. Stoch. Models Bus. Ind. **37**, 961–972 (2021). https://doi.org/10.1002/asmb.2642
21. Li, C., Wang, B.: Fisher linear discriminant analysis. CCIS Northeastern University, 6 (2014)
22. Hearst, M.A., Dumais, S.T., Osuna, E., Platt, J., Scholkopf, B.: Support vector machines. IEEE Intell. Syst. Appl. **13**(4), 18–28 (1998)
23. Comelli, A., Stefano, A., Benfante, V., et al.: Normal and abnormal tissue classification in positron emission tomography oncological studies. Pattern Recogn. Image Anal. **28**, 106–113 (2018). https://doi.org/10.1134/S1054661818010054
24. Kramer, O., Kramer, O.: K-nearest neighbors. In: Dimensionality Reduction with Unsupervised Nearest Neighbors, pp. 13–23 (2013)
25. Michael, E., Ma, H., Li, H., Qi, S.: An optimized framework for breast cancer classification using machine learning. Biomed. Res. Int. **2022**, 2022. https://doi.org/10.1155/2022/8482022
26. Adebiyi, M.O., Arowolo, M.O., Mshelia, M.D., Olugbara, O.O.: A linear discriminant analysis and classification model for breast cancer diagnosis. Appl. Sci. **12**, 11455 (2022). https://doi.org/10.3390/app122211455
27. Comelli, A., et al.: A smart and operator independent system to delineate tumours in Positron Emission Tomography scans. Comput. Biol. Med. **1**(102), 1–15 (2018). https://doi.org/10.1016/j.compbiomed.2018.09.002. Epub 2018 Sep 8 PMID: 30219733
28. Laudicella, R., et al.: [68Ga]DOTATOC PET/CT radiomics to predict the response in GEP-NETs undergoing [177Lu]DOTATOC PRRT: the "Theragnomics" concept. Cancers **14**, 984 (2022). https://doi.org/10.3390/cancers14040984
29. Gerbasi, A., et al.: DeepMiCa: automatic segmentation and classification of breast MIcroCAlcifications from mammograms. Comput. Methods Programs Biomed. **235**, 107483 (2023). https://doi.org/10.1016/j.cmpb.2023.107483
30. Wang, L.: Mammography with deep learning for breast cancer detection. Front. Oncol. **14**, 1281922 (2024). https://doi.org/10.3389/fonc.2024.1281922
31. Thirumalaisamy, S., et al.: Breast cancer classification using synthesized deep learning model with metaheuristic optimization algorithm. Diagnostics **13**, 2925 (2023). https://doi.org/10.3390/diagnostics13182925

Variational Radiomics for Active Surfaces

Rosario Corso[1(✉)] , Haoyue Chen[2] , Albert Comelli[3] ,
and Anthony Yezzi[2]

[1] Dipartimento di Matematica e Informatica, Università degli Studi di Palermo, Via
Archirafi 34, Palermo, Italy
`rosario.corso02@unipa.it`
[2] School of Electrical and Computer Engineering, Georgia Institute of Technology,
Atlanta, GA, USA
`{hchen765,anthony.yezzi}@gatech.edu`
[3] Ri.MED Foundation, Via Bandiera 11, 90133 Palermo, Italy
`acomelli@fondazionerimed.com`

Abstract. We show how radiomics features can be integrated
directly into the data driven terms of a variational active surface using
a novel region based strategy in which the segmentation goal will be
to match learned values of various radiomics features both inside and
outside the region of interest to be localized. We do not merely utilize a
one-time, up-front radiomics feature calculation to generate preprocessed
inputs for a subsequent variational active surface. Rather, we exploit the
variational properties of the radiomics features themselves in the active
surface formulation. We demonstrate the formulation of the corresponding energy functional as well as its gradient flow PDE for an illustrative
selection of common region based radiomics features.

Keywords: Image Segmentation · Radiomics · Active Contours ·
Active Surfaces

1 Introduction

1.1 Active Contours and Surfaces

Over the past decades, active contour and active surface methods have received
a significant amount of attention as a flexible method for image segmentation
in which a dynamic contour (or surface in 3D) evolves towards the boundaries
of a nearby object of interest. Variational active contour methods in particular employ a robust mathematical formulation where the contour evolution is
expressed as a continuous gradient flow in the form of a partial differential equation derived via the *Calculus of Variations* to minimize a given *Energy Functional*. The energy functional in turn is a cost function designed to yield smaller
values for preferred contour configurations. The designed energy functional will
typically include both data driven terms, such as local edge strength along the
contour points or global statistics computed inside and out side the contour,

E. Rodolà et al. (Eds.): ICIAP 2025 Workshops, LNCS 16169, pp. 210–221, 2026.
https://doi.org/10.1007/978-3-032-11317-7_18

as well as regularizing terms, such as penalties on excessive total arclength or curvature.

Several variational active contours methods have been proposed including edge-driven models (with some of the earliest examples being the original *Snakes* method and its variants in [7,15], their more geometric counterparts, *Geodesic Active Contours*, in [5,25]), as well as region based models [20,21,26] including the highly popular Chan-Vese technique [6] which specializes the more general Mumford-Shah model [19,22,23], and even hybrid strategies intended to be tunable between local and global feature sensitiviy [18]. Active surfaces, the 3D counterpart of active contours, have become especially prominent in medical imaging applications and can even be extended into 4D for time-evolving 3D data (for example in cardiac imaging [1,16]). For a more in-depth coverage of the topic of active contours we suggest [3] as well as some surveys [8,9].

1.2   Radiomics

Radiomics is an emerging interdisciplinary field that embraces medicine, informatics and statistics with the aim to provide quantitative information from medical images and support doctors decisions about diagnosis, prognosis and specific treatments [10,17]. With numerical values in hands, it is possible to have important information that may not be recognizable through a traditional visualization. Specifically, the steps of a typical radiomics analysis consist of image acquisition and pre-processing, extraction of features from regions of interest after a segmentation procedure, feature reduction/selection and classification.

Common radiomics features include first-order statistics, shape properties, and descriptors of co-occurrence matrix or other type of matrices. Moreover, *layered* radiomics features can be obtain as well by computing standard radiomics features on the output of various preprocessing filters applied to the original input data (such as wavelet transforms or gradient filtering) [2,4,24]. In order to standardize the feature set and ensure reproducibility, a project called Image Biomarker Standardization Initiative (IBSI) [24,27] was created.

1.3   Contribution: Variational Radiomics for Active Contours/Surfaces

In this work we show how radiomics features can be integrated directly into the data driven terms of a variational active contour or surface using a novel region based strategy in which the segmentation goal will be to match learned values of various radiomcs features both inside and outside the region of interest to be localized. We will demonstrate the full formulation of the energy functional as well as the gradient flow PDE for a illustrative selection of common region based radiomics features. Note that by incorporating the radiomics features directly into the energy functional, our method exploits not only the values of the radiomics features themselves, but also their own variational properties in connection with the evolving contour.

While the vast body of existing work on active contour frameworks can readily be leveraged by introducing preprocessing inputs which depend upon an upfront, one-time calculation of radiomics features (see [12] for a successful recent approach, and [8] for a review of preprocessing-style methods involving general features), we rather seek a framework which can directly incorporate dynamical changes in region-dependent radiomics features inside and outside the evolving contour as it deforms during the optimization process in order to more precisely localize transition boundaries between the exterior and interior features. Such a strategy requires continual recomputation of the radiomics features along with the evolving contour so that both respond to feedback from the other along the path toward the converged optimal solution.

The framework we will present here will be quite general regarding both the choice of radiomics features as well as the dimension of the image data, and may be used to leverage a large class of radiomics based features. While we will specifically develop and demonstrate the strategy in this initial work for 3D active surfaces on 3D image data with a limited number of illustrative examples of region-dependent radiomics features, the mathematical expressions and numerical implementations will be similar for active contours on 2D image data where the same as well as many other radiomics features may be equally exploited.

2 Method

2.1 Energy Functional Term Based on Radiomics Features

In this section we outline a general framework to exploit region based radiomics features and their variational properties with respect to an active surface in the energy functional. The specific choice of the radiomics features chosen here are only meant to be illustrative of a more general strategy in which region dependent radiomics features, together with a choice of distance measure to compare differences in learned feature values versus detected feature values, are used to drive an active surface specifically based on the variational behavior of the radiomics features themselves. Since radiomics is a field related mainly to 3D images, we illustrate the approach for active surfaces, but the active contour version for 2D images can be formulated with obvious modifications.

Given a radiomics feature Y with region-dependent value y, computed within the interior region R_{int} an active surface S, we define the energy of S as

$$E(S) = E\big(y(S)\big) = \frac{(y - \eta)^2}{2\Sigma}, \tag{1}$$

where η and Σ are the mean and variance of the feature Y calculated a-priori over a set of 3D training image data and Regions Of Interest (ROIs). The two parameters η and Σ are considered fixed (independent of the active surface S). The defined energy is small if the value of Y is near its mean, with the nearness scale depending upon the variance Σ. Note, $\dfrac{y - \eta}{\Sigma^{\frac{1}{2}}}$ represents a normalized feature, the so-called *z-score* of y. Hence, the idea behind (1) is that minimizing E

implies moving the surface S towards an a surface whose interior region yields a radiomics feature value y similar to the 'reference' surfaces whose radiomics features were measured and exploited in training.

The variation δE of E with respect to a varation δS of the active surface can be expressed in terms of the varations δy of the radiomics features y as follows

$$\delta E = \frac{y - \eta}{\Sigma}\, \delta\, y = \frac{y - \eta}{\Sigma} \underbrace{\int_S f_Y\,(\delta S \cdot N)\, dS}_{\delta y},$$

where f_Y represents the normal sensitivity of the region dependent feature Y with respect to a surface perturbation δS and where $N(s)$ denotes the outward unit normal for S.

A contribution from exterior region R_{ext} outside S can be also taken into account. Specifically, if Y_{int} and Y_{ext} are features R_{int} and R_{ext} and with values y_{int} and y_{ext}, respectively, we can define the surface energy

$$E(y_{int}, y_{ext}) = b_{int}\frac{(y_{int} - \eta_{int})^2}{2\Sigma_{int}} + b_{ext}\frac{(y_{ext} - \eta_{ext})^2}{2\Sigma_{ext}}, \tag{2}$$

where b_{int} and b_{ext} are non-negative biases. Therefore,

$$\delta E = b_{int}w_{int}\delta y_{int} - b_{ext}w_{ext}\delta y_{ext}$$

$$= b_{int}w_{int}\int_S f_{int}(s)(\delta S \cdot N)\, dS - b_{ext}w_{ext}\int_S f_{ext}(s)(\delta S \cdot N)\, dS$$

with $w_{int} = \Sigma_{int}^{-1}(y_{int} - \eta_{int})$ and $w_{ext} = \Sigma_{ext}^{-1}(y_{ext} - \eta_{ext})$. In conclusion,

$$\nabla_S E(s) = (b_{int}w_{int}(s)f_{int}(s) - b_{ext}w_{ext}(s)f_{ext}(s))N(s).$$

As common, it is possible to add a regularization term in (2) based on a penalty on the surface area of S.

2.2   List of Radiomics Features

As an illustrative example, we consider the following radiomics features computed over the *volume* V of the region R either inside or outside S

$$V = \int_R dx$$

(i.e. R is either R_{int} or R_{ext}) and where I denotes the pointwise image intensity.

$$\text{Mean} \qquad \mu = \frac{1}{V} \int_R I \, dx$$

$$\text{Variance} \qquad \sigma^2 = \frac{1}{V} \int_R (I - \mu)^2 \, dx$$

$$\text{Standard Deviation} \qquad \sigma$$

$$\text{Skewness} \qquad \beta = \frac{1}{V} \int_R \left(\frac{I - \mu}{\sigma} \right)^3 dx$$

$$\text{Kurtosis} \qquad \kappa = \frac{1}{V} \int_R \left(\frac{I - \mu}{\sigma} \right)^4 dx$$

$$\text{Energy} \qquad \xi = \int_R I^2 \, dx$$

$$\text{Root Mean Squared} \qquad \text{rms} = \sqrt{\frac{\xi}{V}}$$

along with the 10th Percentile, Median (50th Percentile), 90th Percentile of the intensity inside R.

2.3   Examples of Shape Gradients

The shape gradients (variational derivatives with respect to the surface S) of the radiomics features described above may be computed as follows. First we define $J = \int_R I \, dx$ and $Q = \int_R I^2 \, dx$ and obtain

$$\delta V = \int_S (\delta S \cdot N) \, dS, \quad \delta J = \int_S I(\delta S \cdot N) \, dS, \quad \delta Q = \int_S I^2(\delta S \cdot N) \, dS,$$

Thus, the first variations of the mean, variance, and standard deviation are

$$\delta\mu = \frac{\delta J - \mu \delta V}{V} = \int_S \frac{I - \mu}{V}(\delta S \cdot N) \, dS,$$

$$\delta\sigma^2 = \frac{\delta Q - \frac{Q}{V}\delta V}{V} - 2\mu\delta\mu = \int_S \frac{(I - \mu)^2 - \sigma^2}{V}(\delta S \cdot N) \, dS,$$

$$\delta\sigma = \frac{\delta\sigma^2}{2\sigma} = \int_S \frac{\sigma}{2} \frac{(\frac{I-\mu}{\sigma})^2 - 1}{V}(\delta S \cdot N) \, dS.$$

which may be plugged into (2) to obtain the corresponding gradient flows for S.

3   Experiments

To assess the proposed method we conducted several tests on a private dataset made of 59 CT images regarding the anatomical district of the left heart ventricle and corresponding ground truths segmented by a radiologist. We divided the dataset in a training set of 47 images and a test set of remaining 12 images, following a random partition with ratio about 4:1. Then, we extracted the features with PyRadiomics software [11], conform to the Image Biomarker Standardization Initiative, on the images of the training set. Finally, we calculated the mean and variance of each feature and used this fixed value in the surface energy expression during the tests.

As one can see in Fig. 1a the part external to the ROI (left ventricle) is heterogeneous, thus the exterior feature values differ a lot from an image to another one. To consider a more homogeneous region, we restrict the domain of the image to a fixed dilation of the initial mask. This means that in the training phase, when the feature averages and variances are calculated, the interior region is the ground truth region and the exterior region is the difference between a dilation and the original ground truth (Fig. 1b shows an example of the regions). In the test phase, the interior is the region inside the evolving surface and the exterior region is the difference between a fixed dilation of the initial mask and the evolving mask. The dilation is made uniformly to all cases.

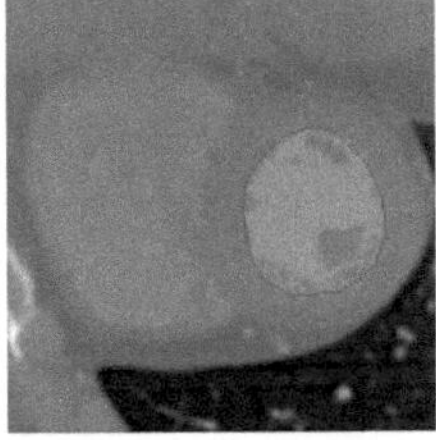

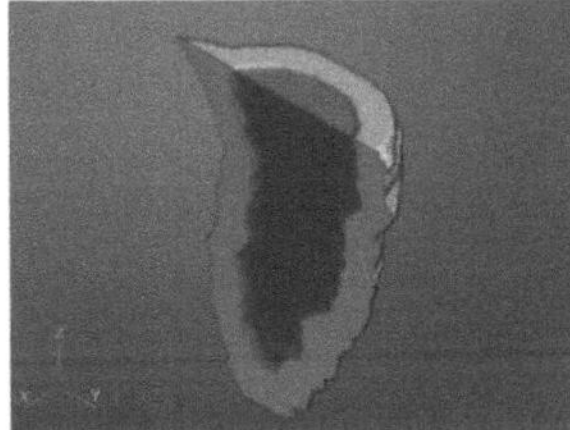

(a) A slice of a CT image of the dataset showing the region of interest surrounded by a red boundary.

(b) An example of internal region (dark gray), corresponding to the ground truth, and exterior region (light gray). The sections are made for a better view.

Fig. 1. Two figures related to our experiments.

As typical, the results of active surface segmentation depend on the initial configuration. To start with initial surfaces close enough to the ground truths, we considered masks obtained by thresholding inside a bounding box. In particular, we first selected the threshold value that produced masks with the best overlapping with the ground truths of the training set. Then, using this value we applied the thresholding on the images of the test set to create the initial masks (we also performed a final filling of the holes).

Moreover, the number of iterations was set equal for each test. The regularization term sometimes improved the results, but with small differences. Therefore, here we decided to show the outcomes of the tests without regularization term.

Concerning the evaluation, we made use of the following classical metrics to rate the performance of the segmentation method:

Dice similarity coefficient

$$DSC(S_1, S_2) = \frac{2|R_1 \cap R_2|}{|R_1| + |R_2|},$$

Average surface distance

$$ASD(S_1, S_2) = \frac{\sum_{s_1 \in S_1} \min_{s_2 \in S_2} d(s_1, s_2) + \sum_{s_2 \in S_2} \min_{s_1 \in S_1} d(s_1, s_2)}{|S_1| + |S_2|},$$

where R_1 is the region (with discretized boundary S_1) segmented by the active surface method, R_2 is the ground truth region (with boundary discretized S_2), $d(\cdot, \cdot)$ is the Euclidean distance and the brackets $|\cdot|$ indicate the voxel number. In this paper, the *DSC* is expressed in percentage, while the *ASD* in mm. The average *DSC* and *ASD* of the initial masks were 90.86% and 1.74 mm, respectively.

Table 1. Performance results considering a single feature on the interior or exterior region and a thresholding mask as initial surface. The names in bold indicate the features that improved the initial segmentation according to the Dice coefficient.

Feature of the interior region	Feature of the exterior region	Dice coefficient	Average distance
Kurtosis		$92.00 \pm 2.81\%$	1.48 ± 0.52
Standard Deviation		$91.87 \pm 3.83\%$	1.52 ± 0.84
Skewness		$91.80 \pm 2.81\%$	1.51 ± 0.49
Variance		$91.72 \pm 4.17\%$	1.57 ± 0.91
10th Percentile		$90.69 \pm 3.14\%$	1.75 ± 0.61
Energy		$74.38 \pm 10.67\%$	5.79 ± 2.57
Volume		$72.91 \pm 24.61\%$	21.99 ± 58.07
Mean		$70.34 \pm 31.50\%$	4.46 ± 3.65
Root Mean Squared		$65.78 \pm 31.21\%$	24.63 ± 59.88
Median		$59.28 \pm 34.20\%$	27.18 ± 58.99
90th Percentile		$30.61 \pm 24.01\%$	29.41 ± 57.81
	90th Percentile	$91.01 \pm 2.91\%$	1.64 ± 0.56
	Root Mean Squared	$90.80 \pm 3.20\%$	1.75 ± 0.65
	Standard Deviation	$90.52 \pm 3.59\%$	1.78 ± 0.65
	Variance	$90.46 \pm 3.64\%$	1.79 ± 0.65
	Mean	$89.57 \pm 2.92\%$	1.94 ± 0.59
	Energy	$88.28 \pm 5.92\%$	2.24 ± 1.12
	Kurtosis	$85.74 \pm 8.35\%$	2.71 ± 1.71
	Median	$83.04 \pm 9.25\%$	3.34 ± 1.88
	Skewness	$82.91 \pm 9.04\%$	3.30 ± 1.92
	Volume	$81.33 \pm 15.08\%$	3.73 ± 3.26
	10th Percentile	$74.56 \pm 21.43\%$	4.33 ± 2.66

Table 2. Performance results considering a single feature on each of the interior and the exterior region. Only the best eight combinations are shown. The optimal bias ratio depends on the choice of features and it was varied between some values of different scale (see also Table 3).

Feature of the interior region	Feature of the exterior region	Dice coefficient	Average distance
Kurtosis	90th Percentile	92.35 ± 2.45%	1.41 ± 0.44
Standard Deviation	90th Percentile	92.28 ± 3.33%	1.48 ± 0.82
Variance	90th Percentile	92.17 ± 3.76%	1.50 ± 0.88
Skewness	90th Percentile	92.14 ± 2.55%	1.44 ± 0.45
Kurtosis	Variance	92.10 ± 2.95%	1.46 ± 0.56
Standard Deviation	Standard Deviation	92.08 ± 3.58%	1.48 ± 0.80
Skewness	Variance	92.06 ± 2.75%	1.45 ± 0.47
Variance	Standard Deviation	92.03 ± 3.94%	1.51 ± 0.88

Table 3. Performance results considering Kurtosis as internal feature and 90th Percentile as external feature and different values of bias ratio $\frac{b_{ext}}{b_{int}}$. The optimal value is indicated with the symbol (∗).

Bias ratio	Dice coefficient	Average distance
10^{-4}	91.96 ± 2.76%	1.49 ± 0.52
10^{-3}	91.88 ± 2.70%	1.51 ± 0.49
10^{-2}	91.83 ± 2.60%	1.52 ± 0.47
(∗) 10^{-1}	92.35 ± 2.45%	1.41 ± 0.44
10^{0}	91.35 ± 2.29%	1.58 ± 0.42
10^{1}	90.96 ± 2.40%	1.65 ± 0.44
10^{2}	90.93 ± 2.33%	1.66 ± 0.43
10^{3}	91.21 ± 2.55%	1.60 ± 0.49
10^{4}	91.18 ± 2.76%	1.61 ± 0.53

Table 1 contains the segmentation results applying the method with one only feature (on the interior or in the exterior region) different for any test. The first and second numbers in the columns of evaluation metrics represent the average and the standard deviation of the values over the test set, respectively. Four features (kurtosis, standard deviation, skewness and variance) on the internal region improved the initial segmentation. About the external part, only one feature (90th percentile) slightly improved the initial segmentation.

The method gives a further enhancement with the simultaneous contributions of features on both regions. Tables 2 and 3 show some of these cases.

To visualize the surface evolution and the convergence of the feature values we collected the segmentation and values at the beginning and after 200, 400, 800 iterations in Table 4.

Table 4. The pictures show the evolution of the surface an image (through two different slices). The chosen features were Kurtosis for the interior region and 90th Percentile for the exterior region, with a bias ratio equals to 10^{-1}. No regularization term was considered. To compare the final segmentation, the ground truth is visualized in the last row. In the last column it is possible to see that, increasing the iterations, the feature values tend to approach the corresponding means (3.90 for Kurtosis and 227.75 for 90th Percentile), calculated in the training phase.

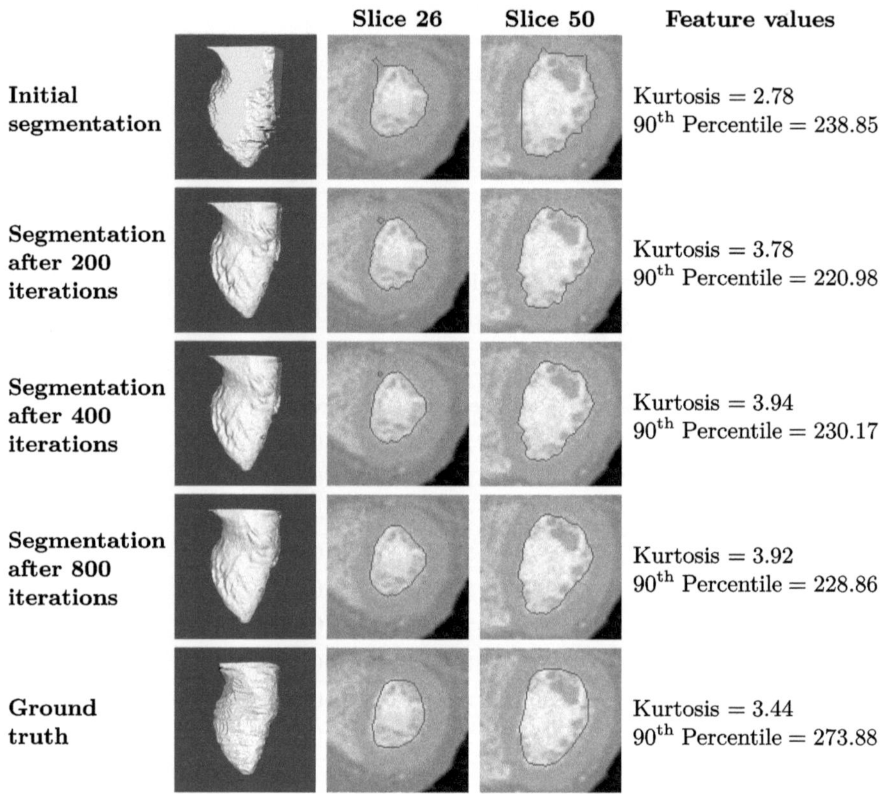

	Slice 26	Slice 50	Feature values
Initial segmentation			Kurtosis = 2.78 90th Percentile = 238.85
Segmentation after 200 iterations			Kurtosis = 3.78 90th Percentile = 220.98
Segmentation after 400 iterations			Kurtosis = 3.94 90th Percentile = 230.17
Segmentation after 800 iterations			Kurtosis = 3.92 90th Percentile = 228.86
Ground truth			Kurtosis = 3.44 90th Percentile = 273.88

4 Conclusion

In this paper we showed how it is possible to directly integrate radiomics into level-set active contour/surface framework and presented a case study. As one can see in Table 1 the choice of features is crucial.

We achieved the goal by considering single features on the interior and exterior region. Ideally, more features give more information, although this is not the general situation and strictly depends on how the features interact between them [14]. In an attempt to extend the setting to multiple features in each region, and then to further improve the results, we are currently considering the Mahalanobis distance [13], a statistic metric giving a unique number from multiple information. The higher the distance is, the more unlikely a datum comes from a given distribution.

To recall the definition, let Y be a probability distribution on $\mathbb{R}^k$ with mean η and non-singular covariance matrix Σ. The number

$$\mathsf{d}(y) = \sqrt{(y - \eta)^T \Sigma^{-1}(y - \eta)}$$

is called the *Mahalanobis distance* of a vector $y \in \mathbb{R}^k$ from η with respect to Y.

Given a vector of features Y_{int} with value $y_{int} = (y_{int}^1, \ldots, y_{int}^n)$ in R_{int} and a vector of features Y_{ext} with value $y_{ext} = (y_{ext}^1, \ldots, y_{ext}^m)$ in R_{ext}, we define the energy of S as

$$E(S) = E(y_{int}, y_{ext}) = \frac{b_{int}}{2}\mathsf{d}_{int}(y_{int})^2 + \frac{b_{ext}}{2}\mathsf{d}_{ext}(y_{ext})^2, \tag{3}$$

denoting with d_{int} and d_{ext} the Mahalanobis distances with respect to interior and exterior features Y_{int} and Y_{ext}, respectively. Note that when only one feature is considered in each region, the energy expression in (3) reduces exactly to (2). Writing

$$\delta y_{int} = \int_S f_{int}(s)(\delta S \cdot N)\, dS, \qquad \delta y_{ext} = \int_S f_{ext}(s)(\delta S \cdot N)\, dS,$$

where $f_{int}(s) = (f_{int}^1(s), \ldots, f_{int}^n(s))$ and $f_{ext}(s) = (f_{ext}^1(s), \ldots, f_{ext}^m(s))$, we have

$$\delta E = b_{int}w_{int} \cdot \delta y_{int} - b_{ext}w_{ext} \cdot \delta y_{ext}$$

$$= b_{int}w_{int}\int_S f_{int}(s)(\delta S \cdot N)\, dS - b_{ext}w_{ext}\int_S f_{ext}(s)(\delta S \cdot N)\, dS$$

with $w_{int} = \Sigma_{int}^{-1}(y_{int} - \eta_{int})$ and $w_{ext} = \Sigma_{ext}^{-1}(y_{ext} - \eta_{ext})$. Thus,

$$\nabla_S E(s) = (b_{int}w_{int}(s) \cdot f_{int}(s) - b_{ext}w_{ext}(s) \cdot f_{ext}(s))N(s).$$

The idea to make use of the Mahalanobis distance is still in progress, indeed some tests suggest the need of understanding how to choose appropriate features. For instance, we took together two of the best features for the interior region, kurtosis and skewness, and with the Mahalanobis formulation we obtained an average Dice coefficient equals to 91.37%, which is below the values of both single features.

Acknowledgments. R.C. is a member of the "Gruppo Nazionale per l'Analisi Matematica, la Probabilità e le loro Applicazioni" (INdAM). A. Y. and H. C. were supported by Army Research Office grant W911NF-22-1-0267.

References

1. Abufadel, A., Yezzi, A., Schafer, R.: 4D segmentation of cardiac data using active surfaces with spatiotemporal shape priors. Springer Berlin Heidelberg (2008)
2. Au, R.C., Tan, W.C., Bourbeau, J., Hogg, J.C., Kirby, M.: Impact of image pre-processing methods on computed tomography radiomics features in chronic obstructive pulmonary disease. Phys. Med. Biol. **66**(24) (2021). https://doi.org/10.1088/1361-6560/ac3eac
3. Blake, A., Isard, M.: Active contours: the application of techniques from graphics, vision, control theory and statistics to visual tracking of shapes in motion. Springer Science & Business Media (2012)
4. Carré, A., Klausner, G., Edjlali, M., Lerousseau, M., et al.: Standardization of brain MR images across machines and protocols: bridging the gap for MRI-based radiomics. Sci. Rep. **10**(1) (2020). https://doi.org/10.1038/s41598-020-69298-z
5. Caselles, V., Kimmel, R., Sapiro, G.: Geodesic active contours. Int. J. Comput. Vision **22**, 61–79 (1997). https://doi.org/10.1023/A:1007979827043
6. Chan, T., Vese, L.: Active contours without edges. IEEE Trans. Image Process. **10**, 266–277 (2001). https://doi.org/10.1109/83.902291
7. Cohen, L.D.: On active contour models and balloons. CVGIP: Image Understanding **53** (1991). https://doi.org/10.1016/1049-9660(91)90028-N
8. Corso, R., Khan, F., Yezzi, A., Comelli, A.: Features for active contour and surface segmentation: a review. Arch. Comput. Methods Eng. (2025). https://doi.org/10.1007/s11831-025-10300-0
9. Cremers, D., Rousson, M., Deriche, R.: A review of statistical approaches to level set segmentation: integrating color, texture, motion and shape. Int. J. Comput. Vision **72**, 195–215 (2007)
10. Gillies, R.J., Kinahan, P.E., Hricak, H.: Radiomics: images are more than pictures, they are data. Radiology **278**(2), 563–577 (2016). https://doi.org/10.1148/radiol.2015151169
11. Griethuysen, J.J.V., et al.: Computational radiomics system to decode the radiographic phenotype. Cancer Res. **77** (2017). https://doi.org/10.1158/0008-5472.CAN-17-0339
12. Guo, Z., et al.: RFLSE: joint radiomics feature-enhanced level-set segmentation for low-contrast SPECT/CT Tumour images. IET Image Proc. **18**(10), 2715–2731 (2024). https://doi.org/10.1049/ipr2.13130
13. Applied Multivariate Statistical Analysis. Springer, Cham (2019). https://doi.org/10.1007/978-3-030-26006-4
14. Iguyon, I., Elisseeff, A.: An introduction to variable and feature selection. J. Mach. Learn. Res. **3** (2003)
15. Kass, M., Witkin, A., Terzopoulos, D.: Snakes: active contour models. Int. J. Comput. Vision **1**, 321–331 (1988). https://doi.org/10.1007/BF00133570
16. Kohlberger, T., Cremers, D., Rousson, M., Ramaraj, R., Funka-Lea, G.: 4D shape priors for a level set segmentation of the left myocardium in spect sequences. In: Proceedings Medical Image Computing and Computer-Assisted Intervention, pp. 92–100. IEEE (2006)
17. Lambin, P., Rios-Velazquez, E., Leijenaar, R., Carvalho, S., et al.: Radiomics: extracting more information from medical images using advanced feature analysis. Eur. J. Cancer **48**(4), 441–446 (2012). https://doi.org/10.1016/j.ejca.2011.11.036

18. Li, H., Yezzi, A.: Local or global minima: flexible dual-front active contours. IEEE Trans. Pattern Anal. Mach. Intell. **29** (2007). https://doi.org/10.1109/TPAMI.2007.250595
19. Mumford, D., Shah, J.: Optimal approximations by piecewise smooth functions and associated variational problems. Commun. Pure Appl. Math. **42**, 577–685 (1989). https://doi.org/10.1002/cpa.3160420503
20. Paragios, N., Deriche, R.: Geodesic active regions for supervised texture segmentation. In: Proceedings of the IEEE International Conference on Computer Vision. vol. 2 (1999). https://doi.org/10.1109/iccv.1999.790347
21. Ronfard, R.: Region-based strategies for active contour models. Int. J. Comput. Vision **13** (1994). https://doi.org/10.1007/BF01427153
22. Tsai, A., Yezzi, A., Willsky, A.: Curve evolution implementation of the Mumford-shah functional for image segmentation, denoising, interpolation, and magnification. IEEE Trans. Image Process. **10**, 1169–1186 (2001). https://doi.org/10.1109/83.935033
23. Vese, L.A., Chan, T.F.: A multiphase level set framework for image segmentation using the Mumford and shah model. Int. J. Comput. Vision **50** (2002). https://doi.org/10.1023/A:1020874308076
24. Whybra, P., Zwanenburg, A., Andrearczyk, V., Schaer, R., et al.: The image biomarker standardization initiative: standardized convolutional filters for reproducible radiomics and enhanced clinical insights. Radiology **310**(2) (2024). https://doi.org/10.1148/radiol.231319
25. Yezzi, A., Kichenassamy, S., Kumar, A., Olver, P., Tannenbaum, A.: A geometric snake model for segmentation of medical imagery. IEEE Trans. Med. Imaging **16**(2), 199–209 (1997)
26. Yezzi, A., Tsai, A., Willsky, A.: A fully global approach to image segmentation via coupled curve evolution equations. J. Visual Commun. Image Represent. **13** (2002). https://doi.org/10.1006/jvci.2001.0500
27. Zwanenburg, A., Vallières, M., Abdalah, M.A., Aerts, H.J., et al.: The image biomarker standardization initiative: standardized quantitative radiomics for high-throughput image-based phenotyping. Radiology **295**(2), 328–338 (2020). https://doi.org/10.1148/radiol.2020191145

Radiomic Analysis of Mouse Magnetic Resonance Images and Correlation of Features Extracted to Vascular Defects in Sickle Cell Disease: Preclinical Applications

Viviana Benfante[1]([✉]) [iD], Liana Hatoum[2,3] [iD], Hannah Song Lee[3,5,6] [iD], Edward A. Botchwey[2,3], Manu O. Platt[2,3,5,6] [iD], and Albert Comelli[4] [iD]

[1] Department of Health Promotion, Mother and Child Care, Internal Medicine and Medical Specialties, Molecular and Clinical Medicine, University of Palermo, 90127 Palermo, Italy
vivianabenfante3@gmail.com
[2] Interdisciplinary Bioengineering Graduate Program, Georgia Institute of Technology, Atlanta, GA, USA
[3] Wallace H. Coulter Department of Biomedical Engineering, Georgia Institute of Technology and Emory University, Atlanta, GA, USA
[4] Ri.MED Foundation, Via Bandiera 11, 90133 Palermo, Italy
[5] NIH Center for Biomedical Engineering Technology Acceleration, Bethesda, MD, USA
[6] National Institute of Biomedical Imaging and Bioengineering, National Institutes of Health, Bethesda, MD, USA

Abstract. Radiomics analysis was used in this study to examine whether it could be possible to distinguish the carotid arteries of sickle cell disease mice (SS) from heterozygous controls (AS) using contrast-free carotid magnetic reso-nance imaging (MRA) images. A total of 66 MRA scans of one-month-old mice and 64 scans of three-month-old mice were used. Using PyRadiomics software, PyRadiomics preprocessing parameters critical for reproducibility include: version 3.0 (IBSI-compliant) 112 radiomic features were extracted from segmented carotid ar-tery images, followed by radiomic feature selection, K-fold splitting of the dataset for cross-validation, and training and validation of a predictive mod-el. At 1 month of age, four radiomic features (Original glcm Imc1; Original firstorder Minimum; Original shape MinorAxisLength; Original shape Ma-jorAxisLength) yielded an accuracy of 74% (AUROC of 0.72; 95% C.I. be-tween 0.60 and 0.84 with a p value of 0.032). At three months, a single fea-ture (Original shape MeshVolume) achieved an accuracy of 76% (AUROC of 0.67; 95% CI between 0.61 and 0.87; p-value of 0.0020). This study shows the efficacy of radiomics in distinguishing arterial features between SCD and control mice. A noninvasive and translational method may be able to assess arterial alterations in SCD for prognostic purposes in the manage-ment of this disease also in humans.

E. Rodolà et al. (Eds.): ICIAP 2025 Workshops, LNCS 16169, pp. 222–232, 2026.
https://doi.org/10.1007/978-3-032-11317-7_19

Keywords: Magnetic Resonance Angiography · Sickle Cell Disease · Texture Analysis · Radiomics · Translational · Machine Learning

1   Introduction

Sickle cell disease (SCD), also known as sickle cell anemia, is a group of inherited diseases that affect hemoglobin (Hb), the primary protein responsible for carrying oxygen to red blood cells. Red blood cells with sickle cell disease have an abnormal shape; instead of being disc-shaped, they are crescent-shaped. Because the red blood cells are sickle-shaped, their movement through the blood vessels creates obstructions to blood flow as a result of a mutation in Hb [11].

The prevalence of SCD in the United States is approximately 100,000, with the prevalence being much higher in sub-Saharan Africa, India, and the Middle East; however, in Central Europe, the prevalence of SCD has increased dramatically in recent years [10].

SCD can be accompanied by mild to severe symptoms, such as anemia, paleness, yellow eyes, chronic and acute pain, dark urine, failure to thrive, inflamed hands and feet, and episodes of full-blown strokes and clinically silent cerebral infarctions [13].

Clinical diagnosis has made some advances, including the use of transcranial Doppler (TCD) analysis and hydroxyurea therapy; however, early screening is not yet capable of reducing the risk of potentially irreversible vascular damage leading to stroke. Identifying new potential imaging SCD biomarkers that indicate early stroke risk is challenging.

Radiomics focuses on the quantitative analysis of medical imaging in order to identify parameters that go beyond typical interpretations of radiology and provide insight into diagnostic or prognostic results [6,12]. Radiomics, through normalization and segmentation processes, extrapolates image information in terms of texture, shape, and intensity distributions of tissues [1,4,5]. These features can then be integrated into predictive models [7] for diagnosis and prognosis by utilizing artificial intelligence algorithms based on machine learning.

In MRI data sets obtained without and with contrast augmentation, radiomic data showed a high rate of accuracy in identifying high-risk aneurysms and could be used as a tool for quantifying aneurysm wall enhancement. Although high-risk atherosclerotic plaques [14,18] as well as subtle differences in plaque morphology or risk of ruptured aneurysms are often difficult to detect visually [15], the use of radiomic analysis is capable of detecting them with a high degree of accuracy.

It is widely accepted that the most effective imaging technique for both high quality soft tissue structures and clear vascular delineation is magnetic resonance imaging and angiography (MRI/MRA). As opposed to computed tomography (CT) and positron emission tomography (PET), it does not use ionizing radiation to perform the procedure. Therefore, it is the most appropriate technique for pediatric patients with SCD who require frequent monitoring over time.

Through radiomics, biomarkers imperceptible to the human eye could be detected earlier in medical images, enabling more accurate risk classification and timely treatment.

In this study, we propose to use radiomics analysis on MR images of SCD mice. We examined the discriminating efficacy of SS vs AS genotypes of SCD mouse models through extracted features.

2   Materials and Methods

2.1   Dataset

Around 25 Townes mice (B6;129-Hbatm1(HBA) Tow Hbbtm2 (HBG1, HBB*) Tow/Hbbtm3 (HBG1, HBB) Tow/J) were used to create a humanized sickle cell anemia model. The following genotypes were considered in this study: homozygous for the β-globin S mutation (SS) and heterozygous for the same littermate (AS) in both males and females. Hemoglobin was analyzed by needle-prick sampling of the paw pad to phenotype each mouse for sickle hemoglobin (HbS) and normal human hemoglobin (HbA). The animals were housed in temperature-controlled rooms with a 12-hour light-dark cycle and had *ad libitum* access to food and water. All animal handling procedures during the experiments were in accordance with the Institutional Animal Care and Use Committee (IACUC) protocols at the Georgia Institute of Technology.

All mice underwent *in vivo* magnetic resonance angiography imaging using a 7T MRI system (PharmaScan, BRUKER) with a Doty Scientific radiofrequency coil (internal diameter = 25 mm) and without contrast agent. Animals were anesthetized by isoflurane inhalation and maintained at 1–2%. The mice were placed in a prone position, covered with a warm water blanket to maintain body temperature, and their vital signs were continuously monitored throughout the imaging process.

2.2   MR Image Acquisition and Segmentation

MRI was performed with a three-dimensional Time of Flight (TOF) sequence with a repetition time of 14 ms, an echo time of 2.6 ms, a flip angle of 20°, a slice thickness of 20 mm, a field of view of $20 \times 20 \times 20$ mm, and a resolution of $256 \times 256 \times 128$ pixels. The MRI acquisition time was approximately 13 min. Angiograms were obtained by generating maximum intensity projections (MIPs) of the raw data.

Segmentation and three-dimensional reconstruction of the carotid arteries was carried out using medical image processing software Mimics (Materialise, Leuven, Belgium). In order to make the images as clear as possible, a minimum threshold was set for each case, and the regions of interest were independently segmented using a tool. During the procedure, noise and vessel cleaning were performed, and only the common carotid arteries were smoothed and retained. Centerline measurements of the cross-sectional area along the entire artery were performed with a resolution of 0.15 mm. To avoid bias in the results, 0.5 mm at the carotid bifurcation were removed from the arterial analysis. Figure 1 shows segmented and reconstructed arteries in MRI images of mice between 1 month and 3 months (AS genotype versus SS genotype).

2.3   Radiomics Workflow

A total of 66 MRI images of carotid arteries from 1-month-old mice ($n = 28$ AS; $n = 38$ SS) and 64 MRI images from 3-month-old mice ($n = 33$ AS; $n = 31$ SS) were analyzed. A total of 25 mice were scanned at both 1 month and 3 months of age. First, MR imaging was performed, then the region of interest was contoured, and 3D reconstruction of the carotid arteries' shape was performed.

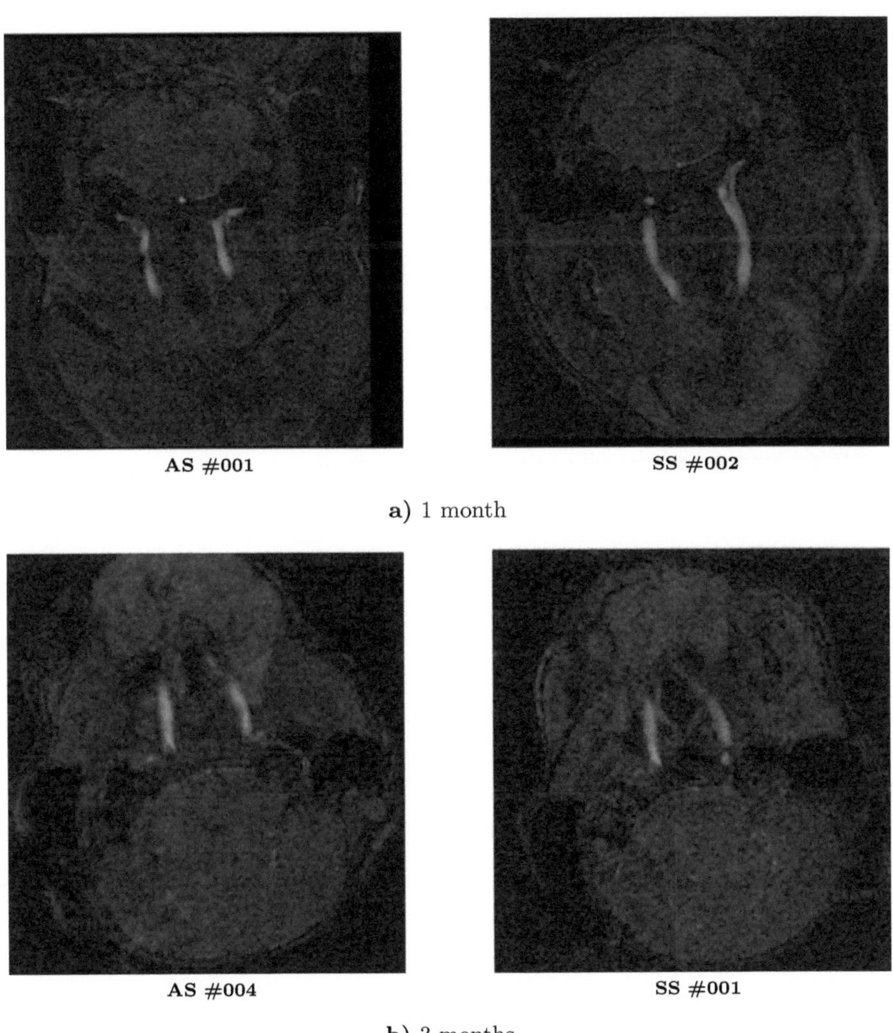

Fig. 1. An example of segmented and reconstructed arteries in MRI images of mice aged at a) 1 month, AS genotype, case 001, and SS genotype, case 002; and at b) 3 months, AS genotype, case 004, and SS genotype, case 001.

Then, a radiomics analysis workflow was set up. The radiomics workflow consists of steps *i*)–*iii*), as follows:

- Radiomics feature extraction;
- Radiomics feature selection;
- K-fold cross-validation and prediction model.

2.3.1 Radiomic Feature Extraction Radiomic feature extraction was conducted using PyRadiomics (3.0), compliant with the Image Biomarker Standardization Initiative (IBSI) [8]. Starting with each segmentation, the volume of each carotid artery was used to automatically extract 112 radiomic features. Among these, third or higher-order texture features were also extracted, in addition to first and second-order ones. In particular, the first-order texture features provide information related to the distribution of the gray level within the ROI, without taking into account the spatial relationships between voxels; the second-order ones take into account the spatial relationships between voxels; and the third-order ones evaluate the spatial relationship be-tween three or more voxels. From the analysis of the gray level histogram, based on mean, variance, skewness, kurtosis and percentiles, the texture parameters were calculated. Contrast, correlation, sum of squares, inverse difference moment [IDM], mean sum, sum of variance, sum of entropy, difference of variance, and difference of entropy were measured for the calculation of the co-occurrence matrix. The run length matrix considered four directions, including run length nonuniformity [RLN], normalized gray level nonuniformity [GLNN], length run emphasis [LRE], and short run emphasis [SRE]. Further details about radiomic features can be found in [17].

2.3.2 Radiomics Feature Selection An operator-independent hybrid statistical method was used to select extracted radiomic features. A discriminant analysis was used to apply the predictive mod-el to the conducted analysis. The most discriminant features along with the re-duction of redundancy among radiomic features with a high correlation were obtained through an innovative sequential mixed descriptive-inferential ap-proach [2] described as follows. The point-biserial correlation coefficients be-tween features and dichotomous variables were calculated, using the reference standard equal to 1 or 0. Subsequently, the features in decreasing order based on the absolute value of the point-biserial correlation coefficient were sorted. A logistic regression analysis was performed by adding one column at a time. Cy-clically, a comparison between the current p-value and the p-value of the previ-ous iteration was performed, until the p-value of the current iteration decreased compared to the previous iteration.

2.3.3 K-Fold Cross-Validation and Predictive Model The predictive model was established as in [2]. The most discriminating features were used for training the predictive model. This was done to compare the carotid arteries in homozygous sickle cell anemia (SS) mice and heterozygous sickle cell anemia

(AS) mice. It was necessary to use the gold standard to provide training input for the classifier. For this step the classification of carotid arteries in homo-zygous sickle cell anemia (SS) mice and heterozygous sickle cell anemia (AS) mice by an experienced researcher was taken as input. For the training and vali-dation phase, the K-fold strategy was adopted. Then, the cases were divided into K-folds. Each fold was used as a validation set and the remaining sets as training sets, for the validation phase. Empirically, a k=5 was considered (K range: 5–15, range: 5), based on a trial and error process. At the end of the training phase, through discriminant analysis it was possible to classify the newly detected radiomic texture. Receiver operating characteristic (ROC) curves with 95% confidence intervals (95% CI) and areas under the ROC curve (AUROC) were calculated to estimate the performance of the most discriminant radiomic features detected on carotid artery images. Sensitivity, specificity and accuracy were calculated. The computing system used to provide computational and statistical analysis was implemented in the Matlab R2019a simulation environment (Math-Works, Na-tick, MA, USA), running on an Intel Core i7 2.5 GHz computer, an iMac 1600 MHz DDR3 16 GB and OS X Sierra. This powerful computational environment facilitated the efficient processing and analysis of complex radiomic data.

3   Results

3.1   Radiomics Analyses of Mice One Month Old: Identification of Four Features that Distinguish SS from AS

A number of 112 radiomic features were extracted from the segmentation of each volume of the common carotid arteries of 1-month-old and 3-month-old mice, respectively. Radiomic feature extraction was performed through PyRadiomics. Normalize was set to True and BinWidth to 0.8 for the extraction of radiomic features, the others were kept by default.

Following the selection step, 1-month-old mice were identified by a group of features reported in Table 1

Table 1. Discriminant radiomic features extracted from carotid artery images of one-month-old mice.

Discriminant Radiomics Features	P-value (<0.05)
Original `Gray-Level Co-occurrence Matrix(glcm)` Imc1	0.0420
Original `firstorder` Minimum	0.0411
Original shape `MinorAxisLength`	0.0126
Original shape `MajorAxisLength`	0.0025

Combined, these features provide robust diagnostic performance for discriminat-ing between SS and AS, with an AUROC of 0.72; 95% C.I. between 0.60 and 0.84 with a p value of 0.032, sensitivity of 68.03%, specificity of 82.16%, accuracy of 74.22% and a Positive Predictive Value % (PPV) of 84.31%. The ROC curve for the combination of the four most discriminated radiomic features in mice at 1 month of age is shown in Fig. 2.

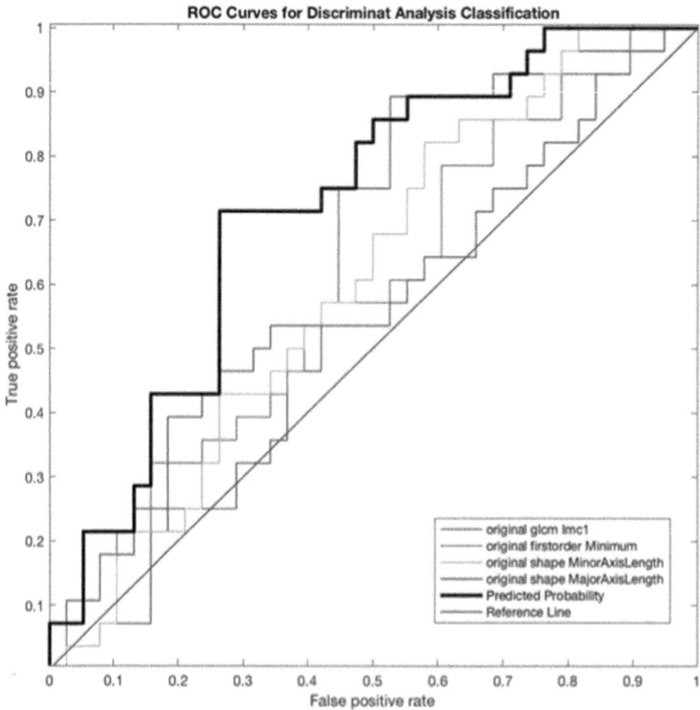

Fig. 2. A ROC curve was plotted for the four characteristics that were analyzed in the cases of one month of age. The curve shows the following features that are most discriminatory: GLCM IMC1 original, First Order Minimum, MinorAxisLength, MajorAxisLength).

3.2 Three Month Old AS and SS Phenotype can be Distinguished by One Radiomic Feature

MR images of three-month-old mice were examined using the radiomics workflow described in Sect. 2.3. Table 2 reports that a single first-order texture feature, the `MeshVolume` shape, was discriminant in separating the two phenotypes, AS and SS.

With an AUROC of 0.67; 95% CI between 0.61 and 0.87; p-value of 0.0020, sensitivity of 67.22%, specificity of 83.88%, accuracy of 76.03%, and positive predictive value (PPV) of 78.57%, this feature is promising for distinguishing AS and SS phenotypes in 3-month-old cases. Figure 3 illustrates the ROC curve for this feature.

Table 2. Discriminant radiomic feature of AS and SS phenotypes in three-month-old mouse cases.

Discriminant Radiomics Features	P-value (<0.05)
Original shape `MeshVolume`	0.0020

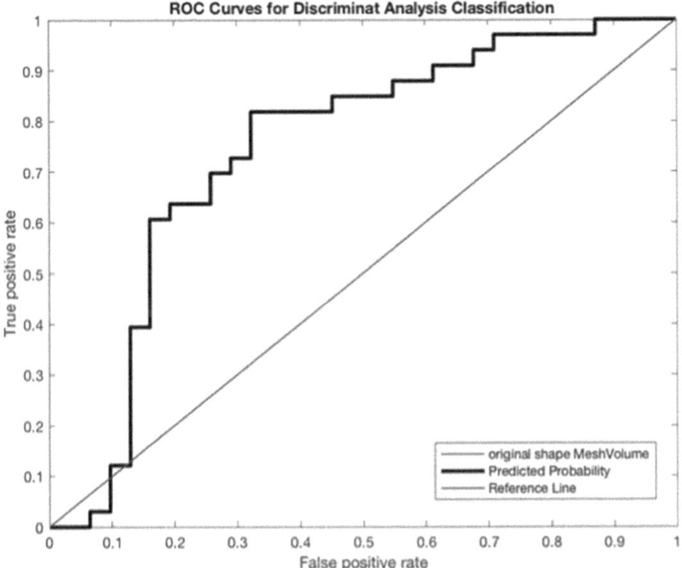

Fig. 3. ROC curve related to radiomic feature, MeshVolume shape, discriminant to discriminate AS and SS phenotypes in three-month-old cases.

4   Discussion

In our study, we propose a radiomic analysis of preclinical magnetic resonance angiography (MRA) images of SCD mice with carotid artery damage. In this work, radiomics was used to investigate its discriminatory power in detecting quantita-tive features in images to differentiate two different disease phenotypes, namely homozygous AS cases and heterozygous SS cases. This also allowed estimates of disease progression from an earlier age (one month) to a later age (3 months of age). At one month of age, a combination of four radiomic features achieved an accuracy of 74%. This suggests that these features capture early vascular changes in SCD. At three months of age, a single feature, mesh volume, provided slightly higher predictive performance (76% accuracy), despite higher variability in dis-ease severity. Although the AUROC obtained in our study shows good perfor-mance, it is not particularly high. This can be attributed to the variability in arte-rial remodeling and disease severity among individual SCD mice. Some limitations should be considered regarding the dataset. For example, one is related to the small dataset, which has been expanded by including both female and male mice. In the future, it would be useful to stratify the data to assess the presence of any differences due to the sex of the animals. Indeed, in the context of SCD, it should also be considered that male SCD sub-jects are more susceptible to mortality and are more susceptible to abdominal-thoracic complications compared to female subjects [16]. Therefore, these preliminary results will be the basis for extending the radi-omics analysis to larger samples to validate the proposed method, and subse-quently investigate its pre- and post-therapeutic treatment approach. Further-more, another consideration should be made regarding the murine species, since in the present work Townes mice affected by SCD were used. Consequently, it will be necessary in the near future to test the validated radiomic approach on patient images to verify its applicability in the clinical field. Two genotypes used in this study were SS and AS mice. Future studies could also include other differ-ent genotypes representing both benign and malignant conditions of the de-scribed pathology. Angio-MR imaging, together with radiomics and advanced machine learning techniques, could allow us to perform less invasive assessments of vas-cular changes in preclinical SCD models. Interesting is the contribution that radiomics can give to the study and characterization of vascular magnetic res-onance angiography images in large arteries such as the carotids in the context of SCD. Such evidence may be useful not only from a diagnostic point of view, but also from a therapeutic one, since, knowing the signaling pathways and the pathways responsible for arterial wall remodeling, it would be possible to inter-vene early at the therapeutic level [3], before irreparable damage such as stroke occurs [9,18]. Although radiomics is already used in fields such as cardiology for the characterization of plaques and prediction of vascular damage, the work we propose on cases of SCD, not yet present in the state of the art, could revolu-tionize the current screening methods, significantly impacting lethal scenarios and further complications.

5   Conclusion

In our study, we demonstrated that radiomics analysis can be used to detect quantitative features that can serve as prognostic and predictive biomarkers in the future, in images of the carotid arteries of mice with SCD, to discrimi-nate across types of SCD genotypes. Incorporating radiomics analysis into MR angiography has the potential to improve the detection of features impercep-tible to the human eye and address the challenges associated with subjective interpretation. By extracting quantitative features from MR images, radiomics analysis provides an objective and systematic approach to discriminate between different SCD geno-types in mouse models, complementing invasive genotyping based on different omics approaches such as genomics. These results suggest that radiomics analysis has a significant performance in the field of vascular imaging detection, providing a complementary and reliable method to overcome the lim-itations of subjective interpretation of MR images as well as time consumption.

References

1. Ali, M., et al.: Prostate cancer detection: performance of radiomics analysis in multiparametric MRI. In: International Conference on Image Analysis and Processing (2024). https://doi.org/10.1007/978-3-031-51026-7_8
2. Barone, S., et al.: Hybrid descriptive-inferential method for key feature selection in prostate cancer radiomics. Appl. Stoch. Models Bus Ind. **37**, 961–972 (2021). https://doi.org/10.1002/asmb.2642
3. Basirinia, G., et al.: Theranostic approaches for gastric cancer: an overview of in vitro and in vivo investigations. Cancers (Basel) **16**, 3323 (2024). https://doi.org/10.3390/cancers16193323
4. Benfante, V., et al.: Grading and staging of bladder tumors using radiomics analysis in magnetic resonance imaging. In: International Conference on Image Analysis and Processing (2024). https://doi.org/10.1007/978-3-031-51026-7_9
5. Cairone, L., et al.: Robustness of radiomics features to varying segmentation algorithms in magnetic resonance images. In: International Conference on Image Analysis and Processing (2022). https://doi.org/10.1007/978-3-031-13321-3_41
6. Canfora, I., et al.: A predictive system to classify preoperative grading of rectal cancer using radiomics features. In: International Conference on Image Analysis and Processing (2022). https://doi.org/10.1007/978-3-031-13321-3_38
7. Giaccone, P., Benfante, V., Stefano, A., Cammarata, F., Russo, G., Comelli, A.: Pet images atlas-based segmentation performed in native and in template space: a radiomics repeatability study in mouse models. In: International Conference on Image Analysis and Processing (2022). https://doi.org/10.1007/978-3-031-13321-3_31
8. van Griethuysen, J., et al.: Computational radiomics system to decode the radiographic phenotype. Cancer Res. **77**, e104–e107 (2017). https://doi.org/10.1158/0008-5472.CAN-17-0339
9. Hyacinth, H., et al.: Association of sickle cell trait with ischemic stroke among African Americans. JAMA Neurol. **75**, 802 (2018). https://doi.org/10.1001/jamaneurol.2018.0571
10. Kavanagh, P., Fasipe, T., Wun, T.: Sickle cell disease. JAMA **328**, 57 (2022). https://doi.org/10.1001/jama.2022.10233
11. Kunz, J., Tagliaferri, L.: Sickle cell disease. Transfus. Med. Hemother. **51**, 332–344 (2024). https://doi.org/10.1159/000540149
12. Lambin, P., et al.: Radiomics: the bridge between medical imaging and personalized medicine. Nat. Rev. Clin. Oncol. **14**, 749–762 (2017). https://doi.org/10.1038/nrclinonc.2017.141
13. Light, J., Boucher, M., Baskin-Miller, J., Winstead, M.: Managing the cerebrovascular complications of sickle cell disease: current perspectives. J Blood Med **14**, 279–293 (2023). https://doi.org/10.2147/JBM.S383472
14. Lin, A., et al.: Radiomics-based precision phenotyping identifies unstable coronary plaques from computed tomography angiography. J. Am. Coll. Cardiol. Cardiovasc. Imaging **15**, 859–871 (2022). https://doi.org/10.1016/j.jcmg.2021.11.016
15. Lyu, J., Xu, Z., Sun, H., Zhai, F., Qu, X.: Machine learning-based CT radiomics model to discriminate the primary and secondary intracranial hemorrhage. Sci. Rep. **13**, 3709 (2023). https://doi.org/10.1038/s41598-023-30678-w
16. Meddings, Z., Rundo, L., Sadat, U., Zhao, X., Teng, Z., Graves, M.: Robustness and classification capabilities of MRI radiomic features in identifying carotid plaque vulnerability. Br. J. Radiol. **97**, 1118–1124 (2024). https://doi.org/10.1093/bjr/tqae057

17. Song, H., et al.: Sickle cell anemia mediates carotid artery expansive remodeling that can be prevented by inhibition of JNK (c-jun N-terminal kinase). Arterioscler. Thromb. Vasc. Biol. **40**, 1220–1230 (2020). https://doi.org/10.1161/ATVBAHA.120.314045
18. Yang, B., et al.: Comparison of ruptured intracranial aneurysms identification using different machine learning algorithms and radiomics. Diagnostics **13**, 2627 (2023). https://doi.org/10.3390/diagnostics13162627

Radiomic Analysis of MR and PET Images of Mice to Identify Possible Microstructural Changes Associated with Early Molecular Changes During Nigral Neuron Deafferentation

Viviana Benfante[1,6]([✉]) [iD], Sara Belloli[2,3]([✉]) [iD], Gaia Faustini[4] [iD], Arianna Bellucci[4] [iD], Rosa Maria Moresco[3,5] [iD], and Albert Comelli[6] [iD]

[1] Department of Health Promotion, Mother and Child Care, Internal Medicine and Medical Specialties, Molecular and Clinical Medicine, University of Palermo, 90127 Palermo, Italy
vivianabenfante3@gmail.com

[2] CNR, Institute of Bioimaging and Biological Complex Systems (IBSBC), 20054 Segrate, Italy
sara.belloli@cnr.it

[3] IRCCS San Raffaele Scientific Institute, Nuclear Medicine Unit, 20132 Milano, Italy

[4] Department of Molecular and Translational Medicine, University of Brescia, 25121 Brescia, Italy

[5] Bicocca University, Department of Medicine and Surgery, Monza, Italy

[6] Ri.MED Foundation, Via Bandiera 11, 90133 Palermo, Italy

Abstract. This study aimed to evaluate the efficacy of radiomics analysis applied to MRI and PET-CT imaging for the early detection of degenerative changes in SYN120 transgenic mice. These mice, which express human C-terminally truncated α-synuclein under the control of the rat tyrosine hydroxylase promoter on an α-synuclein null background, exhibit key neuropathological features characteristic of Parkinson's disease, including progressive α-synuclein aggregate deposition and degeneration of nigrostriatal dopaminergic neurons. Radiomics analysis was conducted on SYN120 transgenic mice and α-synuclein null control mice (OlaHsd) using T2-weighted MRI scans with multiple sequences and PET-CT imaging with three distinct radiotracers: [18F]FDG to assess glucose metabolism, [18F]VC701 to evaluate microglial activation, and [18F]FP-CIT to examine dopamine transporter (DAT) integrity. Imaging was performed at three developmental stages 5, 8, and 10 months to facilitate the identification of early disease biomarkers. Following acquisition, PET and CT images were reconstructed and automatically co-registered. A total of 26 brain regions were delineated using a standardized brain atlas. A three-step radiomics pipeline was applied to extract 93 quantitative features from each defined region. By 10 months of age, the dorsal striatum demonstrated increased metabolic heterogeneity, as indicated by variations in [18F]FDG signal intensity. Whole-brain radiomic analysis further revealed a progressive and diffuse increase in glucose

E. Rodolà et al. (Eds.): ICIAP 2025 Workshops, LNCS 16169, pp. 233–242, 2026.
https://doi.org/10.1007/978-3-032-11317-7_20

metabolism across multiple brain regions. Notably, alterations in DAT expression were most prominent at 5 months in several regions, including the dorsal striatum, whereas signs of neuroinflammation emerged at later time points.

Keywords: Parkinson's disease · PET-CT · MR · Texture Analysis · Radiomics

1   Introduction

Parkinson's disease (PD) is the second most common neurodegenerative disease, characterized by the gradual destruction of dopaminergic neurons in the pars Compacta of the substantia nigra. This condition results in a dopamine deficit in the dorsal striatum and other motor and nonmotor dysfunctions [14]. It has been shown that abnormal aggregates of α-synuclein (α-syn), such as Lewy bodies (LB) and Lewy neurites, and glial cell inclusions are involved in various types of sporadic neurodegenerative diseases called α-synucleinopathies, including idiopathic PD, dementia with Lewy bodies (DLB), multiple system atrophy (MSA), pure autonomic failure (PAF), and REM sleep behavior disorder (RBD) [7]. These types of neurodegenerative diseases are investigated taking advantage of representative preclinical models [1,6,10,11,19]. In the current state of the art, a number of studies are underway to identify new imaging biomarkers to allow early detection of PD, which remains an incurable disease. Monitoring the early phase of nigrostriatal dopaminergic degeneration and the resulting early neuroinflammation could help with the diagnostic process, but most importantly, it is essential for the development of an effective treatment. Despite recent discoveries, the biological mechanism that regulates the correlation between α-Syn aggregates and nigrostriatal synapses in the early phase of PD is still unknown. Currently, there is no suitable imaging biomarker for the definitive and early diagnosis of α-Syn aggregates. However, dopaminergic deficits detected with SPECT and PET radiotracers can support clinical diagnosis. α-Syn aggregates may be a suitable imaging biomarker in synucleinopathies, but there is no way to directly measure whether and to what extent it reduces aggregated α-Syn in the brain. At present, the assessment of these α-Syn aggregates in Parkinson's disease patients using PET radiotracers is limited by off-target binding or poor specificity [17,20]. The field of radiomics, a branch of science based on artificial intelligence, in-volves analyzing biomedical images for quantification of parameters, called ra-diomic features, that are normally invisible to the naked eye [8,18]. Through the normalization and segmentation of images, radiomics provides image information in terms of tissue texture, shape, and intensity distributions [6]. Machine learning-based artificial intelligence algorithms can then be utilized to incorporate these features into predictive models for diagnosis [2] and prognosis [9]. As a result of the enormous difficulty in utilizing current knowledge to detect efficient biomarkers [21] not only from PET images [12], but also from MRI images [5], we propose a new approach that utilizes radiomics analysis as a means of obtaining additional insight into the possible changes in microstructure associat-ed with early molecular alterations in Parkinson's disease. Using MRI and PET images of transgenic (tg) mice for the C-terminally truncated form (1–120) of human α-Syn (SYN120 tg) and control mice C57 black 6 (C57BL/6J-OlaHsd), a comparison between groups of animals of various ages (5, 8 and 10 months) was made. As a readout of the results of this study, a new non-invasive [4] and sensitive diagnostic tool for Parkinson's disease may be developed and translated into the clinic [3].

2 Materials and Methods

2.1 Animals and Study Design

C57BL/6J mice (Charles River, Wilmington, MA), C57BL/6JOlaHsd mice carrying a spontaneous deletion of the α-Syn gene (Harlan Olac, Bicester, UK) and SYN120 transgenic [23] mice were used at 5, 8 and 10 months of age. Animals were maintained under a 12-h light-dark cycle at a room temperature (RT) of 22 °C and had food and water ad libitum. At 5 months of age, the animals underwent T2-weighted MRI with different sequences to obtain information on brain volume, neuronal degeneration and white matter integrity. Subsequently, the same group of animals underwent PET-CT with three different radiotracers: [18F]-FDG, [18F]-FP-CIT and [18F]-VC701, to investigate brain metabolism (neuronal damage), dopamine transporter (DAT) and neuroinflammation, respectively. The different imaging studies were separated by at least one day of rest to avoid prolonged fasting and stress for the animals. The same scheme was repeated at 8 and 10 months of age. All animal handling procedures during the experiments were in accordance with Directive 2010/63/EU of the European Parliament and of the Council of 22 September 2010 on the protection of laboratory animals and the National Guide for Research on the Care and Use of Laboratory Animals approved by Brescia (Min. Aut. n. 261/2020), San Raffaele Hospital (Min. Aut. n. 800/2020-PR) and the University of Ferrara (Min. Aut. 21-2020-PR). Furthermore, all efforts to minimize animal suffering and the 3Rs have been made.

2.2 MRI and PET-CT Imaging and Reconstruction

MRI acquisitions were performed using a 7 T preclinical MRI scanner (Bruker, BioSpec 70/30 USR, Paravision 5.0) equipped with gradients of 450/675 mT/m (slew-rate: 3400–4500 T/m/s; rise-time: 140 μs). MRI scans included 2D T2-weighted images with high resolution Turbo spin-echo technology (TR = 3600 ms, TE = 45 ms, FOV = 14×14 mm, thickness = 0.6 mm, matrix size = 192×192, means = 10) and DTI-EPI (Diffusion Tensor Imaging with spin-echo EPI) sequences (TR = 3500 ms and TE = 32 ms, FOV = 20×20 mm, thickness = 0.75 mm, matrix size = 170×170, means = 1 and repeats = 2). Animals were positioned prone under gas anesthesia (isoflurane and oxygen 1:1) and acquired for a total of 1 h with the above sequences. DTI images were analyzed using

Bruker's software Paravision (version 6.0.1), by manually drawing 8 regions of interest (ROIs: Dorsal striatum, Ventral striatum, Motor cortex Somatosensory cortex, Thalamus, Hippocampus and Visual cortex) and extracting values of Mean Diffu-sivity (MD) and Fractional Anisotropy (FA), as previously reported [24]. For comparison, radiomic analysis has been performed using the 27-region template, but extracting data from the same 8 regions with the exclusion of corpus callosum, which was not foreseen in the template. For PET-CT, the following radiotracers have been used: [18F]-FP-CIT (3.5 MBq), [18F]-FDG (3.9 MBq) and [18F]-VC701 (4.4 MBq), administered to mice via their tail veins. Before undergoing the [18F]-FDG scan, mice were fasted over-night. All scans were performed 1 h after the radiopharmaceutical injection, except for [18F]-VC701 where the uptake time was 2 h. CT and PET scans were performed using X-Cube and β-Cube (Molecubes, Gent, BE), respectively. All mice were first anesthetized with isoflurane and medical air (1:1) in the prone position with the brain centered in the field of view (FOV). During the examination, the main vital parameters were monitored and stabilized. The CT scan had the following characteristics: examination duration: 4 min, X-ray beam duration: 90 s, kVp: 40, current: 400 μA, rotation time: 60 s, angular views: 960. Following each CT acquisition, the bed was moved to the PET scanner for a 20-minute static acquisition. A standard Image Space Reconstruction Algorithm (ISRA) was used to recon-struct CT images, whereas PET images were reconstructed by iterative ordered subset expectation maximization (OSEM), with 30 iterations and a 400×400 voxel size. Figure 1 shows an example of co-registration of MR images with an atlas of 26 brain regions on T2- w brain sequences. CT and PET images were automatically co-registered using Molecubes software. After that, the co-registered images were processed using a mouse brain atlas tool of 27 regions and SPM12. In order to deter-mine the uptake of regional tracers. Radioactivity in each brain region was ex-pressed as the percentage of injected dose per g of tissue (%ID/g) and normalized for a control region (parietal cortex) in case of [18F]-FP-CIT or for the global mean in case of [18F]-VC701 and [18F]-FDG to obtain [(%ID/g/%ID/g)]. A comparison was also made between the uptake values on the right and left sides of each region.

2.3   Radiomics Analysis Workflow

Radiomics workflow was divided into the following steps: first, all MRI scans of the mice were co-registered with an atlas of 27 brain regions; following that, the normalized PET images were co-registered with the Turbo Rare MRI sequences (high resolution T2-weighted) that had already been co-registered with the atlas; finally, through the PyRadiomics software [13] the radiomic features were extracted from the MRI and PET images of each group of mice (SYN120 tg and the C57BL/6JOlaHsd controls). A total of 93 radiomic features were extracted from each ROI. After excluding shape radiomic features that remained unchanged since the MRI and PET images were co-registered with the Atlas, various feature classes have been extracted, including first-order, second-order, and

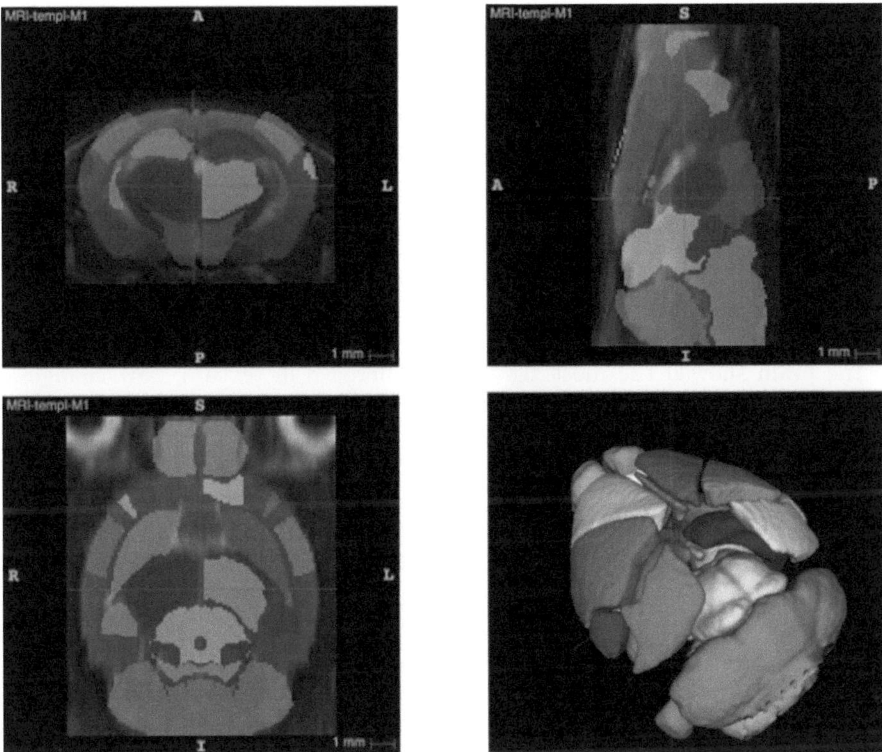

Fig. 1. Co-registration of mouse MR images with an atlas of 26 brain regions on a T2-w brain se-quence.

higher-order statistics (texture matrices). In the different experimental studies, 5 to 10 animals were used for each experimental group.

2.4   Statistical Analyses

To evaluate the performance of the radiomic method used in this comparison study, the unpaired T-test was applied as a statistical method, through which it is possible to compare the means of two independent groups. Radiomic features with a p-value lower than 0.05 were considered significantly different. Thus, it was then possible to identify the percentage of statistically significant radiomic features between the tg mouse groups and the OLA controls for each of the 27 brain regions considered, excluding the background.

3   Results

Radiomic analysis applied to the study of early pre-degenerative alterations associated with α-Syn accumulation in the nigrostriatal neurons of SYN120 tg mice at 5, 8 and 10 months of age provided information on synucleinopathy progression in an early non-symptomatic phase [24]. Due to the absence of symptoms, MR imaging and PET-CT imaging with three different radiotracers allowed identification of early structural/functional disease biomarkers preceding motor and cognitive signs. A radiomics workflow provides a framework for quantitative analyses of bio-medical images, especially when naked eye comparisons do not yield meaningful results, thus becoming an almost indispensable tool [6,25]. According to the state of the art, radiomics can be used to predict PD [18]. Analysis of DTI-MRI images of SYN120 tg mice compared to controls revealed significant changes in signal texture, which evolved differentially over time. This is evident from Fig. 2, where the hippocampus and ventral striatum of the brain showed the most significant difference in texture, 25% change at 5 months (hip-pocampus) and 24.11% change at 8 months (ventral striatum), respectively. No important variations have been observed in cortical regions at the time points considered. To date, Volume of interests (VOIs) analysis of DTI sequences did not reveal any significant differ-ences between the two genotypes over time. In addition to structural information provided by MR, the mice were subjected to functional imaging by [18F]-FP-CIT, [18F]-FDG and [18F]-VC701 PET-CT, to detect dopamine transporter (DAT), brain glucose metabolism and microglial cell activation, respectively. Radiomic analysis of [18F]-FDG images between SYN120 tg mice and control mice suggests that metabolic heterogeneity increases progressively up to ten months of age in the majority of brain regions and earlier in the striatum, olfactory bulb, midbrain, hippocampus and motor and visual cortex (Fig. 3a). In particular, in the dorsal striatum the percentage change in signals was maximum at 22.8% at 10 months of age (Fig. 3a). For comparison, data on the same 27 region template using SPM12 software failed to

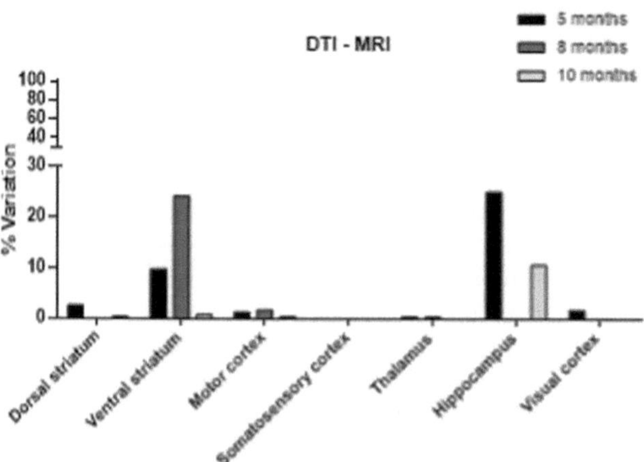

Fig. 2. Comparison of regional brain texture (DTI-MRI) between SYN120 tg and C57BL/6JOlaHsd mice over time. Bars represent the mean of 5 animals per group.

evidence significant differences between SYN120 tg and control mice at any time point (Fig. 3b-d). A much more complex picture emerged from radiomics analysis of [18F]-FP-CIT and [18F]-VC701 PET images (Fig. 4a and b). Regarding dopamine transporter markers, % Variation between brain regions of SYN120 tg and control animals appeared very marked at the early stage (5 months) in dorsal striatum, olfactory bulbs, anterior cingulate and hippocampus; ventral striatum and somatosensory cortex was involved later while midbrain (a region containing substantia nigra) has a constant variation rate over time (Fig. 4a). Regarding microglia activa-tion, the somatosensory cortex exhibited higher % variation between the two genotypes at the earlier time (8.9% at 5 months), together with cerebellum and hindbrain (39.3 and 43.7%, respectively), while variation in the other brain re-gions appeared later (Fig. 4b).

4   Discussion

This study combined two in vivo imaging methods, namely preclinical MRI and PET-CT in transgenic PD mouse models of human α-Syn, with radiomics analysis to identify quantitative changes associated with radiomic features in an early presymptomatic disease phase. The results are encouraging and seem

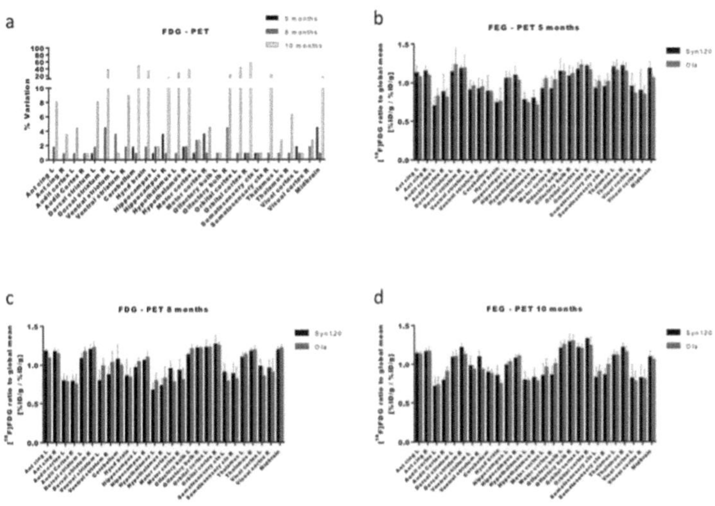

Fig. 3. Comparison of regional brain [18F]-FDG uptake between SYN120 tg and C57BL/6JOlaHsd mice over time. a) % Variation after radiomic analysis at 5, 8 and 10 mounts of age. b) VOIs analysis of [18F]-FDG images using SPM software in 5-month-old transgenic and control animals. Bars represent the mean ± SD of 5 animals per group. c) VOIs analysis of [18F]-FDG images using SPM software in 8 month old transgenic and control animals. Bars represent the mean ± SD of 5 animals per group. d) VOIs analysis of [18F]-FDG images using SPM software in 10 month old transgenic and control animals. Bars represent the mean ± SD of 5 animals per group.

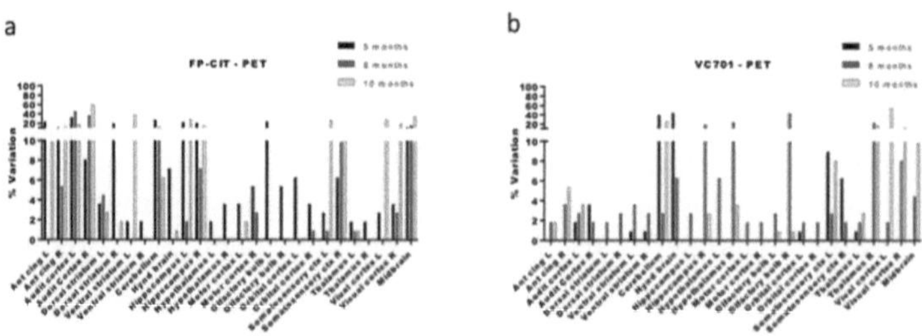

Fig. 4. Comparison of 93 radiomic features between brain regions of SYN120 tg and C57BL/6JOlaHsd mice over time. Bars represent the mean of 5 animals per group. a) % Variation after radiomic analysis of PET-[18F]FP-CIT images. b) % Variation after radiomic analysis of PET-[18F]VC701 images.

to be related to disease progression, compared to C57BL/6JOlaHsd controls. The radiomics analysis of DTI-MRI found progressive and significant differences between PD and control mice. Radiomics applied to morphological MR imaging has shown more predictive results than functional PET imaging. This may encourage increased use of this type of analysis in the early stages of PD [15,16,22]. In parallel, it should be emphasized that some significant and progressive dif-ferences between SYN120 tg and C57BL/6JOlaHsd mice emerged from texture analysis through radiomics of FDG PET-CT images. However, cerebral glucose metabolism highlighted by [18F]-FDG distribution could be associated with radiomic variations at the level of the whole brain and not in distinct regions, even if it is progressive into the disease. Radiomics analysis of [18F]-FP-CIT and [18F]-VC701 signals showed a more complex pattern and the differences appeared region specific and with a different timing. These findings must be confirmed with tissue molecular analysis and immunohistochemistry.

5   Conclusion

Our study demonstrated the preclinical applicability of radiomics analysis to MRI and PET-CT imaging for the early identification of the development of subtle dopaminergic functional alterations, neuroinflammatory alterations as well as metabolic alterations well before the onset of frank deafferentation of nigral neurons in α-Syn transgenic mouse models. Implementing radiomics analysis could improve the detection of structural alteration as those deriving from pathological α-Syn aggregates deposition, indicative of major neurodegenerative disorders. This will address the challenges associated with the subjective interpretation of MRI and PET-CT images. Using radiomics analysis, quantitative features are extracted from morphological and metabolic images, providing a systematic and objective method for early research on novel imaging biomarkers. The most effective therapeutic path can be determined at an early stage by identifying and monitoring future pathological changes and, therefore, guiding the most effective therapeutic path.

References

1. Aerts, H.J.W.L., et al.: Decoding tumour phenotype by noninvasive imaging using a quantitative radiomics approach. Nat. Commun. **5**, 4006 (2014). https://doi.org/10.1038/ncomms5006

2. Ali, M., et al.: Prostate cancer detection: Performance of radiomics analysis in multiparametric MRI (2024). https://doi.org/10.1007/978-3-031-51026-7_8

3. Alongi, P., et al.: Radiomics analysis of brain [18F]FDG PET/CT to predict Alzheimer's disease in patients with amyloid pet positivity: a preliminary report on the application of SPM cortical segmentation, pyradiomics and machine-learning analysis. Diagnostics **12**, 933 (2022). https://doi.org/10.3390/diagnostics12040933

4. Basirinia, G., et al.: Theranostic approaches for gastric cancer: An overview of in vitro and in vivo investigations. Cancers (Basel) **16**, 3323 (2024). https://doi.org/10.3390/cancers16193323

5. Benfante, V., et al.: Grading and staging of bladder tumors using radiomics analysis in magnetic resonance imaging (2024). https://doi.org/10.1007/978-3-031-51026-7_9

6. Cairone, L., et al.: Robustness of radiomics features to varying segmentation algorithms in magnetic resonance images (2022). https://doi.org/10.1007/978-3-031-13321-3_41

7. Calabresi, P., Mechelli, A., Natale, G., Volpicelli-Daley, L., Di Lazzaro, G., Ghiglieri, V.: Alpha-synuclein in Parkinson's disease and other Synucleinopathies: from overt neurodegeneration back to early synaptic dysfunction. Cell Death Dis. **14**, 176 (2023). https://doi.org/10.1038/s41419-023-05672-9

8. Caminiti, S.P., et al.: Dopaminergic connectivity reconfiguration in the dementia with Lewy bodies continuum. Parkinsonism Relat. Disord. **108**, 105288 (2023). https://doi.org/10.1016/j.parkreldis.2023.105288

9. Canfora, I., et al.: A predictive system to classify preoperative grading of rectal cancer using radiomics features (2022). https://doi.org/10.1007/978-3-031-13321-3_38

10. Faustini, G., et al.: Synapsin III gene silencing redeems alpha-synuclein transgenic mice from Parkinson's disease-like phenotype. Mol. Ther. **30**, 1465–1483 (2022). https://doi.org/10.1016/j.ymthe.2022.01.021

11. Gao, V., Briano, J.A., Komer, L.E., Burré, J.: Functional and pathological effects of -synuclein on synaptic snare complexes. J. Mol. Biol. **435**, 167714 (2023). https://doi.org/10.1016/j.jmb.2022.167714

12. Giaccone, P., Benfante, V., Stefano, A., Cammarata, F.P., Russo, G., Comelli, A.: Pet images atlas-based segmentation performed in native and in template space: A radiomics repeatability study in mouse models (2022). https://doi.org/10.1007/978-3-031-13321-3_31

13. van Griethuysen, J.J.M., et al.: Computational radiomics system to decode the radiographic phenotype. Can. Res. **77**, e104–e107 (2017). https://doi.org/10.1158/0008-5472.CAN-17-0339

14. Guatteo, E., Berretta, N., Monda, V., Ledonne, A., Mercuri, N.B.: Pathophysiological features of nigral dopaminergic neurons in animal models of Parkinson's disease. Int. J. Mol. Sci. **23**, 4508 (2022). https://doi.org/10.3390/ijms23094508

15. Jian, Y., et al.: Prediction of cognitive decline in Parkinson's disease based on MRI radiomics and clinical features: a multicenter study. CNS Neurosci. Ther. **30** (2024). https://doi.org/10.1111/cns.14789

16. Jiang, H., et al.: Radiomics incorporating deep features for predicting Parkinson's disease in 123i-Ioflupane SPECT. EJNMMI Phys. **11**, 60 (2024). https://doi.org/10.1186/s40658-024-00651-1

17. Korat, Å., et al.: Alpha-synuclein pet tracer development—an overview about current efforts. Pharmaceuticals **14**, 847 (2021). https://doi.org/10.3390/ph14090847

18. Meiser, J., Weindl, D., Hiller, K.: Complexity of dopamine metabolism. Cell Commun. Signal. **11**, 34 (2013). https://doi.org/10.1186/1478-811X-11-34

19. Schulz-Schaeffer, W.J.: The synaptic pathology of -synuclein aggregation in dementia with Lewy bodies, Parkinson's disease and Parkinson's disease dementia. Acta Neuropathol. **120**, 131–143 (2010). https://doi.org/10.1007/s00401-010-0711-0

20. Smith, R., et al.: The -synuclein pet tracer [18F] ACI-12589 distinguishes multiple system atrophy from other neurodegenerative diseases. Nat. Commun. **14**, 6750 (2023). https://doi.org/10.1038/s41467-023-42305-3

21. Stefano, A.: Challenges and limitations in applying radiomics to pet imaging: possible opportunities and avenues for research. Comput. Biol. Med. **179**, 108827 (2024). https://doi.org/10.1016/j.compbiomed.2024.108827

22. Sun, J., et al.: Identification of Parkinson's disease and multiple system atrophy using multimodal PET/MRI radiomics. Eur. Radiol. **34**, 662–672 (2023). https://doi.org/10.1007/s00330-023-10003-9

23. Svoboda, K., Yasuda, R.: Principles of two-photon excitation microscopy and its applications to neuroscience. Neuron **50**, 823–839 (2006). https://doi.org/10.1016/j.neuron.2006.05.019

24. Tofaris, G.K., et al.: Pathological changes in dopaminergic nerve cells of the substantia Nigra and olfactory bulb in mice transgenic for truncated human -synuclein(1–120): implications for Lewy body disorders. J. Neurosci. **26**, 3942–3950 (2006). https://doi.org/10.1523/JNEUROSCI.4965-05.2006

25. Tupe-Waghmare, P., Rajan, A., Prasad, S., Saini, J., Pal, P.K., Ingalhalikar, M.: Radiomics on routine T1-weighted MRI can delineate Parkinson's disease from multiple system atrophy and progressive supranuclear palsy. Eur. Radiol. **31**(11), 8218–8227 (2021). https://doi.org/10.1007/s00330-021-07979-7

Automated Prostate Gland Detection in MRI Using YOLOv8: A Deep Learning Approach for Early Diagnosis

Akhtar Rasool[1] , Muhammad Ali[1(✉)] , Leonardo Salvaggio[1,2] ,
Viviana Benfante[3] , and Albert Comelli[1]

[1] Ri.MED Foundation, Via Bandiera 11, 90133 Palermo, Italy
`amuhammad@Fondazionerimed.com`
[2] Department of Biomedicine, Neuroscience and Advanced Diagnostics (BiND), University of Palermo, Palermo, Italy
[3] Department of Health Promotion, Mother and Child Care, Internal Medicine and Medical Specialties, Molecular and Clinical Medicine, University of Palermo, 90127 Palermo, Italy

Abstract. Accurate identification of the prostate gland on magnetic resonance imaging (MRI) is a critical step in the diagnostic and treatment planning workflow for prostate cancer. Manual delineation, however, remains time-consuming and prone to interobserver variability. This study investigates the applicability of YOLOv8, a state-of-the-art real-time object detection model, for automatic prostate gland detection in axial T2-weighted MRI slices. A dataset of 71 patients consisting of 1019 annotated prostate slices were used, encompassing heterogeneous imaging conditions and glandular anatomies. YOLOv8 was trained using advanced preprocessing techniques, anchor-free detection, and optimized loss functions to improve localization precision and inference speed. The model achieved a mean Average Precision (mAP@0.5) of 0.98, with peak precision and recall values of 1.00 and 0.99, respectively. The F1-score reached 0.97, indicating a strong balance between detection accuracy and sensitivity. Although the confusion matrix revealed a mild obstacle in background separation, the overall results demonstrate that YOLOv8 offers robust, efficient, and clinically promising performance for automated prostate gland detection, facilitating faster and more consistent MRI interpretation.

Keywords: Prostate · magnetic resonance imaging · YOLOv8 · Average precision · axial T2 weighted MRI slices

1 Introduction

Prostate cancer (PCa), a highly prevalent malignancy in male populations, requires early diagnosis to improve clinical management. Magnetic Resonance

E. Rodolà et al. (Eds.): ICIAP 2025 Workshops, LNCS 16169, pp. 243–254, 2026.
https://doi.org/10.1007/978-3-032-11317-7_21

Imaging (MRI) is central to modern diagnostic workflows, enabling high resolution visualization critical for identifying tumor sites and evaluating their aggressiveness potential [13].

Accurate interpretation of prostate MRI examinations remains critically contingent upon radiologist expertise, introducing inherent subjectivity and substantial time requirements. This dependency creates significant operational constraints in clinical practice [12], where diagnostic efficiency and reliability are paramount for time-sensitive therapeutic decisions. To address this limitation, we explore the applicability of YOLOv8, a real-time object detection framework specifically optimized to balance speed and accuracy in non-medical domains. In contrast to region-based or segmentation-focused approaches [5,19], YOLOv8 employs a single-stage architecture that facilitates rapid inference without compromising detection performance. Although the framework has shown efficacy in tasks such as lung nodule identification and breast cancer screening, its potential utility in the context of prostate cancer detection has not yet been thoroughly investigated [2].

This study aims to evaluate the effectiveness of YOLOv8 in prostate gland detection using a data set of 1019 annotated MRI slices. Specifically, we focus on two core challenges: (1) ensuring clinical applicability through real-time image processing for immediate diagnostic support, and (2) achieving robust detection across heterogeneous imaging conditions, including variations in contrast and glandular anatomy. By evaluating YOLOv8's performance in this context, we highlight its potential to bridge the gap between high-performance detection and practical clinical integration in prostate cancer diagnostics.

2 Materials and Methods

2.1 Population

We retrospectively reviewed the radiological reports of 155 consecutive patients who underwent multiparametric prostate MRI examinations at our hospital between June 1, 2019, and January 31, 2023. MRI examinations were performed for at least one of the following clinical indications: the appearance of prostatic symptoms such as urinary flow obstruction, lower urinary tract symptoms, haematuria; abnormal findings at transrectal ultrasound and/or digital rectal examination; an elevated prostate-specific antigen (PSA) level greater than 2.5 ng/mL; follow-up after radiation therapy; or a PSA level exceeding 0.2 ng/mL in patients previously treated with radical prostatectomy for prostate cancer. Exclusion criteria included incomplete MRI examinations due to intolerance, discomfort, or claustrophobia (n = 12); history of radical prostatectomy (n = 21); previous transurethral resection of the prostate (TURP) (n = 28); pelvic radiotherapy (n = 17). Inclusion criteria for the selected cases were the availability of axial T2-weighted images acquired with a standardized protocol and adequate image quality for segmentation purposes. Adequate image quality was defined as complete coverage of the prostate gland in the axial plane, absence of severe motion or susceptibility artifacts, and clear visibility of the prostatic capsule margins on T2-weighted images. MRI studies with inadequate image quality for segmentation purpose $(n = 6)$ were also excluded. After applying these criteria, a total of 71 prostate MRI studies were included in the final analysis. The presence or histological nature of prostate lesions was not considered for this study, as the aim was the automated detection of the prostate gland, independent of disease status.

2.2 MRI Technique

All MRI examinations were performed using a 1.5T MRI scanner (Achieva, Philips Healthcare, Best, The Netherlands) equipped with a 16-channel pelvic phased-array coil (HD Torso XL). A standardized imaging protocol was applied to all patients to ensure consistency. The routine clinical prostate MRI protocol included T2-weighted turbo spin-echo sequences acquired in axial, coronal, and sagittal planes covering the prostate and seminal vesicles, as well as axial T1-weighted turbo spin-echo images. Diffusion-weighted imaging (DWI) was also performed using b-values of 0, 700, and $1400 \, s/mm^2$, with subsequent generation of apparent diffusion coefficient (ADC) maps. Following the acquisition of unenhanced sequences, a dynamic contrast-enhanced (DCE) MRI was conducted by administering gadoteric acid (Gd-DOTA, Dotarem, Guerbet) at a dose of 1 mmol/kg of body weight, injected at a rate of 3 mL/s, followed by a 20–30 mL saline flush at the same rate. Axial three-dimensional spoiled gradient echo volumetric images were then acquired to assess prostate perfusion. For the purpose of this study, only axial T2-weighted images were used for both manual segmentation [10,14] and automatic detection of the prostate gland.

2.3 Architecture of YOLOv8 Model

This study employs YOLOv8, an advanced object detection algorithm, as the core model for developing a reliable and efficient early detection system for prostate gland. YOLOv8 was selected due to its substantial improvements over previous iterations in the YOLO series, particularly in terms of accuracy, real-time processing capability, and adaptability to various computational environments. The detection pipeline begins with preprocessing, where input images are resized to a fixed resolution and pixel values are normalized. The image is then divided into a grid, with each cell representing a potential object location (Fig. 1) [20]. The dataset was split into 70% training, 30% validation.

A deep convolutional neural network extracts feature maps from the input, which are used to predict multiple bounding boxes per grid cell, along with associated class probabilities and confidence scores. Non-maximum suppression is applied to remove redundant and overlapping detections, retaining only the most confident results [3]. Finally, post-processing steps eliminate low-confidence detections and consolidate overlapping boxes that correspond to the same object. This comprehensive and modular pipeline allows YOLOv8 to maintain high detection accuracy while ensuring efficient processing, making it well-suited for deployment in medical diagnostic applications.

YOLOv8 introduces several architectural advancements to enhance performance: A CSPDarknet53 backbone using Cross-Stage Partial connections to enhance gradient flow and reduce redundancy [18]. A C2f neck module, replacing the traditional FPN, which improves the fusion of semantic and spatial features—critical for detecting small lesions [21]. A multi-scale detection head with dynamic anchor assignment and refined loss functions for precise localization [7]. Integration of Spatial Pyramid Pooling Fast (SPPF) for enhanced multi-scale context understanding. Training enhancements such as mosaic augmentation and mixed precision to improve generalizability.

Additionally, YOLOv8 supports configurable variants (YOLOv8-C/D/E, enabling users to balance performance and computational efficiency.While YOLOv8 achieves state-of-the-art results on COCO and VOC benchmarks and supports real-time inference on edge devices, it remains underexplored for prostate MRI diagnostics an opportunity this study aims to address [15].

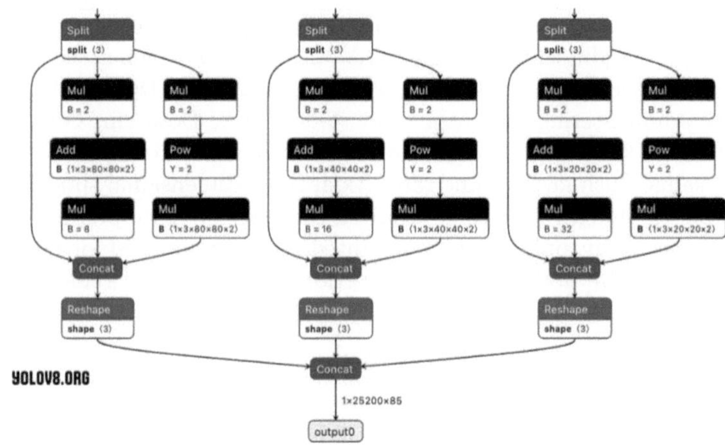

Fig. 1. YOLOv8 Architecture.

2.4 Evaluation Matrics

To evaluate the performance of the YOLOv8 model, we used mean Average Precision at IoU 0.5 (mAP@0.5), a standard metric in object detection tasks. It measures the average precision across all recall levels, providing a balanced view of accuracy and sensitivity. A prediction is considered correct if the IoU between the predicted and ground truth bounding boxes is at least 0.5.

mAP@0.5 is especially relevant for medical imaging, where accurate localization is crucial. In this study, it was calculated during both training and validation to monitor model performance. Additional metrics such as precision, recall, F1-score, and confusion matrix were also used to ensure a comprehensive evaluation.

3   Results

In this study, the YOLOv8 model was trained on a prostate MRI dataset for 50 epochs. Rather than directly detecting and classifying tumors [6], the model was designed to identify the prostate region in which a tumor may be located. This approach aims to support future segmentation procedure by minimizing background interference and focusing on the relevant anatomical area [17]. Performance was evaluated using metrics such as bounding box loss, precision-recall, recall-confidence, and F1-score. The results highlight the model's confidence in accurately identifying the prostate [16], thereby facilitating potential future detection [11] and classification of prostate cancer lesions while excluding non-relevant tissues.

3.1   Experimental Design

Prostate gland detection using the YOLOv8n algorithm begins with annotated MRI scans outlining gland boundaries (Fig. 2). Images are preprocessed through normalization, resizing, and enhancement, with annotations converted to bounding boxes. The model, configured for a single-class custom dataset, uses pretrained weights and is trained for 50 epochs with a batch size of 16 and image size of 320. AdamW optimizer with a 0.002 learning rate is employed, along with augmentations like horizontal flips, random erasing, and auto-augmentation. Deterministic training ensures reproducibility. Training progress is tracked using loss and mAP, while validation performance is assessed using precision, recall, and F1-score. Non-Maximum Suppression refines predictions, and medical expert input ensures clinical relevance.

3.2   Performance Evaluation

The performance evaluation was conducted using the Prostate MRI dataset, which contains images with a resolution of 320×320, to assess the effectiveness of our model. YOLOv8 was trained on this dataset for 25 epochs, and its performance was analyzed using a range of metrics, including accuracy, recall-confidence, precision-confidence, precision-recall, F1-score, and the confusion matrix. The Precision-Recall (PR) curve for prostate gland detection using the YOLOv8 model demonstrates outstanding performance. The curve remains close to the top of the graph across nearly the entire recall range, indicating consistently high precision. This suggests that the model generates very few false positives—when it detects a prostate gland, it is highly likely to be correct. A sharp drop in precision occurs only near the maximum recall value, which is typical in high-performing models attempting to capture every possible true positive, sometimes at the cost of introducing more false positives. The reported mean Average Precision at an Intersection over Union threshold of 0.5 (mAP@0.5) is 0.98, highlighting the model's exceptional accuracy in detecting prostate gland (Fig. 3). Since the curve pertains to a single class—prostate—the high precision and recall reinforce the model's robustness and reliability for this specific

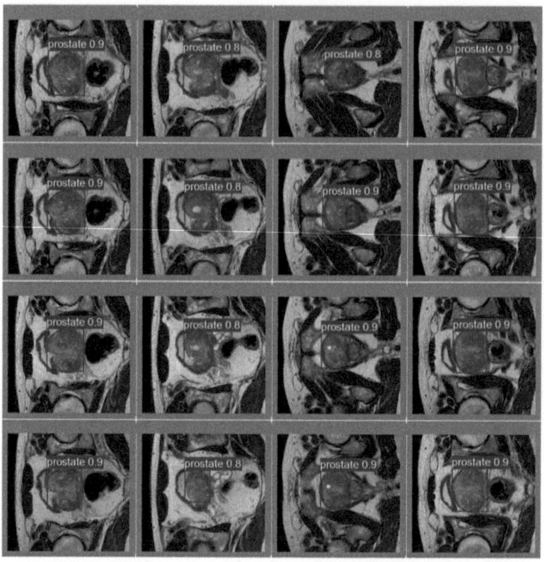

Fig. 2. Prostate gland detection.

detection task. Overall, the curve reflects a highly effective detection model that holds strong potential for integration into clinical workflows, particularly where accurate and early identification of prostate is essential for the detection of the lesions.

The Precision-Confidence Curve illustrates the relationship between prediction confidence and precision in detecting prostate gland. As shown in the graph, precision improves steadily with increasing confidence scores, reaching its peak at a confidence threshold of approximately 0.823, where the model achieves a perfect precision score of 1.00 (Fig. 4). This indicates that at higher confidence levels, the model's predictions are highly reliable, with virtually no false positives. At lower confidence thresholds, the precision is slightly lower—starting around 0.6—but quickly rises, suggesting that while the model is capable of identifying prostate gland, it becomes significantly more accurate as confidence increases. This curve is particularly useful in clinical settings where precision is critical, as it helps determine an optimal confidence threshold that balances detection with the need to minimize false alarms. The strong performance shown in this plot reinforces the effectiveness of YOLOv8 in delivering accurate, predictions in prostate gland detection.

The F1-Confidence Curve shows how the model's F1-score—a harmonic mean of precision and recall—varies with the prediction confidence level. The curve peaks at a confidence threshold of 0.454, where the F1-score reaches 0.97, indicating optimal model performance at this point (Fig. 5). This threshold represents the best trade-off between precision and recall, where the model maintains both high sensitivity (true positive rate) and high accuracy in its predictions. As

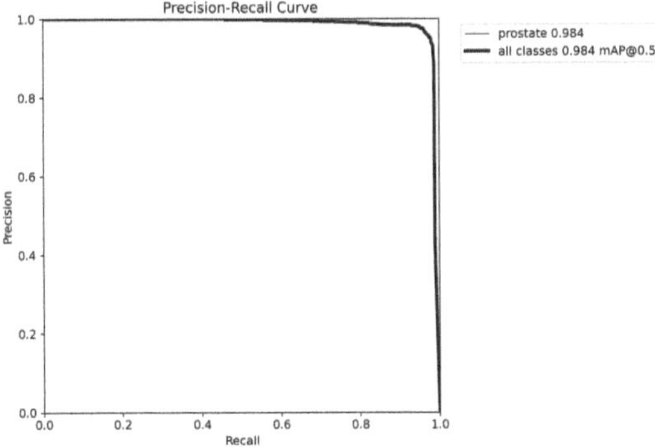

Fig. 3. The figure shows a Precision-Recall Curve with a high mean Average Precision (mAP@0.5) of 0.984, indicating excellent model performance. The curve remains near the top-left corner, reflecting strong precision and recall balance across all classes.

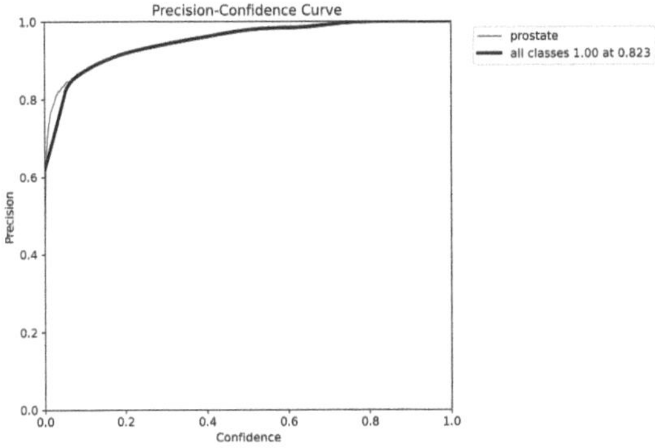

Fig. 4. The Precision-Confidence Curve shows the relationship between precision (0.0–0.8) and confidence thresholds (0.0–1.0) for a classification model. At a confidence threshold of 0.823, the model achieves perfect precision (1.00) across all classes.

the confidence level increases beyond this point, the F1-score begins to decline sharply, suggesting that while the precision may continue to improve, recall drops significantly—fewer true positives are captured, which affects the overall balance. On the other hand, at very low confidence values, the F1-score is also reduced, likely due to a higher number of false positives. The curve demonstrates that the model performs exceptionally well over a wide range of confidence levels, with consistently high F1-scores before the sharp decline. This performance highlights the model's robustness and reliability in detecting prostate gland, especially when calibrated to the optimal confidence threshold.

The Recall-Confidence Curve demonstrates that the model maintains high recall—close to 1.0—across a wide range of low to moderate confidence thresholds. This indicates that, at lower confidence levels, the model is highly sensitive and successfully identifies nearly all true positive cases of prostate gland.

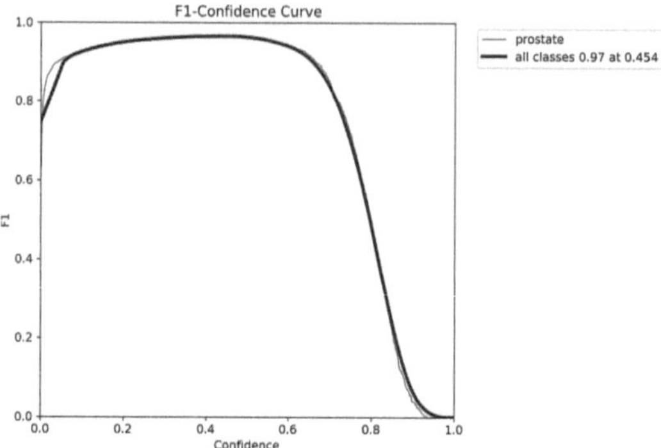

Fig. 5. The F1-Confidence Curve shows performance metrics across confidence thresholds, with Prostate achieving 0.97 F1-score at 0.454 confidence for all classes. The curve ranges from 0.0 to 1.0, displaying varying F1-scores (0.0 to 0.8) at different confidence levels.

However, as the confidence threshold increases beyond approximately 0.6, recall begins to drop sharply. At higher confidence levels, the model becomes more conservative in its predictions, prioritizing precision at the expense of missing some true positives. Notably, the model achieves a recall of 0.99 at a confidence level of 0.000, highlighting that even when minimal confidence is required, it captures nearly all relevant detections (Fig. 6). This is a typical trade-off in object detection systems: as confidence increases, the model becomes stricter in what it considers a valid detection, which reduces the number of false positives but also increases the number of missed true positives. Overall, this curve confirms that YOLOv8 is highly effective in identifying prostate gland when sensitivity is prioritized. The ideal confidence threshold would need to be carefully selected based on the clinical context—lower thresholds for screening scenarios where missing a case is critical, and higher thresholds when false positives must be minimized

The confusion matrix shows that the model performs well in identifying the "prostate" class, with high precision (98.82%) and good recall (93.32%). It correctly predicts 1007 prostate slices but misses 72, and misclassifies 12 background instances as prostate. The F1-score of 96% indicates balanced performance. However, the absence of true negatives suggests a potential class imbalance or weak background detection, highlighting the need for improvement in identifying background regions (Fig. 7).

4   Discussion

Prostate cancer is one of the most common cancers in men and requires early and accurate detection for effective treatment. MRI offers detailed imaging of the prostate gland [1,9], and machine learning models like YOLOv8 are increasingly used to automatically detect prostate regions and cancerous lesions, enhancing

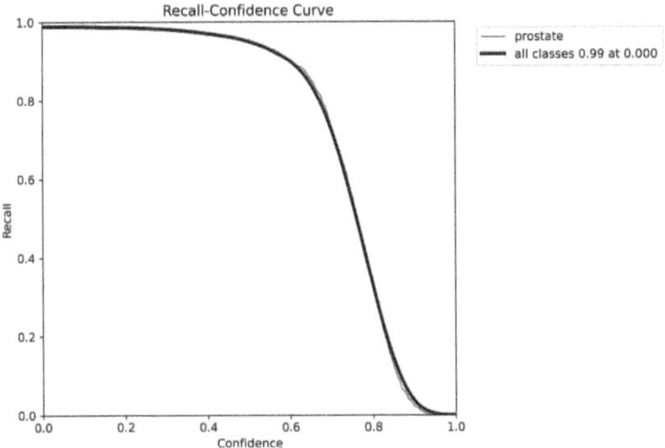

Fig. 6. The Recall-Confidence Curve shows how recall varies with confidence thresholds, with recall decreasing as confidence increases. For all classes, a high confidence score of 0.99 is achieved at a very low threshold (0.000).

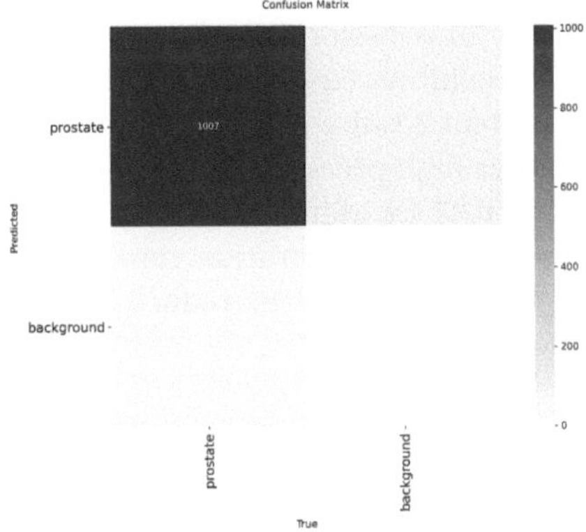

Fig. 7. Confusion Matrix for prostate detection.

diagnostic precision and speed This study demonstrates the strong potential of the YOLOv8 model for accurate and efficient detection of the prostate gland in annotated MRI scans. The primary objective was to evaluate the model's performance in automatically identifying and localizing the prostate gland, a critical anatomical structure in diagnostic and treatment planning for various urological conditions, including prostate cancer [4,8]. This study demonstrates the strong potential of the YOLOv8 model for accurate and efficient detection of prostate gland from annotated MRI scans. Through systematic training over 50 epochs using a high-resolution dataset, the model achieved consistently high performance across multiple key evaluation metrics, including mean Average Precision (mAP), precision, recall, and F1-score.

The Precision-Recall curve remained near the upper boundary across most of the recall range, indicating that YOLOv8 maintains high precision without significantly compromising sensitivity. The model's reported mAP@0.5 of 0.98

further confirms its robustness and reliability in the domain of medical imaging. Additionally, the Precision-Confidence curve revealed a precision of 1.0 at a confidence threshold of 0.823, suggesting that high-confidence predictions are almost always correct—a crucial feature for clinical deployment, where minimizing false positives is critical.

The F1-Confidence curve identified an optimal threshold at 0.454, where the model achieved its peak F1-score of 0.97, reflecting a strong balance between precision and recall. This behavior was consistent across a range of confidence levels, emphasizing the model's reliability under varying conditions. Similarly, the Recall-Confidence curve demonstrated high sensitivity at lower thresholds, with recall values approaching 1.0. Although recall decreased at higher confidence levels, this trade-off between sensitivity and specificity allows for adaptable calibration depending on clinical requirements.

Importantly, the model's predictions were validated in collaboration with medical experts, ensuring that its outputs are not only statistically sound but also clinically meaningful. This adds a significant layer of credibility to the study and supports the feasibility of real-world deployment.

In summary, YOLOv8 exhibits excellent performance in detecting prostate gland on MRI, with consistent results across a variety of confidence thresholds. Its ability to adapt for either high precision or high sensitivity through threshold tuning makes it a flexible tool for different diagnostic settings. These findings highlight YOLOv8's readiness for integration into clinical workflows, where it could significantly aid in early detection, reduce diagnostic subjectivity, and ultimately improve patient outcomes.

References

1. Agnello, L., Comelli, A., Ardizzone, E., Vitabile, S.: Unsupervised tissue classification of brain MR images for voxel-based morphometry analysis. Int. J. Imaging Syst. Technol. **26**(2), 136–150 (2016). https://doi.org/10.1002/ima.22168
2. Ali, M.L., Zhang, Z.: The YOLO framework: a comprehensive review of evolution, applications, and benchmarks in object detection. Computers **13**(12), 336 (2024). https://doi.org/10.3390/computers13120336
3. Ali, M., et al.: Applications of artificial intelligence, deep learning, and machine learning to support the analysis of microscopic images of cells and tissues. J. Imaging **11**(2), 59 (2025). https://doi.org/10.3390/jimaging11020059
4. Ali, M., et al.: Prostate cancer detection: performance of radiomics analysis in multiparametric MRI. In: Foresti, G.L., Fusiello, A., Hancock, E. (eds.) ICIAP 2023. LNCS, vol. 14366, pp. 83–92. Springer, Cham (2023). https://doi.org/10.1007/978-3-031-51026-7_8
5. Alongi, P., et al.: Radiomics analysis of brain [18f] fdg pet/ct to predict Alzheimer's disease in patients with amyloid pet positivity: a preliminary report on the application of SPM cortical segmentation, pyradiomics and machine-learning analysis. Diagnostics **12**(4), 933 (2022). https://doi.org/10.3390/diagnostics12040933
6. Alongi, P., et al.: Artificial intelligence applications on restaging [18F]FDG PET/CT in metastatic colorectal cancer: a preliminary report of morpho-functional radiomics classification for prediction of disease outcome. Appl. Sci. **12**(6), 2941 (2022). https://doi.org/10.3390/app12062941

7. Bai, L., Xu, W.H.: Improved printed circuit board defect detection scheme. Sci. Rep. **15**(1), 2389 (2025). https://doi.org/10.1038/s41598-025-85245-2

8. Basirinia, G., et al.: Theranostic approaches for gastric cancer: an overview of in vitro and in vivo investigations. Cancers **16**(19), 3323 (2024). https://doi.org/10.3390/cancers16193323

9. Benfante, V., et al.: Grading and staging of bladder tumors using radiomics analysis in magnetic resonance imaging. In: Foresti, G.L., Fusiello, A., Hancock, E. (eds.) ICIAP 2023. LNCS, vol. 14366, pp. 93–103. Springer, Cham. (2023). https://doi.org/10.1007/978-3-031-51026-7_9

10. Cairone, L., et al.: Robustness of radiomics features to varying segmentation algorithms in magnetic resonance images. In: Mazzeo, P.L., Frontoni, E., Sclaroff, S., Distante, C. (eds.) ICIAP 2022. LNCS, vol. 13373, pp. 462–472. Springer, Cham (2022). https://doi.org/10.1007/978-3-031-13321-3_41

11. Canfora, I., et al.: A predictive system to classify preoperative grading of rectal cancer using radiomics features. In: Mazzeo, P.L., Frontoni, E., Sclaroff, S., Distante, C. (eds.) ICIAP 2022. LNCS, vol. 13373, pp. 431–440. Springer, Cham (2022). https://doi.org/10.1007/978-3-031-13321-3_38

12. Comelli, A., et al.: A smart and operator independent system to delineate Tumours in positron emission tomography scans. Comput. Biol. Med. **102**, 1–15 (2018). https://doi.org/10.1016/j.compbiomed.2018.09.002

13. Correia, E.T.d.O., Baydoun, A., Li, Q., Costa, D.N., Bittencourt, L.K.: Emerging and anticipated innovations in prostate cancer mri and their impact on patient care. Abdominal Radiol. **49**(10), 3696–3710 (2024). https://doi.org/10.1007/s00261-024-04423-4

14. Giaccone, P., Benfante, V., Stefano, A., Cammarata, F.P., Russo, G., Comelli, A.: Pet images atlas-based segmentation performed in native and in template space: a radiomics repeatability study in mouse models. In: Mazzeo, P.L., Frontoni, E., Sclaroff, S., Distante, C. (eds.) ICIAP 2022. LNCS, vol. 13373, pp. 351–361. Springer, Cham (2022). https://doi.org/10.1007/978-3-031-13321-3_31

15. Liu, J., Liu, X., Chen, H., Luo, S.: MDD-YOLOv8: a multi-scale object detection model based on yolov8 for synthetic aperture radar images. Appl. Sci. (2076-3417) **15**(4) (2025). https://doi.org/10.3390/app15042239

16. Piras, A., et al.: Artificial intelligence and statistical models for the prediction of radiotherapy toxicity in prostate cancer: A systematic review. Appl. Sci. **14**(23), 10947 (2024). https://doi.org/10.3390/app142310947

17. Russo, G., et al.: Feasibility on the use of radiomics features of 11 [c]-met PET/CT in central nervous system Tumours: preliminary results on potential grading discrimination using a machine learning model. Curr. Oncol. **28**(6), 5318–5331 (2021). https://doi.org/10.3390/curroncol28060444

18. Salamone, G., et al.: Vacuum-assisted wound closure with mesh-mediated fascial traction achieves better outcomes than vacuum-assisted wound closure alone: a comparative study. World J. Surg. 1–8 (2017). https://doi.org/10.1007/s00268-017-4354-3

19. Salvaggio, G., et al.: Deep learning network for segmentation of the prostate gland with median lobe enlargement in t2-weighted MR images: comparison with manual segmentation method. Curr. Probl. Diagn. Radiol. **51**(3), 328–333 (2022). https://doi.org/10.1067/j.cpradiol.2021.06.006

20. Sharma, A., Kumar, V., Longchamps, L.: Comparative performance of yolov8, YOLOv9, YOLOv10, YOLOV11 and faster R-CNN models for detection of multiple weed species. Smart Agricu. Technol. **9**, 100648 (2024). https://doi.org/10.1016/j.atech.2024.100648
21. Vernuccio, F., et al.: Diagnostic performance of qualitative and radiomics approach to parotid gland tumors: which is the added benefit of texture analysis? Br. J. Radiol. **94**(1128), 20210340 (2021). https://doi.org/10.1259/bjr.20210340

Artificial Intelligence-Assisted Image Segmentation in Scratch Wound Healing Assays Using Threshold and Level Set Methods

Viviana Benfante[1]($\boxtimes$) iD, Muhammad Ali[2] iD, Farhan Khan[2], Sebastiano Piana[1], Alessandro Sperandeo[1], Anthony Yezzi[3] iD, and Albert Comelli[2] iD

[1] Pharmaceutical Factory, La Maddalena S.P.A., Via San Lorenzo Colli, 312/d, 90146 Palermo, Italy
vivianabenfante3@gmail.com
[2] Ri.MED Foundation, Via Bandiera 11, 90133 Palermo, Italy
[3] School of Electrical and Computer Engineering, Georgia Institute of Technology, Atlanta 30332, GA, USA

Abstract. In wound healing assays, accurate segmentation is essential to quantifying cell migration. Manual methods are labor-intensive and subject to individual interpretation. This study presents a robust, automated image analysis framework that integrates entropy- based thresholding techniques in conjunction with level set-based contour analyses to segment scratch wound regions in microscopy images. PC3 Prostate cancer cells were treated with Indole-3-carbinol (I3C) and Docetaxel, and the method demonstrated consistent accuracy across all treatment conditions. Shanbhag entropy-driven thresholding achieved Dice scores up to 0.98 as well, particularly excelling in treated samples. Level set models, including Local Binary Fitting (LBF) and Distance Regularized Level Set Evolution (DRLSE), provided robustness against intensity variation, with DSCs ranging from 0.99 to 0.97. Experimental results show drug-induced inhibition of wound closure, validating both the segmentation pipeline and its application to pharmacological screening. The framework offers a scalable solution for high-throughput wound healing analysis.

Keywords: Scratch assay · Cell Migration · Image Segmentation · Level Set Methods · Prostate cancer

1 Introduction

Cell migration is a crucial process under physiological and pathological conditions, including tissue regeneration, immune responses, and cancer metastasis [5]. Scratch wound healing assay is a widely adopted in vitro technique to assess collective cell migration by creating a controlled gap in a cell monolayer and observing its closure over time [15]. However, traditional methods of analyzing

E. Rodolà et al. (Eds.): ICIAP 2025 Workshops, LNCS 16169, pp. 255–265, 2026.
https://doi.org/10.1007/978-3-032-11317-7_22

these assays are labor-intensive and subject to subjective variability, limiting their scalability and throughput, especially for high-content screenings [14].

Advancements in image processing and artificial intelligence have opened up new possibilities for automating the segmentation and quantification of the region in wound scratch assays [1]. Accurate segmentation remains challenging due to uneven illumination, heterogeneous cell morphology, and non-uniform backgrounds typically present in microscopy images. To address these issues, this study introduces a robust, adaptive image segmentation framework that integrates entropy-based with level set-based active contour models [21].

This study aims to evaluate the inhibitory effects of Docetaxel (DOC) [20] and Indole-3-carbinol (I3C) [10], both individually and in combination, on the migration of PC-3 prostate cancer cells using an automated image analysis framework [22]. By integrating in MATLAB with advanced level set models specifically Local Binary Fitting (LBF) and Distance Regularized Level Set Evolution (DRLSE) the proposed method enables accurate and reproducible segmentation of wound areas across multiple time points. This approach improves the reliability of wound quantification and facilitates precise assessment [12] of treatment-induced changes in cancer cell migration dynamics.

2 Materials and Methods

2.1 Cell Culture

PC3 human prostate cancer cell line was provided by the Gynecological Oncology Research Unit of the European Institute of Oncology under the direction of Dr. Ugo Cavallaro. Cells were cultured in F-12 medium supplemented with 10% fetal bovine serum, 4 mM L-glutamine, and 1% penicillin-streptomycin. Cells were cultured at 37°C in an atmosphere of 95% air and 5% CO_2.

2.2 Scratch Wound Healing Migration Assay

PC-3 prostate cancer cells (0.3×10^6) were added to 6-well microtiter plates with complete culture medium. After 24 h of seeding, only 60 µM I3C dissolved in 0.5% DMSO was added to one well for 24 h, while a pretreatment of 60 µM I3C dissolved in 0.5% DMSO was added for 24 h followed by treatment with 1.8 nM DOC dissolved in 0.5% DMSO for 24 h in another well. Control samples were left untreated. At the end of the treatments, 48 and 24 h later, respectively (90% cell confluence), the cell layer in the well was scratched and wounded by scraping with a p10 pipette tip. Subsequently, the cells were washed with phosphate-buffered saline to wash away the cellular waste and the wounded cultures were incubated in serum-free culture medium [16].

2.3 Semi-automatic Methods for Image Segmentation

From 1 to 6 h post-scratch and 24 h post-scratch, images were taken of each wound using the fully integrated EVOS M5000 Imaging System (ThermoFisher

Scientific). Using a 4x fluorite magnification, images in RGB were acquired with a resolution of 2048×1536 pixels without artifacts, 0.13 numerical aperture (NA) and 10.58 mm as working distance (WD). Images were acquired of all three samples to be compared: a) PC3 incubated with I3C (alone) for 24 h, b)PC3 pre-treated with I3C for 24 h and then treated with DOC for 24 h and, c)PC3 not administered with any compound (control). To establish a gold standard, each image was manually delineated using the polygon selection tool. As a result, the algorithm could later be validated. Area measurements were extracted from the segmented regions to quantify changes in wound size over time, providing a robust metric for assessing cell migration.

To segment and quantify wound regions in a time series of scratch assay images, we employed an enhanced thresholding pipeline that includes Shanbhag entropy-based method.

2.3.1 Thresholding Methods

2.3.1.1 ROI Detection via Edge Projection. Let $E(x, y)$ denote the edge intensity at pixel (x, y), obtained from the Sobel edge detection of the grayscale image. The vertical edge profile is computed as a column-wise summation:

$$P(y) = \sum_{x=1}^{W} E(x, y) \tag{1}$$

where W is the image width. This profile, $P(y)$, is then inverted and smoothed using a moving average filter to reduce noise and enhance the prominence of the scratch gap. The scratch center is identified as the location of the maximum value in the smoothed inverted profile.

To extract the Region of Interest (ROI) containing the scratch and its boundary regions, a dynamic expansion strategy is employed. Starting from the identified scratch center, the profile is scanned upwards and downwards until the edge activity drops below a calculated threshold T, defined as:

$$T = \mu_P + \alpha \sigma_P \tag{2}$$

where μ_P and σ_P are the mean and standard deviation of the smoothed profile, respectively, and $\alpha = 0.4$ is an empirically chosen parameter. The upper and lower bounds of the ROI are determined based on this condition, and a small padding is added on both sides to ensure that the migrating cell fronts are included within the extracted ROI.

2.3.1.2 Shanbhag Thresholding. We also implemented Shanbhag entropy-based global thresholding method [25]. This method computes the optimal threshold t by maximizing the sum of entropies between the foreground and background:

$$H(t) = -\sum_{i=0}^{t} \frac{p_i}{P_t} \log\left(\frac{p_i}{P_t}\right) - \sum_{i=t+1}^{L} \frac{p_i}{1 - P_t} \log\left(\frac{p_i}{1 - P_t}\right) \tag{3}$$

where p_i represents the normalized histogram and $P_t = \sum_{i=0}^{t} p_i$.

The scratch area A_{scratch} was estimated from the number of background pixels N_b, scaled by the square of the physical pixel size s (in μm), extracted from image metadata:

$$A_{\text{scratch}} = N_b \cdot s^2 \quad [\mu m^2] \tag{4}$$

2.3.1   Level Set Methods To refine the segmentation further, we employed two level set-based active contour models: the Local Binary Fitting (LBF) model [18] and the Distance Regularized Level Set Evolution (DRLSE) model [19].

The level set function ϕ was initialized from image intensity via thresholding:

$$\phi_0(x) = c_0 \cdot (1 - 2 \cdot \mathbb{I}(I(x) < T)) \tag{5}$$

Here, $T = \frac{\max(I) - \min(I)}{2}$ is the threshold, $\mathbb{I}$ is the indicator function, and c_0 is a constant defining the signed distance scale.

2.3.2.1 Local Binary Fitting (LBF) Model. The *Local Binary Fitting (LBF)* model [18] is a region-based active contour method that handles intensity inhomogeneity using local image statistics. It minimizes the following energy:

$$
\begin{aligned}
E_{\text{LBF}}(\phi) =& \lambda_1 \int_\Omega |I(x) - f_1(x)|^2 K_\sigma(x - y) H(\phi(y)) \, dy \\
&+ \lambda_2 \int_\Omega |I(x) - f_2(x)|^2 K_\sigma(x - y)[1 - H(\phi(y))] \, dy \\
&+ \nu \int_\Omega |\nabla H(\phi(x))| \, dx
\end{aligned}
\tag{6}
$$

where: $I(x)$ is the image intensity; $f_1(x)$, $f_2(x)$ are local means inside and outside the contour; $\phi(x)$ is the level set function; $H(\cdot)$ and $\delta(\cdot)$ are regularized Heaviside and Dirac delta functions; K_σ is a Gaussian kernel; λ_1, λ_2, ν are weighting parameters.

The kernel K_σ provides spatial weighting, enhancing detection under uneven lighting.

Gradient descent evolution is given by:

$$\frac{\partial \phi}{\partial t} = \delta(\phi) \left[-\lambda_1 e_1(x) + \lambda_2 e_2(x) \right] + \nu \text{div} \left(\frac{\nabla \phi}{|\nabla \phi|} \right) \tag{7}$$

with:

$$e_1(x) = \int_\Omega K_\sigma(x - y)|I(y) - f_1(x)|^2 dy, \quad e_2(x) = \int_\Omega K_\sigma(x - y)|I(y) - f_2(x)|^2 dy$$

This flow balances local fitting and contour smoothness for robust boundary detection.

2.3.2.2 Distance Regularized Level Set Evolution (DRLSE). The *Distance Regularized Level Set Evolution (DRLSE)* model by Li et al. [19] improves level set stability by incorporating a distance regularization term, eliminating the need for reinitialization and maintaining the signed distance property during evolution.

The total energy functional is:

$$E_{\text{DRLSE}}(\phi) = \mu R(\phi) + \lambda L(\phi) + \alpha A(\phi) \tag{8}$$

where $R(\phi)$ enforces distance regularity; $L(\phi)$ attracts the contour to edges; $A(\phi)$ controls area-based motion; μ, λ, and α are weights.

These terms are defined as:

$$R(\phi) = \int_{\Omega} \frac{1}{2}(|\nabla\phi| - 1)^2 dx, \quad L(\phi) = \int_{\Omega} \delta(\phi)|\nabla\phi|g(x)\, dx, \quad A(\phi) = \int_{\Omega} H(\phi)g(x)\, dx \tag{9}$$

with edge indicator:

$$g(x) = \frac{1}{1 + |\nabla(G_\sigma * I)(x)|^2} \tag{10}$$

where G_σ is a Gaussian kernel and $*$ denotes convolution.

The gradient descent flow for minimizing the energy is:

$$\frac{\partial\phi}{\partial t} = \mu\left(\Delta\phi - \text{div}\left(\frac{\nabla\phi}{|\nabla\phi|}\right)\right) + \lambda\delta(\phi)\text{div}\left(g(x)\frac{\nabla\phi}{|\nabla\phi|}\right) + \alpha g(x)\delta(\phi) \tag{11}$$

where $\Delta\phi$ is the Laplacian, $\text{div}(\cdot)$ the divergence, $\delta(\phi)$ the regularized Dirac delta, and $\nabla\phi$ the gradient of ϕ. This evolution balances regularization, edge attraction, and area control for robust contour evolution.

3   Results

Figure 1 through 3 provide visual comparisons of segmentation outputs at 0 h and 24 h for each method. The overlays and binary masks confirm accurate scratch area delineation [11] and allow for clear tracking of cell movement over time. The Shanbhag thresholding methods provided faster, computationally efficient segmentation, while level set methods yielded smoother contours and were more robust to intensity inhomogeneities. The performance of various image segmentation techniques was evaluated to quantify wound closure in scratch tests under three experimental conditions: untreated, treated with I3C, and treated with both I3C and DOC. Segmentation accuracy was assessed using Dice scores, and changes in scratch area were tracked over 24 h. Figure 4(a) illustrate the temporal progression of scratch closure measured via Shanbhag thresholding respectively, and level set methods (LBF and DRLSE, Fig. 4(b,c). Across all segmentation methods: untreated cells showed a significant reduction in scratch area, indicating active cell migration and wound closure. Treated cells (I3C and I3C + DOC) exhibited reduced closure rates, reflecting inhibited cell migration likely due to drug effects. Among treated samples, the combination therapy further

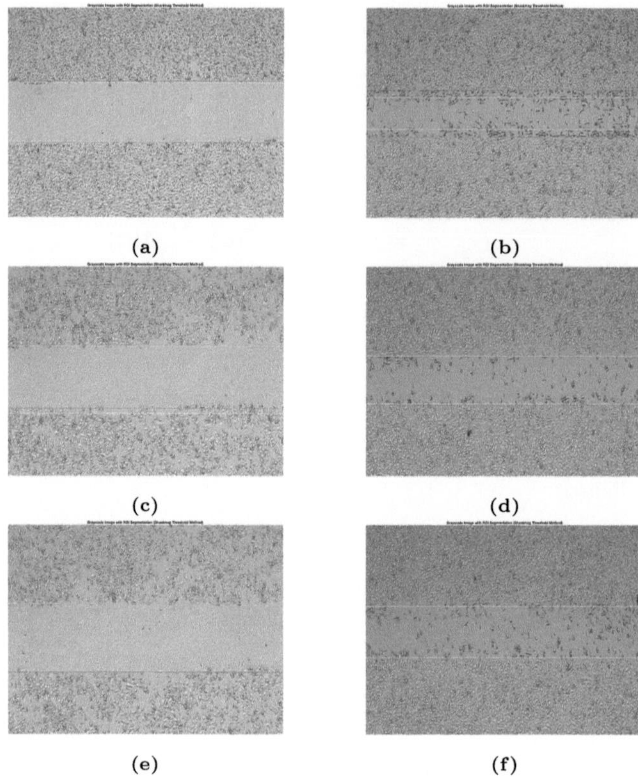

Fig. 1. Scratch assay segmentation results using the Shanbhag thresholding method. The grayscale images represent original wound healing assay frames. Red overlays denote segmented cell regions within the scratch area, extracted using shanbhag thresholding method. Green horizontal lines mark the automatically detected region of interest (ROI) corresponding to the scratch, while yellow lines indicate the ground truth boundaries manually annotated for evaluation. Subfigures (a) and (b) show untreated cell samples (NT PC3) at 0 h and 24 h, respectively; (c) and (d) correspond to treated cell sample (PC3+I3C) at 0 h and 24 h; and (e) and (f) represent treated cell samples (PC3+I3C+DOC) at 0 h and 24 h. (Color figure online)

suppressed wound closure compared to I3C alone. Table 1 presents Dice scores for three segmentation methods applied to the different experimental conditions. All methods achieved comparable performance, with scores ranging between 0.99 and 0.96 across conditions. Level set methods (LBF, DRLSE) demonstrated consistent accuracy [23] across treated and untreated samples (Fig. 1).

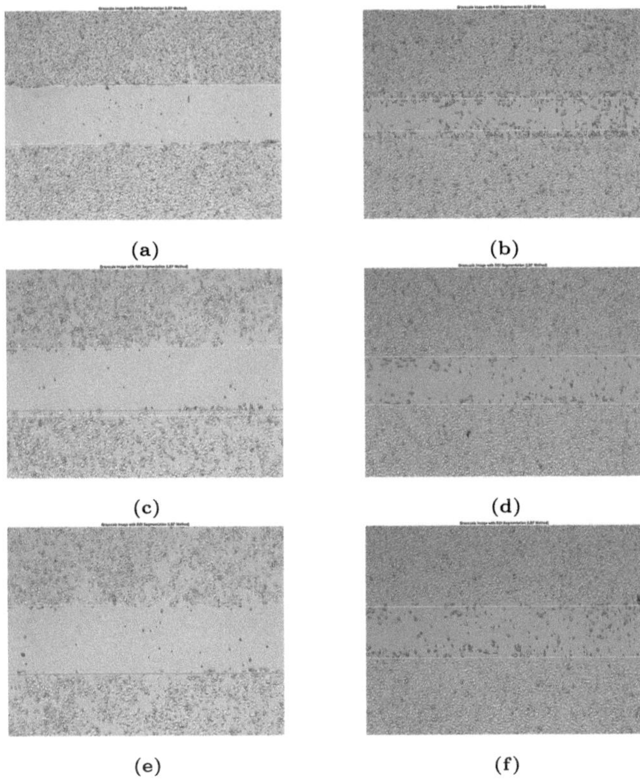

Fig. 2. Scratch assay segmentation results using the Local Binary Fitting (LBF) level set method. Subfigures (a) and (b) show untreated cell sample (NT PC3) at 0 h and 24 h, respectively; (c) and (d) correspond to treated cell sample (PC3+I3C) at 0 h and 24 h; and (e) and (f) represent treated cell sample (PC3+I3C+DOC) at 0 h and 24 h.

Table 1. Comparison of Dice scores for scratch area segmentation among treated (PC3+I3C, and PC3+I3C+DOC) and untreated (NT PC3) samples

Methods	NT PC3	PC3 + I3C	PC3+I3C+DOC
Shanbhag threshold method	0.976 ± 0.037	0.935 ± 0.226	0.940 ± 0.227
Local binary fitting level set method	0.976 ± 0.031	0.987 ± 0.010	0.992 ± 0.004
Distance regularized level set evolution	0.972 ± 0.030	0.984 ± 0.010	0.980 ± 0.009

4   Discussion

The experimental results show the efficacy of the proposed segmentation framework for analyzing scratch wound healing assays under challenging imaging conditions [17]. The combination of entropy-based thresholding, and level set methods [9,26] enabled accurate and reproducible quantification of wound areas across

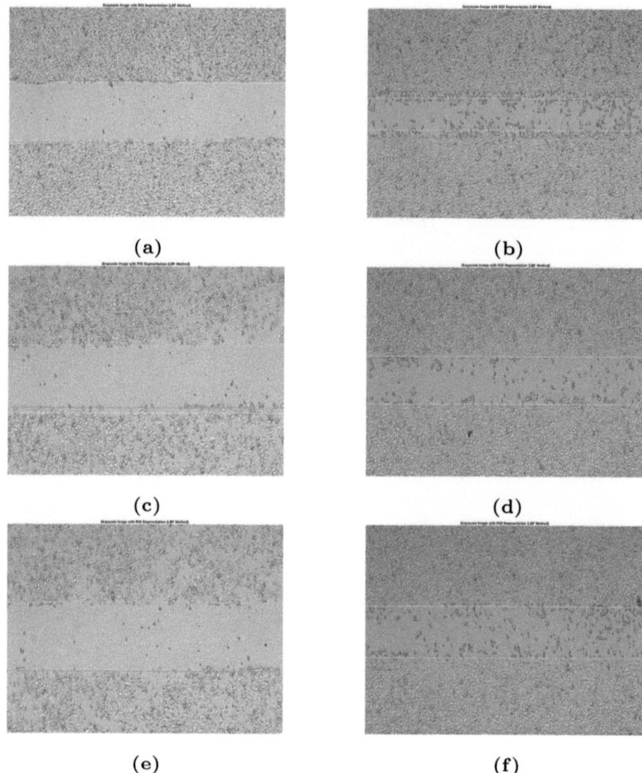

Fig. 3. Scratch assay segmentation results using the Distance Regularized Level Set Evolution (DRLSE) method. Subfigures (a) and (b) show untreated cell samples (NT PC3) at 0 h and 24 h, respectively; (c) and (d) correspond to treated cell samples (PC3+I3C) at 0 h and 24 h; and (e) and (f) represent treated cell samples (PC3+I3C+DOC) at 0 h and 24 h.

multiple time points (0 to 24h) and treatment conditions [2]. Shanbhag thresholding methods demonstrated fast and computationally efficient performance, with Shanbhag slightly outperforming others in untreated samples based on Dice scores. These methods are particularly suitable for high-throughput screening where rapid segmentation is critical. However, they were somewhat sensitive to illumination variability and uneven background intensity. In contrast, the level set-based models—Local Binary Fitting (LBF) and Distance Regularized Level Set Evolution (DRLSE)—offered enhanced robustness against intensity inhomogeneity, producing smoother segmentation boundaries. LBF, in particular, effectively adapted to local image characteristics, while DRLSE preserved contour regularity without requiring reinitialization, reducing computational overhead.

Functionally, wound closure analysis showed clear drug-induced inhibition of PC-3 cell migration. Untreated cells exhibited consistent wound healing, while treatment with I3C and its combination with DOC markedly suppressed closure

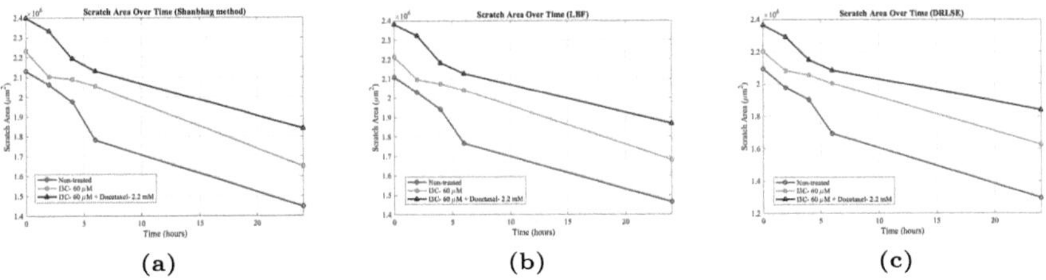

(a)   (b)   (c)

Fig. 4. These graphs show scratch area changes over time (024 h) using three image analysis methods: (a) Shanbhag threshold, (b) LBF, and (c) DRLSE. Scratch closure was assessed under three conditions: untreated (red), I3C (60 µM , green), and I3C + DOC (60 µM + 1.8 nM, blue). Untreated cells showed greater scratch closure, indicating higher cell migration, while both drug treatments reduced closure, suggesting inhibitory effects.

rates. These observations align with known cytotoxic properties of the compounds and validate the utility of automated segmentation [4, 8, 13] in pharmacological assessment

5   Conclusion

A robust and adaptable image analysis pipeline is presented in this study for quantifying the rate of wound healing associated with cut wounds based on the migratory behavior of tumor cell lines. This pipeline integrates entropy-based thresholding with advanced level set models approach achieved high segmentation accuracy and was effective in quantifying wound closure dynamics in response to polyphenolic and drug treatments [6].

By automating scratch area estimation, the proposed method reduces manual labor and subjective bias in image analysis [3, 7], enabling scalable, high-content screening of cell migration behavior. Future work will explore integration with deep learning [1, 24] models to further enhance segmentation generalizability and extend the framework to 3D and time-lapse imaging modalities.

Acknowledgment. This work was supported by the following grant and project: "RADIATIONS" under the PNRR—Next Generation EU, Mission 4, Component 2—Investment 1.5—Cascading Call—Ecosystem of Innovation "THE—Tuscany Health Ecosystem"—code ECS00000017—CUP B83C22003930001, to Viviana Benfante, Alessandro Sperandeo, and Albert Comelli;

References

1. Ali, M., et al.: Applications of artificial intelligence, deep learning, and machine learning to support the analysis of microscopic images of cells and tissues. J. Imaging **11**(2), 59 (2025). https://doi.org/10.3390/jimaging11020059

2. Ali, M., et al.: Prostate cancer detection: Performance of radiomics analysis in multiparametric MRI. In: International Conference on Image Analysis and Processing, pp. 83–92. Springer (2023). https://doi.org/10.1007/978-3-031-51026-7_8

3. Ali, M., Benfante, V., Di Raimondo, D., Laudicella, R., Tuttolomondo, A., Comelli, A.: A review of advances in molecular imaging of rheumatoid arthritis: From in vitro to clinic applications using radiolabeled targeting vectors with technetium-99m. Life **14**(6), 751 (2024). https://doi.org/10.3390/life14060751

4. Alongi, P., et al.: Artificial intelligence applications on restaging [18f] fdg pet/ct in metastatic colorectal cancer: A preliminary report of morpho-functional radiomics classification for prediction of disease outcome. Appl. Sci. **12**(6), 2941 (2022). https://doi.org/10.3390/app12062941

5. Azeem, M., et al.: Assessing anticancer, antidiabetic, and antioxidant capacities in green-synthesized zinc oxide nanoparticles and solvent-based plant extracts. Heliyon **10**(14) (2024). https://doi.org/10.1016/j.heliyon.2024.e34073

6. Basirinia, G., et al.: Theranostic approaches for gastric cancer: an overview of in vitro and in vivo investigations. Cancers **16**(19), 3323 (2024). https://doi.org/10.3390/cancers16193323

7. Benfante, V., et al.: Grading and staging of bladder tumors using radiomics analysis in magnetic resonance imaging. In: International Conference on Image Analysis and Processing, pp. 93–103. Springer (2023). https://doi.org/10.1007/978-3-031-51026-7_9

8. Cairone, L., et al.: Robustness of radiomics features to varying segmentation algorithms in magnetic resonance images. In: International Conference on Image Analysis and Processin, pp. 462–472. Springer (2022). https://doi.org/10.1007/978-3-031-13321-3_41

9. Canfora, I., et al.: A predictive system to classify preoperative grading of rectal cancer using radiomics features. In: International Conference on Image Analysis and Processing, pp. 431–440. Springer (2022). https://doi.org/10.1007/978-3-031-13321-3_38

10. Centofanti, F., et al.: Synthetic methodologies and therapeutic potential of indole-3-carbinol (i3c) and its derivatives. Pharmaceuticals **16**(2), 240 (2023). https://doi.org/10.3390/ph16020240

11. Comelli, A., et al.: Tissue classification to support local active delineation of brain tumors. In: Annual Conference on Medical Image Understanding and Analysis, pp. 3–14. Springer (2019). https://doi.org/10.1007/978-3-030-39343-4_1

12. Corso, R., Khan, F., Yezzi, A., Comelli, A.: Features for active contour and surface segmentation: a review. Archives of Computational Methods in Engineering, pp. 1–27 (2025). https://doi.org/10.1007/s11831-025-10300-0

13. Giaccone, P., Benfante, V., Stefano, A., Cammarata, F.P., Russo, G., Comelli, A.: Pet images atlas-based segmentation performed in native and in template space: a radiomics repeatability study in mouse models. In: International Conference on Image Analysis and Processing, pp. 351–361. Springer (2022). https://doi.org/10.1007/978-3-031-13321-3_31

14. Grada, A., Otero-Vinas, M., Prieto-Castrillo, F., Obagi, Z., Falanga, V.: Research techniques made simple: analysis of collective cell migration using the wound healing assay. J. Investig. Dermatol. **137**(2), e11–e16 (2017). https://doi.org/10.1016/j.jid.2016.11.020

15. Jonkman, J.E., et al.: An introduction to the wound healing assay using live-cell microscopy. Cell Adhes. migration **8**(5), 440–451 (2014). https://doi.org/10.4161/cam.36224

16. Kauanova, S., Urazbayev, A., Vorobjev, I.: The frequent sampling of wound scratch assay reveals the opportunity window for quantitative evaluation of cell motility-impeding drugs. Front. Cell Develop. Biol. **9**, 640972 (2021). https://doi.org/10.3389/fcell.2021.640972

17. Laudicella, R., et al.: [68ga] dotatoc pet/ct radiomics to predict the response in gep-nets undergoing [177lu] dotatoc prrt: "The theragnomics" concept. Cancers **14**(4), 984 (2022). https://doi.org/10.3390/cancers14040984

18. Li, C., Kao, C.Y., Gore, J.C., Ding, Z.: Implicit active contours driven by local binary fitting energy. In: Proceedings of the 2007 IEEE Conference on Computer Vision and Pattern Recognition (CVPR), pp. 1–7. IEEE (2007). https://doi.org/10.1109/CVPR.2007.383014

19. Li, C., Xu, C., Gui, C., Fox, M.D.: Distance regularized level set evolution and its application to image segmentation. IEEE Trans. Image Process. **19**(12), 3243–3254 (2010). https://doi.org/10.1109/TIP.2010.2069690

20. Montero, A., Fossella, F., Hortobagyi, G., Valero, V.: Docetaxel for treatment of solid tumours: a systematic review of clinical data. Lancet Oncol. **6**(4), 229–239 (2005). https://doi.org/10.1016/S1470-2045(05)70094-2

21. Mouritzen, M.V., Jenssen, H.: Optimized scratch assay for in vitro testing of cell migration with an automated optical camera. Journal of visualized experiments: JoVE (138), 57691 (2018). https://doi.org/10.3791/57691

22. Piras, A., et al.: Artificial intelligence and statistical models for the prediction of radiotherapy toxicity in prostate cancer: a systematic review. Appl. Sci. **14**(23), 10947 (2024). https://doi.org/10.3390/app142310947

23. Russo, G., et al.: Feasibility on the use of radiomics features of 11 [c]-met pet/CT in central nervous system tumours: Preliminary results on potential grading discrimination using a machine learning model. Curr. Oncol. **28**(6), 5318–5331 (2021). https://doi.org/10.3390/curroncol28060444

24. Salvaggio, G., et al.: Deep learning network for segmentation of the prostate gland with median lobe enlargement in t2-weighted mr images: comparison with manual segmentation method. Curr. Probl. Diagn. Radiol. **51**(3), 328–333 (2022). https://doi.org/10.1067/j.cpradiol.2021.06.006

25. Shanbhag, A.G.: Utilization of information measure as a means of image thresholding. Graph. Models Image Proc. **56**(5), 414–419 (1994). https://doi.org/10.1006/cgip.1994.1037

26. Vernuccio, F., et al.: Diagnostic performance of qualitative and radiomics approach to parotid gland tumors: which is the added benefit of texture analysis? Br. J. Radiol. **94**(1128), 20210340 (2021). https://doi.org/10.1259/bjr.20210340

Unsupervised MR to CT Synthesis: A Comparative Analysis of Explicit Structural Constrained Adversarial Learning and Adversarial Diffusion Models

Luca Cruciata[1,2], Muhammad Ali[2], Salvatore Contino[1], Viviana Benfante[3], Roberto Pirrone[1], and Albert Comelli[2]

[1] Department of Engineering, University of Palermo, Viale delle Scienze, Palermo, 90128 Palermo, Italy

[2] Ri.MED Foundation, Via Bandiera 11, 90133 Palermo, Italy
acomelli@fondazionerimed.com

[3] Department of Health Promotion, Mother and Child Care, Internal Medicine and Medical Specialties, Molecular and Clinical Medicine, University of Palermo, 90127 Palermo, Italy

Abstract. This study employs deep learning algorithms to minimize ionizing radiation exposure during radiotherapy planning while preserving essential diagnostic information. We utilized CycleGAN+U-Net and SynDiff models to generate synthetic computed tomography (sCT) images from T1-weighted MRI scans of 37 patients exhibiting a spectrum of neurological conditions, ranging from mild cognitive impairment to severe Alzheimer's disease. Model development and evaluation employed a rigorous k-fold cross-validation strategy. The dataset was partitioned, reserving 4 subjects as an independent test set. The remaining 33 subjects were utilized in a five-fold cross-validation process, systematically rotating subsets for training and validation to optimize model generalizability and performance assessment. From the comparison of the performances of the two applied models, SynDiff obtained higher segmentation accuracy values of ME (1066.92), MAE (1066.92), MSE (1474882.42), RMSE (1212.81), PSNR (34.67) and SSIM (0.001) than Cycle Gan + Unet and can be used to generate accurate sCT from brain MRI images. A key advantage of this AI-based sCT generation is the significant reduction in patient ionizing radiation exposure. This approach facilitates the preferential use of MRI, leveraging its superior soft-tissue contrast to provide essential information for detecting, monitoring, and diagnosing intracranial pathologies

Keywords: MRI · sCT · Deep Learning · Cycle Gan · Unet · SynDiff · Brain

1   Introduction

The scientific community has shown particular interest in replacing computed tomography (CT) with magnetic resonance imaging (MRI). MRI provides safer and more precise planning, especially for patients who require long-term care and wish to minimize the risk of side effects associated with radiation exposure.

Traditional radiation therapy (RT) relies on CT scans that provide diagnostic images for accurate dose calculation. These images allow precise targeting of the radiation beam using linear accelerators (linacs), which helps to reduce the ability of diseased cells to grow and reproduce [15]. However, exposure to ionizing radiation poses significant health risks, especially for patients undergoing repeated imaging. As the dose of X-rays increases, so does the frequency of cellular damage and the incidence of cancerous neoplasms.

Recent research has focused on using artificial intelligence (AI) to combine information from multiple imaging modalities [2]. Compared to CT, MRI offers superior contrast and soft tissue delineation [1,3,9]. The notion that all steps in the RT chain may be performed using MRI has grown in popularity [11]. This is due to the ability to derive CT-equivalent information from MRI images, enabling dose calculation during treatment planning and DRR-based patient positioning, which helps minimize radiation exposure to healthy tissues.

MRI-derived synthetic computed tomography (sCT) images are an active area of research. Wong et al. [10] presented a practical application of sCT in 2020 for lymphoma treatment using intensity-modulated radiation therapy (IMRT). Traditionally, total marrow and lymphoid irradiation are based on CT images. However, MRI scans are typically easier and faster to segment, thanks to their superior ability to highlight soft tissues.

Accurate sCT generation from MRI is also vital for PET attenuation correction in PET/MRI hybrid systems [7,8,13,18]. Several automatic sCT generation methods exist in the literature, and they can be classified into three main categories:

1. **Tissue segmentation-based approaches:** These involve segmenting MRI voxels into tissues such as air, fat, soft tissue, and bone, followed by mass density assignment to synthesize CT images [4,6,14,16].
2. **Atlas-based approaches:** These use image registration to align a target MRI to an atlas image, enabling image warping to generate an sCT [14,16].
3. **Learning-based approaches:** These employ statistical learning or model fitting to map MRI voxel intensities to CT values, including the use of deep learning techniques [14,16].

In recent years, deep learning has enabled diagnostic systems to plan radiotherapy starting solely from MRI images. In this context, our goal was to compare two deep learning models for generating sCT images from MRI: i) Cycle-GAN + UNet and ii) SynDiff.

These models were evaluated using a dataset of T1-weighted MRI and CT brain scans from 37 individuals, some with mild brain deterioration and others with advanced Alzheimer's disease [17].

2   Materials and Methods

2.1   Dataset

This study utilized the CERMEP-IDB-MRXFDG dataset, which includes T1-weighted MRI brain scans and CT images acquired from 37 subjects. These subjects presented varying degrees of cerebral deterioration, ranging from mild atrophy to clinically diagnosed Alzheimer's disease. All data were anonymized in full compliance with privacy regulations and data protection protocols.

To ensure anatomical consistency across subjects, co-registration techniques were employed to align the same brain regions within each individual's imaging data. Beyond classical co-registration, the dataset was also processed in Montreal Neurological Institute (MNI) space. Within this space, each image was encoded into gray and white matter probability maps, allowing each voxel to be assigned a probabilistic classification corresponding to specific brain tissue types.

All volumes underwent a rigorous preprocessing pipeline, which included normalization, intensity standardization, and background removal. These steps ensured that the dataset was of high quality, homogeneous across subjects, and suitable for further computational analysis and deep learning model training.

2.2   MRI and CT Acquisition and Reconstruction

MRI sequences were obtained on a Siemens Sonata 1.5 T scanner. Three-dimensional anatomical T1-weighted sequences (MPRAGE) were acquired in sagittal orientation (TR = 2400 ms, TE = 3.55 ms, inversion time = 1000 ms, flip angle = $8°$). The images were reconstructed into a $160 \times 192 \times 192$ matrix with voxel dimensions of $1.2 \times 1.2 \times 1.2$ mm^3 (axial field of view = 230.4 mm). Sagittal Fluid-Attenuated Inversion Recovery (FLAIR) images (TR = 6000 ms, TE = 354 ms, inversion time = 2200 ms, flip angle = $180°$) were acquired with a $176 \times 196 \times 256$ matrix and a voxel size of $1.2 \times 1.2 \times 1.2$ mm^3 (axial field of view = 307.2 mm).

CT data were acquired on a Siemens Biograph mCT64 scanner. Low-dose CT images for attenuation correction were acquired with a tube voltage of 100 keV and reconstructed into a $512 \times 512 \times 233$ matrix with a voxel size of $0.6 \times 0.6 \times 1.5$ mm^3 (axial field of view = 349.5 mm) [17].

2.3   Cycle Gan + Unet

The CycleGAN + U-Net model [12] proposes an innovative approach to generating computed tomography (CT) images from magnetic resonance imaging (MRI) and is capable of working without paired data. This is particularly useful in clinical settings, since aligned MRI and CT scans are often difficult, costly, or impossible to obtain. The method overcomes this limitation by incorporating the principle of cyclic consistency, typical of CycleGANs, enabling the network to learn structural mappings between domains rather than relying solely on pixel-wise translation.

The architecture is based on the classical Generative Adversarial Network (GAN), composed of a generator and a discriminator. The generator, implemented as a ResNet with 9 residual convolutional blocks, extracts relevant features from MRI images to generate corresponding synthetic CT images. The discriminator, a PatchGAN, evaluates the realism of synthetic images by focusing on small local patches, enhancing the ability to capture fine structural and textural details [5].

To ensure stable training and maintain compatibility with other architectures, several modifications were introduced in the generator:

- Downsampling and upsampling for spatial coherence to restore original dimensions and avoid artifacts, constraining output values in the range $[-1, 1]$.
- Reduction in the number of residual blocks (limited to 9) to prevent feature flattening and loss of structural information.
- Skip connections to preserve domain mapping fidelity.
- Instance normalization for improved training stability and output quality.

The discriminator follows the PatchGAN design, with each block consisting of:

- Convolutional layer (kernel 4×4, stride 2, padding 1),
- Normalization (BatchNorm or InstanceNorm, except in the first layer),
- LeakyReLU activation (slope ~ 0.2).

A distinguishing feature of this model is the addition of an anatomical consistency constraint. A U-Net is used as a structural feature extractor to enforce anatomical coherence between the original MRI and the synthetic CT, using a dedicated structural loss. When available, segmentation masks of anatomical structures are also used to guide the generator.

The total loss function is defined as:

$$\mathcal{L}_{\text{total}} = \mathcal{L}_{\text{cyc}} + \mathcal{L}_{\text{shape}} + \mathcal{L}_{\text{MI}} + \mathcal{L}_{\text{GAN}} \tag{1}$$

The cycle consistency loss is:

$$\mathcal{L}_{\text{cyc}}(G, F) = \|F(G(I_{\text{MR}})) - I_{\text{MR}}\|_1 + \|G(F(I_{\text{CT}})) - I_{\text{CT}}\|_1 \tag{2}$$

where I_{MR} and I_{CT} are input MR and CT images, respectively, and $F(G(I_{\text{MR}}))$ and $G(F(I_{\text{CT}}))$ are their reconstructions after domain translation.

The mutual information (MI) loss is defined as:

$$\mathcal{L}_{\text{MI}} = \sum_{y \in I_{\text{MR}}} \sum_{x \in I_{\text{MR}}} p(x, y) \log \left(\frac{p(x, y)}{p(x)p(y)} \right) \tag{3}$$

where $p(x)$ and $p(y)$ are the marginal distributions of I_{MR} and $G(I_{\text{MR}})$ respectively, and $p(x, y)$ is the joint distribution.

Cycle Gan + Unet model was implemented using the hyperparameters reported in the Table 1.

Table 1. CycleGAN + U-Net hyperparameter setup (unpaired training).

Hyperparameter	Value
G_{LR} (Generator Learning Rate)	2×10^{-4}
D_{LR} (Discriminator Learning Rate)	2×10^{-4}
Image Size	256
G_{Filter} (Generator Filters)	64
D_{Filter} (Discriminator Filters)	64
λ (Cycle Consistency Weight)	10
Batch Size	16
Input Channels	1
Output Channels	1

2.4 SynDiff

The SynDiff model [19] proposes a novel approach to the unsupervised style-transfer of medical images between different diagnostic modes, combining diffusion models and classical adversarial techniques. In our application, this methodology is utilized for the transition from the MRI to the CT domain. This method generates images with coherent anatomical structures and intensity values that accurately represent real values. Furthermore, the model works without corresponding image pairs. This is particularly useful in the medical field, where obtaining annotated data for each modality can be expensive or impractical.

Conditional Diffusion Model. The SynDiff model [19] uses a conditional diffusion process to model the target image distribution. The diffusion process gradually adds noise to the target image, while the inverse process attempts to reconstruct the original image by removing the noise conditioned on the source image.

The diffusion process can be described as follows:

$$q(x_t \mid x_{t-1}) = \mathcal{N}\left(x_t; \sqrt{1 - \beta_t}\, x_{t-1}, \beta_t I\right) \tag{4}$$

where x_t represents the image at time step t, β_t is the added noise rate, and $\mathcal{N}$ denotes a normal distribution. The reverse process is approximated by a neural model which predicts the added noise, allowing for reconstruction of the original image.

Adversarial Loss. To ensure realistic synthetic images, SynDiff incorporates adversarial loss using a discriminator D that attempts to distinguish between real and synthetic images. The generator G (the diffusion model) attempts to fool the discriminator. The adversarial loss can be expressed as:

$$\min_{G_\phi^A, G_\phi^B, G_\theta^A, G_\theta^B} \max_{D_\phi^A, D_\phi^B, D_\theta^A, D_\theta^B} L_{adv} \tag{5}$$

where

$$
\begin{aligned}
L_{adv} = {}& \lambda_{2\phi}\Big[\mathbb{E}_{q(y^B|x_0^A)} \log D_\phi^B(y^B) + \mathbb{E}_{p_\phi^B(y^B|x_0^A)} \log\big(1 - D_\phi^B(G_\phi^B(x_0^A))\big) \\
& + \mathbb{E}_{q(y^A|x_0^B)} \log D_\phi^A(y^A) + \mathbb{E}_{p_\phi^A(y^A|x_0^B)} \log\big(1 - D_\phi^A(G_\phi^A(x_0^B))\big)\Big] \\
& + \lambda_{2\theta}\Big[\mathbb{E}_{q(y^B|x_0^A)} \log D_\theta^B(y^B) + \mathbb{E}_{p_\theta^B(y^B|x_0^A)} \log\big(1 - D_\theta^B(G_\theta^B(x_0^A))\big) \\
& + \mathbb{E}_{q(y^A|x_0^B)} \log D_\theta^A(y^A) + \mathbb{E}_{p_\theta^A(y^A|x_0^B)} \log\big(1 - D_\theta^A(G_\theta^A(x_0^B))\big)\Big] \quad (6)
\end{aligned}
$$

assuming that x is a real image and $\tilde{x}$ is a synthetic image generated by G.

Cycle Consistency Loss. To preserve overall anatomical features during translation, SynDiff implements cyclic consistency loss. This means an image translated from domain A to domain B and then back to A should look similar to the original image. The cyclic consistency loss is given by:

$$
\begin{aligned}
L_{cyc} = \mathbb{E}\Big[& \lambda_{1\phi}\big(\|x_0^A - \hat{x}_0^A\|_1 + \|x_0^B - \hat{x}_0^B\|_1\big) \\
& + \lambda_{1\theta}\big(\|x_0^A - \check{x}_0^A\|_1 + \|x_0^B - \check{x}_0^B\|_1\big)\Big]
\end{aligned}
\quad (7)
$$

where

- $\hat{x}_0^A = G_\phi^A(\tilde{y}^B)$ synthetic data generated in domain A from a sample in domain B,
- $\hat{x}_0^B = G_\phi^B(\tilde{y}^A)$ synthetic data generated in domain B from a sample in domain A,
- x_0^A, x_0^B are original images from domains A and B,
- $\lambda_{1\theta}$ weight of the non-diffusion module,
- $\lambda_{1\phi}$ weight of the diffusion module.

The SynDiff model was implemented with a batch size of 2, learning rate of 0.0001, $\lambda = 0.5$ as relative weight for the cycle coherence loss, $\mu = 0.5$ as relative weight for the gradient penalty, $T = 1000$, a step size $k = 250$, and a total of $T/k = 4$ diffusion passes. The lower and upper bounds on the noise variance schedule were chosen as $\beta_{min} = 0.1$ and $\beta_{max} = 2$.

2.5   Training

In order to reduce the overall number of samples and allow the model to focus only on slices containing relevant information, slices with zero information content, typically found at the beginning and end of three-dimensional acquisitions, were removed. The 3D volumes were then split into slices, addressing the problem from a two-dimensional (2D) perspective. Data leakage, i.e., contamination between the training and test sets with images that have temporal or spatial correlations (as in the case of video sequences), was prevented by avoiding random division. For this purpose, the dataset was partitioned by extracting a test set consisting of 4 samples, representing about 10% of the total of 37 patients, to be used strictly outside the K-Fold validation procedure. The k-fold cross-validation strategy is typically used when only a few images are available. Accordingly, MRI and computed tomography studies are divided into k-folds. Specifically, the 33 patients were stratified into k-folds as follows: the dataset was divided into 5 subgroups of patients and each holdout method was repeated 5 times for both models. At each round, each subgroup of the five was considered as the test set and the other 4 as the training set. Cross-contamination between the test and training sets was avoided. The training of the networks was conducted following a standard approach, in which the discriminator (D) and generator (G) update steps are alternated. Stochastic gradient descent (SGD) on minibatches was used for optimization, in con-junction with the Adam algorithm. This scheme allows frequent updating of the weights, facilitating a more efficient fit to the data and improving the ability of the networks to learn from the information provided. Each model was trained for up to 150 epochs. An early stopping procedure was also implemented, which interrupts the training in the event that the model performance does not show a significant improvement (change threshold of 1×10^{-4}) for 50 consecutive epochs. This mechanism was introduced to reduce the risk of overfitting and optimize the use of computational resources.A high-end HPC system equipped with a GPU (NVIDIA RTX 8000 with 48 GB of RAM, 4608 CUDA Cores) was used to train all networks and run inference.

2.6   Evaluation Metrics

Evaluation of the generating capability of each algorithm was performed using the following six error metrics commonly employed in medical image processing: ME, MAE, MSE, RMSE, PSNR, and SSIM. These metrics quantify the discrepancy between the synthesized image and the baseline image in terms of numerical and perceptual fidelity.

Mean Error (ME) provides an estimate of the mean deviation between the original and generated image pixels. The error sign indicates the direction of the deviation (overestimation or underestimation). It is formally defined as:

$$\mathrm{ME} = \frac{1}{N} \sum_{i=1}^{N} (x_i - \hat{x}_i) \tag{8}$$

where N represents the total number of pixels, x_i the pixel value of the original image, and $\hat{x}_i$ the corresponding value in the generated image.

Mean Absolute Error (MAE) considers the mean difference in absolute terms, providing a nondirectional measure of error:

$$\mathrm{MAE} = \frac{1}{N} \sum_{i=1}^{N} |x_i - \hat{x}_i| \tag{9}$$

Mean Squared Error (MSE) quadratically penalizes larger differences, resulting in particular sensitivity to outliers. It is defined as:

$$\mathrm{MSE} = \frac{1}{N} \sum_{i=1}^{N} (x_i - \hat{x}_i)^2 \tag{10}$$

Root Mean Squared Error (RMSE) represents the square root of the MSE and provides an indication of the mean absolute error in Hounsfield Units (HU). Being derived from the MSE, it retains sensitivity to large errors but presents a more direct interpretation in terms of the original physical unit.

Peak Signal-to-Noise Ratio (PSNR) is a logarithmic metric expressed in decibels (dB) that evaluates the ratio between the maximum signal intensity value possible for a pixel and the reconstruction error (expressed through the RMSE). It is given by:

$$\mathrm{PSNR} = 20 \cdot \log_{10} \left(\frac{\mathrm{MAX}_I}{\sqrt{\mathrm{MSE}}} \right) \tag{11}$$

where MAX_I represents the maximum possible value a pixel can assume (e.g., 4095 for 12-bit CT images). Higher values indicate higher perceptual quality of the generated image.

Structural Similarity Index (SSIM) assesses structural similarity between two images by combining three components: luminance, contrast, and texture. The metric is dimensionless and varies in the range $[-1, 1]$, where 1 represents full identity. It is calculated as:

$$\mathrm{SSIM}(x, y) = \frac{(2\mu_x\mu_y + c_1)(2\sigma_{xy} + c_2)}{(\mu_x^2 + \mu_y^2 + c_1)(\sigma_x^2 + \sigma_y^2 + c_2)} \tag{12}$$

where μ_x, μ_y are the local means of the two images, σ_x^2, σ_y^2 are the variances, and σ_{xy} the covariance. The terms c_1 and c_2 are empirical stabilization constants, typically defined as $c_1 = (k_1 L)^2$ and $c_2 = (k_2 L)^2$, with L being the intensity dynamic range and k_1, k_2 parameters less than 1 (e.g., 0.01 and 0.03 respectively).

3   Results

We applied the approaches of accurate sCT image generation from a collection of T1-weighted MRI brain scans images, acquired from 37 subjects and shown a comparison of the original CT images of patient #006 with the corresponding sCT images obtained with SynDiff and Cycle GAN + Unet in slices #070 and #140, respectively Fig. 1. Table 2 shows the performance evaluation using the k-fold strategy, highlight-ing the best metrics for the different models. The results refer to the test set on the 4 patients [0006, 0012, 0025, 0032]. Cycle Gan + Unet showed a ME of -345.692, a MAE of 496.278, a MSE of 426909.472, a RMSE of 650.01, a PSNR of 40.116 and a SSIM of 0.126 best performed, while SynDiff achieved a ME of 1066.92, a MAE of 1066.92, a MSE of 1474882.42, a RMSE of 1212.81, a PSNR of 34.67 and a SSIM of 0.001.

Table 2. Performance of DL networks on the 4 patients [0006, 0012, 0025, 0032].

Fold	Model	ME	MAE	MSE	RMSE	PSNR	SSIM
Fold 1	Cycle GAN + UNet	1066.92	1066.92	1474882.43	1212.81	34.67	0.001
	SynDiff	-345.64	496.74	427681.40	650.55	40.11	0.124
Fold 2	Cycle GAN + UNet	1066.92	1066.92	1474882.43	1212.81	34.67	0.001
	SynDiff	-345.89	497.59	427501.55	650.46	40.11	0.120
Fold 3	Cycle GAN + UNet	1066.92	1066.92	1474882.38	1212.81	34.67	0.001
	SynDiff	-345.64	494.70	425473.56	648.91	40.13	0.125
Fold 4	Cycle GAN + UNet	1066.92	1066.92	1474882.43	1212.81	34.67	0.001
	SynDiff	-345.64	495.75	426315.80	649.60	40.12	0.124
Fold 5	Cycle GAN + UNet	1066.92	1066.92	1474882.43	1212.81	34.67	0.001
	SynDiff	-345.65	496.61	427575.05	650.53	40.11	0.137
Mean	Cycle GAN + UNet	1066.92	1066.92	1474882.42	1212.81	34.67	0.001
	SynDiff	-345.69	496.28	426909.47	650.01	40.12	0.126

4   Discussion and Conclusions

Conventional radiotherapy (RT) relies on CT scans that provide diagnostic images that accurately calculate the radiation dose to be administered to patients. However, exposure to ionizing radiation can pose a serious health risk, especially if patients need to undergo multiple imaging sessions. In recent years, research has focused on using artificial intelligence (AI) to combine information from different imaging techniques. This is mainly because it is convenient to have a method that can extract CT-equivalent information from MRI images for dose calculation during treatment planning and patient positioning. Such MRI-based CT-equivalent images are called pseudo-CT or synthetic CT (sCT).

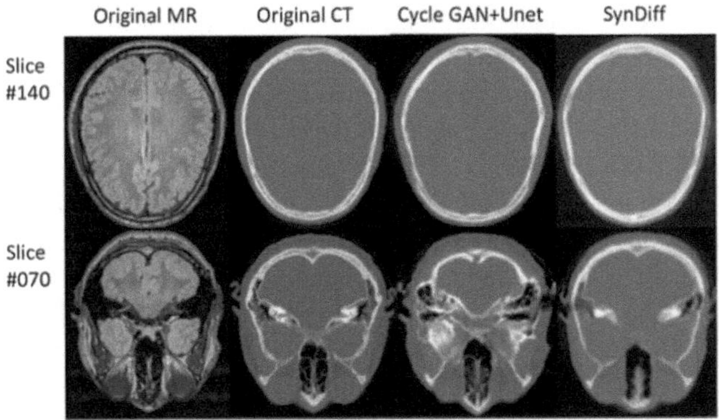

Fig. 1. Comparison of the original MRI and CT images of patient #006 with the corresponding synthetic CT (sCT) images obtained using SynDiff and Cycle GAN + UNet for slices #070 and #140, respectively.

Our study was based on the biomedical application of DP methods, including Cycle Gan + Unet and SynDiff models. The Cycle Gan + Unet model is based on the classical GAN structure, which is composed of a generator and a discriminator. For extracting the relevant features from the MRI image and generating the corresponding synthetic CT image, the generator is a ResNet with 9 convolutional residual blocks. The discriminator is a PatchGAN, which evaluates the authenticity of the synthetic image by operating on small local patches, resulting in a more accurate capture of structural and textural details. In addition, U-Net network was introduced, whose function is to compare and force the anatomical coherence between the original MR images and the corresponding synthetic CT images, through a specific structural loss. The SynDiff model is unsupervised style-transfer of medical images between different diagnostic modes, combining diffusion models and classical adversarial techniques. This method generates images with coherent anatomical structures and intensity values that accurately represent real values. Therefore, we applied DL models to improve approaches for generating accu-rate sCT from MRI images. These comparisons were applied to a collection of T1 MRI brain scans and CT images, acquired from 37 subjects [17]. The dataset was divided by extracting a test set composed of 4 samples, equal to about 10% of the total of 37 patients, to be used exclusively outside the K-Fold validation proce-dure. Consequently, MR and CT studies are divided into k folds. In particular, the 33 patients were stratified in five-fold as follows: the dataset was divided into 5 subgroups of patients and each holdout method was repeated 5 times both for each models. At each round, each subgroup of the five was considered as a test set and the other 4 as training set. Cross-contamination was avoided between test and training sets. According to our analysis, SynDiff achieved higher ME, MAE, MSE, RMSE, PSNR and SSIM than Cycle Gan + Unet. Therefore, these data demonstrated that SynDiff can be used to achieve accurate sCT generation from brain MR images com-pared to Cycle Gan + Unet method. This study demonstrated the feasibility and validity/effectiveness of the SynDiff model for generating accurate sCT images from brain MRI scans, outperforming the CycleGAN+U-Net approach based on quantitative metrics (ME = 1066.92, MAE = 1066.92, MSE = 1474882.42, RMSE = 1212.81, PSNR = 34.67, SSIM = 0.001). However, a key limitation is the relatively small sample size of 37 patients, with only 4 reserved for independent testing. The observed performance of SynDiff, while promising within this specific cohort, might not fully represent its accuracy across broader populations with diverse anatomical variations or pathologies. Future validation on larger, independent datasets is essential to confirm the robustness and clinical applicability of this method. Despite this limitation, our results contribute valuable insights into the potential of advanced diffusion-based models for MRI-to-CT synthesis in radiotherapy planning. In conclusion, our study revealed how DL algorithms can be used to achieve accurate sCT generation from brain MRI images, leveraging SynDiff analyses and outperforming the Cycle Gan + Unet method enable fast and accurate analysis to support the medical doctor.

References

1. Agnello, L., Comelli, A., Ardizzone, E., Vitabile, S.: Unsupervised tissue classification of brain MR images for voxel-based morphometry analysis. Int. J. Imaging Syst. Technol. **26**, 136–150 (2016). https://doi.org/10.1002/ima.22168
2. Ali, M., et al.: Applications of artificial intelligence, deep learning, and machine learning to support the analysis of microscopic images of cells and tissues. J Imaging **11**, 59 (2025). https://doi.org/10.3390/jimaging11020059
3. Ali, M., et al.: Prostate cancer detection: Performance of radiomics analysis in multiparametric MRI (2024). https://doi.org/10.1007/978-3-031-51026-7_8
4. Alongi, P., et al.: Radiomics analysis of brain [18f]fdg pet/ct to predict Alzheimer's disease in patients with amyloid pet positivity: A preliminary report on the application of spm cortical segmentation, pyradiomics and machine-learning analysis. Diagnostics **12**, 933 (2022). https://doi.org/10.3390/diagnostics12040933
5. Canfora, I., et al.: A predictive system to classify preoperative grading of rectal cancer using radiomics features (2022). https://doi.org/10.1007/978-3-031-13321-3_38
6. Comelli, A., et al.: Development of a new fully three-dimensional methodology for Tumours delineation in functional images. Comput. Biol. Med. **120**, 103701 (2020). https://doi.org/10.1016/j.compbiomed.2020.103701
7. Comelli, A., Stefano, A., Benfante, V., Russo, G.: Normal and abnormal tissue classification in positron emission tomography oncological studies. Pattern Recognit Image Anal. **28**(1), 106–113 (2018). https://doi.org/10.1134/S1054661818010054
8. Comelli, A., et al.: Tissue classification to support local active delineation of brain tumors (2020). https://doi.org/10.1007/978-3-030-39343-4_1
9. Comelli, A., et al.: A smart and operator independent system to delineate Tumours in positron emission tomography scans. Comput. Biol. Med. **102**, 1–15 (2018). https://doi.org/10.1016/j.compbiomed.2018.09.002
10. Dong, X., et al.: Synthetic CT generation from non-attenuation corrected pet images for whole-body pet imaging. Phys. Med. Biol. **64**, 215016 (2019). https://doi.org/10.1088/1361-6560/ab4eb7

11. Edmund, J., Nyholm, T.: A review of substitute CT generation for MRI-only radiation therapy. Radiat. Oncol. **12**, 28 (2017). https://doi.org/10.1186/s13014-016-0747-y
12. Ge, Y., et al.: Unpaired MR to CT synthesis with explicit structural constrained adversarial learning, pp. 1096–1099 (2019). https://doi.org/10.1109/ISBI.2019.8759529
13. Hsu, S.H., Cao, Y., Huang, K., Feng, M., Balter, J.: Investigation of a method for generating synthetic CT models from MRI scans of the head and neck for radiation therapy. Phys. Med. Biol. **58**, 8419–8435 (2013). https://doi.org/10.1088/0031-9155/58/23/8419
14. Jin, C.B., et al.: Deep CT to MR synthesis using paired and unpaired data. Sensors **19**, 2361 (2019). https://doi.org/10.3390/s19102361
15. Laudicella, R., et al.: [68ga]dotatoc pet/CT radiomics to predict the response in GEP-nets undergoing [177lu]dotatoc prrt: The theragnomics concept. Cancers (Basel) **14**, 984 (2022). https://doi.org/10.3390/cancers14040984
16. Lei, Y., et al.: Magnetic resonance imaging-based pseudo computed tomography using anatomic signature and joint dictionary learning. J. Med. Imaging **5**(3), 034001–034001 (2018). https://doi.org/10.1117/1.JMI.5.3.034001
17. Mérida, I., et al.: CERMEP-IDB-MRXFDG: a database of 37 normal adult human brain [^{18}F]FDG PET, T1 and FLAIR MRI, and CT images available for research. EJNMMI Res. **11**(1), 1–10 (2021). https://doi.org/10.1186/s13550-021-00830-6
18. Nie, D., Cao, X., Gao, Y., Wang, L., Shen, D.: Estimating CT image from MRI data using 3d fully convolutional networks (2016). https://doi.org/10.1007/978-3-319-46976-8_18
19. Unsupervised medical image translation with adversarial diffusion models: Özbey, M., Dalmaz, O., Dar, S., Bedel, H., Özturk, å, Güngör, A., Çukur, T. IEEE Trans. Med. Imaging **42**, 3524–3539 (2023). https://doi.org/10.1109/TMI.2023.3290149

Paravertebral Space Muscles Segmentation in Patients with Oral Squamous Cell Carcinoma: A Preliminary Step for Radiomics Studies

Paolo Buscemi[1], Gaspare Centineo[2], Muhammad Ali[3(✉)] (iD),
Viviana Benfante[4] (iD), Antonio Lo Casto[1], and Albert Comelli[3] (iD)

[1] Department of Biomedicine, Neuroscience and Advanced Diagnostics (BiND),
University of Palermo, Palermo, Italy
[2] Department of Electronics and Telecommunications, Politecnico Di Torino,
Turin, Italy
[3] Ri.MED Foundation, Via Bandiera 11,90133Palermo, Italy
amuhammad@Fondazionerimed.com
[4] Department of Health Promotion, Mother and Child Care, Internal Medicine and
Medical Specialties, Molecular and Clinical Medicine, University of Palermo,
90127 Palermo, Italy

Abstract. The aim of this study is to present a deep learning (DL) algorithm for accurate C3 paravertebral space muscle delineation in magnetic resonance (MR) images of patients with oral squamous cell carcinoma (OSCC). This study tries to enhance the methodologies employed by medical professionals in radiomics by incorporating operator-independent segmentation techniques. Such approaches are essential for accurately identifying target regions and developing reliable predictive models.

We investigated a deep learning model C-ENet, using head and neck MR scans from 146 patients, either pre-surgical or during follow-up. The model's segmentation performance was assessed by comparing its output to expert-defined ground truth annotations.

Our findings indicate that the C-ENet model can achieve effective segmentation of C3 paravertebral space muscles, reaching a Dice Similarity Coefficient (DSC) of 67.41%. In summary, the study confirms that deep learning can be successfully utilized for fast, radiologist-free segmentation of the C3 paravertebral region, enabling consistent, user-independent outputs for downstream radiomics applications.

Keywords: magnetic resonance · oral squamous cell carcinoma · C3 paravertebral space muscle · C-ENet · Segmentation

1 Introduction

Head and neck malignant tumors represent approximately 3% of all malignant tumors in Italy [23]. Similar to trends reported across Europe, the annual incidence

E. Rodolà et al. (Eds.): ICIAP 2025 Workshops, LNCS 16169, pp. 278–288, 2026.
https://doi.org/10.1007/978-3-032-11317-7_24

of head and neck cancers in Italy is approximately 18 cases per 100,000 inhabitants. About 93% of these are epithelial tumors, and 90% are malignant squamous cell carcinomas [24]. MRI is referral technique in the evaluation of oral cavity cancer, because it allows to evaluate the deep extension of tumor, invasion of adjacent bone structures, and locoregional lymph node metastasis. Furthermore it ensure a great accuracy in the evaluation of depth of invasion (DOI), that is a fondamental parameter in the local stagin of oral cavity cancer according to the 8th edition of AJCC Cancer Staging [23,24]. Malnutrition involves the depletion of energy stores, protein reserves, and essential nutrients, ultimately impairing health and contributing to increased morbidity and mortality [32]. Among chronic diseases, cancer shows the highest prevalence of malnutrition, which negatively impacts quality of life, healthcare resource utilization, and survival outcomes, frequently limiting access to pharmacological or surgical interventions [7,27,30].

Oral cavity squamous cell carcinomas (OSCC) represent the second most frequent malignancy linked to malnutrition, preceded only by gastrointestinal cancers [31]. This association is not solely related to tumor staging, but also to early-onset dysphagia, the side effects of radiotherapy such as xerostomia, and the consequences of surgical resection, which frequently result in impaired chewing and swallowing function [26]. Despite its impact, cancer-related malnutrition remains underdiagnosed, underestimated, and insufficiently addressed in daily clinical management [28].

Suitable methods for measuring the amount of lean body mass are necessary. Recently, an alternative technique was described using MRI to measure the volume of paravertebral space muscles at C3 level [1,12]. This approach may present some considerable advantages. In fact, an MRI of the neck is always available in these patients, even at follow-up, but it would be useful to have an automatic tool for the segmentation of this area to make this approach easy to use [9,14,36].

Deep learning (DL) techniques have enabled the automation of numerous tasks across various domains. From a functional standpoint, DL leverages advanced algorithms to generate synthetic imagery, recognize patterns of behavior, classify visual data, and perform image segmentation [16,18]. Among the most commonly employed DL architectures in medical imaging are convolutional neural networks (CNNs), a class of feed-forward networks particularly suited for processing modalities such as computed tomography (CT), magnetic resonance imaging (MRI), and contrast-enhanced ultrasound (CEUS) [17].

Within this framework, our goal was to implement CNNs–specifically, a customized Efficient Neural Network (C-ENet) architecture to enhance the precision of image segmentation, support object detection, and improve classification accuracy in view of future radiomics applications [5,6,33].

2   Materials and Methods

2.1   Patients

All patients provided written consent for the use of their data for scientific purposes. The local Ethical Committee Palermo 1 approved the study protocol (reference #13/2024 of the 22nd May 2024), and the same approval also covered the clinical study previously published [12]. The present analysis addresses a distinct methodological aim. This study was conducted in accordance with the Declaration of Helsinki.

Patients who underwent MR Head and Neck examinations for OSCC at the University Hospital Policlinico "P. Giaccone" in Palermo (Italy) between January 2023 and February 2024 were retrospectively identified from our institutional electronic medical records and included in this study. Finally, 146 patients were consecutively enrolled. In these patients, MR examinations of head and neck were performed at the time of diagnosis or as follow-up imaging.

2.2   MR Examination

Images and reports of MRI were retrospectively analyzed. Briefly, T1 sequences with and without fat saturation and intravenous gadolinium, T2, and STIR sequences on the axial and coronal planes with a maximum slice thickness of 4 mm were acquired using a 1.5T and 3T MRI device (Ingenia 3.0 T, Philips Healthcare, Best, The Netherlands; Achieva 1.5 T, Philips Healthcare, Best, The Netherlands) and stored in our Picture Archiving and Communication System (PACS).

2.3   Manual Segmentation

One board-certified diagnostic radiologist with at least 20 years of experience in Head and Neck imaging reviewed the MR images. The radiologist was blinded to the patients' clinical characteristics.

The C3 muscle analysis was performed by drawing ROIs along the profile of posterior paraspinal muscles in five slices at the C3 vertebra level on MR axial T2 images using the "3D Slicer" software [20], obtaining a volume. Sternocleido-mastoid muscles were excluded from the segmentation area.

2.4   The Proposed Deep Learning Model

C-ENet is a tailored neural network derived from the ENet architecture, known for its balance of fast inference speed and high accuracy, with detailed architectural insights available in [29]. The primary customization in C-ENet (custom-ENet) involves modifying ENet's output layer from $128 \times 64 \times 64$ to $256 \times 64 \times 64$. This structural adjustment has been shown to improve the average Dice Similarity Coefficient (DSC) and reduce performance variability relative to the original configuration.

2.5   Loss Functions

A common challenge in biomedical image segmentation arises when the region of interest occupies a much smaller volume compared to the surrounding anatomical structures. As a result, the majority of voxels are labeled as background, leading to a class imbalance that can negatively affect the performance of deep learning (DL) models.

To mitigate this imbalance, specialized loss functions can be employed. One effective solution is the Tversky loss, introduced in [35], which incorporates both the Dice Similarity Coefficient and a penalty-based strategy to enhance segmentation accuracy:

$$T(\alpha, \beta) = \frac{\sum_{i=1}^{N} p_{0i} g_{0i}}{\sum_{i=1}^{N} p_{0i} g_{0i} + \alpha \sum_{i=1}^{N} p_{0i} g_{1i} + \beta \sum_{i=1}^{N} p_{1i} g_{0i}} \tag{1}$$

where p_{0i} is the probability that voxel i is inside the target, and p_{1i} is the probability that voxel i is outside the target. The ground truth training label g_{0i} is 1 for target and 0 for background, and vice versa for the label g_{1i}.

By adjusting the parameters α and β, the trade-off can be controlled between false positives and false negatives. To obtain the Dice similarity coefficient, α and β must be set to 0.5.

Setting $\beta > 0.5$ places more emphasis on false negatives in the slices with small target regions.

2.6 Training

The deep learning approach utilized in this study handled an MRI dataset from 146 patients, employing a 5-fold cross-validation strategy to manage the limited image availability; this involved dividing the dataset into five patient subgroups, with each round using one subgroup as the test set and the remaining four as the training set, meticulously avoiding cross-contamination. To augment the training data and prevent overfitting, the input training set underwent various transformations including random rotations (up to 20°), X and Y translations (up to 0.2), shearing (10°), horizontal flips, and zooming (0.1), with data normalization performed by subtracting the mean and dividing by the standard deviation for each MRI study [21]. Key learning parameters for the C-ENet, derived from an initial set of 23 patients, included an Adam optimizer with a learning rate of 0.0001, a consistent batch size of 60 slices, and a Tversky loss function set with $\alpha = 0.3$ and $\beta = 0.7$. All models were trained for a maximum of 100 epochs, with an automatic early stopping criterion halting training if the loss did not decrease for 10 consecutive epochs, all executed on a high-end HPC system featuring a Dell Inc. Precision 7760 workstation equipped with an NVIDIA RTX 5000 GPU with 16 GB of RAM and CPU Intel® Xeon® W-11955M x 16 cores with 128 GB..

2.7 Evaluation Metrics

The performance of the proposed approaches was based on the Dice similarity coefficient (DSC) (which measures the spatial overlap between the reference volume and the segmented volume), sensitivity, volume overlap error (VOE),

Relative volume difference (RVD), and Precision calculated as mean, standard deviation, and confidence interval.

In general, a Dice similarity coefficient equal to 0% means no overlap, while a value of 100% indicates perfect overlap.

$$\text{DSC} = \frac{2 \times TP}{2 \times TP + FP + FN} \times 100\% \tag{2}$$

where TP, FP, and FN are the number of true positives, false positives, and false negatives respectively.

Precision, is defined as:

$$\text{Precision} = \frac{TP}{TP + FP} \times 100\% \tag{3}$$

Finally, VOE, RVD, Recall and F1-score are defined respectively as:

$$\text{VOE} = 1 - \frac{TP}{TP + FP + FN} \tag{4}$$

$$\text{RVD} = \pm \frac{|A| - |B|}{|B|} \tag{5}$$

$$\text{Recall} = \frac{TP}{TP + FN} \tag{6}$$

$$\text{F1-score} = \frac{2 \cdot TP}{2 \cdot TP + FP + FN} \tag{7}$$

in which A is the prediction and B is the ground truth.

A one-way ANOVA test was applied to test the significance of performance of C-ENet.

3   Results

We applied the proposed segmentation approach (C-ENet) to analyze C3 paravertebral muscle morphology in OSCC patients. Table 1 and Fig. 1 reports the performance evaluation based on a 5-fold cross-validation strategy. The C-ENet model achieved an average Dice Similarity Coefficient (DSC) of $67.41 \pm 7.70\%$, reflecting the challenging nature of the dataset and the variability across patients. Figures 2 and 3 shows the profiles of training DSC and Tversky loss function for one fold. The C-ENet model reaching a training DSC of nearly 0.8 in less than 26 epochs. The complexity and perfomance of the C-ENet deep model are rispectivily 793,917 Nummber of Parameters Trainable, 11,426 Number of Parameters non Trainable, 11 MB Size on Disk, 3.20 s on CPU and 1.28 on GPU Inference Times/Dataset.

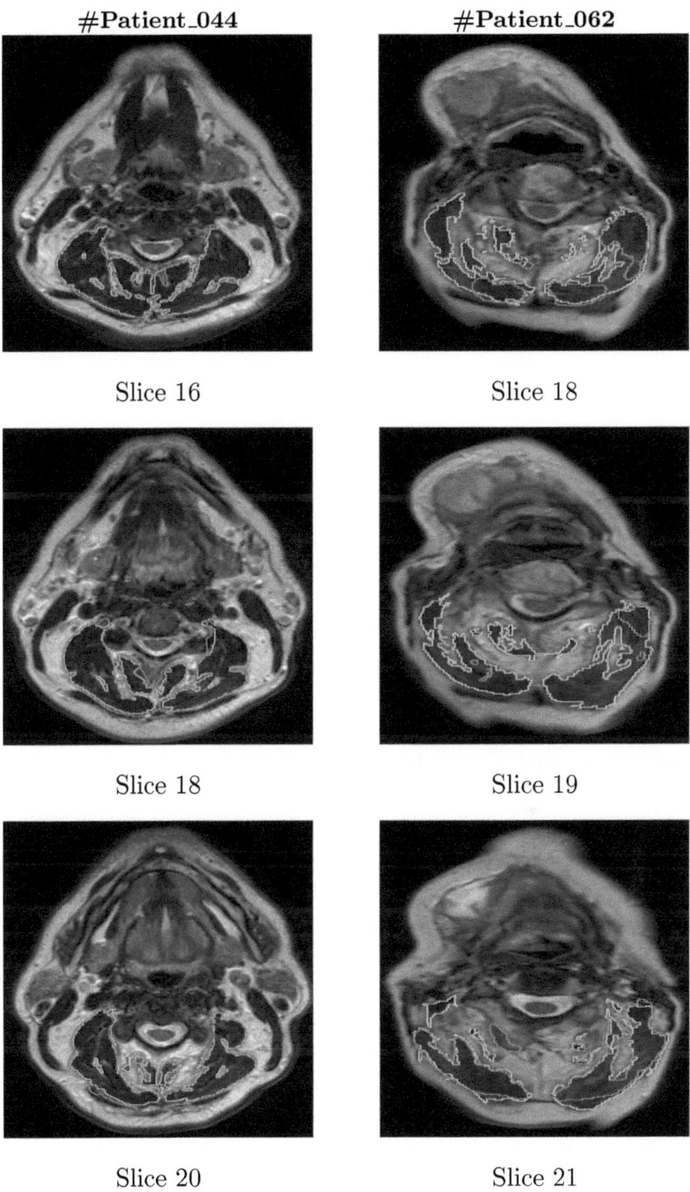

Fig. 1. Comparison of segmentation performance in Fold-3 between the best-performing case (**#Patient_044**) and the worst-performing case (**#Patient_062**). Ground truth (Yellow) and C-ENet predictions (Red)(Color figure online)

Table 1. Performance of C-ENet

Fold	DICE %	VOE %	VD %	Prec.	Rec.	F1
Fold 1	70.81 ± 15.34	43.41 ± 15.14	31.23 ± 24.17	65.54 ± 13.80	79.50 ± 19.82	70.81 ± 15.34
Fold 2	71.17 ± 15.20	43.17 ± 13.78	45.04 ± 25.44	61.45 ± 14.90	86.24 ± 19.02	71.17 ± 15.20
Fold 3	71.62 ± 12.71	42.74 ± 14.95	51.12 ± 39.58	65.92 ± 14.17	85.06 ± 18.74	71.62 ± 12.71
Fold 4	53.69 ± 16.60	61.63 ± 14.85	29.75 ± 24.26	62.00 ± 12.65	51.03 ± 20.89	53.69 ± 16.60
Fold 5	69.77 ± 9.58	45.62 ± 11.13	65.66 ± 47.64	58.25 ± 12.51	90.53 ± 9.05	69.77 ± 9.58
Media	**67.41 ± 7.70**	**47.31 ± 8.08**	**44.56 ± 14.88**	**62.63 ± 3.17**	**78.47 ± 15.84**	**67.41 ± 7.70**

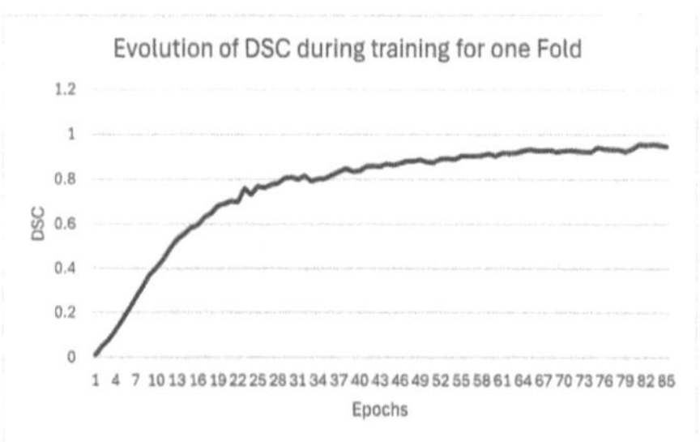

Fig. 2. Evolution of training DSC for C-ENet

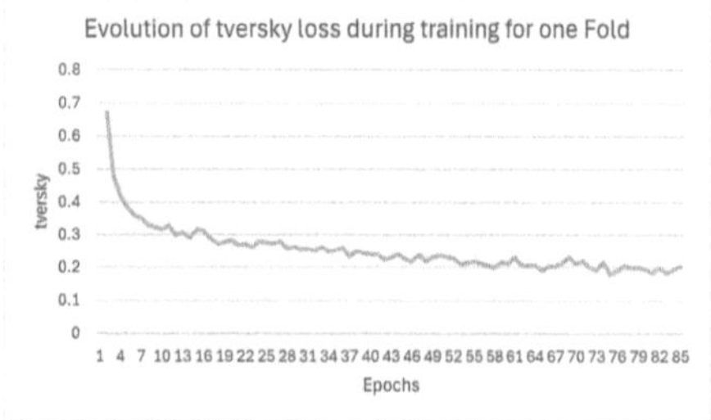

Fig. 3. Evolution of training Tversky loss for C-ENet

4 Discussion and Conclusions

The management of oncologic patients presents significant challenges during hospitalization and, potentially, even greater complexity following discharge. For specific subgroups, such as those affected by head and neck cancers, the implementation of structured nutritional assessments and systematic monitoring programs is strongly recommended [25]. Given the difficulties in obtaining traditional malnutrition indicators in all patients, especially during hospital stays, there is a growing need for alternative diagnostic approaches [12] which may complement or substitute bioelectrical impedance analysis (BIA), a method that can have notable limitations in this clinical context [10].

Indeed, patients undergoing surgery for OSCC frequently require advanced reconstructive techniques, often involving the use of skin and muscle flaps harvested from the limbs. These procedures can interfere with proper electrode placement, affecting BIA reliability. Moreover, post-operative conditions such as localized edema of the neck and upper limbs, commonly observed after lymph node dissection, may further distort BIA measurements [13]. While CT and DEXA currently represent reference methods for body composition analysis, their routine use remains limited due to accessibility constraints.

In this study, we applied segmentation techniques on MR images to assess the volume of paravertebral muscles at the C3 level [2]. This approach offers distinct advantages: neck MRI scans are routinely available for these patients, including during follow-up [23,24]. Furthermore, as reported in previous research [12].

paravertebral muscles serve as a stable anatomical reference and are less susceptible to atrophic changes related to physical inactivity, unlike lumbar or lower limb muscles, which may be affected in bedridden or sedentary patients [11,37]. However, in OSCC patients–especially those undergoing radical or modified radical neck dissections–the sternocleidomastoid muscle is often resected in advanced disease stages. For this reason, we excluded this muscle from our analysis, thereby establishing a segmentation protocol applicable to a wider range of clinical scenarios, including longitudinal follow-up. Additionally, we segmented the paravertebral space muscles of the neck using five axial slices at the C3 level, enabling the calculation of muscle volume. Transitioning from two-dimensional to three-dimensional segmentation is expected to enhance measurement accuracy [8,12].

Indeed, the automatic characterization of paravertebral space muscles in MR images may be a support in the real calculation of features, in order to avoid manual characterizations and resulting mistakes [15]. Deep learning (DL) techniques have demonstrated superior performance in generating more precise and consistent synthetic images [3], performing image classification, and particularly enhancing segmentation tasks compared to traditional manual methods [22]. Specifically, convolutional neural networks (CNNs) have emerged as a promising tool in the field of medical imaging [4].

In this study, we focused on biomedical applications of CNN architectures, including UNet and C-ENet models. The latter is an advancement built upon the ENet framework, known for its balance between accuracy and rapid inference. Our modification to ENet involved increasing the output dimension from $128 \times 64 \times 64$ to $256 \times 64 \times 64$ in C-ENet, which led to improved Dice Similarity Coefficient (DSC) scores and reduced variability in segmentation outcomes.

We applied these DL models to enhance image segmentation of magnetic resonance (MR) scans from a cohort of patients diagnosed with oral squamous cell carcinoma (OSCC). The dataset was retrospectively collected at our institution, comprising 146 patients. Images were divided into training and testing groups, and a five-fold cross-validation approach was adopted to manage the relatively small sample size. For each fold, one subset served as the test set while the remaining four subsets constituted the training data, ensuring no data leakage between training and testing.

The C-ENet model yielded encouraging segmentation results, achieving an average DSC of $67.41 \pm 7.70\%$. This performance reflects the inherent complexity and heterogeneity within the dataset. While this DSC is somewhat lower than values reported in other contexts, it nonetheless confirms the practical applicability of the method.

Despite the limited cohort size, our findings validate the potential of the C-ENet architecture for reliable image segmentation, with DSC scores approaching 67% on average.

In summary, our investigation demonstrates that DL models, particularly C-ENet, can be effectively utilized for automated segmentation of paravertebral muscles. These methods provide fast, reproducible, and operator-independent analyses without the need for radiologist involvement [19,34]. The clinical implementation of such algorithms for automatic delineation and segmentation of paravertebral muscles at the C3 vertebral level could offer valuable tools for assessing sarcopenia and support future radiomics research.

References

1. Agnello, L., Comelli, A., Ardizzone, E., Vitabile, S.: Unsupervised tissue classification of brain MR images for voxel-based morphometry analysis. Int. J. Imaging Syst. Technol. **26**(2), 136–150 (2016). https://doi.org/10.1002/ima.22168
2. Ali, M., et al.: Applications of artificial intelligence, deep learning, and machine learning to support the analysis of microscopic images of cells and tissues. J. Imaging **11**(2), 59 (2025). https://doi.org/10.3390/jimaging11020059
3. Ali, M., et al.: Prostate cancer detection: Performance of radiomics analysis in multiparametric MRI. In: Image Analysis and Processing – ICIAP 2023 Workshops, Lecture Notes in Computer Science, vol. 14366, pp. 83–92 (2024). https://doi.org/10.1007/978-3-031-51026-7_8
4. Ali, M., Benfante, V., Raimondo, D.D., Laudicella, R., Tuttolomondo, A., Comelli, A.: A review of advances in molecular imaging of rheumatoid arthritis: from in vitro to clinic applications using radiolabeled targeting vectors with technetium-99m. Life **14**(6), 751 (2024). https://doi.org/10.3390/life14060751
5. Alongi, P., et al.: Radiomics analysis of brain [18f]fdg pet/CT to predict alzheimer's disease in patients with amyloid pet positivity: A preliminary report on the application of spm cortical segmentation, pyradiomics and machine-learning analysis. Diagnostics **12**(4), 933 (2022). https://doi.org/10.3390/diagnostics12040933
6. Alongi, P., et al.: Artificial intelligence applications on restaging [18f]fdg pet/ct in metastatic colorectal cancer: A preliminary report of morpho-functional radiomics classification for prediction of disease outcome. Appl. Sci. **12**(6), 2941 (2022). https://doi.org/10.3390/app12062941
7. Basirinia, G., et al.: Theranostic approaches for gastric cancer: an overview of in vitro and in vivo investigations. Cancers **16**(19), 3323 (2024). https://doi.org/10.3390/cancers16193323
8. Belavý, D., Miokovic, T., Armbrecht, G., Felsenberg, D.: Hypertrophy in the cervical muscles and thoracic discs in bed rest? J. Appl. Physiol. **115**, 586–589 (2013). https://doi.org/10.1152/japplphysiol.00376.2013
9. Benfante, V., Salvaggio, G., Ali, M., Comelli, A.: Grading and staging of bladder tumors using radiomics analysis in magnetic resonance imaging. In: Image Analysis and Processing – ICIAP 2023 Workshops, Lecture Notes in Computer Science, vol. in press, pp. 93–103 (2024). https://doi.org/10.1007/978-3-031-51026-7_9
10. de Bree, R., Meerkerk, C., Halmos, G., Mäkitie, A., Homma, A., Rodrigo, J.e.a.: Measurement of sarcopenia in head and neck cancer patients and its association with frailty. Front Oncol **12**, 884988 (2022). https://doi.org/10.3389/fonc.2022.884988
11. Bril, S., van Beers, M., Chargi, N., Carrillo Minulina, N., Smid, E.E.A.: Skeletal muscle mass at c3 is a strong predictor for skeletal muscle mass at l3 in sarcopenic and non-sarcopenic patients with head and neck cancer. Oral Oncol **122**, 105558 (2021). https://doi.org/10.1016/j.oraloncology.2021.105558

12. Buscemi, P., et al.: Nutritional factors and survival in a cohort of patients with oral cancer. Front. Nutr. **12**, 1530460 (2025). https://doi.org/10.3389/fnut.2025.1530460

13. Buscemi, S., Blunda, G., Maneri, R., Verga, S.: Bioelectrical characteristics of type 1 and type 2 diabetic subjects with reference to body water compartments. Acta Diabetol. **35**, 220–223 (1998). https://doi.org/10.1007/s005920050135

14. Cairone, L., et al.: Robustness of radiomics features to varying segmentation algorithms in magnetic resonance images. In: Image Analysis and Processing – ICIAP 2022 Workshops, Lecture Notes in Computer Science, vol. in press, pp. 462–472 (2022). https://doi.org/10.1007/978-3-031-13321-3_41

15. Canfora, I., et al.: A predictive system to classify preoperative grading of rectal cancer using radiomics features. In: Image Analysis and Processing – ICIAP 2022 Workshops, Lecture Notes in Computer Science, vol. 13373, pp. 431–440 (2022). https://doi.org/10.1007/978-3-031-13321-3_38

16. Cederholm, T., et al.: Glim criteria for the diagnosis of malnutrition - a consensus report from the global clinical nutrition community. Clin. Nutr. **38**, 1–9 (2019). https://doi.org/10.1016/j.clnu.2018.08.002

17. Chen, C., et al.: Deep learning for cardiac image segmentation: a review. Front Cardiovasc Med (2020). https://doi.org/10.3389/fcvm.2020.00025

18. Comelli, A., Stefano, A., Benfante, V., Russo, G.: Normal and abnormal tissue classification in positron emission tomography oncological studies. Pattern Recognit Image Anal. **28**(1), 106–113 (2018). https://doi.org/10.1134/S1054661818010054

19. Comelli, A., Stefano, A., Russo, G., Sabini, M.G.: A smart and operator independent system to delineate tumours in positron emission tomography. Comput. Meth. Programs Biomed. **165**, 1–9 (2018). https://doi.org/10.1016/j.compbiomed.2018.09.002

20. Fedorov, A., et al.: 3d slicer as an image computing platform for the quantitative imaging network. Magn. Reson. Imaging **30**, 1323–1341 (2012). https://doi.org/10.1016/j.mri.2012.05.001

21. Giaccone, P., Benfante, V., Stefano, A., Cammarata, F.P., Russo, G., Comelli, A.: Pet images atlasbased segmentation performed in native and in template space: a radiomics repeatability study in mouse models. In: Image Analysis and Processing – ICIAP 2022 Workshops, Lecture Notes in Computer Science, vol. 13373, pp. 351–361 (2022). https://doi.org/10.1007/978-3-031-13321-3_31

22. Groendahl, A., Skjei Knudtsen, I., Huynh, B., Mulstad, M.e.a.: A comparison of methods for fully automatic segmentation of tumors and involved nodes in pet/ct of head and neck cancers. Phys Med Biol **66** (2021). https://doi.org/10.1088/1361-6560/abe553

23. International agency for research on cancer: world health organization. Global Cancer Observatory (2024). https://gco.iarc.fr/. Accessed 10 Apr 2024

24. Istituto superiore di sanità, sistema nazionale linee guida: aprile (2022). https://www.iss.it/documents/20126/8403839/LG493-AIOM-Head-and-neck. Accessed 10 Apr 2024

25. Langius, J., Zandbergen, M., Eerenstein, S., van Tulder, M., Leemans, C., Kramer, M.: Effect of nutritional interventions on nutritional status, quality of life and mortality in patients with head and neck cancer receiving (chemo)radiotherapy: a systematic review. Clin. Nutr. **32**, 671–678 (2013). https://doi.org/10.1016/j.clnu.2013.06.012

26. Martinovic, D., et al.: Nutritional management of patients with head and neck cancer—a comprehensive review. Nutrients **15**, 1864 (2023). https://doi.org/10.3390/nu15081864

27. Matsui, R., Rifu, K., Watanabe, J., Inaki, N., Fukunaga, T.: Impact of malnutrition as defined by the glim criteria on treatment outcomes in patients with cancer: A systematic review and meta-analysis. Clin. Nutr. **42**, 615–624 (2023). https://doi.org/10.1016/j.clnu.2023.02.019

28. Muscaritoli, M., et al.: Espen practical guideline: clinical nutrition in cancer. Clin. Nutr. **40**, 2898–2913 (2021). https://doi.org/10.1016/j.clnu.2021.02.005

29. Paszke, A., Chaurasia, A., Kim, S., Culurciello, E.: Enet: a deep neural network architecture for real-time semantic segmentation (2016). https://arxiv.org/abs/1606.02147, arXiv:1606.02147

30. Piras, A., et al.: Artificial intelligence and statistical models for the prediction of radiotherapy toxicity in prostate cancer: a systematic review. Appl. Sci. **14**(23), 10947 (2024). https://doi.org/10.3390/app142310947

31. Pressoir, M., et al.: Prevalence, risk factors and clinical implications of malnutrition in french comprehensive cancer centres. Br. J. Cancer **102**, 966–971 (2010). https://doi.org/10.1038/sj.bjc.6605578

32. Ravasco, P., Monteiro-Grillo, I., Marques Vidal, P., Camilo, M.: Impact of nutrition on outcome: a prospective randomized controlled trial in patients with head and neck cancer undergoing radiotherapy. Head Neck **27**, 659–668 (2005). https://doi.org/10.1002/hed.20221

33. Russo, G., et al.: Feasibility on the use of radiomics features of 11[C]-met pet/ct in central nervous system tumours: Preliminary results on potential grading discrimination using a machine learning model. Curr. Oncol. **28**(6), 5318–5331 (2021). https://doi.org/10.3390/curroncol28060444

34. Salamone, G., et al.: Vacuum-assisted wound closure with mesh-mediated fascial traction achieves better outcomes than vacuum-assisted wound closure alone: a comparative study. World J. Surg. (1), 1–8 (2017). https://doi.org/10.1007/s00268-017-4354-3

35. Salehi, S.S.M., Erdogmus, D., Gholipour, A.: Tversky Loss Function for Image Segmentation Using 3D Fully Convolutional Deep Networks. In: Wang, Q., Shi, Y., Suk, H.-I., Suzuki, K. (eds.) MLMI 2017. LNCS, vol. 10541, pp. 379–387. Springer, Cham (2017). https://doi.org/10.1007/978-3-319-67389-9_44

36. Salvaggio, G., et al.: Deep learning network for segmentation of the prostate gland with median lobe enlargement in t2-weighted mr images: Comparison with manual segmentation method. Curr. Probl. Diagn. Radiol. **51**(3), 328–333 (2022). https://doi.org/10.1067/j.cpradiol.2021.06.006

37. Zwart, A.T., et al.: Skeletal muscle mass and sarcopenia can be determined with 1.5-T and 3-T neck MRI scans, in the event that no neck CT scan is performed. Eur. Radiol. **31**(6), 4053–4062 (2020). https://doi.org/10.1007/s00330-020-07440-1

Preliminary Study Exploring the Potential of ML Models Applied to Radiomic Features Extracted from PSMA-PET: Prediction of the PRIMARY and E-PSMA Scores in the Evaluation of Primary PCa Lesions

Viviana Benfante[1], Pierpaolo Purpura[1], Costanza Longo[2], Albert Comelli[3(✉)], and Pierpaolo Alongi[1,4]

[1] Advanced Diagnostic Imaging—INNOVA Project, Department of Radiological Sciences, A.R.N.A.S. Civico, Di Cristina e Benfratelli Hospitals, P.zza N. Leotta 4, 90127 Palermo, Italy
[2] Nuclear Medicine Unit, A.R.N.A.S. Civico Di Cristina e Benfratelli Hospitals, P.zza N. Leotta 4, 90127 Palermo, Italy
[3] Ri.MED Foundation, Via Bandiera 11, 90133 Palermo, Italy
acomelli@fondazionerimed.com
[4] Department of Biomedicine, Neuroscience and Advanced Diagnostics (BiND), University of Palermo, Palermo, Italy

Abstract. We explored the performance potential of radiomics analysis of PSMA-PET images for predicting response to PSMA-targeted therapies in high-grade prostate cancer (PCa). Radiomics analysis was performed on 18F-PSMA-1007 PET images of 18 patients with histologically confirmed or verified disease at 2-year follow-up. The radiomics workflow involved segmenting regions of interest (ROIs) from images, extracting features from ROIs, selecting features using a hybrid descriptive-inferential approach. A hybrid descriptive-inferential approach combines data summarization with statistical testing. Descriptive statistics characterize the dataset (e.g., mean, SD), while inferential methods assess significance and generalizability. This dual approach provides both insight and evidence-based conclusions.

Among the 112 radiomics features, `original_gray level co-occurrence matrix (glcm)_DifferenceEntropy` and `original_first-order_Entropy` performed exemplary in predicting E-PSMA scores from primary PCa lesions with AUROC 100%, $p < 0.001$. The `original_Gray Level Run Length Matrix (glrlm)_LongRunLowGrayLevelEmphasis` feature showed suboptimal performance in providing PRIMARY scores with AUROC 70.8%, $p = 0.02$.

Radiomic analysis of PSMA PET images has clinical application in predicting internationally validated PSMA-PET scoring systems in PCa staging, establishing a roadmap for increased efficiency in predicting PRIMARY score in TM PCa staging.

Keywords: EANM Guidelines · PSMA for Prostate Cancer · PSMA-PET · Guidelines for Prostate Cancer · PET/CT · Radiomics · Segmentation · Machine Learning · Predictive Model · Texture Analysis

1 Introduction

Currently, PCa ranks as the second most prevalent cancer type in men and the fifth most prominent cause of cancer-related death. PCa is characterized by multiple occurrences and several novel therapeutic approaches have been included in the toolbox [9,11]. Artificial intelligence is widely used in cancer imaging, performing three main clinical tasks: tumor detection, characterization, and monitoring. Detection refers to localizing objects of interest within biomedical images, characterization refers to identifying particular biomarkers, and monitoring refers to determining the prognosis of the disease. Molecular imaging plays a crucial role in the development of PCa, and prostate-specific membrane antigen (PSMA) is abundant on the majority of PCa cells [5]. In patients with primary intermediate/high-risk PCa, PSMA-PET imaging significantly enhanced disease staging when compared with computed tomography (CT) scans or multiparametric MRIs [2,22,24], thereby bypassing unnecessary cross-sectional imaging and bone scintigraphy. As a result of the potential controversy in regards to the suitability of PSMA-PET in clinical settings [18], PSMA-PET is currently under investigation. Radiomics is an advanced technique which uses sophisticated mathematics and statistics to determine quantitative parameters, called radiomic features, from biomedical imaging, in support of medical treatment decisions [21]. By employing Deep Learning (DL) and Machine Learning (ML) algorithms [23], radiomics has shown significant potential for optimizing diagnostic [25], prognostic, and predictive outcomes in a variety of cancer types, including PCa [3,8]. To improve PCa diagnosis and staging accuracy, several scoring systems have been developed. In particular, the PRIMARY score has significantly influenced the clinical reporting of PSMA-PET results. Recent research shows that artificial intelligence (AI)-based approaches, such as ML, DL, and convolutional neural networks (CNN), can be used to automate PSMA-PET metastasis detection, predict response to PSMA-targeted therapies to detect intraprostatic cancer [20]. This preliminary study investigated the feasibility for applying machine learning algorithms to radiomic features extracted from PSMA-PET to identify primary PCa classified using the PRIMARY score for PCa staging.

2 Materials and Methods

In this retrospective analysis, we considered 18 patients with PCa, referred to the Nuclear Medicine Unit of the Civic Hospital A.R.N.A.S. of Palermo for diagnosis staging and restaging of 18F-PSMA-1007 digital PET/CT (PSMA-PET/CT). Each participant in the study provided informed consent. Patients were included for histologically confirmed disease or verified with a 2-year follow-up. Radiomic analysis was performed on 18F-PSMA-1007 PET images of all 18 patients (18 prostatic lesions). The manual segmentation procedure was performed in agreement by a radiologist and a nuclear medical doctor both with 15 years of experience on the prostate. ITK-SNAP 4.2.2 was used to segment the lesions (tumor, node, and bone metastasis) [12]. A total of 112 radiomic features were extracted using Pyradiomics 3.1.0 software [19]. Finally, a hybrid machine learning model was applied to select most discriminant radiomic features. In this way, the most discriminant radiomic features were identified to predict the PRIMARY score and E-PSMA score in primary lesions.

2.1 18F-PSMA-1007 PET/CT Technique

According to the standard 18F-PSMA-1007 PET/CT protocol used at our institution, scans were performed following international clinical recommendations [17]. The clinical protocol included a full-body PET scan (6–8 beds, 2–3 min per bed position), followed by intravenous administration of 3–4 MBq/kg body mass of 18F-PSMA-1007 for 90–120 min (late scans may help in identifying lesions located near the ureter or bladder) [1] and a low-dose CT (120 kV, 80–120 mA) co-registered without contrast. PET images (voxel size 256 × 256) were processed with CT-assisted attenuation correction. 3D reconstruction was based on the Ordered Subset Expectation Maximization algorithm with two sequential processes. Two nuclear medicine physicians qualitatively assessed the scans. Based on the expert medical doctor assessment of the inclusion criteria, 18F-PSMA-1007 PET/CT positivity was verified by the assessors using PRIMARY Score Criteria and E-PSMA.

2.2 Manual Segmentation and Radiomics Features Extraction and Analysis

Volumetric segmentation was performed with ITK-SNAP 4.2.2 software [18,26]. In particular, a radiologist and a nuclear medical doctor both with 15 years of experience on the prostate, assessed and segmented PET tumors by agreement. An elevated PET signal intensity was defined as hyperintense when the tumor signal intensity exceeded that of non-tumor tissue. SUVmax was used as a PET value to recognize the most active lesions for global disease status assessment. The same volume was converted into the same region of CT image for morphological feature extraction. Subsequently, 112 radiomic features were extracted using Pyradiomics from the aforementioned VOI from each tumor in PET and CT images, in turn [10,19]. This software is an open-source scientific computing language that can be in-stalled on any computer system. In addition to providing a flexible analysis plat-form, PyRadiomics also provides a back-end interface that enables automated data processing, feature definition, and batch processing. The extracted radiomic features belong to the following classes: (a) shape features, focusing on geometric aspects, such as shape and volume; (b) statistical features, including first-order statistical features (histogram-based) derived from ROI intensity values, and (c) higher-order statistical features (texture) designed for quantifying texture perception and spatial intensity within a VOI. Normalize was configured to True and BinWidth to 0.13 for radiomic features extraction with Pyradiomics, the others were retained by default. The Fig. 1 shows an example of PET imaging segmentation at the level of 18F-PSMA uptake in prostate tumor.

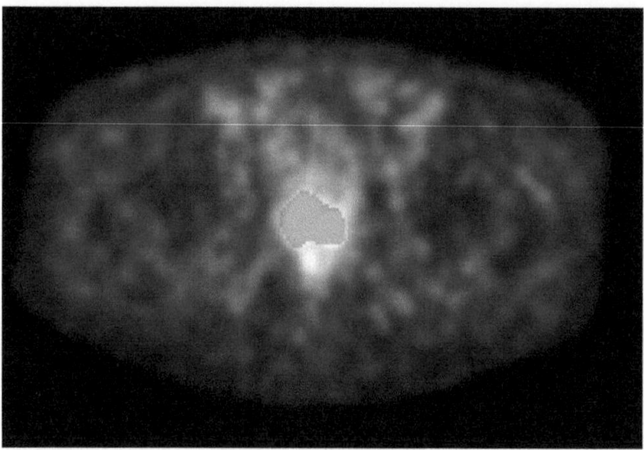

Fig. 1. PET image of the prostate with segmented tumor region (in red) showing elevated 18F-PSMA uptake. The segmentation was performed manually using ITK-SNAP by experienced clinicians, and the resulting volume of interest was used for subsequent radiomic analysis. (Color figure online)

2.3   Hybrid Radiomics Features Selection and Machine Learning Predictive Model

In this study, an innovative descriptive-inferential technique was used to identify discriminatory features and decrease their redundancy. After the selection step, a predictive model was created employing discriminant analysis. Once the training phase was complete, the model was able to classify new PET lesions. Using a k-fold cross-validation strategy, the data was randomly divided into training and validation sets [15]. In the next step, one fold served as a validation set, whereas the other folds as training sets. Using folds, the training and validation sets maintained the same patient status percentage. We determined that k = 5 is the best value for our analysis using the trial and error method. In order to increase the accuracy of the results and, therefore, the robustness of the method, the k-fold process was repeated five times.

2.4 Predictive Performance

As part of the evaluation of the performance of the predictive method [6], the sensitivity, specificity, positive predictive value (PPV), accuracy, receiver operating characteristic (ROC), and areas under the receiver operating characteristic (AUROC) with 95% confidence intervals (CI) were calculated for identifying PRIMARY and E-PSMA scores for PCa staging using radiomic features extracted from PSMA-PET. For the computational statistical analysis, we used the Matlab R2019a simulation environment (MathWorks, Natick, MA, USA), running on an Intel Core i7 2.5GHz computer, 16 GB of iMac 1600 DDR3 MHz memory, and OS X Sierra. Radiomic data were processed and analyzed efficiently using this powerful computational environment.

3 Results

18 prostatic lesions were selected from 18 PCa patients included in the study. After ROI segmentation, the classification model analysis was divided into:

- radiomic feature extraction;
- most discriminant radiomic feature selection;
- machine learning model training and validation using k-fold strategy;
- evaluation performance [4];

3.1 Performance of Radiomics

For primary prostatic lesions, excellent performances for the prediction of E-PSMA expression grades were evidenced by the following two radiomic features: `original_glcm_Difference Entropy` and `original_firstorder_Entropy`. These radiomic features gave sensitivity 100%, specificity 90.5%, AUROC 100%, $p < 0.01$ (Fig. 2a).

For predicting the E-PSMA interpretation scale, `original_Gray Level Dependence Matrix (gldm)_LargeDependenceLowGrayLevelEmphasis` was significant but with limited sensitivity (36.7%) despite a high specificity of 87.8%, (AUROC 77.3%, $p = 0.04$), as shown in Fig. 2b.

In the prediction of PRIMARY scores, suboptimal performance was supported by the radiomic feature `original_glrlm_LongRunLow GrayLevelEmphasis`, with a sensitivity of 52.6%, specificity of 86.9%, AUROC of 70.8%, $p = 0.03$ (Fig. 2c).

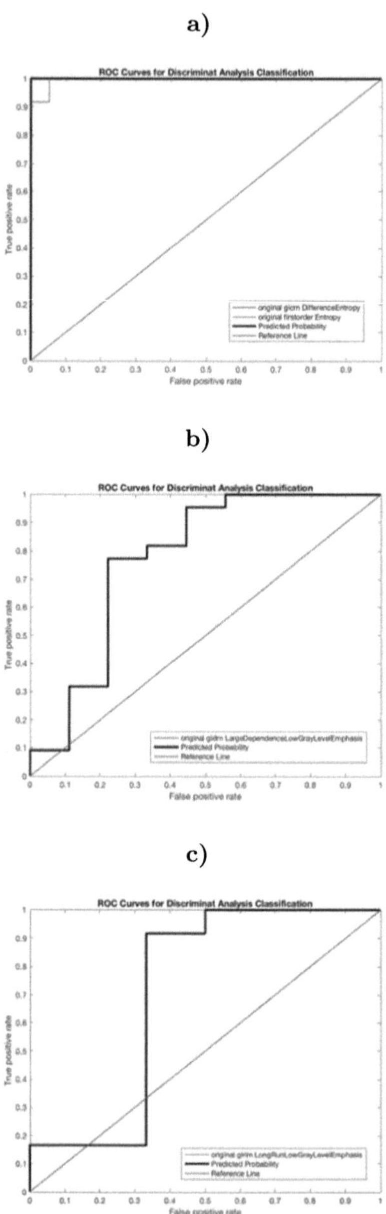

Fig. 2. ROC for the radiomic model obtained for the prediction of E-PSMA expression grades: (a) in primary prostatic lesions (`original_glcm_DifferenceEntropy` and `original_firstorder_Entropy` features); (b) for the E-PSMA interpretation scale (`original_gldm_LargeDependenceLowGrayLevelEmphasis` feature); (c) prediction of PRIMARY scores (`original_glrlm_LongRunLowGrayLevelEmphasis` radiomic feature).

4   Discussion

PSMA-PET staging can better delineate localized PCa and improve treatment. Radiomics applied to prostate 18F-PSMA-PET may help improve tumor localization, significantly enhancing diagnostic confidence of patients with PCa and relapses thanks to the additional information provided by radiomic features extracted from prostate tumor.

A computational device can measure tumor size, prostate volume, and radiomic features directly [16]. Thus, objective results can be obtained independently of the competence of the nuclear medicine doctor as well as independently of the operator. This will support medical operators in predicting international scoring systems on PSMA-PET in PCa diagnosis.

In this work, the performance of radiomics analysis was evaluated with 18F-PSMA-1007 PET images of 18 patients with histologically confirmed or verified PCa disease with 2-year follow-up (18 prostate lesions). The results of our analysis of primary lesions identified the radiomic features `original_glcm_DifferenceEntropy` and `original_firstorder_Entropy` as significantly discriminating for E-PSMA expression grades. GLCM quantifies the frequency of voxels with the same intensity at a predetermined distance along a fixed direction.

For the E-PSMA interpretation scale, `original_gldm_Large DependenceLowGrayLevelEmphasis` was highly discriminant in specificity, but poor in sensitivity. GLDM quantifies the number of voxel segments whose intensities in a specific axis are equal.

For primary prostatic lesions, two features (`original_glcm _DifferenceEntropy` and `original_firstorder_Entropy`) showed excellent performance in predicting E-PSMA expression grades (sensitivity 100%, specificity 90.5%, AUROC 100%, $p = 0.00$). For the prediction of the E-PSMA interpretation scale, `original_gldm_LargeDependenceLowGrayLevelEmphasis` was significant but with limited sensitivity (36.7%) despite high specificity (87.8%, AUROC 77.3%, $p = 0.04$).

The `original_glrlm_LongRunLowGrayLevelEmphasis` feature showed suboptimal performance in predicting PRIMARY scores (sensitivity 52.6%, specificity 86.9%, AUROC 70.8%, $p = 0.02$). GLRLM quantifies consecutive voxels with the same intensity in fixed directions.

As a result of these findings, radiomic analysis of PSMA PET images could be a promising method for detecting the presence of a high-grade PCa in the prostate gland [10,13]. Significant AUC and high sensitivity provide evidence that this methodology could be helpful as a yy tool to diagnose and assess the disease. Due to its reasonable specificity and accuracy, radiomic analysis may help decrease false positives [7,14]. A major limitation of this retrospective study is the small sample size (n = 18 lesions), which may lead to overfitting due to the high number of extracted radiomic features (n = 112). This limits the generalizability of the results, and the findings should be considered preliminary.

Additionally, the inclusion of only patients with confirmed histological diagnosis or 2-year follow-up–verified disease may introduce a selection bias. Validation on larger, more diverse cohorts is needed to confirm these results.

5   Conclusion

In this study, we assessed the potential clinical value of combining radiomics with machine learning algorithms in the prediction of primary PCa classification following PRIMARY score and E-PSMA Criteria for PCa patients underwent PSMA-PET. Implementing radiomics analysis may improve the detection of PCa and reduce the possibility of subjective interpretation of PSMA-PET images. By extracting quantitative features from PSMA-PET images, radiomics analysis provides an objective and systematic approach to disease identification and classification, potentially improving accuracy and reducing interpretative variability. These results suggest that radiomics analysis predicts PRIMARY score in primary lesions. However, further future analyses involving larger datasets and prospective multicenter studies will be essential.

References

1. Afshar-Oromieh, A., et al.: Pet imaging with a [68ga]gallium-labelled PSMA ligand for the diagnosis of prostate cancer: biodistribution in humans and first evaluation of Tumour lesions. Eur. J. Nucl. Med. Mol. Imaging **40**, 486–495 (2013). https://doi.org/10.1007/s00259-012-2298-2
2. Agnello, L., Comelli, A., Ardizzone, E., Vitabile, S.: Unsupervised tissue classification of brain MR images for voxel-based morphometry analysis. Int. J. Imaging Syst. Technol. **26**, 136–150 (2016). https://doi.org/10.1002/ima.22168
3. Ali, M., et al.: Applications of artificial intelligence, deep learning, and machine learning to support the analysis of microscopic images of cells and tissues. J Imaging **11**, 59 (2025). https://doi.org/10.3390/jimaging11020059
4. Ali, M., et al.: Prostate cancer detection: performance of radiomics analysis in multiparametric MRI (2024) (2024). https://doi.org/10.1007/978-3-031-51026-7_8
5. Ali, M., Benfante, V., Di Raimondo, D., Laudicella, R., Tuttolomondo, A., Comelli, A.: A review of advances in molecular imaging of rheumatoid arthritis: from in vitro to clinic applications using radiolabeled targeting vectors with technetium-99m. Life **14**, 751 (2024). https://doi.org/10.3390/life14060751
6. Alongi, P., et al.: Radiomics analysis of brain [18F]FDG PET/CT to predict Alzheimer's disease in patients with amyloid pet positivity: a preliminary report on the application of SPM cortical segmentation, pyradiomics and machine-learning analysis. Diagnostics **12**, 933 (2022). https://doi.org/10.3390/diagnostics12040933
7. Alongi, P., et al.: Prostate-specific membrane antigen-PET/CT may result in stage migration in prostate cancer: performances, quantitative analysis, and potential criticism in the clinical practice. Nucl. Med. Commun. **45**, 622–628 (2024). https://doi.org/10.1097/MNM.0000000000001850
8. Alongi, P., et al.: Artificial intelligence applications on restaging [18F]FDG PET/CT in metastatic colorectal cancer: a preliminary report of morpho-functional radiomics classification for prediction of disease outcome. Appl. Sci. **12**, 2941 (2022). https://doi.org/10.3390/app12062941

9. Basirinia, G., et al.: Theranostic approaches for gastric cancer: an overview of in vitro and in vivo investigations. Cancers (Basel) **16**, 3323 (2024). https://doi.org/10.3390/cancers16193323

10. Benfante, V., et al.: Grading and staging of bladder tumors using radiomics analysis in magnetic resonance imaging (2024). https://doi.org/10.1007/978-3-031-51026-7_9

11. Bergengren, O., et al.: 2022 update on prostate cancer epidemiology and risk factors–a systematic review. Eur. Urol. **84**, 191–206 (2023). https://doi.org/10.1016/j.eururo.2023.04.021

12. Cairone, L., et al.: Robustness of radiomics features to varying segmentation algorithms in magnetic resonance images (2022). https://doi.org/10.1007/978-3-031-13321-3_41

13. Canfora, I., et al.: A predictive system to classify preoperative grading of rectal cancer using radiomics features (2022). https://doi.org/10.1007/978-3-031-13321-3_38

14. Ceci, F., et al.: E-PSMA: the EANM standardized reporting guidelines v1.0 for PSMA-PET. Eur. J. Nucl. Med. Mol. Imaging **48**(5), 1626–1638 (2021). https://doi.org/10.1007/s00259-021-05245-y

15. Comelli, A., Stefano, A., Benfante, V., Russo, G.: Normal and abnormal tissue classification in positron emission tomography oncological studies. Pattern Recogn. Image Anal. **28**(1), 106–113 (2018). https://doi.org/10.1134/S1054661818010054

16. Corso, R., Comelli, A., Salvaggio, G., Tegolo, D.: New parametric 2D curves for modeling prostate shape in magnetic resonance images. Symmetry (Basel) **16**, 755 (2024). https://doi.org/10.3390/sym16060755

17. Fendler, W., et al.: PSMA PET/CT: joint EANM procedure guideline/SNMMI procedure standard for prostate cancer imaging 2.0. Eur. J. Nucl. Med. Mol. Imaging **50**, 1466–1486 (2023). https://doi.org/10.1007/s00259-022-06089-w

18. Giaccone, P., Benfante, V., Stefano, A., Cammarata, F., Russo, G., Comelli, A.: Pet images atlas-based segmentation performed in native and in template space: a radiomics repeatability study in mouse models. Int. Conf. Image Anal. Process. (2022). https://doi.org/10.1007/978-3-031-13321-3_31

19. van Griethuysen, J., et al.: Computational radiomics system to decode the radiographic phenotype. Cancer Res. **77**, e104–e107 (2017). https://doi.org/10.1158/0008-5472.CAN-17-0339

20. Piras, A., et al.: Artificial intelligence and statistical models for the prediction of radiotherapy toxicity in prostate cancer: a systematic review. Appl. Sci. **14**, 10947 (2024). https://doi.org/10.3390/app142310947

21. Russo, G., et al.: Feasibility on the use of radiomics features of 11[c]-MET PET/CT in central nervous system Tumours: preliminary results on potential grading discrimination using a machine learning model. Curr. Oncol. **28**, 5318–5331 (2021). https://doi.org/10.3390/curroncol28060444

22. Salamone, G., et al.: Vacuum-assisted wound closure with mesh-mediated fascial traction achieves better outcomes than vacuum-assisted wound closure alone: a comparative study. World J. Surg. (1), 1–8 (2017). https://doi.org/10.1007/s00268-017-4354-3

23. Salvaggio, G., et al.: Deep learning network for segmentation of the prostate gland with median lobe enlargement in t2-weighted MR images: comparison with manual segmentation method. Curr. Probl. Diagn. Radiol. **51**, 328–333 (2022). https://doi.org/10.1067/j.cpradiol.2021.06.006

24. Stefano, A.: Challenges and limitations in applying radiomics to pet imaging: possible opportunities and avenues for research. Comput. Biol. Med. **179**, 108827 (2024). https://doi.org/10.1016/j.compbiomed.2024.108827
25. Vernuccio, F., et al.: Diagnostic performance of qualitative and radiomics approach to parotid gland tumors: which is the added benefit of texture analysis? Br. J. Radiol. **94**(1128), 20210340 (2021). https://doi.org/10.1259/bjr.20210340
26. Yushkevich, P., et al.: User-guided 3D active contour segmentation of anatomical structures: significantly improved efficiency and reliability. Neuroimage **31**, 1116–1128 (2006). https://doi.org/10.1016/j.neuroimage.2006.01.015

Prostate MRI Segmentation: A Comparative Analysis of U-Net and E-Net Deep Learning Models

Gaspare Centineo[1], Leonardo Salvaggio[2,3], Syed Ibrar Hussain[3], Muhammad Ali[3], Kristen Meiburger[1], Filippo Molinari[1], and Albert Comelli[3(✉)]

[1] Biolab - Department of Electronics and Telecommunications, Politecnico Di Torino, Turin, Italy
[2] Department of Biomedicine, Neuroscience and Advanced Diagnostics (BiND), University of Palermo, Palermo, Italy
[3] Ri.MED Foundation, Via Bandiera 11, 90133 Palermo, Italy
acomelli@fondazionerimed.com

Abstract. Accurate segmentation of the prostate gland in magnetic resonance imaging (MRI) is important for the early detection and treatment of prostate cancer. Manual segmentation is time-consuming and prone to interobserver variability. This study evaluates and compares two deep learning models, U-Net and E-Net, for automated prostate segmentation using axial T2-weighted MRI scans. Both models were trained and validated using a five-fold cross-validation approach, and tested on a hold-out dataset. Performance was assessed using metrics such as Dice Similarity Coefficient (DSC), precision, recall, and surface distance measures. E-Net achieved superior results with a mean DSC of 81.61%, outperforming U-Net's 77.53%, while also demonstrating significantly lower computational complexity and faster inference times. Despite the performance gap, testing showed statistically significant difference between the models. These findings highlight the potential of lightweight deep learning architectures, particularly E-Net, for accurate and efficient prostate segmentation in clinical settings.

Keywords: Prostate Cancer · Deep Learning · MRI · U-Net · E-Net · Medical Image Segmentation

1 Introduction

Prostate cancer (PCa) remains one of the most frequently diagnosed malignancies among men worldwide, with a significant public health challenge due to increasing incidence and mortality rate [7]. This growing burden demands the urgent need for an accurate and early diagnosis to support effective treatment strategies and improve survival outcomes. Magnetic Resonance Imaging (MRI), particularly multiparametric MRI (mpMRI), has become a cornerstone

E. Rodolà et al. (Eds.): ICIAP 2025 Workshops, LNCS 16169, pp. 299–309, 2026.
https://doi.org/10.1007/978-3-032-11317-7_26

in PCa detection, providing superior anatomical detail and tumor characterization. Radiomics [5] and quantitative image analysis have further improved the diagnostic capabilities of mpMRI, but their efficacy is highly dependent on accurate segmentation of the prostate and cancerous regions [16] .

Manual segmentation remains the clinical standard, but is labor intensive, time-consuming, and prone to interobserver variability. To address these limitations, deep learning-based approaches have emerged as powerful tools for automated prostate segmentation. Among these, convolutional neural network (CNN) architectures such as U-Net and E-Net have demonstrated exceptional performance in capturing complex anatomical structures and enhancing segmentation accuracy [12]. These models offer improved reproducibility, reduced processing time, and facilitate real-time integration into radiology workflows [8].

This paper focuses on evaluating the performance and clinical applicability of U-Net and E-Net architectures in automated segmentation of prostate anatomy and MRI, with the aim of supporting a more precise diagnosis and personalized treatment planning for patients with PCa.

2   Materials and Methods

2.1   Population

The dataset used for this work consists of prostate MRI scans acquired between June 1, 2019, and January 31, 2023, at the Azienda Ospedaliera Universitaria Policlinico (AOUP) "Paolo Giaccone", under the University of Palermo. A total of 71 prostate MRI images were selected, all with a Prostate Imaging Reporting and Data System (PI-RADS) score of 3 or higher. The dataset was further categorized into two groups based on histological analysis from fusion biopsy: Group 1 (n = 45; 61.6%) included prostate tissue with histological features consistent with neoplastic changes, while Group 2 (n = 28; 38.4%) comprised prostate tissue showing non-neoplastic conditions, such as prostatitis and prostatic intraepithelial neoplasia. This dataset provides a diverse range of prostate tissue types, enabling the development of segmentation models aimed at accurately delineating the prostate in both healthy and altered tissue regions.

2.2   MRI Technique

MRI examinations were performed using a 1.5T MRI scanner (Achieva, Philips Healthcare, Best, The Netherlands) with a 16-channel pelvic phased array coil (HD Torso XL). The imaging protocol was kept constant for all patients. Standard clinical prostate MRI acquisition included T2-weighted turbo spin-echo sequences in axial, coronal, and sagittal projections of the prostate and seminal vesicles, as well as axial T1-weighted turbo spin-echo images. In addition, diffusion-weighted imaging (DWI) with b-values of 0, 700, and 1400 s/mm was performed, subsequently processed to generate apparent diffusion coefficient (ADC) maps. After the acquisition of unenhanced images, patients were administered gadoteric acid (Gd-DOTA, Dotarem, Guerbet) at a dose of 1 mmol/kg of body weight, with an infusion rate of 3 mL/s, followed by 20–30 mL of saline solution at the same rate. Subsequently, axial three-dimensional spoiled gradient-echo interpolated volumetric images were acquired to capture prostate perfusion via dynamic contrast-enhanced MRI (DCE). For the purpose of this study, axial T2-weighted images were mainly used.

2.3   U-Net and E-Net

U-Net and E-Net Architectures To enhance segmentation performance, several modifications were introduced to the original U-Net architecture. Notably, all 3×3 convolutional layers were replaced with larger 5×5 kernels to expand the network's receptive field, which, in our experience, led to improved segmentation accuracy. Each convolutional layer was followed by a dropout layer with a 10% rate to mitigate overfitting and enhance generalization. Unlike the original U-Net, which does not apply padding, zero-padding was employed in our version to preserve spatial dimensions between input and output feature maps. The original U-Net utilized a 32×32 feature map with 1024 channels at the deepest layer of the encoder. In contrast, our implementation began with an input resolution of 512×512 and 32 filters in the initial convolutional block, doubling the number of filters after each max-pooling operation until reaching 256 filters at a resolution of 64×64. Here, "feature maps" refer to the outputs of convolutional layers that capture meaningful patterns in the data, while "max pooling" refers to a common downsampling operation in CNNs that reduces spatial dimensions by selecting the maximum value from non-overlapping sub-regions—typically halving the height and width of the input [13].

E-Net, on the other hand, was originally designed for rapid inference and high accuracy in applications such as augmented reality and automotive environments. Its architecture is built upon residual blocks, each composed of three convolutional layers: a 1×1 projection for dimensionality reduction, a standard convolution, and a 1×1 expansion layer, all followed by batch normalization. E-Net integrates multiple convolution types within an encoder-decoder framework for image segmentation, including asymmetric convolutions—implemented as sequential 5×1 and 1×5 layers—to reduce parameter count and computational load. While a standard 5×5 convolution contains 25 parameters, the equivalent asymmetric convolution only requires 10, making the network more efficient. Additionally, E-Net begins with a single initial block and incorporates multiple variations of a bottleneck layer—an architecture that forces the model to compress features through repeated downsampling, thereby encouraging the learning of the most salient and task-relevant information in the input [9].

2.4   Loss Functions

Various loss functions have been proposed in the literature to optimize deep learning-based medical image segmentation models. One of the most widely used is the Dice Similarity Coefficient (DSC), which quantifies the overlap between predicted segmentation masks and ground truth annotations. Mathematically, the DSC represents the harmonic mean of false positives (FPs) and false negatives (FNs), treating both error types equally. To provide greater flexibility in weighting these errors, the Tversky loss, derived from the Tversky index, was introduced [11]. The Tversky index is defined as:

$$S(P, G; \alpha, \beta) = \frac{|P \cap G|}{|P \cap G| + \alpha |P \setminus G| + \beta |G \setminus P|} \tag{1}$$

P and G denote the predicted and ground truth label sets, respectively, while α and β are weighting parameters that control the relative penalties for false positives and false negatives. The Tversky loss function can then be formulated as:

$$T(\alpha, \beta) = \frac{\sum_{i=1}^{N} p_{0i} g_{0i}}{\sum_{i=1}^{N} p_{0i} g_{0i} + \alpha \sum_{i=1}^{N} p_{0i} g_{1i} + \beta \sum_{i=1}^{N} p_{1i} g_{0i}} \tag{2}$$

In this equation, p_{0i} is the predicted probability that voxel i belongs to the target region (e.g., prostate), and p_{1i} is the probability that it belongs to the background. The ground truth labels g_{0i} and g_{1i} are binary indicators for foreground and background, respectively. By adjusting the values of α and β, one can fine-tune the loss function to emphasize either precision or recall.

For example, setting $\alpha = \beta = 0.5$ recovers the standard Dice similarity coefficient (DSC), while enforcing $\alpha + \beta = 1$ yields a general form of the F-score. Choosing $\beta > 0.5$ biases the model toward recall, making it especially useful in medical imaging tasks where missing foreground regions (false negatives) is more critical than over-segmentation.

2.5   Training Phase

The training of the deep learning models and the subsequent inference phases were performed on a Dell Inc. Precision 7760 workstation equipped with an NVIDIA RTX 5000 GPU with 16 GB of RAM and CPU Intel® Xeon® W-11955M x 16 cores with 128 GB. The optimization of the learning rates for each architecture was conducted experimentally: a learning rate of 0.0001 was adopted for the ENet network, while it was set to 0.00001 for the UNet network. In both cases, the Adam optimizer and a batch size set to twenty four slices per iteration were used. The training of the models was carried out for a maximum of 100 epochs, with the use of an early stopping criterion in case the loss function showed no improvement for at least 10 consecutive epochs.

Data Training. In deep learning-based medical image analysis, data is typically partitioned into three subsets: training, validation, and testing. The testing set, often referred to as the hold-out set, remains completely excluded from the training process and is used solely for final performance evaluation. When data availability is limited as is common in medical imaging a k-fold cross-validation approach is often employed to enhance reliability. In this study, we applied a five-fold cross-validation strategy due to the restricted dataset size [4]. For each network architecture, five models were trained by iteratively using four folds for training and the remaining fold for validation. Given that the networks were designed for 2D segmentation, individual MRI slices were extracted from each study and used as inputs to the models.

Model performance was assessed by averaging the results across the five validation folds.

To reduce overfitting and improve model generalization, data augmentation was applied during training. Each input slice was randomly transformed using six augmentation techniques: rotation, translation (in both x and y directions), shearing, horizontal flipping, and zooming. In line with standard deep learning preprocessing practices, pixel-wise normalization was also performed. For each fold, the mean and standard deviation of pixel intensities were computed using the training data, and these values were used to normalize the input slices prior to model training.

An initial subset of 16 patient studies was used to empirically determine optimal learning rates. The Adam optimizer was employed with a learning rate of 0.0001 for E-Net and 0.00001 for U-Net. All models were trained using a batch size of 8 slices and the Tversky loss function with 0.3 and 0.7. Training was allowed to continue for a maximum of 100 epochs, with early stopping implemented based on training loss: if no improvement was observed over 10 consecutive epochs, training was halted. All experiments were conducted on an NVIDIA RTX 5000 laptop card, equipped with 16GB RAM

2.6 Data Analysis

To evaluate the performance of the automatic segmentation models, a set of commonly used shape-based performance metrics was computed for each clinical case, as established in the literature. The metrics included precision, recall, Dice Similarity Coefficient (DSC), volume overlap error (VOE), volumetric difference (RVD). For each metric, the mean, standard deviation (std), and 95% confidence interval (CI) were reported. To determine statistically significant differences in segmentation performance across different network architectures, a Two-tailed paired samples t-test was conducted on the dice similarity coefficient (DSC). A p-value of less than 0.05 was considered indicative of statistical significance [14].

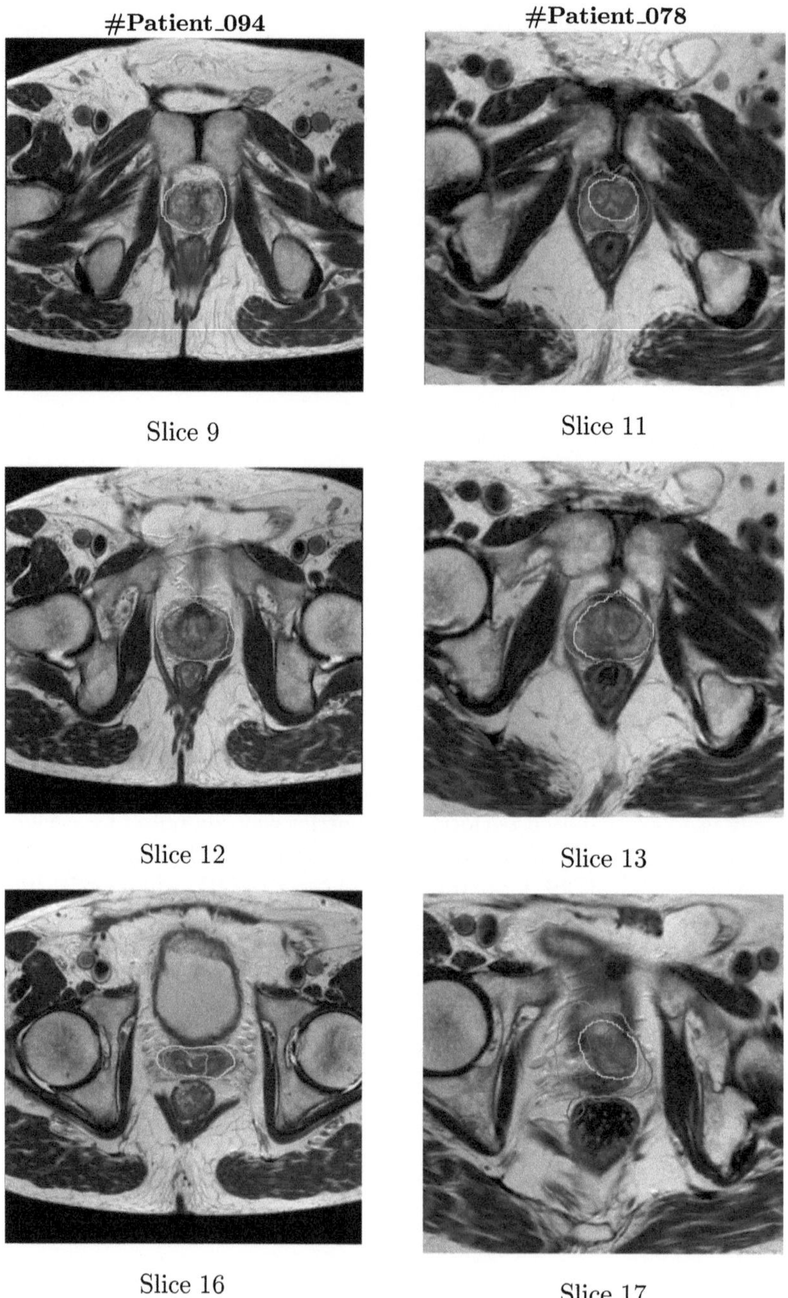

Fig. 1. Comparison of segmentation performance in Fold-5 between the best-performing case (**#Patient_094**) and the worst-performing case (**#Patient_078**). Ground truth (Yellow), U-Net predictions (Blue) and E-Net predictions (Red).(Color figure online)

3   Results

The proposed segmentation methods were applied to automatically identify and delineate the prostate in MRI scans. We compared the performance of these two well-established models: ENet and UNet. Table 1 and Fig. 1 present the quantitative evaluation results obtained using a k-fold cross-validation strategy. The ENet model achieved the best performance, with a Dice Similarity Coefficient (DSC) of $81.61 \pm 2.80\%$, outperforming UNet, which reached a DSC of $77.53 \pm 2.59\%$. Figure 2 and Fig. 3 show the profiles of training DSC and Tversky loss function for one fold. Both DSC and Tversky loss profiles indicate that the ENet model converges much faster than both the UNet. This suggests that training could be obtained faster with ENet. The fact that the training loss of the UNet model is lower than ENet suggests the presence of overfitting, although the number of UNet filters in each layer was reduced.

Table 2 and Table 3 show respectively the two-tailed paired samples t-test on the Dice Similarity Coefficient (DSC) and the comparison of computational complexity and performance for each model. This suggests that the two networks show statistically different DSC results with a t-value of 2.89 and a p-value of $0.045 < 0.05$, demonstrating that E-Net is more performant in terms of DSC. Furthermore, as reported in Table 4, the E-Net model has a number of trainable parameters significantly lower than the U-Net model (362,992 vs. over 1.9 million in U-Net), which makes E-Net faster in inference both with CPU and GPU (2.18 s CPU and 1.88 s GPU vs. 11.61 s CPU and 4.66 s GPU in U-Net), and therefore applicable in real-time on clinical systems.

Table 1. Performance comparison between E-Net and U-Net

E-Net

	DICE(%)	VOE(%)	RVD(%)	PREC.(%)	REC.(%)	F1 (%)
Fold 1	78.04 ± 17.12	33.39 ± 18.8	21.16 ± 19.07	79.2 ± 8.36	80.46 ± 22.55	78.04 ± 17.12
Fold 2	78.17 ± 14.75	33.81 ± 16.83	56.33 ± 79.86	69.76 ± 18.78	93.94 ± 4.33	78.17 ± 14.75
Fold 3	85.74 ± 3.74	24.78 ± 5.40	13.66 ± 12.06	87.27 ± 6.08	85.84 ± 10.51	85.74 ± 3.74
Fold 4	83.00 ± 6.10	28.61 ± 8.66	28.15 ± 21.81	75.65 ± 10.77	93.60 ± 5.52	83.00 ± 6.10
Fold 5	83.12 ± 4.58	28.62 ± 6.58	26.07 ± 17.94	78.35 ± 10.84	90.80 ± 8.18	83.12 ± 4.58
Media	$\mathbf{81.61 \pm 2.8}$	$\mathbf{29.84 \pm 3.006}$	$\mathbf{29.07 \pm 10.9}$	$\mathbf{78.04 \pm 4.27}$	$\mathbf{88.928 \pm 4.62}$	$\mathbf{81.61 \pm 2.8}$

U-Net

	DICE(%)	VOE(%)	RVD(%)	PREC.(%)	REC.(%)	F1 (%)
Fold 1	71.56 ± 21.44	40.56 ± 22.41	24.33 ± 25.76	83.44 ± 9.67	66.63 ± 25.51	71.56 ± 21.44
Fold 2	79.33 ± 7.49	33.66 ± 9.69	22.88 ± 12.86	78.9 ± 9.55	82.15 ± 13	79.33 ± 7.79
Fold 3	80.02 ± 4.77	33.05 ± 6.5	17.97 ± 10.02	89.25 ± 3.01	73.04 ± 8	80.02 ± 4.77
Fold 4	77.04 ± 11.37	36.17 ± 12.79	20.08 ± 17.39	85.63 ± 6.32	72.7 ± 16.38	77.04 ± 11.37
Fold 5	79.74 ± 3.27	33.56 ± 4.59	28.55 ± 16.41	74.65 ± 9.96	87.94 ± 8.69	79.74 ± 3.27
Media	$\mathbf{77.53 \pm 2.59}$	$\mathbf{35.4 \pm 2.37}$	$\mathbf{22.76 \pm 2.98}$	$\mathbf{82.37 \pm 4.47}$	$\mathbf{76.49 \pm 6.84}$	$\mathbf{77.53 \pm 2.59}$

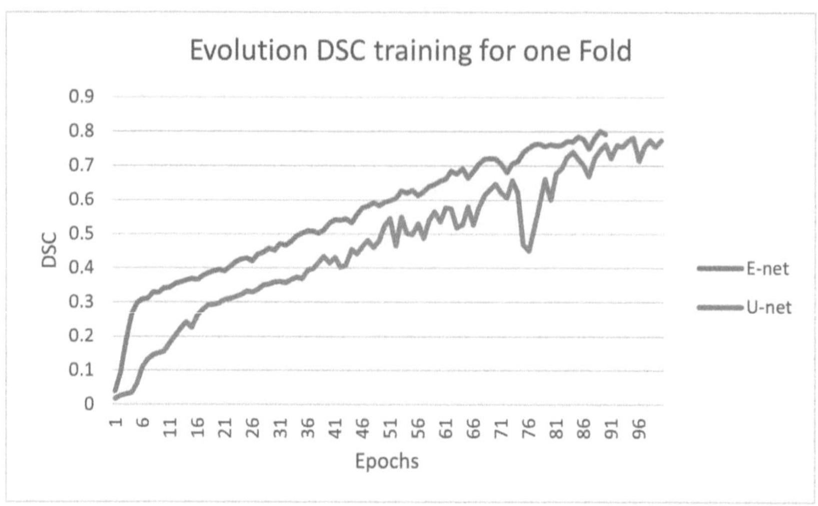

Fig. 2. Evolution of training DSC for both Networks.

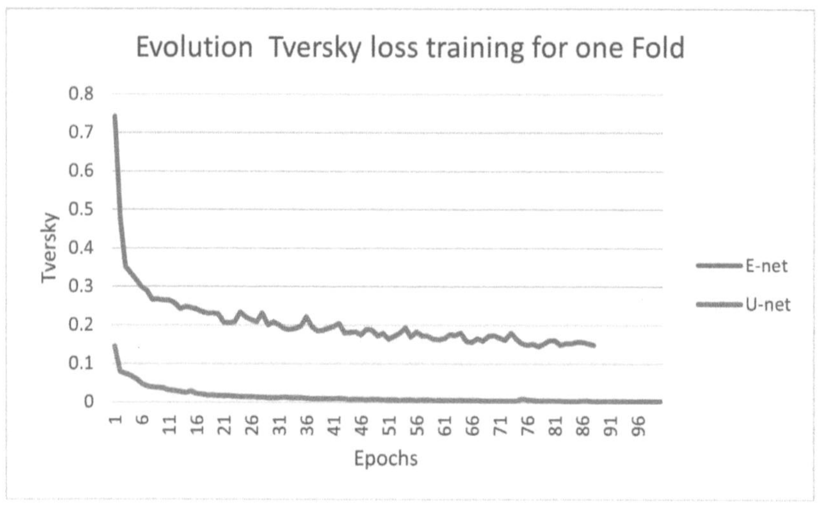

Fig. 3. Evolution of training Tversky loss for both Networks.

Table 2. Two-tailed paired samples t-test on the dice similarity coefficient (DSC) showed statistical differences between segmentation methods.

Two-tailed paired samples t-test	t-Value	p-Value
E-Net vs U-Net	2.88463755	0.044803672

Table 3. Segmentation metrics (DSC, precision, recall) comparing U-Net vs. E-Net

Model Name	Number of Parameters		Size on Disk	Inference Times/Dataset	
	Trainable	**Non-Trainable**		**CPU**	**GPU**
E-Net	362,992	8352	5.8 MB	2.18 s	1.88 s
U-Net	5,403,874	0	65.0 MB	11.61 s	4.66 s

4   Discussion and Conclusions

Accurate segmentation of the prostate gland in MRI scans is a crucial step in the diagnosis, treatment planning, and monitoring of prostate cancer [2]. Multi-parametric MRI (mpMRI) provides rich anatomical and functional information, making it the gold standard for prostate imaging [6]. However, manual segmentation is time-consuming, subjective, and prone to variability [15]. Deep learning approaches have emerged as powerful tools to automate this process, improving consistency and efficiency in clinical workflows [1,3]. In this study, we compared the performance of two deep learning architectures U-Net and E-Net for the task of automated prostate segmentation in MRI. Both models demonstrated promising results, with E-Net slightly outperforming U-Net in most evaluation metrics, including the Dice Similarity Coefficient (81.61% vs. 77.53%), precision, and recall. Despite these numerical differences, statistical analysis via Two-tailed paired samples t-test indicates significant difference between the models (p = 0.0448), suggesting that the two networks show statistically different DSC results demonstrating that E-Net is more performant as DSC.

One of the key advantages of E-Net lies in its computational efficiency [10]. With significantly fewer trainable parameters (362,992 vs. over 1.9 million in U-Net), E-Net achieved faster inference times and required less storage, making it better suited for integration into clinical workflows or deployment in resource-constrained environments. This efficiency, combined with high segmentation accuracy, highlights E-Net as a viable solution for real-time prostate imaging analysis.

On the other hand, U-Net, despite its heavier architecture and longer inference time, remains a strong baseline, particularly when modified with larger convolutional kernels to enhance the receptive field. These enhancements helped narrow the performance gap with E-Net, confirming U-Net's continued relevance in medical image segmentation tasks.

The robustness of both models was further supported by consistent performance across five-fold cross-validation and a separate hold-out test set. The use of advanced loss functions such as the Tversky index also contributed to the networks' ability to balance false positives and false negatives—an essential aspect in clinical settings where missing a lesion can be far more detrimental than over-segmentation.

In conclusion, both U-Net and E-Net offer effective solutions for automatic segmentation of the prostate in MRI, with E-Net providing a better trade-off between performance and computational cost. These findings support the integration of lightweight deep learning models into radiological practice to enhance diagnostic precision, reduce workload, and improve patient outcomes in prostate cancer care. Future work may explore multi-modal data integration, 3D volumetric segmentation, and domain adaptation techniques to further improve segmentation robustness and generalizability.

References

1. Ali, M., et al.: Applications of artificial intelligence, deep learning, and machine learning to support the analysis of microscopic images of cells and tissues. J. Imaging **11**(2) (2025). https://doi.org/10.3390/jimaging11020059
2. Ali, M., et al.: Prostate cancer detection: performance of radiomics analysis in multiparametric MRI. In: Image Analysis and Processing – ICIAP 2023 Workshops, Lecture Notes in Computer Science, vol. 14366, pp. 83–92. Springer (2024). https://doi.org/10.1007/978-3-031-51026-7_8
3. Alongi, P., et al.: Radiomics analysis of brain [^{18}f]fdg pet/ct to predict alzheimer's disease in patients with amyloid pet positivity: a preliminary report on the application of spm cortical segmentation, pyradiomics and machine-learning analysis. Diagnostics **12**(4) (2022). https://doi.org/10.3390/diagnostics12040933
4. Ameen, Y.A., Badary, D.M., Abonnoor, A.E.I., Hussain, K.F., Sewisy, A.A.: Which data subset should be augmented for deep learning? a simulation study using urothelial cell carcinoma histopathology images. BMC Bioinf. **24**(1), 75 (2023). https://doi.org/10.1186/s12859-023-05199-y
5. Canfora, I., et al.: A predictive system to classify preoperative grading of rectal cancer using radiomics features. In: Image Analysis and Processing – ICIAP 2022 Workshops, Lecture Notes in Computer Science, vol. 13373, pp. 431–440. Springer (2022). https://doi.org/10.1007/978-3-031-13321-3_38
6. Corso, R., Comelli, A., Salvaggio, G., Tegolo, D.: New parametric 2d curves for modeling prostate shape in magnetic resonance images. Symmetry **16**(6) (2024). https://doi.org/10.3390/sym16060755
7. Litwin, M.S., Tan, H.J.: The diagnosis and treatment of prostate cancer: a review. JAMA **317**(24), 2532–2542 (2017). https://doi.org/10.1001/jama.2017.7248
8. Molière, S., et al.: Reference standard for the evaluation of automatic segmentation algorithms: quantification of inter observer variability of manual delineation of prostate contour on mri. Diagn. Interv. Imaging **105**(2), 65–73 (2024). https://doi.org/10.1016/j.diii.2023.08.001
9. Paszke, A., Chaurasia, A., Kim, S., Culurciello, E.: Enet: a deep neural network architecture for real-time semantic segmentation. arXiv preprint arXiv:1606.02147 (2016). https://doi.org/10.48550/arXiv.1606.02147
10. Salamone, G., et al.: Vacuum-assisted wound closure with mesh-mediated fascial traction achieves better outcomes than vacuum-assisted wound closure alone: a comparative study. World J. Surg. , 1–8 (2017). https://doi.org/10.1007/s00268-017-4354-3
11. Salehi, S.S.M., Erdogmus, D., Gholipour, A.: Tversky loss function for image segmentation using 3d fully convolutional deep networks. In: International workshop on machine learning in medical imaging, pp. 379–387. Springer (2017). https://doi.org/10.1007/978-3-319-67389-9_44

12. Salvaggio, G., et al.: Deep learning network for segmentation of the prostate gland with median lobe enlargement in t2-weighted mr images: Comparison with manual segmentation method. Curr. Probl. Diagn. Radiol. **51**(3), 328–333 (2022). https://doi.org/10.1067/j.cpradiol.2021.06.006
13. Sarsembayeva, T., Mansurova, M., Abdildayeva, A., Serebryakov, S.: Enhancing u-net segmentation accuracy through comprehensive data preprocessing. J. Imaging **11**(2), 50 (2025). https://doi.org/10.3390/jimaging11020050
14. Sherer, M.V., et al.: Metrics to evaluate the performance of auto-segmentation for radiation treatment planning: a critical review. Radiother. Oncol. **160**, 185–191 (2021). https://doi.org/10.1016/j.radonc.2021.05.003
15. Turkbey, B., et al.: Fully automated prostate segmentation on MRI: comparison with manual segmentation methods and specimen volumes. Am. J. Roentgenol. **201**(5), W720–W729 (2013). https://doi.org/10.2214/AJR.12.9712
16. Zhou, W., et al.: Interpretable multiparametric MRI radiomics-based machine learning model for preoperative differentiation between benign and malignant prostate masses: a diagnostic, multicenter study. Front. Oncol. **15**, 1541618 (2025). https://doi.org/10.3389/fonc.2025.1541618

A Template-Independent Method
for Parkinson's Diagnosis
from Handwritten Patterns

Lorenzo Putzu[1(✉)] , Roberta Angioni[2], and Andrea Loddo[2]

[1] Department of Electrical and Electronic Engineering, University of Cagliari,
Cagliari, Italy
`lorenzo.putzu@unica.it`
[2] Department of Mathematics and Computer Science, University of Cagliari,
Cagliari, Italy
`r.angioni7@studenti.unica.it` , `andrea.loddo@unica.it`

Abstract. Parkinson's Disease (PD) is a chronic and progressive neurodegenerative disorder that primarily impairs motor function. In recent years, advancements in deep learning (DL) have opened promising avenues for early, non-invasive diagnosis of PD through the analysis of handwriting patterns. This study investigates the use of Convolutional Neural Network (CNN) and Vision Transformers (ViT) models for classifying hand-drawn spiral and meander images collected from both healthy individuals and PD patients. We introduce a robust preprocessing pipeline tailored to enhance diagnostic features and implement a patient-exclusive data partitioning strategy to ensure clinical validity, an aspect often overlooked in prior studies. Our experiments evaluate multiple DL architectures under different fine-tuning strategies, and results show that the proposed preprocessing leads to noticeable improvements in classification accuracy. Despite more rigorous and realistic evaluation protocols, our models achieve performance levels comparable to or exceeding those in existing literature, highlighting both the effectiveness and clinical applicability of our approach.

1 Introduction

Parkinson's disease (PD) is a progressive neurodegenerative disorder that primarily affects motor control, causing symptoms such as tremor, rigidity, bradykinesia, and postural instability [14,18]. Affecting nearly 1% of people over the age of 60, PD presents a significant global health concern as life expectancy rises. Early and accurate diagnosis is challenging due to the subtle onset and subjectivity of traditional assessments [14].

Recent sensor and computational advances now allow for quantitative, objective assessment through kinematic activity detection—the measurement and analysis of movement data [13,21]. In particular, the integration of kinematic activity detection with advanced machine learning (ML) and deep learning techniques enables more accurate, objective, and early diagnosis and monitoring of

E. Rodolà et al. (Eds.): ICIAP 2025 Workshops, LNCS 16169, pp. 310–321, 2026.
https://doi.org/10.1007/978-3-032-11317-7_27

Parkinson's disease, addressing limitations of traditional clinical assessment and supporting scalable tools for remote and real-world management [9].

Artificial intelligence (AI) has significantly impacted medical diagnostics, providing advanced tools for the analysis and interpretation of complex clinical data. In the context of neurodegenerative diseases, including PD, drawing and handwriting tasks, which require fine motor coordination, offer an insightful, non-invasive method for early diagnosis through image classification [9,13].

In this work, we explore the application of CNN and ViT models for classifying hand-drawn spiral and meander patterns produced by both healthy individuals and PD patients. We propose an ad hoc pre-processing procedure tailored for this kind of image and aimed at homologating the images of spirals and meanders to a common handwritten template. We then compare several DL models through a series of controlled experiments, analysing the impact of the proposed pre-processing procedure.

The rest of his work is organised as follows. Section 2 provides a literature review of PD diagnosis, including challenges of traditional diagnosis and existing automated methodologies. Section 3 describes the materials and methods used in this work. Section 4 presents our experimental evaluation, the experimental setup and the obtained results. Conclusions and future directions are discussed in Sect. 5.

2   Related Work

PD is a progressive neurodegenerative disorder that affects over 10 million individuals worldwide and remains difficult to diagnose, especially in its early stages [14,18]. Clinical diagnosis traditionally relies on expert evaluation and scales such as the Unified PD Rating Scale (UPDRS), which are inherently subjective and often inconsistent, particularly in detecting early or mild motor symptoms [13].

In recent years, advances in AI have enabled the development of automated, objective methods for assessing PD. These methods typically extract features from various patient behaviours or biometric signals. For example, gait analysis has received considerable attention, employing wearable sensors or camera-based systems to capture spatial and temporal characteristics of walking patterns. Deep learning techniques are increasingly used to classify gait phases, predict joint trajectories, and detect anomalies like freezing of gait (FOG) [13,15,21].

Similarly, voice and speech signals, which reflect non-motor symptoms like hypophonia and speech monotony, have been used to train models for PD detection. ML classifiers have shown promising accuracy using acoustic and phonation features [19]. Other approaches include motion capture from videos (e.g., Open-Pose) [15], or multi-modal fusion techniques that combine handwriting, speech, and sensor data [17].

Among these strategies, handwriting analysis stands out as a particularly non-invasive, accessible, and interpretable method, suitable for both in-clinic and remote assessments. Tasks such as drawing spirals or meanders are commonly used to elicit motor impairment, including tremors, rigidity, and bradykinesia, which manifest visibly in the trace [11,12].

Several recent studies have proposed DL models for handwriting-based PD diagnosis. Jiang et al. [10] introduced a convolutional architecture with an attention mechanism tailored to detect tremor-induced irregularities, achieving 96.5% accuracy. Kamran et al. [11] combined multiple handwriting datasets and applied transfer learning to reach 99.22% accuracy. Wang et al. [20] used a coordinate attention Swin Transformer to enhance feature localization, achieving 92.68% accuracy across two datasets. Other works have applied Convolutional Autoencoders (CAE) [12] or hybrid CNN-ViT models [23], demonstrating that handwriting images, especially spirals and waves, can serve as powerful markers for PD diagnosis.

However, many of these studies treat different drawing types (spirals, meanders) separately, apply favourable data splits that may include images from the same patient in both training and testing sets, or rely on private datasets, making direct comparison difficult. Our study addresses these limitations by introducing a preprocessing procedure that enables unified analysis of different handwriting templates. Moreover, we adopted a patient-based data partitioning strategy to better reflect clinical use cases.

3   Materials and Methods

3.1   Dataset

Our methodology has been based on the images obtained from the publicly available NewHandPD dataset [16]. According to the authors, all participants provided informed consent at the time of data collection, and the dataset was collected in accordance with institutional ethical standards. The data used in our experiments consists solely of anonymised handwriting images, with no personally identifiable information. This dataset includes data from 66 participants, divided into two cohorts: Healthy (35 individuals) and PD patients (31 individuals). It also includes basic demographic metadata for all participants, such as age, gender, and handedness (right-handed or left-handed). These metadata enable basic cohort characterisation and help control for potential biases related to handedness or age distribution.

Each participant was instructed to complete 9 static drawing tasks: 4 spiral drawings, 4 meanders and 1 circular movement (without a specific template). In total, the NewHandPD dataset contains 594 labelled images. Image filenames follow a structured naming convention to facilitate access. For example, sp1-H1.jpg refers to the first spiral image drawn by Healthy participant 1, while sp1-P1.jpg indicates the same exam for Patient 1.

To ensure consistency and improve diagnostic relevance, we restricted our analysis to images that included a predefined visual template (spirals or meanders). Prior studies have demonstrated that template-guided drawings yield higher discriminative power, as they provide a structured reference that highlights motor impairments typical of PD, unlike freehand tasks, which introduce greater variability and ambiguity [4]. Consequently, we excluded images lacking a visible template, resulting in a curated subset of 528 samples, composed of 280 images from healthy individuals and 248 from PD patients.

Importantly, to ensure experimental reproducibility and clinical reliability, we implemented a patient-exclusive partitioning strategy for training and testing sets. Unlike several prior studies that used random splitting procedures, or even leave-one-out cross-validation schemes, without accounting for subject-level separation (thus risking data leakage and overly optimistic results) [4], we ensured that no individual contributed data to both sets. This approach prevents overfitting to patient-specific traits and aligns more closely with real-world diagnostic deployment, where a model is applied to entirely unseen patients. Specifically, we selected participants with non-overlapping IDs: the first 24 and last 11 healthy subjects (IDs 28–38) for training and testing, respectively, and the first 22 and last 9 PD patients (IDs 24–32) for training and testing, respectively. Given that each participant contributed 4 spirals and 4 meanders, this split yielded a training set of 368 images (192 healthy, 176 PD) and a testing set of 160 images (88 healthy, 72 PD).

3.2   Methodology

We designed a dedicated preprocessing pipeline aimed at normalising and unwrapping both spiral and meander drawings into a comparable linear format. This step was crucial to avoid the classifier learning to distinguish between template types (spiral vs. meander) instead of focusing on the quality and fidelity of the traced paths, which is the most meaningful feature for identifying motor impairments related to PD from hand-drawn traces. To analyse patients' drawings (blue pen trace) superimposed on predefined black templates (spirals or meanders), we developed a multi-stage image processing pipeline. The goal was to extract, linearise, and reformat the hand-drawn traces for subsequent quantitative analysis and classification.

Template Extraction and Skeletonisation. Each input image, containing both the printed black template and the overlaid blue pen trace, was first converted to the HSV colour space. A mask was applied to isolate the black template by thresholding pixels within a specified HSV range, explicitly excluding blue ink. The largest connected component in the resulting binary mask was extracted and assumed to correspond to the full template region. To enhance robustness and standardise comparison across samples, a binarised reference template (either spiral or meander) was resized to match the bounding box of the detected template in the input image and then superimposed onto the mask. The skeleton of the resulting template region was computed using morphological skeletonisation.

Path Extraction from Skeleton. We extracted a continuous, ordered path from the skeleton for image unwrapping. This was achieved by modelling the skeleton as an undirected graph where each skeleton pixel was a node connected to its 8-neighbour pixels. The longest geodesic path between any two nodes in the graph was selected as the reference path. This ensured that the most extended traceable route within the template structure was followed, even in cases with slight occlusions or noise.

Image Unwrapping Along the Path. The greyscale version of the original image was sampled orthogonally along the extracted path using bilinear interpolation. At each point along the path, a line perpendicular to the tangent direction was sampled with a fixed width (100 pixels, half of the average distance between template lines), generating a rectangular unwrapped image. This process effectively transformed the 2D curvilinear drawing into a linear representation suitable for alignment and comparison, which results in our unwrapped-only image version.

Standardisation and Reformatting. To ensure all images had consistent dimensions for downstream processing, the unwrapped image was vertically cropped (if necessary) to a fixed height. The cropped image was then rearranged into a more compact format by splitting it horizontally into blocks and reassembling them side by side for CNN compatibility. The block height was automatically selected to minimise the aspect ratio difference between height and width. The result of this step is our unwrapped-boxed image version.

The pseudocode of the whole preprocessing procedure is reported in Algorithm 1 and the results for two sample images, one meander and one spiral, are shown in Fig 1.

Algorithm 1. Preprocessing Pipeline for Template-Based Hand-Drawn Images

---

Require: Image I containing a visible drawing template
1: **Segmentation**: Extract only the black regions of the template using HSV thresholding $\rightarrow T$
2: **Normalization**: Align T with a standard version of the template $\rightarrow T_{norm}$
3: **Morphological cleaning**: Remove noise and keep the largest connected component $\rightarrow T_{clean}$
4: **Skeletonisation**: Compute the skeleton of $T_{clean} \rightarrow S$
5: **Path extraction**: Find the longest path in the skeleton $S \rightarrow P$
6: **Unwrapping**: Unwrap the original image along the path $P \rightarrow I_{unwrap}$
7: **Height normalization**: Crop if necessary I_{unwrap} to a fixed height
8: **Reformatting**: Split I_{unwrap} horizontally for CNN input $\rightarrow I_{unwrap-box}$
9: **return** $I_{unwrap}, I_{unwrap-box}$

---

4 Experimental Evaluation

This section describes the models and experimental settings, the main results of our experimental evaluation and a comparison with alternative existing methods.

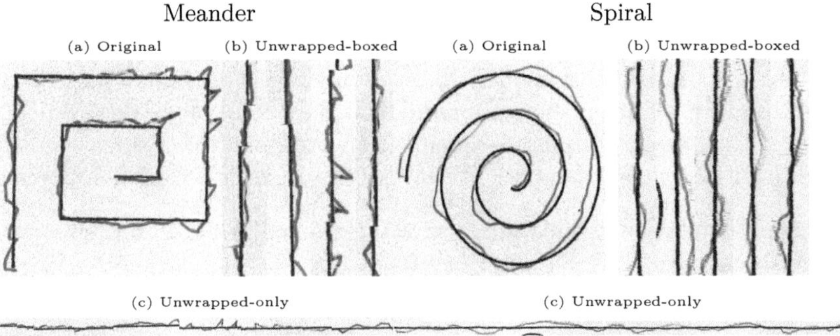

Fig. 1. Sample images of Meander (left) and Spiral (right) with the corresponding (a) original images, (b) unwrapped and boxed image, and (c) unwrapped-only image.

4.1 Experimental Setup

To assess the impact of the proposed preprocessing procedure in a robust and architecture-agnostic manner, we evaluated multiple DL models through a series of controlled experiments. Specifically, to reduce the risk of results being biased by the particular characteristics of a single architecture, we selected well-established yet structurally diverse models from the major families of DL architectures. These include a residual network (ResNet-152), a densely connected network (DenseNet-121), and a Vision Transformer (ViT). Notably, even among the CNN-based models, we deliberately chose architectures with different connectivity patterns, skip connections in ResNet versus dense connections in DenseNet, to better capture the variability in model behaviour across architectural designs. Detailed descriptions of each architecture and its configurations are provided below.

ResNet-152, belonging to the ResNet family of architectures, comprises 152 layers and is emblematic of residual learning principles [6]. These architectures incorporate skip connections or recurrent units between convolution and pooling layer blocks, coupled with batch normalisation post blocks [8].

DenseNet-121 is part of the DenseNet family, introduced to address the constraint in traditional CNNs where the number of layers (L) equals the number of connections. This architecture introduces $L(L + 1)/2$ connections, wherein the output of each layer serves as input to all subsequent layers. The number of filters in each convolution layer dynamically varies based on the growth rate parameter, k, thereby influencing the overall parameter count; k is set to 32 in the original work [7].

ViT stands as the pioneering transformer-based architecture proposed for CV tasks [5]. It leverages a self-attention mechanism, supplanting traditional convolution layers with self-attention layers. Images are partitioned into non-overlapping patches of fixed size and processed sequentially through transformer layers, thus furnishing spatial information to the model.

The DL architectures used in this study were initially pre-trained on the large-scale ImageNet dataset [3], a standard benchmark for natural image recognition tasks. To adapt these models to our domain-specific binary classification task, we retained all layers except for the final fully connected layer, which was replaced with a newly initialised classification head suitable for distinguishing between healthy and PD samples. We investigated two fine-tuning strategies:

- **last**: only the final classification layer was retrained, while all preceding layers were frozen;
- **full**: all layers of the network were retrained end-to-end.

Training was conducted for a maximum of 100 epochs, with early stopping applied based on validation loss. To define the validation set, we randomly sampled 10% of the training data, and training was terminated if no improvement in validation loss was observed for 10 consecutive epochs. We used a batch size of 32, the Stochastic Gradient Descent (SGD) optimiser, and a learning rate of 0.0001. To mitigate overfitting, we incorporated a dropout layer with a rate of 0.2 and applied L2 weight decay with a coefficient of 0.00001. Additionally, given the limited size of the training set, we implemented a data augmentation strategy that included horizontal and vertical flipping, centre cropping, and colour jittering to increase variability and improve model generalisation.

4.2 Experimental Results

The results obtained using the selected DL architectures on the NewHandPD dataset are summarised in Table 1, reporting the performance in terms of Accuracy (Acc.), Precision (Prec.), Recall (Rec.), and F1-score (F1). To evaluate the effectiveness of our preprocessing strategy, we compare three image configurations: the original images, the unwrapped-only version, and the unwrapped-boxed version, which includes both unwrapping and spatial cropping. It is important to note that the unwrapped-only format was used exclusively with the ViT architecture, due to its flexible input handling. In contrast, applying this representation to CNNs would have required resizing procedures that could significantly distort the visual structure of the drawings, thereby undermining the integrity of the input data. As can be observed, among the evaluated architectures, DenseNet-121 consistently achieved the highest performance across different image type configurations and fine-tuning strategies, emerging as the most effective model overall. Contrary to initial expectations, the ViT model yielded the lowest average performance, regardless of image type or training strategy, suggesting a limited ability to capture discriminative features in this specific context. This underperformance can be attributed to several factors. First, ViTs generally require large-scale datasets to fully exploit their self-attention mechanisms; in data-scarce scenarios, such as the one considered in this work, they are more prone to overfitting or undergeneralisation. Second, the patch-based tokenisation used in ViTs may not align well with the fine-grained, continuous nature of handwriting, particularly for spatially elongated or sparse content such as spirals and meanders. This representation can fragment key visual structures and hinder effective feature modelling. Third, unlike CNNs, ViTs lack strong inductive biases, such as locality and translation equivariance, which are especially advantageous for learning repetitive and structured patterns typical of handwriting tasks.

Table 1. Results obtained with the different DL models and different fine-tuning strategies over the three image configurations: original, unwrapped-boxed and unwrapped-only. Best results for each DL model are highlighted in bold.

Image Type	Fine-tuning	DL Model	Acc.	Prec.	Rec.	F1
Original	last	ResNet-152	91.88	92.13	91.88	91.82
Original	full	ResNet-152	91.25	91.59	91.25	91.18
Unwrapped-boxed	last	ResNet-152	92.50	92.51	92.50	92.49
Unwrapped-boxed	full	ResNet-152	**93.12**	**93.12**	**93.12**	**93.12**
Original	last	DenseNet-121	92.50	92.86	92.50	92.44
Original	full	DenseNet-121	91.88	92.33	91.88	91.80
Unwrapped-boxed	last	DenseNet-121	90.00	85.33	90.00	88.68
Unwrapped-boxed	full	DenseNet-121	**95.62**	**95.78**	**95.62**	**95.61**
Original	last	ViT	86.25	86.49	86.25	86.23
Original	full	ViT	80.00	84.48	80.00	78.82
Unwrapped-boxed	last	ViT	86.25	89.00	86.25	85.77
Unwrapped-boxed	full	ViT	**93.12**	**93.62**	**93.12**	**93.06**
Unwrapped-only	last	ViT	90.00	91.54	90.00	89.79
Unwrapped-only	full	ViT	82.50	86.72	82.50	81.59

Regarding the fine-tuning strategies, the full fine-tuning approach generally outperformed the strategy where only the final layer was retrained. The only notable exception was the ViT model, which showed an inverse trend in two out of three image type configurations. As for the proposed preprocessing pipeline, results confirm the effectiveness of the unwrapped-boxed variant, which led to consistent and, in some cases, significant improvements in classification performance. On the other hand, the unwrapped-only version produced slightly lower scores, but still outperformed the original images when evaluated with the ViT model. Given that this representation was only tested on a single ViT architecture, it cannot be definitively ruled out, especially considering its favourable performance relative to the baseline (original images).

These findings suggest that future work should explore alternative Vision Transformer architectures that are better suited to processing fine-grained and spatially elongated inputs, such as handwriting images. Additionally, exploring hybrid or multi-phase fine-tuning strategies may help mitigate data scarcity issues and enhance the model's ability to capture task-relevant patterns in transformer-based frameworks.

4.3   Comparison with the Literature

Comparing our results with those reported in previous studies is inherently challenging due to several factors. First, not all authors employed the same dataset; in many cases, custom or private datasets were used, making direct comparison of performance metrics unreliable. Moreover, even among studies that utilised the same dataset (such as NewHandPD), there are significant discrepancies in data partitioning strategies. For example, some works trained and tested their models exclusively on specific subsets of the data, such as spiral or meander drawings only, rather than combining both, which reduces task complexity. Furthermore, many studies adopted data partitioning schemes that are highly favourable but not clinically realistic. These include randomly splitting individual samples across training and testing sets, which can lead to the same patient being present in both sets, thereby inflating performance due to patient-specific bias. Some even adopted leave-one-out cross-validation, which, while exhaustive, suffers from the same limitation unless explicitly done at the patient level.

Despite these discrepancies, we have included a comparative summary table (Table 2) reporting the most recent studies that utilised the NewHandPD dataset. The table includes, for each study, the reported classification accuracy, as it remains the most widely used metric in the literature, the type of images employed (e.g., spirals, meanders, or both), and the partitioning strategy adopted, including the proportion of data used for training. This provides a contextualised and transparent view of how our results relate to the current state of the art.

Table 2. Comparison of existing ML/DL approaches on the NewHandPD dataset.

Reference	Image Type	Split Type	Split Size	Method	Acc.
[11] (2021)	Spiral	Random	90/10	ResNet-50	88.46
[11] (2021)	Meander	Random	90/10	AlexNet	92.31
[11] (2021)	Spiral+Meander	Random	90/10	ResNet-50	97.73
[2] (2022)	Spiral	Random	75/25	Custom CNN	61.5
[2] (2022)	Meander	Random	75/25	Custom CNN	67.8
[22] (2023)	Not Specified	Random	80/20	Siamese-CNN	92.86
[1] (2023)	Not Specified	Random	80/20	DL+KNN	95.29
[20] (2023)	Not Specified	Random	70/30	Ensemble	97.02
[4] (2024)	Spiral+Meander	Random	Leave-one-out	CNN+SVM	95.45
This Work	Spiral+Meander	Patient based	≈ 70/30	DenseNet-121	**95.62**

As can be seen, all the other authors used a random splitting procedure, which does not ensure that data from the same individual is never shared between the training and testing sets. To the best of our knowledge, this is the only study that explicitly enforces this clinically sound evaluation protocol on the NewHandPD dataset. This strategy more accurately reflects real-world diagnostic scenarios, in which models are expected to generalise to entirely unseen patients, making our performance estimates more realistic and clinically meaningful. Despite the stricter and more clinically realistic evaluation conditions applied in our study, our method achieved performance that is comparable to, or in some cases surpasses, that of prior work, as shown in Table 2. This result underscores not only the effectiveness of our preprocessing pipeline and DL models but also their robustness to variability in drawing templates. Together, these findings strengthen the case for the clinical validity and practical applicability of our approach in real-world diagnostic settings.

5   Conclusion

This work presented a comprehensive study on the automatic classification of PD using DL models trained on static handwriting images from the NewHandPD dataset. Our primary goal was to evaluate the effectiveness of modern architectures, both CNNs and ViTs, in distinguishing between healthy individuals and patients based on hand-drawn spiral and meander patterns.

A central contribution of this study lies in the preprocessing pipeline specifically designed to normalise and enhance the visual structure of different drawing templates, such as spirals and meanders. This approach allows the model to handle these two distinct types of images within a unified framework, enabling joint training and evaluation. In contrast, previous works often treated these drawing types separately, limiting generalizability and requiring dedicated models for each.

Another important methodological contribution is the adoption of a patient-based partitioning strategy, ensuring that no individual contributes images to both the training and testing sets. This setup more closely reflects real-world clinical conditions and prevents overly optimistic results caused by patient-specific features leaking into both subsets.

Our experiments demonstrate that DenseNet-121 consistently achieves the best classification performance, and the introduced preprocessing strategy shows clear advantages in most settings. Even under stricter evaluation conditions, our models reach performance levels comparable to or exceeding those reported in the literature.

These results confirm the clinical validity and robustness of our approach, particularly in handling variability across different drawing templates and in ensuring reliable generalisation to unseen patients. Although the dataset lacks detailed clinical annotations—such as medication status, disease stage, or motor sub-scores—which limits stratified analysis by disease severity, it does include relevant demographic information such as age, sex, and handedness. Combined with the presence of both spirals and meanders, this demographic diversity contributes to a more heterogeneous and ecologically valid evaluation scenario, supporting the general applicability of our method in realistic diagnostic settings.

Future research should explore specialised visual transformer architectures better suited for elongated or structured image formats, as well as multi-phase fine-tuning strategies that could enhance model adaptation while preserving generalisation.

Acknowledgments. This work was partially supported by the project SERICS (PE00000014) under the NRRP MUR program funded by the EUNGEU.

References

1. Abdullah, S.M., et al.: Deep transfer learning based parkinson's disease detection using optimized feature selection. IEEE Access **11**, 3511–3524 (2023). https://doi.org/10.1109/ACCESS.2023.3233969
2. Biswas, S., Kaur, N., Seeja, K.R.: Early detection of parkinson's disease from hand drawings using CNN and LSTM. In: AIST 2022 - 4th International Conference on Artificial Intelligence and Speech Technology (2022). https://doi.org/10.1109/AIST55798.2022.10065159
3. Deng, J., Dong, W., Socher, R., Li, L.J., Li, K., Fei-Fei, L.: ImageNet: a large-scale hierarchical image database . In: 2009 IEEE Computer Society Conference on Computer Vision and Pattern Recognition Workshops (CVPR Workshops), pp. 248–255. IEEE Computer Society, Los Alamitos (2009). https://doi.org/10.1109/CVPR.2009.5206848
4. Dong, S., Liu, J., Wang, J.: Diagnosis of parkinson's disease based on hybrid fusion approach of offline handwriting images. IEEE Signal Process. Lett. (2024). https://doi.org/10.1109/LSP.2024.3496579
5. Dosovitskiy, A., et al.: An image is worth 16x16 words: transformers for image recognition at scale. In: 9th International Conference on Learning Representations, ICLR 2021, Virtual Event, Austria, May 3-7, 2021. OpenReview.net (2021)
6. He, K., Zhang, X., Ren, S., Sun, J.: Deep residual learning for image recognition. In: 2016 IEEE Conference on Computer Vision and Pattern Recognition (CVPR), pp. 770–778 (2016)
7. Huang, G., Liu, Z., van der Maaten, L., Weinberger, K.Q.: Densely connected convolutional networks. In: 2017 IEEE Conference on Computer Vision and Pattern Recognition, CVPR 2017, Honolulu July 21-26, 2017, pp. 2261–2269. IEEE Computer Society (2017)
8. Ioffe, S., Szegedy, C.: Batch normalization: accelerating deep network training by reducing internal covariate shift. In: Bach, F.R., Blei, D.M. (eds.) Proceedings of the 32nd International Conference on Machine Learning, ICML 2015, Lille, France, 6-11 July 2015. JMLR Workshop and Conference Proceedings, vol. 37, pp. 448–456. JMLR.org (2015)
9. Islam, M.A., Majumder, M.Z.H., Hussein, M.A., Hossain, K.M., Miah, M.S.: A review of machine learning and deep learning algorithms for parkinson's disease detection using handwriting and voice datasets. Heliyon **10**, e25469 (2024). https://doi.org/10.1016/J.HELIYON.2024.E25469
10. Jiang, X., Yu, H., Yang, J., Liu, X., Li, Z.: A new network structure for parkinson's handwriting image recognition. Med. Eng. Phys. **139**, 104333 (2025). https://doi.org/10.1016/J.MEDENGPHY.2025.104333
11. Kamran, I., Naz, S., Razzak, I., Imran, M.: Handwriting dynamics assessment using deep neural network for early identification of parkinson's disease. Futur. Gener. Comput. Syst. **117**, 234–244 (2021). https://doi.org/10.1016/J.FUTURE.2020.11.020

12. Kasab, M., Masri, N.A., Moghrabi, R.A.: Detection of parkinson's disease through spiral drawings using convolutional autoencoders. In: 2024 IEEE International Conference on Smart Systems and Power Management, IC2SPM 2024, pp. 1–5 (2024). https://doi.org/10.1109/IC2SPM62723.2024.10841337

13. Khan, A., Galarraga, O., Garcia-Salicetti, S., Vigneron, V.: Deep learning for quantified gait analysis: a systematic literature review. IEEE Access (2024). https://doi.org/10.1109/ACCESS.2024.3434513

14. Khan, A.U., Akram, M., Daniyal, M., Zainab, R.: Awareness and current knowledge of parkinson's disease: a neurodegenerative disorder. Int. J. Neurosci. **129**, 55–93 (2019). https://doi.org/10.1080/00207454.2018.1486837

15. Pecoraro, P.M., Marsili, L., Espay, A.J., Bologna, M., di Biase, L.: Computer vision technologies in movement disorders: a systematic review. Move. Disorders Clinic. Practice (2025). https://doi.org/10.1002/MDC3.70123

16. Pereira, C.R., Weber, S.A., Hook, C., Rosa, G.H., Papa, J.P.: Deep learning-aided parkinson's disease diagnosis from handwritten dynamics. In: Proceedings - 2016 29th SIBGRAPI Conference on Graphics, Patterns and Images, SIBGRAPI 2016, pp. 340–346 (2017). https://doi.org/10.1109/SIBGRAPI.2016.054

17. Santhosh, K., Dev, P.P., A., B.J., Lynton, Z., Das, P., Ghaderpour, E.: A modified gray wolf optimization algorithm for early detection of parkinson's disease. Biomed. Signal Process. Control **109**, 108061 (2025). https://doi.org/10.1016/J.BSPC.2025.108061

18. Sveinbjornsdottir, S.: The clinical symptoms of parkinson's disease. J. Neurochemis., 318–324 (2016). https://doi.org/10.1111/JNC.13691

19. Tanveer, M., Rashid, A.H., Kumar, R., Balasubramanian, R.: Parkinson's disease diagnosis using neural networks: survey and comprehensive evaluation. Inf. Process. Manag. **59**, 102909 (2022). https://doi.org/10.1016/J.IPM.2022.102909

20. Wang, N., et al.: A coordinate attention enhanced swin transformer for handwriting recognition of parkinson's disease. IET Image Proc. **17**, 2686–2697 (2023). https://doi.org/10.1049/IPR2.12820

21. Watts, J., Niethammer, M., Khojandi, A., Ramdhani, R.: Machine learning model comparison for freezing of gait prediction in advanced parkinson's disease. Front. Aging Neurosci. **16** (2024). https://doi.org/10.3389/FNAGI.2024.1431280

22. Zhao, A., Wu, H., Chen, M., Wang, N.: A spatio-temporal siamese neural network for multimodal handwriting abnormality screening of parkinson's disease. Int. J. Intell. Syst. **2023**, 9921809 (2023). https://doi.org/10.1155/2023/9921809

23. Özdemir, E.Y., Özyurt, F.: Elasticnet-based vision transformers for early detection of parkinson's disease. Biomed. Signal Process. Control **101**, 107198 (2025). https://doi.org/10.1016/J.BSPC.2024.107198

2nd International Workshop on Computer Vision for Environment Monitoring and Preservation (CVEMP 2025)

Workshop

2nd International Workshop on Computer Vision for Environment Monitoring and Preservation (CVEMP 2025)

In conjunction with ICIAP 2025—Rome, Italy, September 15–19, 2025

Workshop Organization

Organizers

Paolo Russo — Sapienza University of Rome, Italy
Fabiana Di Ciaccio — University of Florence, Italy
Diego Marcos — Inria Université Côte d'Azur, France

Program Committee

Irene Amerini — Sapienza University of Rome, Italy
Ananthu Aniraj — Inria University of Montpellier, France
Roberto Beraldi — Sapienza University of Rome, Italy
Benjamin Bourel — Inria University of Montpellier, France
Silvio Del Pizzo — Parthenope University of Naples, Italy
Maxime Fromholtz — Inria University of Montpellier, France
Pietro Fusco — University of Campania Luigi Vanvitelli, Italy
Dino Ienco — Inria University of Montpellier, France
Roberto Interdonato — Cirad, UMR TETIS, France
Luca Maiano — Sapienza University of Rome, Italy
Pietro Manganelli Conforti — Sapienza University of Rome, Italy
Andrea Masiero — University of Florence, Italy
Nizar Massouh — BonsXAI
Erica Nocerino — University of Sassari, Italy
Lorenzo Papa — Sapienza University of Rome, Italy
Gabriele Proietti Mattia — Sapienza University of Rome, Italy
Marta Sanzari — Sapienza University of Rome, Italy
Lorenzo Stacchio — University of Bologna, Italy
Salvatore Venticinque — University of Campania Luigi Vanvitelli, Italy

Leveraging Annotation Efficiency Strategies for AI-Driven Detection and Segmentation of Fruits, Flowers, and Diseased Leaves

Samral Tahirli[1](✉), Federico Frontali[1], Serena Baiocco[2], and Silvia Zuffi[3]

[1] Awentia s.r.l., Imola, Italy
{samral.tahirli,federico.frontali}@awentia.com
[2] Terremerse Soc. Coop., Ravenna, Italy
sbaiocco@terremerse.it
[3] IMATI-CNR, Milan, Italy
silvia.zuffi@cnr.it
https://www.awentia.com , https://terremerse.it

Abstract. In this work, we present a scalable approach for the automatic detections and segmentation of fruits, flowers and diseased leaves in RGB images. While many general methods for detection and segmentation demonstrate high performance, they underperform in domain-specific scenarios as they have not been trained for specific and fine-grain applications like the ones we address here. General-purpose methods can be adapted to specific tasks if large, annotated datasets are available for training. However, creating such data is labor-intensive, requiring capturing plants under a variety of conditions as well as domain specific knowledge for image annotation. To address this challenge, we propose a methodology that combines fine-tuning of general-purpose vision models with efficient annotation techniques, including pseudo-labeling and data augmentation. Evaluation using COCO API metrics demonstrates high detection and segmentation accuracy with reduced annotation overhead. Our approach offers a scalable and efficient solution for automated plant monitoring, supporting early disease detection, resource optimization, and improved crop health assessment in precision agriculture.

1 Introduction

In agriculture, the analysis of flowers and fruits in different stages of growth is crucial to monitor crop development, predict harvest time, and optimize resource allocation. Similarly, early detection of leaf diseases plays a vital role in preventing disruptive outbreaks. Traditional manual inspection is labor-intensive and subjective, calling for automated, scalable, and cost-effective alternatives. Precision agriculture integrates advanced sensing and machine learning techniques to optimize crop management and improve agricultural productivity. Among these technologies, computer vision has emerged as a powerful tool for fruit detection and disease diagnosis in plants. The ability to automatically analyze plant

Fig. 1. Images collected under various conditions, illustrating the challenges for the detection and segmentation of fruits, flowers and leaves.

conditions from RGB images would help farmers to take data-driven decisions, improve water management, and mitigate crop loss due to diseases.

Recent advances in AI-based object detection and segmentation methods have the potential to enable the accurate recognition of fruits and diseased leaves under varying environmental conditions. However, challenges arise due to the specificity of the tasks, which require discrimination among infection variations, occlusions, intra-class variability in fruit appearance, and disease symptom similarity. Addressing these challenges requires robust models capable of distinguishing between healthy and infected leaves and between fruit growth stages in complex scenes (Fig. 1). Recent general-purpose detection and segmentation methods [7,10] [?, ?] while highly effective in many zero-shot problems, lack specificity to many relevant problems in precision agriculture. How to fill this gap is the goal of our study.

In this work, we develop a semi-supervised pipeline that fine-tunes pre-trained CV models for the segmentation and detection of fruits and flowers at multiple growth stages, and leaves exhibiting early signs of disease. To obtain a system with high accuracy and robustness, the proposed method integrates state-of-the-art models with a novel strategy for training annotation management. Additionally, data augmentation is employed to improve model robustness against domain shifts, ensuring reliable detection and segmentation under varying environmental conditions. This combination of strategies enables the system to accurately segment flowers, fruits, and diseased leaves at different growth stages, addressing key challenges such as intra-class variability, occlusions, and symptom similarity in plant diseases.

We illustrate the effectiveness of our approach with experiments comparing pre-trained models with models fine-tuned on newly annotated data, demonstrating high detection and segmentation accuracy with reduced annotation overhead. We apply our pipeline to the problem of disease detection and fruits and flower segmentation, but our approach is general, and can be applied to other problems in precision agriculture that require creating specialized models with limited manual labor.

Fig. 2. Examples of target images, including fruits (left) and leaves with diseases.

2   Related Work

Advancements in precision agriculture have increasingly incorporated computer vision and machine learning techniques to enhance crop management. A comprehensive review by Ali *et al.* [1] discusses non-destructive methods for plant disease detection, highlighting the efficacy of image processing and spectroscopy-based approaches.

Deep learning (DL) models have been pivotal in automating fruit detection and disease diagnosis. Xiao *et al.* [18] provide an extensive overview of DL applications in fruit detection and recognition, addressing challenges such as limited high-quality datasets and the need for lightweight models. Similarly, a recent study reviews DL techniques in plant disease detection, emphasizing the role of convolutional neural networks (CNNs) in accurate disease identification [14].

The labor-intensive nature of manual annotation has led to the exploration of semi-supervised learning (SSL) methods to improve annotation efficiency. Pseudo-labeling, where a model generates labels for unlabeled data, has been shown to enhance predictive performance in plant disease diagnosis [5]. In agricultural contexts, SSL frameworks have been applied to tasks like weed detection, demonstrating improved model performance with reduced labeled data requirements [8]. Additionally, Zhang *et al.* [19] introduce "Wheat Teacher," a semi-supervised wheat head detector combining pseudo-labeling and consistency regularization, achieving metrics comparable to fully supervised models using only a fraction of labeled data. Arshad *et al.* [2] propose an Assisted Few-Shot Learning approach to improve vision-language models (VLMs) ability in image classification tasks with scarce annotated data by optimizing the selection of input examples. Ciarfuglia *et al.* [3] address the problem of detecting and segmenting grapes by training with scarce and noisy labelled data. They exploit video data, and leverage 3D Structure from Motion (SfM) techniques to generate additional training labels from already labeled samples. Milioto *et al.* [9] exploit vegetation indices for training a CNN-based segmentation method for the semantic segmentation of sugar beet plants, weeds, and background in crop fields. Roggiolani *et al.* [12] address the problem of joint semantic, plant instance, and leaf instance segmentation of crop fields from RGB data. They propose a single CNN architecture that simultaneously addresses these tasks by exploiting their underlying hierarchical structure, By leveraging hierarchical relationships between these segmentation tasks, the model learns from limited annotated data more effectively.

To further alleviate annotation burdens, the development of pre-trained models on plant-specific datasets has been proposed. Dong *et al.* [4] construct a large-scale plant disease dataset to pre-train models, resulting in improved diagnostic accuracy and reduced training times. Weyler *et al.* [16] introduce PhenoBench, a dataset and benchmarks for semantic segmentation in the agricultural domain. The dataset includes dense, pixel-wise annotations for crop (sugar beet) and weed instances, as well as detailed leaf-level annotations. Images are captured with unmanned aerial vehicles (UAVs) and present limited occlusion and clear background separation, while here we address more challenging scenarios (Fig. 2).

Recent advances in vision transformers (ViTs) and self-supervised representation learning have shown strong potential in agricultural applications. Dasom *et al.* [13] propose a transformer-based approach in detecting small fruits, making it highly suitable for practical use in real cases. Similarly, Yuzhi *et al.* [15] explore contrastive self-supervised learning for plant disease classification, significantly reducing the need for annotated samples by leveraging unlabeled data. Active learning approaches have also been explored to minimize annotation costs. A study by Xue *et al.* [17] applies meta-learning strategies to field-acquired cereal crop images, achieving strong recognition performance with very limited labeled data. These strategies align with our goals of annotation efficiency and domain-specific generalization.

3   Method

This work explores leveraging semi-supervised learning to streamline annotation for object detection and addresses domain generalization challenges in agriculture. Pseudo-labeling is the primary semi-supervised learning method to effectively utilize unlabeled data. Additionally, data augmentation and pre-training with external data enhance domain generalization, making the model more robust to domain shifts. For our approach, an adapted zero-shot segmentation framework based on SAM [7] is used, while customized YOLO [10][?] models were selected as the foundation for our detection pipeline.

3.1   Annotation Management Pipeline

The flowchart (Fig. 3) illustrates the decision process for image labeling and model training, based on whether the target object is considered common or uncommon. Common objects, which we define as those frequently found in public datasets, such as grapes or apples, are typically supported by reliable pretrained models. In these cases, a pretrained detection model, such as Faster R-CNN [11], is used to produce preliminary bounding boxes, which are subsequently verified and refined. Uncommon objects are manually labeled.

The bounding boxes are then imported into CVAT (Computer Vision Annotation Tool) [6] for verification and correction. In this phase, the wrong bounding boxes are removed, the missed detections are manually added, and the inaccurate bounding boxes are refined.

After this process, which goal is to ensure accuracy, we identify the task type, either object detection (bounding boxes) or object segmentation (pixel-level masks). In case of detection, the verified bounding boxes are used to train or fine-tune a detection model. For segmentation, an adapted zero-shot segmentation framework based on SAM [7] is used to generate initial masks, which are then reviewed and manually corrected before training a segmentation model tailored for the real agricultural case. The outcome is a newly fine-tuned model optimized for detecting or segmenting common objects.

If the object is uncommon—such as a specialized early stage of special fruits and rare plant diseases—manual labeling is required. This involves collecting a certain amount of images, and using the CVAT annotation tool to draw bounding boxes or segmentation masks manually, depending on the task. Once annotations are finalized, the next step is determining whether the task is detection or segmentation.

For detection, the manually labeled bounding boxes are used to train a detection model which is customized for real-world agricultural scenarios. For segmentation, an adapted zero-shot segmentation framework based on SAM [7] or similar tools can assist in mask generation, which is then refined and finalized in CVAT before training a segmentation model. The result is a new finetuned model tailored to detecting or segmenting uncommon objects.

In summary, the workflow follows a structured approach: if the object is common, leverage pretrained models, validate annotations, and train a detection or segmentation model as needed. If the object is uncommon, manually label the dataset, verify the annotations, and proceed with training. This method ensures efficient annotation and model training, optimizing detection and segmentation for a wide range of applications.

4   Experiments

This section presents the training details and data augmentation strategies applied to the custom model, along with a comparison between the pretrained and custom model results. It also includes a discussion of the data-centric ablation study and the evaluation of its outcomes on a distributed test dataset. For this study, we use the YOLOX-M model as a custom object detection framework. The training phase is fundamental, relying on a custom dataset that captures real-world farming conditions. This dataset includes images from various growth stages, different lighting conditions, and occlusions caused by leaves and farming equipment. In this study, the dataset is specifically curated for fruit detection, with a focus on grape detection, ensuring the model learns to identify grapes accurately under diverse environmental scenarios. (Fig. 2).

The dataset consists of images and videos captured using cameras mounted on tractors as they move through the field. These images are continuously

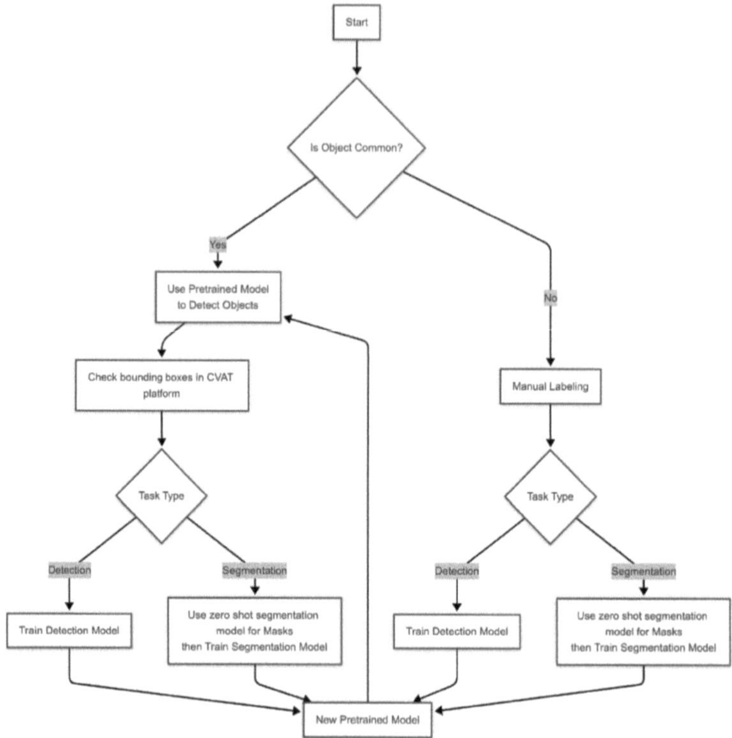

Fig. 3. Flowchart describing the process for finetuning detection and segmentation networks.

recorded and stored in a database. Annotation is performed using a combination of automated and manual methods. A pre-trained model is utilized to generate initial labels for certain parts of the dataset, while the remaining annotations are manually verified and corrected to ensure accuracy.

The YOLOX model was trained on $2 \times$ NVIDIA L4 GPUs using mixed-precision training with float16, which significantly improved computational efficiency. A batch size of 32 was used, with a learning rate set to 0.01. The total training time ranged from 72 to 96 h, depending on the specific run and preprocessing overhead. These configurations were selected through empirical tuning to ensure an optimal balance between performance and stability. The dataset used in this study is not publicly available due to data privacy restrictions.

A iterative training approach was adopted: initial training on this curated dataset, followed by fine-tuning with additional unseen data to enhance generalization and using our model as pretrained model to label more dataset. The model's performance was evaluated using newly collected field data, ensuring its robustness across diverse scenarios.

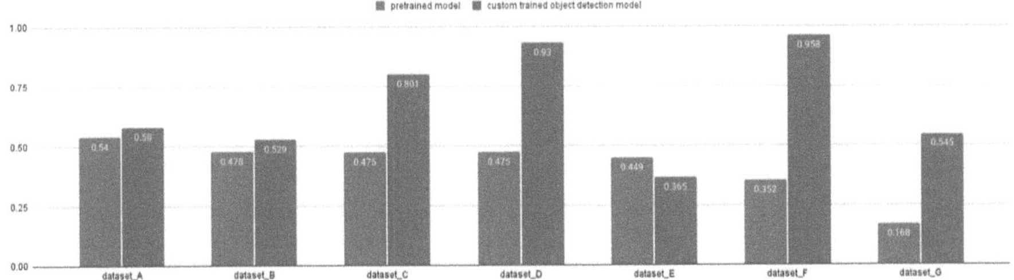

Fig. 4. Performance of pre-trained and custom-trained object detection models on unseen grapefruit data.

Comparison with Baseline . A comparative analysis was conducted between the custom trained object detection model and a pre-trained alternative to establish a performance benchmark. The pre-trained model's predictions served as a reference for assessing the custom trained object detection model's effectiveness on unseen data. Figure 4 reports performance for pre-trained and custom models.

The plot in Fig. 4 is intended to highlight insight by including both pre-trained model results and our own trained model results side by side on a variety of datasets. As pretrained models are typically trained on large-scale datasets, comparing them with our model allows us to better understand its strengths and weaknesses. This comparison helps evaluate how well our model generalizes and where further improvements may be needed.

The evaluation of the two models was conducted using different datasets to evaluate their generalization capabilities across various scenarios. All datasets focus on white grapes, but differ across dimensions such as distance from the camera (close or far), lighting conditions (shadow or sunlight), image quality (clear or blurry), and object visibility (occluded or not occluded). Datasets A and B consist of close-range, shadowed scenes with clear but partially occluded grapes, simulating challenging indoor or shaded environments. Dataset C captures far-distance, shadowed grapes that are clear and fully visible, while Dataset D introduces sunlight exposure with no occlusion. Dataset E presents the most challenging conditions, with close-range, sunlit, blurry, and occluded grapes, testing the model's robustness to noise and partial visibility. Lastly, Datasets F and G offer a mix of close/far distance and varying lighting and occlusion, enabling a comprehensive assessment of model generalization across multiple capture conditions.

Evaluations utilized the COCO metric, specifically Mean Average Precision (mAP), to quantify detection accuracy. These results highlight that it is possible to obtain predicted bounding boxes for common objects using a pre-trained model, making it a practical initial approach to reduce the need for manual labeling. However, while the pre-trained model provided reliable predictions, the custom trained object detection model exhibited superior confidence scores and fewer mispredictions.

A data-centric ablation study was conducted to assess the impact of training data composition on model performance. Starting with a baseline dataset, we trained an initial model (called 1st model here), which served as our reference. We then incrementally added two additional datasets in separate experiments—resulting in the 2nd model and the 3rd model - to observe performance differences. Results are presented in Fig 6. This step-by-step approach allowed us to evaluate how each dataset contributes to overall model generalization. The observed trends provide meaningful insights into the influence of dataset diversity and scale. It is also important to note that there is a distributional difference between the datasets used for training and those used in the ablation study, further emphasizing the relevance of evaluating performance under varied conditions. After the data-centric ablation study, a model comparison experiment was conducted with seven different datasets, each of which was intended to reflect various real-world conditions that influence model performance. The datasets were divided according to four major variables: fruit kind (black or white), object distance from the camera (close or far), movement speed (fast or slow), and environmental lighting (sun or shadow). For instance, Dataset A consisted of close, fast black fruit in shadow, and Dataset B consisted of close, slow black fruit in sunlight. Dataset C consisted of close, slow white fruit in sunlight conditions so that the model's ability to distinguish between fruit colors under constant lighting could be tested. Dataset D consisted of far, fast black fruit in shadow, testing the model's resistance under low-light and far conditions. Dataset E also contained far and fast black fruit but in sunlight, the difference from Dataset D being the lighting conditions. Dataset F was far, slow black fruit in sunny conditions, while Dataset G was far, slow white fruit, again in sunlight. This systematically distributed dataset structure allowed for a detailed analysis of the interaction between environmental conditions and object attributes in influencing model detection accuracy and generalizability. The performance comparison of the three model versions demonstrates a stepwise enhancement in detection accuracy with increasing data-centric enhancements. Starting from the baseline (1st model), accuracy ranged from 50.00 percent to 81.58 percent across datasets, with Dataset A and Dataset B having relatively good starting points. The 2nd model introduced noticeable improvements, with remarkable improvement for Dataset B (from 81.40% to 93.02%) and Dataset D, which reached a perfect 100% accuracy. The 3rd model, with additional data included, kept improving performance in most cases—Dataset G, for instance, went up from 58.33% to 83.33%, and Dataset F went up from 71.43% to 85.71%. While Dataset C, with white fruit in stable sunlight, was lower in overall accuracy, it still shows comparative improvement between models (from 63.64% to 72.73%). These results highlight that data sets that involved long distances and high-velocity movement, especially under challenging lighting conditions, were most favorably affected by the model enhancements. Overall, the data-driven evolution between the 1st and 3rd model resulted in more robust generalization across environmental as well as object-related conditions.

As part of the training strategy, data augmentation techniques were implemented to improve model generalization under varying real-world conditions. These include geometric transformations (random rotations, scaling, flips), photometric adjustments (brightness, contrast, hue variation), and task-specific augmentations such as simulated occlusions and synthetic leaf discoloration. Additionally, advanced augmentation methods such as Mosaic Augmentation, which combines four images into one to enhance object context and scale diversity, and MixUp Augmentation, which blends pairs of images and labels to improve regularization and robustness, were applied. These augmentations are applied consistently in both training phases—initial supervised training and later fine-tuning—helping to mitigate domain shifts and boost model robustness.

Figures 5 shows predictions with the pre-trained object detection model and custom model, allowing for a visual assessment of their performance.

Fig. 5. Comparison of pre-trained model's (middle) and custom trained object detection model's (bottom) on unseen images.

5   Limitations and Future Work

While the proposed pipeline demonstrates strong performance in reducing annotation costs and improving segmentation accuracy, several limitations remain.

Label	Dataset	1st_model	2nd_model	3rd_model
close-fast-shadow-black	Dataset A	81.58	92.11	92.11
close-slow-sun-black	Dataset B	81.40	93.02	95.35
close-slow-sun-white	Dataset C	63.64	72.73	69.09
far-fast-shadow-black	Dataset D	77.78	100.00	100.00
far-fast-sun-black	Dataset E	50.00	66.67	66.67
far-slow-sun-black	Dataset F	71.43	85.71	85.71
far-slow-sun-white	Dataset G	58.33	83.33	83.33

Fig. 6. Detection Accuracy Across Datasets for Three Model Versions.

First, although data augmentation and pretraining mitigate domain shift, drastic environmental variations (e.g., different seasons, extreme lighting, or unseen crop species) may still degrade performance. Addressing this fully may require domain adaptation techniques or continual learning strategies.

Second, pseudo-labeling, while effective for common objects, can introduce label noise, especially in cluttered scenes or under occlusion. We observe that noise accumulation may impact performance unless followed by careful human verification and confidence-based filtering.

Lastly, our pipeline currently operates on RGB imagery only. Although sufficient for many tasks, disease detection—particularly in early or asymptomatic stages—could benefit significantly from multimodal inputs. Future work will explore integrating thermal and multispectral data to improve sensitivity and robustness in detecting plant stress and disease.

6   Conclusion

In conclusion, this research aims to enhance object detection efficiency by integrating semi-supervised learning and domain generalization techniques. By minimizing reliance on extensive labeled datasets, the proposed approach enables models to generalize more effectively across diverse environments. Through a structured workflow that optimally leverages pretrained models for common objects and manual annotation for uncommon ones, the methodology ensures high-quality training data and robust model performance. The combination of automated labeling assistance, verification in CVAT, and tailored training strategies advances the state-of-the-art in object detection. Ultimately, this research contributes to the development of more adaptable and scalable object detection systems capable of performing reliably across varying domains and real-world conditions.

References

1. Ali, M.M., Bachik, N.A., Muhadi, N.A., Tuan Yusof, T.N., Gomes, C.: Non-destructive techniques of detecting plant diseases: A review. Physiol. Mol. Plant Pathol. **108**, 101426 (2019). https://doi.org/10.1016/j.pmpp.2019.101426
2. Arshad, M.A., et al.: Assisted few-shot learning for vision-language models in agricultural stress phenotype identification. In: Adaptive Foundation Models: Evolving AI for Personalized and Efficient Learning (2024). https://openreview.net/forum?id=WutpswD3ea
3. Ciarfuglia, T.A., Motoi, I.M., Saraceni, L., Fawakherji, M., Sanfeliu, A., Nardi, D.: Weakly and semi-supervised detection, segmentation and tracking of table grapes with limited and noisy data. Comput. Electr. Agricul. **205**, 107624 (2023). https://doi.org/10.1016/j.compag.2023.107624
4. Dong, X., et al.: Pddd-pretrain: A series of commonly used pre-trained models support image-based plant disease diagnosis. Plant Phenomics **5**, 0054 (2023). https://doi.org/10.34133/plantphenomics.0054
5. Ferreira, R.E.P., Lee, Y.J., Dórea, J.R.R.: Using pseudo-labeling to improve performance of deep neural networks for animal identification. Sci. Rep. **13**(1), 13875 (2023). https://doi.org/10.1038/s41598-023-40977-x
6. Intel: CVAT: Computer vision annotation tool (2020). https://github.com/opencv/cvat
7. Kirillov, A., et al.: Segment anything. arXiv:2304.02643 (2023)
8. Li, R., et al.: A semi-supervised diffusion-based framework for weed detection in precision agricultural scenarios using a generative attention mechanism. Agriculture **15**(4) (2025). https://doi.org/10.3390/agriculture15040434
9. Milioto, A., Lottes, P., Stachniss, C.: Real-time semantic segmentation of crop and weed for precision agriculture robots leveraging background knowledge in CNNs. In: Proceedings of the IEEE International Conference on Robotics and Automation (ICRA). pp. 2229–2235 (2018). https://doi.org/10.1109/ICRA.2018.8460962. https://arxiv.org/abs/1709.06764
10. Redmon, J., Divvala, S., Girshick, R., Farhadi, A.: You only look once: Unified, real-time object detection. In: Proceedings of the IEEE Conference on Computer Vision and Pattern Recognition (CVPR), pp. 779–788 (2016)
11. Ren, S., He, K., Girshick, R., Sun, J.: Faster R-CNN: Towards real-time object detection with region proposal networks. In: Advances in Neural Information Processing Systems (NeurIPS), pp. 91–99 (2015)
12. Roggiolani, G., Sodano, M., Guadagnino, T., Magistri, F., Behley, J., Stachniss, C.: Hierarchical approach for joint semantic, plant instance, and leaf instance segmentation in the agricultural domain. In: Proceedings of the IEEE International Conference on Robotics and Automation (ICRA), pp. 1–7 (2023). https://doi.org/10.1109/ICRA48891.2023.10160918. https://arxiv.org/abs/2210.07879
13. Seo, D., Lee, S.K., Kim, J.G., Oh, I.S.: High-precision peach fruit segmentation under adverse conditions using Swin transformer. J. Name **XX**(YY), ZZZ–ZZZ (2024). DOI_here, URL_here
14. Upadhyay, A., et al.: Deep learning and computer vision in plant disease detection: a comprehensive review of techniques, models, and trends in precision agriculture. Artif. Intell. Rev. **58**(3), 92 (2025). https://doi.org/10.1007/s10462-024-11100-x
15. Wang, Y., et al.: Classification of plant leaf disease recognition based on self-supervised learning. Agronomy **14**(3), 500 (2024). https://doi.org/10.3390/agronomy14030500

16. Weyler, J., et al.: PhenoBench—a large dataset and benchmarks for semantic image interpretation in the agricultural domain. arXiv preprint arXiv:2306.04557 (2023). https://arxiv.org/abs/2306.04557
17. Wu, X., Deng, H., Wang, Q., Lei, L., Gao, Y., Hao, G.: Meta-learning shows great potential in plant disease recognition under few available samples. Plant J. **114** (2023). https://doi.org/10.1111/tpj.16176
18. Xiao, F., Wang, H., Xu, Y., Zhang, R.: Fruit detection and recognition based on deep learning for automatic harvesting: An overview and review. Agronomy **13**(6) (2023). https://doi.org/10.3390/agronomy13061625
19. Zhang, R., Yao, M., Qiu, Z., Zhang, L., Li, W., Shen, Y.: Wheat teacher: a one-stage anchor-based semi-supervised wheat head detector utilizing pseudo-labeling and consistency regularization methods. Agriculture **14**(2) (2024). https://doi.org/10.3390/agriculture14020327

A Deep Learning Model to Predict GNSS from InSAR Data

Michelangelo Caretto[1], Antonio Alliegro[1], Rosario Milazzo[1],
Osmari Aponte[2], Andrea Gatti[2], Eugenio Realini[2], Lia Morra[1],
and Tatiana Tommasi[1(✉)]

[1] Politecnico di Torino, Turin, Italy
`tatiana.tommasi@polito.it`
[2] Geomatics Research and Development srl, Lomazzo, Italy

Abstract. Accurate monitoring of ground deformation is crucial for hazard mitigation, infrastructure management, and environmental protection. Interferometric Synthetic Aperture Radar (InSAR) and Global Navigation Satellite System (GNSS) are two complementary geospatial technologies whose integration relies predominantly on physical modeling and geometric transformations for fusion. This paper introduces a novel deep learning model that predicts GNSS-like three-dimensional ground displacements at InSAR measurement locations, using weak supervision from spatially sparse GNSS data. Our approach leverages a Dynamic Graph Convolutional Neural Network (DGCNN) backbone to model spatial dependencies among localized InSAR-derived features, effectively calibrating InSAR measurements to correct for viewing geometry limitations. The proposed method is evaluated in an area in the Netherlands affected by induced seismicity and ground subsidence across different experimental scenarios, with a particular focus on predicting ground deformations in time windows not experienced at training time.

1 Introduction

Interferometric Synthetic Aperture Radar (InSAR) and Global Navigation Satellite System (GNSS) are powerful geospatial techniques that serve as the most relevant source of high-precision deformation data, which are critical for geotechnical engineering, seismology, volcanology, and infrastructure health monitoring [18,20,21]. Specifically, GNSS works on the basis of antennas installed on the ground that capture signals from orbiting satellites and provide three-dimensional ground displacement with millimeter-level precision at discrete points and continuously over time. However, technical issues like power loss can cause interruptions in data collection, and monitoring large areas requires several dozen receivers, which adds to the system's complexity and cost. On the other hand, InSAR provides extensive spatial coverage and can detect large-scale surface changes, but it is limited by its one-dimensional Line-Of-Sight (LOS) viewing geometry and by the temporal resolution of satellite acquisitions, which is determined by revisit intervals over the area of interest ranging from 6 to 42

E. Rodolà et al. (Eds.): ICIAP 2025 Workshops, LNCS 16169, pp. 339–351, 2026.
https://doi.org/10.1007/978-3-032-11317-7_29

days, depending on the satellite mission [9]. Moreover, the monitoring precision of InSAR data is easily affected by phase unwrapping issues due to temporal and spatial decorrelation, as well as by orbital and atmospheric errors [11].

Considering their complementarity, a synergy between these technologies has the potential to improve the reliability and comprehensiveness of ground deformation monitoring, offering a more detailed and timely understanding of terrain variations. Several studies proposed using the differences between GNSS and InSAR data to correct the latter, employing surface fitting models and accounting for spatial distribution errors via tailored physics-based approaches [12,17,19], but deep learning techniques remain largely unexplored in this context.

In this work we formalize a first deep learning model that takes InSAR ground displacement measurements as input and produces as output their calibrated version, aligned with what a GNSS receiver at the same location would have observed. Given the spatial sparsity of GNSS data, the model is designed to operate in a weakly supervised setting.

In the following, Sect. 2 describes the data used to train and validate the proposed model; Sect. 3 provide details on the backbone architecture and learning objective of the proposed deep neural network model; Sect. 4 evaluates the performance of the proposed method, with a physics-based approach as an upper bound reference. Finally, conclusions are drawn in Sect. 5.

2   Data Nature and Sources

InSAR is a sophisticated method utilized for mapping ground deformation through radar images captured from satellites orbiting the Earth. This technology enables the monitoring of ground shifts even during adverse weather conditions and at night. It operates by transmitting microwave signals toward the Earth's surface, where they are reflected and recorded to generate high-resolution Synthetic Aperture Radar (SAR) images. InSAR combines two SAR images of the same area taken at different times. By measuring the difference in phase between the two images, it is possible to create an interferogram, which reveals patterns indicating ground deformation or topography. The phase information is initially wrapped, so phase unwrapping is performed to convert it into a continuous signal, allowing accurate measurement of ground displacement. This unwrapped phase is translated into LOS displacement, which detects movements such as subsidence, uplift, or horizontal shifts (see Fig. 1, Left).

The main limitation of InSAR is that the resulting motion measurements are restricted to the one-dimensional LOS viewing geometry. However, surface motions (resulting from subsurface deformation processes) generally occur in the three spatial dimensions (*i.e.* east (E), north (N), and up (U)). This means that the InSAR analysis of a stack of SAR images is not able to fully capture the magnitude and direction of surface motions [14].The viewing geometry of the satellite LOS is defined by the incidence angle θ (the angle between the local zenith and the looking vector of the satellite) and the satellite heading α. In what

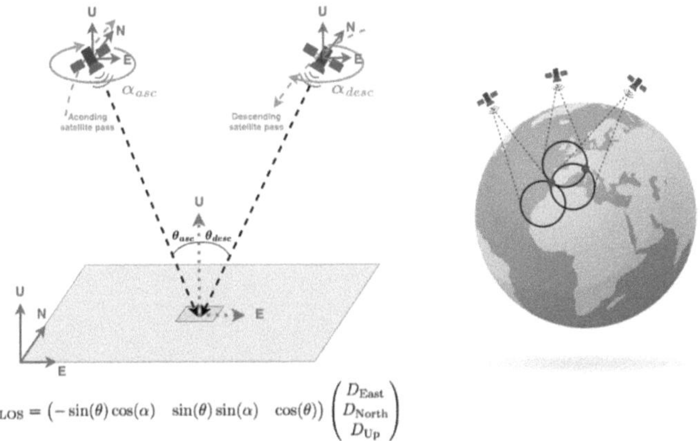

$$D_{\text{LOS}} = \begin{pmatrix} -\sin(\theta)\cos(\alpha) & \sin(\theta)\sin(\alpha) & \cos(\theta) \end{pmatrix} \begin{pmatrix} D_{\text{East}} \\ D_{\text{North}} \\ D_{\text{Up}} \end{pmatrix}$$

Fig. 1. InSAR viewing geometry for Line-Of-Sight (LOS) measurements on ascending and descending satellite passes. **Right**:An image illustrating trilateration: the GNSS receiver lies on the perimeter of the first satellite's coverage area. With a second satellite, its location narrows to one of two intersection points. A third satellite pinpoints the 2D location, and a fourth (not shown) allows calculation of elevation.

follows, we describe surface motions in terms of displacement. A LOS displacement D_{LOS} is composed of the 3D displacement components $D_{East}, D_{North}, D_{Up}$ [7,8].

GNSS consists of a constellation of satellites orbiting the Earth in very specific trajectories. For global coverage, it is estimated that a constellation requires 18 to 30 satellites. Navigation satellites provide orbit information and accurate timing (and other services) to radio receivers specifically designed to capture and decode the content of the signal message. The position on or near most of the Earth's surface can be calculated using a mathematical process known as trilateration [10]. The absolute position of an object is its latitude, longitude, and elevation above mean sea level and can be determined when the signal from four (or more) GNSS satellites is received at the same time [6].

The signal from a single satellite provides a general location of a point around the perimeter of a circular area that covers approximately 35% of the Earth's surface a huge area [4]. Note that the location is not within the area of coverage, but on the perimeter because the distance to the satellite is a known length, and anywhere inside the perimeter will be a different, shorter distance. When a second satellite can be seen, the coverage of that satellite will overlap part of the first satellite's coverage. This means that the GNSS receiver is at one of the two points of intersection of the perimeters of the coverage areas. When a third satellite can be seen, the point of intersection of all three coverage area perimeters will be the location of the GNSS receiver. That is, an accurate two-dimensional (longitude and latitude coordinates) position of the GNSS receiver on the Earth's surface (see Fig. 1, Right). When a fourth satellite can be seen,

Fig. 2. Partial overview of the studied region of Groningen, including the GNSS stations (points shown in red) and InSAR data (points shown in gray). (Color figure online)

elevation or altitude can be determined with trigonometry using the longitude-latitude coordinate and the additional "line" to the fourth satellite.

Datasets. For our work we focus on the area of Groningen (Netherlands). This region is known for natural gas extraction which has led to induced seismicity, ground deformation and subsidence. Monitoring these dynamics accurately is critical for public safety, infrastructure planning, and policymaking.

The GNSS dataset for this area is distributed by the Nevada Geodetic Laboratory [5]. It includes 50 stations, their geographical coordinates, and daily time series of displacement measures in three directions (East, North, and Up) with up to 60 timestamps for each station. The dataset was pre-processed by removing common-mode errors [13,22] and detrending for plate-tectonic motion, thereby isolating localized deformation signals. The centrally located GNSS station *TJUC* (see Fig. 2) served as reference, and the resulting time series were smoothed with a robust moving-average filter to enhance the signal-to-noise ratio and clarify deformation trends.

We consider the InSAR data for the same area of Groningen provided by the European Ground Motion Service (EGMS) of the European Environment Agency (EEA): the whole dataset consists of 1.7×10^6 measurement points, also known as *Persistent Scatterers Interferometry* (PSI). It delivers precise and up-to-date information on ground motion across Europe [2] as part of the Copernicus Land Monitoring Service (CLMS). The dataset is composed of ground motion time series obtained from descending and ascending orbits of Copernicus Sentinel-1. Specifically, every sample consists of direction displacement values in the satellite's LOS for ascending and descending orbits, including point coordinates (x, y, z), values summarizing the average displacement ($V_{\mathrm{LOS}}, A_{\mathrm{LOS}}$), quality measures (RMSE, amplitude dispersion, temporal coherence), and attributes that describe the LOS between the satellite and the ground point (incidence angle (θ) and track angle (ϕ)), as well as a set of cumulative displacement measurements (D_{LOS}), each with a specified date and time. The InSAR dataset spans from January 2018 to December 2022, with a total of 526 timestamps for ascending orbits (divided into 2 orbits: 015, 088) and 538 timestamps for descending orbits (divided into 2 orbits: 037, 139). The data provides dense spatial coverage across the monitored area, making it suitable for large-scale ground motion analysis. For our study we focus on a subset corresponding to a single ascending orbit (Orbit 15). After this selection, 259 timestamps remain available.

3   Algorithm Description

Input and Output Specification. The input to our deep neural network consists of an $N \times 6$ array, where each row represents a 3D point and concatenates its spatial coordinates $P^N_{i=1} = [x, y, z]_i$ to the InSAR measurements $S^N_{i=1} = [\alpha, \theta, D_{\text{LOS}}]_i$, including the azimuth angle (α), the incidence angle (θ), and the Line-Of-Sight displacement (D_{LOS}). The azimuth angle (α) is computed as the satellite's track angle (ϕ) plus 90 ∘ to the right, accounting for the fact that Sentinel-1 is a right-looking satellite, and it is clamped within a $0359°$ range. The neural network produces as output an $N \times 3$ array, where each row corresponds to the estimated corrected displacement in the east, north, and up directions $\hat{\delta}_i = [\hat{D}_{East}, \hat{D}_{North}, \hat{D}_{Up}]_i$ for the respective i-th input InSAR measurement point. The training supervision for the neural network is provided by GNSS measurements, organized as an $M \times 6$ array, where M represents the number of GNSS stations. It holds $M << N$, reflecting the lower spatial density of GNSS stations compared to InSAR measurement points. Each row is defined by concatenating the spatial coordinates $G^M_{j=1} = [x, y, z]_j$ to the ground truth measurements $\delta_j = [D_{East}, D_{North}, D_{Up}]_j$.

InSAR Data Structuring. The full InSAR dataset consists of 1.7×10^6 measurement points spanning a large geographic area. Feeding all measurements into the neural network is both computationally infeasible and suboptimal, as many points are spatially distant and thus weakly correlated. To address this, during training, the dataset is partitioned into local subsets (patches) using a spherical query ball grouping mechanism [1]. Grouping is performed based on the 3D spatial coordinates P_i of each InSAR measurement, with pairwise distances computed using the Haversine formula [15] to accurately reflect geodesic distances on the Earth's surface. Patch centers, referred to as anchor points, are selected using Farthest Point Sampling (FPS) to ensure uniform spatial coverage across the dataset. Each patch consists of all points within a spherical neighborhood (radius of 100 m) around its anchor. Overlapping patches are allowed, so individual InSAR points may be included in multiple patches. Compared to k-nearest neighbors (k-NN), the spherical query ball approach is less sensitive to outliers, as it excludes spatially distant or isolated points that may otherwise be included due to the fixed neighbor count. This patch-based strategy enables the

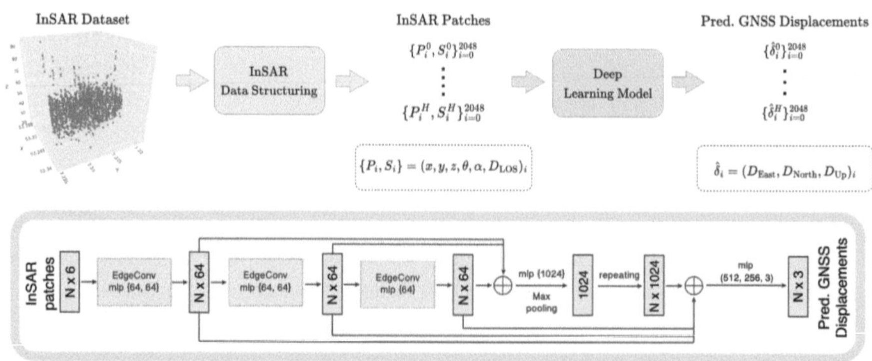

Fig. 3. Top: Overview of the pipeline used to train the proposed deep learning model. The InSAR data are structured in patches which are provided as input to the model. The model predicts the GNSS displacement that would have been observed if a GNSS station were located at the same InSAR input position. The model's parameters are optimized with the goal of minimizing the loss in Eq. 1. **Bottom**: Model architecture. The EdgeConv outputs serve as local descriptors and are aggregated to the global descriptor for each point ($N \times 1024$). The symbol $\oplus$ represents concatenation.

network to learn localized mappings from InSAR measurements to 3D GNSS displacement vectors, while keeping training computationally tractable.

Deep neural Network Training and Inference. For training the deep neural network, we use InSAR patches containing $N = 2048$ measurement points (see the top part of Fig. 3). The goal is to predict, for each point, the GNSS displacement that would have been observed if a GNSS station were located at the same position. To address the sparse supervision provided by $M = 50$ GNSS measurements, we define a mapping strategy by associating each instance i only with its nearest neighbor $j^*(i) = \arg\min_j \|P_i - G_j\|_2$. The training loss function is computed as the weighted squared difference between the network's prediction and the displacement measured at the nearest GNSS station for the same timestamp. The weighting is introduced to rescale the loss contribution proportionally to the inverse of the distance between the InSAR point and its nearest GNSS station. More formally, the loss can be written as follows:

$$\mathcal{L} = \frac{1}{N} \sum_{i=1}^{N} W_i(\hat{\delta}_i - \delta_{j^*(i)})^2 \, , \tag{1}$$

where the weighting function follows a cosine decay profile, truncated beyond a certain threshold ($D_t = 100\,\mathrm{m}$):

$$W_i = \begin{cases} \frac{1}{2}\left(1 + \cos\left(\pi \frac{\|P_i - G_{j^*(i)}\|_2}{D_t}\right)\right) & \text{if } \|P_i - G_{j^*(i)}\|_2 \leq D_t \\ 0 & \text{if } \|P_i - G_{j^*(i)}\|_2 > D_t \, . \end{cases} \tag{2}$$

At inference time, the deep neural network may consider spatial regions disjoint from those observed during training (*i.e.* $P_{test} \neq P_{i=1}^N$). Alternatively, the location can be the same observed during training but at different timestamps with corresponding variations in the InSAR measures (*i.e.* $S_{test} \neq S_{i=1}^N$). Given the input sample $[P_{test}, S_{test}]$, the model predicts the corresponding $\hat{\delta}_{test}$ correction for unseen timestamps either within the same time window of the training data or for different/future time windows. Note that in this phase we directly feed the query points, (*i.e.* the InSAR measurements for which 3D GNSS displacements are to be predicted, into the network without applying the patching strategy. Since test points are typically localized within a small spatial region, they can be processed efficiently in a single forward pass.

Backbone Architecture and Implementation Details. To design the architecture of our deep learning model we take inspiration from the Dynamic Graph Convolutional Neural Network (DGCNN) [16] (see the bottom part of Fig. 3). This architecture is specifically tailored to effectively aggregate feature information from spatially proximate measurements, where proximity is defined based on a spatial k-NN graph. We use $k = 12$, considering the Haversine distance [15] to calculate the distance between geographical coordinates. Each *EdgeConv* module takes as input a tensor of shape $N \times f$, computes edge features for each point (row in the tensor) by applying a multi-layer perceptron (mlp) with the number of layer neurons defined as $\{a_1, a_2, \ldots, a_n\}$, and generates a tensor of shape $N \times k \times a_n$, which reduces to $N \times a_n$ after pooling among neighboring edge features. EdgeConv outputs are local feature descriptors: by stacking multiple modules the network progressively considers a broader spatial context, while the shortcut connections allow it to keep information about the local spatial dependencies, enhancing the model's representational capacity. At last, three fully-connected layers of dimension $\{512, 256, 3\}$ project the internal representation to three-dimensional pointwise measures. Dropout with keep probability of 0.5 is used in the last two fully-connected layers. All layers include LeakyReLU. The network is trained using the Stochastic Gradient Descent (SGD) with a base learning rate of 0.1, which is gradually decayed during training with a cosine annealing rule to ensure stable convergence. Training is conducted over 300 epochs on a single NVIDIA RTX A5000 GPU.

4   Experimental Analysis

For our experimental evaluation we focus on fixed regions around specific GNSS stations, with the model trained and tested on the same InSAR points but at different time windows. Specifically, we consider two settings: *Intermediate* and *Future Window*. In the first case, the model is trained on the entire temporal range, except for a continuous 20% segment from the middle of the timeline, which is used for testing. In the second case, the model is trained on the initial percentage (90% and 50%) of available timestamps and evaluated on future, unseen timestamps (respectively 10% and 50%). The evaluation also runs on the training windows, considering that only a subset of the available timestamps in those time ranges were actually used for training, thus the remaining ones can be considered for testing. We mainly focus on a 100 m radius area centered on

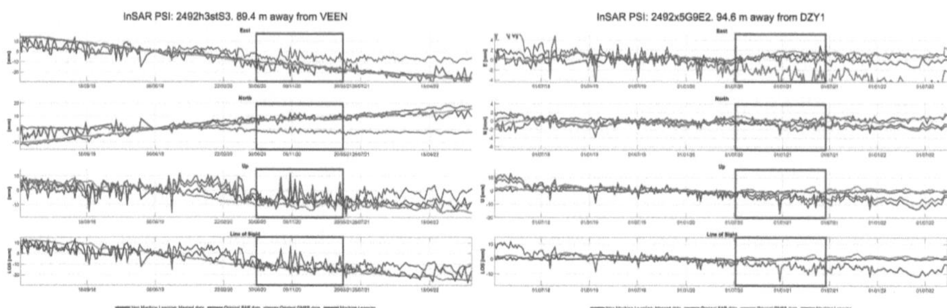

Fig. 4. Intermediate Window setting. **Left**: VEEN station, **Right**: DZY1 station. The learning phase runs on 60 evenly spaced timestamps selected from the initial 50% and the last 30% periods. The central 20% time segment is withheld during training, and GNSS 3D displacement is predicted within this excluded window (red rectangle) as well as on the unused timestamps in the training period. Our approach is indicated as *Machine Learning* (red line).(Color figure online)

the *VEEN* and *DZY1* GNSS stations. The first was chosen due to the challenging nature of the time series at that station: the InSAR data and the GNSS ground-truth measurements follow divergent trends, suggesting that accurately predicting one from the other is inherently difficult. The second station was randomly picked among the remaining sites.

A Non-machine Learning Baseline. The proposed deep learning approach models spatial dependencies from the data and does not require GNSS measurements at test time. An alternative *non-Machine Learning* (non-ML) based method to obtain a ground displacement at a certain point in space consists of leveraging physical models to integrate GNSS and InSAR measurements through geometric transformations and error-weighted fusion [3]. Specifically, GNSS and InSAR data are synchronized by aligning their timestamps and establishing a common reference time. For each InSAR point located within a 100 m radius of a GNSS station, the incidence (θ) and azimuth (α) angles are used to rotate the GNSS 3D measurements so that one of the axes is aligned with the InSAR LOS. This allows all subsequent operations to be performed directly on the LOS. An error-weighted least-squares solution is applied to fuse these measurements, and out-of-LOS components are derived from the pre-processed GNSS data. Finally, the resulting InSAR deformation field is rotated back into the East, North, Up coordinate system. This strategy can be considered as an upper bound for our deep learning model, as it has access at test time to the GNSS data of the closest stations to the prediction point.

Intermediate Window Setting. This setting mimics a realistic operational scenario in which GNSS recordings are temporarily unavailable, for example due to station failure or maintenance downtime. The left part of Fig. 4 depicts the obtained results for the VEEN station. Notably, the proposed deep learning model is not exposed to any data within the intermediate gap window dur-

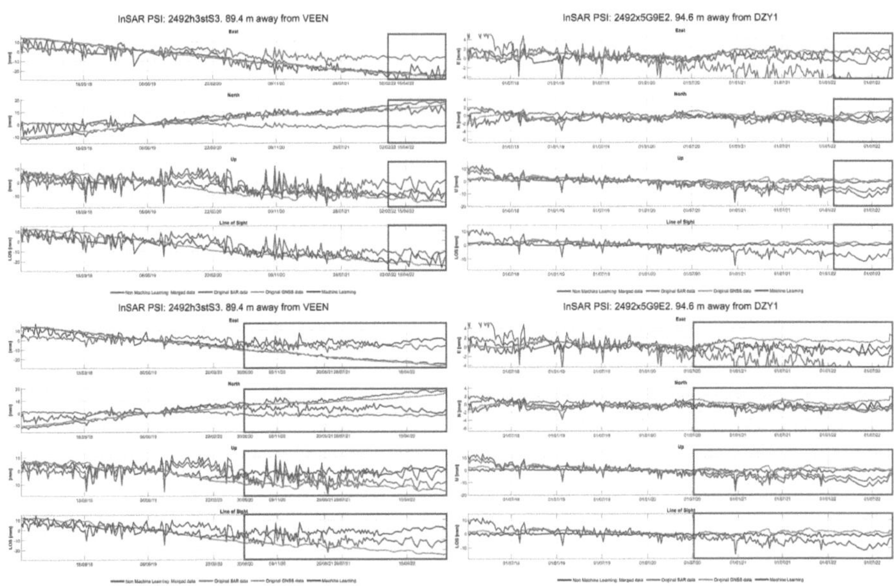

Fig. 5. Future Window 10% (**top row**) and 50% (**bottom row**) for VEEN (**left column**) and DZY1 (**right column**). The deep learning model is trained on 60 timestamps within the first 90% or 10% of the time range and predicts GNSS 3D displacement over the remaining timestamp in the training window as well as in the future window (red rectangle). Our approach is indicated as *Machine Learning* (red line).(Color figure online)

Table 1. Discrepancy (RMSE) between our deep learning and the non-ML baseline approach calculated in the Future Window 50%.

Timestamps	East [mm]	North [mm]	Up [mm]	LOS [mm]
30	10.35	5.94	14.36	15.60
60	8.93	5.49	12.20	12.98
90	6.86	4.43	9.79	9.78

ing training, yet its predictions (red line) remain well aligned with the GNSS ground truth (green line) across the entire timeline and for all the components. The non-ML baseline (blue line) appears close to the GNSS for the East and North components, and to the InSAR data in the Up component with an upward diverging behavior towards the end of the training window. In the LOS component, the baseline (blue line) coincides with the InSAR measurement (and hides the corresponding gray line). Similar observations can be done from the results at the DZY1 station (right part of Fig. 4), although there is a general lower variability of the results in the intermediate test window.

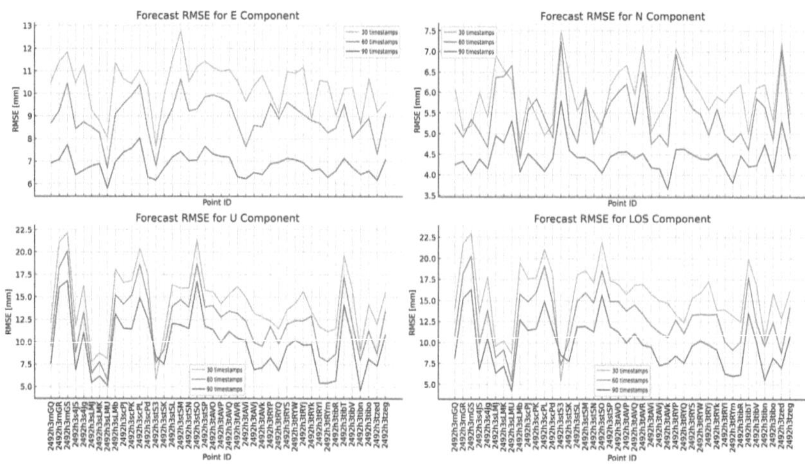

Fig. 6. RMSE per InSAR Point ID (we report the alphanumeric ID of each PSI) for different displacement components (E, N, U, LOS) within the Future Window 50%, when using varying numbers of training timestamps.

Future Window Setting. The top left part of Fig. 5 shows the results for the VEEN station when 90% of the time frame is used for training. Within the 10% future window, the predictions of our deep learning method show a general alignment with the ground truth GNSS data for all the components. Further insights can be obtained by comparing the results of our approach in the 10% future window with what it predicted in the same ending time range in the previously considered intermediate window scenario. Indeed, in that case, the final 10% range of interest was part of the training time period. The similarity in the performance indicates that, regardless of the availability of timestamps at training time, in this window the model performance remains consistent. The results are analogous also for the DZY1 station as indicated by the top right part of the figure. We remind that the non-ML baseline merges the GNSS and InSAR data having always access to all the measures (no distinction between training and test), thus the results for this method are exactly the same already observed in the intermediate window scenario.

The bottom part of Fig. 5 presents a more challenging setting where the model is required to predict displacements in a time window nearly two and a half years into the future with respect to the training one. Our deep learning model (red line) closely tracks the GNSS signal on the test timestamps within the training period, but its predictive accuracy decreases in later timestamps. All the East, North, and Up components reveal a gradual departure from the GNSS signal (green line), with predictions increasingly resembling the raw InSAR input (gray line) rather than the true ground GNSS displacements when moving into the future. For the DZY1 station the output of the machine learning method (red line) remains more consistent with the GNSS data (green line) during the forecast period, demonstrating better alignment in all displacement components compared to what was happening for VEEN. The issue appears less evident on DZY1 station than on the VEEN one. Interestingly, for VEEN in the Up direction the deep learning and non-ML results appear aligned, and the same happens for the East and North components for DZY1.

Impact of Training Timestamps on Model Performance. To evaluate how the number of timestamps used in the training phase affects the model's performance, we conduct an additional analysis using the *Future Window 50%* experimental setup. Specifically, we test configurations with 30, 60, and 90 uniformly distributed timestamps within the training period and evaluate the root mean square difference (we indicate it as RMSE where the E stands for *error*) between our deep learning model and a non-ML baseline in the future 50% time window. Figure 6 shows the results averaged over the future window for each of the displacement components East (E), North (N), Up (U), and Line-of-Sight (LOS), at 42 PSI (Persistent Scatterers Interferometry) points close to the VEEN station. The results are further summarized in Table 1 by averaging over all the PSIs. The numbers indicate that the discrepancy between the deep learning approach and the physics-based solution reduces by increasing the number of observed timestamps at training time. In summary, these findings confirm that the prediction capabilities of our model benefit significantly from higher training data density, particularly when predicting over unseen temporal intervals.

5   Conclusions

Our work focuses on the design of a deep learning model that learns how to calibrate InSAR measurements using sparse GNSS observations during training. At test time, given a new InSAR measure, the model predicts the GNSS displacement that would have been observed if a GNSS station were located at the same point in space and time as the input InSAR. We framed it as a weakly supervised problem and designed a graph neural network that was has been trained and evaluated on publicly available data in the area of Groningen (Netherlands). As a baseline, we adopted a physics-based model that merges the available InSAR and GNSS data at inference time. As the latter is not available for the deep learning model, the non-machine learning baseline can be considered as a reference upper bound.

The experimental analysis on two PSI (close to VEEN and DZY1 stations) showed differences in trend between our deep learning predictions and those of the baseline. However, a final overall assessment on 42 PSI (in the VEEN area) in the challenging future 50% scenario that requires predicting displacements nearly two and a half years into the future, showed that the proposed deep learning model becomes progressively more coherent with the baseline when increasing the number of observed timestamps during training. These findings confirm the importance of training data density for accurate and stable predictions.

This study has been limited to models trained and tested on regions centered around individual GNSS stations. While the learning and inference sets were temporally but not spatially disjoint in the current setup, this is not a constraint imposed by the model, which is in principle capable of generalizing to entirely unseen areas. This is a relevant direction for future work, along with extending the analysis to additional datasets that capture landslide-prone terrains or areas affected by seismic faults and earthquake-induced deformation. Our approach can serve as a first step toward deep learning based data-driven models for ground deformation calibration, potentially paving the way for broader integration of InSAR and GNSS data in operational monitoring systems.

Acknowledgments. This study was carried out within NODES - Nord Ovest Digitale E Sostenibile, funded by the European Union - NextGenerationEU, Mission 4 Component 2 - ECS00000036 - CUP E13B22000020001.

References

1. Pointnet++: Deep hierarchical feature learning on point sets in a metric space. In: NeurIPS (2017)
2. European Environment Agency (EEA). Ground Motion Service (EGMS) (2023). https://land.copernicus.eu/pan-european/european-ground-motion-service
3. Aponte, O., Gatti, A., Realini, E.: Reconstruction of 3d deformation from GNSS and insar data: A case study in groningen using EGMS data. In: EGU General Assembly (2025)
4. Blewitt, G.: Global navigation satellite system (GNSS) and satellite navigation explained (2024). www.advancednavigation.com/tech-articles/global-navigation-satellite-system-gnss-and-satellite-navigation-explained/
5. Blewitt, G., Hammond, W., Kreemer, C.: Harnessing the GPS data explosion for interdisciplinary science. Eos **99**(2), e2020943118 (2018)
6. Enge, P.K.: The global positioning system: Signals, measurements, and performance. Int. J. Wirel. Inf. Netw. **1**(2) (1994)
7. Fialko, Y., Simons, M., Agnew, D.: The complete (3-d) surface displacement field in the epicentral area of the 1999 mw7. 1 hector mine earthquake, California, from space geodetic observations. Geophys. Res. Lett. **28**(16) (2001)
8. Fuhrmann, T., Garthwaite, M.C.: Resolving three-dimensional surface motion with insar: constraints from multi-geometry data fusion. Rem. Sen. **11**(3) (2019)
9. Imperatore, P., Pepe, A., Sansosti, E.: High performance computing in satellite SAR interferometry: a critical perspective. Rem. Sens. **13**(23), 4756 (2021)
10. Kaplan, E., Hegarty, C.: Understanding GPS/GNSS: Principles and Applications, 3rd edn. Inc, Artech House (2017)
11. Lee, J.C., Shirzaei, M.: Novel algorithms for pair and pixel selection and atmospheric error correction in multitemporal insar. Rem. Sens. Envir. **286** (2023)
12. Neely, W.R., Borsa, A.A., Silverii, F.: Ginsar: A CGPS correction for enhanced insar time series. IEEE Trans. Geosci. Rem. Sens. **58**(1) (2020)
13. Nikolaidis, R.: Observation of Geodetic and Seismic Deformation with the Global Positioning System. University of California, San Diego (2002)
14. Rosen, P.A., Hensley, S., Joughin, I.R., Li, F.K., Madsen, S.N., Rodriguez, E., Goldstein, R.M.: Synthetic aperture radar interferometry. Proc. IEEE **88**(3), 333–382 (2000)
15. Van Brummelen, G.: Heavenly mathematics: The forgotten art of spherical trigonometry. Princeton University Press (2017)

16. Wang, Y., Sun, Y., Liu, Z., Sarma, S.E., Bronstein, M.M., Solomon, J.M.: Dynamic graph CNN for learning on point clouds. ACM TOG (2019)
17. Weiss, J.R., et al.: High-resolution surface velocities and strain for Anatolia from sentinel-1 insar and GNSS data. Geophys. Res. Lett. **47**(17) (2020)
18. Xu, K., Gan, W., Wu, J., Hou, Z.: A robust method for 3-d surface displacement fields combining GNSS and single-orbit insar measurements with directional constraint from elasticity model. GPS Solut. **26**(2) (2022)
19. Xu, X., Sandwell, D.T., Klein, E., Bock, Y.: Integrated sentinel-1 insar and GNSS time-series along the san Andreas fault system. J. Geophys. Res. Solid Earth **126**(11) (2021)
20. Yan, H., Dai, W., Xie, L., Xu, W.: Fusion of GNSS and insar time series using the improved stre model: applications to the san Francisco bay area and southern California. J. Geodesy **96**(7) (2022)
21. Zhang, S., et al.: Two-dimensional deformation monitoring for spatiotemporal evolution and failure mode of Lashagou landslide group, northwest china. Landslides **20**(2), 447–459 (2023)
22. Zheng, K., Zhang, X., Sang, J., Zhao, Y., Wen, G., Guo, F.: Common-mode error and multipath mitigation for subdaily crustal deformation monitoring with high-rate GPS Observations. GPS Sol. **25**, 1–15 (2021)

Enhancing Camouflaged Object Detection via Diffusion Model Augmentation

Henry O. Velesaca[1,2] and Angel D. Sappa[1,3(✉)]

[1] ESPOL Polytechnic University, Campus Gustavo Galindo, Guayaquil, Ecuador
hvelesac@espol.edu.ec
[2] Software Engineering Department, Research Center for Information and
Communication Technologies (CITIC-UGR), University of Granada, 18071
Granada, Spain
[3] Computer Vision Center, 08193 Bellaterra, Spain
sappa@ieee.org

Abstract. This paper proposes a novel approach to improve camouflaged object detection (COD) techniques using a data augmentation strategy based on a diffusion model. COD represents a significant challenge in computer vision, being crucial in applications such as military surveillance, medical image analysis, and ecological studies. Although recent advances in deep learning have improved COD performance, they heavily rely on well-annotated datasets, which are particularly difficult to obtain in the context of camouflaged objects. The proposed method addresses the problem of data sparsity while maintaining crucial features of camouflaged objects and their relationships with the environment. Incorporating a diffusion model into the data augmentation pipeline shows improvement in the performance of the COD models, which varies between the different techniques and the proposed evaluation metrics.

Keywords: Camouflaged object detection · Computer vision · Diffusion models · Pest detection

1 Introduction

Camouflaged Object Detection (COD) represents a significant challenge in computer vision, where the goal is to identify objects that are deliberately or naturally blended with their surroundings [5]. This task is particularly crucial in various applications, including military surveillance, medical image analysis, and ecological studies [17]. Unlike traditional object detection, COD faces unique challenges due to the inherent similarity between target objects and their backgrounds, making it difficult for conventional detection methods to achieve satisfactory results [28]. Recent advances in deep learning have significantly improved COD performance [31]. However, these improvements are heavily dependent on large-scale and well-annotated datasets, which are particularly challenging to obtain in the context of camouflaged objects due to the time-consuming nature

© The Author(s), under exclusive license to Springer Nature Switzerland AG 2026
E. Rodolà et al. (Eds.): ICIAP 2025 Workshops, LNCS 16169, pp. 352–364, 2026.
https://doi.org/10.1007/978-3-032-11317-7_30

of data collection and annotation [14]. This limitation often leads to models that underperform in real-world scenarios or fail to generalize across different camouflage patterns and environments.

To address these challenges, data augmentation has emerged as a promising solution [13]. Traditional augmentation techniques, such as geometric transformations and color adjustments, while useful, often fail to capture the complex nature of camouflage patterns and their interaction with the environment [30]. This limitation has motivated the exploration of more sophisticated approaches to data augmentation.

In recent years, diffusion models have demonstrated remarkable capabilities in generating high-quality, diverse images [19]. These models, which learn to gradually denoise images through an iterative process, have shown particular promise in maintaining structural coherence and semantic relationships in generated images [20]. However, their application to the specific challenge of COD data augmentation remains largely unexplored. This paper proposes a methodology to improve COD techniques using data augmentation with diffusion models. Our method addresses the data scarcity problem while maintaining the crucial characteristics of camouflaged objects and their relationships with their surroundings. By incorporating diffusion models into the data augmentation pipeline, we can generate synthetic yet realistic training samples that enhance the robustness and generalization capabilities of COD models.

The manuscript is organized as follows. Section 2 presents work related to camouflaged object detection and data augmentation using classical and diffusion model approaches. Section 3 presents the proposed methodology to improve camouflaged object detection using specialized techniques for this task. Experimental results, comparisons, and discussions with different COD approaches to validate the proposed methodology are presented in Sect. 4. Finally, conclusions are presented in Sect. 5.

2 Background

This section summarizes some of the most relevant techniques related to the camouflaged object detection techniques and data augmentation approaches.

2.1 Camouflaged Object Detection Approaches

Camouflaged Object Detection has emerged as a crucial research field in computer vision, with significant advances in recent years. The early work of Qin et al. [18] on salient object detection significantly influences the development of COD techniques. In 2021, Fan et al. [5] establish the fundamental foundations for concealed object detection, providing a robust theoretical framework for this challenging problem. In 2022, Chen et al. [3] propose a context-aware cross-level fusion-based approach, improving the accuracy of camouflaged object identification, while Chen et al. [4] introduce a boundary-guided network that enhances the precision in detecting edges and features of camouflaged objects. In the same year, Liu et al. [16] address the modeling of aleatoric uncertainty in camouflaged object detection. The field continues to evolve in 2023 with significant contributions such as the high-resolution iterative feedback network presented by Hu et al. [9], and the edge-aware network for COD presented by Sun et al. [24]. In the same year, Ji et al. [11] introduce an innovative deep gradient learning technique that significantly improves the efficiency of camouflaged object detection. More recently, in 2024, Yang et al. [29] expand the field towards specific applications with PlantCamo, focusing on camouflage detection in plants.

These advances contribute to creating more robust and accurate systems for camouflaged object detection, setting new standards in the field and paving the way for future research and practical applications.

2.2   Data Augmentation Approaches

Data augmentation has become a fundamental technique in deep learning, especially in computer vision tasks, where the availability of labeled data may be limited. Classic augmentation techniques include geometric transformations such as rotation, flipping, cropping, and rescaling, as well as color space modifications such as changes in brightness, contrast, and saturation [22]. These basic transformations help artificially increase the size of the training dataset and improve the generalization capabilities of models. A popular tool that has revolutionized the implementation of these techniques is Albumentations [1], a Python library that provides an efficient and optimized interface for applying data augmentation transformations. This library has become a standard in the computer vision community due to its superior performance and wide range of available transformations.

Recent research uses synthetic data and domain adaptation to combat data scarcity. Virtual environments effectively train deep learning models for camera pose estimation [2] and image dehazing [21]. Synthetic datasets aid instance segmentation [26] and agricultural applications like corn kernel classification [23], reducing annotation costs. This approach, combined with traditional augmentation, improves model performance while decreasing reliance on manual labeling.

In a significant advancement in the field of deep learning and data augmentation, Islam et al. [10] present DiffuseMix, a groundbreaking data augmentation technique that preserves labels using diffusion models. This method addresses one of the fundamental challenges in machine learning: generating synthetic data that maintains the integrity of the original labels. DiffuseMix leverages the power of diffusion models to create realistic and meaningful variations of the training data while ensuring that semantic labels remain consistent and accurate. This approach not only improves the diversity of the training dataset but also helps prevent overfitting and improves the robustness of deep learning models. The technique demonstrates promising results on several benchmark datasets, establishing itself as a valuable contribution to the computer vision and machine learning community.

3   Proposed Methodology

This section details the different stages followed to carry out the proposed methodology. Figure 1 shows the overall pipeline of the proposed methodology.

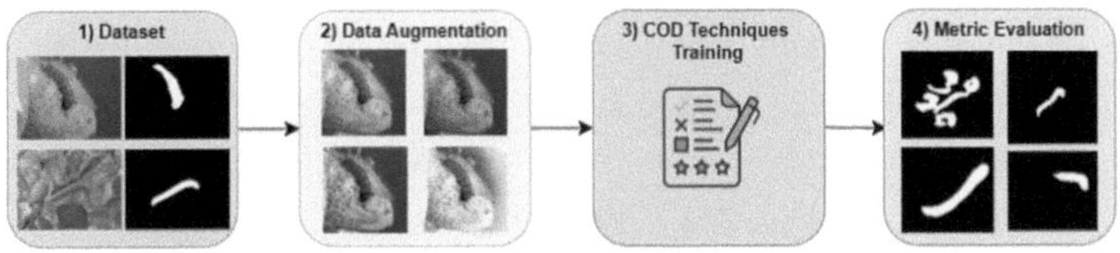

Fig. 1. Overall pipeline of the proposed methodology.

3.1   Data Augmentation

For the proposed work, two data augmentation approaches are used to enrich and diversify the training data set. These approaches are carefully selected to maximize the variability and quality of the training data.

The first approach corresponds to data augmentation using the classical method, specifically Albumentations [1], which include geometric and photometric transformations of the images, such as random rotations, horizontal and vertical flips, brightness and contrast adjustments, random cropping, scale changes, the addition of Gaussian noise and moderate Gaussian blur. These traditional transformations help improve the robustness of the model against basic variations that may occur in real scenarios. For the second data augmentation approach, a more advanced technique based on diffusion models is used, specifically, the work DiffuseMix proposed by Islam et al. [10]. This methodology leverages the generative power of diffusion models to create new synthetic samples that preserve important semantic features, generate realistic variations while maintaining the original labels, introduce significant diversity into the dataset, produce examples that cover cases underrepresented in the original dataset, and maintain coherence between the generated image and its corresponding segmentation mask.

3.2   SOTA COD Techniques

To validate the proposed methodology, SOTA COD techniques are first selected using various state-of-the-art surveys [6,15,28,31] as references. After reviewing the best-performing techniques and available codes, nine camouflaged object detection techniques are selected for fine-tuning using two data augmentation approaches: the classical method Albumentations [1] and the diffusion model-based technique DiffuseMix [10]. It is worth noting that DiffuseMix is not evaluated for segmentation tasks in its original paper.

Table 1. Distinctive features of the evaluated camouflage techniques.

Technique	Year	Image Size (px)	Backbone	#Param. (M)
BASNet [18]	2019	256×256	ResNet-34 [8]	87.06
SINet-v2 [5]	2021	352×352	Res2Net-50 [7]	24.93
BGNet [4]	2022	416×416	Res2Net-50 [7]	77.80
C^2F-Net [3]	2022	352×352	Res2Net-50 [7]	26.36
OCENet [16]	2022	352×352	ResNet-50 [8]	58.17
EAMNet [24]	2023	384×384	Res2Net-50 [7]	30.51
DGNet [11]	2023	352×352	EfficientNet [25]	8.30
HitNet [9]	2023	352×352	PVTv2 [27]	25.73
PCNet [29]	2024	352×352	PVTv2 [27]	27.66

The nine SOTA COD techniques evaluated are: BASNet [18], SINet-v2 [5], BGNet [4], C^2F-Net [3], OCENet [16], EAMNet [24], DGNet [11], HitNet [9], and PCNet [29]. Table 1 presents a comparative analysis of their architectural characteristics. Models from 2019 to 2024 are considered. Each model is trained for 150 epochs, and a batch size of 16 is considered. Techniques such as SINet-v2, C^2F-Net, OCENet, DGNet, HitNet, and PCNet use input images of 352×352 pixels, while BASNet, BGNet, and EAMNet use sizes of 256×256, 416×416, and 384×384 pixels, respectively. This dimensional consistency among most models indicates a standardized approach in the field.

The backbone architectures demonstrate significant diversity in their design choices. Res2Net50 serves as the backbone for four techniques (SINet-v2, BGNet, C^2F-Net, and EAMNet). Meanwhile, PCNet and HitNet implement PVT-V2 (Pyramid Vision Transformer V2), and DGNet utilizes EfficientNet, showcasing various architectural approaches to COD tasks. In addition, BASNet and OCENet use ResNet-34 and ResNet-50, respectively. The computational complexity, measured in terms of parameter count, varies substantially across these models. DGNet exhibits the most efficient architecture with 8.30 million parameters, while BASNet, BGNet, and OCENet represent the most complex design with 87.06, 77.80, and 58.17 million parameters, respectively. The remaining models (SINet-v2, C^2F-Net, EAMNet, HitNet, PCNet) maintain 24–31 million parameter counts, suggesting an optimal complexity range for COD applications.

3.3 Metric Evaluation

This study employs five widely recognized evaluation metrics to evaluate Camouflaged Object Detection (COD) performance. These metrics provide comprehensive assessment criteria for analyzing detection accuracy and effectiveness across different models. The Structure-measure (S_α), weighted F-measure (F_β^w), Mean Absolute Error (M), E-measure (E_ϕ), and F-measure (F_β). The S_α metric quantifies the structural similarity between prediction and ground truth maps.

The F_β^w represents an enhanced evaluation metric that extends the traditional F_β by incorporating spatial weights, providing a better assessment of segmentation quality with emphasis on boundary accuracy and location-based importance of detected pixels. The M metric focuses on pixel-level error evaluation between the normalized prediction and ground truth. The E_ϕ metric simultaneously evaluates the global and local accuracy of COD based on human visual perception mechanisms. The F_β provides a synthetic measure that considers both precision and recall components. For both F-measure and E-measure metrics, different scores can be obtained according to different precision-recall pairs. This leads to the computation of mean F-measure (F_β^{mean}) and maximum F-measure (F_β^{max}). Similarly, the E-measure utilizes maximum and mean variants, denoted as E_ϕ^{mean} and E_ϕ^{max}, which are also employed as evaluation metrics.

On the other hand, to evaluate the quality of the images obtained with both approaches, we use the work proposed by [12]. The evaluation metrics comprise three fundamental aspects: the Reconstruction Fidelity Score ($S_{R_f}^Q$), which measures the ability to preserve the original structure and content of the image, where a value close to 1 indicates excellent conservation of the original features; the Boundary Visibility Score (S_b^Q), which evaluates the quality and definition of the edges in the generated image, where higher values indicate better preservation of object contours; and finally, the Combined Score (S_α^Q), which acts as a comprehensive metric that balances both the reconstruction fidelity and the edge quality, providing an overall assessment of the quality camouflage, where a value close to 1 represents optimal performance in the three metrics.

4   Results and Discussion

This section presents experimental results of the proposed methodology. Model performance evaluation uses the assessment metrics detailed in Sect. 3.3.

4.1   Dataset

To carry out the different experiments in this work, a dataset on the cotton bollworm [17] is used, representing a comprehensive collection of specialized images focusing on one of the most destructive pests of agriculture that poses a significant threat to cotton crops worldwide. Created by Meng et al. [17], this dataset is a vital resource for researchers, specifically designed to address the complex challenges of detecting these elusive pests in various environmental conditions and varied scenes. The dataset, spanning a wide range of visual data capturing cotton bollworms in different contexts and environments, has been made publicly available to the scientific community through the Kaggle platform[1], enabling further research and development in pest detection and the agricultural context. The original dataset consists of 856 images for training, 161 images for validation, and 56 images for testing. While the dataset augmented with (AwA) [1] and

---

[1] https://www.kaggle.com/datasets/kexinmeng1/the-dataset-of-cotton-bollworms.

(*AwD*) [10] results in 4633 and 4406 images respectively. Figure 2 shows examples of the images that are part of the dataset augmented with the DiffuseMix technique [10].

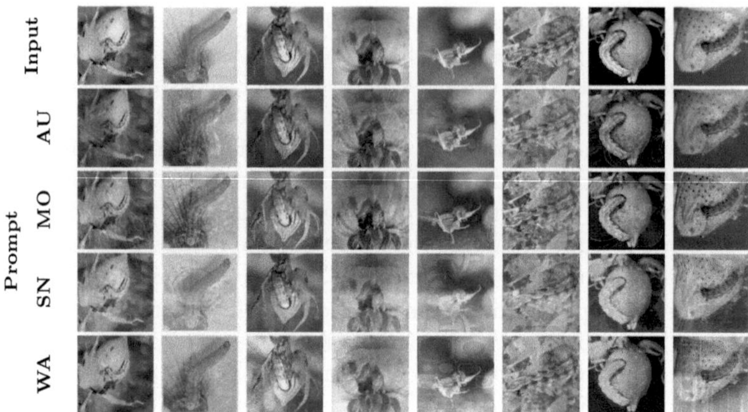

Fig. 2. Examples of images generated using DiffuseMix [10]. Prompt: Watercolor Art (WA), Snowy (SN), Autumn (AU), Mosaic (MO).

Table 2. Evaluation of the quality of camouflaged images according to [12] .The first two rows show the absolute median evaluation for three metrics for augmentation using Albumentations (*AwA*) and DiffuseMix (*AwD*) techniques. The last row shows the improvement (%) when using the DiffuseMix technique.

Data Augmentation	$S_{R_f}^{Q}$ ↑	S_b^{Q} ↑	S_α^{Q} ↑
AwA [1]	0.631	0.529	0.586
AwD [10]	0.637	0.539	0.603
Improv.(%)	+0.06%	+1.0%	+1.7%

4.2 Quantitative Results

Before evaluating each COD model's performance, the quality of the images generated with the augmentation techniques for the different data sets is measured as a first step. Table 2 shows in the first two rows the evaluation results of the metrics proposed by [12] on the images obtained with the data augmentation techniques. The last row shows the percentage difference between both data augmentation techniques, where it can be seen that the data augmentation technique using DiffuseMix presents better performance in all metrics. Based on

Table 3. Evaluation metrics for each COD technique according to the metrics described in Sect. 3.3. The best three performing results are highlighted in red (first), blue (second), and green (third) respectively.

Technique	$S_\alpha \uparrow$	$F_\beta^w \uparrow$	$M \downarrow$	$E_\phi^{adp} \uparrow$	$E_\phi^{mean} \uparrow$	$E_\phi^{max} \uparrow$	$F_\beta^{adp} \uparrow$	$F_\beta^{mean} \uparrow$	$F_\beta^{max} \uparrow$
BASNet [18]	0.7860	0.6839	0.0277	0.8938	0.8811	0.8890	0.7085	0.7073	0.7235
SINet-v2 [5]	0.8588	0.7979	0.0184	0.9430	0.9571	0.9727	0.7837	0.8086	0.8392
BGNet [4]	0.8573	0.7964	0.0185	0.9588	0.9532	0.9625	0.8264	0.8323	0.8527
C^2F-Net [3]	0.8399	0.7184	0.0214	0.9368	0.9329	0.9441	0.7910	0.7989	0.8215
OCENet [16]	0.8598	0.8042	0.0164	0.9421	0.9543	0.9647	0.7809	0.8115	0.8416
EAMNet [24]	0.8627	0.8136	0.0174	0.9576	0.9611	0.9686	0.8164	0.8320	0.8570
DGNet [11]	0.8565	0.8005	0.0163	0.9468	0.9585	0.9666	0.7827	0.8096	0.8377
HitNet [9]	0.8422	0.7978	0.0178	0.9601	0.9663	0.9724	0.7991	0.8060	0.8244
PCNet [29]	0.8463	0.7885	0.0184	0.9342	0.9471	0.9642	0.7691	0.7924	0.8311
BASNet [18]	0.7761	0.6739	0.0269	0.9064	0.8726	0.9065	0.6992	0.7009	0.7113
SINet-v2 [5]	0.8612	0.8123	0.0167	0.9493	0.9594	0.9689	0.7960	0.8199	0.8466
BGNet [4]	0.8666	0.8237	0.0165	0.9698	0.9600	0.9708	0.9356	0.8431	0.8584
C^2F-Net [3]	0.8487	0.7346	0.0212	0.9422	0.9509	0.9635	0.7931	0.8191	0.8450
OCENet [16]	0.8548	0.7961	0.0179	0.9494	0.9528	0.9646	0.7831	0.8100	0.8352
EAMNet [24]	0.8617	0.8180	0.0166	0.9524	0.9608	0.9674	0.8073	0.8303	0.8552
DGNet [11]	0.8645	0.8068	0.0164	0.9454	0.9492	0.9597	0.7927	0.8196	0.8403
HitNet [9]	0.8542	0.8206	0.0157	0.9704	0.9673	0.9713	0.8298	0.8318	0.8416
PCNet [29]	0.8550	0.7949	0.0176	0.9425	0.9479	0.9535	0.7891	0.8057	0.8275

The first nine rows are grouped under *AwA [1]*; the last nine rows are grouped under *AwD [10]*.

Table 4. Percentage difference between the results obtained for each COD technique and each data augmentation approach shown in Table 3. Positive values indicate improvement in the diffusion model-based approach, negative values indicate a deterioration in performance. The last row shows the improvement (%) of the DiffuseMix concerning the Albumentations technique.

Technique	ΔS_α	ΔF_β^w	ΔM	ΔE_ϕ^{adp}	ΔE_ϕ^{mean}	ΔE_ϕ^{max}	ΔF_β^{adp}	ΔF_β^{mean}	ΔF_β^{max}
BASNet [18]	-0.99	-1.00	-0.08	+1.26	-0.85	+1.75	-0.93	-0.64	-1.22
SINet-v2 [5]	+0.24	+1.44	-0.17	+0.63	+0.23	-0.38	+1.23	+1.13	+0.74
BGNet [4]	+0.93	+2.73	-0.20	+1.10	+0.68	+0.83	+10.92	+1.08	+0.57
C^2F-Net [3]	+1.62	-0.02	+0.54	+1.80	+1.94	+0.21	+0.27	+2.02	+2.35
OCENet [16]	-0.50	-0.81	+0.15	+0.73	-0.15	-0.01	+0.22	-0.15	-0.64
EAMNet [24]	-0.10	+0.44	-0.08	-0.08	-0.52	-0.03	-0.12	-0.17	-0.18
DGNet [11]	+0.80	+0.63	+0.01	+0.01	-0.14	-0.93	+0.69	+1.00	+0.26
HitNet [9]	+1.20	+2.28	-0.21	+1.03	+0.10	-0.11	+3.07	+2.58	+1.72
PCNet [29]	+0.87	+0.64	-0.08	+0.83	+0.08	+0.08	-1.07	+1.33	-0.36
Improv.(%)	+0.37	+0.54	0.00	+0.50	-0.27	+1.50	+0.31	+1.06	+0.24

these results, it is observed that the diffusion model achieves an improvement in the preservation of the structure and original content of the image ($S_{R_f}^{Q}$ +0.06%), a better definition of edges and contours (S_b^{Q} +1.0%) and consequently, a better overall balance between fidelity and visual quality of the image (S_α^{Q} +1.7%).

After evaluating the quality of the datasets, the next step is to compare the results between data augmentation using Albumentations [1] and DiffuseMix [10] techniques to improve the COD task. Table 3 shows the results of this evaluation. Overall, DiffuseMix exhibits superior performance across most metrics. BGNet with DiffuseMix achieves the best results in several key metrics, including S_α (0.8666), F_β^w (0.8237), F_β^{adp} (0.9356), F_β^{mean} (0.8431), and F_β^{max} (0.8584). Hit-Net with DiffuseMix also stood out, achieving the best performance in the M metric (0.0157, where lower values are better) and in the E_ϕ^{adp} (0.9704) and E_ϕ^{mean} (0.9673) metrics. Among the results obtained when the augmentation is performed with Albumentations, EAMNet and SINet-v2 show notable performance, with SINet-v2 reaching the highest E_ϕ^{max} (0.9727) across all comparisons. DGNet using Albumentations achieves the second-best result in the M metric (0.0163). Interestingly, implementing the diffusion model generally improves the performance of the base techniques, particularly in precision and edge detection metrics.

On the other hand, Table 4 shows the comparative analysis between augmentation with Albumentations [1] and augmentation with DiffuseMix [10]; it reveals interesting patterns in performance differences across various COD techniques. HitNet and BGNet demonstrate the most substantial improvements when implemented with the DiffuseMix technique, showing notable enhancements across all evaluation metrics. Interestingly, some techniques like BASNet show a slight deterioration in performance with the DiffuseMix approach, with decreases ranging from -0.64% to -1.22% across various metrics. The median difference performance comparison between both strategies indicates that DiffuseMix technique generally outperforms Albumentations technique, with positive differences across S_α, F_β^w, E_ϕ^{adp}, E_ϕ^{max}, F_β^{adp}, F_β^{mean}, and F_β^{max} metrics except for M, which remains neutral (0.00%) value and E_ϕ^{mean} with a negative difference of -0.27%. The most substantial improvements are observed in F_β^{mean} and E_ϕ^{max}, with gains of 1.50% and 0.91% respectively. The diffusion model approach improves performance, though its effectiveness varies across techniques and metrics.

4.3 Qualitative Results

Figure 3 compares ground truth (GT) masks and predicted masks across different techniques. In these visualizations, three distinct regions are highlighted: white areas represent successful matches between GT and predicted masks, indicating accurate detection; red areas denote false positive regions (over-segmentation), where the model incorrectly predicts camouflaged objects in areas not marked in the GT; and green areas indicate false negative regions (miss-segmentation), representing areas where the model fails to detect camouflaged objects present in

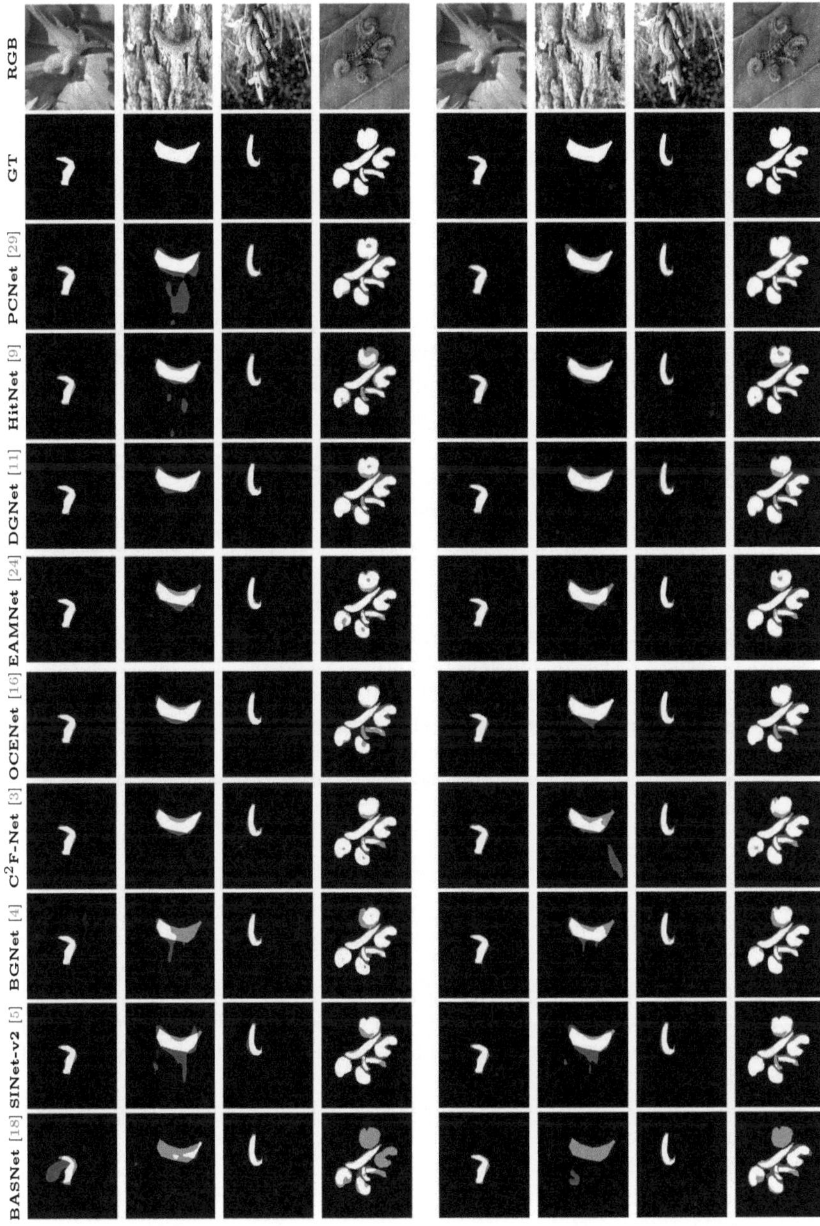

Fig. 3. Prediction results of nine state-of-the-art COD techniques trained with augmentation with Albumentations [1] (*col 1st to 4th*) and DiffuseMix [10] (*col 5th to 8th*). White areas represent successful matches between GT and predicted masks; red areas denote false positive regions (over-segmentation); and green areas indicate false negative regions (miss-segmentation).(Color figure online)

the GT. Visual analysis reveals that COD techniques trained with the Albumentations method tend to produce more false positives (red) and false negatives (green) compared to their counterparts trained with diffusion model-augmented datasets. This observation suggests that the diffusion model-based approach leads to more precise and accurate object detection, with fewer instances of both over-segmentation and missed detections.

5   Conclusions

Experimental results demonstrate that the diffusion model-based data augmentation technique outperforms the classical method, showing an improvement in the quality of the camouflaged image detection reflected in an increase in the S_α^Q metric ($+1.7\%$), which shows an overall better balance between reconstruction fidelity score and boundary visibility score. On the other hand, a comparison shows an improvement in the metrics using DiffuseMix, which indicates that the diffusion model technique generally outperformes the classical method, with positive improvements across most metrics (S_α, F_β^w, E_ϕ^{adp}, E_ϕ^{max}, F_β^{adp}, F_β^{mean}, and F_β^{max}), the most substantial being in F_β^{mean} and E_ϕ^{max}, with gains of 1.50% and 0.91% respectively.

Acknowledgements. This work was supported in part by Grant PID2021-128945NB-I00 funded by MCIN/AEI/10.13039/501100011033 and by "ERDF A way of making Europe", in part by the Air Force Office of Scientific Research Under Award FA9550-24-1-0206. The second author acknowledges the support of the Generalitat de Catalunya CERCA Program to CVC's general activities, and the Departament de Recerca i Universitats from Generalitat de Catalunya with reference 2021 SGR01499.

References

1. Buslaev, A., Iglovikov, V.I., Khvedchenya, E., Parinov, A., Druzhinin, M., Kalinin, A.A.: Albumentations: fast and flexible image augmentations. Information **11**(2), 125 (2020)
2. Charco, J.L., Sappa, A.D., Vintimilla, B.X., Velesaca, H.O.: Camera pose estimation in multi-view environments: from virtual scenarios to the real world. Image Vision Comput. **110**, 104182 (2021)
3. Chen, G., Liu, S.J., Sun, Y.J., Ji, G.P., Wu, Y.F., Zhou, T.: Camouflaged object detection via context-aware cross-level fusion. Trans. Circuits Syst. Video Technol. **32**(10), 6981–6993 (2022)
4. Chen, T., Xiao, J., Hu, X., Zhang, G., Wang, S.: Boundary-guided network for camouflaged object detection. Knowl. Based Syst. **248**, 108901 (2022)
5. Fan, D.P., Ji, G.P., Cheng, M.M., Shao, L.: Concealed object detection. Trans. Patt. Analy. Mach. Intell. **44**(10), 6024–6042 (2021)
6. Fan, D.P., Ji, G.P., Xu, P., Cheng, M.M., Sakaridis, C., Van Gool, L.: Advances in deep concealed scene understanding. Visual Intell. **1**(1), 16 (2023)
7. Gao, S.H., Cheng, M.M., Zhao, K., Zhang, X.Y., Yang, M.H., Torr, P.: Res2net: a new multi-scale backbone architecture. Trans. Patt. Analy. Mach. Intell. **43**(2), 652–662 (2019)

8. He, K., Zhang, X., Ren, S., Sun, J.: Deep residual learning for image recognition. In: Conf. on Computer Vision and Pattern Recognition, pp. 770–778 (2016)

9. Hu, X., et al.: High-resolution iterative feedback network for camouflaged object detection. In: Conf. on Artificial Intelligence. vol. 37, pp. 881–889 (2023)

10. Islam, K., Zaheer, M.Z., Mahmood, A., Nandakumar, K.: Diffusemix: label-preserving data augmentation with diffusion models. In: Conf. on Computer Vision and Pattern Recognition, pp. 27621–27630 (2024)

11. Ji, G.P., Fan, D.P., Chou, Y.C., Dai, D., Liniger, A., Van Gool, L.: Deep gradient learning for efficient camouflaged object detection. Mach. Intell. Res. **20**(1), 92–108 (2023)

12. Lamdouar, H., Xie, W., Zisserman, A.: The making and breaking of camouflage. In: Int. Conf. on Computer Vision, pp. 832–842 (2023)

13. Le, M.Q., Tran, M.T., Le, T.N., Nguyen, T.V., Do, T.T.: Unveiling camouflage: a learnable fourier-based augmentation for camouflaged object detection and instance segmentation. arXiv (2023)

14. Li, X., et al.: Camouflaged object detection with counterfactual intervention. Neurocomputing **553**, 126530 (2023)

15. Liang, Y., Qin, G., Sun, M., Wang, X., Yan, J., Zhang, Z.: A systematic review of image-level camouflaged object detection with deep learning. Neurocomputing **566**, 127050 (2024)

16. Liu, J., Zhang, J., Barnes, N.: Modeling aleatoric uncertainty for camouflaged object detection. In: Winter Conference on Applications of Computer Vision, pp. 1445–1454 (2022)

17. Meng, K., Xu, K., Cattani, P., Mei, S.: Camouflaged cotton bollworm instance segmentation based on pvt and mask r-cnn. Comput. Electron. Agricul. **226**, 109450 (2024)

18. Qin, X., Zhang, Z., Huang, C., Gao, C., Dehghan, M., Jagersand, M.: Basnet: boundary-aware salient object detection. In: Conf. on Computer Vision and Pattern Recognition (2019)

19. Rombach, R., Blattmann, A., Lorenz, D., Esser, P., Ommer, B.: High-resolution image synthesis with latent diffusion models. In: Conference on Computer Vision and Pattern Recognition, pp. 10684–10695 (2022)

20. Saharia, C., et al.: Photorealistic text-to-image diffusion models with deep language understanding. Adv. Neural. Inf. Process. Syst. **35**, 36479–36494 (2022)

21. Sappa, A.D., Suárez, P.L., Velesaca, H.O., Carpio, D.: Domain adaptation in image dehazing: Exploring the usage of images from virtual scenarios. In: Int. Conf. on Computer Graphics, Visualization, Computer Vision and Image Processing, pp. 85–92 (2022)

22. Shorten, C., Khoshgoftaar, T.M.: A survey on image data augmentation for deep learning. J. Big Data **6**(1), 1–48 (2019)

23. Suárez, P.L., Velesaca, H.O., Carpio, D., Sappa, A.D.: Corn kernel classification from few training samples. Artif. Intell. Agriculture **9**, 89–99 (2023)

24. Sun, D., Jiang, S., Qi, L.: Edge-aware mirror network for camouflaged object detection. In: Int. Conf. on Multimedia and Expo, pp. 2465–2470. IEEE (2023)

25. Tan, M., Le, Q.: Efficientnet: Rethinking model scaling for convolutional neural networks. In: Int. Conf. on Machine Learning, pp. 6105–6114. PMLR (2019)

26. Velesaca, H.O., Suárez, P.L., Carpio, D., Sappa, A.D.: Synthesized image datasets: Towards an annotation-free instance segmentation strategy. In: International Symposium Advances in Visual Computing, pp. 131–143. Springer (2021)

27. Wang, W., et al.: Pvt v2: Improved baselines with pyramid vision transformer. Computational Visual Media **8**(3), 415–424 (2022)

28. Xiao, F., et al.: A survey of camouflaged object detection and beyond. arXiv (2024)
29. Yang, J., Wang, Q., Zheng, F., Chen, P., Leonardis, A., Fan, D.P.: Plantcamo: plant camouflage detection. arXiv (2024)
30. Yang, L., et al.: Diffusion models: a comprehensive survey of methods and applications. ACM Comput. Surv. **56**(4), 1–39 (2023)
31. Zhong, J., Wang, A., Ren, C., Wu, J.: A survey on deep learning-based camouflaged object detection. Multimedia Syst. **30**(5), 268 (2024)

Seeing the Heat: Vision Transformers for Spatiotemporal Temperature Prediction

Paolo Russo[1]([envelope]) [iD] and Fabiana Di Ciaccio[2] [iD]

[1] Department of Computer, Control and Management Engineering, Sapienza University of Rome, Rome, Italy
paolo.russo@diag.uniroma1.it
[2] Deparment of Civil and Environmental Engineering, University of Florence, Florence, Italy
fabiana.diciaccio@unifi.it

Abstract. Forecasting land surface temperature (LST) is a key task in environmental monitoring. While Transformer-based architectures have recently shown promise in modeling temperature time series, it remains unclear whether representing geophysical signals as raw sequences or as visual encodings offers better predictive accuracy. In this work, we investigate this question by comparing two autoregressive architectures: a standard sequence-based Transformer operating on raw LST values, and a video-based model that leverages visual encodings of temperature fields on a pre-trained encoder.

To this end, we convert hourly LST grids from the Copernicus Land Monitoring Service and the ERA5 datasets into color-coded heatmaps using OpenCV to further train a TimeSformer-based model that performs spatiotemporal attention across the resulting video-like sequences. The visual backbone extracts structured features from both space and time, which are then passed to an autoregressive decoder to forecast the next 72 h of temperature evolution. Our framework is evaluated on a multi-year dataset covering the region of Florence (Italy), and compared against a previously validated Transformer model trained directly on numerical signals.

Experimental results show that the vision-based model achieves competitive performance with respect to the numeric baseline. The study highlights the potential of vision transformers for environmental forecasting tasks, bridging computer vision and climate modeling.

1 Introduction

Accurate and reliable forecasting of land surface temperature (LST) is becoming increasingly critical in the context of global climate change, urbanization, and environmental management. Temperature fluctuations and extreme heat events have significant impacts not only on ecosystems and agriculture but also on urban infrastructure, energy demand, and human health [20]. Particularly in

E. Rodolà et al. (Eds.): ICIAP 2025 Workshops, LNCS 16169, pp. 365–376, 2026.
https://doi.org/10.1007/978-3-032-11317-7_31

urban and peri-urban areas, the phenomenon of Urban Heat Islands (UHIs) exacerbates temperature extremes, creating microclimatic variations that are difficult to capture and predict with coarse-resolution global models.

Traditionally, environmental forecasting relies on numerical weather prediction (NWP) models, which are grounded in physical laws and simulate atmospheric processes through systems of partial differential equations [1]. While powerful, these models often require intensive computational resources and may suffer from limitations in spatial resolution, data latency, and adaptability to hyperlocal conditions. With the advent of large-scale Earth observation programs such as Copernicus, researchers now have access to high-frequency and spatially resolved LST data from satellite platforms, opening up to new opportunities for data-driven forecasting approaches.

Recent advances in Deep Learning (DL) have shown strong potential for modeling complex spatiotemporal phenomena from such datasets [6]. Transformer architectures, originally developed for natural language processing, have demonstrated state-of-the-art performance in time series modeling, weather prediction, and environmental forecasting. In particular, autoregressive Transformer models can learn long-range dependencies and capture sequential dynamics in structured data, making them suitable for predicting temperature trends across space and time. However, these models typically operate on raw numerical representations of gridded data, potentially neglecting spatial patterns that are visually informative but not explicitly encoded in tabular formats.

At the same time, the field of computer vision has seen significant advances with the advent of Vision Transformer-based architectures (ViT) [8], which extend the self-attention paradigm from natural language processing to image and video understanding. The ViT introduced a new way of processing visual data by modeling image patches as sequences, leading to the use of high-performing architectures in vision tasks without relying on convolutions. Building on this foundation, models such as ViViT [3] and TimeSformer [4] are specifically designed to capture both spatial and temporal dependencies within visual sequences.

In particular, TimeSformer adopts a factorized self-attention mechanism that decomposes attention across spatial and temporal dimensions: its flexibility and effectiveness have been demonstrated in a wide range of domains, from human action recognition and video classification to scientific tasks such as cloud motion prediction, rainfall nowcasting, and Earth observation sequence modeling. These models, often pre-trained on large-scale video datasets such as Kinetics-400, have shown strong transfer learning capabilities, making them attractive candidates for tasks involving dynamic environmental processes.

Despite their success, a fundamental question remains open in the literature: is it more effective to model geophysical time series using traditional Transformer architectures that operate directly on raw numerical sequences, or to convert these data into visual representations and process them using video-based architectures? Both approaches are prevalent, yet comparative studies remain scarce.

In this work, we aim to address this question by directly comparing a sequential Transformer model operating on raw LST values with a Vision Transformer architecture applied to the same data converted into a sequence of images. Specifically, each hourly LST field is transformed into a color-coded heatmap using OpenCV's colormap functions, effectively creating a visual frame that preserves spatial temperature patterns. These heatmaps are then fed to a TimeSformer-based encoder to extract spatiotemporal features through factorized attention. The encoded representations are passed to an autoregressive Transformer decoder that generates future numerical temperature predictions step-by-step. This hybrid formulation combines the strengths of vision-based modeling i.e. spatial awareness and structured pattern recognition, with the temporal reasoning capabilities of autoregressive sequence modeling, offering a novel perspective on environmental data forecasting. Moreover, the vision approach enables the exploitation of pre-trained visual features, with promising generalization capabilities with respect to Transformer models that need to be trained from scratch.

Our approach is evaluated on a multi-year dataset constructed by combining hourly LST fields from the Copernicus Land Monitoring Service (CLMS) and near-surface temperature data from the ERA5 reanalysis. The study focuses on a geographically and climatically diverse area around Florence, Italy. We design a series of experiments to compare our method against a numerical baseline model, an encoder-decoder Transformer trained directly on raw temperature values instead of visual inputs, as proposed in previous work [7].

By analyzing the effectiveness of vision-based modeling in this context, we aim to address several research questions: (1) Can visual encoding of temperature sequences improve forecast accuracy over purely numerical deep models? (2) How well do TimeSformer-based architectures capture spatiotemporal patterns in environmental data? (3) What is the role of pre-trained visual features on the final accuracy?

The contributions of this paper are as follows:

- We introduce a novel framework that combines heatmap image encoding of LST data with a TimeSformer-based Transformer model for autoregressive temperature forecasting.
- We provide a comparative analysis against deep numerical sequence models on identical data splits, quantifying the advantages of vision-based learning.

The rest of the paper is organized as follows: Sect. 2 reviews relevant literature on LST forecasting, Deep Learning approaches, and vision-based environmental models. Section 3 presents the proposed methodology, including dataset preprocessing, model architecture, and visual encoding strategies. Section 4 describes the experimental setup and discusses the the results with a comparison of the two modeling paradigms. Finally, Sect. 5 concludes the paper and outlines future research directions.

2 Related Work

LST forecasting has traditionally relied on numerical weather prediction models that solve physical equations to predict future states [1]. While NWP provides the backbone for operational weather and climate forecasts, these physics-based models can be computationally intensive and are limited by their resolution [22]. To complement such models, data-driven approaches have emerged for LST and near-surface air temperature prediction [19]. Early studies employed statistical models (e.g. regression, time series) and classical machine learning, but recent work increasingly focuses on Deep Learning to capture the complex spatiotemporal patterns of surface temperature dynamics. For example, An and Kim [2] demonstrated that convolutional neural networks (CNNs) trained on climate simulation data significantly improved the skill of seasonal U.S. surface temperature forecasts, outperforming both linear regression baselines and dynamical climate models for lead times up to 6 months. Recurrent neural networks (RNNs) and their variants have shown particular success in LST forecasting tasks by modeling temporal dependencies in climate data [21]. Long Short-Term Memory (LSTM) networks are especially popular for climate time series because they address the long-term dependency problem and can capture nonlinear relationships in sequential data [14]. Jyostina et al. proposed a hybrid data-driven framework for daily LST forecasting that combined an LSTM with empirical mode decomposition to better capture multi-scale temporal patterns [12]. Kartal and Sekertekin compared a multilayer perceptron, a standard LSTM, and a convolutional LSTM (ConvLSTM) for one-step-ahead LST prediction using MODIS satellite data [13]. In their setup, each day's LST is treated as a 2D image (map) and fed into a ConvLSTM to model both the temporal sequence and the spatial neighborhood influences. Similarly, Maddu et al. [16] developed a hybrid LSTMBiLSTM model to predict daily surface temperature in several Indian coastal cities, using meteorological variables (wind, humidity, etc.) as inputs. Their bidirectional recurrent model captured past and future context, attaining high predictive skill in those urban time series.

The success of Transformers in natural language and vision has encouraged their adoption in geospatial analytics. Vision Transformers bypass convolution by splitting images into patches and using self-attention to model global relationships. In remote sensing, ViT models and their variants have been applied to tasks ranging from land-use classification to change detection [23]. Beyond static imagery, spatiotemporal transformer models are gaining traction for environmental data: researchers have explored video-transformer architectures like ViViT, TimeSformer, and VideoMAE for processing sequences of Earth observations [11]. These models use attention mechanisms to capture both spatial structures in each frame and temporal dependencies across frames, an appealing capability for climate and remote sensing applications that deal with image time series. For example, a Temporal ViT model by Heidarianbaei et al. was designed to extract features from Satellite Image Time Series, enabling improved multisensor land cover classification by concurrently learning spatial and temporal patterns [10]. Gao et al. introduced Earthformer, a spacetime transformer with "cuboid" attention, to predict Earth system variables. Earthformer achieved state-of-the-art results on benchmarks like precipitation nowcasting (treating radar maps as video frames) and even forecasting the El Niño Southern Oscillation index [9]. Another milestone is the Pangu-Weather model, which employs a transformer-based neural network to perform global medium-range weather forecasting. Pangu-Weather was able to predict 10-day future weather at 0.25° resolution with accuracy rivaling the ECMWF numerical model, demonstrating the potential of pure AI models in operational forecasting [5]. Likewise, a recent transformer model [15] for severe weather nowcasting by Küçük et al. uses geostationary satellite images as input and predicts future high-resolution radar reflectivity fields up to 2 h ahead. This model leverages the transformer's ability to integrate information across multiple image channels (IR, lightning etc.) to infer precipitation evolution, and showed robust performance for rapidly developing storms. These examples illustrate how attention-based visual models are being tailored to environmental data, from classifying static remote sensing scenes to predicting the evolution of complex geophysical fields. The introduction of large-scale pretraining is another trend: for instance, Qin et al. propose TiMo, a foundation model for satellite image time series that uses a hierarchical vision transformer with masked autoencoding for pretraining [18]. Such approaches, inspired by VideoMAE and other self-supervised transformers, aim to learn general spatiotemporal representations that can be fine-tuned for various Earth monitoring tasks.

A key step in applying computer vision models to environmental forecasting is to represent the geophysical data in image or video format. Many environmental variables naturally lend themselves to grid-based image representations: for example, a land surface temperature field over a region can be viewed as a grayscale image or a color-coded heatmap, where pixel intensity or color corresponds to temperature. For instance, for precipitation nowcasting, it is common to encode radar reflectivity or satellite cloud images as 2D grids of pixel values and apply video prediction models [17]. The rationale for such visual encoding is twofold: (1) it enables the use of well-established deep vision architectures (and even pre-trained weights) on scientific data, and (2) it preserves the spatial structure of the data, allowing convolution or attention layers to detect meaningful geospatial features (fronts, clusters, urban hotspots) automatically.

In parallel, Transformer-based models have revolutionized the modeling of sequential and spatiotemporal data across a wide range of domains. TimeSformer in particular introduces a factorized attention mechanism that separately handles spatial and temporal dimensions, allowing it to scale to longer video sequences while retaining interpretability [4].

Our work is positioned at the intersection of these developments. It bridges heatmap-based encoding of satellite-derived LST grids and vision transformer modeling for spatiotemporal forecasting, extending the autoregressive paradigm of previous numerical studies. Specifically, we employ a TimeSformer model that applies factorized attention across space and time, effectively capturing temporal dependencies across frames as well as spatial temperature patterns within each frame. Unlike traditional models that predict scalar values, we leverage full visual frames, facilitating both pixel-wise interpretation and a more accurate representation of localized temperature anomalies over time.

3   Methodology

3.1   Dataset and Preprocessing

The datasets used in this study are the Hourly Land Surface Temperature Global (Version 2.0) product from the Copernicus Land Monitoring Service (CLMS)[1] and the ERA5 hourly reanalysis dataset[2] from the Copernicus Climate Data Store. These datasets are employed in two separate experimental settings to evaluate the performance of our approach on different types of temperature measurements.

The CLMS product provides hourly LST observations at approximately 5 km spatial resolution, derived from geostationary satellite sensors. The ERA5 dataset provides hourly 2-meter air temperature at a coarser spatial resolution of approximately 31 km ($0.25° \times 0.25°$), generated through data assimilation and numerical weather prediction modeling.

From these datasets, we extracted a 5×5 pixel spatial grid covering a 25km $\times$ 25km area and 131km $\times$ 131km area respectively of the region of Florence and its surroundings.

Data from both the datasets has been converted to Celsius degrees and normalized between 0 and 1 range for both the baseline Transformer model and the vision-based model. The time range considered spans from January 19, 2021, to the end of 2024, yielding over 30,000 hourly frames. For evaluation, one month from each year was selected (July 2021, October 2022, January 2023, and April 2024) covering all four seasons to capture high variability and extreme temperatures.

In order to keep a fair comparison with the experiment conducted in the baseline Transformer model [7], data has been divided and processed with the same procedure. In particular, each training sample consists of a 10-day window: the first 7 d (168 h) are used as input, while the subsequent 3 d (72 h) form the prediction target. Missing values in CLMS were handled via linear interpolation, as higher-order methods produced artifacts, while ERA5 data, already gap-free by design, required no linear interpolation. However, as the employed Timesformer pre-trained model requires a fixed number of 64 frames as input, we truncated the input data in order to feed just the last 64 heatmaps out of the 168 full sequence.

To enable the processing by TimeSformer vision model, each LST grid was converted to a heatmap image using JET colormap through `applyColorMap()` OpenCV's function. This step transforms scalar temperature values into color-coded representations, preserving spatial gradients. Two different examples of heatmaps obtained for the CLMS are shown in Fig. 1.

---

[1] https://land.copernicus.eu/en/products/temperature-and-reflectance/hourly-land-surface-temperature-global-v2-0-5km.

[2] https://cds.climate.copernicus.eu/datasets/reanalysis-era5-single-levels?tab=overview.

(a) Heatmap of LST field (morning, summer).   (b) Heatmap of LST field (morning, winter).

Fig. 1. Examples of heatmap representations of LST fields generated from the CLMS data using OpenCV. Colors tending toward red correspond to higher temperatures. (Color figure online)

3.2 Proposed Model Architecture

Our model is designed to forecast land surface temperature (LST) over a 72-hour horizon based on the preceding 64 h of satellite-derived temperature fields, encoded as color heatmaps. The architecture follows a video prediction pipeline centered on the TimeSformer model, a spatiotemporal vision transformer specifically crafted for learning from video sequences. The framework follows an encoder-decoder model and is composed of three main components: image tokenization and embedding, the TimeSformer backbone encoder, and an autoregressive Transformer decoder for instant-by-instant forecasting.

Each hourly temperature field is first transformed into a 2D heatmap image using the OpenCV's `applyColorMap()` function and the JET colormap, and resized to the pre-trained resolution of 224×224 pixels. The resulting RGB image is then divided into fixed-size, non-overlapping square patches (16×16 pixels). Each patch is flattened and projected into a vector space via a linear embedding layer. Positional encodings are added to retain spatial information lost during flattening. The tokenized sequence representing a single image thus becomes the input to the TimeSformer.

TimeSformer [4] is a video-based Vision Transformer that factorizes attention into spatial and temporal dimensions. This factorization makes the model computationally efficient and more interpretable for video-like environmental data. Specifically, it applies:

– *Spatial Attention:* for each frame independently, attention is computed across all patches, allowing the model to focus on significant temperature structures (e.g., hotspots, gradients).

– *Temporal Attention:* once spatial patterns are extracted, the model computes attention across corresponding patches across the entire input sequence (i.e., the same spatial position across time), enabling it to model temporal evolution at each location.

This dual attention design enables TimeSformer to capture complex dependencies both within each heatmap (spatial coherence) and across heatmaps (temporal progression). Unlike traditional CNN-RNN hybrids, this approach avoids inductive biases about locality and enables long-range modeling in both space and time.

Once the 64-hour input sequence is encoded via TimeSformer, an autoregressive Transformer decoder is used to generate the next 72 frames. At each prediction step $t+i$, the decoder receives as input the previously predicted value $\hat{x}_{t+i-1}$ and the encoded context, producing the next forecast $\hat{x}_{t+i}$. This setup preserves temporal causality and enables the decoder to iteratively refine its predictions based on previously generated outputs, improving continuity and realism in the generated temperature sequences. It has to be noted that we used the same decoder module which has been employed by the baseline Transformer method, which incorporates a straightforward Transformer model as encoder.

Training Strategy. The model is trained to minimize the Mean Absolute Error (L1 loss). Empirically, L1 loss was found to produce sharper, more accurate results than RMSE, especially in preserving the structure of localized anomalies. While the baseline model has been trained for 30 epochs, due to computational power constraints the vision-based model has been trained for 10 epochs. The whole experimental setup has been written in Python while using Pytorch framework, and the code is made available online[3].

An illustration of the overall pipeline is provided in Fig. 2.

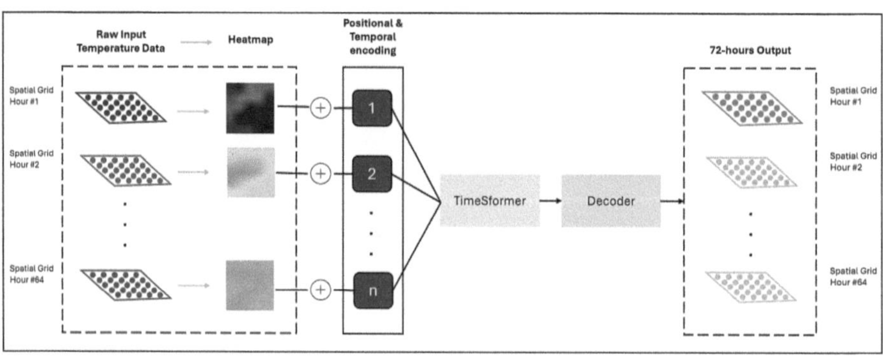

Fig. 2. Overview of the proposed method: LST grids are converted to heatmaps and processed temporally using TimeSformer before generating autoregressive predictions.

---

[3] https://github.com/engharat/tempformer_forecast_2025.

The model is trained using a combination of L1 (mean absolute error) and perceptual losses. Empirically, L1 loss provided better generalization and robustness against outliers, achieving a lower MAE compared to L2 (RMSE-based) loss. Training was performed using Adam optimizer with early stopping to prevent overfitting. All images were normalized to the [0, 1] range before encoding.

This pipeline enables the transformation of satellite time series into an image-sequence learning task, leveraging the power of vision-based models for environmental forecasting.

4 Experiments and Results

Both the baseline and vision-based models were trained independently on the CLMS and ERA5 datasets using the same data splits and preprocessing pipeline described in Sect. 3. Training was performed using the Adam optimizer with an initial learning rate of 1×10^{-4} and a batch size of 1 for the vision model and 32 for the baseline one. The baseline model has been trained for 30 epochs, while the vision model has been fine-tuned for 10 epochs due to the high computational requirements. The models were trained on a single NVIDIA 4090 GPU.

To ensure a fair comparison, the same input sequence length (64 frames) and forecasting horizon (72 h) were maintained for both models. Input images were resized to 224×224 pixels and normalized to the [0,1] range. The loss function for both models was the Mean Absolute Error (MAE), which showed improved sharpness and structure preservation compared to RMSE in preliminary tests. We experimented with added noise on the input signals as well as data augmentation techniques, without measuring any increase of performance.

Table 1 reports the Mean Absolute Error (MAE) scores obtained on both CLMS and ERA5 datasets.

Table 1. MAE comparison between baseline and proposed model.

Dataset	Baseline MAE	Vision Model MAE
CLMS	0.758	**0.741**
ERA5	**0.408**	0.419

The proposed TimeSformer-based vision model achieved a lower MAE than the numerical Transformer baseline on the CLMS dataset, with a difference of error equal to 0.017. This indicates that the spatial information encoded visually in the heatmaps and the pre-trained visual features have an impact to model accuracy.

Conversely, the baseline model outperformed the vision model on the ERA5 dataset, where spatial resolution is coarser and benefits from pre-existing numerical consistency. Here, the vision model yielded 0.011 bigger error than the baseline model, suggesting that visual modeling is more effective when high-resolution spatial structure is present, while lower-resolution, not interpolated data may benefit more from raw signals approaches.

These results support the hypothesis that the usefulness of vision-based modeling is context-dependent, and its advantages are most prominent when the input data contains spatial heterogeneity that can be captured by attention mechanisms across patches.

While the visual model demonstrated a measurable improvement on the CLMS dataset, several limitations and directions for enhancement emerge from the analysis. First, further investigation is needed to assess the role of missing or incomplete values in the input grids, as these may affect the quality and consistency of spatial patterns extracted by the vision encoder. Similarly, the impact of training hyperparameters, especially the number of epochs, should be systematically evaluated, as longer training may improve convergence, particularly for models initialized with pre-trained weights. For this reason, we cannot rule out that training the vision model (which is currently trained for 10 epochs) up to 30 epochs could produce further increase of performances.

Moreover, we hypothesize that the vision-based architecture could yield stronger results when applied to broader geographic regions. The high-resolution visual features learned through pretraining may better capture spatial heterogeneity in larger spatial contexts than in small patches. Exploring the application of the TimeSformer to larger domains may therefore amplify its advantages over traditional sequence models.

Architecturally, the current framework could benefit from more efficient or hierarchical transformer backbones, such as the Swin Transformer, which offer better scalability and inductive biases for structured visual inputs. Additionally, ensemble strategies may help improve robustness, especially in conditions with high temporal volatility or during extreme temperature events.At the same time, the role of the colormap function warrants further investigation. Although the JET colormap was chosen due to its widespread use, it cannot be ruled out that alternative mappings may yield higher accuracy.

A promising research strategy could be the integration of visual and numerical streams into hybrid or multimodal transformer architectures. Such approaches could leverage the complementary strengths of raw temporal signals and visual encodings. Techniques such as wavelet-based decomposition or attention fusion may further enhance the model's capacity to localize transient anomalies and long-term trends. Finally, increasing the spatial resolution of the prediction grid and extending validation to additional regions and seasonal conditions will help verify generalizability and uncover potential limitations of the current formulation.

5 Conclusion

This work presented a comparative study between two Transformer-based approaches for land surface temperature forecasting: an autoregressive Transformer operating on raw temperature values, and a vision-based model that leverages spatially encoded LST heatmaps and a TimeSformer encoder. By conducting parallel experiments on two distinct datasets-CLMS and ERA5-we aimed to evaluate the effectiveness of visual versus signal modeling paradigms for environmental forecasting tasks.

Our findings indicate that the vision-based approach provides better performance on high-resolution satellite data from CLMS, demonstrating its ability to capture spatial structures and local anomalies through attention-based visual representations. In contrast, the numerical Transformer performs better on the ERA5 reanalysis dataset, likely due to its coarser resolution and smoother spatiotemporal dynamics, which favor direct modeling of raw values.

These results underscore how the characteristics of the input data, such as resolution, noise, and spatial coherence, play a crucial role in determining the most appropriate modeling strategy. Vision-based models appear particularly well suited for high-resolution, spatially detailed datasets, while numerical models remain competitive on lower-resolution or more consistent data.

Several directions for future improvement emerge from this study. Enhancing the visual encoding process, experimenting with lightweight or hierarchical vision backbones, and conducting extended evaluations over larger spatial areas could improve both accuracy and efficiency. Hybrid architectures that combine raw and visual inputs, as well as ensemble methods, may further boost robustness and generalization. Additionally, attention interpretability tools could provide valuable insights into the model's decision-making process.

References

1. Al-Yahyai, S., Charabi, Y., Gastli, A.: Review of the use of numerical weather prediction (NWP) models for wind energy assessment. Renew. Sustain. Energy Rev. **14**(9), 3192–3198 (2010)
2. An, Y., Kim, H.: Improving the predictability of the us seasonal surface temperature with convolutional neural networks trained on cesm2 lens. J. Geophy. Res. Atmos. **129**(15), e2024JD040961 (2024)
3. Arnab, A., Dehghani, M., Heigold, G., Sun, C., Lučić, M., Schmid, C.: Vivit: a video vision transformer. In: Proceedings of the IEEE/CVF international conference on computer vision, pp. 6836–6846 (2021)
4. Bertasius, G., Wang, H., Torresani, L.: Is space-time attention all you need for video understanding? In: ICML. vol. 2, p. 4 (2021)
5. Bi, K., Xie, L., Zhang, H., Chen, X., Gu, X., Tian, Q.: Pangu-weather: a 3d high-resolution model for fast and accurate global weather forecast. arXiv preprint arXiv:2211.02556 (2022)
6. Conforti, P.M., Russo, P., Di Ciaccio, F.: Advancing sea surface temperature forecasting with deep learning techniques on copernicus data: an application in the gulf of trieste. In: 2024 IEEE International Workshop on Metrology for the Sea; Learning to Measure Sea Health Parameters (MetroSea), pp. 83–88. IEEE (2024)
7. Di Ciaccio, F., Russo, P., EI, P., Angelini, R., Tucci, G.: Towards sustainable heritage conservation: Transformer-based temperature forecasting in the city of florence. International Archives of the Photogrammetry, Remote Sensing and Spatial Information Science (2025)

8. Dosovitskiy, A., et al.: An image is worth 16×16 words: transformers for image recognition at scale. arXiv preprint arXiv:2010.11929 (2020)

9. Gao, Z., et al.: Earthformer: exploring space-time transformers for earth system forecasting. Adv. Neural. Inf. Process. Syst. **35**, 25390–25403 (2022)

10. Heidarianbaei, M., Kanyamahanga, H., Dorozynski, M.: Temporal vit-u-net tandem model: enhancing multi-sensor land cover classification through transformer-based utilization of satellite image time series. ISPRS Annals Photogr. Remote Sens. Spatial Inf. Sci. **10**, 169–177 (2024)

11. Jiao, L., et al.: Transformer meets remote sensing video detection and tracking: a comprehensive survey. IEEE J. Select. Topics Appl. Earth Observ. Remote Sens. **16**, 1–45 (2023)

12. Jyostna, B., et al.: Multiscale rainfall forecasting using a hybrid ensemble empirical mode decomposition and LSTM model. Model. Earth Syst. Environ. **11**(2), 1–15 (2025)

13. Kartal, S., Sekertekin, A.: Prediction of modis land surface temperature using new hybrid models based on spatial interpolation techniques and deep learning models. Environ. Sci. Pollut. Res. **29**(44), 67115–67134 (2022)

14. Kim, K., Kim, D.K., Noh, J., Kim, M.: Stable forecasting of environmental time series via long short term memory recurrent neural network. IEEE Access **6**, 75216–75228 (2018)

15. Küçük, Ç., Giannakos, A., Schneider, S., Jann, A.: Transformer-based nowcasting of radar composites from satellite images for severe weather. Artif. Intell. Earth Syst. **3**(2), e230041 (2024)

16. Maddu, R., Vanga, A.R., Sajja, J.K., Basha, G., Shaik, R.: Prediction of land surface temperature of major coastal cities of india using bidirectional LSTM neural networks. J. Water and Climate Change **12**(8), 3801–3819 (2021)

17. Prudden, R., et al.: A review of radar-based nowcasting of precipitation and applicable machine learning techniques. arXiv preprint arXiv:2005.04988 (2020)

18. Qin, X., et al.: Timo: spatiotemporal foundation model for satellite image time series. arXiv preprint arXiv:2505.08723 (2025)

19. Reichstein, M., et al.: Deep learning and process understanding for data-driven earth system science. Nature **566**(7743), 195–204 (2019)

20. Singh, N., Singh, S., Mall, R.: Urban ecology and human health: implications of urban heat island, air pollution and climate change nexus. In: Urban ecology, pp. 317–334. Elsevier (2020)

21. Suleman, M.A.R., Shridevi, S.: Short-term weather forecasting using spatial feature attention based LSTM model. IEEE Access **10**, 82456–82468 (2022)

22. Tang, Y., Lean, H.W., Bornemann, J.: The benefits of the met office variable resolution NWP model for forecasting convection. Meteorol. Appl. **20**(4), 417–426 (2013)

23. Yao, J., Zhang, B., Li, C., Hong, D., Chanussot, J.: Extended vision transformer (exvit) for land use and land cover classification: a multimodal deep learning framework. IEEE Trans. Geosci. Remote Sens. **61**, 1–15 (2023)

Workshop on Computer Vision and Generative Models for Medical Imaging (CVGMMI 2025)

Workshop

Workshop on Computer Vision and Generative Models for Medical Imaging (CVGMMI 2025)

In conjunction with ICIAP 2025—Rome, Italy, September 15–19, 2025

Workshop Organization

Chairs and Organizers

Daniele Ravì (Chair)	University of Messina, Italy and University College London, UK
Massimo Villari	University of Messina, Italy
Lemuel Puglisi	University of Catania, Italy and University College London, UK and Queen Square Analytics, UK
Giulio Minore	University College London, UK

Program Committee

Daniele Ravì	University of Messina, Italy
Massimo Villari	University of Messina, Italy
Lemuel Puglisi	University of Catania, Italy
Giulio Minore	University College London, UK
Giovanni Lonia	University of Messina, Italy
Alec Sargood	University College London, UK
Mattia Litrico	University of Catania, Italy
Francesco Guarnera	University of Catania, Italy

Supporting Cardiac MRI Generation with High Fidelity Digital Twin

Giulio Del Corso[1]([✉])[iD], Claudia Caudai[1][iD], Roberto Verzicco[2,3][iD], Francesco Viola[2][iD], and Sara Colantonio[1][iD]

[1] Institute of Information Science and Technologies "A. Faedo", National Research Council of Italy, Pisa, Italy
giulio.delcorso@isti.cnr.it
[2] Gran Sasso Science Institute (GSSI), L'Aquila, Italy
[3] University of Rome - Tor Vergata, Rome, Italy

Abstract. The generation of cardiac cine-MRI for both healthy and pathological cases is a highly relevant topic for simulation, training applications, and the creation of high-quality synthetic datasets. Despite recent advances in generative techniques, cardiac MRI remains particularly challenging due to the difficulty of producing realistic anatomical contexts and time sequences that adhere to the physical constraints imposed by blood flow dynamics. However, the emergence of high-fidelity computational models of the human heart paves the way for generating image sequences conditioned on the shapes and positions of cardiac chambers, as defined by the underlying physiological model. In this proof of concept, we present an artificial neural network derived from medical image deformation models. The network is trained on 360 cine-MRI volumes from the public M&M dataset and is designed to combine existing MRI sequences with conditional information generated by a high-fidelity cardiac digital twin, enabling the generation of both short-axis 3D MRI sequences.

Keywords: Cardiac Modeling · Video to Video Generation · Digital Twins · Deep Learning · Physical-Informed Generative AI

1 Introduction

Generative artificial intelligence (GenAI) models are emerging as a promising solution to major challenges in healthcare, including data scarcity, privacy concerns, and bias [15]. Generative approaches can generate synthetic data that closely mimic real-world datasets. This allows researchers and practitioners to overcome the challenges associated with accessing and using sensitive, high-quality, curated patient data. Additionally, they can promote fairness and equity by addressing data availability gaps for underrepresented patient subgroups or disease conditions and by providing high-resolution data in low-resource settings [23]. Among various state-of-the-art generative models—including Variational Autoencoders, Diffusion Models, and Multimodal Foundational Models—adversarial approaches have shown a notable ability to enhance realism [13].

E. Rodolà et al. (Eds.): ICIAP 2025 Workshops, LNCS 16169, pp. 381–392, 2026.
https://doi.org/10.1007/978-3-032-11317-7_32

These methods have been employed to synthesize medical imaging datasets across multiple anatomical regions, including endoscopy [18], brain MRI [25], and cardiac ultrasound [7].

Among the most advanced non-invasive techniques for evaluating cardiac function, cine-cardiac magnetic resonance imaging (MRI) is widely regarded as the gold standard. However, the application of GenAI to the synthesis of 3D cine-MRI images remains relatively unexplored, primarily due to the complexity of cardiac dynamics—especially in pathological cases. Some generation are derived directly from knowledge distillation from a large language model, incorporating it into a pre-trained latent diffusion mode equipped with a spatio-temporal attention mechanism [16]. The difficulty of generating realistic cine-MRI images can be partially mitigated by incorporating conditional information—for example, guiding image synthesis through a multi-tissue semantic segmentation module. This approach helps reinforce the generation of semantically consistent and realistic images by leveraging tissue labels of anatomical structures surrounding the heart [1]. Alternative approaches include the integration of deformable image registration strategies, such as those leveraging a variational recurrent artificial neural network (ANN) to achieve spatio-temporal registration, as implemented in DragNet [27].

However, the challenge of generating cine-MRI sequences that are perceived as realistic by medical experts —particularly in pathological cases— has yet to be fully addressed. In fact, a recent survey on advanced generative methods found that the majority (80.8%) of generated images were rated poorly by healthcare professionals, primarily due to anatomically incorrect artifacts, while 78.1% of the images were deemed unsuitable even for use in medical education [24]. While motion artifacts can be addressed using a posteriori adversarial methods [28], ensuring anatomical coherence throughout the sequence remains an exceptionally challenging task for conventional generative AI approaches.

In this paper, we present a *proof-of-concept* demonstrating how a high-fidelity cardiac numerical model (i.e., a *digital twin*) can support the generation of realistic cine-MRI sequences from small datasets. Specifically, a version of a partially annotated public database (M&M [17]) is adapted to the task at hand, and a segmentation model based on MRI deformation is introduced to infer the missing annotations. A high-fidelity multi-chamber cardiac numerical model [26] is then introduced to generate sequences of segmentation masks that characterize the motion and compression of the ventricular chambers and myocardium in both healthy and pathological cases. Finally, a second VoxelMorph-based deep model is adapted to learn smooth deformations between segmentation masks of consecutive frames, which are then applied to generate volumetric and cine-MRI sequences consistent with the underlying physics.

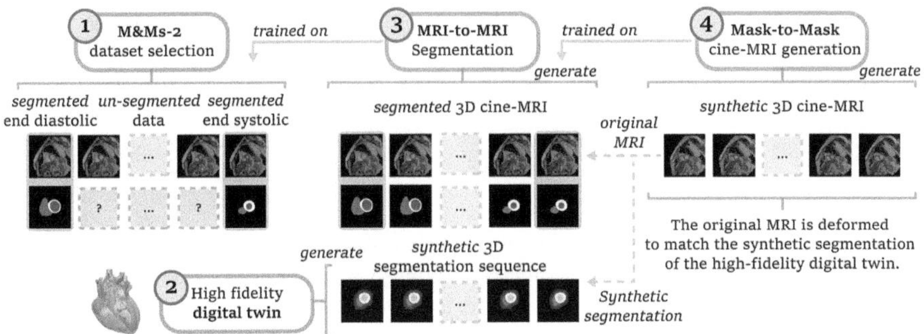

Fig. 1. Roadmap of the presented work: (1) Selection of the available dataset and identification of segmented ED/ES volumes; (2) Definition of a digital twin to generate synthetic segmentation sequences; (3) Training of an MRI-to-MRI deformation model and segmentation of the dataset; (4) Training of a Mask-to-Mask deformation model on the labelled dataset and integration with the digital twin to produce synthetic MRI sequences.

2 Methods

The proposed method demonstrates how a state-of-the-art cardiac digital twin can be used—together with a dataset of real cine-MRI sequences and a deformation model—to generate synthetic cine-MRI sequences. The digital twin simulates the heart's compression and decompression dynamics over a specified time frame and under defined physiological conditions. A sequence of segmentation masks is then extracted from the simulated heart using a ray casting procedure. These segmentation masks are transformed into a synthetic cine-MRI by: (1) sampling a real cine-MRI sequence from the dataset, (2) deforming the segmentation of the real cine-MRI to match the deformation of the synthetic sequence at each time step, and (3) applying the resulting deformation field to the original cine-MRI to generate a realistic, synthetic cine-MRI sequence.

As illustrated in Fig. 1, the presented strategy is based on: **(1)** cleaning and augmenting a dataset of cardiac cine-MRI (Sect. 2.1); **(2)** introducing a cardiac digital twin to generate physically reliable mask sequences (Sect. 2.2); **(3)** formally introducing a deformation model along with an appropriate training scheme (Sect. 2.3); **(4)** training a deformation MRI-to-MRI model on that dataset to define a coherent sequence of 3D masks (Sect. 2.4); and **(5)** training a mask-to-mask deformation model on the generated sequences of 3D masks (Sect. 2.5).

2.1 Cardiac Video Dataset and Data Augmentation

While several datasets containing MRI volumes exist, most are not suitable for our specific application, as they lack ventricular masks or horizontal (short-axis)

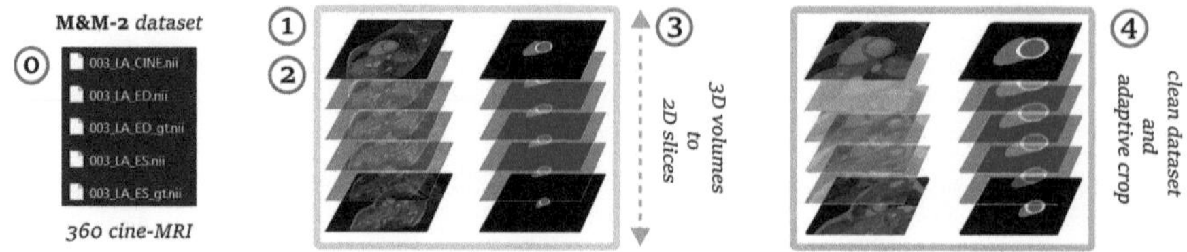

Fig. 2. The training and testing dataset is derived from the M&M challenge and comprises 360 cine-MRI scans. The sequences undergo a five-step procedure to clean the data, crop the images, and generate paired ESED MRIs and corresponding segmentation masks.

views of the heart. Among them, the UK Biobank (UKBB) provides a large-scale dataset of annotated images—over 4,875 subjects and 93,500 pixelwise annotated images [21]. However, it requires specific authorization for access and is therefore not as freely available as the other datasets presented here.

The datasets considered for this work include the *Cardiac MRI Dataset* with 33 patients [2], 45 cine-MRI scans from the *Cardiac Atlas Project* [20], the 389 cine-MRI of OCMR dataset [6], and the ACDC-MICCAI 2017 challenge dataset with 150 patients [5]. However, we decided to focus on the more extensive and freely available *M&M2* dataset [17], which includes 360 cine-MRI studies with End Diastolic and End Systolic segmentations, as well as both long-axis and short-axis views. The ACDC dataset is a subset of M&M2 and was therefore excluded to avoid duplication. The OCMR dataset, developed for image reconstruction, contains only 53 fully sampled volumes, while the other datasets differ in acquisition procedures and image types, making proper integration challenging.

As illustrated in Fig. 2, using **(0)** the 360 videos from the *M&Ms-2* challenge [17], we applied a rigorous data augmentation procedure to maximize the available information for training and testing our method. In particular, **(1)** we isolated the *End Diastolic* (ED) and *End Systolic* (ES) phases—3D volumes for which corresponding 3D masks were annotated by medical experts. Therefore, for each of the 360 original patients, we derived two annotated volumes (ES and ED), resulting in a total of 720 samples [augmentation factor: ×2]. To train a deformation model, **(2)** both ES → ED and ED → ES transitions can be used, resulting in an additional twofold data augmentation [augmentation factor×2, 1440 3D samples]. The volumes are characterized by varying scales and resolutions; therefore, to standardize the datasets, adaptive cropping along the xy-plane and rescaling to 120×120 pixels using bilinear interpolation are applied. Formally, the adaptive crop must ensure that ES and ED volumes are cropped identically for each slice of the 3D volume. To achieve this, we defined the cropping region as the minimal bounding box that includes the masks of both ES and ED 2D slices, extended by 5 pixels in each direction. To train a 3D deep ANN, the 1,440 samples are typically insufficient. Therefore, **(3)** we split each volume into its paired ES-ED 2D slices, significantly increasing the dataset

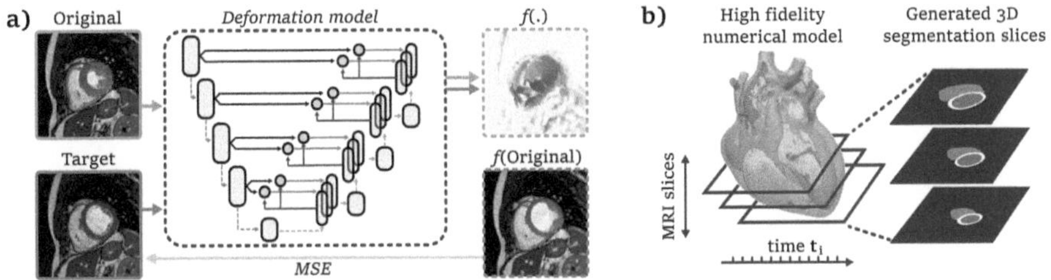

Fig. 3. a) Schematic of the deformation model. The original and target images are provided to the deformation model, which outputs a deformation map $f(\cdot)$. This deformation map is then applied to the original image. The quality of the deformation is evaluated, among other metrics, using the mean squared error (MSE) between the target image and the deformed original: MSE (Target, f (Original)). **b)** High-fidelity cardiac numerical model capable of generating 3D segmentation slices for each time point t_i.

size from 1,440 3D pairs to 16,256 2D pairs (+1028%). The dataset must be cleaned **(4)** to ensure each pair is suitable for training and testing. In particular, each slice pair must include labeled masks. This is necessary because the labeling process focuses on the left and right ventricles, while the MRI volumes also contain atria and additional slices outside the region of interest. In addition, in some cases the number of annotated slices in the ED and ES phases does not match due to differences in the annotation process, and these pairs must also be removed. This leads to a reduction in the available data to a total of *9,400 samples* (−42%), which represents the final size of our dataset. The aforementioned strategy can potentially lead to data leakage, as 2D slices are taken from the same patient and, moreover, the ES → ED and ED → ES pairs from the same patient are strongly correlated. Therefore, from this point onward, every split between training, validation, and test sets implicitly assumes patient-level stratification, ensuring that all slices and pairs from a given patient belong to the same subset.

Proper *on-the-fly data augmentation* that preserves the physical properties underlying the generated images can significantly improve the performance of the trained model [9]. Therefore, a synthetic data augmentation strategy [factor ×48] is applied each time a batch is sampled for training. Each pair of 2D images undergoes random transformations, including horizontal flipping [×2], rotation by 0◦/90◦/180◦/270◦ [×4], white noise addition at 0% or 5% [×2], and random zooming that preserves full mask visibility in both images [×3]. This results in a total of 451,200 augmented 2D pairs to be used for training and testing purposes.

2.2 GPU Accelerated Cardiac Digital Twin

The proposed generation of cine-MRI sequences (Sect. 2.5) relies on the ability to produce realistic temporal sequences of ventricular chamber and myocardium positions, along with their corresponding compression and decompression dynamics. The drastic increase in computational power has led in recent years to a proliferation of cardiac electrophysiology models and their coupling with structural solvers [11]. As shown in Fig. 3 b), a state-of-the-art model of cardiac fluid dynamics was introduced by Viola et al. [26]. The model, equipped with GPU acceleration, integrates realistic electrophysiology with a comprehensive fluid dynamics framework based on the immersed boundary method for solving the NavierStokes equations. It enables the simulation of fine-scale fluid behavior and the evaluation of ventricular wall deformations, including those associated with pathological conditions. Once the simulation parameters are defined—such as heart size and morphology, patient sex, heart rate, and the presence of pathologies (including, but not limited to, arrhythmias, myocardial infarction, and bundle branch blocks)—the cardiac digital twin generates a physically realistic sequence of deformations of the atrial and ventricular chambers. By selecting fixed time points and defining a sequence of secant planes through the ventricular chambers—corresponding to the slice planes of an MRI scan— it is possible, via a ray casting procedure [22], to determine the affiliation of each point in the slice to a specific chamber (LV/RV) or to the surrounding myocardium. This procedure enables the generation of a synthetic sequence of patient-specific cardiac segmentations, which can be used to produce individual 2D/3D MRI frames/volumes or even full 3D cine-MRI cardiac sequences.

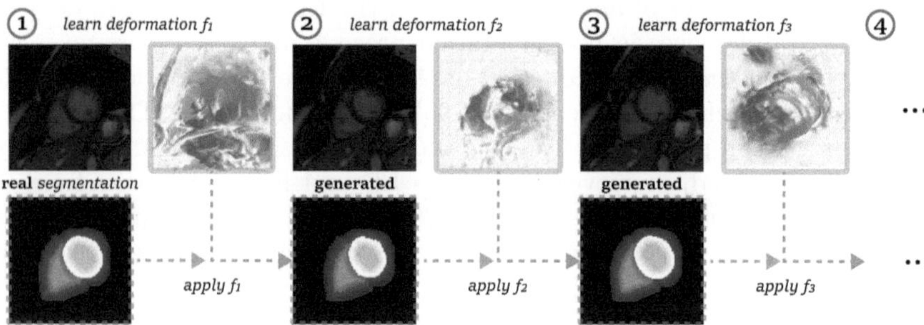

Fig. 4. Segmentation via deformation of the cine-MRI sequence is performed as follows. Given two consecutive 2D MRI frames, $\mathrm{MRI}_i \rightarrow \mathrm{MRI}_{i+1}$, the trained deformation model learns the deformation $f_i : \mathrm{MRI}_i \rightarrow \mathrm{MRI}_{i+1}$. This deformation $f_i(\cdot)$ is then iteratively applied to the corresponding segmentation mask according to $\mathrm{Seg}_{i+1} = f_i(\mathrm{Seg}_{i+1})$.

2.3 Deformation Model and Training Scheme

The primary artificial neural network architecture used for both segmentation (Sect. 1) and MRI generation (Sect. 2) is a deformation-based model inspired by VoxelMorph —a state-of-the-art image registration framework specifically designed for aligning medical images [3].

As illustrated in Fig. 3 a), the model takes a pair of images (an Original/Source image and a Target/Fixed one) and, using a U-Net architecture (featuring increasing layer complexity and skip connections), produces a 2D or 3D dense deformation field $f(\cdot)$ that warps the moving image to align it with the fixed one (f(Original) $\approx$ Target). During training, the standard loss function typically combines an image similarity term (e.g., MSE - mean squared error) with a regularization term that promotes smoothness in the deformation field ($\ell^2(\cdot)$). Extensions that balance these two components or compute image similarity on associated warped structures (e.g., segmentation masks) are investigated in this work. The implementation is based on the open-source PyTorch version of VoxelMorph available on GitHub (VoxelMorph), released under the Apache-2.0 License. All subsequent code was developed in native PyTorch (PyTorch).

All subsequent model training implicitly assumes stratification and a homogeneous subdivision among the training (80%), validation (10%), and test (10%) sets. Performance scores are reported only on the test set to provide unbiased estimates of generalization capability. Training is performed for a maximum of 200 epochs using an adaptive triangular learning rate schedule (warm-up from $1 \cdot e^{-3}$ to $1 \cdot e^{-2}$, plateau at ending epoch down to $1 \cdot e^{-5}$), with early stopping based on a patience of 10 epochs on the validation set. A batch size of 20 (2D slices) is used to stabilize the training process. Training is conducted with GPU acceleration on an *NVIDIA GeForce RTX 3060* (6 GB GDDR6), while other computations are performed on an *AMD Ryzen 7 5800H* (8 cores, 16 threads, 3.2 GHz) with 16 GB of RAM.

2.4 Step I: Segmentation via Deformation

The selected dataset includes only 3D segmentations of the End-Systolic and End-Diastolic phases. However, to apply a proper mask-to-mask deformation model for generating a smooth video sequence, we need the full video sequence of annotations (i.e., Left Ventricle, Right Ventricle, and Myocardium). While the application of a state-of-the-art classical U-Net-based annotator can produce high-quality segmentation masks, it is essential to ensure temporal consistency between the End-Systolic (ES) and End-Diastolic (ED) intermediate annotations. Therefore, we trained a 2D VoxelMorph with underlying U-Net shape (encoder: [32,32,32,32], decoder: [32,32,32,32,32,16]) model on the available ES/ED couples which underwent the data augmentation procedure. We modified classical reconstructiojn loss $\mathcal{L} = \ell^2$ (**field**) $+$ MSE (**reconstruction**)by adding the influence ($\lambda = 0.01$) on the regularization part ℓ^2 to identify smoother fields and by applying a focal reconstruction loss on MSE $\mathcal{F}(\cdot)$. The focal reconstruction loss allows the network to place greater emphasis on reconstructing the central part of the image, which is, by design, centered on the barycenter of the segmented ventricles. In addition, the MSE is computed on both the 2D MRI slice and the corresponding 2D segmentation mask, encouraging lower reconstruction error for deformations that correctly align the relevant ventricular components (i.e., LV chamber, RV chamber, and Myocardium). In summary, the total loss is defined as:

$$\mathcal{L} = \lambda \cdot \ell^2 \left(\textbf{field}\right) + \frac{(1 - \lambda)}{2} \cdot \left(\text{MSE}\left(\mathcal{F}\left(\textbf{MRI}\right)\right) + \text{MSE}\left(\textbf{Mask}\right)\right) \tag{1}$$

The trained model takes as input two MRI slices (original and target) along with the original segmentation mask. From the two MRI slices, a deformation map (i.e., the spatial transform f) is computed and applied to the original segmentation mask to generate the segmentation of the target image.

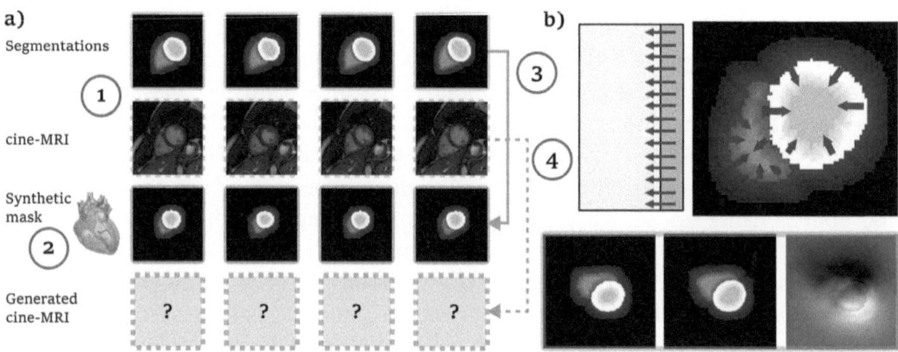

Fig. 5. a) Mask-to-mask model that adapts an existing cine-MRI sequence to a synthetic sequence of LV/RV positions; b) Segmentation smoothing to prevent MRI collapse and unrealistic deformations.

2.5   Step II: Mask to Mask/MRI Deformation

The main model aims to generate a cardiac cine-MRI sequence from an original cine-MRI sequence (together with its segmentation masks) and a synthetic mask sequence produced by a high-fidelity numerical model. As illustrated in Fig. 5 a), the model takes as input an existing 3D cine-MRI sequence with corresponding segmentation masks (**1**) and an additional synthetic segmentation mask sequence generated by the digital twin (**2**). The deformation model learns a mask-to-mask deformation between the real and synthetic segmentations, producing a sequence of deformation maps $f_i(\cdot)$ (**3**). These deformation maps are then applied to the original cine-MRI sequence to generate a realistic sequence of 3D MRI volumes whose chamber positioning and deformation match those of the synthetic masks (**4**). To train this model, we used the dataset generated in the previous step and optimized the loss defined in Eq. 1. Unlike the previous model, this one takes as input two segmentation masks rather than MRI images, while still evaluating the loss on the subsequent MRI frame. However, the original VoxelMorph learns a continuous deformation $\mathcal{F}$ field by operating on pixel-level, highly differentiated images. Therefore, applying a continuous deformation to a

Table 1. Results of different model performances evaluated on an independent set.

Model	Tested on	Training	Validation	Test
[1] *Segmentation*	MRI-to-MRI	0.0009	0.0012	0.0015
[2] *Generation/Deformation*	Mask-to-Mask	0.0013	0.0017	0.0018

binary mask —or, as in this case, to multi-class masks— can result in physically unrealistic deformations. With reference to Fig. 5 b), the deformation method has no incentive to modify pixels that share the same label in both the source and target images. Consequently, the deformation mask exhibits highly discontinuous behavior, characterized by minimal or no changes inside the chambers and myocardium, and a pronounced collapse of the regions outside the segmentation toward the edges of the target image. A simple yet effective solution to induce more realistic deformation is to modify the input masks by introducing smooth gradients within each region (LV, RV, Myocardium). This approach preserves a high contrast between individual segmentation values to maintain the integrity of the main structures, while adding edge-focused deformation to encourage a deformation vector field that yields more realistic images.

3   Numerical Examples

As shown in Table 1, four main families of experiments were conducted to quantitatively assess the effectiveness of the proposed methods. Specifically, with reference to model [1] (the segmentation-by-deformation model described in Sect. 2.4), trained using the loss defined in Eq. 1 —which incorporates both the mean squared error (MSE) on the MRI and the ability to accurately deform the underlying segmentation mask— the model achieves excellent performance on the test set, with error rates approaching 0.0015. While alternative methods —such as attention-based U-Net architectures applied directly to the second MRI— may yield higher segmentation scores numerically, the proposed approach ensures greater temporal consistency across the segmentation sequence. This consistency translates into video generation with fewer visual artifacts, as perceived by human observers.

Referring again to Table 1, the second model [2]—the mask-to-mask model designed to generate MRI sequences, as described in Sect. 2.5—was evaluated on the same unseen data. The model achieved a combined loss of 0.0018, demonstrating its ability to accurately deform the smooth segmentation mask. The ability to generate realistic MRI images solely from the deformation of the segmentation mask is illustrated in Fig. 6. Here, the original and target images —selected as end-systolic (ES) and end-diastolic (ED) frames of the same slice for clarity— are compared against f(original), which represents the original image deformed using the displacement field *derived from the segmentation masks*, rather than from the MRI images themselves.

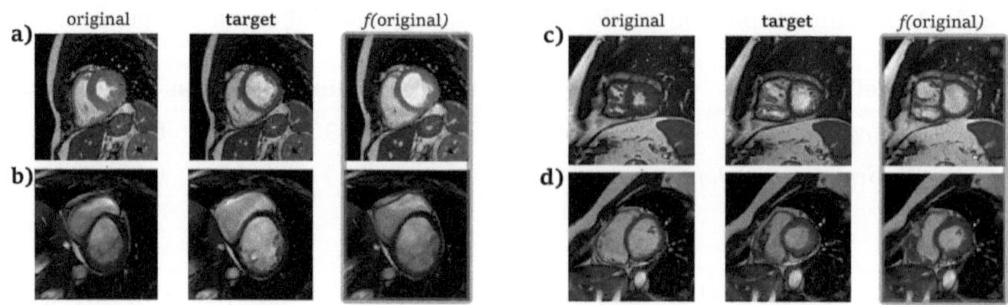

Fig. 6. Examples of deformations learned from segmentation masks. The model takes as input two (smoothed) segmentation masks and applies the learned deformation to the corresponding MRIs.

4   Conclusion

In this paper, we presented an adaptation of a state-of-the-art Deep Deformation Model [3] to generate realistic cardiac cine-MRI sequences, using conditional information derived from a high-fidelity digital twin of the human heart [26]. Despite the extremely limited amount of training data (only 360 available cine-MRI scans [17]), the conditional information provided by the digital replica is sufficiently informative to enable the generation of realistic volumes and plausible activation sequences. This *proof-of-concept* demonstrates the potential of physics-informed numerical models for generating realistic images and videos, even in data-scarce contexts that typically constrain the performance of generative AI. Potential applications therefore range from educational uses of generative models [19], to the creation of uncertainty aware synthetic datasets — crucial for developing and testing innovative methodologies [8,10]— and even to the refinement of existing measurement procedures based on generative AI, such as biventricular measurements compared with conventional k-space segmented cine images [12].

The main limitations of this work lie in the quantity and quality of the data used. In particular, the absence of intermediate segmentations between end-systole (ES) and end-diastole (ED) necessitates the use of intermediate segmentation models (see Sect. 2.4), potentially leading to error propagation during training. In addition, the need to switch from 3D (volumes) to 2D (slides) representations (see Sect. 2.1) and operate at the level of individual temporal frames limits the model's potential and introduces minor artifacts that may be perceptible to the human eye. In this context, integrating the generated videos with adversarial mechanisms —such as CycleGAN [13], Pix2Pix image-to-image translation [14], or video-to-video conditional translation [4]— could be a promising direction to enhance temporal coherence and overall visual quality. A second limitation arises when using simple deformation patterns: while the generated images exhibit realistic positioning of the ventricular chambers and myocardium, they lack corresponding texture variations. Given that many cardiac pathologies manifest through both abnormal chamber motion and altered tissue texture, this limitation restricts applicability to healthy subjects or to pathologies that do not significantly affect MRI texture, such as conduction disorders involving the His bundle or Purkinje fibers. Future work will therefore focus on the potential to locally synthesize disease-specific texture characteristics, such as those associated with tissue necrosis or scar formation.

References

1. Al Khalil, Y., Amirrajab, S., Lorenz, C., Weese, J., Pluim, J., Breeuwer, M.: On the usability of synthetic data for improving the robustness of deep learning-based segmentation of cardiac magnetic resonance images. Med. Image Analy. **84**, 102688 (2023)
2. Andreopoulos, A., Tsotsos, J.K.: Efficient and generalizable statistical models of shape and appearance for analysis of cardiac mri. Med. Image Analy. **12**(3), 335–357 (2008)
3. Balakrishnan, G., Zhao, A., Sabuncu, M., Guttag, J., Dalca, A.V.: Voxelmorph: a learning framework for deformable medical image registration. IEEE TMI: Trans. Med. Imaging **38**, 1788–1800 (2019)
4. Bansal, A., Ma, S., Ramanan, D., Sheikh, Y.: Recycle-GAN: unsupervised video retargeting. In: Proceedings of the European Conference On Computer Vision (ECCV), pp. 119–135 (2018)
5. Bernard, O., et al.: Deep learning techniques for automatic mri cardiac multi-structures segmentation and diagnosis: is the problem solved? IEEE Trans. Med. Imaging **37**(11), 2514–2525 (2018)
6. Chen, C., et al.: Ocmr (v1. 0)–open-access multi-coil k-space dataset for cardiovascular magnetic resonance imaging. arXiv preprint arXiv:2008.03410 (2020)
7. Chen, T., et al.: Ultrasound image-to-video synthesis via latent dynamic diffusion models. In: International Conference on Medical Image Computing and Computer-Assisted Intervention, pp. 764–774. Springer (2024)
8. Del Corso, G., Colantonio, S., Caudai, C.: Shedding light on uncertainties in machine learning: formal derivation and optimal model selection. J. Franklin Inst., 107548 (2025)
9. Del Corso, G., et al.: Facial landmark identification and data preparation can significantly improve the extraction of newborns' facial features. In: 2024 IEEE 18th International Conference on Automatic Face and Gesture Recognition (FG), pp. 1–7. IEEE (2024)
10. Del Corso, G., Volpini, F., Caudai, C., Moroni, D., Colantonio, S.: Descriptor: not-a-database of synthetic shapes benchmarking dataset (nada-synshapes). IEEE Data Descriptions (2025)
11. DeSimone, A., et al.: Segregated algorithms for the numerical simulation of cardiac electromechanics in the left human ventricle. Math. Mechanobiol. Cetraro, Italy **2018**, 81–116 (2020)
12. Ghanbari, F., et al.: Free-breathing, highly accelerated, single-beat, multisection cardiac cine MRI with generative artificial intelligence. Radiol. Cardiothoracic Imaging **7**(2), e240272 (2025)
13. Harms, J., et al.: Paired cycle-gan-based image correction for quantitative cone-beam computed tomography. Med. Phys. **46**(9), 3998–4009 (2019)
14. Henry, J., Natalie, T., Madsen, D.: Pix2pix GAN for image-to-image translation. Res. Gate Publ. 1–5 (2021)

15. Ibrahim, M., et al.: Generative ai for synthetic data across multiple medical modalities: a systematic review of recent developments and challenges. Comput. Biol. Med. **189**, 109834 (2025)
16. Liu, C., Yuan, X., Yu, Z., Wang, Y.: Texdc: text-driven disease-aware 4d cardiac cine MRI images generation. In: Proceedings of the Asian Conference on Computer Vision, pp. 3005–3021 (2024)
17. Martín-Isla, C., et al.: Deep learning segmentation of the right ventricle in cardiac mri: the m&ms challenge. IEEE J. Biomed. Heal. Inf. **27**(7), 3302–3313 (2023)
18. Osuala, R., et al.: medigan: a python library of pretrained generative models for medical image synthesis. J. Med. Imaging **10**(6), 061403–061403 (2023)
19. Preiksaitis, C., Rose, C.: Opportunities, challenges, and future directions of generative artificial intelligence in medical education: scoping review. JMIR Med. Educ. **9**, e48785 (2023)
20. Radau, P., Lu, Y., Connelly, K., Paul, G., Dick, A.J., Wright, G.A.: Evaluation framework for algorithms segmenting short axis cardiac MRI. MIDAS J. (2009)
21. Raisi-Estabragh, Z., Harvey, N.C., Neubauer, S., Petersen, S.E.: Cardiovascular magnetic resonance imaging in the uk biobank: a major international health research resource. European Heart J. Cardiovascular Imaging **22**(3), 251–258 (2021)
22. Ray, H., Pfister, H., Silver, D., Cook, T.A.: Ray casting architectures for volume visualization. IEEE Trans. Visualization Comput. Graph. **5**(3), 210–223 (1999)
23. Sizikova, E., et al.: Synthetic data in radiological imaging: current state and future outlook. BJR—Artifi.Intell. **1**(1), ubae007 (05 2024)
24. Temsah, M.H., et al.: Art or artifact: evaluating the accuracy, appeal, and educational value of ai-generated imagery in dall e 3 for illustrating congenital heart diseases. J. Med. Syst. **48**(1), 54 (2024)
25. Tudosiu, P.D., et al.: Realistic morphology-preserving generative modelling of the brain. Nat. Mach. Intell. **6**(7), 811–819 (2024)
26. Viola, F., Del Corso, G., De Paulis, R., Verzicco, R.: Gpu accelerated digital twins of the human heart open new routes for cardiovascular research. Sci. Reports **13**(1), 8230 (2023)
27. Zakeri, A., et al.: Dragnet: learning-based deformable registration for realistic cardiac mr sequence generation from a single frame. Med. Image Anal. **83**, 102678 (2023)
28. Zhang, L., et al.: Motion artifact removal in coronary ct angiography based on generative adversarial networks. European Radiol. **33**(1), 43–53 (2023)

GeMM-GAN: A Multimodal Generative Model Conditioned on Histopathology Images and Clinical Descriptions for Gene Expression Profile Generation.

Francesca Pia Panaccione[(✉)] [iD], Carlo Sgaravatti[(✉)] [iD], and Pietro Pinoli [iD]

DEIB - Dipartimento Elettronica, Informazione e Bioingegneria, Politecnico di Milano, Milan, Italy
{francescapia.panaccione,carlo.sgaravatti,pietro.pinoli}@polimi.it

Abstract. Biomedical research increasingly relies on integrating diverse data modalities, including gene expression profiles, medical images, and clinical metadata. While medical images and clinical metadata are routinely collected in clinical practice, gene expression data presents unique challenges for widespread research use, mainly due to stringent privacy regulations and costly laboratory experiments. To address these limitations, we present GeMM-GAN, a novel Generative Adversarial Network conditioned on histopathology tissue slides and clinical metadata, designed to synthesize realistic gene expression profiles. GeMM-GAN combines a Transformer Encoder for image patches with a final Cross Attention mechanism between patches and text tokens, producing a conditioning vector to guide a generative model in generating biologically coherent gene expression profiles. We evaluate our approach on the TCGA dataset and demonstrate that our framework outperforms standard generative models and generates more realistic and functionally meaningful gene expression profiles, improving by more than 11% the accuracy on downstream disease type prediction compared to current state-of-the-art generative models. Code will be available at: https://github.com/francescapia/GeMM-GAN.

Keywords: Generative AI · Deep Learning · Computer Vision · Generative Adversarial Networks · Multimodal Learning

1 Introduction

The integration of heterogeneous biomedical data, ranging from clinical records and histopathology images to gene expression profiles, has the potential to transform our understanding of disease mechanisms and enable more personalized healthcare. Yet, in practical settings, the availability of these data modalities is highly uneven. Clinical metadata and histopathology slides are routinely collected across institutions, whereas transcriptomic profiles remain limited due to

E. Rodolà et al. (Eds.): ICIAP 2025 Workshops, LNCS 16169, pp. 393–404, 2026.
https://doi.org/10.1007/978-3-032-11317-7_33

their cost, privacy implications, and lack of standardization [11]. Most medical centers can readily collect tissue slides and corresponding whole slide images (WSIs), but may lack the infrastructure and expertise to perform RNA sequencing experiments for gene expression analysis. However, gene expression analysis is often essential for guiding personalized treatment selection. For example, PAM50-guided treatment decisions in breast cancer have led to significant improvements in patient survival outcomes by enabling better risk stratification and optimal therapy selection for each tumor subtype [12].

Deep generative models offer a promising solution for addressing this data gap. Recent progress in this domain has demonstrated their ability to model complex, high-dimensional biological data and produce realistic, privacy-preserving synthetic gene expression profiles [1]. However, prior approaches to transcriptomic generation have typically relied on simplified inputs (either low-dimensional vectors or pathology images alone) failing to fully capture the rich context available in real-world clinical settings [9,18]. Other approaches [16,20], instead, model the problem as a prediction task, thus predicting the gene expression profile from histopathology images, without any generative modelling, missing the opportunity to leverage cross-modal relationships for synthetic data generation.

To overcome these limitations, we introduce **GeMM-GAN**, a Multimodal Generative Adversarial Network conditioned both on histopatology images and clinical descriptions of the patients to generate biologically coherent gene expression profiles *in silico*. To the best of our knowledge, we are the first addressing the problem of conditioning a generative model of gene expressions on both images and text. Our method builds on prior work on gene expression prediction from Whole Slide Images (WSIs), leveraging patch-based preprocessing via Otsu thresholding [16] and Transformer-based modelling of image embeddings [20]. We furthermore study how to integrate the textual modality into this architecture, by leveraging Feature-wise Linear Modulation (FiLM) [15] of the patch embeddings and by desinging a Cross-Attention mechanism to extract meaningful information from both modalities to effectively condition our generative model. We then leverage a Wasserstein Generaive Adversarial Network with Gradient Penalty (WGAN-GP) [5] to generate gene expression profiles.

We evaluate GeMM-GAN using standard metrics from the generative modelling literature, such as distributional alignment and downstream predictive performance, to demonstrate its ability to produce realistic, functionally meaningful gene expression profiles.

Our contribution can be summarized as:

- We introduce a novel multimodal framework for the generation of gene expression profiles conditioned jointly on histopathology images and clinical descriptions.
- We explore cross-modal fusion strategies (FiLM and cross-attention) for integrating textual and visual information into a shared latent representation.

– We show that our model can generate gene expression profiles that preserve biologically meaningful patterns and can support downstream predictive tasks, improving current state-of-the-art performance.

2   Related Work

Recent efforts at the intersection of computational pathology and transcriptomics have focused primarily on the prediction of gene expression values from histopathology images using deep learning. For example, Schmauch et al. [16] and Wang et al. [19] developed models to predict RNA-Seq profiles directly from WSIs. Similarly, Chlis et al. [4] predicted the expression profile of every cell in an imaging flow cytometry experiment. Although effective, these approaches are inherently discriminative, aiming to map visual features to molecular outputs. They do not address the underlying limitations of data scarcity and privacy.

Parallel research streams have explored the development of generative models for transcriptomics. For instance, Vinas et al. [18], Lacan et al. [9], and Panaccione et al. [14] evaluated GANs for generating synthetic omics data. Yet, these approaches operate within the molecular domain, without incorporating imaging or clinical modalities. Conversely, image generation from transcriptomics has been investigated by Tariq et al. [2]. using diffusion models. However, this approach does not address the challenge of augmenting transcriptomic datasets, and the multimodal framework remains less comprehensive compared to ours.

To the best of our knowledge, no existing approach generates gene expression profiles conditioned simultaneously on histopathology images and structured clinical metadata. Our framework, GeMM-GAN, is the first to integrate these heterogeneous modalities as conditioning signals in a generative model for transcriptomic synthesis.

3   Problem Formulation

In this work, we address the problem of generating biologically plausible gene expression profiles conditioned on multimodal inputs comprising Whole Slide Images (WSIs) and clinical descriptions of the patient. Specifically, let $x_g \in \mathbb{R}^g$ be a gene expression profile with g genes, $\mathbf{I} \in \mathbb{R}^{W \times H \times 3}$ a WSI, where W and H are the sizes of the slide and $x_t \in \mathcal{T}$ the textual description. As a WSI is usually characterized by very high dimensions for W and H, we denote with $x_p \in \mathbb{R}^{P \times P \times 3}$ a single patch of the WSI, and thus represent the WSI as a set of patches $\{x_p^{(i)}\}$. Our goal is to learn a conditional generative model $G(x_g|\{x_p^{(i)}\}, x_t)$ that captures the joint distribution of gene expression patterns given the visual and textual context. More in detail, we aim to learn a model that enables:

– Conditional generation: synthesizing gene expression profiles consistent with provided image and text inputs.
– Multimodal conditioning: integrating visual features and textual semantics to inform gene-level patterns.

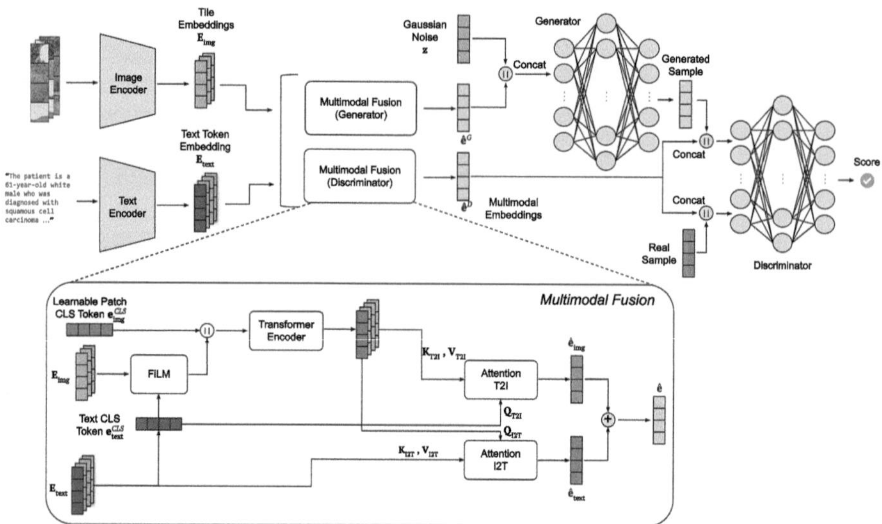

Fig. 1. The architecture of our generative model. The input patches and textual descriptions are embedded with the Image and Text Encoder, respectively. These embeddings are combined in the Multimodal Fusion network to condition a WGAN-GP.

– Biological fidelity: producing outputs that reflect realistic and biologically coherent expression levels, potentially useful for downstream tasks such as disease classification.

4 Method

At a high level, GeMM-GAN consists of three main steps, as depicted in Fig. 1. We first extract patches from tissue slides, embed each patch with an *Image Encoder* and embed the clinical description with a *Text Encoder*. These embeddings are then fused in a *Multimodal Fusion* network that produces as output a single multimodal embedding capturing information from both modalities. Finally, the multimodal embedding is used as a conditioning vector for a Wasserstein GAN with Gradient Penalty (WGAN-GP) [5], which generates gene expression profiles. All these steps are jointly trained in an end-to-end manner.

4.1 Single-Modal Networks

Our generative model is conditioned on the WSI patches $\{\mathbf{x_p}^{(i)}\}$ and the textual description $\mathbf{x_t}$. To enable multimodal conditioning, we project each input modality into a latent space having the same dimension d. For the textual modality, we tokenize and embed the input using the *Text Encoder*, that outputs an embedding matrix $\mathbf{E_{text}} \in \mathbb{R}^{M \times d}$, where M is the maximum number of tokens allowed for the textual description. The first row of this matrix is the Text CLS token $\mathbf{e}_{\mathbf{text}}^{\mathbf{CLS}} \in \mathbb{R}^d$, which is directly modelled by the *Text Encoder*. For the imaging modality, we randomly sample a subset $\{\mathbf{x_p}^{(i_j)}\}_{j=0}^{N}$ of N patches, which are processed by the *Image Encoder*, producing as output an embedding matrix $\mathbf{E_{img}} \in \mathbb{R}^{N \times d}$. As the *Image Encoder* process each patch separately, there is no natural CLS token inside $\mathbf{E_{img}}$. Thus, we use a learnable vector $\mathbf{e}_{\mathbf{img}}^{\mathbf{CLS}} \in \mathbb{R}^d$ as a surrogated Patch CLS token for the image modality.

4.2   Multimodal Fusion

Our Multimodal Fusion network takes as input the patch embeddings $\mathbf{E_{img}}$ and the text tokens $\mathbf{E_{text}}$, producing as output a joint representation $\hat{\mathbf{e}} \in \mathbb{R}^d$:

$$\hat{\mathbf{e}} = MultimodalFusion(\mathbf{E_{img}}, \mathbf{E_{text}}). \tag{1}$$

We first modulate the patch embeddings exploiting Feature-wise Linear Modulation (FiLM) [15], leveraging the Text CLS Token to condition the image features. Specifically, FiLM learns two functions $\gamma : \mathbb{R}^d \to \mathbb{R}^d$ and $\beta : \mathbb{R}^d \to \mathbb{R}^d$ that are used to compute a feature-wise affine transformation of each patch embedding:

$$FiLM(\mathbf{E_{img}}, \mathbf{e}_{\mathbf{text}}^{\mathbf{CLS}}) = \gamma(\mathbf{e}_{\mathbf{text}}^{\mathbf{CLS}}) \odot \mathbf{E_{img}}_{i,c} + \beta(\mathbf{e}_{\mathbf{text}}^{\mathbf{CLS}}). \tag{2}$$

The result is a set of modulated patch embeddings, denoted by $\mathbf{E}_{\mathbf{img}}^{\mathbf{FiLM}} \in \mathbb{R}^{N \times d}$, which emphasized features that are more relevant, according to the textual description. We then prepend the learnable Patch CLS Token $\mathbf{e}_{\mathbf{img}}^{\mathbf{CLS}}$ to the modulated patch embeddings and input these to a Transformer Encoder [17]. This enables us to capture relationships between different patches of the WSI and, at the same time, summarize the most relevant patterns of the WSI into the updated Patch CLS Token. Indeed, WSIs might contain patches that refer to the diagnosed disease and other patches that are not relevant. We denote with $\mathbf{E}_{\mathbf{img}}' \in \mathbb{R}^{(N+1) \times d}$ the output of the Transformer Encoder, having as first row the updated Patch CLS Token, denoted by $\mathbf{e}_{\mathbf{img}}^{\mathbf{CLS}'}$.

Finally, to capture relationships between the two input modalities, we use a bidirectional Cross-Attention mechanism exploiting two Multi-Head Attention:

$$\hat{\mathbf{e}}_{img} = MultiHeadAttention_{T2I}(Q = \mathbf{e}_{\mathbf{text}}^{\mathbf{CLS}}, K = \mathbf{E}_{\mathbf{img}}', V = \mathbf{E}_{\mathbf{img}}'), \tag{3}$$

$$\hat{\mathbf{e}}_{text} = MultiHeadAttention_{I2T}(Q = \mathbf{e}_{\mathbf{img}}^{\mathbf{CLS}'}, K = \mathbf{E_{text}}, V = \mathbf{E_{text}}), \tag{4}$$

where *T2I* and *I2T* stands for *Text2Image* and *Image2Text*, respectively. The Text CLS Token attends over the visual tokens to obtain $\hat{\mathbf{e}}_{img}$, while the updated Patch CLS Token attends over the textual embeddings to produce $\hat{\mathbf{e}}_{text}$. The final multimodal embedding is given by the sum of both: $\hat{\mathbf{e}} = \hat{\mathbf{e}}_{text} + \hat{\mathbf{e}}_{img}$.

4.3   Generative Model

As generative model, we adopt the WGAN-GP architecture [5], which remains the state-of-the-art for transcriptomic data generation due to its ability to handle high-dimensional inputs with stable training dynamics. While diffusion-based models have recently begun to emerge in this domain [8], they are still in early exploratory stages and not yet widely adopted. Our choice of WGAN-GP is further supported by empirical comparisons, where alternative models such as VAEs are consistently underperformed in terms of sample quality and distributional fidelity.

We condition the WGAN-GP on the output of the Multimodal Fusion network. To obtain the two conditioning vectors (one for the generator and one for the discriminator) to be concatenated with both real and fake gene expression profiles, we use the same Multimodal Fusion network, with different learnable parameters θ_G and θ_D:

$$\hat{\mathbf{e}}^{\mathbf{G}} = MultimodalFusion^G(\mathbf{E_{img}}, \mathbf{E_{text}}; \theta_G), \tag{5}$$

$$\hat{\mathbf{e}}^{\mathbf{D}} = MultimodalFusion^D(\mathbf{E_{img}}, \mathbf{E_{text}}; \theta_D), \tag{6}$$

The Generator, a Multi-Layer Perceptron (MLP), takes a noise vector $\mathbf{z} \sim \mathcal{N}(0, I_d)$ and a multimodal embedding $\hat{\mathbf{e}}^{\mathbf{G}}$ to produce a gene expression profile $\mathbf{x_g^{gen}}$. The Discriminator, also an MLP, receives $\hat{\mathbf{e}}^{\mathbf{D}}$ along with real and generated profiles, and learns to distinguish between them. This conditional setup guides the generator to produce biologically meaningful outputs semantically consistent with the input tissue slides and textual descriptions.

5   Implementation Details

5.1   Data

The input data consist of paired histopathology whole-slide images (WSIs), clinical metadata, and matched gene expression profiles, all retrieved from TCGA public repository[1] and focused on nineteen different tumor types. Gene expression profiles, obtained from RNA sequencing, are quantified using FPKM (Fragments Per Kilobase of transcript per Million mapped reads), providing normalized measures of gene activity specific to the disease context. WSIs are ultra-high-resolution images of tissue sections, which are computationally infeasible to analyze in full, and are therefore typically subdivided into smaller tiles for downstream processing (Fig. 2). Clinical metadata is available in JSON format and includes patient-specific information such as demographics, cancer subtype, and treatment history, all related to the disease condition. We convert this structured metadata into concise case summaries using a quantized instruction-tuned language model. We used a version of Llama3-8B fine-tuned on medical data [2].

---

[1] https://www.cancer.gov/ccg/research/genome-sequencing/tcga.
[2] https://huggingface.co/ContactDoctor/Bio-Medical-Llama-3-8B.

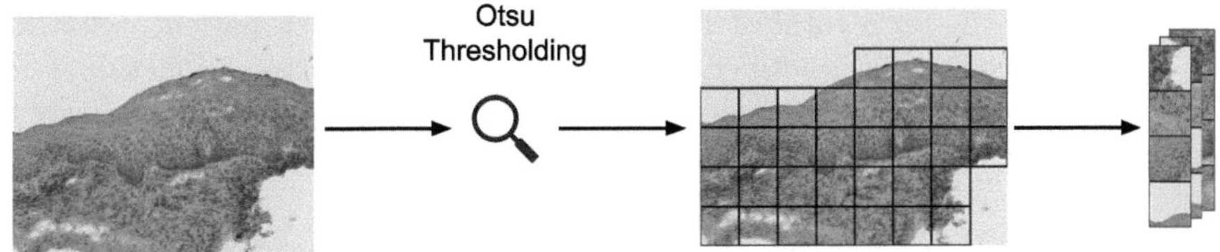

Fig. 2. Preprocessing of the WSIs with Otsu Thresholding to extract valid image patches that contain at least 20% of tissue content.

Irrelevant fields are removed, and the remaining data is serialized into prompts, resulting in 200-word descriptions that capture disease site, demographics, and experimental conditions. This step is essential to align the TCGA input format with the intended use case, where users may input a WSI and write a text description to obtain a plausible gene expression.

5.2 Preprocessing

To prepare data for multimodal generation (Fig. 2), we segment tissue regions from pathology slides using Otsu Thresholding [13] and extract high-resolution tiles of size $P = 256$, keeping only those tiles with more than 20% of tissue content. For the gene expression data, we remove genes with more than 90% missing values and apply standard score normalization (z-score).

5.3 Embedding Models and Hyperparameters

We use UNI [3] as an Image Encoder, which offers a pretrained Vision Transformer for histopathology tiles. As a Text Encoder, we exploit Clinical Modern-BERT [10][3][4]. As these two models have a different embedding size, *i.e.* 1024 for UNI and 768 for Clinical ModernBERT, we add a linear projection layer to the outputs of the two models to align the embedding sizes to $d = 256$. In practice, as these models are already pretrained on a high amount of data, we train only the linear projection layer and freeze the pretrained layers. For the Generator and Discriminator, we use two MLPs having two hidden layers of size 256. To train our model, we select $N = 256$ random patches at each training step.

6 Experiments

We test our method on the TCGA dataset, that provides both tissue slides and clinical descriptions of each patient, associated with a gene expression profile derived from RNA-seq. For storage and computational purposes, we select only a subset of TCGA by keeping only the samples having a tissue slide of dimension

---

[3] https://huggingface.co/Simonlee711/Clinical_ModernBERT.

[4] Both UNI and Clinical ModernBERT are accessible through Huggingface.

less than 100 MB, for a total of 1944 clinical cases. For each patient, following the preprocessing steps described in Sect. 5.2, we obtain expression values for 18,868 genes, where each gene corresponds to a column in the tabular matrix to be generated.

6.1   Evaluation Metrics

We evaluate our approach with three classes of metrics: (i) *unsupervised metrics*, (ii) *detectability* and (iii) *utility*.

Unsupervised Metrics

Precision and Recall [7] are metrics used to assess the quality and relevance of the generated data with respect to the real data. Given the real dataset $\mathbf{X} = \{x_1, \ldots, x_n\}$ and the generated dataset $\hat{\mathbf{X}} = \{\hat{x}_1, \ldots, \hat{x}_m\}$, the first step is to compute the manifolds for the real and generated samples $H_{\mathbf{X}}^t$ and $H_{\hat{\mathbf{X}}}^t$, respectively. This is done by forming a sphere around each point of the two datasets with a radius corresponding to the euclidean distance to t-th closest neighbor within the same dataset. A binary function $f(h, H)$ is then defined to evaluate whether the sample h falls within the manifold H.

Precision and recall can be computed as follows:

$$\text{precision}_t(\mathbf{X}, \hat{\mathbf{X}}) = \frac{1}{|\hat{\mathbf{X}}|} \sum_{\hat{\mathbf{x}} \in \hat{\mathbf{X}}} f(\hat{\mathbf{x}}, H_{\mathbf{X}}^t), \tag{7}$$

$$\text{recall}_t(\mathbf{X}, \hat{\mathbf{X}}) = \frac{1}{|\mathbf{X}|} \sum_{\mathbf{x} \in \mathbf{X}} f(\mathbf{x}, H_{\hat{\mathbf{x}}}^t). \tag{8}$$

We compute precision and recall considering the 10-th nearest neighbors.

Structural Coherence. To assess structural quality, we compute the mean squared error (C. MSE) between real and synthetic gene-gene correlation matrices. Lower C. MSE values indicate better preservation of biologically meaningful co-expression patterns.

Supervised Metrics

Detectability. The detectability measures how well a classfication model can distinguish generated from real samples. In our experiments, we use Logistic Regression and Multi-Layer Perceptron as classifiers and Accuracy and F1-Score as figures of merit. In particular, low values for Accuracy and F1-Score indicate high similarity between real and generated samples.

Utility. The utility measures the performance of models trained on downstream tasks. In our experiments, we train classification models on generated data to predict the disease type or the primary site given a gene expression profile and test these models on real data. Higher performance (accuracy and F1-Score) indicates a more realistic data generation.

6.2   Competitors

We compare our results with three different benchmarks: a Vanilla WGAN-GP, that is a WGAN-GP without any conditioning, a Conditional WGAN-GP, conditioned on two categorical variables, namely the disease type and the primary site, and a Conditional Variational Autoencoder (CVAE), conditioned on the same variables of the Conditional WGAN-GP. Please note that we cannot perform an utility evaluation on the Vanilla WGAN-GP as this is unconditional. Differently, by conditioning the other baselines on the disease type and the primary site, we can use the conditional variables as labels of the generated data to train the classification models.

6.3   Experimental Setup

We performed our experiments on a machine with the AMD Ryzen 1950X CPU, with 128 GB of RAM, using two NVIDIA A6000 GPUs with 48 GB of VRAM. We employed PyTorch[5] version 2.6.0 to define our model and perform the experimental evaluation, and CUDA version 12.4 for model training on the GPU devices. We perform an 80/20 train-test split, ensuring that the histopathology images, clinical metadata, and corresponding gene expression profiles are divided consistently across sets. We train all the models using a latent dimension $d = 256$ for the noise and with a batch size of 64.

6.4   Main Results

As shown in Table 1, our method demonstrates superior recall performance (0.8190) while maintaining balanced precision-recall characteristics. Although precision (0.7103) is lower than conditional WGAN-GP (0.8436), the higher recall indicates successful capture of dataset variability, effectively avoiding mode collapse common in GAN-based approaches [6]. Most notably, our method achieves exceptional gene correlation preservation with MSE of 0.0007, representing 98.2% of improvement over the Conditional WGAN-GP (0.0038). This dramatic improvement demonstrates our model's ability to preserve biologically meaningful gene relationships.

Detectability analysis reveals that our synthetic data is significantly harder to distinguish from real data using LR, indicating high-quality synthesis despite genomic data complexity. However, sophisticated classifiers like MLPs maintain high detectability, presenting an area for future investigation.

Our method consistently outperforms baselines across all utility metrics. For disease type classification, we achieve 16.9% improvement in RF accuracy and 11.1% improvement in F1 score. Primary site classification shows even greater improvements: 13.4% in RF accuracy and 10.0% in F1 score. These improvements stem from our model's powerful embedding-based conditioning, which generates semantically meaningful synthetic data that maintains relevant biological patterns for downstream applications.

---

[5] https://pytorch.org/.

Table 1. Results of our models on the TCGA dataset, compared with three different benchmarks: a WGAN-GP (*Vanilla WGAN-GP*), a WGAN-GP (*Cond. WGAN-GP*) and CVAE, both conditioned on disease type and primary site. Mean and standard deviation across 10 generation runs are reported. Best results are highlighted in bold.

Metric	Vanilla WGAN-GP	Cond. WGAN-GP	CVAE	Proposed
Unsupervised Metrics				
Precision ↑	0.8282(0.0000)	**0.8436(0.0051)**	0.4892(0.0015)	0.7103(0.0047)
Recall ↑	0.6477(0.0031)	0.7372(0.0038)	0.7956(0.0131)	**0.8190(0.0068)**
C. MSE ↓	0.0095(0.0012)	0.0038(0.0001)	0.2837(0.0027)	**0.0007(0.0000)**
Detectability				
LR (Acc.) ↓	0.8459(0.0142)	0.8801(0.0014)	0.8931(0.0077)	**0.5236(0.0179)**
LR (F1) ↓	0.8662(0.0107)	0.8925(0.0119)	0.9003(0.0070)	**0.5444(0.0286)**
MLP (Acc.) ↓	0.9887(0.0040)	0.9904(0.0019)	0.9992(0.0006)	**0.9797(0.0141)**
MLP (F1) ↓	0.9886(0.0040)	0.9905(0.0019)	0.9992(0.0006)	**0.9794(0.0145)**
Utility - Disease Type				
RF (Acc.) ↑	–	0.6367(0.0403)	0.5725(0.0196)	**0.7449(0.0127)**
RF (F1) ↑	–	0.7422(0.0287)	0.6698(0.0165)	**0.8249(0.0087)**
MLP (Acc.) ↑	–	0.9262(0.0048)	0.9031(0.0007)	**0.9354(0.0072)**
MLP (F1) ↑	–	0.9247(0.0053)	0.8803(0.0006)	**0.9322(0.0072)**
Utility - Primary Site				
RF (Acc.) ↑	–	0.6118(0.0244)	0.4228(0.0075)	**0.6936(0.0061)**
RF (F1) ↑	–	0.6800(0.0201)	0.4726(0.0049)	**0.7478(0.0090)**
MLP (Acc.) ↑	–	0.8254(0.0091)	0.6782(0.0176)	**0.8436(0.0069)**
MLP (F1) ↑	–	0.8094(0.0070)	0.5735(0.0222)	**0.8264(0.0072)**

6.5   Ablation Study

To evaluate the contribution of both text and image modalities to the generative process, we conducted an ablation study comparing different conditioning strategies. In particular, we investigated whether our proposed image encoding strategy provides a more informative signal than simpler alternatives, such as the mean of the patch embeddings or the use of a CLS token for the input text. The results in Table 2 highlight that image-based conditioning plays a crucial role in balancing precision and recall, as demonstrated by the significantly more stable values across all image-informed variants. In contrast, conditioning solely on text yields a high precision (0.9513) but markedly lower recall (0.5205), suggesting that it leads to more selective but less comprehensive generations. From a utility perspective, image conditioning consistently yields higher accuracy and F1 scores, indicating that it carries richer semantic information for downstream tasks. Our full model achieves the highest overall performance in both utility prediction and detectability metrics, confirming its robustness and practical utility. While we acknowledge the trade-offs between unsupervised and supervised met-

Table 2. Ablation study. We compare our conditional generative models with different strategies for the conditioning vector of the WGAN-GP: (i) the mean of the patch embeddings (*Mean Image*), (ii) the CLS token embedding of the input text (*Text CLS Token*), (iii) a Transformer Encoder for the patch embeddings (*Patch Transformer*), (iv) our model without the final Cross Attention (*FiLM*), (v) our model without FiLM (*Cross Attention*). Results are the mean of 10 generation runs.

Method	Unsupervised			Detectability (LR)		Utility (RF)	
	Prec.	Recall	C. MSE	Acc.	F1	Acc.	F1
Mean Image	0.8128	0.7026	0.0014	0.5449	0.5730	0.7167	0.8011
Text CLS Token	**0.9513**	0.5205	0.0045	0.5571	0.5797	0.5000	0.6330
Patch Transformer	0.8500	0.7821	0.0013	0.5692	0.5772	0.7333	0.8158
FiLM	0.7910	**0.8821**	0.0015	0.5853	0.5967	0.6513	0.7555
Cross Attention	0.7000	0.8346	0.0011	0.5526	0.5524	0.7256	0.8112
Ours	0.7103	0.8190	**0.0007**	**0.5236**	**0.5444**	**0.7449**	**0.8249**

rics, our approach proves to be the most effective in producing realistic and functionally valuable samples.

7 Conclusions

This work introduced GeMM-GAN, the first multimodal generative framework that conditions gene expression synthesis on both histopathology images and clinical metadata. The model achieves strong performance across standard evaluation metrics, producing statistically realistic and biologically coherent gene expression profiles.

Through a detailed ablation study, we quantified the contribution of each modality to the generative process, revealing the crucial role of histopathology images. These findings demonstrate that tissue morphology encodes rich semantic information, which significantly enhances the biological fidelity of the generated profiles. This analysis also improves the interpretability of our model by clarifying how visual and textual inputs influence the generation pipeline.

Future work will extend this framework to generate histopathology images from gene expression profiles and clinical data. This bidirectional capability would enable researchers to simulate how genetic changes appear in tissue samples, advancing our understanding of disease mechanisms and supporting the development of cross-modal AI tools for precision medicine.

References

1. van Breugel, B., Liu, T., Oglic, D., van der Schaar, M.: Synthetic data in biomedicine via generative artificial intelligence. Nat. Rev. Bioeng. **2**(12), 991–1004 (2024)

2. Carrillo-Perez, F., et al.: Rna-to-image multi-cancer synthesis using cascaded diffusion models. bioRxiv pp. 2023–01 (2023)
3. Chen, R.J., et al.: Towards a general-purpose foundation model for computational pathology. Nat. Med. (2024)
4. Chlis, N.K., Rausch, L., Brocker, T., Kranich, J., Theis, F.J.: Predicting single-cell gene expression profiles of imaging flow cytometry data with machine learning. Nucleic Acids Res. **48**(20), 11335–11346 (2020)
5. Gulrajani, I., Ahmed, F., Arjovsky, M., Dumoulin, V., Courville, A.C.: Improved training of wasserstein gans. Adv. Neural Inf. Process. Syst. **30** (2017)
6. Kossale, Y., Airaj, M., Darouichi, A.: Mode collapse in generative adversarial networks: an overview. In: 2022 8th International Conference on Optimization and Applications (ICOA), pp. 1–6. IEEE (2022)
7. Kynkäänniemi, T., Karras, T., Laine, S., et al.: Improved precision and recall metric for assessing generative models. In: Advances in Neural Information Processing Systems. Curran Associates Inc. (2019)
8. Lacan, A., André, R., Sebag, M., Hanczar, B.: In silico generation of gene expression profiles using diffusion models. bioRxiv, pp. 2024–04 (2024)
9. Lacan, A., Sebag, M., Hanczar, B.: GAN-based data augmentation for transcriptomics: survey and comparative assessment. Bioinformatics **39**(Supplement_1), i111–i120 (2023)
10. Lee, S.A., Wu, A., Chiang, J.N.: Clinical modernbert: an efficient and long context encoder for biomedical text. arXiv preprint arXiv:2504.03964 (2025)
11. Liu, H., Zhang, Y., Luo, J.: Contrastive learning-based histopathological features infer molecular subtypes and clinical outcomes of breast cancer from unannotated whole slide images. Comput. Biol. Med. **170**, 107997 (2024)
12. Ohnstad, H.O., et al.: Prognostic value of pam50 and risk of recurrence score in patients with early-stage breast cancer with long-term follow-up. Breast Cancer Res. **19**, 1–12 (2017)
13. Otsu, N., et al.: A threshold selection method from gray-level histograms. Automatica **11**(285–296), 23–27 (1975)
14. Panaccione, F.P., Mongardi, S., Masseroli, M., Pinoli, P.: Biogan: enhancing transcriptomic data generation with biological knowledge. Bioengineering **12**(6), 658 (2025)
15. Perez, E., Strub, F., De Vries, H., Dumoulin, V., Courville, A.: Film: visual reasoning with a general conditioning layer. In: Proceedings of the AAAI Conference on Artificial Intelligence. vol. 32 (2018)
16. Schmauch, B., et al.: A deep learning model to predict rna-seq expression of tumours from whole slide images. Nat. Commun. **11**(1), 3877 (2020)
17. Vaswani, A.: Attention is all you need. Adv. Neural Inf. Process. Syst. (2017)
18. Viñas, R., Andrés-Terré, H., Liò, P., Bryson, K.: Adversarial generation of gene expression data. Bioinformatics **38**(3), 730–737 (2022)
19. Wang, A.T., Dhruba, S.R., Wang, K., Campagnolo, E.M., Shulman, E.D., Ruppin, E.: Deep learning inference of cell type-specific gene expression from breast tumor histopathology. bioRxiv, pp. 2025–05 (2025)
20. Zheng, Y., et al.: Digital profiling of cancer transcriptomes from histology images with grouped vision attention. BioRxiv, pp. 2023–09 (2024)

A Comparative Evaluation of Diffusion Based Networks for Multiple Sclerosis Lesion Segmentation

Alessia Rondinella[1]([✉])[iD], Francesco Guarnera[1][iD], Alessandro Ortis[1][iD], Elena Crispino[2][iD], Giulia Russo[2][iD], Francesco Pappalardo[2][iD], and Sebastiano Battiato[1][iD]

[1] Department of Mathematics and Computer Science, University of Catania, Catania, Italy
{alessia.rondinella,francesco.guarnera,alessandro.ortis,
sebastiano.battiato}@unict.it
[2] Department of Drug and Health Sciences, University of Catania, Catania, Italy
{elena.crispino,giulia.russo,francesco.pappalardo}@unict.it

Abstract. Semantic segmentation of Multiple Sclerosis (MS) lesions in longitudinal MRI is essential for tracking disease progression. This study evaluates how well various deep learning segmentation models, commonly used in medical imaging, generalize when integrated into a diffusion model framework. We conduct extensive experiments testing different architectural configurations and inference strategies to identify optimal setups for MS lesion segmentation. In particular, we examine how combining outputs from multiple diffusion time steps affects performance and robustness. Results show that certain backbone architectures significantly enhance performance, and that appropriate inference strategies can further improve segmentation accuracy. These findings highlight the potential of diffusion-based methods for clinical MS analysis and offer guidance on model and inference selection to achieve reliable lesion segmentation from MRI scans. Our work contributes to the growing body of research applying generative models to medical imaging, especially in the context of progressive diseases like MS.

Keywords: Multiple Sclerosis · Denoising Diffusion Models · Lesion segmentation

1 Introduction

Multiple Sclerosis (MS) is a chronic inflammatory disease of the Central Nervous System, characterized by demyelination and axonal damage that lead to diverse and progressive neurological symptoms [5,13]. Early diagnosis is challenging due to the complex brain anatomy and the subtle nature of initial lesions, often resulting in delayed treatment and irreversible damage.

E. Rodolà et al. (Eds.): ICIAP 2025 Workshops, LNCS 16169, pp. 405–416, 2026.
https://doi.org/10.1007/978-3-032-11317-7_34

Magnetic Resonance Imaging (MRI) plays a key role in early MS detection. A typical longitudinal protocol includes FLAIR, T1-w, T2-w, and PD-w sequences, each offering unique contrast. FLAIR images are especially effective at revealing periventricular lesions, while T1-w helps detect atrophy, T2-w highlights inflammatory regions, and PD-w improves visualization of spinal cord lesions [4,12]. However, interpreting such multimodal data remains time-consuming and complex, underscoring the need for automated, robust tools for lesion detection and monitoring.

This work extends the MSSegDiff framework initially introduced by Rondinella et al. [22], a diffusion-based model for MS lesion segmentation. Our contributions include:

1. Architectural Variations: Evaluation of alternative segmentation backbones within MSSegDiff to assess performance and generalizability.
2. Enhanced Inference Module: Analysis of different inference strategy for aggregating outputs across timesteps, enhancing segmentation quality.
3. Validation on a MSLesSeg Dataset [8]: Benchmarking on a newly heterogeneous dataset collected across different scanners, also used in the *ICPR 2024 MSLesSeg Competition* [21].
4. Open Release of Code: Full public release of our implementation to foster transparency and community engagement at https://github.com/alessiarondinella/MSSegDiff.

Through these contributions, we demonstrate that the constructed architecture and proposed inference strategies lead to substantial improvements in segmentation performance. Notably, the MSSegDiff framework achieved high accuracy in segmenting MS lesions on the new dataset, despite the inherent variability in the acquisition settings. This highlights the practical utility of our approach for real-world clinical applications.

The rest of the paper is organized as follows: Sect. 2 reviews related work, Sect. 3 details our method, Sect. 4 reports results, and Sect. 5 concludes the study.

2   Related Work

Recent advances in medical image segmentation, have leveraged a variety of deep learning architectures to enhance accuracy and efficiency. In the last few years, several notable approaches have emerged, each contributing to the field with unique methodologies and improvements. One significant development is the application of transformer-based models in medical imaging. The Vision Transformer (ViT), introduced in [6] has been adapted for segmentation tasks with promising results. Following this, the Swin Transformer, proposed in [14], has shown superior performance by utilizing hierarchical feature extraction with shifted windows, allowing for efficient global context modeling. Following these works, numerous studies have proposed modifying the backbone structures of

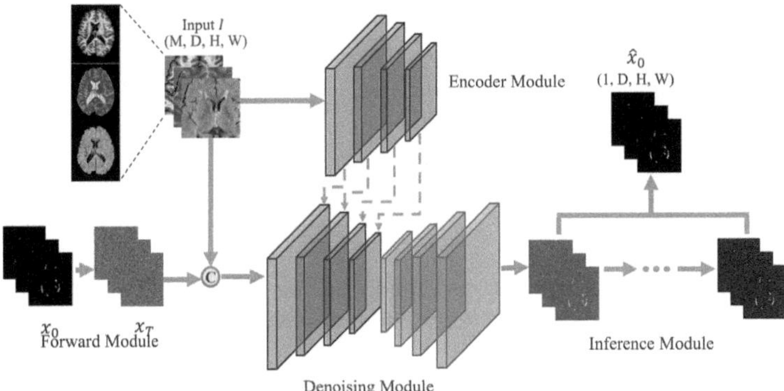

Fig. 1. Overview of the MSSegDiff [22] architecture. The architecture consists of four key modules: (1) the Forward Module, which implements the forward diffusion process by gradually adding Gaussian noise to the ground-truth segmentation mask; (2) the Encoder Module, which captures discriminative features from the channel-wise concatenated volumetric MRI sequences ('C' in the figure represents the channel-wise concatenation); (3) the Denoising Module, responsible for the reverse diffusion process, progressively removing the noise to produce a clean segmentation mask; and (4) the Inference Module, which combines predictions from each timestep to generate the final segmentation mask.

medical image segmentation models by incorporating transformers, such as SwinUNet [1], a hybrid model combining Swin Transformer and U-Net that achieves robust performance in multi-organ segmentation task.

Another noteworthy approach is the use of convolutional neural networks (CNNs) with attention mechanisms [24]. Authors in [17] proposed an Attention U-Net, integrates spatial attention gates to highlight relevant features, improving segmentation accuracy. This model has inspired the development of various attention-based architectural adaptations aiming to enhance segmentation performance, as done in [20] which propose a Fully Convolutional DenseNet with attention blocks for MS lesion segmentation in 2D images and in [9] where authors propose an Attention u-Net for the same purpose. Moreover, diffusion models have gained traction in the field of medical image analysis. These models, originally introduced for generative tasks, have been adapted for image segmentation by incorporating noise-injection and denoising processes to improve robustness and accuracy of prediction results.

Authors in [26], proposed a segmentation model based on Diffusion Probabilistic Models (DDPM) [11] with a dynamic conditional encoding, which aims to learn segmentation by conditioning with the image prior. The same authors in [27] that integrate transformer into a diffusion segmentation model with various conditional techniques over the denoising network to perform multi-organ segmentation.

These advancements highlight the ongoing evolution of segmentation techniques, with recent models focusing on integrating attention mechanisms, leveraging transformer architectures, and employing diffusion processes to achieve state-of-the-art performance in medical image segmentation tasks.

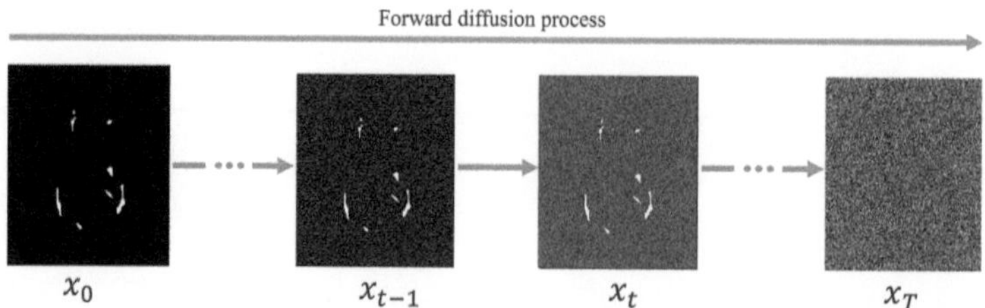

Fig. 2. This image depicts the forward module. The MS lesion segmentation mask is gradually perturbed with Gaussian noise, for a number of timesteps T, until it becomes completely noisy.

3 Overview of the MSSegDiff Architecture

MSSegDiff [22] is designed to learn the denoising process of MS lesion segmentation masks from volumetric, multimodal MRI data. As depicted in Fig. 1, the overall pipeline comprises four main components: Forward Module, Encoder Module, Denoising Module, and Inference Module. Noise is initially added to the ground-truth segmentation masks, and the model learns to reverse this process, gradually reconstructing accurate segmentation outputs. The architecture integrates a U-Net backbone with an encoder-decoder structure, receiving as input both the MRI volume I (T1-w, T2-w, FLAIR with dimension $M \times D \times H \times W$) and the noisy segmentation mask x_T generated via the forward diffusion process.

The Forward Module (Fig. 2) implements the forward diffusion process, which involves introducing T steps of Gaussian noise to the input image x_0. This process gradually corrupts the original ground-truth x_{GT} until the signal is completely destroyed. It is modeled as a fixed Markov chain where each step is defined as:

$$q(x_t|x_{t-1}) = \mathcal{N}(x_t; \sqrt{1 - \beta_t}x_{t-1}, \beta_t I). \tag{1}$$

Here, x_t is the corrupted image at time t, $\mathcal{N}$ is the normal distribution with mean $\sqrt{1 - \beta_t}x_{t-1}$ and variance $\beta_t I$. Based on the original DDPM paper [11], we set $T = 1000$ and adopted a linear schedule for β_t from 1e−4 to 0.02, enabling a gradual and controlled corruption process.

The Encoder Module (EM) processes the MRI data by concatenating T1-w, T2-w, and FLAIR sequences along the channel dimension. This multi-channel input I provides complementary information: FLAIR captures periventricular inflammation, T1-w offers anatomical contrast, and T2-w identifies edema. The encoder extracts rich 3D spatial and modality-aware features, which are essential for segmenting MS lesions. Since the encoder shares its architecture with the U-Net's downsampling path, its output can be seamlessly integrated into the main architecture via skip connections.

The Denoising Module (DM) takes the noisy segmentation mask x_T and iteratively reconstructs a clean mask $\hat{x}_0$. It follows a probabilistic schedule, applying small corrections in early steps and more substantial adjustments later. Unlike single-step models (e.g., GANs), this gradual process ensures higher fidelity and consistency with lesion structures. The encoder's features condition the denoising path, reducing variability and improving robustness.

At test time, the Inference Module performs T denoising steps using the Denoising Diffusion Implicit Models (DDIM) method [23]. The predicted segmentation mask $\hat{x}_0$ is refined progressively, with multiple intermediate outputs that can be combined to reduce uncertainty and enhance accuracy. Figure 3 illustrates an example output.

The integration of these components enables MSSegDiff to perform high-quality, multimodal, 3D MS lesion segmentation with robust denoising capabilities, leveraging both the structural and functional contrasts present in the MRI volumes.

3.1   Dataset, Training Setup and Evaluation

The experiments were conducted using a subset of the ISBI 2015 dataset [2], which includes 21 MRI scans from 5 patients annotated by two raters. Only T1-w, T2-w, and FLAIR sequences were used, as MS lesions are more visible in these modalities. Each volume underwent rigid registration to MNI space, brain extraction, and non-uniformity correction. To ensure consistency across scans, preprocessing involved MONAI-based transformations including intensity normalization, foreground cropping, and padding. Data augmentation (random crops of $96 \times 96 \times 96$, flips, scaling, intensity shifts) helped improve generalization given the small dataset size.

Training and inference were performed in PyTorch [18] and MONAI [3] on a single NVIDIA A100 GPU using AdamW optimizer (LR=1e4, weight decay=1e3) and a Cosine Annealing learning rate schedule [15] with warmup. The loss function combines Dice, Binary Cross Entropy, and Mean Squared Error to compare the predicted segmentation $\hat{x}_0$ with the ground-truth mask x_{GT}:

$$Loss(\hat{x}_0, x_{GT}) = Loss_{DSC} + Loss_{BCE} + Loss_{MSE} \qquad (2)$$

Inference used a sliding window approach (stride = 0.5) to reduce memory load and ensure full coverage.

Evaluation followed a Leave-One-Subject-Out Cross-Validation (LOSO-CV) with five folds, each using 3 patients for training, 1 for validation, and 1 for testing, ensuring that no temporal data leakage occurred across folds. This method contrasts with some state-of-the-art approaches ([9,19,20]), where configurations involve using the entire patient for model training, reserving only one time point for testing. The model evaluation process entailed a comparison between the predicted segmentation masks and the provided ground-truth data annotated by Rater 1. Performance was assessed using Dice Score (DSC), True Positive Rate (TPR), Positive Predictive Value (PPV), Lesion False Positive Rate (LFPR), Lesion True Positive Rate (LTPR), Absolute Volume Difference (AVD), and Average Symmetric Surface Distance (ASSD), providing both voxel-wise and lesion-wise evaluation of segmentation accuracy.

Table 1. Results obtained from the comparative studies performed considering different model configurations. The table reports the mean evaluation metrics across all folds.

Model	Mean on 5 Fold						
	DSC↑	TPR↑	PPV↑	LTPR↑	LFPR↓	AVD↓	ASSD↓
MSSegDiff	**0.7526**	**0.7617**	**0.7700**	0.6652	0.2782	**0.2251**	**0.7699**
MSSegDiff+EncoderSegResNet	0.7167	0.7103	0.7656	0.6467	**0.2698**	0.2765	1.1371
MSSegDiff+SegResNet	0.7140	0.7485	0.7602	0.6237	0.2705	0.3800	1.1888
MSSegDiff+SwinUNETR	0.7093	0.7364	0.7345	**0.6654**	0.3261	0.3409	0.9624
MSSegDiff+MultiEncoder	0.7024	0.7598	0.6890	0.6244	0.3275	0.4096	1.2689

4   Comprehensive Evaluation of Comparative Studies

This chapter presents the main extensions introduced in this paper wrt [22]. Sections 4.1 explore modifications to the MSSegDiff backbone, evaluating their effect on performance. Section 4.2 introduces and assesses alternative inference strategies for aggregating predictions across the diffusion steps. Lastly, Sect. 4.3 reports on the validation of MSSegDiff on the MSLesSeg dataset [8], demonstrating the model's robustness in real-world clinical scenarios and underscoring the dataset's relevance for the research community.

4.1   Performance of Different Configuration Models

We explore different configurations of the MSSegDiff architecture, depending on which backbone architectures are used as the Encoder Module and Denoising Module. We chose to use network architectures based on the U-Net model in all configurations due to their proven capability to produce accurate segmentation maps, even with limited input data. This is particularly important in medical imaging, where access to large datasets is often restricted.

State-of-the-art medical image segmentation often incorporates custom modifications to basic architectures to enhance performance. Each configuration is trained for 1200 epochs. Following the completion of training, we employ the validation set to select the model obtained at the epoch with the highest validation Dice Score. For our studies we use the appropriately customized implementation of architectures available in MONAI.

The experiments were carried out on ISBI2015 dataset using three representative and recent network structure in medical image segmentation, namely BasicUNet [7], SegResNet [16] and SwinUNETR [10].

Table 1 reports the segmentation performance of the tested configurations.

We first evaluated the baseline MSSegDiff model [22] using the BasicUNet architecture with added squeeze-and-attention (SA) [29] layers customized for 3D images. Here, the BasicUNet encoder serves as the EM, while the full attention-enhanced BasicUNet acts as the DM. SA layers were introduced after each block,

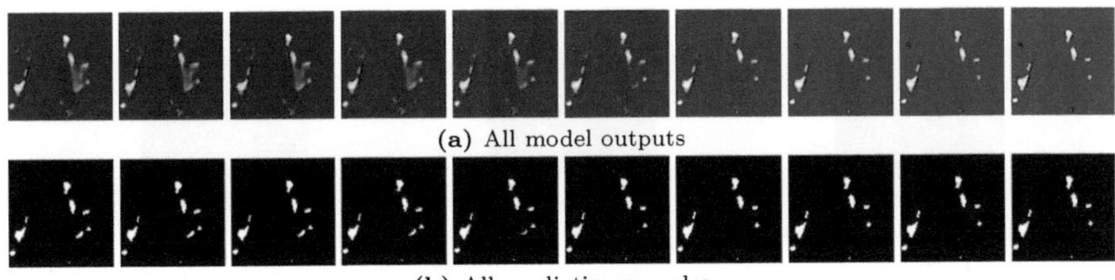

(a) All model outputs

(b) All prediction samples

Fig. 3. Image shows 3 a model outputs and 3 b the prediction of $\hat{x}_0$ through $T = 10$ steps.

in both upsampling and downsampling paths of the backbone. This configuration achieved strong results, demonstrating the benefits of attention mechanisms in focusing on relevant features.

Next, MSSegDiff+EncoderSegResNet replaced the EM with SegResNet's encoder while keeping the DM unchanged. Despite SegResNet's strengths in feature extraction, this change did not yield notable improvements over the BasicUNet-based baseline.

We then tested MSSegDiff+SegResNet, employing SegResNet for both EM and DM to fully leverage its capabilities. However, performance remained comparable to the previous configuration, suggesting limited added value within this diffusion-based framework.

The MSSegDiff+SwinUNETR configuration aimed to evaluate SwinUNETR, a transformer-based model known for capturing long-range dependencies effectively. Despite its theoretical advantages, it performed worse than the other setups—likely due to transformer limitations with small medical imaging datasets.

Finally, MSSegDiff+MultiEncoder explored splitting the EM into three modality-specific encoders whose outputs were aggregated before integration with the DM. This design aimed to better exploit multimodal MRI information but ultimately resulted in the lowest Dice score in our tests, indicating that this added complexity did not translate into better segmentation performance.

4.2 Performance of Different Inference Methods

This section expands on the enhanced inference strategies introduced in this paper. We evaluate the performance of various inference methods applied to the MSSegDiff framework, focusing on the fusion of segmentation masks generated across multiple diffusion steps. At test time, the diffusion model iterates through T steps using the DDIM method [23]. Each step generates an increasingly refined segmentation mask. As described in Sect. 3 DDIM is able to produce refined masks over iterations, so we decided to use $T = 10$ to test the generalization capabilities of the model. Figure 3 shows an example of prediction masks obtained from each iteration of DDIM with $T = 10$ step. Leveraging the insight

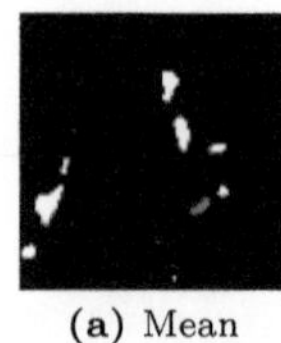 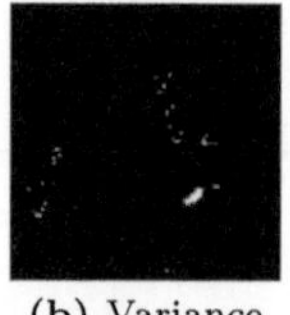 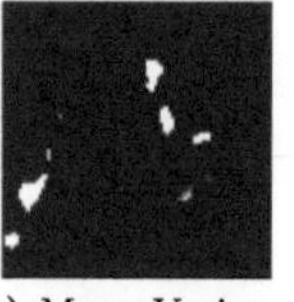

(a) Mean   (b) Variance   (c) Mean-Variance

Fig. 4. Example of segmentation masks obtained by 4 a the mean of the masks obtained from each iteration, 4 b their variance, and 4 c the difference between the mean and the variance.

that with an increasing number of testing steps, the prediction becomes progressively more accurate and the prediction uncertainty decreases, we decided to assess various inference methods. This methods leverage the fusion of segmentation masks obtained from each iteration. Merging the predictions generated at each iteration can ensure more robust segmentation results compared to using only the final segmentation, as done in traditional generative tasks. As the first method, we decided to calculate the mean of the segmentation masks obtained at each step to generate the final segmentation mask. Additionally, we calculated the variance of the output predictions. Figure 4 a shows an example of the mean mask obtained, while Fig. 4 b illustrates the variance. From the variance image, it is noticeable that the pixels most challenging to predict are those delineating the boundaries of MS lesions. As an additional inference method, we decided to compute the difference between the mean and the variance. An example of the result is shown in Fig. 4 c.This stems from the insight that is particularly challenging for our models to accurately predict the boundaries of the lesions. These boundaries are already difficult to identify due to the small and complex structure of the lesions, making precise segmentation a significant challenge for the network. To further enhance the fusion process, we decided to use two additional methods. The first method, originally proposed in [28] and utilized in [22], introduces a Step-Uncertainty based Fusion (SUF) module to fuse the segmentation masks based on the number of steps and the prediction uncertainty. The uncertainty is estimated by performing multiple forward passes through the diffusion model. Each step produces different outputs, generated from different random noise, which are then used to calculate the uncertainty map.

The second method we tested is the Simultaneous Truth and Performance Level Estimation (STAPLE) algorithm, proposed in [25], a widely used method in medical imaging also employed in [26,27]. STAPLE is a voting algorithm that creates a final segmentation mask by performing pixel-by-pixel voting of all the segmentation masks. It has been demonstrated that the accuracy of STAPLE improves with an increasing number of segmentation masks considered for voting. In Table 2 we report the results of the different inference methods in terms of the mean Dice score, evaluated over all folds (best results are denoted as bold). From the results, it can be seen that the inference methods Mean, Mean-var, and SUF achieve similar results, showing consistent performance across different network configurations. In contrast, the STAPLE algorithm exhibits poor

Table 2. Comparison of inference methods based on mean Dice score across all folds.

Model	Mean DSC on 5 Fold			
	Mean	Mean-var	SUF	STAPLE
MSSegDiff	**0.7517**	**0.7495**	**0.7526**	**0.6690**
MSSegDiff+EncoderSegResNet	0.7133	0.7246	0.7167	0.6381
MSSegDiff+SegResNet	0.7149	0.7169	0.7140	0.5655
MSSegDiff+SwinUNETR	0.7073	0.7108	0.7093	0.6246
MSSegDiff+MultiEncoder	0.7023	0.7049	0.7025	0.5511
Mean	0,7179	**0,7213**	0,7190	0,6097

performance across all network configurations. A possible reason for this could be the low number of predictions used for the voting process. A higher number of predictions might lead to better results. Additionally, it is evident that the MSSegDiff configuration provides the highest Dice scores across all inference methods, confirming it as the best configuration. Specifically, the SUF inference method achieves the highest Dice score in this configuration. However, the SUF method does not perform better in all configurations, while Mean-var method achieves a Dice score of 0.7213, the highest across all network configurations, slightly surpassing the SUF method. Given the fact that the best network configuration MSSegDiff obtains the highest Dice scores and achieves the best performance using SUF as the inference method, we decided to use the SUF method and report the results in Table 1 using this inference method.

4.3   Evaluation on a New Dataset

This section presents the results obtained by training the MSSegDiff architecture on a newly collected, heterogeneous dataset, called MSLesSeg, proposed for the MSLesSeg challenge [21]. This dataset represents a significant resource for evaluating the performance of segmentation models in challenging, real-world scenarios. This comprehensively annotated dataset contains labeled MRI scans acquired from 75 patients. The dataset includes scans captured at multiple timepoints: 50 patients with one timepoint, 15 with two, and 10 with three or more. In total, 115 MRI series were acquired, each containing three modalities: T1-weighted (T1-w), T2-weighted (T2-w), and Fluid-Attenuated Inversion Recovery (FLAIR). Lesion annotations were meticulously performed by experts, leveraging FLAIR for lesion delineation and T1-w/T2-w for contextual characterization.

To evaluate the MSSegDiff framework, we utilized the training/test split defined by the MSLesSeg challenge. This dataset's diversity, stemming from scans acquired on different MRI scanners with field strengths of 1.5T and 3T, poses a unique challenge for segmentation models, making it an ideal benchmark for robust validation. Our results, summarized in Table 2, demonstrate the congruence of MSLesSeg dataset with established datasets like ISBI-2015.

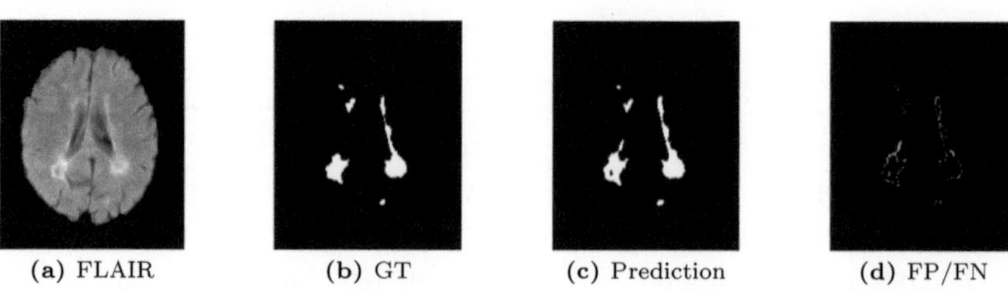

(a) FLAIR   (b) GT   (c) Prediction   (d) FP/FN

Fig. 5. Example of a segmented mask obtained by training the proposed architecture on MSLesSeg dataset [8]. The image shows a FLAIR slice 5 a of a test patient, its ground-truth mask 5 b the prediction 5 c and 5 d the False Positive (in green) and the False Negative (in red) pixels.

Specifically, MSSegDiff achieved a maximum Dice score of 0.7067 on the test set, comparable to ISBI results, despite the absence of a unified acquisition protocol.

An example of segmentation predictions on a patient from the MSLesSeg dataset is shown in Fig. 5, illustrating the model's ability to accurately segment lesions under varying conditions. Despite the overall strong performance, qualitative inspection of the predictions revealed consistent challenges in precisely delineating lesion boundaries, especially in cases with very small or diffuse lesions. This issue is particularly evident in the MSLesSeg dataset, which includes scans with high variability in acquisition protocols and subtle lesion appearances, making boundary prediction more uncertain and error-prone in these challenging cases.

5   Conclusion

In this study, we enhanced the MSSegDiff framework [22] by improving its architecture, refining inference strategies, and expanding validation methods, with the full implementation released openly to encourage transparency and community engagement. Our experiments showed that while advanced architectures are promising, the traditional U-Net-based configuration remains the most effective, especially when combined with attention mechanisms that boost segmentation accuracy. The SUF inference strategy consistently delivered high-quality results by better managing prediction uncertainty. Validation on the heterogeneous MSLesSeg dataset demonstrated the model's strong generalization, achieving competitive Dice scores even with varied acquisition protocols and scanner types. This confirms MSSegDiff's suitability for real-world clinical applications and highlights the MSLesSeg dataset's value as a benchmark for future work. These segmentation improvements have direct clinical relevance by enabling more accurate and consistent lesion quantification over time, which is critical for early diagnosis, monitoring disease progression, and evaluating treatment response. Improved delineation of lesion boundaries can support radiologists and neurologists in making more informed decisions about initiating or adjusting therapies, potentially leading to better patient outcomes and personalized treatment planning. Our results also indicate that transformer-based backbones tend to underperform in this context, likely due to their higher data requirements and the limited training samples available in medical imaging. Addressing this limitation will require exploring advanced data augmentation, semi-supervised or transfer learning strategies, and developing larger, standardized multi-center datasets to fully leverage the potential of these architectures, while future work should also focus on improving computational efficiency and further enhancing robustness across diverse acquisition settings to support broader clinical adoption.

References

1. Cao, H., et al.: Swin-unet: unet-like pure transformer for medical image segmentation. In: European Conference on Computer Vision, pp. 205–218. Springer (2022)
2. Carass, A., et al.: Longitudinal multiple sclerosis lesion segmentation: resource and challenge. Neuroimage **148**, 77–102 (2017)
3. Cardoso, M.J., et al.: Monai: an open-source framework for deep learning in healthcare. arXiv preprint arXiv:2211.02701 (2022)
4. Chong, A., Chandra, R.V., Chuah, K., Roberts, E., Stuckey, S.: Proton density mri increases detection of cervical spinal cord multiple sclerosis lesions compared with t2-weighted fast spin-echo. Am. J. Neuroradiol. **37**(1), 180–184 (2016)
5. Coles, A.: Alastair compston, alasdair coles. Lancet **372**, 1502–1517 (2008)
6. Dosovitskiy, A., et al.: An image is worth 16x16 words: transformers for image recognition at scale. arXiv preprint arXiv:2010.11929 (2020)
7. Falk, T., et al.: U-net: deep learning for cell counting, detection, and morphometry. Nat. Methods **16**(1), 67–70 (2019)
8. Guarnera, F., et al.: Mslesseg: baseline and benchmarking of a new multiple sclerosis lesion segmentation dataset. Sci. Data **12**(1), 1–10 (2025)
9. Hashemi, M., Akhbari, M., Jutten, C.: Delve into multiple sclerosis (ms) lesion exploration: a modified attention u-net for ms lesion segmentation in brain mri. Comput. Biol. Med. **145**, 105402 (2022)
10. Hatamizadeh, A., Nath, V., Tang, Y., Yang, D., Roth, H.R., Xu, D.: Swin unetr: swin transformers for semantic segmentation of brain tumors in MRI images. In: International MICCAI Brainlesion Workshop, pp. 272–284. Springer (2021)
11. Ho, J., Jain, A., Abbeel, P.: Denoising diffusion probabilistic models. Adv. Neural. Inf. Process. Syst. **33**, 6840–6851 (2020)
12. Janardhan, V., Suri, S., Bakshi, R.: Multiple sclerosis: hyperintense lesions in the brain on nonenhanced t1-weighted mr images evidenced as areas of t1 shortening. Radiology **244**(3), 823–831 (2007)
13. Lassmann, H.: Multiple sclerosis pathology. Cold Spring Harbor Perspect. Med. **8**(3), a028936 (2018)
14. Liu, Z., et al.: Swin transformer: Hierarchical vision transformer using shifted windows. In: Proceedings of the IEEE/CVF International Conference on Computer Vision, pp. 10012–10022 (2021)
15. Loshchilov, I., Hutter, F.: Sgdr: stochastic gradient descent with warm restarts. arXiv preprint arXiv:1608.03983 (2016)
16. Myronenko, A.: 3d MRI brain tumor segmentation using autoencoder regularization. In: Brainlesion: Glioma, Multiple Sclerosis, Stroke and Traumatic Brain Injuries: 4th International Workshop, BrainLes 2018, Held in Conjunction with MICCAI 2018. Revised Selected Papers, Part II 4, pp. 311–320. Springer (2019)

17. Oktay, O., et al.: Attention u-net: learning where to look for the pancreas. arXiv preprint arXiv:1804.03999 (2018)
18. Paszke, A., et al.: Pytorch: an imperative style, high-performance deep learning library. Adv. Neural Inf. Process. Syst. **32** (2019)
19. Raab, F., Wein, S., Greenlee, M., Malloni, W., Lang, E.: A multimodal 2d convolutional neural network for multiple sclerosis lesion detection. Authorea Preprints (2023)
20. Rondinella, A., et al.: Boosting multiple sclerosis lesion segmentation through attention mechanism. Comput. Biol. Med. **161**, 107021 (2023)
21. Rondinella, A., et al.: Icpr 2024 competition on multiple sclerosis lesion segmentation—methods and results. In: International Conference on Pattern Recognition, pp. 1–16. Springer (2024)
22. Rondinella, A., et al.: Enhancing multiple sclerosis lesion segmentation in multimodal mri scans with diffusion models. In: 2023 IEEE International Conference on Bioinformatics and Biomedicine (BIBM), pp. 3733–3740 (2023). https://doi.org/10.1109/BIBM58861.2023.10385334
23. Song, J., Meng, C., Ermon, S.: Denoising diffusion implicit models. arXiv preprint arXiv:2010.02502 (2020)
24. Vaswani, A., et al.: Attention is all you need. Adv. Neural Inf. Process. Syst. **30** (2017)
25. Warfield, S.K., Zou, K.H., Wells, W.M.: Simultaneous truth and performance level estimation (staple): an algorithm for the validation of image segmentation. IEEE Trans. Med. Imaging **23**(7), 903–921 (2004)
26. Wu, J., et al.: Medsegdiff: Medical image segmentation with diffusion probabilistic model. In: Medical Imaging with Deep Learning, pp. 1623–1639. PMLR (2024)
27. Wu, J., Ji, W., Fu, H., Xu, M., Jin, Y., Xu, Y.: Medsegdiff-v2: diffusion-based medical image segmentation with transformer. In: Proceedings of the AAAI Conference on Artificial Intelligence. vol. 38, pp. 6030–6038 (2024)
28. Xing, Z., Wan, L., Fu, H., Yang, G., Zhu, L.: Diff-unet: a diffusion embedded network for volumetric segmentation. arXiv preprint arXiv:2303.10326 (2023)
29. Zhong, Z., et al.: Squeeze-and-attention networks for semantic segmentation. In: Proceedings of the IEEE/CVF Conference on Computer Vision and Pattern Recognition, pp. 13065–13074 (2020)

Latent Space Synergy: Text-Guided Data Augmentation for Direct Diffusion Biomedical Segmentation

Muhammad Aqeel[1]([✉])(iD), Maham Nazir[2](iD), Zanxi Ruan[1](iD), and Francesco Setti[1](iD)

[1] Department of Engineering for Innovation Medicine, University of Verona, Strada le Grazie 15, Verona, Italy
`muhammad.aqeel@univr.it`
[2] Department of Computer Science, Beihang University, Beijing, China

Abstract. Medical image segmentation suffers from data scarcity, particularly in polyp detection where annotation requires specialized expertise. We present SynDiff, a framework combining text-guided synthetic data generation with efficient diffusion-based segmentation. Our approach employs latent diffusion models to generate clinically realistic synthetic polyps through text-conditioned inpainting, augmenting limited training data with semantically diverse samples. Unlike traditional diffusion methods requiring iterative denoising, we introduce direct latent estimation enabling single-step inference with $T\times$ computational speedup. On CVC-ClinicDB, SynDiff achieves 96.0% Dice and 92.9% IoU while maintaining real-time capability suitable for clinical deployment. The framework demonstrates that controlled synthetic augmentation improves segmentation robustness without distribution shift. SynDiff bridges the gap between data-hungry deep learning models and clinical constraints, offering an efficient solution for deployment in resource-limited medical settings.

Keywords: Medical Image Segmentation · Diffusion Model · Polyp Detection · Text-Guided Synthesis

1 Introduction

Biomedical image segmentation plays a critical role in modern healthcare, enabling precise diagnosis and treatment planning [14,22]. In gastrointestinal endoscopy, automated polyp segmentation has emerged as a particularly important application with the potential to improve colorectal cancer screening accuracy and reduce missed detection rates during real-time procedures. However, data scarcity represents the most fundamental bottleneck limiting the development of robust medical segmentation systems. Medical datasets are constrained

---

M. Aqeel and M. Nazir—Equal contribution.

E. Rodolà et al. (Eds.): ICIAP 2025 Workshops, LNCS 16169, pp. 417–428, 2026.
https://doi.org/10.1007/978-3-032-11317-7_35

by privacy regulations, costly expert annotation, and time-intensive boundary delineation [15,18]. In polyp segmentation, this challenge is compounded by significant morphological diversity, requiring extensive annotated examples that current public datasets cannot provide [5]. Traditional data augmentation in medical imaging relies on geometric transformations [22,23]. While providing some benefit, these approaches cannot generate new pathological variations needed for robust model generalization. Recent GAN-based synthesis methods have shown promise but suffer from limited controllability and mode collapse [28].

The emergence of diffusion probabilistic models has revolutionized image generation [13,17]. Text-conditioned diffusion models enable semantically-guided data augmentation through clinical descriptions [21]. However, traditional diffusion-based segmentation requires computationally intensive multi-step inference, making it impractical for clinical deployment [27,29]. Latent diffusion models (LDMs) address these computational limitations by operating in compressed latent spaces [21]. Their text-conditioning capabilities present an opportunity to create clinically-informed augmentation strategies, where domain expertise guides the generation of diverse pathological variations [19].

In this paper, we present SynDiff, a framework that addresses medical data scarcity through text-guided synthetic data augmentation while maintaining computational efficiency for practical deployment. Our approach leverages latent diffusion models to generate diverse synthetic polyp images guided by clinical descriptions, effectively expanding training datasets with semantically meaningful variations. We integrate this with a direct latent estimation technique enabling single-step segmentation inference, eliminating the computational burden of iterative denoising while preserving performance. Our contributions are:

- A text-guided data augmentation framework using latent diffusion models to address medical data scarcity through semantically-controlled synthetic polyp generation.
- An efficient single-step segmentation approach that maintains competitive performance while dramatically reducing computational requirements compared to multi-step diffusion methods.
- Comprehensive evaluation demonstrating that synthetic data augmentation preserves segmentation quality (96.0% Dice, 92.9% IoU) while expanding dataset diversity for improved model robustness.

2   Related Work

Medical image segmentation has evolved significantly with deep learning approaches, with U-Net [22] establishing the encoder-decoder paradigm that remains foundational for medical tasks. Recent advances include transformer-based architectures like TransUNet [7] and UNETR [12] that capture long-range dependencies, and specialized polyp segmentation methods such as PraNet [10] and SANet [26]. Despite these architectural innovations, all approaches fundamentally require extensive annotated datasets to achieve robust performance, a constraint that significantly limits practical clinical deployment.

Traditional augmentation through geometric transformations fails to capture pathological diversity [5]. Alternative approaches address data scarcity through meta-learning [1,4] and self-supervised refinement [2,3], which enhance model robustness by learning better representations from limited data. In contrast, synthetic generation methods directly create new training samples: GANs suffer from mode collapse and training instability [8], while recent diffusion-based approaches like DiffBoost [30] demonstrate superior controllability but lack integration with downstream tasks. Our work bridges this gap by combining controllable synthesis with task-specific optimization.

Diffusion models have emerged as powerful generative frameworks, with medical applications explored in MedSegDiff [28] and MedSegDiff-V2 [27] that treat segmentation as conditional generation through iterative denoising. Latent Diffusion Models [21] reduced computational overhead by operating in compressed latent spaces while enabling text-conditioning for semantic control. However, existing approaches require multi-step inference processes that are computationally prohibitive for clinical deployment [29], while single-step inference methods explored in natural image domains [25] remain underexplored for medical segmentation.

While progress has been made in individual components—synthetic medical data generation, efficient diffusion inference, and medical segmentation—no prior work integrates these elements to address both data scarcity and computational efficiency simultaneously. Our work fills this gap by combining text-guided synthetic data generation with single-step diffusion segmentation, enabling enhanced training diversity while maintaining practical deployment efficiency.

3   SynDiff Framework

Our proposed approach, SynDiff, inspired by [11,16], integrates text-guided synthetic data generation with single-step diffusion segmentation to address the dual challenges of limited training data and computational efficiency in biomedical image segmentation as shown in Fig. 1. The framework operates in two distinct phases: synthetic data generation using Stable Diffusion XL (SDXL) [20] for text-guided inpainting, followed by end-to-end training of a single-step segmentation model that processes both real and synthetic data.

3.1   Latent Diffusion Model

Both the data generation and segmentation components operate within a latent diffusion framework to reduce computational overhead. For the segmentation pipeline, we employ a trainable vision encoder τ_θ to encode an image $C \in \mathbb{R}^{H \times W \times 3}$ into its latent representation $z_c = \tau_\theta(C)$. For segmentation maps, we

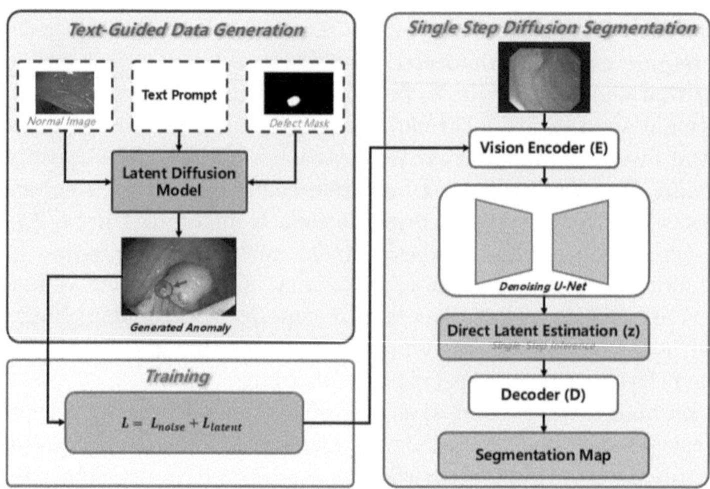

Fig. 1. Overview of the SynDiff framework. (Left) Offline text-guided data generation: Stable Diffusion XL (SDXL) inpainting takes a normal endoscopic image, clinical text prompt, and binary mask to generate synthetic polyp images with corresponding ground truth masks. (Right) Single-step segmentation pipeline: input images are encoded through a trainable vision encoder, processed by a denoising U-Net for direct latent estimation, and decoded to produce segmentation masks in a single inference step.

leverage a pre-trained autoencoder with encoder E and decoder D for perceptual compression. Given a segmentation map $X \in \mathbb{R}^{H \times W \times 3}$ in pixel space, the encoder produces a latent representation $z = E(X)$, and the decoder recovers the segmentation map as $\hat{X} = D(z)$, where $z \in \mathbb{R}^{h \times w \times c}$. For the data generation pipeline, SDXL operates in its latent space where the normal image, mask, and text prompt are jointly processed—the text conditions the diffusion process through cross-attention while the mask guides spatial inpainting. Both pipelines operate entirely in compressed latent spaces, significantly reducing computational requirements while preserving essential structural information.

3.2 Text-Guided Synthetic Data Generation

To address medical data scarcity, we implement an offline synthetic data generation process using SDXL inpainting conditioned on clinical text descriptions. The generation process takes three inputs: a normal endoscopic image i_n randomly sampled from available normal cases, a text description d_s specifying desired polyp characteristics such as "small sessile polyp with irregular surface texture," and a binary mask m_s indicating the spatial region where the polyp should be synthesized. Within the SDXL latent diffusion framework, these inputs are jointly processed: the text prompt conditions the generation through cross-attention mechanisms while operating with frozen pre-trained weights to generate a synthetic image $i_s = \text{SDXL}_{\text{inpaint}}(i_n, d_s, m_s)$ containing a realistic polyp within the masked region. The binary mask m_s simultaneously serves as the ground truth segmentation label for the generated synthetic image i_s. We generate 100 synthetic samples using 50 diverse text prompts describing various polyp morphologies, sizes, and surface characteristics, augmenting our training dataset with approximately 20% additional data.

3.3   Direct Latent Estimation for Single-Step Segmentation

Our segmentation component introduces a direct latent estimation strategy that enables single-step inference, achieving theoretical computational speedup of $T\times$ compared to traditional diffusion approaches requiring T denoising steps. During training, the latent representation of a segmentation map z_0 undergoes forward diffusion by adding Gaussian noise for t timesteps to obtain:

$$z_t = \sqrt{\bar{\alpha}_t} z_0 + \sqrt{1 - \bar{\alpha}_t} n \tag{1}$$

where n is random Gaussian noise and $\bar{\alpha}_t$ controls the noise schedule. The denoising U-Net $f(\cdot)$ is trained to estimate the noise as:

$$\tilde{n} = f(z_t, z_c) \tag{2}$$

where z_c is the conditioning image's latent representation. The key innovation lies in directly estimating the clean latent z_0 from the noisy latent z_t in a single step through:

$$\tilde{z}_0 = \frac{1}{\sqrt{\bar{\alpha}_t}}(z_t - \sqrt{1 - \bar{\alpha}_t} \tilde{n}) \tag{3}$$

This enables a dual supervision strategy where we minimize both noise prediction error:

$$\mathcal{L}_{\text{noise}} = ||n - \tilde{n}||_1 \tag{4}$$

and direct latent estimation error:

$$\mathcal{L}_{\text{latent}} = ||z_0 - \tilde{z}_0||_1 \tag{5}$$

The combined loss function is:

$$\mathcal{L} = \mathcal{L}_{\text{noise}} + \lambda \mathcal{L}_{\text{latent}} \tag{6}$$

with $\lambda = 1$, enabling the model to learn both denoising and direct prediction capabilities simultaneously. The dual supervision strategy is particularly effective for segmentation tasks due to the discrete nature of segmentation masks, which allows for more stable latent predictions and direct optimization of boundary precision—critical for accurate polyp delineation. By leveraging the structural simplicity of binary masks, our approach achieves single-step inference while maintaining segmentation quality, making it suitable for real-time clinical deployment.

3.4   Latent Fusion and Training Protocol

To integrate image features with segmentation information efficiently, we implement concatenation-based latent fusion rather than computationally expensive cross-attention mechanisms. The latent representations of the conditioning image z_c and segmentation map are concatenated along the channel dimension before processing by the denoising U-Net. Our trainable vision encoder τ_θ shares the same architecture as the segmentation encoder E but remains trainable to adapt pre-trained natural image features to medical imaging characteristics. During training, the vision encoder τ_θ and denoising U-Net $f(\cdot)$ are updated through backpropagation, while the autoencoder components (E, D) and SDXL inpainting model remain frozen with pre-trained weights. The training proceeds with random timestep sampling $t \sim \text{Uniform}(1, 1000)$ during forward diffusion, ensuring the model learns to handle varying noise levels for robust single-step inference.

3.5   Inference and Computational Efficiency

During inference, the framework processes input medical images through a single forward pass, eliminating the iterative sampling required by traditional diffusion methods. An input image is encoded through the trainable vision encoder τ_θ to obtain z_c, which conditions the denoising U-Net to directly estimate the segmentation latent $\tilde{z}_0$ at a fixed timestep $t = 50$ (empirically determined for optimal quality-efficiency balance). The estimated latent is decoded through the frozen decoder D to produce the final segmentation mask. This single-step approach reduces computational complexity from $O(T \times \text{U-Net forward pass})$ to $O(1 \times \text{U-Net forward pass})$, making the method suitable for real-time clinical deployment while maintaining competitive segmentation accuracy on both real and synthetic training data.

4   Experimental Results

4.1   Dataset and Evaluation Metrics

We evaluate SynDiff on the CVC-ClinicDB [6] dataset, consisting of 550 RGB colonoscopy images with 488 training and 62 test images. Each image includes precise polyp boundary annotations for binary segmentation. To ensure statistical reliability, we report Dice Coefficient (DC) and Intersection over Union (IoU) metrics with mean $\pm$ standard deviation across 5-fold cross-validation. Additional boundary quality metrics include Hausdorff Distance at 95th percentile (HD95) and Normalized Surface Distance (NSD) for clinical precision assessment, as accurate boundary delineation is crucial for surgical planning in clinical practice.

4.2   Implementation Details

Text-Guided Data Generation: We implement SDXL [20] inpainting using the Diffusers library with carefully crafted clinical prompts including "sessile polyp with irregular surface texture," "pedunculated polyp on mucosal fold," and "flat adenomatous lesion" while employing negative prompts "smooth healthy tissue, normal colon wall" to ensure realistic synthesis. The inpainting process utilizes binary masks derived from original dataset annotations, ensuring synthetic polyps are placed in anatomically plausible locations.

Training: Models are trained for 100,000 steps using AdamW optimizer (learning rate $= 1 \times 10^{-5}$, batch size $= 4$) on NVIDIA RTX 4090. We use a KL-regularized autoencoder with $8\times$ downsampling ($256 \times 256 \rightarrow 32 \times 32 \times 4$ latent space), with all components initialized from pre-trained Stable Diffusion weights.

Inference: Single-step prediction at fixed timestep $t = 50$ with Gaussian noise concatenated to image latents for direct mask estimation, eliminating iterative sampling required by traditional diffusion approaches.

Table 1. Polyp segmentation performance comparison on CVC-ClinicDB dataset.

Method	Dice (%)	IoU (%)	HD95 (mm)	NSD (%)
SSFormer [24]	94.4 ± 0.4	89.9 ± 0.6	12.3 ± 2.1	87.2 ± 1.8
Li-SegPNet [23]	92.5 ± 0.5	86.0 ± 0.7	15.6 ± 3.2	84.1 ± 2.3
Diff-Trans [9]	95.4 ± 0.3	92.0 ± 0.4	8.7 ± 1.5	89.8 ± 1.2
SDSeg [16]	95.8 ± 0.2	92.6 ± 0.3	7.9 ± 1.3	90.4 ± 1.1
SynDiff	$\mathbf{96.0 \pm 0.3}$	$\mathbf{92.9 \pm 0.5}$	$\mathbf{7.2 \pm 1.1}$	$\mathbf{91.1 \pm 1.0}$

4.3   Quantitative Results

We compare SynDiff against established polyp segmentation approaches representing different architectural paradigms. SSFormer [24] employs a pyramid transformer architecture for multi-scale feature extraction, while Li-SegPNet [23] utilizes lightweight separable convolutions for efficient segmentation. Among diffusion-based methods, Diff-Trans [9] combines diffusion models with transformer architectures for iterative refinement, whereas SDSeg [16] implements stable diffusion for medical image segmentation but requires multiple denoising steps.

Table 1 presents a comprehensive performance comparison on CVC-ClinicDB, demonstrating the effectiveness of our integrated approach. SynDiff achieves competitive performance with $96.0 \pm 0.3\%$ Dice coefficient and $92.9 \pm 0.5\%$ IoU, substantially outperforming traditional CNN-based architectures with a 1.6% Dice improvement over SSFormer (94.4% Dice) and a 3.5% improvement over Li-SegPNet (92.5% Dice). When compared to recent diffusion-based approaches, our method achieves marginal but consistent improvements over SDSeg (95.8% Dice) and Diff-Trans (95.4% Dice) while offering significant computational advantages through single-step inference.

The boundary quality metrics reveal particularly encouraging results with HD95 of 7.2 ± 1.1mm and NSD of 91.1 ± 1.0%, indicating superior edge preservation crucial for clinical applications. These improvements in boundary accuracy are clinically significant, as precise polyp delineation directly impacts diagnostic confidence and treatment planning decisions. The consistent low standard deviations across all metrics demonstrate robust performance across cross-validation folds, which is essential for reliable clinical deployment.

Table 2. Ablation study on key components of SynDiff framework.

Configuration	Dice (%)	IoU (%)
Baseline (Real data only)	93.7 ± 0.4	89.6 ± 0.6
+ Text-guided augmentation	96.0 ± 0.3	92.9 ± 0.5
+ Multi-step inference ($T = 50$)	96.1 ± 0.3	93.0 ± 0.4
+ Frozen vision encoder	95.2 ± 0.4	91.8 ± 0.6
+ Noise loss only ($\lambda = 0$)	95.4 ± 0.3	92.1 ± 0.5
+ Traditional augmentation	94.8 ± 0.4	90.7 ± 0.6

4.4   Ablation Study

Table 2 demonstrates a systematic analysis of component contributions. Text-guided augmentation provides the most significant improvement (+2.3% Dice over baseline), while the comparison between single-step and multi-step inference shows minimal performance difference (96.0% vs 96.1% Dice), validating our core hypothesis that single-step inference maintains quality while reducing computational requirements. The trainable vision encoder contributes meaningfully

Table 3. Impact of different augmentation strategies on segmentation performance.

Augmentation Type	Synthetic Samples	Dice (%)	IoU (%)
None	0	93.7 ± 0.4	89.6 ± 0.6
Traditional (rotation, flip)	-	94.8 ± 0.4	90.7 ± 0.6
GAN-based	100	95.2 ± 0.3	91.8 ± 0.5
Text-guided (20)	20	94.1 ± 0.4	90.2 ± 0.6
Text-guided (50)	50	95.1 ± 0.3	91.5 ± 0.5
Text-guided (100)	**100**	**96.0 $\pm$ 0.3**	**92.9 $\pm$ 0.5**
Text-guided (200)	200	95.6 ± 0.4	92.1 ± 0.6

(+0.8% over frozen), demonstrating the importance of domain-specific adaptation. The dual loss formulation also proves beneficial, as noise-only supervision reduces performance to 95.4% Dice.

4.5 Data Augmentation Analysis

Figure 2 shows synthetic polyp samples generated through text-guided inpainting, demonstrating morphological diversity across sizes, shapes, and textures. The generated samples exhibit clinically plausible characteristics with realistic color variations and anatomically consistent placement. Table 3 evaluates different quantities of synthetic data added to our full training set of 488 real scans. The optimal performance is achieved with 100 synthetic samples (approximately 20% augmentation), where insufficient augmentation (20–50 samples) limits model exposure to variability, while excessive synthetic data (200 samples) introduces slight performance degradation due to distribution shift. Notably, even minimal text-guided augmentation (20 samples) shows comparable performance to traditional methods, highlighting the quality of our synthetic generation. Our text-guided approach significantly outperforms traditional geometric

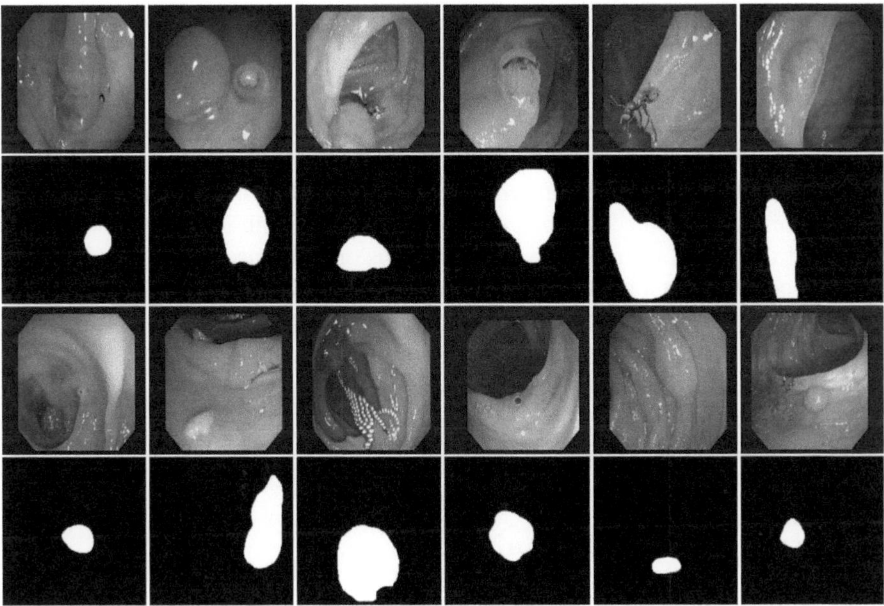

Fig. 2. Synthetically generated endoscopic images with corresponding binary segmentation masks. The first and third rows display synthetic colonoscopy images created through our text-guided generation approach, while the second and fourth rows present their corresponding binary masks. White regions in the masks indicate the synthetic anomalous tissue generated through the inpainting process. This augmented dataset enhances the diversity of training samples for our single-step diffusion segmentation model.

augmentation (+1.2% Dice) and GAN-based synthesis (+0.8% Dice), validating the effectiveness of semantically-controlled generation. The consistent improvement across both Dice and IoU metrics indicates that text-guided synthesis enhances both overall overlap and boundary delineation accuracy.

4.6 Computational Efficiency Analysis

Our single-step approach achieves theoretical complexity reduction from $O(T)$ to $O(1)$ compared to multi-step diffusion methods. Table 4 demonstrates practical efficiency gains, with SynDiff completing inference in 0.08 s compared to 1.8-2.3 s for existing diffusion methods, representing a 22–28× speedup. This efficiency improvement stems from our direct latent estimation strategy that eliminates iterative denoising while maintaining competitive accuracy, making the method suitable for real-time clinical applications.

Table 4. Inference efficiency comparison across diffusion segmentation methods.

Method	Inference Steps	Time (s)
Diff-Trans	50	1.8
SDSeg	100	2.3
SynDiff	**1**	**0.08**

5 Conclusion

We introduced SynDiff, a framework that addresses data scarcity and computational efficiency in medical image segmentation through text-conditioned synthetic data augmentation and single-step diffusion inference. On CVC-ClinicDB, our method achieved a Dice coefficient of $96.0 \pm 0.3\%$ and HD95 of 7.2 ± 1.1 mm. The direct latent estimation approach reduced inference time from 1.8-2.3 s to 0.08 s while maintaining segmentation accuracy. Our analysis revealed that augmenting with 100 synthetic samples optimizes performance, with text-guided generation outperforming conventional augmentation methods. These findings demonstrate the potential of combining semantic text guidance with efficient diffusion inference for practical medical imaging applications.

6 Future Work and Limitations

While SynDiff demonstrates competitive performance, limitations include evaluation on a single dataset, reliance on carefully engineered text prompts for realistic synthesis, and computational requirements that remain higher than traditional CNN approaches. Future work should prioritize multi-dataset validation across diverse imaging modalities, integration of clinical expert feedback into synthetic data generation, and exploration of the framework's effectiveness on other medical imaging tasks beyond polyp segmentation to establish broader clinical applicability.

Acknowledgements. This study was carried out within the PNRR research activities of the consortium iNEST (Interconnected North-Est Innovation Ecosystem) funded by the European Union Next-GenerationEU (Piano Nazionale di Ripresa e Resilienza (PNRR) Missione 4 Componente 2, Investimento 1.5 D.D. 1058 23/06/2022, ECS_00000043).

References

1. Aqeel, M., Sharifi, S., Cristani, M., Setti, F.: Meta learning-driven iterative refinement for robust anomaly detection in industrial inspection. In: European Conference on Computer Vision, pp. 445–460. Springer (2024)
2. Aqeel, M., Sharifi, S., Cristani, M., Setti, F.: Self-supervised learning for robust surface defect detection. In: International Conference on Deep Learning Theory and Applications (2024)
3. Aqeel, M., Sharifi, S., Cristani, M., Setti, F.: Self-supervised iterative refinement for anomaly detection in industrial quality control. In: International Joint Conference on Computer Vision, Imaging and Computer Graphics Theory and Applications (2025)
4. Aqeel, M., Sharifi, S., Cristani, M., Setti, F.: Towards real unsupervised anomaly detection via confident meta-learning. In: Accepted to Proceedings of the IEEE/CVF International Conference on Computer Vision (2025)
5. Baranchuk, D., Rubachev, I., Voynov, A., Khrulkov, V., Babenko, A.: Label-efficient semantic segmentation with diffusion models. arXiv preprint arXiv:2112.03126 (2021)
6. Bernal, J., Sánchez, F.J., Fernández-Esparrach, G., Gil, D., Rodríguez, C., Vilariño, F.: Wm-dova maps for accurate polyp highlighting in colonoscopy: Validation vs. saliency maps from physicians. Comput. Med. Imaging graph. **43**, 99–111 (2015)
7. Chen, J., et al.: Transunet: Transformers make strong encoders for medical image segmentation. arXiv preprint arXiv:2102.04306 (2021)
8. Choi, E., Biswal, S., Malin, B., Duke, J., Stewart, W.F., Sun, J.: Generating multi-label discrete patient records using generative adversarial networks. In: Machine Learning for Healthcare Conference, pp. 286–305. PMLR (2017)
9. Chowdary, G.J., Yin, Z.: Diffusion transformer u-net for medical image segmentation. In: International Conference on Medical Image Computing and Computer-Assisted Intervention, pp. 622–631. Springer (2023)
10. Fan, D.P., et al.: Pranet: Parallel reverse attention network for polyp segmentation. In: International Conference on Medical Image Computing and Computer-assisted Interventionm, pp. 263–273. Springer (2020)
11. Girella, F., Liu, Z., Fummi, F., Setti, F., Cristani, M., Capogrosso, L.: Leveraging latent diffusion models for training-free in-distribution data augmentation for surface defect detection. In: International Conference on Content-Based Multimedia Indexing (CBMI) (2024)
12. Hatamizadeh, A., et al.: Unetr: Transformers for 3d medical image segmentation. In: Proceedings of the IEEE/CVF Winter Conference on Applications of Computer Vision, pp. 574–584 (2022)
13. Ho, J., Jain, A., Abbeel, P.: Denoising diffusion probabilistic models. Adv. Neural. Inf. Process. Syst. **33**, 6840–6851 (2020)

14. Isensee, F., Jaeger, P.F., Kohl, S.A., Petersen, J., Maier-Hein, K.H.: nnu-net: a self-configuring method for deep learning-based biomedical image segmentation. Nat. Methods **18**(2), 203–211 (2021)

15. Jha, D., et al.: Kvasir-seg: A segmented polyp dataset. In: International conference on multimedia modeling, pp. 451–462. Springer (2019)

16. Lin, T., Chen, Z., Yan, Z., Yu, W., Zheng, F.: Stable diffusion segmentation for biomedical images with single-step reverse process. In: Medical Image Computing and Computer Assisted Intervention – MICCAI 2024, pp. 656–666. Springer Nature Switzerland, Cham (2024)

17. Nichol, A.Q., Dhariwal, P.: Improved denoising diffusion probabilistic models. In: International Conference on Machine Learning, pp. 8162–8171. PMLR (2021)

18. Orlando, J.I., et al.: Refuge challenge: a unified framework for evaluating automated methods for glaucoma assessment from fundus photographs. Med. Image Anal. **59**, 101570 (2020)

19. Peebles, W., Xie, S.: Scalable diffusion models with transformers. In: Proceedings of the IEEE/CVF International Conference on Computer Vision, pp. 4195–4205 (2023)

20. Podell, D., et al.: Sdxl: Improving latent diffusion models for high-resolution image synthesis. arXiv preprint arXiv:2307.01952 (2023)

21. Rombach, R., Blattmann, A., Lorenz, D., Esser, P., Ommer, B.: High-resolution image synthesis with latent diffusion models. In: Proceedings of the IEEE/CVF Conference on Computer Vision and Pattern Recognition, pp. 10684–10695 (2022)

22. Ronneberger, O., Fischer, P., Brox, T.: U-net: Convolutional networks for biomedical image segmentation. In: Medical image computing and computer-assisted intervention–MICCAI 2015: 18th International Conference, Munich, Germany, October 5-9, 2015, Proceedings, Part III 18, pp. 234–241. Springer (2015)

23. Sharma, P., Gautam, A., Maji, P., Pachori, R.B., Balabantaray, B.K.: Li-segpnet: Encoder-decoder mode lightweight segmentation network for colorectal polyps analysis. IEEE Trans. Biomed. Eng. **70**(4), 1330–1339 (2022)

24. Shi, W., Xu, J., Gao, P.: Ssformer: A lightweight transformer for semantic segmentation. In: 2022 IEEE 24th International Workshop on Multimedia Signal Processing (MMSP), pp. 1–5. IEEE (2022)

25. Song, Y., Dhariwal, P., Chen, M., Sutskever, I.: Consistency models (2023)

26. Wei, J., Hu, Y., Zhang, R., Li, Z., Zhou, S.K., Cui, S.: Shallow attention network for polyp segmentation. In: Medical Image Computing and Computer Assisted Intervention–MICCAI 2021: 24th International Conference, Strasbourg, France, September 27–October 1, 2021, Proceedings, Part I 24, pp. 699–708. Springer (2021)

27. Wu, J., et al.: Medsegdiff: Medical image segmentation with diffusion probabilistic model. In: Medical Imaging with Deep Learning (2024)

28. Wu, J., Ji, W., Fu, H., Xu, M., Jin, Y., Xu, Y.: Medsegdiff-v2: Diffusion-based medical image segmentation with transformer. In: Proceedings of the AAAI Conference on Artificial Intelligence (2024)

29. Xing, Z., Wan, L., Fu, H., Yang, G., Zhu, L.: Diff-unet: A diffusion embedded network for volumetric segmentation. arXiv preprint arXiv:2303.10326 (2023)

30. Zhang, Z., et al.: Diffboost: Enhancing medical image segmentation via text-guided diffusion model. IEEE Trans. Med. Imaging (2024)

Benchmarking GANs, Diffusion Models, and Flow Matching for T1w-to-T2w MRI Translation

Andrea Moschetto[1]([✉]), Lemuel Puglisi[1][iD], Alec Sargood[3],
Pierluigi Dell'Acqua[2], Francesco Guarnera[1][iD], Sebastiano Battiato[1][iD],
and Daniele Ravì[2][iD]

[1] Università degli Studi di Catania, Catania, CT 95124, Italy
`andrea.moschetto@studium.unict.it`
[2] Università degli Studi di Messina, Messina, ME 98122, Italy
[3] University College London, London, LDN WC1E 6BT, UK

Abstract. Magnetic Resonance Imaging (MRI) enables the acquisition of multiple image contrasts, such as T1-weighted (T1w) and T2-weighted (T2w) scans, each offering distinct diagnostic insights. However, acquiring all desired modalities increases scan time and cost, motivating research into computational methods for cross-modal synthesis. To address this, recent approaches aim to synthesize missing MRI contrasts from those already acquired, reducing acquisition time while preserving diagnostic quality. Image-to-image (I2I) translation provides a promising framework for this task. In this paper, we present a comprehensive benchmark of generative models—specifically, Generative Adversarial Networks (GANs), diffusion models, and flow matching (FM) techniques—for T1w-to-T2w 2D MRI I2I translation. All frameworks are implemented with comparable settings and evaluated on three publicly available MRI datasets of healthy adults. Our quantitative and qualitative analyses show that the GAN-based Pix2Pix model outperforms diffusion and FM-based methods in terms of structural fidelity, image quality, and computational efficiency. Consistent with existing literature, these results suggest that flow-based models are prone to overfitting on small datasets and simpler tasks, and may require more data to match or surpass GAN performance. These findings offer practical guidance for deploying I2I translation techniques in real-world MRI workflows and highlight promising directions for future research in cross-modal medical image synthesis. Code and models are publicly available at https://github.com/AndreaMoschetto/medical-I2I-benchmark

Keywords: Image-to-Image Translation · Magnetic Resonance Imaging · Generative AI · Flow Matching · Diffusion Models · Generative Adversarial Networks

E. Rodolà et al. (Eds.): ICIAP 2025 Workshops, LNCS 16169, pp. 429–440, 2026.
https://doi.org/10.1007/978-3-032-11317-7_36

1   Introduction

Magnetic Resonance Imaging (MRI) is one of the most powerful and widely used medical imaging techniques. A unique strength of MRI lies in its ability to generate multiple image contrasts—such as T1-weighted (T1w), T2-weighted (T2w), and FLAIR images—each emphasizing different tissue characteristics. For example, T1w images are useful for highlighting fat, while T2w images are more effective at detecting fluid. As a result, multiple scan types are often required for a single patient, which can increase both the duration and cost of the MRI examination. To address these challenges, there is increasing interest in methods that can reduce the number of scans needed without sacrificing diagnostic value. One promising approach is image-to-image (I2I) translation, where computational models estimate missing image contrasts from those already acquired, for example, synthesizing a T2w image starting from a T1w one. By synthesizing additional modalities from existing scans, I2I translation has the potential to accelerate MRI workflows, reduce patient burden, and lower healthcare costs. Furthermore, I2I techniques can help address situations where certain contrasts are missing due to time and cost constraints or incomplete acquisitions, ensuring that clinicians and downstream models still have access to the necessary imaging information for accurate diagnosis and analysis.

Recent advances in generative artificial intelligence have significantly enhanced the generation of synthetic MRI contrasts and have emerged as powerful tools for I2I translation. These methods offer a data-driven approach to synthesizing missing MRI modalities from available ones. Among them, Generative Adversarial Networks (GANs)—such as Pix2Pix [13]—have been widely adopted for their ability to produce high-resolution and realistic outputs by learning mappings between paired image domains in an adversarial manner. More recently, diffusion models [10] have demonstrated superior performance in generating high-fidelity images by iteratively refining noise into structured outputs. Another promising direction is Flow Matching (FM) [15], which models transformations between distributions through learned continuous dynamics.

Despite significant advancements in generative modeling for MRI contrast translation, a comprehensive and fair comparison of recent methods is still lacking. In this paper, we benchmark state-of-the-art generative models for synthesizing T2w 2D axial slices from corresponding T1w slices within a unified experimental framework. To our knowledge, this is the first direct comparison of GANs, diffusion models, and FMs for this task, all implemented with the same U-Net backbone [18]. We present both quantitative and qualitative evaluations highlighting the strengths and limitations of each approach.

Our study aims to identify the model that achieves the best image fidelity, structural consistency, and computational efficiency simultaneously, providing clear guidance for deploying I2I translation in real-world MRI applications.

2   Related Works

Medical I2I translation has been explored using both supervised [2] and self-supervised [12] learning approaches, with recent advances increasingly favoring generative models. Early studies in generative I2I translation predominantly focused on GANs, which use a generatordiscriminator architecture to synthesize realistic images through adversarial training. Within this framework, researchers have tried to enhance translation quality by optimizing input representations and network architectures. For example, [14,21] proposed a conditional GAN for T2w image synthesis from T1w inputs. Specifically, in [14] the authors analyze the impact of different input resolutions and dimensionalities. Similarly, Dey et al. [8] introduced MTSR-MRI, a GAN-based framework that jointly performs super-resolution and modality translation in a lower-dimensional embedding space.

Diffusion models have emerged as a powerful alternative in generative modeling. Building on the limitations of GANs, these new models provide greater training stability and excel at producing high-fidelity, diverse images by gradually reversing a learned noise process. This advancement has set a new standard for image synthesis, enabling more reliable and realistic results [9]. Although these methods have shown success on 2D natural images, a major challenge lies in extending them to the 3D volumetric medical domain while maintaining computational efficiency. Choo et al. [7] tackle this challenge by introducing a slice-consistent 2D diffusion model for CT-to-MRI translation. Their approach incorporates style conditioning and inter-slice trajectory alignment to preserve 3D anatomical coherence, while avoiding direct computation on full 3D volumes. Similarly, MC-IDDPM [17] employs a diffusion process for synthetic CT generation from MRI, using a shifted-window transformer V-Net (Swin-VNet) to produce high-quality synthetic CT scans aligned with MRI anatomy.

In recent years, FM models [15] have been proposed as a generalized framework that enables more efficient training and sampling by learning continuous dynamics to map between source and target data distributions. Several studies have successfully applied flow matching to a range of image-related tasks, including medical image synthesis [22], image restoration [16], and natural I2I translation [6]. However, there is limited research specifically focused on applying flow matching to medical I2I translation.

3 Preliminaries

We consider the task of medical I2I translation using a dataset of paired T1w and T2w axial slices, denoted as $\{(x^{(i)}, y^{(i)})\}_{i=1}^{N}$, where each $x^{(i)}, y^{(i)} \in \mathbb{R}^{H \times W}$ represents the central axial slice from the T1w and T2w volumes of the i-th subject. The objective is to generate the T2w slice $y^{(i)}$ given the corresponding T1w slice $x^{(i)}$, which can be framed as learning a conditional distribution $p(y \mid x = x^{(i)})$. To approximate this distribution, we train a generative model G_θ, parameterized by θ, on the available paired dataset. We evaluate three distinct families of generative models—GANs, diffusion models, and FMs—which we describe in the sections below.

3.1 Pix2Pix

The Pix2Pix model [13] approaches I2I translation as a supervised learning task using a conditional GAN. The generator G_θ^{p2p}, with parameters θ, learns to translate a T1w slice $x^{(i)}$ into a synthetic T2w slice $\hat{y}^{(i)} = G_\theta^{p2p}(x^{(i)})$, while a discriminator D_ϕ, with parameters ϕ, is trained simultaneously to distinguish between real pairs $(x^{(i)}, y^{(i)})$ and fake pairs $(x^{(i)}, \hat{y}^{(i)})$. The training objective combines an adversarial loss with a pixel-wise reconstruction loss:

$$\mathcal{L}_{p2p} = \mathbb{E}_{x^{(i)}, y^{(i)}} \left[\log D_\phi(x^{(i)}, y^{(i)}) \right] + \mathbb{E}_{x^{(i)}} \left[\log \left(1 - D_\phi(x^{(i)}, G_\theta^{p2p}(x^{(i)})) \right) \right] \\ + \lambda \| y^{(i)} - G_\theta^{p2p}(x^{(i)}) \|_1, \tag{1}$$

where λ controls the trade-off between realism and fidelity to the ground truth.

3.2 Conditional Diffusion Model

Diffusion models [10] are generative models that learn to reverse a Markovian noising process applied to the data. In our setting, the forward process incrementally corrupts a target T2w slice, $y_0^{(i)} = y^{(i)}$, with Gaussian noise over T steps. At each step t, the noisy image $y_t^{(i)}$ is sampled from:

$$q(y_t^{(i)} \mid y_{t-1}^{(i)}) = \mathcal{N}(y_t^{(i)}; \sqrt{1 - \beta_t}\, y_{t-1}^{(i)}, \beta_t I), \tag{2}$$

where β_t follows a predefined variance schedule. After T steps, the image is fully transformed into noise, i.e., $y_T^{(i)} \sim \mathcal{N}(0, I)$. The goal of the model is to learn the reverse process that transforms the noise sample $y_T^{(i)}$ back into a realistic T2w image $y_0^{(i)}$. The reverse process $y_t^{(i)} \to y_{t-1}^{(i)}$ can be expressed by the transition probability:

$$p(y_{t-1}^{(i)} \mid y_t^{(i)}; \epsilon) = \mathcal{N}\left(y_{t-1}^{(i)}; \frac{1}{\sqrt{\alpha_t}} \left(y_t^{(i)} - \frac{\beta_t}{\sqrt{1 - \bar{\alpha}_t}} \epsilon \right), \tilde{\beta}_t I \right) \tag{3}$$

where α_t, $\bar{\alpha}_t$, and $\tilde{\beta}_t$ are values derived from the variance schedule β_t and are defined in [10]. Here, $\epsilon \sim \mathcal{N}(0, I)$ is the noise used to obtain $y_t^{(i)}$ from the original image $y_0^{(i)}$ during the forward diffusion process.

A neural network $G_\theta(y_t^{(i)}, t)$ is trained to predict the noise ϵ by minimizing the following loss:

$$\mathcal{L}_\epsilon = \mathbb{E}_{t, y_0^{(i)}, \epsilon \sim \mathcal{N}(0, I)} \left[\left\| \epsilon - G_\theta(y_t^{(i)}, t) \right\|^2 \right]. \tag{4}$$

In essence, the model learns to reconstruct a clean image by progressively removing noise, step by step, using a learned denoising function.

To adapt diffusion models for conditional generation, we incorporate the T1w slice $x^{(i)}$ into the denoising process, and modify the loss term to include the conditioning:

$$\mathcal{L}_{cdm} = \mathbb{E}_{t,y_0^{(i)},\epsilon\sim\mathcal{N}(0,I)} \left[\left\|\epsilon - G_\theta^{cdm}(y_t^{(i)}, t; x^{(i)})\right\|^2\right]. \tag{5}$$

Specifically, we explore two conditioning strategies: (i) concatenating $x^{(i)}$ with the noisy input $y_t^{(i)}$ along the channel dimension at every denoising step, which we refer to as "Concat. Diffusion"; and (ii) employing a ControlNet [23] module, trained on top of a pre-trained, frozen unconditional diffusion backbone. Specifically, the ControlNet injects conditioning into the decoder layers of the U-Net.

3.3   Flow Matching

Given a source (known) distribution p_0 and a target (unknown) distribution p_1, the goal of FM is to define a time-dependent flow ψ_t that transports a sample $z_0 \sim p_0$ such that $\psi_1(z_0) \sim p_1$. This flow is governed by a time-dependent velocity field u_t through the ordinary differential equation (ODE):

$$\frac{d}{dt}\psi_t(z_0) = u_t(\psi_t(z_0)) \quad z_0 \sim p_0 \quad t \in [0,1] \tag{6}$$

If u_t is known, the flow ψ_t can be obtained by numerically solving this ODE. In practice, u_t is often unknown or intractable. FM addresses this by learning u_t using a neural network G_θ^{fm}. In its simplest form, FM assumes we have access to a target sample $z_1 \sim p_1$ during training, allowing us to condition the flow entirely on this known endpoint and predetermine where a sample z_0 should be transported. This allows us to define a linear path:

$$z_t = \psi_t(z_0) = (1 - t)z_0 + tz_1, \tag{7}$$

with a constant, conditional velocity field:

$$u_t(\psi_t(z_0) \mid z_1) = \frac{dz_t}{dt} = z_1 - z_0. \tag{8}$$

The model is trained to match the conditional velocity field along this path. This leads to the Conditional Flow Matching (CFM) objective:

$$\mathcal{L}_{cfm} = \mathbb{E}_{t,\,z_0\sim p_0,\,z_1\sim p_1} \left[\left\|G_\theta^{fm}(z_t, t) - (z_1 - z_0)\right\|^2\right]. \tag{9}$$

Although this setup requires access to z_1 during training, the objective induces the same gradient as the original FM loss involving the true velocity field u_t [15]. As a result, minimizing $\mathcal{L}_{cfm}$ leads to learning correct flow dynamics that generalize to unseen samples at inference time.

We explore two conditioning strategies for employing an FM model to generate a T2w axial slice $y^{(i)}$ from its corresponding T1w slice $x^{(i)}$. The first strategy, which we refer to as *Concat. FM*, initializes the source distribution as $p_0 = \mathcal{N}(0,I)$ and conditions the generation process by concatenating the T1w slice $x^{(i)}$ with the intermediate sample z_t along the channel dimension before regressing the velocity field. The second strategy, which we refer to as *Direct FM*, uses the T1w slice $x^{(i)}$ directly as the source sample z_0, learning a mapping to the corresponding T2w slice $y^{(i)}$.

4   Experimental Design

This section describes our proposed benchmark study comparing multiple generative models for T1w-to-T2w I2I translation, covering datasets, preprocessing and experimental protocol.

4.1   Datasets

We train and evaluate the models on paired T1w and T2w structural MRI scans from three publicly available datasets: IXI (560 subjects), the Human Connectome Project (HCP, 1002 subjects), and the Cambridge Centre for Ageing and Neuroscience (CamCAN, 633 subjects). All participants across these datasets were healthy adults, with an average age of 41.4±17.7 years and 54% of subjects being female.

4.2   Preprocessing

We process all MRI data through a standardized pipeline designed to ensure consistency, anatomical alignment, and intensity normalization across subjects and imaging modalities. The preprocessing steps include:

- **Bias field correction:** We correct intensity inhomogeneities using the N4ITK algorithm [20], which mitigates spatially varying signal artifacts and enhances tissue contrast.
- **Brain extraction:** We perform the skull-stripping procedure with SynthStrip [11] to isolate brain tissue from non-brain structures, reducing irrelevant variability and focusing subsequent processing on the relevant anatomy.
- **Spatial normalization:** We affinely register each scan to the MNI152 $1\,\mathrm{mm}^3$ template using the Advanced Normalization Tools (ANTs) [3], ensuring all images conform to a common anatomical space.
- **Intensity normalization:** We apply WhiteStripe [19] normalization independently to T1w and T2w images, referencing white matter regions to standardize intensity distributions and reduce inter-subject and inter-scan variability.
- **Slice extraction:** For each subject, we extract the central axial slice from both T1w and T2w volumes, and pad it to 224×192 pixels.

Table 1. Quantitative results for the T1w-to-T2w I2I translation task. Metrics are reported as mean ± standard deviation across the test set for SSIM, MSE, and PSNR.

Method	SSIM ↑	MSE ↓	PSNR ↑
ControlNet [23]	0.363 ± 0.057	0.3887 ± 0.1427	14.800 ± 10.498
Concat. Diffusion [10]	0.469 ± 0.037	0.5110 ± 0.2157	15.892 ± 9.168
Concat. FM [15]	0.715 ± 0.007	0.0097 ± 0.0000	20.326 ± 1.143
Direct FM [15]	0.732 ± 0.001	0.0106 ± 0.0000	20.011 ± 1.361
Pix2Pix [13]	**0.862 ± 0.001**	**0.0054 ± 0.0000**	**22.915 ± 3.428**

4.3   Experimental Protocol

We develop and evaluate all generative models within a unified experimental framework to facilitate fair and comprehensive comparison. Key aspects include:

- **Model architecture:** Each model employs a 2D U-Net encoder-decoder backbone with skip connections between corresponding layers. This architecture utilizes two residual blocks at each resolution level, with the number of feature channels progressing from 32, 64, 64, and 64 across its levels. It also integrates attention mechanisms at the latter two resolution levels.
- **Data splits:** We partition the dataset into training (80%), validation (5%), and testing (15%) sets, with strict subject-level separation to prevent data leakage and ensure unbiased evaluation.
- **Training details:** We train all networks with a batch size of 6, selected to optimize GPU memory usage. Training proceeds for up to 300 epochs, with early stopping based on validation loss to prevent overfitting.
- **Evaluation metrics:** We assess model performance using the Structural Similarity Index (SSIM), Mean Squared Error (MSE), and Peak Signal-to-Noise Ratio (PSNR) between the ground-truth and predicted T2w scan. These metrics quantify fidelity, structural preservation, and noise characteristics.

4.4   Implementation Details

We implement all models using the MONAI framework [5], which facilitates reproducibility. For diffusion-based models, we adopt a standard training configuration with $T = 1000$ denoising steps and a linear variance schedule ranging from $\beta_1 = 1 \times 10^{-4}$ to $\beta_T = 2 \times 10^{-2}$. Ancestral sampling is used during inference. ControlNet is trained and evaluated using the same diffusion schedule and sampling procedure. For the FM-based models, we employ the Euler ODE solver with 300 integration steps. In the Pix2Pix loss, we set $\lambda = 100$. We perform all training and experiments on a Tesla T4 GPU with 16GB of VRAM.

5   Results

In our experiments, we conduct three types of analyses: i) a quantitative evaluation focusing on comparing the different approaches in terms of SSIM, MSE, and

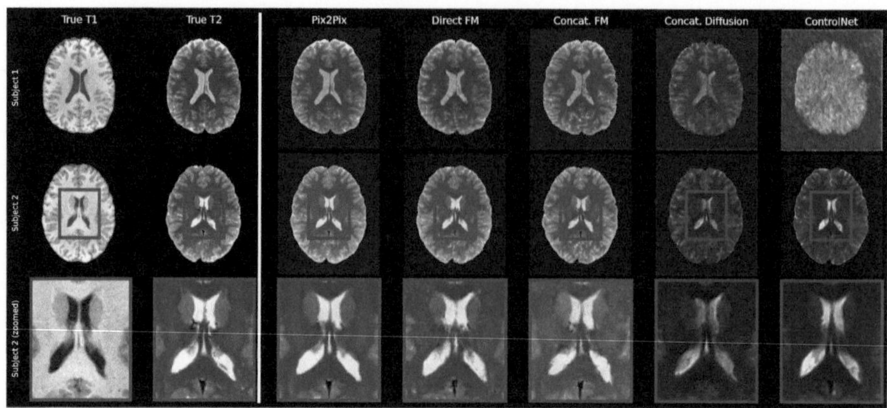

Fig. 1. Visual comparison of predictions from different methods across two randomly selected test subjects. The first and second columns display the input T1w and ground-truth T2w slices, respectively. The remaining columns show the predictions from the evaluated generative models. Rows one and two correspond to the two different subjects, while row three displays a zoomed-in region from the second subject, highlighting anatomical details within the brain MRI.

PSNR performance; ii) a qualitative assessment involving visual comparison of the results; and iii) a comparison of resource allocation with respect to execution time, memory usage, and the number of parameters.

5.1   Quantitative Comparison

Table 1 presents the quantitative results of our benchmark experiment. Pix2Pix demonstrably outperforms all other evaluated methods, achieving the highest SSIM (0.862), lowest MSE (0.0054), and highest PSNR (22.915), collectively indicating superior preservation of structural information, anatomical details, and image quality. FM approaches show moderate but notable performance: Direct FM achieves a slightly higher SSIM (0.732) compared to Concat. FM (0.715), indicating reasonable reconstruction capabilities for both. However, Concat. FM demonstrates marginally better performance in terms of MSE (0.0097 versus 0.0106). PSNR values are comparable for both variants, with Concat. FM at 20.326 and Direct FM at 20.011. Conversely, diffusion-based models show significant performance limitations; the Concat. Diffusion model yields an SSIM of only 0.469, a substantially elevated MSE of 0.5110, and a PSNR of 15.892, indicating considerable challenges in maintaining anatomical fidelity and structural consistency. ControlNet demonstrates the least accurate performance, obtaining the lowest SSIM (0.363), the highest MSE (0.3887), and the lowest PSNR (14.800), suggesting inconsistent output quality and unreliable synthesis capabilities.

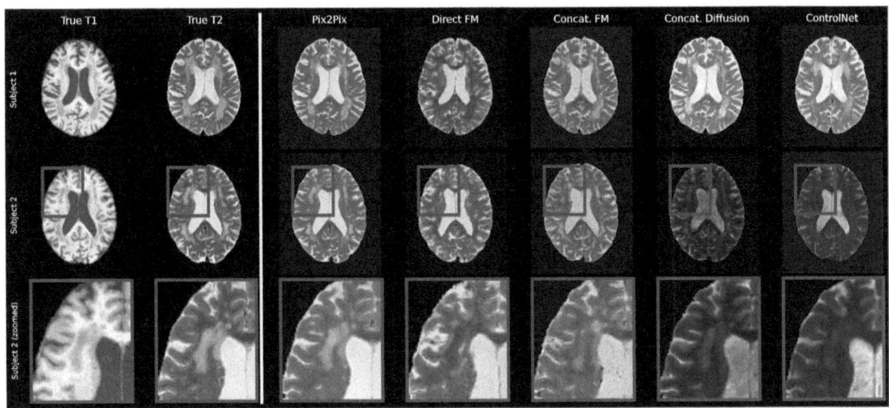

Fig. 2. Visual comparison of predictions from different methods across two test subjects that show white matter lesions. The first and second columns display the input T1w and ground-truth T2w slices, respectively. The remaining columns show the predictions from the evaluated generative models. Rows one and two correspond to the two different subjects, while row three displays a zoomed-in region from the second subject, highlighting an area containing white matter lesions.

5.2 Qualitative Assessment and Visual Analysis

This section provides a visual assessment of the evaluated methods for two case studies, examining (i) the structural fidelity of the predicted output and (ii) the ability to transfer white matter lesions from T1w to T2w scans.

Evaluating Structural Fidelity. We visually compare the predicted T2w scans to the ground-truth images in Fig. 1, observing distinct qualitative patterns that confirm our quantitative findings. Pix2Pix consistently produces the most clinically acceptable and realistic synthetic images, demonstrating sharp anatomical boundaries, appropriate tissue contrast differentiation, and minimal noise artifacts. The generated images closely approximate ground truth T2w scans in terms of gray matterwhite matter contrast and overall neuroanatomical structure preservation. Flow matching methods demonstrate reasonable structural consistency, though they exhibit slightly increased noise levels and reduced sharpness compared to Pix2Pix. In contrast, diffusion-based models produce images with noticeable artifacts, blurred boundaries, and inconsistent tissue intensities. ControlNet exhibits the most severe qualitative limitations, frequently failing to synthesize coherent T2w images with acceptable anatomical fidelity.

Evaluating White Matter Lesions Transfer. The evaluation of subjects with pronounced white matter lesions, as shown in Fig. 2, yields critical insights into each method's capability to transfer pathological features. Pix2Pix maintains superior performance, accurately translating lesion visibility and contrast

Table 2. Computational efficiency comparison of generative models for the T1w-to-T2w I2I task. Metrics reported include the number of parameters, inference time (Inf. Time), and inference memory usage (Inf. Memory).

Method	#Params. ↓	Inf. Time ↓	Inf. Memory ↓
ControlNet [23]	3,255,745	57.87 s	12.42 MB
Concat. Diffusion [10]	2,328,737	41.33 s	8.88 MB
Concat. FM [15]	2,328,737	11.94 s	8.88 MB
Direct FM [15]	2,328,449	11.61 s	8.88 MB
Pix2Pix [13]	2,328,449	0.05 s	8.88 MB

characteristics. FM methods, notably Direct FM, sometimes perform lesion inpainting rather than faithful reconstruction, indicating a potential misinterpretation of lesions as artifacts. Both diffusion-based models consistently fail to accurately represent pathological features, as they are not able to transfer the lesion accurately.

5.3   Resources Allocation at Inference Time

Table 2 presents a comparative overview of the computational resources required by each generative model at inference time, including the number of parameters, computational time, and memory usage. Among the evaluated methods, Pix2Pix is the most resource-efficient, requiring only 2,328,449 parameters for the generator network, an extremely fast inference time of 0.05 s, and a low memory footprint of 8.88 MB. In contrast, ControlNet and the standard diffusion model are notably less efficient. ControlNet demands the highest number of parameters (3,255,745), the longest inference time (57.87 s), and the largest memory usage (12.42 MB). Concat Diffusion, while slightly more efficient than ControlNet, still lags behind both the flow matching and GAN-based approaches. Both flow matching variants (Concat. FM and Direct FM) offer a balanced compromise. They match Pix2Pix in memory usage and parameter count, but have inference times (around 1112 s) that are significantly longer than Pix2Pix, though much faster than diffusion and ControlNet. Overall, Pix2Pix is clearly the best method in terms of computational efficiency, combining the fastest inference, lowest memory use, and smallest parameter count. This makes it highly suitable for real-time or resource-constrained clinical applications.

6   Conclusion

The synthesis of T2w MRI from T1w scans marks a significant step forward in medical imaging. This approach addresses key clinical challenges by reducing scan times, lowering costs, and improving access to comprehensive imaging. T1w images are best for anatomical detail, while T2w images excel at identifying fluid and pathology. Reliable cross-modal synthesis could decrease the need for multiple MRI acquisitions while maintaining diagnostic quality.

Our benchmarking results show that Pix2Pix, a GAN-based approach, consistently outperforms both FM and diffusion-based methods across all evaluated metrics. Although prior work has reported that flow-based models can surpass GANs when trained on large-scale datasets [9], recent studies [1,4] have highlighted their tendency to overfit and memorize the training distribution under low-data regimes or simplified tasks. In our setting which is characterized by relatively small datasets and low-dimensional 2D axial slices, this overfitting behavior was particularly pronounced in flow-based models. Therefore, our findings may not generalize to other settings involving larger datasets, higher-dimensional data, or more complex tasks.

For this reason, future work could investigate model performance across varying dataset sizes and a range of data dimensionalities—from low-resolution to high-resolution 3D MRI—as well as evaluate the generalizability of these models to out-of-distribution data. Finally, expanding the scope of this work by including other modalities (e.g., CT, PET) will aid in developing future methods and broaden the applicability of I2I frameworks. Such extensions could open new avenues for clinical and research use.

Acknowledgement. The work of Francesco Guarnera has been supported by MUR in the framework of PNRR PE0000013, under project "Future Artificial Intelligence Research FAIR". This work is partially supported by the Horizon Europe "Open source deep learning platform dedicated to Embedded hardware and Europe" project (Grant Agreement, project 101112268 - NEUROKIT2E)

References

1. Akbar, M.U., Wang, W., Eklund, A.: Beware of diffusion models for synthesizing medical images–a comparison with GANS in terms of memorizing brain MRI and chest x-ray images. Mach. Learn. Sci. Tech. **6**(1), 015022 (2025)
2. Alkan, C., Cocjin, J., Weitz, A.: Magnetic resonance contrast prediction using deep learning. Google Scholar (2016)
3. Avants, B.B., Epstein, C.L., Grossman, M., Gee, J.C.: Symmetric diffeomorphic image registration with cross-correlation: evaluating automated labeling of elderly and neurodegenerative brain. Med. Image Anal. **12**(1), 26–41 (2008)
4. Bertrand, Q., Gagneux, A., Massias, M., Emonet, R.: On the closed-form of flow matching: generalization does not arise from target stochasticity. arXiv preprint arXiv:2506.03719 (2025)
5. Cardoso, M.J., et al.: Monai: an open-source framework for deep learning in healthcare. arXiv preprint arXiv:2211.02701 (2022)
6. Chadebec, C., Tasar, O., Sreetharan, S., Aubin, B.: LBM: latent bridge matching for fast image-to-image translation. arXiv preprint arXiv:2503.07535 (2025)
7. Choo, K., Jun, Y., Yun, M., Hwang, S.J.: Slice-consistent 3d volumetric brain CT-to-MRI translation with 2d brownian bridge diffusion model. In: International Conference on Medical Image Computing and Computer-Assisted Intervention, pp. 657–667. Springer (2024)

8. Dey, A., Ebrahimi, M.: MTSR-MRI: Combined modality translation and super-resolution of magnetic resonance images. In: Medical Imaging with Deep Learning, pp. 743–757. PMLR (2024)

9. Dhariwal, P., Nichol, A.: Diffusion models beat GANS on image synthesis. Adv. Neural. Inf. Process. Syst. **34**, 8780–8794 (2021)

10. Ho, J., Jain, A., Abbeel, P.: Denoising diffusion probabilistic models. Adv. Neural. Inf. Process. Syst. **33**, 6840–6851 (2020)

11. Hoopes, A., Mora, J.S., Dalca, A.V., Fischl, B., Hoffmann, M.: Synthstrip: skull-stripping for any brain image. Neuroimage **260**, 119474 (2022)

12. Huang, Y., Zheng, F., Sun, X., Li, Y., Shao, L., Zheng, Y.: Generalized brain image synthesis with transferable convolutional sparse coding networks. In: ECCV, pp. 183–199. Springer (2022)

13. Isola, P., Zhu, J.Y., Zhou, T., Efros, A.A.: Image-to-image translation with conditional adversarial networks. In: Proceedings of the IEEE conference on computer vision and pattern recognition, pp. 1125–1134 (2017)

14. Kawahara, D., Nagata, Y.: T1-weighted and t2-weighted MRI image synthesis with convolutional generative adversarial networks. Rep. Pract. Oncol. Radiother. **26**(1), 35–42 (2021)

15. Lipman, Y., Chen, R.T., Ben-Hamu, H., Nickel, M., Le, M.: Flow matching for generative modeling. arXiv preprint arXiv:2210.02747 (2022)

16. Martin, S., Gagneux, A., Hagemann, P., Steidl, G.: PNP-flow: plug-and-play image restoration with flow matching. arXiv preprint arXiv:2410.02423 (2024)

17. Pan, S., et al.: Synthetic CT generation from MRI using 3d transformer-based denoising diffusion model. Med. Phys. **51**(4), 2538–2548 (2024)

18. Ronneberger, O., Fischer, P., Brox, T.: U-Net: convolutional networks for biomedical image segmentation. In: Navab, N., Hornegger, J., Wells, W.M., Frangi, A.F. (eds.) MICCAI 2015. LNCS, vol. 9351, pp. 234–241. Springer, Cham (2015). https://doi.org/10.1007/978-3-319-24574-4_28

19. Shinohara, R.T., et al.: Statistical normalization techniques for magnetic resonance imaging. NeuroImage: Clinical **6**, 9–19 (2014)

20. Tustison, N.J., et al.: N4itk: improved n3 bias correction. IEEE Trans. Med. Imaging **29**(6), 1310–1320 (2010)

21. Vaidya, A., Stough, J.V., Patel, A.A.: Perceptually improved t1-t2 MRI translations using conditional generative adversarial networks. In: Medical Imaging 2022: Image Processing. vol. 12032, pp. 505–511. SPIE (2022)

22. Yazdani, M., Medghalchi, Y., Ashrafian, P., Hacihaliloglu, I., Shahriari, D.: Flow matching for medical image synthesis: Bridging the gap between speed and quality. arXiv preprint arXiv:2503.00266 (2025)

23. Zhang, L., Rao, A., Agrawala, M.: Adding conditional control to text-to-image diffusion models. In: Proceedings of the IEEE/CVF international conference on computer vision, pp. 3836–3847 (2023)

Human-Object Interaction: Integrating Egocentric and Exocentric Perspectives (HOI-EGO-EXO 2025)

Workshop

Human-Object Interaction: Integrating Egocentric and Exocentric Perspectives (HOI-EGO-EXO 2025)

In conjunction with ICIAP 2025—Rome, Italy, September 15–19, 2025

Workshop Organization

Organizers

Matteo Dunnhofer	University of Udine, Italy
Giovanni Maria Farinella	University of Catania, Italy
Emanuele Frontoni	University of Macerata, Italy
Antonino Furnari	University of Catania, Italy
Christian Micheloni	University of Udine, Italy
Marina Paolanti	University of Macerata, Italy

Program Committee

Adriano Mancini	Polytechnic University of Marche, Italy
Paolo Sernani	University of Macerata, Italy
Lorenzo Stacchio	University of Macerata, Italy
Rocco Pietrini	Polytechnic University of Marche, Italy
Emanuele Balloni	Polytechnic University of Marche, Italy
Zaira Manigrasso	University of Udine, Italy
Moritz Nottebaum	University of Udine, Italy
Devis Salierno	University of Udine, Italy
Rosario Leonardi	University of Catania, Italy
Michele Mazzamuto	University of Catania, Italy
Francesco Ragusa	University of Catania, Italy
Rosario Forte	University of Catania, Italy

A Benchmark of Egocentric Scene Graph Prediction Methods for Understanding Human-Object Interactions

Asfand Yaar[(✉)], Ivan Rodin, Giovanni Maria Farinella, and Antonino Furnari

Department of Mathematics and Computer Science, University of Catania,
Catania, Italy
{asfand.yaar,ivan.rodin,antonino.furnari,giovanni.farinella}@unict.it

Abstract. Egocentric videos captured by wearable cameras provide a rich, first-person view of human-object interactions, offering unique insights into how people perform everyday activities. However, understanding long-form egocentric video remains challenging due to the complex temporal structure and dynamic nature of this data. In this work, we address the task of Egocentric Action Scene Graph (EASG) generation, which constructs structured graphs that represent hands, objects, and their semantic relationships over time. Unlike previous methods that rely on ground-truth object annotations, we evaluate models in two settings: in the presence of ground-truth object annotations, and in the absence of such annotations, where algorithms rely on predicted objects. We evaluate multiple methods, including transformer-based and fully connected models, for predicting temporally coherent and semantically rich scene graphs. The benchmarked approaches capture the hierarchical and compositional aspects of first-person behavior, potentially enabling improved interpretation and generalization for downstream tasks such as action anticipation and video summarization.

Keywords: Egocentric Vision · Scene Graph Generation · Human-Object Interaction · Temporal Video Understanding · Graph-Based Action Representation

1 Introduction

Wearable cameras have made it possible to widely capture egocentric videos that provide a close, first-person view of how people interact with objects and their surroundings. Unlike traditional third-person footage, egocentric videos capture the wearer's hands and field of view, offering detailed evidence of visual attention, manipulation patterns, and task-relevant interactions with objects and the environment. This rich perspective has proven valuable across numerous applications, including action anticipation [2,5,18], video summarization [3], and episodic memory retrieval [5]

E. Rodolà et al. (Eds.): ICIAP 2025 Workshops, LNCS 16169, pp. 445–456, 2026.
https://doi.org/10.1007/978-3-032-11317-7_37

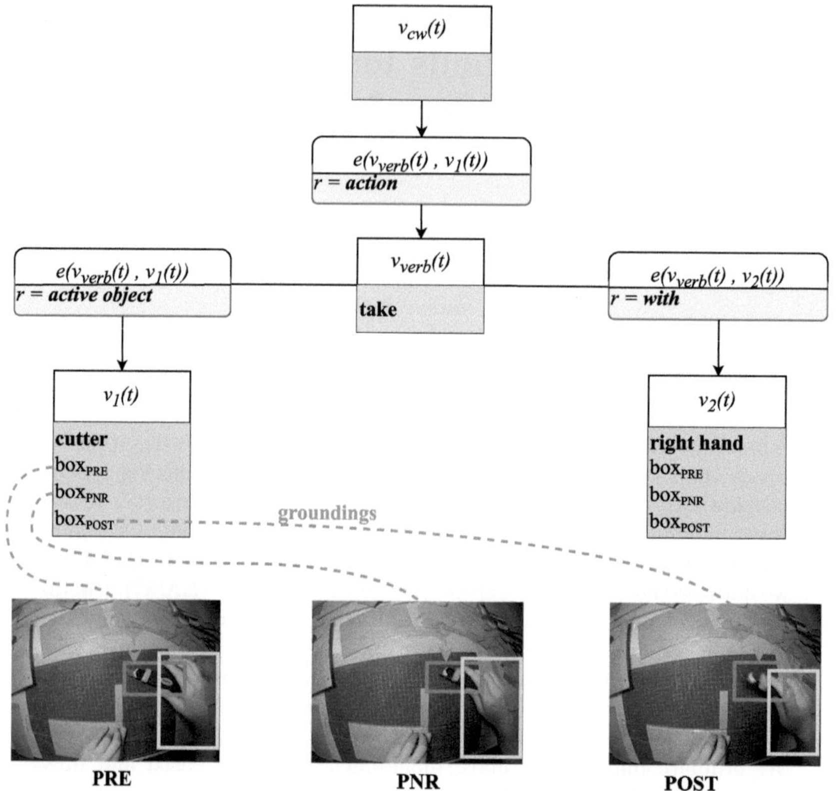

Fig. 1. Egocentric Action Scene Graph (EASG) as a time-varying directed graph G(t), where nodes represent the camera wearer, action verb, or objects, and edges define their relationships.

Despite these advances, understanding the full complexity of long-form egocentric video remains a significant challenge. First-person videos often span minutes or hours and contain multiple, interleaved activities occurring across diverse physical contexts. Existing annotation schemes [10,15] whether verbnoun pairs, object bounding boxes, or short, isolated action segments tend to focus on individual actions or brief time intervals . While useful for many tasks, these short-range representations fail to capture the temporal continuity and compositional structure that characterize extended sequences of behavior. Consequently, models trained on such limited annotations struggle to generalize beyond narrow segments or to reason effectively about the relationships between actions and objects over longer timescales.

Scene graphs offer a promising framework for structured and interpretable modeling of visual content by representing entities as nodes and their semantic relationships as edges. In static images and third-person video, scene graphs have facilitated tasks such as visual question answering, relationship detection, and spatio-temporal reasoning. However, applying scene graph techniques to egocentric, long-form video introduces unique challenges, including frequent occlusions due to rapid head movements, recurrent object reappearances, and the need to track hands and tools as they move in and out of view. Furthermore, the dynamic nature of first-person data requires scene graphs that evolve over time, continuously updating node attributes and edge relationships to reflect ongoing actions.

In this work, we tackle the problem of Egocentric Action Scene Graph (EASG) generation, as formalized in [19]. Given a continuous egocentric video, our goal is to construct a sequence of scene graphs where nodes correspond to hands and objects, edges encode semantic relations such as *from* or *with* and each graph is grounded to a specific temporal segment (see Fig. 1 and Fig. 2). This representation captures the hierarchical and compositional structure of first-person behavior by encoding not only which objects are present, but also how they are used, in what order, and in relation to specific actions. We benchmark several methods for EASG generation, comparing transformer based models with fully connected graph prediction layers. Notably, prior work [19] assumed access to ground-truth object annotations at test time. We extend our evaluation to the more realistic setting where objects must be predicted from given frames. Our experiments show that, despite the challenges, it is feasible to generate temporally coherent and semantically rich scene graphs directly from egocentric video streams. By advancing methods for EASG generation, we aim to support more interpretable and generalizable approaches to egocentric video understanding. We believe that EASGs have the potential to improve downstream tasks such as action forecasting and video summarization and open new avenues for research in egocentric vision, bridging the gap between low-level perception and high-level behavior understanding. Overall, our contributions are as follows:

- We establish a comprehensive benchmarking framework for Egocentric Action Scene Graph generation, comparing transformer-based and fully connected models under unified experimental settings.
- We extend evaluation protocols by removing the reliance on ground-truth object annotations at test time, instead assessing model performance using predicted objects.
- We provide an extensive experimental analysis on the EASG-Ego4D [19] egocentric video dataset, offering insights into model performance, limitations, and promising directions for future research.

2   Related Work

First Person Vision (FPV) systems, enabled by wearable cameras, offer a unique egocentric perspective for capturing human-object interactions in real-world settings. These systems often rely on deep learning to detect actions like item pickups and to identify the objects involved, which is crucial for applications such as action anticipation [2,5,18], video summarization [3], and episodic memory retrieval [5].

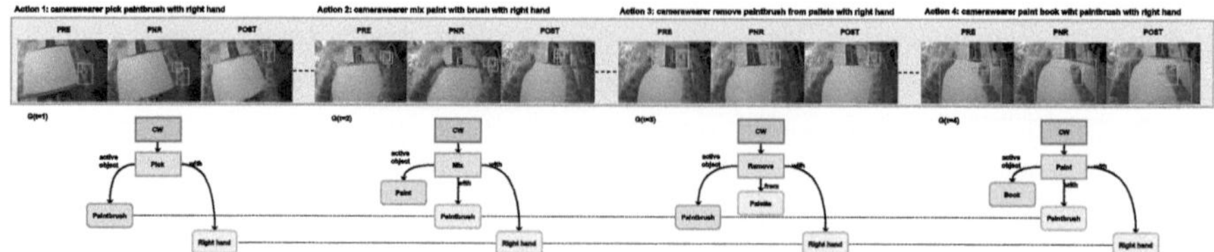

Fig. 2. Egocentric Action Scene Graphs *(G(t))* represent dynamic, time-varying interactions, with action verbs (blue nodes), active objects (green nodes), other objects (yellow nodes), and the camera wearer (grey nodes). Edges denote relationships, while dashed lines illustrate temporal progression, forming a long-form representation of the egocentric video.

Graph-Based Representations in Visual Data. Graph-based methods have become popular for modeling complex visual relationships. They have been successfully applied to image synthesis [6,7], video representation learning [22], object detection in videos [24], and person re-identification [23]. Transformer-based models further enable generating structured scene graphs from images [11], while dynamic scene graph generation has shown promise in capturing spatial and temporal relations for video understanding [1,4,10,15]. Despite this progress, few works have addressed scene graph generation specifically for egocentric videos.

Dynamic Scene Graphs. Scene graphs represent visual entities as nodes and their relationships as edges, providing a structured way to understand complex scenes. In video, dynamic scene graphs extend this idea by modeling how these relationships evolve over time. Prior work on dynamic scene graphs mainly focuses on short-range object and action relations [10,15], often in third-person views. However, egocentric video presents unique challenges like rapid head movements, occlusions, and frequent object reappearances, which demand graph representations that capture long-term, evolving human-object interactions. Rodin et al. [19] introduced the EASG task to fill this gap by generating temporally grounded graphs capturing hands, objects, and their interactions from first-person videos. We extend this work by benchmarking multiple methods, including transformer-based and fully connected models, and by evaluating settings where object instances must be predicted rather than given.

Scene Graphs in Egocentric Video. While scene graph generation has been widely studied in general video analysis and autonomous driving perception [9,12], its use in egocentric FPV remains limited. Existing works often target specific tasks like audio-video diarization [13,14] or embodied navigation [21], or Ego-Topo [16] models scene topology for long-term understanding. These approaches tend to focus on particular applications and do not fully capture the breadth of human-object interactions in egocentric video. Our work addresses

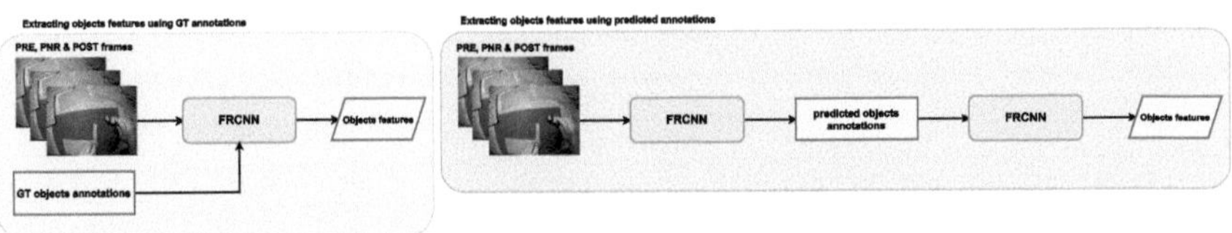

Fig. 3. Objects features extraction from PRE, PNR, and POST frames using pre-trained Faster-RCNN [17], based on both ground-truth and predicted annotations.

EASG tasks tailored for FPV, designed to model detailed human-object interactions across varied scenarios. This representation supports real-time action recognition and enhances long-term understanding of egocentric activities, such as object manipulation in daily tasks and shopping environments

3 Poblem Definition and Evaluation Protocol

We tackle the EASG generation task to model human-object interactions in egocentric video. Our approach focuses on capturing both the spatial extents of actions and the semantic relationships between the camera wearer and the objects involved. Following [19], each EASG graph $G(t)$ corresponds to an egocentric action and is constructed using information from three key temporal frames: Precondition (PRE), Point of No Return (PNR), and Postcondition (POST). The graph is composed of several types of nodes: the camera wearer node $v_{cw}(t)$, which represents the person performing the action; the verb node $v_{verb}(t)$, representing the specific action being performed; and object nodes $V_{obj}(t)$, representing all relevant objects involved, including the camera wearer's hands. Edges $E(t)$ between these nodes capture relationships such as *to* or *from*, representing both direct interactions and contextual connections (see Fig. 1). By connecting these graphs along multiple frames, we obtain a long-term description of the scene, which can support downstream tasks (see Fig. 2).

To generate these graphs, we evaluate multiple methods, including transformer-based models that leverage self-attention to capture temporal dependencies and fully connected (FC) layers as a baseline. Additionally, we explore variants that incorporate bounding box coordinates to assist object detection and relationship modeling, improving the precision of node and edge predictions. We evaluate our proposed method on the EASG-Ego4D [19] dataset under two primary evaluation setups: *With Constraint*, where each scene graph is restricted to a single verb-object relationship, and *No Constraint*, where the graph structure permits multiple relationships without limitation. To evaluate the robustness of our approach, we conduct experiments in two scenarios. The first uses ground-truth object annotations, following previous work [19], where object features are extracted from the PRE, PNR, and POST frames using accurate bounding box coordinates (see Fig. 3). The second scenario relies on predicted object detections from a pre-trained detector, introducing challenges like detection errors and inaccurate bounding boxes. Features are then extracted

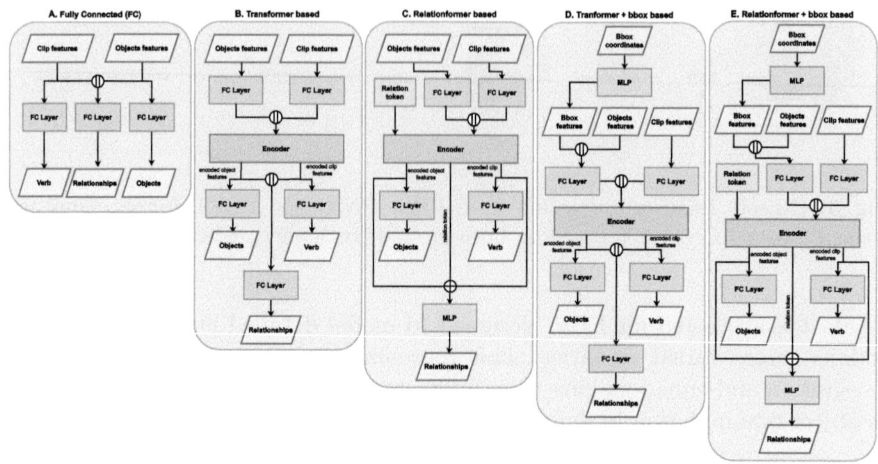

Fig. 4. Overview of different methods for Egocentric Action Scene Graph generation and their respective processing pipelines.

based on these predicted locations. This second setting makes our approach more practical and applicable to real-world settings where such annotations are not accessible.

To assess the quality of the generated graphs, we evaluate the framework across three tasks: *Edge Classification (Edge Cls)*, which involves predicting the correct relationships between verbs and objects; *Scene Graph Classification (SG Cls)*, which evaluates the accurate prediction of object categories and relationships; and *Egocentric Action Scene Graph Classification (EASG Cls)*, which measures the ability to construct the entire EASG, including all nodes and edges. Performance is measured using standard Recall@K metrics (R@10, R@20, R@50). Training is carried out for 10 epochs using the Adam optimizer [8] with a learning rate of 10^{-3}, following the experimental setup adapted from [19]. A random guessing baseline is also included for reference.

4   Compared Methods

We evaluate the performance of several methods to understand the impact of architectural choices and input features on egocentric scene graph generation. Each model is designed to predict object classes, verbs, and verb-object relationships, but they differ in how they process input features and encode relationships and spatial context.

Fully Connected (FC): This model serves as a baseline and relies solely on Fully Connected (FC) layers (see Fig. 4A). It takes as input object features and clip-level features. The object features, with a shape of $N \times 1024$ (where N is the number of objects), are extracted from the PRE, PNR, and POST frames using a pre-trained Faster R-CNN model [17], as illustrated in Fig. 3. The clip-level features, with a shape of 1×2304, are provided as part of the EASG-Ego4D dataset [19]. The list of all verbs, relations, and object nouns in EASG-Ego4D dataset are shown in Table 3. To predict object classes, the object features are passed through a dedicated FC layer. Similarly, verb prediction is performed by passing the clip-level features through another FC layer. For relationship prediction, the object and clip features are concatenated and passed through an additional FC layer. This provides a valuable point of comparison for evaluating the impact of more sophisticated structured, attention-based methods.

Transformer Based: The Transformer-based model replaces the FC-only setup with a transformer encoder, leveraging the self-attention to capture global contextual dependencies (see Fig. 4B). Object features and clip-level features are first passed through separate FC layers to project them into a shared feature space. These features are then concatenated and processed by the transformer encoder. The encoder output is then split and fed into two separate FC layers to predict object classes and verb category. For verbobject relationship prediction, the encoded objects features and encoded clip-level features are concatenated and processed by an additional FC layer. While this model is not explicitly designed for relational reasoning, it serves as a strong baseline due to its expressive power and flexible, general-purpose architecture.

Relationformer Based: This model is a transformer variant designed to capture verbobject and objectobject interactions in visual scenes (see Fig. 4C). It introduces a learnable relation token between each pair of encoded clip-level and object features, forming structured triplets of the form (verb, relation token, object), following the approach proposed in [20]. These triplets are passed through a Multi-Layer Perceptron (MLP) composed of two linear layers, with a ReLU activation and a dropout layer applied in between, to predict the relationship between the verb and the object. This design enables the model to explicitly attend to relational cues, improving its capacity to capture complex interaction dynamics in egocentric settings.

Transformer + bbox: This model extends the Transformer baseline by incorporating objects bounding box coordinates as spatial features (see Fig. 4D). To align the spatial information with the object features, the bounding box coordinates are first passed through an MLP. This MLP consists of two linear layers with a ReLU activation and dropout in between, and it projects the raw coordinates into the same feature space as the object features. The resulting spatial embeddings are then concatenated with the corresponding object features. Both the enriched object features and the clip-level features are then passed through separate FC layers to bring them into a shared feature space. These representations are concatenated and processed by the transformer encoder, enabling the model to reason jointly over visual and spatial cues. This added spatial grounding enhances the model's ability to localize objects and more accurately infer their interactions.

Relationformer + bbox: This model combines the Relationformer baseline with objects bounding box features to incorporate both relational structure and

Table 1. Comparison of different methods for Egocentric Action Scene Graph generation using ground-truth object annotations, evaluated under With Constraint and No Constraint settings. Results are reported as Recall at 10, 20, and 50 (R@10, R@20, R@50). Bold values indicate the best performance, while underlined values represent the second-best.

Method	With Constraint									No Constraint									Average
	Edge Cls			SG Cls			EASG Cls			Edge Cls			SG Cls			EASG Cls			
	R@10	R@20	R@50	R@10	R@20	R@50	R@10	R@20	R@50	R@10	R@20	R@50	R@10	R@20	R@50	R@10	R@20	R@50	
Random Guess	8.0	8.0	8.0	0.2	0.4	1.0	0.0	0.0	0.0	36.5	_72.6_	_99.9_	0.3	0.5	1.0	0.0	0.0	0.0	13.1
A. FC based	60.4	60.4	60.4	41.4	44.3	50.6	14.3	16.4	17.9	94.4	**99.8**	100	_51.6_	58.2	62.4	14.7	18.3	**20.9**	52.0
B. Transformer	66.3	66.3	66.3	_45.9_	49.0	55.6	**15.7**	_19.4_	_22.4_	_97.4_	**99.8**	100	**53.7**	_60.5_	_64.4_	**15.6**	**19.1**	_20.3_	**57.4**
C. Relationformer	**70.1**	**70.1**	**70.1**	_45.9_	**49.8**	_56.5_	13.0	16.6	19.6	_97.4_	**99.8**	100	51.4	60.1	63.8	12.3	15.6	16.8	_57.2_
D. Transformer + bbox	65.2	65.2	65.2	34.8	39.1	48.3	9.3	12.5	15.8	96.9	**99.8**	100	39.3	50.1	57.3	8.8	11.4	13.0	49.0
E. Relationformer + bbox	_69.8_	_69.8_	_69.8_	**46.0**	_49.5_	**57.1**	_15.6_	**19.7**	**23.0**	**97.5**	**99.8**	100	**53.7**	**61.1**	**64.8**	_15.5_	_18.9_	20.2	56.2

spatial context (see Fig. 4E). The bounding box coordinates are processed following the same procedure as in the Transformer + bbox model, projected into the object feature space using a MLP and then concatenated with the corresponding object features. The enriched object features and clip-level features are then passed through separate FC layers to project them into a shared feature space, after which they are concatenated. These combined features, along with learnable relation tokens, are then fed into the transformer encoder as in the Relationformer baseline. By integrating both spatial grounding and structured relational reasoning, this variant aims to improve the modeling of egocentric interactions and enhance predictive performance across object, verb, and relationship prediction tasks.

5   Results and Comparison

Table 1 presents the results for all three prediction tasks, Edge Classification, Scene Graph Classification, and Egocentric Action Scene Graph Classification, across various baseline models using ground-truth object annotations. When provided with ground-truth annotations, transformer-based models outperform the FC baseline, demonstrating the advantage of self-attention mechanisms for capturing structured relationships in egocentric scenes. Among all baseline models, the Transformer achieves the best overall performance, especially in the *No Constraint* setting, reinforcing the effectiveness of attention-based mechanisms for modeling complex interactions. The Relationformer model follows closely with the second-best average score. It uses the a relation token that allows the model to explicitly represent and reason about interactions between verb and object, further improving relationship predictions. Moreover, adding bounding box coordinates provides the model with explicit spatial information, helping it understand where objects are located and how they relate to each other. This enhances its accuracy in predicting interactions by focusing attention on relevant parts of the scene. The FC baseline serves as a solid reference point but is consistently outperformed by attention-based models across all tasks, highlighting the benefits of contextual modeling through self-attention mechanisms.

Table 2. Performance comparison of various methods for Egocentric Action Scene Graph generation using predicted objects, evaluated under With Constraint and No Constraint settings. Metrics reported include Recall at 10, 20, and 50 (R@10, R@20, R@50). Bold values indicate the best performance, while underlined values represent the second-best.

Method	With Constraint									No Constraint									Average
	Edge Cls			SG Cls			EASG Cls			Edge Cls			SG Cls			EASG Cls			
	R@10	R@20	R@50	R@10	R@20	R@50	R@10	R@20	R@50	R@10	R@20	R@50	R@10	R@20	R@50	R@10	R@20	R@50	
Random Guess	8.0	8.0	8.0	0.2	0.4	1.0	0.0	0.0	0.0	36.5	<u>72.6</u>	<u>99.9</u>	0.3	0.5	1.0	0.0	0.0	0.0	13.1
A. FC based	**42.5**	**42.5**	**42.5**	**34.1**	**35.9**	**41.2**	**7.6**	**11.2**	**14.4**	99.1	**99.9**	100	**42.9**	**64.5**	<u>81.5</u>	**9.9**	**14.0**	**20.1**	**48.0**
B. Transformer	25.1	25.1	25.1	10.7	15.1	19.1	2.7	3.9	5.2	99.1	**99.9**	100	24.8	50.5	79.0	4.7	9.5	18.3	<u>39.9</u>
C. Relationformer	22.9	22.9	22.9	7.4	11.4	16.0	0.7	1.4	3.2	<u>99.2</u>	**99.9**	100	26.0	<u>61.0</u>	80.3	3.2	7.1	15.7	39.0
D. Transformer + bbox	22.9	22.9	22.9	5.5	9.0	15.3	1.5	1.8	2.7	**99.3**	**99.9**	100	6.1	55.2	76.5	2.3	8.8	15.7	37.1
E. Relationformer + bbox	29.9	<u>29.9</u>	<u>29.9</u>	<u>17.4</u>	<u>21.0</u>	<u>25.1</u>	<u>4.2</u>	<u>5.7</u>	<u>7.6</u>	**99.3**	**99.9**	100	<u>33.5</u>	51.9	**81.8**	<u>6.1</u>	<u>10.3</u>	<u>18.7</u>	37.3

Table 2 shows the results for all three prediction tasks, Edge Classification, Scene Graph Classification, and Egocentric Action Scene Graph Classification, across various baseline models using predicted object annotations. When evaluated with predicted objects instead of ground-truth, all models experience a performance drop, underlining the importance of reliable object detection in this task. Interestingly, in this more challenging setting, the FC baseline outperforms transformer-based models. This outcome suggests that FC models are more robust to noisy or imperfect detections, whereas transformer-based models are more affected by errors in object detection. To evaluate the robustness of our approach, we experiment with two settings. The first uses ground-truth object annotations to extract features from the PRE, PNR, and POST frames (see Fig. 3). The second uses predicted object detections from a pre-trained detector, introducing real-world challenges such as detection noise and inaccurate bounding boxes. Despite these challenges, our method exhibits encouraging performance across all evaluation metrics, even without access to ground-truth object annotations at inference time, highlighting its robustness under realistic conditions. This evaluation is therefore more reflective of practical deployment scenarios. Future work should prioritize this more realistic evaluation protocol to advance the field.

Table 3. The list of all verbs, relations, and object nouns in EASG-Ego4D dataset.

Verbs	add, adjust, align, apply, arrange, attach, beat, bend, break, bring, bring-out, brush, carry, carry-out, carry-up, carve, change, check, chop, clean, clean-off, clear, climb, close, collect, connect, cover, crumple, cut, cut-off, cut-out, detach, dip, dip in, disconnect, divide, drag, drill, drive, drop, drop-out, drop up, dry, dust, empty, examine, fasten, feel, fetch, fill-up, fit, fix, flap, flip, fold, force, glue, grab, grasp, grip, hammer, hang, hit, hold, hold-up, insert, inspect, iron, join, keep, knead, knit, lay, leave, lift, lift-up, loose, loosen, loosen out, losse, lower, mark, measure, mix, mount, move, move-off, move-up, open, operate, pack, paint, pass, peel, pet, pick, pick-out, pick-up, place, place down, plaster, play, point, position, pour, pour down, pour-in, pour-off, pour-out, press, pull, pull-out, push, push-down, push-in, put, put-away, put-down, put-in, put-off, put-on, put-out, raise, read, release, remove, reposition, rest, return, rinse, roll, rotate, rub, sand, scan, scoop, scoop-out, scrap, scrape, scratch, screw, screw-in, scrub, search, separate, seperate, set, sew, shake, shape, shave, shift, shuffle, slice, slide, smoothen, soak, spin, split, spray, spread, spread out, sprinkle, squeeze, squeeze out, stick, stir, store, straighten, straighten-out, streche, stretch, sweep, swing, swirl, switch, switch-off, take, take-off, take-out, take-up, tap, taste, tear-off, test, throw, throw-away, tie, tight, tighten, tilt, touch, transfer, trim, turn, turn off, turn out, turn over, twist, unfold, unhang, unlock, unplug, unscrew, untangle, untie, untighten, unwrap, uproot, use, wash, water, wear, wet, wipe, wipe-off, withdraw, wrap
Relations	active object, around, from, in, inside, into, on, onto, out, through, to, towards, under, up, verb, with
Objects	both hands, left hand, right hand, adapter, apron, art, bag, bar, basket, battery, bed, belt holder, bench, bicycle, bicycle wheel, bike, bike part, bin, board, bobbin, bolt, book, booklet, bookshelf, bottle, bowl, bowls, box, boxer, brake, brake shoe pack, branch, bread, brick, broom, brush, bucket, bulb, bunch, button, cabbage, cabinet, cable, cables, caliper, camera, can, cap, car, car part, carburetor, card, cardboard, carpet, carton, case, casing, cello, cement, chaff, chain, chair, charger, chips, chisel, chopping board, clamp, cleaner, clip, cloth, clothe, clothes, clothing material, compartment, computer, connector, container, control, cooker, cord, cot, counter, countertop, cover, cracker, craft, crumbs, cup, cupboard, cutter, cutting board, debris, derailleur, desk, detergent, dirt, dish, dog, door, dough, dough strip, dough strips, drawer, dress, drill, drill bit, driller, drink, driver, drum, dust, dustbin, dustpan, egg, engine, fabric, fabrics, faucet, fence, file, filter, finger, floor, flour, flower, foam, food, fork, fridge, fuel, furniture, gasket, gauge, gear, generator, glass, glasses, glove, glue, gouge, grass, grater, grease, grinder, ground, guitar, hammer, hand, handle, hanger, heap, hoe, hoes, holder, hose, ice, ice cubes, ice tray, insulator, iron, iron box, ironbox, ironing board, jack, jacket, jar, jug, keg, key, keyboard, knife, knob, knot, ladder, laptop, layer, leaf, leg, lever, lid, lift, light, liquid, lock, machine, manual, marker, mask, mat, matchstick, material, measure, meat, metal, metal board, milk, mirror, mixer, mixture, mop stick, motorbike, motorcycle, mouse, mouth, mower, mug, multimeter, nail, napkin, needle, net, newspaper, note, nozzle, nut, nylon, oil, onion, oven, pack, pad, paddle, paint, paint brush, paintbrush, palette, pan, pants, paper, papers, part, pastry, pedal, peel, peeler, pen, pencil, phone, photo, picture, piece, piece of cloth, pieces, pile, piler, pin, pipe, pizza, plank, plant, planter, plastic wraps, plate, plates, platform, plier, pliers, plug, pocket, pole, polythene, pot, potato, pruner, pruning sheer, pump, purse, rack, rag, rail, railing, rake, refrigerator, rim, ring, rod, rod metal, roller, room, root, rope, ropes, rubber, ruler, sachet, salt, sand, sandpaper, sauce, saucer, saw, scaffold, scale, scarf, scissors, scoop, scooter, scourer, scraper, scrapper, screw, screwdriver, seat, seed, sellotape, serviette, shaft, shear, shears, sheet, shelf, shelve, shirt, short, side, sieve, sink, sink faucet, slab, smartphone, soap, sock, socket, soil, spanner, spatula, spice, sponge, spoon, spring, stack, stairs, stand, steel, stick, stool, stove, strand, string, sugar, switch, table, table cloth, tablet, tag, tank, tap, tape, terminal, thread, tie, timber, timer, tin, tire, tissue, tomato, toolbox, tools, top, torch, towel, toy, train, trash, tray, trey, trimmer, trolley, trouser, trowel, trunk, tub, tube, tyre, umbrella, vacuum cleaner, valve, vase, vice, vine, waist, wall, wallet, wardrobe, washer, water, water hose, weed, wheel, windshield, wipe, wiper, wire, wire cutter, wires, with, wood, wood plank, wooden block, wooden stand, workbench, worktop, wrapper, wrench, yarn, yeast

6   Conclusion

In this work, we evaluated a range of methods for generating Egocentric Action Scene Graphs (EASGs), providing a benchmark for structured modeling of human-object interactions in first-person videos. Our experiments demonstrate that transformer-based architectures, especially those enriched with spatial inputs (e.g., bounding box coordinates) and relational mechanisms (e.g., relation tokens), achieve superior performance across various tasks. Including spatial cues enhances object localization and contextual grounding, while relation tokens facilitate more expressive modeling of verb-object relationship through triplet reasoning. Our experiments, conducted under both ground-truth and predicted objects settings, confirm the robustness and generalizability of these structured models. Qualitative analyses further illustrate their ability to capture fine-grained and context-aware interactions in visually complex egocentric scenes. These findings highlight the value of structured representations and inductive biases in egocentric video understanding and establish a solid foundation for future research in long-horizon activity modeling, graph-based perception, and downstream applications such as assistive intelligence and embodied AI systems.

Acknowledgements. This research has been funded by the European Union - Next Generation EU, Mission 4 Component 1 CUP E53D23008280006 - Project PRIN 2022 EXTRA-EYE, and FAIR PNRR MUR Cod. PE0000013 - CUP: E63C22001940006.

References

1. Cong, Y., Liao, W., Ackermann, H., Rosenhahn, B., Yang, M.Y.: Spatial-temporal transformer for dynamic scene graph generation. In: Proceedings of the IEEE/CVF International Conference on Computer Vision, pp. 16372–16382 (2021)
2. Damen, D., et al.: Scaling egocentric vision: The epic-kitchens dataset. In: Proceedings of the European Conference on Computer Vision (ECCV), pp. 720–736 (2018)
3. Del Molino, A.G., Tan, C., Lim, J.H., Tan, A.H.: Summarization of egocentric videos: a comprehensive survey. IEEE Trans. Hum. Mach. Syst. **47**(1), 65–76 (2016)
4. Feng, S., Mostafa, H., Nassar, M., Majumdar, S., Tripathi, S.: Exploiting long-term dependencies for generating dynamic scene graphs. In: Proceedings of the IEEE/CVF Winter Conference on Applications of Computer Vision, pp. 5130–5139 (2023)
5. Grauman, K., et al.: Ego4d: Around the world in 3,000 hours of egocentric video. In: Proceedings of the IEEE/CVF Conference on Computer Vision and Pattern Recognition, pp. 18995–19012 (2022)
6. Herzig, R., Bar, A., Xu, H., Chechik, G., Darrell, T., Globerson, A.: Learning canonical representations for scene graph to image generation. In: Vedaldi, A., Bischof, H., Brox, T., Frahm, J.-M. (eds.) ECCV 2020. LNCS, vol. 12371, pp. 210–227. Springer, Cham (2020). https://doi.org/10.1007/978-3-030-58574-7_13
7. Johnson, J., Gupta, A., Fei-Fei, L.: Image generation from scene graphs. In: Proceedings of the IEEE Conference on Computer Vision and Pattern Recognition, pp. 1219–1228 (2018)

8. Kingma, D.P., Ba, J.: Adam: A method for stochastic optimization. arXiv preprint arXiv:1412.6980 (2014)

9. Kochakarn, P., De Martini, D., Omeiza, D., Kunze, L.: Explainable action prediction through self-supervision on scene graphs. In: 2023 IEEE International Conference on Robotics and Automation (ICRA), pp. 1479–1485. IEEE (2023)

10. Li, R., Zhang, S., He, X.: Sgtr: End-to-end scene graph generation with transformer. In: proceedings of the IEEE/CVF Conference on Computer Vision and Pattern Recognition, pp. 19486–19496 (2022)

11. Lu, Y., et al.: Context-aware scene graph generation with seq2seq transformers. In: Proceedings of the IEEE/CVF International Conference on Computer Vision, pp. 15931–15941 (2021)

12. Malawade, A.V., Yu, S.Y., Hsu, B., Kaeley, H., Karra, A., Al Faruque, M.A.: Roadscene2vec: a tool for extracting and embedding road scene-graphs. Knowl.-Based Syst. **242**, 108245 (2022)

13. Min, K.: Intel labs at ego4d challenge 2022: a better baseline for audio-visual diarization. arXiv preprint arXiv:2210.07764 (2022)

14. Min, K.: Sthg: Spatial-temporal heterogeneous graph learning for advanced audio-visual diarization. arXiv preprint arXiv:2306.10608 (2023)

15. Nag, S., Min, K., Tripathi, S., Roy-Chowdhury, A.K.: Unbiased scene graph generation in videos. In: Proceedings of the IEEE/CVF Conference on Computer Vision and Pattern Recognition, pp. 22803–22813 (2023)

16. Nagarajan, T., Li, Y., Feichtenhofer, C., Grauman, K.: Ego-topo: environment affordances from egocentric video. In: Proceedings of the IEEE/CVF Conference on Computer Vision and Pattern Recognition, pp. 163–172 (2020)

17. Ren, S., He, K., Girshick, R., Sun, J.: Faster r-CNN: towards real-time object detection with region proposal networks. Adv. Neural Inf. Process. Syst. **28** (2015)

18. Rodin, I., Furnari, A., Mavroeidis, D., Farinella, G.M.: Predicting the future from first person (egocentric) vision: a survey. Comput. Vis. Image Underst. **211**, 103252 (2021)

19. Rodin, I., Furnari, A., Min, K., Tripathi, S., Farinella, G.M.: Action scene graphs for long-form understanding of egocentric videos. In: Proceedings of the IEEE/CVF Conference on Computer Vision and Pattern Recognition, pp. 18622–18632 (2024)

20. Shit, S., et al.: Relationformer: a unified framework for image-to-graph generation. In: European Conference On Computer Vision, pp. 422–439. Springer (2022)

21. Singh, K.P., Salvador, J., Weihs, L., Kembhavi, A.: Scene graph contrastive learning for embodied navigation. In: Proceedings of the IEEE/CVF International Conference on Computer Vision, pp. 10884–10894 (2023)

22. Wang, X., Gupta, A.: Videos as space-time region graphs. In: Proceedings of the European Conference on Computer Vision (ECCV), pp. 399–417 (2018)

23. Wu, Y., Bourahla, O.E.F., Li, X., Wu, F., Tian, Q., Zhou, X.: Adaptive graph representation learning for video person re-identification. IEEE Trans. Image Process. **29**, 8821–8830 (2020)

24. Yuan, Y., Liang, X., Wang, X., Yeung, D.Y., Gupta, A.: Temporal dynamic graph LSTM for action-driven video object detection. In: Proceedings of the IEEE International Conference on Computer Vision, pp. 1801–1810 (2017)

A Real-Time System for Egocentric Hand-Object Interaction Detection in Industrial Domains

Antonio Finocchiaro$^{(\boxtimes)}$, Alessandro Sebastiano Catinello,
Michele Mazzamuto, Rosario Leonardi, Antonino Furnari,
and Giovanni Maria Farinella

Department of Mathematics and Computer Science, University of Catania,
Catania, Italy
`1000006272@studium.unict.it`, `ale.catinello.c@gmail.com`,
`michele.mazzamuto@phd.unict.it`,
`{rosario.leonardi,antonino.furnari,giovanni.farinella}@unict.it`
`https://iplab.dmi.unict.it/fpv/`

Abstract. Hand-object interaction detection remains an open challenge in real-time applications, where intuitive user experiences depend on fast and accurate detection of interactions with surrounding objects. We propose an efficient approach for detecting hand-objects interactions from streaming egocentric vision that operates in real time. Our approach consists of an action recognition module and an object detection module for identifying active objects upon confirmed interaction. Our Mamba model with EfficientNetV2 as backbone for action recognition achieves 38.52% p-AP on the ENIGMA-51 benchmark at 30fps, while our fine-tuned YOLOWorld reaches 85.13% AP for hand and object. We implement our models in a cascaded architecture where the action recognition and object detection modules operate sequentially. When the action recognition predicts a contact state, it activates the object detection module, which in turn performs inference on the relevant frame to detect and classify the active object.

Keywords: Hand-Object Interaction · Online Video Understanding · Wearable Vision

1 Introduction

Hand-Object Interaction (HOI) detection is a fundamental challenge, particularly when both high accuracy and low latency are required. In industrial scenarios, such as those involving laboratory tools, detecting these interactions becomes crucial, as discussed in [6]. A straightforward method for HOI detection in 2D frames involves detecting hands and objects and applying an Intersection over

A. Finocchiaro and A.S. Catinello—These authors share first authorship

Union (IoU) threshold to infer contact based on bounding box overlap. While state-of-the-art models exist for HOI detection, [6,11], they are typically computationally intensive and slow. For this reason, we propose a simple yet effective approach that addresses these limitations. Our method leverages the real-time capabilities of detectors like YOLOWorld [3] to identify active objects. Despite its simplicity, the approach proves to be both effective and lightweight. Moreover, when coupled with an Action Recognition (AR) module, it becomes even more efficient by avoiding per-frame inference. Finally, we developed an HOI framework which leverages the Meta Quest 3 device. In this implementation, Mamba [5] was used as the AR module, while YOLOWorld was employed as the object detector to handle active object retrieval analyzing the overlap between detected hands and objects.

1.1 Problem Formulation

As previously defined, the core challenge of this work is the real-time detection of HOIs in egocentric video streams captured in industrial environments. The problem involves taking streaming video as input and producing localized "events" indicating object interactions as output. To be useful in industrial scenarios, the system must perform in real-time. The performance of the AR module is evaluated using point-level AP (p-AP), as in [12]. The OD module is evaluated using standard metrics, such as average precision (AP), recall, and harmonic mean (HM). HOI metrics are adopted from [6]. Both modules take RGB frames as input. The OD module predicts hand coordinates, hand side, active object coordinates, and object class based on the overlap between bounding boxes. The system behavior depends on the AR module's output, resulting in the following cases. For example, if the AR module predicts a contact but no overlap is detected, the OD module outputs no contact. Conversely, if the AR module predicts a contact and the OD module detects an overlap, the contact state provided by the AR module is used alongside the information retrieved by the OD module.

2   Related Works

The practical application of HOI is central to developing effective Mixed Reality (MR) systems for user assistance on wearable devices. Recent work, for example, demonstrates how to support industrial workers by inferring interactions from 2D object detectors and standard MR features, thereby highlighting the need for efficient, real-time models [7]. Egocentric HOI detection focuses on identifying interactions with objects from a first-person perspective, requiring diverse annotation types ranging from basic bounding boxes to multimodal labels including hand detection, contact state estimation, active object identification, segmentation masks, and temporal interaction boundaries. Early work [11] framed HOI detection as a multitask problem involving hand/object localization and contact-state inference, which was extended to egocentric settings through the

MECCANO dataset for industrial applications [10] and the EgoHOS benchmark for pixel-level segmentation [15]. The current standard was established with the EPIC-KITCHENS VISOR dataset [4], unifying contact-state prediction, bounding box detection, and instance segmentation in a comprehensive Hand-Object Segmentation task. Recent advances have embraced transformer architectures adapted from third-person HOI detection [17] to capture long-range dependencies and contextual relationships between hands, objects, and scene elements through global-local correlation mechanisms. Simultaneously, synthetic data approaches have emerged to address limited annotation availability, with works like HOI-Synth [6] demonstrating effectiveness across established benchmarks (VISOR, EgoHOS, ENIGMA-51) through systematic domain adaptation strategies. Despite these advances, existing approaches face significant computational challenges, particularly for real-time applications. Most current methods employ end-to-end architectures that jointly optimize object detection and interaction understanding, leading to complex models with substantial inference overhead. Transformer-based approaches, while effective at capturing contextual relationships, introduce quadratic computational complexity that limits their deployment in resource-constrained environments. Similarly, multi-modal integration strategies often require sophisticated fusion mechanisms that further increase computational demands. These limitations highlight the need for efficient architectures that can maintain high accuracy while enabling real-time performance. The challenge of achieving real-time video understanding is not unique to HOI and is a central focus in the related field of Online Action Detection (OAD). OAD seeks to identify an action in a video stream as early as possible from partial observations, making predictions without access to future frames. Recent advancements are largely dominated by Transformer-based architectures like OadTR [14] and TeSTra [16] due to their strength in processing long sequences. Alongside these, alternative architectures have demonstrated competitive performance while addressing efficiency concerns. For instance, the RNN-based MiniRoad [1] offers similar results with a smaller memory footprint, and emerging models have shown success by replacing Transformer components with Mamba-based architectures [2], highlighting the active exploration of different architectural backbones for this real-time task.

In contrast to these approaches, our method addresses both computational and annotation complexity challenges by decomposing egocentric HOI detection into two lightweight components: object detection and contact state estimation. This decomposition enables independent optimization of each module, reducing inference time while requiring minimal annotations compared to comprehensive frameworks demanding segmentation masks, active object vectors, and temporal boundaries [6]. Our approach achieves computational efficiency and practical scalability for real-time applications without sacrificing fine-grained interaction understanding.

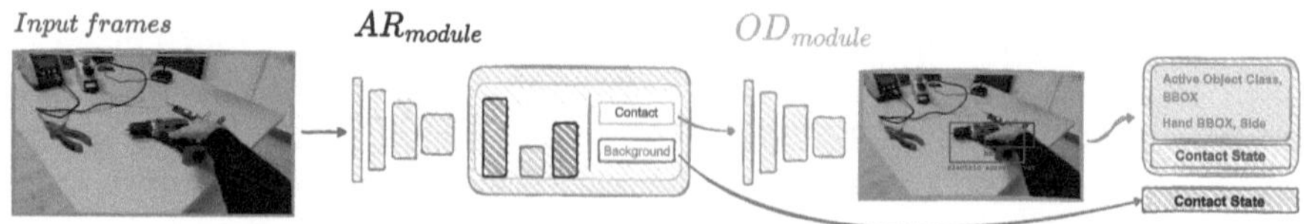

Fig. 1. Schema of the proposed approach. The AR module predicts the contact state online. Whenever a contact is detected, the object detection module identifies the active object class by evaluating the Intersection over Union (IoU) between objects and hands.

3   Methodologies

In this section, we present the pipeline and its components. We detail the integration of the OD and AR modules, describing each component's role and its interaction with the overall framework.

3.1   Pipeline Overview and Component Details

We consider two pipelines for detecting hand-object interactions: a baseline and our proposed methodology, illustrated in Fig. 1. Both pipelines employ an OD module to localize hands and objects, but differ fundamentally in how interaction is inferred. The baseline relies on a simple spatial heuristic: after detecting hands and objects using OD, it applies an IoU threshold to infer contact. In contrast, our approach introduces an AR module that first detects potential interaction events. Upon a positive prediction, the AR module activates the OD stage to localize the relevant hand and object. While Fig. 1 illustrates only our proposed pipeline, we describe both approaches in detail below, including their respective architectures, training procedures, and integration strategies.

Action Recognition using Mamba. As AR module, we adopted a Mamba model, which is a modified version of TeSTra [16], with the main goal to predict the action occurring in each frame. The model predicts a single "contact" action in an online fashion, discriminating it from the background class. The contact class represents those time instants where the hand has just touched something

Our architecture includes a sequence of 3 Mamba blocks observing the previous 3 s of frames. These mamba blocks are preceded by a linear projection of the input features and followed by a classification layer that outputs the classification score, which is followed by a softmax function. We extracted the frame-wise feature using DINOv2 [8] and trained the model using these features as input. Since DINOv2 is not suitable for a real-time scenario, we also tested the EfficientNetV2 model [13], distilled from DINOv2. In particular, we fine-tuned EfficientNetV2 with DINOv2's features as ground truth to force the similarity between the two architectures, with the use of the MSE as loss function.

Object Detection with YOLO. To detect hands and active objects in the scene, we implemented an OD module. We fine-tuned different OD models, such as RT-DETR and YOLOWorld, and selected the best-performing model. Since we needed a fast and simple solution for online processing, we classified an object as active based on its IoU value with the hand having the highest confidence score in the scene, with the threshold being an hyperparameter. We limited the recognized hands to the two with the highest confidence scores. Additionally, we distinguished left from right hands by considering the position of the bounding box centroid relative to the center of the image's x-axis.

4   Experimental Settings and Results

We trained our models on the merge of the ENIGMA-51 [9] training and validation set. This egocentric dataset provides videos captured in an industrial scenario, including 45,505 RGB frames, 12,597 interaction frames annotated with 14,036 interactions, and 9,342 active objects. We used the dataset to train the RT-DETR and YOLO models, using its fixed taxonomy. We used the ENIGMA-51 test set to evaluate the individual modules. Meanwhile, for the HOI performance evaluation, we used a subset of the ENIGMA-51 test set involved in [6]. This subset contains several ENIGMA-51 clips with a different split. Hence, we filtered out the clips used for training. All experiments were performed on an NVIDIA Tesla V100S 32GB.

4.1   Action Recognition Settings and Results

We formulate the AR task with the constraint to predict the single frame in which the action occurs rather than the segment that defines the start and the end of the action. To accomplish this, we trained our Mamba model for 100 epochs, with an AdamW optimizer, a learning rate of 1e-5 and, a batch size of 32. The high framerate (30 fps) of the dataset significantly increase the inherent class imbalance of the annotations. To address this, only during training, we implemented a targeted downsampling strategy, reducing the video framerate from the original 30fps to 4 fps while ensuring all frames with positive annotations were preserved. This approach was complemented by the use of a Focal Loss function, configured to assign a higher weight to the positive class. This combined strategy facilitates a stable training process and, critically, allows the recurrent nature of the Mamba architecture to be leveraged for effective inference on the original high-framerate videos, achieving robust performance.

We evaluated our AR models by measuring the p-AP, as in [12]. Following the metric formulation, given a set of time thresholds, for each ground truth frame, we calculate the nearest prediction greedily favoring high confidence prediction. If the prediction has the same class as the ground truth and the time distance between the two is lower than the time threshold, the prediction is matched to the ground truth. Any unmatched prediction will count as a false positive, while unmatched ground truth instances count as false negatives. To leverage

Table 1. Comparison of various backbones with Mamba. p-AP results are reported as percentages. Execution Time is reported in seconds (it refers to the time that the complete model needs to process a single frame), model Size in Mb. Best results in bold and second-best results underlined.

Configuration	p-AP@4fps(%)↑	p-AP@30fps(%)↑	Execution Time(s)↓	Size(Mb) ↓
Mamba + DINOv2	<u>66.28</u>	**40.25**	0.06	<u>448.0</u>
Mamba + EfficientNetV2	61.32	24.72	**0.007**	446.0
Mamba + EfficientNetV2 Distilled	**68.37**	<u>38.52</u>	**0.007**	446.0

the powerful feature representations of DINOv2 within a real-time system, we adopted a knowledge distillation strategy. The large DINOv2 model serves as the "teacher". While EfficientNetV2 acts as the "student". The objective is to align the student's feature space with the teacher's one. To achieve this, we used the Mean Squared Error (MSE) as the training loss. Due to hardware and time constraints, distilling the model on the entire dataset was computationally prohibitive. To obtain a feasible yet representative training environment, we randomly sampled 10% of the frames from the complete dataset. This subset was then split into a final training set and a validation set, which were used to effectively train our model and monitor its performance.

The quantitative results for the Mamba-based action recognition module, detailed in Table 1, highlight a clear trade-off between accuracy and computational efficiency across different backbones.

The configuration using DINOv2 as a feature extractor establishes a strong performance baseline, achieving an p-AP of 66.28% at 4fps and 40.25% at 30fps. However, with an execution time of 0.06 seconds per frame, this model's high latency makes it unsuitable for real-time applications at 30fps.

In contrast, employing the standard EfficientNetV2 backbone drastically improves efficiency, reducing the execution time to just 0.007 s. This gain in speed comes with a significant performance penalty, as the p-AP drops to 61.32% (from 66.28%) at 4fps and to 24.72% (from 40.25%) at 30fps.

The distilled EfficientNetV2 model emerges as the optimal configuration. Notably, it not only improves upon the standard EfficientNetV2 but surpasses all other variants at 4fps, achieving a top p-AP of 68.37%. While its 30fps performance does not exceed the DINOv2 baseline, it demonstrates a substantial recovery of accuracy compared to the non-distilled version. Crucially, it achieves this strong performance while maintaining the real-time execution speed of 0.007 s, confirming that the distillation process was effective in creating a model that successfully balances high accuracy with low computational overhead.

4.2 Object Detection Models Comparison

We tuned and evaluated RT-DETR Large, YOLOv10m, YOLOv8m-world, and YOLOv8x-world. All object detectors were fine-tuned for 200 epochs with 7 epochs of patience, the optimizer was SGD with a learning rate of 0.01 and

Table 2. Comparison of object detection models. For each row, we report Average Precision (AP), Recall, and Harmonic Mean (HM) in percentages, along with Execution Time in seconds, and model Size (in MB). Best results in **bold**, second-best in underlined.

Model	AP(%) ↑	Recall(%) ↑	HM(%) ↑	Execution Time(s) ↓	Size(Mb) ↓
RT-DETR Large	82.9	71.2	76.6	0.070	<u>686.0</u>
YOLOv10m	83.3	70.1	76.1	0.039	754.0
YOLOv8m-world	**85.1**	<u>77.7</u>	<u>81.2</u>	**0.037**	**634.0**
YOLOv8x-world	<u>84.8</u>	**78.9**	**81.7**	<u>0.038</u>	862.0

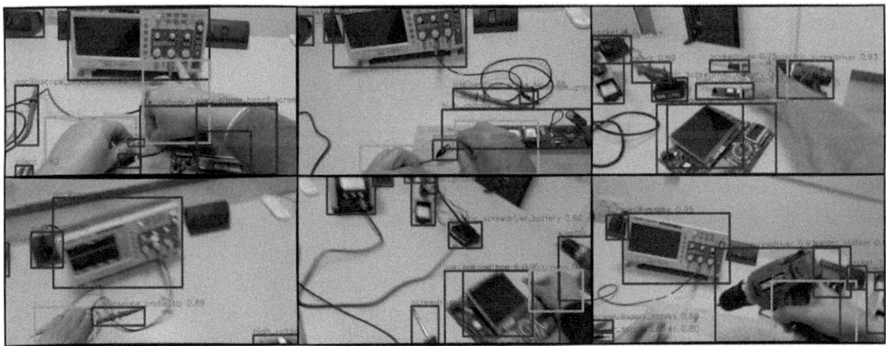

Fig. 2. Representative samples of YOLOWorldM predictions on some validation images of ENIGMA-51. Blue boxes represent objects, green boxes represent hands. (Color figure online)

a momentum of 0.9. The batch size was set to 16, and the input image size to 640x640 without stretching, using padding. The evaluation was done with a confidence threshold of 0.7.

Table 2 reports the results of the evaluated models. Among the evaluated models, YOLOv8m-world achieves the highest AP of 85.1%, making it the most accurate in detecting objects. It is also the most efficient one, with the lowest inference (0.037 s) and the smallest model size (634 MB). On the other hand, while YOLOv8x-world has a slightly lower AP (84.8%), it achieves the highest recall (78.9%), indicating a stronger ability to detect a wider range of objects. This comes at the cost of a slightly larger model size (862 MB) and a slightly longer inference time. The YOLOv10m model delivers a solid AP of 83.3%, but it has a slightly lower recall (70.1%) compared to the YOLOWorld models. While its inference speed (0.039 s) is competitive, its model size of 754 MB places it between YOLOv8m-world and YOLOv8x-world in terms of storage efficiency. Meanwhile, RT-DETR Large, despite achieving a respectable AP of 82.9% and a large size (686.0), has a lower HM value (76.6%) and a slightly higher latency (0.070 s).

Table 3. Comparison of AP for Hand, Hand + State, Hand + Side, and Hand + All across different IoU thresholds. Best results are shown in **bold**, and second-best results are <u>underlined</u>.

IoU Threshold	AP Hand(%) ↑	AP Hand + State(%) ↑	AP Hand + Side(%) ↑	AP Hand + All(%) ↑
0.01	90.86	**47.34**	88.41	**33.90**
0.05	90.86	<u>46.12</u>	88.41	<u>33.52</u>
0.1	90.86	35.86	88.41	28.57
0.2	90.86	23.29	88.41	20.28
0.3	90.86	19.65	88.41	14.12

Figure 2 shows the predictions of some validation images obtained with a confidence score threshold of 0.5.

4.3 Active Object Retrieval Criteria

The OD module is integrated into the system with the purpose of identifying the active object once the AR module has detected a contact. As such, the OD module operates conditionally, relying on the output of the AR module. To associate an active object with the hands in the scene, we adopt a straightforward strategy: among all detected objects, the one with the highest IoU with the hand bounding boxes is selected as the active object, provided the IoU exceeds a predefined threshold. Considering no learning is involved, the IoU threshold is the only parameter that requires tuning in this setup.

The performance of the HOI have been evaluate using:

- **AP Hand**: Average Precision of the hand detections.
- **AP Hand + State**: Average Precision of the hand detections considering the correctness of the hand state.
- **AP Hand + Side**: mean Average Precision of the <hand, active object> detected pairs.
- **AP Hand + All**: combinations of AP Hand, AP Hand + State and AP Hand + Side metrics.

The results reported in Table 3 were computed only on frames where contact has been annotated. Under this assumption, we activate the OD module through an oracle that does not penalize false positives detected by the module. As shown, the best results are obtained with an IoU threshold of 0.01. This low threshold requirement highlights the need for a more advanced method to detect interactions.

Considering that Mamba has no integrated mechanism to retrieve the class of the object, we decided to use YOLOWorld as a model to retrieve the active object in the scene and use Mamba to detect the action. Figure 3 illustrates some ENIGMA-51 test samples. As can be seen, the algorithm returns one object per frame, choosing between objects and hands based on the highest IoU. The frames inside the green boxes indicate successes, where YOLO correctly detects the hand

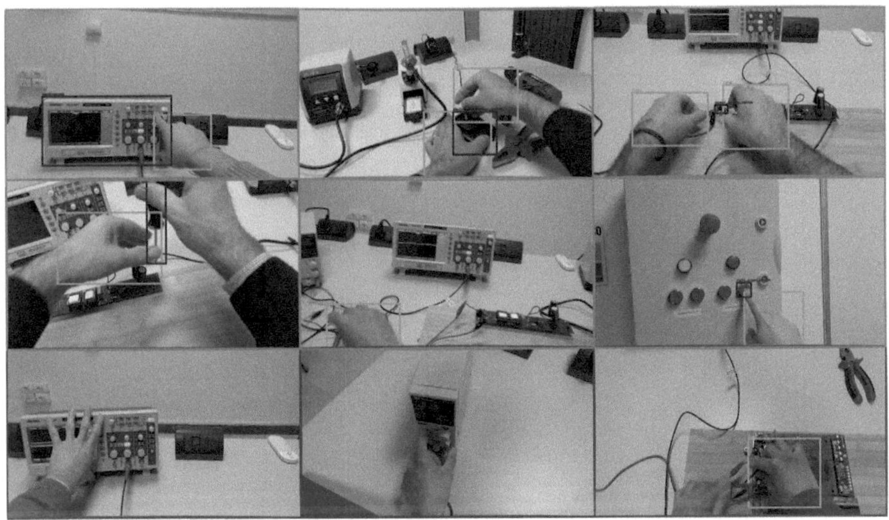

Fig. 3. Representative samples of active object retrieval performed by YOLOWorld. Green frames highlight success cases, red frames indicate failure cases. (Color figure online)

Table 4. Comparison of AP for Hand, Hand+State, Hand+Side, and Hand+All using ORACLE+YOLO and MAMBA-based pipelines. ORACLE+YOLO serves as an upper bound using ground-truth labels, while MAMBA@30/60+YOLO leverages predictions from the past 30 and 60 frames, respectively. Best results are in **bold**, second-best in underlined.

Method	AP Hand(%) ↑	AP Hand + State(%) ↑	AP Hand + Side(%) ↑	AP Hand + All(%) ↑	Execution Time(s)↓	Size(Mb) ↓
ORACLE+YOLO	**90.86**	**47.34**	**88.41**	**33.90**	**0.057**	**634.0**
MAMBA@30+YOLO	36.34	23.22	35.36	14.98	0.064	1080
MAMBA@60+YOLO	45.43	28.85	44.37	19.70	0.064	1080

and the objects, and the active object prediction is correct. Red boxes indicate failures and can be due to various reasons, such as objects or hands not being recognized by YOLO, IoU threshold not being met, or multiple objects close together, leading to incorrect object selection.

4.4 Overall Pipeline Evaluation

We evaluate two strategies for triggering object detection (OD) in the context of hand-object interaction (HOI): a heuristic oracle-based method and a learned approach using the AR module. In the oracle-based setup, we assume perfect knowledge of when a contact occurs and trigger the OD module (YOLOv8m-world) only on frames known to contain contact. Within these frames, the active object is identified using the aforementioned mechanism involving bounding boxes overlap. Despite being driven by a simple heuristic, this method establishes an upper bound on performance, as it leverages ground-truth contact

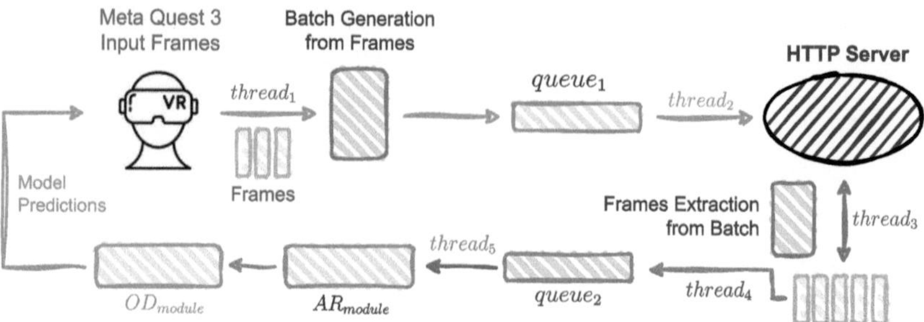

Fig. 4. Representation of the full pipeline, from input frames recording to model predictions back to the Meta Quest 3.

timing. In contrast, the Mamba-based AR module detects contact events over time and autonomously triggers the OD module. For online processing, we apply a temporal windowing strategy that looks only at past frames: if the AR module detects contact within the preceding 30 or 60 frames from the current frame, the OD is triggered for that frame. This past-only approach ensures real-time compatibility while reducing sensitivity to occasional missed detections and better capturing the temporal nature of interactions.

Table 4 compares the results. The first row (ORACLE+YOLO) represents the upper bound. The second and last rows show the results when using our AR model with different temporal windows. Notably, the oracle setting does not penalize false positives from YOLO, as OD is invoked only in frames known to contain contact. This leads to higher reported values compared to the AR-triggered setup, where the OD is invoked based on predicted contact, potentially introducing more uncertainty. A more balanced comparison would require accounting for false positives in the oracle-driven YOLO as well, which would likely narrow the performance gap.

5   Implementation with Meta Quest 3

In this section, we present a full pipeline that leverages a wearable device to capture frames and acts as a practical example. To operate in a real-time fashion, we utilize the Meta Quest 3 headset as the source for video streaming and employ multiple threads to handle processes concurrently. All frames are captured via a Unity-based interface, enabling the user to start and stop the recording using the Meta Quest controller. When the recording begins, the Meta Quest stores up to 60 frames. $thread_1$ is responsible for creating a batch of frames and inserting them into $queue_1$. Subsequently, $thread_2$ retrieves the batch and sends it to an HTTP server. The HTTP server receives the batch, and $thread_3$ unpacks it into 60 individual frames. From this, $thread_4$ extracts the frames, duplicates each one, and send them into $queue_2$, which facilitates the flow of frames to the models. $thread_5$ serves as the model worker for $queue_2$; it extracts frames in groups of 30, sends them to the models for concurrent predictions, and finally returns the results to the Meta Quest 3 for user feedback. The explained pipeline is illustrated in Fig. 4. Regarding the Unity application used to send the frames and visualize user feedback, our work is based on the Android MediaProjector API. This allows us to record the left eye screen of the Meta Quest 3 and access the raw frames. We adopted this method because we believe that an "edge-based" solution like this one is superior to casting. In fact, the alternative is to use the casting option of Meta and than record the browser: this solution, even if it is faster in latency when operating under optimal solution, is not viable if we consider that it is mandatory that both the visor and the computer needs to be in the same network. With our solution, by adopting an HTTP request to an arbitrary endpoint, we can send those frames independently of the network used for both of the elements included.

6   Conclusion

In this work, we tackled the pressing challenge of detecting hand-object interactions from an egocentric perspective in real-time, a critical capability for industrial applications. We introduced an efficient, two-stage approach that decomposes this complex problem into online contact detection and active object identification. Our proposed pipeline first leverages an AR module to determine when contact occurs, which in turn triggers an OD module to identify the specific hand and object involved in the interaction. Our experimental results validate the effectiveness of this methodology. For the AR task, we demonstrated that a Mamba-based model, when paired with a distilled EfficientNetV2 feature extractor, successfully balances high accuracy with the low latency required for real-time applications, processing frames in just $0.007\,\mathrm{s}$. For the OD component, YOLOWorld emerged as the superior model, achieving the highest accuracy (85.13% AP) and fastest inference speed ($0.037s$) among the tested alternatives. Our analysis revealed that a very low IoU threshold was required for optimal performance, even with oracle contact data, highlighting the limitations of inferring interactions from spatial overlap alone. The subsequent performance decrease when combining the Mamba-based Action Recognition module and YOLOWorld is justified, as the system transitions from an idealized oracle to a practical, learned model. The oracle setup represents a theoretical upper bound on performance because it uses ground-truth data to trigger object detection only on frames with guaranteed contact, a method that is not penalized for false positives. Finally, we demonstrated the practical applicability of our work by developing and implementing the entire pipeline on a Meta Quest 3 headset. Code and implementation details are available at: https://github.com/JustCati/iCODE-Server.

Acknowledgements. This research has been funded by the European Union - Next Generation EU, Mission 4 Component 1 CUP E53D23008280006 - Project PRIN 2022 EXTRA-EYE, and FAIR – PNRR MUR Cod. PE0000013 - CUP: E63C22001940006. A. Finocchiaro has been supported by Research Program PIAno di inCEntivi per la Ricerca di Ateneo 2024/2026, project "Multi-Agent Simulator for Real-Time Decision-Making Strategies in Uncertain Egocentric Scenarios" - University of Catania.

References

1. An, J., Kang, H., Han, S.H., Yang, M.H., Kim, S.J.: Miniroad: minimal RNN framework for online action detection. In: Proceedings of the IEEE/CVF International Conference on Computer Vision, pp. 10341–10350 (2023)
2. Chen, G., et al.: Video mamba suite: state space model as a versatile alternative for video understanding. arXiv preprint arXiv:2403.09626 (2024)
3. Cheng, T., Song, L., Ge, Y., Liu, W., Wang, X., Shan, Y.: Yolo-world: real-time open-vocabulary object detection. In: Proceedings of the IEEE/CVF Conference on Computer Vision and Pattern Recognition (2024)
4. Darkhalil, A., et al.: Epic-kitchens visor benchmark: video segmentations and object relations. Adv. Neural. Inf. Process. Syst. **35**, 13745–13758 (2022)
5. Gu, A., Dao, T.: Mamba: linear-time sequence modeling with selective state spaces. arXiv preprint arXiv:2312.00752 (2023)
6. Leonardi, R., Ragusa, F., Furnari, A., Farinella, G.M.: Exploiting multimodal synthetic data for egocentric human-object interaction detection in an industrial scenario. Comput. Vision Image Understanding
7. Mazzamuto, M., Ragusa, F., Resta, A., Farinella, G. M., Furnari, A.: A wearable device application for human-object interactions detection. In: VISIGRAPP (4: VISAPP), pp. 664–671 (2023)
8. Oquab, M., et al.: Dinov2: learning robust visual features without supervision. arXiv preprint arXiv:2304.07193 (2023)
9. Ragusa, F., et al.: Enigma-51: towards a fine-grained understanding of human-object interactions in industrial scenarios. arXiv preprint arXiv:2309.14809 (2023)
10. Ragusa, F., Furnari, A., Livatino, S., Farinella, G.M.: The meccano dataset: understanding human-object interactions from egocentric videos in an industrial-like domain. In: Proceedings of the IEEE/CVF Winter Conference on Applications of Computer Vision, pp. 1569–1578 (2021)
11. Shan, D., Geng, J., Shu, M., Fouhey, D.F.: Understanding human hands in contact at internet scale (2020)
12. Shou, Z., et al.: Online detection of action start in untrimmed, streaming videos. In: Proceedings of the European Conference on Computer Vision (ECCV), pp. 534–551 (2018)
13. Tan, M., Le, Q.: Efficientnetv2: smaller models and faster training. In: International Conference on Machine Learning, pp. 10096–10106. PMLR (2021)
14. Wang, x., et al.: Oadtr: online action detection with transformers. In: Proceedings of the IEEE/CVF International Conference on Computer Vision, pp. 7565–7575 (2021)
15. Zhang, L., Zhou, S., Stent, S., Shi, J.: Fine-grained egocentric hand-object segmentation: Dataset, model, and applications. In: European Conference on Computer Vision, pp. 127–145. Springer (2022)

16. Zhao, Y., Krähenbühl, P.: Real-time online video detection with temporal smoothing transformers. In: European Conference on Computer Vision, pp. 485–502. Springer (2022)
17. Zou, C., et al.: End-to-end human object interaction detection with hoi transformer. In: Proceedings of the IEEE/CVF Conference on Computer Vision and Pattern Recognition, pp. 11825–11834 (2021)

Toward Trustworthy AI in Retail: A Case Study of Vending Machines in Public Environments

Alessandro Galdelli[1], Luigi Di Bello[2], Marco Contigiani[2], Mattia Sospetti[2], Mauro D'Aloisio[2], Valerio Placidi[2], and Rocco Pietrini[3]

[1] Università Politecnica delle Marche, Ancona, Italy
`a.galdelli@staff.univpm.it`
[2] Med Innovations, Porto Recanati, Italy
`{luigi.dibello,marco.contigiani,mattia.sospetti,mauro.daloisio,`
`valerio.placidi}@medinnovations.it`
[3] Universitas Mercatorum, Rome, Italy
`rocco.pietrini@unimercatorum.it`

Abstract. Artificial intelligence (AI) systems are becoming increasingly embedded in physical retail environments, enabling real-time personalisation through face and behaviour analysis. This paper critically examines the ethical implications of one such system: an intelligent vending machine (VM) infrastructure deployed in 30 real-world locations across Italy. This infrastructure uses computer vision and machine learning to adapt promotional content based on inferred age, gender and interaction patterns. While such systems are technically effective and privacy-preserving, they raise important concerns around fairness, transparency, consent, and profiling. Drawing on the EU's Ethics Guidelines for Trustworthy AI and the ALTAI framework, we evaluate the VM system against core ethical principles, particularly those of justice and explicability. Key challenges identified include demographic bias in personalisation, the absence of meaningful user awareness and implicit profiling in public spaces. Based on our findings, we propose practical enhancements to improve fairness-aware design, embedded transparency mechanisms and symbolic consent models. Our work provides a field-tested ethical assessment of AI in physical retail and offers actionable strategies for aligning real-world deployments with responsible AI standards.

Keywords: Artificial Intelligence · Trustworthy AI · Vending Machine · Retail · AI Ethics

1 Introduction

In recent years, the integration of artificial intelligence (AI) technologies into physical retail environments has paved the way for a deeper understanding of consumer behaviour, personalised marketing and data-driven decision-making [3, 4, 11, 12]. Vending machines (VMs) are evolving from passive distribution devices into advanced, interactive systems that can adapt to customers' needs in real time [14]. Using computer vision and machine learning, vending machines can detect individuals, analyse their facial features, estimate demographic attributes such as age and gender, and infer behavioural patterns in order to make product recommendations or inform promotional strategies. These innovations have the potential to enhance user experience and optimise business performance [2].

However, the growing use of such technologies in public and semi-public spaces raises important ethical and regulatory issues [7,10,17]. Integrating AI into VMs introduces new forms of implicit profiling, nudging and automated decision-making, which may affect consumer autonomy, privacy and fairness, often without the user's explicit knowledge or consent. These challenges are particularly pertinent in light of recent policy developments, such as the European Union's General Data Protection Regulation (GDPR) and the Artificial Intelligence Act (AI Act). These acts establish strict guidelines for the lawful, transparent and accountable use of AI, particularly in systems involving biometric categorisation or behavioural prediction.

This paper addresses the ethical implications of smart VMs that use face and behaviour recognition technology. Building on a previous work [2], in which the authors proposed and validated a low-cost edge/cloud architecture for detecting and segmenting shoppers in a large-scale, real-world VM lab across 30 locations in Italy, this paper shifts the focus towards critically examining how such systems align, or fail to align, with principles of Trustworthy AI. Specifically, we evaluate the implications of AI-enabled virtual machine (VM) features in terms of privacy protection, fair demographic classification, explainable machine decisions, and respect for human autonomy. Our analysis is based on a unique dataset comprising over one million consumer interactions. It considers not only the technical capabilities of the system, but also its social, legal and psychological implications. By situating our case study within recognised ethical frameworks, such as the Assessment List for Trustworthy AI (ALTAI)[1], HLEG-AI 2019 [1] and the risk-based classification system of the EU AI Act, we provide practical insights and recommendations for the responsible design and governance of intelligent retail systems.

This paper makes three key contributions. First, we provide a structured ethical analysis of AI-based shopper recognition and adaptation mechanisms in vending environments. Second, we map the system's current design against relevant regulatory and ethical frameworks, highlighting areas of compliance and risk. Third, we propose a set of actionable design recommendations aimed at enhancing transparency, fairness, and accountability in AI-driven retail technologies.

The paper is structured as follows: Sect. 2 provides a review of the literature on AI-based behaviour analysis paying particular attention to the ethical concerns raised in previous works. Section 3 introduces the core ethical principles for trustworthy AI. It then discusses how these principles pertain to AI-driven consumer profiling and interaction in vending contexts. Section 4 offers a critical

---

[1] https://altai.insight-centre.org.

analysis of the real-world system architecture that was previously deployed in 30 vending machines across Italy. Drawing on empirical evidence from over one million shopper interactions, it evaluates the system's alignment with the earlier introduced ethical and legal requirements, with a particular focus on fairness, privacy, and explainability. Section 5 proposes a set of strategic enhancements to strengthen the system's trustworthiness. These include design recommendations to mitigate bias in demographic estimation, improve user transparency and introduce human oversight mechanisms. Finally, Sect. 6 concludes the paper by summarising the main insights and contributions. It considers the wider implications for the ethical deployment of AI in retail and suggests areas for future research, such as user-centred evaluations and the integration of fairness auditing tools.

2   Related Works

This section provides a detailed analysis of literature on the ethics of the analysis of human behaviour, paying particular attention to key studies that have influenced the ongoing discussion on AI ethics. The focus is on three critical areas: privacy, informed consent and data protection.

Surez et al. [16] conducted a thorough review of the peer-reviewed literature on ethical decision-making in behaviour analysis and related health disciplines. Their research identified 55 distinct ethical decision-making frameworks across 60 articles, spanning seven core professional domains, such as medicine and psychology, and 22 specialised subfields, including dentistry and family medicine. Through consensus-driven analysis, the researchers highlighted nine frequently recommended actions within these models. They noted that the vast majority (52) followed a sequential process, while a smaller number (23) incorporated a problem-solving approach. These nine actions correspond closely with the steps set out in the Behaviour Analyst Certification Board's 2020 Code of Ethics, highlighting a shared professional understanding of key ethical practices.

Building on this basis, Contreras et al. [5] emphasised the importance of evidence-based practice (EBP) in improving ethical decision-making in applied behaviour analysis (ABA). Referencing Slocum et al. [15], they highlighted how the EBP model in ABA can provide a robust framework for ethical judgement. They also discussed how integrating EBP aligns with the ethical expectations set out by the Behaviour Analyst Certification Board, and they proposed concrete strategies to help behaviour analysts refine their ethical practices.

Taking a more critical perspective, Wilkenfeld and McCarthy [19] examined the ethical issues surrounding the treatment of Autism Spectrum Disorder (ASD) using Applied Behaviour Analysis (ABA). Despite its widespread endorsement, they argued that ABA may conflict with fundamental bioethical principles such as justice and nonmaleficence. The authors contended that ABA can undermine the autonomy of both children receiving treatment and parents making care decisions, highlighting the ethical tensions that can emerge when clinical efficacy is prioritised without sufficient ethical oversight.

Kelly et al. [8] examined the ethical principles established by the Behaviour Analyst Certification Board (BACB) and explored how certified behaviour analysts can apply these standards to their clinical and professional decision-making processes. They discussed the challenge of communicating these standards to non-specialist audiences, and explained how the BACB has developed a set of complementary ethical principles to support the dissemination of these standards and value-based decision-making. These principles are intended to help practitioners navigate ethical dilemmas and convey their professional norms effectively.

Building upon both EU AI ethics principles and applied computer vision technologies, a more recent contribution further advances this line of research by proposing a methodological framework for designing ethically sound AI systems in retail [18]. The study focuses on the VRAI framework, a deep learning system designed to analyse shopper behaviour, count people, and re-identify individuals, to demonstrate the urgent requirement for fairness and transparency in AI pipelines. The authors emphasise that bias and opacity can severely undermine trust, and propose strategies for embedding ethical requirements directly into system design. These include techniques for mitigating bias and tools for increasing system explainability, thereby promoting technical robustness and ethical integrity. The framework bridges the gap between abstract ethical principles and practical deployment, offering a way to adopt AI more widely and more trustingly in retail environments. Similarly, [13] demonstrated the application of multimodal deep learning approaches for person re-identification in cultural heritage environments, using RGB-D cameras in top-view configurations to analyze visitor behavior while preserving privacy. Their SeSAME system, deployed in museum settings, shares architectural similarities with retail AI systems in terms of edge processing, behavioral analysis, and the challenge of balancing personalization with ethical considerations.

While the existing literature has made significant progress in formalising ethical principles for behaviour analysis and exploring their implications for AI in clinical and controlled environments, there is still limited understanding of how these principles can be applied in practice to AI systems operating autonomously in public and commercial contexts. The few studies that address this issue, such as that of Tiribelli et al. [18], emphasise the importance of embedding fairness and explainability in retail AI systems, yet they do not examine the ethical implications of real-time systems that interact directly with consumers. This paper addresses this underexplored area by offering a detailed ethical analysis of an AI-powered vending machine ecosystem, based on large-scale deployment data and operational experience. Situating our discussion within current European regulatory frameworks and principles for trustworthy AI, we demonstrate how face and behaviour recognition technologies in retail present novel challenges regarding consent, fairness and transparency, and propose practical design strategies for their ethical and responsible use.

3   Ethical Principles for Trustworthy AI in Consumer Behaviour Understanding

As smart VMs become more prevalent in everyday public life, it is not only a regulatory necessity but also a social imperative to embed ethical principles into their design and operation. These AI-powered systems detect and analyse consumer presence, estimate demographic features such as age and gender, and adapt their promotional content accordingly. They operate at the intersection of commercial innovation and individual rights. In such contexts, where physical space, attention and autonomy are limited, ensuring trustworthy AI requires developers and deployers to engage with established ethical frameworks in a concrete, operational manner. To guide this process, we refer to the Ethics Guidelines for Trustworthy AI issued by the European Commission's High-Level Expert Group on AI and the Assessment List for Trustworthy AI (ALTAI). These frameworks articulate five core principles for ethical and responsible AI development [6]:

- **Beneficence:** ensuring that AI promotes individual and societal well-being, including accessibility and inclusive product offerings.
- **Non-maleficence:** minimising the risk of harm, particularly from misclassification or discriminatory responses.
- **Autonomy:** respecting consumers' capacity to act freely, without unknowingly being profiled or manipulated.
- **Justice and fairness:** preventing unfair bias, stereotyping and unjust targeting in system outputs.
- **Explicability:** ensuring that system decisions are understandable, transparent and open to scrutiny.

To begin our ethical analysis, we will first summarise the functional architecture of the intelligent vending machine system under evaluation. As illustrated in Fig. 1, each vending unit is equipped with a low-cost embedded camera that performs real-time people and face detection and analysis to estimate age and gender. Individuals interacting with the VM are assigned a unique, anonymised ID, which is used to associate demographic and behavioural data with vending interactions, such as time spent interacting with the machine, product selection and payment method. This information is processed locally and then transmitted to the cloud, where it is aggregated and used to support adaptive promotional strategies. This architecture enables personalisation without storing biometric images and forms the basis of our fairness and transparency assessments.

While all five principles are relevant, the nature of the intelligent VM system, which is deployed across 30 public and semi-public spaces in Italy, makes two principles particularly salient: Justice and Fairness and Explicability.

The VM system uses face analysis technologies to categorise shoppers by age and gender, with the aim of segmenting behaviour and tailoring product recommendations or promotional offers. While technically efficient, such categorisation models often rely on pre-trained computer vision networks that may

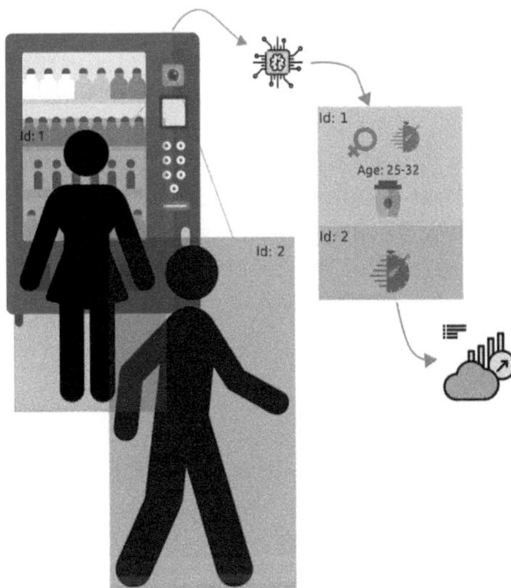

Fig. 1. Overall architecture. The camera on the vending machine (VM) tracks multiple individuals and performs multi-face detection, enabling age and gender classification. The resulting data is processed locally and sent to the cloud for behavior analysis and promotional adaptation.

embed demographic bias, for example by overestimating the age of women or misclassifying darker skin tones. When these outputs are used to trigger specific promotions (e.g. offering a coffee and chocolate combo to older men), there is a risk of reinforcing stereotypes or unintentionally excluding certain demographic groups from favourable offers. Furthermore, fairness should be considered not only in terms of accuracy metrics, but also in relation to equal treatment and access to opportunities. In vending contexts, where consumers have little time or opportunity to negotiate, biased algorithmic outcomes may subtly but persistently influence consumption patterns, trust and perceived inclusion. Therefore, our system must be assessed for group fairness (do all groups benefit equally from adaptive features?) and procedural fairness (were design choices inclusive and transparent?).

In the context of intelligent VMs, explicability is particularly challenging due to the nature of the interaction: brief, screenless and ambient. Users may not realise that they are being analysed, that their demographic attributes are being inferred, or that their behaviour is informing algorithmic decisions. Without clear communication, even a system that is non-intrusive and processes data at the edge, thus protecting privacy in technical terms, may fall short of ethical expectations related to transparency and informed consent.

Thus, explicability must be approached in two layers here:

- **System-level explicability:** the architecture, logic and data flows must be auditable and well documented for regulatory and technical oversight.

- **User-level transparency:** symbolic indicators (e.g. signage to indicate that AI is in use), opt-out opportunities and public-facing materials should explain what the system does, what data is processed and how personalisation occurs.

These two principles, fairness and explicability, form the backbone of our ethical analysis of AI-driven vending systems. In the following sections, we critically assess how our deployed system aligns with these requirements and propose concrete measures for enhancing its trustworthiness, based on both empirical evidence and normative reasoning.

4   Results

In order to evaluate the ethical implications of intelligent vending machines operating in real-world environments, we applied the Trustworthy AI principles, particularly those of fairness and explicability, to a computer vision-based system for shopper analysis that had already been deployed. Installed in 30 vending machine locations across Italy, the system generates behavioural and demographic insights from over one million customer interactions per year. The system is designed to operate in a non-intrusive, privacy-preserving manner. It estimates age and gender using a low-cost camera, tracks stopping behaviour and links consumer profiles with purchasing events to enable adaptive promotional strategies. While the system has demonstrated strong technical performance in real-world scenarios, the ethical audit conducted in this study highlights several important considerations.

4.1   Fairness Implications in Shopper Profiling and Promotion

The system uses a face-based classification pipeline to estimate the age group and gender of shoppers as part of a behavioural segmentation strategy. Although the TinyYOLOv2 [20] and CNN-based classification models [9] used were pre-trained on diverse datasets, their accuracy, measured at 91.2% for age and 94.4% for gender in controlled environments, may not be equally applicable to all ethnicities, lighting conditions or body positions encountered in public vending settings. Therefore, algorithmic bias remains a potential concern, particularly for demographic groups that were underrepresented in the training data.

Promotional campaigns conducted in the original deployment, such as targeting healthy snacks at younger women or combo offers at senior men, demonstrated modest but measurable improvements in sales performance (e.g. a 3.7% category uplift). However, these strategies also revealed ethical trade-offs: if promotional access depends on demographic inference, individuals who are misclassified may be excluded from favourable pricing or tailored offers. More critically, if such patterns persist across locations, the system may contribute to structural bias in access to products or services, even unintentionally. Furthermore, the deployed system did not explicitly address intersectional fairness, which considers the combined effect of multiple attributes such as gender, age, and group behaviour. Future iterations should consider reducing reliance on demographic classifiers in promotional strategies or implementing fairness-aware models that explicitly monitor group-level impacts.

4.2 Explicability and Transparency in Public Interaction

In the evaluated deployment, computer vision processing was conducted entirely at the edge, with no images stored or transmitted to the cloud. While this approach enhances data minimisation and aligns with the privacy-by-design principle of the GDPR, user awareness and system transparency remained limited. There was no dynamic interface to inform consumers that an AI system was in use, nor were there any symbolic indicators to explain how their behaviour might affect pricing or product visibility. This lack of real-time transparency raises questions about informed engagement. Unlike digital interfaces (e.g. e-commerce), physical VMs offer users few options to understand, challenge or opt out of algorithmic personalisation. Interviews with operators and consumers conducted during the pilot phase revealed that many users were unaware of personalisation, and those who noticed behavioural adaptation assumed it was randomised rather than systematically driven by computer vision inference. To meet the ethical standard of explicability, the system would benefit from enhancements such as:

- Physical signage indicating the presence of AI-driven personalisation; QR codes or NFC links to access privacy and explainability information;
- A consent mechanism that distinguishes between anonymous telemetry and behaviourally adaptive responses.

To summarise the findings of our ethical analysis, we present our key observations across four critical dimensions: fairness, transparency, consent and profiling logic. For each dimension, we reflect on the current implementation of the intelligent vending machine system, identify any ethical concerns that have arisen during deployment and propose targeted recommendations for mitigation and improvement. This overview is intended to guide the future development and ongoing governance of AI systems operating in consumer-facing environments, ensuring they align with the principles of Trustworthy AI. The summary is reported in Table 1.

5 Discussions

In this section, we propose targeted enhancements to strengthen the trustworthiness of AI-based retail systems operating in public spaces. The ethical evaluation of our deployed VM system has revealed important opportunities for improvement in terms of fairness, transparency, and user agency. Although the technical architecture, which is based on real-time edge processing and anonymised data handling, demonstrates strong alignment with the principles of privacy by design, the current implementation of the system still leaves gaps in the embedding of ethical requirements into the consumer experience, design choices and long-term governance.

Table 1. Summary of ethical evaluation of the smart vending machine system

Ethical Dimension	Current Practice	Identified Concern	Recommendation
Fairness	Age and gender-based segmentation and targeted promotions	Risk of bias, stereotyping, and exclusion due to misclassification or unbalanced training data	Monitor fairness metrics; reduce reliance on demographic profiling; incorporate behavior-only clustering and alternative criteria
Transparency	No user-facing explanation of AI-based personalization	Low user awareness; no way to verify or interpret system behavior	Add visual signage or symbolic indicators; use QR codes or NFC tags to link to system and privacy information
Consent	Implicit profiling via proximity and interaction	Lack of explicit or informed consent mechanisms	Explore symbolic opt-in models; offer opt-out via mobile interface or signage; comply with data minimization principles
Profiling Logic	Personalization based on combined demographic and behavioral features	Possibility of unfair treatment or exclusion in promotional targeting	Use behavior-only fallback clustering; incorporate rotation/randomization to reduce systemic bias

5.1 Fairness-Aware Design and Monitoring

To address fairness concerns, future iterations of the system should implement bias auditing and fairness monitoring pipelines. This would involve tracking demographic performance disparities (e.g. misclassification rates based on age or gender), evaluating group-based outcomes in terms of promotional access, and adopting fairness-aware machine learning models where applicable. Behaviour-only clustering strategies, which rely on patterns of interaction rather than inferred identity, should be introduced as an alternative to demographic segmentation that preserves privacy. Furthermore, randomly rotating promotional content across groups or conducting A/B tests with fairness constraints can mitigate the risk of systematic exclusion or over-targeting.

5.2 Embedded Transparency Mechanisms

Although technical transparency is ensured through the documentation and auditability of the system pipeline, transparency for users remains limited. To address this issue, we propose integrating low-barrier mechanisms for achieving symbolic and procedural transparency. For example, clear signage could be placed on or near the vending machine to indicate the presence of AI and personalisation algorithms. QR codes or NFC access points that link to explanations in a format that is easy for humans to read, which detail how data is used, what personalisation occurs and how consumers can request more information or raise concerns. Icons or colour-coded prompts to indicate when personalisation is activated in response to behavioural signals. These measures aim to raise awareness without disrupting usability, helping users to feel more in control of the interaction, and ensuring compliance with the EU AI Act's emerging legal standards for explainability.

5.3   Respecting Autonomy and Consent in the Physical Space

In the absence of explicit opt-in interfaces in vending contexts, ensuring meaningful user autonomy requires us to redefine consent in a contextually appropriate way. We propose adopting a model of symbolic or ambient consent, whereby users are informed via signage and can choose to proceed with or avoid interaction. This could be reinforced by offering non-personalised modes of interaction or by giving consumers the opportunity to disable personalisation for a session via simple mechanisms, such as a 'privacy button' or an opt-out QR code. In the long term, retailers and technology providers should consider implementing dynamic consent models, particularly if additional biometric or behavioural features are introduced. These models should include the capacity for users to access their data, understand how it is being processed, and retract their consent if desired.

5.4   Human Oversight and Ethical Governance

Enhancing trustworthiness requires institutional measures that extend beyond the technical level. These measures include:

- appointing an internal ethics officer or compliance lead for AI-based retail technologies;
- Conducting regular algorithmic impact assessments (AIAs) to monitor risks and benefits.
- establishing feedback channels that allow consumers to report concerns or ask questions about the system.

These governance mechanisms align with the 'human oversight' requirement of the EU AI Act and ensure that ethical concerns are continuously addressed as the system evolves and scales.

6   Conclusions and Future Works

As AI technologies become more prevalent in public and commercial spaces, it is crucial that they are designed and deployed in a way that reflects technical efficiency, ethical accountability, and societal trust. In this paper, we examine an intelligent vending machine system that uses face and behaviour analysis to personalise product offerings in real time. Building on our previous technical work, we shifted the focus to a critical ethical analysis based on the principles of Trustworthy AI, particularly fairness and explicability, as defined by leading European and global frameworks. Through evaluating a large-scale, real-world deployment involving over one million consumers across 30 vending machine locations, we identified key ethical risks and challenges. These include the potential for demographic bias, a lack of informed consent and limited user awareness of AI-driven personalisation. Although the system adheres to robust privacy-by-design principles through edge processing and data minimisation, it falls short of meeting expectations regarding fairness in profiling and transparency in user interaction. To address these issues, we have proposed a series of practical enhancements, including fairness-aware monitoring, symbolic transparency mechanisms, ambient consent models and institutional governance measures, such as algorithmic impact assessments. These recommendations aim to operationalise ethical principles in a domain where user interaction is brief, physical and often unconscious, highlighting the unique demands of embedding ethical safeguards into ambient AI systems. This work makes a valuable contribution to the broader discourse on responsible AI, offering a practical, field-tested case study of how ethical frameworks can be applied to intelligent retail technologies. Aligning with ethical principles can be beneficial for fostering long-term consumer trust and neccessary for regulatory compliance. Future research will explore methods for integrating real-time user feedback, assessing perceived fairness and transparency across different demographic groups, and refining personalisation algorithms to better align with ethical goals. As intelligent systems become more autonomous and pervasive, the retail sector must take the lead in establishing standards for ethical, inclusive and human-centred AI.

References

1. Ai, H.: High-level expert group on artificial intelligence. Ethics GuidelinesTrustworthy AI **6** (2019)
2. Allegrino, F., Gabellini, P., Di Bello, L., Contigiani, M., Placidi, V.: The vending shopper science lab: deep learning for consumer research. In: Cristani, M., Prati, A., Lanz, O., Messelodi, S., Sebe, N. (eds.) ICIAP 2019. LNCS, vol. 11808, pp. 307–317. Springer, Cham (2019). https://doi.org/10.1007/978-3-030-30754-7_31
3. Araf, R.I.: A systematic review of data-driven insights in retail: transforming consumer behavior and market trend. Innovatech Eng. J. **1**(01), 46–55 (2024)
4. Cao, L.: Artificial intelligence in retail: applications and value creation logics. Int. J. Retail Distr. Manag. **49**(7), 958–976 (2021)
5. Contreras, B.P., Hoffmann, A.N., Slocum, T.A.: Ethical behavior analysis: Evidence-based practice as a framework for ethical decision making. Behav. Anal. Practice, 1–16 (2021)
6. Jobin, A., Ienca, M., Vayena, E.: The global landscape of AI ethics guidelines. Nat. Mach. Intell. **1**(9), 389–399 (2019)
7. Katirai, A.: Ethical considerations in emotion recognition technologies: a review of the literature. AI and Ethics **4**(4), 927–948 (2024)
8. Kelly, E.M., Greeny, K., Rosenberg, N., Schwartz, I.: When rules are not enough: Developing principles to guide ethical conduct. Behav. Anal. Pract. **14**, 491–498 (2021)

9. Levi, G., Hassner, T.: Age and gender classification using convolutional neural networks. In: Proceedings of the IEEE Conference on Computer Vision and Pattern Recognition Workshops, pp. 34–42 (2015)
10. Mirishli, S.: Ethical implications of AI in data collection: balancing innovation with privacy. arXiv preprint arXiv:2503.14539 (2025)
11. Mirza, J.B., Hasan, M.M., Paul, R., Hasan, M.R., Asha, A.I., et al.: AI-driven business intelligence in retail: Transforming customer data into strategic decision-making tools. AIJMR-Adv. Int. J. Multidis. Res. 3(1) (2025)
12. Pagala, I., Asir, M., Mere, K., Lestari, U.P., Siddiqa, H.: Consumer behavior in the age of ai: The role of personalized marketing and data analytics in shaping purchase decisions. Dinasti Int. J. Educ. Manag. Soc. Sci. 5(6) (2024)
13. Paolanti, M., Pierdicca, R., Pietrini, R., Martini, M., Frontoni, E.: Sesame: Re-identification-based ambient intelligence system for museum environment. Pattern Recogn. Lett. 161, 17–23 (2022)
14. Perfetti, A., Pietrini, R., Scarpi, D., Gistri, G.: Snack dilemma: how vending machines influence choice of virtue and vice foods. J. Retail. Consum. Serv. 87, 104369 (2025)
15. Slocum, T.A., Detrich, R., Wilczynski, S.M., Spencer, T.D., Lewis, T., Wolfe, K.: The evidence-based practice of applied behavior analysis. Behav. Analyst 37, 41–56 (2014)
16. Suarez, V.D., Marya, V., Weiss, M.J., Cox, D.: Examination of ethical decision-making models across disciplines: common elements and application to the field of behavior analysis. Behav. Anal. Pract. 16(3), 657–671 (2023)
17. Sugianto, N., Tjondronegoro, D., Stockdale, R., Yuwono, E.I.: Privacy-preserving AI-enabled video surveillance for social distancing: responsible design and deployment for public spaces. Inf. Technol. People 37(2), 998–1022 (2024)
18. Tiribelli, S., Giovanola, B., Pietrini, R., Frontoni, E., Paolanti, M.: Embedding AI ethics into the design and use of computer vision technology for consumer's behaviour understanding. Comput. Vis. Image Underst. 248, 104142 (2024)
19. Wilkenfeld, D.A., McCarthy, A.M.: Ethical concerns with applied behavior analysis for autism spectrum" disorder". Kennedy Inst. Ethics J. 30(1), 31–69 (2020)
20. Zhou, H., Xiao, Y., Zheng, Z., Yang, B.: Yolov2-tiny target detection system based on fpga platform. In: 2022 3rd International Conference on Big Data, Artificial Intelligence and Internet of Things Engineering (ICBAIE), pp. 289–292. IEEE (2022)

From Smart Assistants to Smart Spaces: AI in Fashion Retail and the Emerging Need for In-Store Behavioral Intelligence

Alessandro Galdelli[1][iD], Luigi Di Bello[2], Marco Contigiani[2], Mattia Sospetti[2][iD],
Mauro D'Aloisio[2], Valerio Placidi[2], and Rocco Pietrini[3(✉)][iD]

[1] Università Politecnica delle Marche, Ancona, Italy
a.galdelli@staff.univpm.it
[2] Med Innovations, Porto Recanati, Italy
{luigi.dibello,marco.contigiani,mattia.sospetti,mauro.daloisio,
valerio.placidi}@medinnovations.it
[3] Universitas Mercatorum, Roma, Italy
rocco.pietrini@unimercatorum.it

Abstract. In the context of the digital transformation of the retail sector, physical fashion stores are increasingly seeking intelligent systems that can provide real-time insights into shopper behaviour. This paper presents an AI-powered monitoring system that has been deployed in a large fashion retail store. It integrates a dense network of Xovis top-view sensors with a real-time analytics. The system uses a temporal proximity-based clustering algorithm to perform fine-grained analysis of customer flow, gender composition, and group segmentation. Over the course of one year, the system recorded and analysed more than 479,000 shopper interactions, revealing temporal patterns, spatial preferences and behavioural trends. The results demonstrate a consistent predominance of female shoppers, a high frequency of individual shopping behaviour and distinct peak hours that align with daily routines and promotional campaigns. The analytics are designed to support operational decision-making and enables managers to optimise store layout, staffing and marketing strategies. The proposed system is a scalable, privacy-conscious solution for behavioural intelligence in physical retail. It offers a foundation for predictive modelling, adaptive merchandising and data-driven retail design. Future developments will explore extended behavioural profiling, automated alerts and integration with business KPIs to enhance strategic decision-making.

Keywords: Retail Analytics · Shopper Behavior Analysis · People Tracking · Fashion Retail

1 Introduction

The fashion retail industry is undergoing a digital transformation driven by artificial intelligence (AI), real-time analytics and ubiquitous sensing technologies [4,12,19,20]. In this evolving landscape, physical retail environments are

E. Rodolà et al. (Eds.): ICIAP 2025 Workshops, LNCS 16169, pp. 482–493, 2026.
https://doi.org/10.1007/978-3-032-11317-7_40

increasingly adopting smart systems to bridge the gap in understanding consumer behaviour in both in-store and online environments [14,27]. Although e-commerce platforms can easily track user actions and preferences, traditional stores still face significant challenges in capturing detailed, structured and actionable data about customer activity. People monitoring technologies play a crucial role in this context, providing retailers with valuable insights into customer behaviour, space utilisation, and engagement patterns within the store [22,23].

One of the most promising technologies in this field is the use of top view 3D cameras, such as Xovis sensors, which enable the precise tracking of people's movements while protecting their privacy. When integrated into a broader AI-powered analytics pipeline, these cameras can support advanced tasks such as gender and group detection, traffic heat mapping, conversion analysis and path optimisation. Despite the availability of sensor-based people counters, literature on integrated, AI-based systems capable of real-time consumer type clustering, adaptive temporal analysis and intuitive visualisation at scale is still limited.

This paper addresses this gap by presenting a comprehensive, AI-driven people monitoring solution that has been implemented in a fashion retail store. The system uses a multi-camera setup with a top view, comprising two entrance cameras and 27 internal Xovis cameras. This is combined with analytics that classifies visitors by gender and group composition (single, couple or group), as well as providing insights into hourly traffic flows. A key innovation of our work is the group clustering algorithm that distinguishes shoppers into meaningful behavioural categories, which are then used to provide personalised spatial and temporal insights. Unlike traditional people counting systems, our approach emphasises the semantic interpretation of in-store presence and flow. This enables business-relevant actions such as optimising layout, staffing and marketing strategies. In order to evaluate the performance and robustness of the proposed system, we conducted a year-long monitoring campaign in a medium-sized fashion retail store. The sensing infrastructure consisted of a network of 29 Xovis top-view stereo vision cameras, two positioned at the entrances and 27 distributed throughout the internal layout of the store, configured to ensure full spatial coverage and minimal blind spots. The raw positional data extracted from the Xovis sensors was processed through a custom AI pipeline incorporating real-time person detection, trajectory stitching and clustering algorithms for group composition analysis. Using proximity thresholds and temporal co-presence windows, the system automatically identified whether individuals were shopping alone, in pairs or in larger groups, enabling dynamic classification of shoppers. Gender classification was derived from embedded camera features combined with confidence-calibrated post-processing filters. The system logged all traffic patterns at hourly, daily and weekly intervals and stored over 480,000 unique detections throughout the year. This large-scale temporal dataset allowed us to identify recurring behavioural trends, such as peak visiting hours (typically between 17:00 and 18:00), a higher female presence (73%), and a predominance of single shoppers (48.7%). Furthermore, longitudinal analysis allowed us to detect seasonal variations and weekday/weekend distribution patterns, providing store managers with granular, evidence-based insights to inform layout design, personnel scheduling and promotional strategies.

The main contributions of this work are as follows: i) the design and the developmentof a sensor-based architecture for in-store monitoring using top-view Xovis cameras, achieving seamless integration across entry and internal spaces. ii) an AI-based clustering model that categorises shoppers into groups (single, couple or group) based on movement and proximity patterns. iii) a real-time analytics that showcases aggregated statistics on gender, group size and the time-based distribution of people, thereby enhancing interpretability for retail stakeholders. iv) a real-world case study in a high-traffic fashion store to demonstrate the system's performance, usability, and potential to support data-driven decision-making in physical retail.

The paper is structured as follows: Section 2 provides a review of the literature on AI-based consumer behaviour monitoring in physical retail environments, focusing particularly on sensor-driven analytics, group detection and AI clustering methods. It also draws attention to the current limitations in terms of scalability, interpretability and privacy-aware implementation. Section 3 describes the methodology adopted in the development of our monitoring system. This includes the top-view, multi-camera architecture based on Xovis sensors; the AI pipeline for person detection, trajectory clustering and gender/group classification; and the analytics framework. The section also details the data collection protocol and the rationale behind shopper clustering strategies. Section 4 presents a comprehensive analysis of the results obtained from one year of continuous in-store monitoring. Based on empirical evidence from almost half a million recorded interactions with shoppers, it discusses trends in gender distribution, group composition and hourly traffic patterns. Particular emphasis is placed on the added value of AI-driven group segmentation and the system's real-time performance in a busy fashion retail environment. Finally, Sect. 5 concludes the paper by summarising the main contributions and impact of the proposed system. It also outlines potential future research areas, such as cross-store benchmarking, consumer consent mechanisms and the adoption of multimodal AI models to enable more nuanced behavioural analysis in retail spaces.

2   Related Works

This section reviews the main technological trends and application domains that are emerging at the intersection of AI and grocery retail. Three dominant lines of development are examined in detail: smart shopping assistants that improve navigation and engagement through real-time support; dynamic pricing systems that optimise product costs based on contextual data; and AI-driven automation platforms that enable seamless in-store experiences through sensor-based interactions. Together, these innovations demonstrate the ambition to embed intelligence into every layer of the physical store. However, they also expose certain limitations, particularly the lack of integrated systems for understanding spatial group behaviour and real-time in-store dynamics.

AI is transforming the in-store grocery shopping experience, introducing new levels of personalisation, operational efficiency and responsiveness. Retailers are adopting AI technologies to meet consumer expectations and remain competitive in an ever-changing market landscape [21]. One key area of application involves smart shopping assistants, which appear in the form of mobile apps, kiosks, and voice-controlled systems. These tools offer real-time navigation, product search, and personalized recommendations, significantly improving the customer journey [1,5,9,10,13,15–17,28,29]. For instance, Amazon's integration of Alexa enables hands-free queries about product locations or recipe suggestions [3], while Walmart's app uses prior purchase history and indoor positioning to streamline store navigation.

AI also underpins dynamic pricing systems, which adjust product prices in real time according to factors such as customer demand, inventory levels, and competitor actions [18,26,33]. These systems use machine learning algorithms to analyze diverse data sources—ranging from foot traffic and purchase history to weather trends and social media activity—to optimize prices. Dynamic pricing allows for both broad and personalized strategies: general adjustments based on stock levels or local events, and tailored offers based on individual preferences or loyalty data. Despite benefits, dynamic pricing must be implemented with care to avoid customer backlash over perceived unfairness [6,26].

Another frontier is fully automated grocery stores powered by AI and sensor-based systems. Technologies like AiFi eliminate traditional checkouts through a combination of computer vision, RFID tagging, and mobile integration [24]. These systems automatically recognize items as customers shop and complete transactions upon exit, enabling a seamless, cashier-less experience [30]. The benefits of AI integration in grocery retail are multifaceted. AI enhances personalization by analyzing customer behavior to offer tailored promotions and product suggestions [8,11,21]. It also supports inventory optimization, predicting demand and minimizing both out-of-stock situations and waste [7,25,34]. Checkout times are reduced thanks to automated systems, and security improves through AI-based surveillance and fraud detect [5,25].

While the existing literature extensively explores AI applications in grocery retail, such as smart shopping assistants, dynamic pricing and automated checkout systems, these developments have largely focused on enhancing the individual consumer experience or operational processes through digital interfaces and personalised services. However, relatively little attention has been paid to real-time analysis of in-store human behaviour using spatial AI and overhead sensing, particularly in non-grocery retail sectors such as fashion. Most AI solutions are device-centric, relying on mobile apps, loyalty programmes or RFID interactions without offering a holistic understanding of in-store dynamics at a population level. Furthermore, few systems address the segmentation of shoppers into behavioural groups (e.g. singles, couples, social clusters) or their temporal and spatial movement patterns throughout the physical space. Recent advances in cultural heritage environments have demonstrated the potential of RGB-D sensing for visitor behavior analysis. The SeSAME system [23] employed deep learning techniques with RGB-D cameras to analyze visitor trajectories and re-identification patterns in museum settings, achieving privacy-preserving behavioral insights through anonymous tracking. While focused on cultural spaces, their approach to spatial movement analysis and visitor flow optimization shares methodological similarities with retail behavioral intelligence, particularly in addressing privacy concerns while extracting actionable insights from human movement patterns. The absence of integration between AI-based behavioural clustering, privacy-aware sensing and visual analytics restricts retailers' ability to optimise store layout, staffing and marketing strategies based on actual customer flow. Our work addresses this gap by introducing an AI-powered monitoring framework for fashion retail. This framework leverages top-view stereo vision sensors and real-time clustering algorithms to capture, classify, and visualise shopper interactions at scale. No customer-side technology or intrusive data collection is required.

3   Materials and Methods

The proposed system was designed to monitor and analyse in-store shopper behaviour in a large fashion retail environment over the course of a full operational year. The aim was to go beyond basic people counting and provide a scalable, transparent and privacy-conscious solution that can segment and visualise customer behaviour using AI-driven clustering models. This section details the system's architecture and implementation across four methodological layers: (i) sensing infrastructure and spatial configuration; (ii) an AI-based pipeline for behavioural classification; (iii) analytics for real-time and longitudinal insights; and (iv) data collection and the experimental protocol. The complete camera deployment is illustrated in Fig. 1, showing the spatial distribution of entrance and internal sensors, as well as the coverage paths across key product and interaction zones. Further details will be given in the following Subsections.

3.1   Sensing Infrastructure and Store Coverage

The physical infrastructure comprises 29 Xovis PC-series stereo vision cameras installed in a top-view configuration, which together cover the entire store layout. The cameras are equipped with processing units that can detect and track people using 3D depth sensing and visual analysis. The sensors were mounted at ceiling height (3.5 m) to ensure an optimal field of view and minimise occlusions caused by interior structures or shelving.

The configuration was as follows:

- Two entrance cameras were positioned above the main doors to monitor foot traffic in and out of the store and establish the ground truth for people flow.
- Twenty-seven internal cameras were deployed to cover strategic zones such as promotional displays, fitting rooms, product-specific areas (e.g. accessories, denim and outerwear) and the cash register area.

Fig. 1. Top-view layout of the fashion retail store with the deployment of 29 Xovis stereo vision cameras (Camera1 – Camera29). The green lines indicate inter-camera coverage zones, ensuring full spatial monitoring throughout the store. (Color figure online)

Each camera operates independently and generates anonymised detection data, including:

A unique anonymised person ID;

- Position in 3D space (X, Y, Z) relative to the store layout;
- Timestamped events for entering/exiting and trajectory coordinates;
- Estimated gender based on body silhouette and gait characteristics.

This sensor data is streamed in real time to a local edge server running a multi-threaded ingestion and synchronisation module, which merges the data streams into a unified spatio-temporal database.

3.2   Shopper Detection, Classification, and Group Clustering

The raw data from each sensor is pre-processed to reconstruct full shopper trajectories:

- Noise filtering is applied to remove ghost detections and non-human movement (e.g., staff repositioning objects) [2].
- Stitching across cameras uses spatial proximity and temporal continuity to reconstruct shopper movement as they pass from one sensor's field of view to another [31].

- A Kalman filter and tracklet scoring mechanism ensures robustness in presence of sensor overlap or temporary occlusion [32].

We apply an additional post-processing refinement:

- Smoothing over trajectory segments to correct for misclassifications caused by pose variation.
- Exclusion rules to remove low-confidence estimates from aggregated counts.

This component is not used for individual targeting but for aggregate analysis and trend detection, in line with privacy and fairness constraints. To segment the population into behavioral categories (*single, couple, group*), we developed a temporal proximity-based clustering algorithm. This approach builds on the following assumptions:

- Shoppers walking together for a prolonged duration and within a short Euclidean distance are likely to be socially connected.
- Synchrony in movement direction and speed further reinforces co-presence classification.

For each timestamp t, we compute:

- The pairwise distance matrix $D_{ij}(t)$ for all detected people.
- A moving window of co-presence duration ΔT to aggregate proximity over time.
- A directional similarity index based on the cosine similarity of movement vectors.

Pairs of shoppers satisfying the following conditions:

$$D_{ij}(t) < \tau_d = 1.5 \text{ meters}$$
$$\Delta T \geq 10 \text{ seconds}$$
$$\cos(\theta_{ij}) > 0.9$$

are merged into clusters.

Clusters are then labeled as:

- **Single**: size $= 1$
- **Couple**: size $= 2$
- **Group**: size ≥ 3

This clustering is executed in real time, and results are logged for further statistical aggregation.

4   Results and Discussions

This section presents the key outcomes of deploying our people monitoring system in a fashion retail store for one year. The results are organised to illustrate the demographic composition of shoppers, their temporal dynamics and their spatial behaviours, all of which were inferred from the AI-powered sensing infrastructure.

4.1 Gender and Group Distribution

Over the monitored period, a total of 37,979 people were detected entering the store, the majority of whom were female shoppers. As shown in Table 1, approximately 73% of entries were identified as female and 27% as male. Clustering by group revealed that almost half (48.78%) of visitors were shopping alone, 41.04% came in pairs (classified as couples) and 10.18% came in larger groups. These statistics support the initial assumption that the store's layout and product segmentation strongly encourage female consumer engagement and individual shopping behaviours.

Table 1. Gender and Group Distribution

Category	Count	Percentage (%)
Female	27,724	73.00
Male	10,254	27.00
Single	18,526	48.78
Couple	15,586	41.04
Group	3,866	10.18

4.2 Temporal Patterns of Traffic Flow

An hourly breakdown of visitor data, averaged across the monitored period, is provided in Table 2. Clear patterns emerged, with traffic peaking between 16:00 and 18:00, which corresponded to the time when people were doing their after-work and post-lunch shopping. Moderate traffic was observed in the morning hours (particularly between 09:00 and 11:00), while visits decreased sharply after 19:00. These findings are consistent with the typical rhythms of urban retail and provide empirical evidence to support optimising staff shifts and promotional activities based on time-of-day segmentation.

4.3 Peak Days and Special Events

Table 3 lists the five days with the highest visitor traffic, along with the gender breakdown. Specifically, on 17 June 2025, during a promotional campaign, a peak of 3,258 entrances was recorded. On all high-traffic days, the number of female visitors consistently outnumbered the number of male visitors, which reinforces the earlier discussed findings. These insights can be used to schedule campaigns more effectively and predict operational stress on key dates.

These results confirm the effectiveness of the system in detecting, classifying, and clustering shopper behavior in real time.

Table 2. Hourly Traffic Distribution

Hour	Average Visitors
09:00	249
10:00	529
11:00	608
12:00	464
13:00	405
14:00	512
15:00	574
16:00	724
17:00	691
18:00	680
19:00	466
20:00	122

Table 3. Top 5 Days by Total Entrances

Date	Total Entrances	Female	Male
2025-06-17	3,258	2,202	1,056
2025-06-22	1,586	1,073	513
2025-06-29	1,643	1,118	525
2025-06-16	1,880	1,304	576
2025-06-19	1,199	812	387

5   Conclusions and Future Works

This paper presents a comprehensive, AI-based system for monitoring shopper behaviour in fashion retail stores, using a network of Xovis top-view sensors. By combining real-time people tracking with a temporal clustering model, we demonstrated the potential of such systems to generate valuable insights into gender composition, group behaviour, temporal dynamics and spatial engagement patterns. The year-long deployment confirmed that the system is technically robust and operationally impactful, supporting informed decisions regarding staff scheduling, layout optimisation and campaign planning. Using explainable clustering based on movement patterns provided meaningful customer segmentation without compromising privacy, thus aligning with ethical design principles.

Based on the results of this study, several areas of future research and development are being considered. One of the main priorities is integrating behavioural data with business-oriented KPIs, such as sales performance and inventory turnover. Linking path trajectories and group segmentation with point-of-sale (POS) and merchandising systems will enable the analysis of the relationship between shopper movement and purchasing behaviour. This will provide store managers with more targeted and actionable intelligence. Another promising area involves incorporating predictive analytics. While the current system is effective for real-time monitoring, it could be further enhanced by incorporating time-series forecasting models. These models would anticipate foot traffic volumes, group compositions and zone-level occupancy based on historical patterns and contextual variables, such as promotions, holidays and weather conditions. Future work will also examine how adaptive store environments can be developed. The aim is to enable visual merchandising strategies that can respond to live occupancy and shopper flow, such as dynamic signage or lighting adjustments, to create more immersive and effective in-store experiences. In terms of behavioural modelling, efforts will be made to extend the system's demographic profiling capabilities, in accordance with privacy-preserving design principles. Attributes such as average dwell time, repeated visits or path efficiency may offer a more detailed behavioural description without resorting to personal identification. Finally, a visual analytics dashboard is being designed to improve usability and decision support. This will include automated alerts, comparative analytics between multiple store locations and customisable views tailored to different user roles, such as marketing managers, store planners or data analysts. These enhancements will transform the system from an observational tool into a strategic asset within the broader framework of intelligent retail.

References

1. Al-Fraihat, D., Alzaidi, M., Joy, M.: Why do consumers adopt smart voice assistants for shopping purposes? A perspective from complexity theory. Intell. Syst. Appl. **18**, 200230 (2023)
2. Alahi, A., Goel, K., Ramanathan, V., Robicquet, A., Fei-Fei, L., Savarese, S.: Social LSTM: Human trajectory prediction in crowded spaces. In: Proceedings of the IEEE Conference on Computer Vision and Pattern Recognition, pp. 961–971 (2016)
3. Aw, E.C.X., Tan, G.W.H., Cham, T.H., Raman, R., Ooi, K.B.: Alexa, what's on my shopping list? Transforming customer experience with digital voice assistants. Technol. Forecast. Soc. Chang. **180**, 121711 (2022)
4. Bieńkowska, J.: The effects of artificial intelligence on the fashion industry—opportunities and challenges for sustainable transformation. Sustain. Devel. (2024)
5. Bruwer, L.A., Madinga, N.W., Bundwini, N.: Smart shopping: the adoption of grocery shopping apps. British Food J. **124**(4), 1383–1399 (2022)
6. Bykadorov, I.: Pricing in dynamic marketing: the cases of piece-wise constant sale and retail discounts. In: Olenev, N., Evtushenko, Y., Khachay, M., Malkova, V. (eds.) OPTIMA 2020. LNCS, vol. 12422, pp. 27–39. Springer, Cham (2020). https://doi.org/10.1007/978-3-030-62867-3_3
7. Chakraborty, D., Altekar, S.: What drives people to use grocery apps? The moderating & mediating role of customer involvement and trust. Indian J. Market. **51**(11), 23–37 (2021)

8. Chakraborty, D., Kar, A.K., Patre, S., Gupta, S.: Enhancing trust in online grocery shopping through generative AI chatbots. J. Bus. Res. **180**, 114737 (2024)

9. Chintala, S.C., Liaukonytė, J., Yang, N.: Browsing the aisles or browsing the app? How online grocery shopping is changing what we buy. Mark. Sci. **43**(3), 506–522 (2024)

10. Dodon, M., Damian, C.: Hbag-smart assistant for healthy shopping. In: 2022 International Conference and Exposition on Electrical And Power Engineering (EPE), pp. 589–592. IEEE (2022)

11. Gooljar, V., Issa, T., Hardin-Ramanan, S., Abu-Salih, B.: Sentiment-based predictive models for online purchases in the era of marketing 5.0: a systematic review. J. Big Data **11**(1), 107 (2024)

12. Goti, A., Querejeta-Lomas, L., Almeida, A., de la Puerta, J.G., López-de Ipiña, D.: Artificial intelligence in business-to-customer fashion retail: a literature review. Mathematics **11**(13), 2943 (2023)

13. Gutiérrez Gómez, F., Abascal-Mena, R.: Chancho Assistant: Smart Shopping Guided by Consumer Habits. In: Stephanidis, C., (ed.) HCI 2018. CCIS, vol. 852, pp. 247–254. Springer, Cham (2018). https://doi.org/10.1007/978-3-319-92285-0_34

14. Knof, M., Stock-Homburg, R., Schurer, J.: How in-store sensor technologies can help retailers to understand their customers: overview on two decades of research. Int. Rev. Retail Distr. Cons. Res. **34**(3), 381–398 (2024)

15. Kothavale, S., Pawar, S., Kankarej, S., Patil, S., Raut, R.: Smart indoor navigation, shopping recommendation & queue less billing based shopping assistant using AI. In: 2021 International Conference on Artificial Intelligence and Machine Vision (AIMV), pp. 1–4. IEEE (2021)

16. Kumar, A., Chakraborty, S., Bala, P.K.: Text mining approach to explore determinants of grocery mobile app satisfaction using online customer reviews. J. Retail. Consum. Serv. **73**, 103363 (2023)

17. Lee, J., Park, G., Jung, H.: Seejang: Smart, easy, and economical offline shopping assist app development through a design thinking process (poster). In: Proceedings of the 17th Annual International Conference on Mobile Systems, Applications, and Services, pp. 602–603 (2019)

18. Li, D., Xin, J.: Deep learning-driven intelligent pricing model in retail: from sales forecasting to dynamic price optimization. Soft Comput. 1–17 (2024)

19. McDonald, S.D.: The internet of things (IoT) revolution: Transforming the fashion supply chain. In: Use of Digital and Advanced Technologies in the Fashion Supply Chain, pp. 167–222. Springer (2025)

20. Mohiuddin Babu, M., Akter, S., Rahman, M., Billah, M.M., Hack-Polay, D.: The role of artificial intelligence in shaping the future of agile fashion industry. Prod. Plan. Control **35**(15), 2084–2098 (2024)

21. Nandhakumar, V., Jyothsna, B., Gnanapriya, S.: Ai-driven produce management and self-checkout system for supermarkets. In: 2023 4th International Conference on Electronics and Sustainable Communication Systems (ICESC), pp. 965–975. IEEE (2023)

22. Paolanti, M., Liciotti, D., Pietrini, R., Mancini, A., Frontoni, E.: Modelling and forecasting customer navigation in intelligent retail environments. J. Intell. Rob. Syst. **91**, 165–180 (2018)

23. Paolanti, M., Pietrini, R., Pietrini, R., Martini, M., Frontoni, E.: SeSAME: Re-identification-based ambient intelligence system for museum environment. Pattern Recogn. Lett. **161**, 17–23 (2022)

24. Prasanna, B.L., Reddy, M.P.P., Kumar, S.S., Mareddy, S.: Rfid-enabled smart shopping cart for enhancing shopping efficiency and preventing theft. In: 2024 5th International Conference on Recent Trends in Computer Science and Technology (ICRTCST), pp. 485–490. IEEE (2024)

25. Rios, J.H., Vera, J.R.: Dynamic pricing and inventory control for multiple products in a retail chain. Comput. Indus. Eng. **177**, 109065 (2023)

26. Riseth, A.N.: Dynamic pricing in retail with diffusion process demand. IMA J. Manag. Math. **30**(3), 323–344 (2019)

27. Roe, M., Spanaki, K., Ioannou, A., Zamani, E.D., Giannakis, M.: Drivers and challenges of internet of things diffusion in smart stores: a field exploration. Technol. Forecast. Soc. Chang. **178**, 121593 (2022)

28. Sherlin Solomi, V., Srujana Reddy, C., Naga Tripura, S.: Smart shopping using embedded based autocart and android app. In: International Conference on Image Processing and Capsule Networks, pp. 704–712. Springer (2022)

29. Trude, A.C., et al.: I don't want an app to do the work for me: a qualitative study on the perception of online grocery shopping from small food retailers. J. Acad. Nutr. Diet. **124**(7), 804–822 (2024)

30. Uganya, G., Nisha, A.S.A., Vinoth, R., Rajah, A.J.L.: Intelligent smart trolley system using arduino. In: 2023 Intelligent Computing and Control for Engineering and Business Systems (ICCEBS), pp. 1–5. IEEE (2023)

31. Ukita, N., Moriguchi, Y., Hagita, N.: People re-identification across non-overlapping cameras using group features. Comput. Vis. Image Underst. **144**, 228–236 (2016)

32. Wojke, N., Bewley, A., Paulus, D.: Simple online and realtime tracking with a deep association metric. In: 2017 IEEE International Conference on Image Processing (ICIP), pp. 3645–3649. IEEE (2017)

33. Yang, X., Zhang, J., Hu, J.Q., Hu, J.: Nonparametric multi-product dynamic pricing with demand learning via simultaneous price perturbation. Eur. J. Oper. Res. **319**(1), 191–205 (2024)

34. Zeeshan, M., Negi, N.S., Markan, D.: Least: The smart grocery app. In: Integrated Emerging Methods of Artificial Intelligence & Cloud Computing, pp. 79–85. Springer (2021)

4th International Workshop on Fine Art Pattern Extraction and Recognition ((FAPER 2025)

Workshop

4th International Workshop on Fine Art Pattern Extraction and Recognition (FAPER 2025)

In conjunction with ICIAP 2025—Rome, Italy, September 15–19, 2025

Workshop Organization

Organizers

Gennaro Vessio	University of Bari Aldo Moro, Italy
Giovanna Castellano	University of Bari Aldo Moro, Italy
Fabio Bellavia	University of Palermo, Italy
Sinem Aslan	University of Milan, Italy
Raffaele Scaringi	University of Bari Aldo Moro, Italy
Nicola Fanelli	University of Bari Aldo Moro, Italy

Program Committee

Elisa H. Barney	Luleå Tekniska Universitet, Sweden
Paola Barra	Parthenope University of Naples, Italy
Eva Cetinic	University of Zurich, Switzerland
Mahdi Champour	University of Hamburg, Germany
Lucia Cipolina Kun	University of Bristol, UK
Marco Fanfani	University of Florence, Italy
Ehsanollah Kabir	Tarbiat Modares University, Iran
Ronak Kosti	Friedrich-Alexander-Universität Erlangen-Nürnberg, Germany
Remo Pareschi	University of Molise, Italy
Albert Ali Salah	Utrecht University, Netherlands
Amanda Wasielewski	Uppsala University, Sweden

Face Detection in Historic Manuscripts A Case Study on the Wenceslas Bible

Heinz Hofbauer[(✉)], Julia Hintersteiner, Manfred Kern, and Andreas Uhl

Paris Lodron University of Salzburg, Salzburg, Austria
{hofbauer,uhl}@cs.sbg.ac.at,
{julia.hintersteiner,manfred.kern}@plus.ac.at
https://www.wavelab.at

Abstract. We want to analyse the faces depicted in the Wenceslas Bible, an illustrated bible from the late 14th century (circa 1390), however, this requires automatic finding of faces in the illustrations of the bible. This is a difficult task due to the fact that we are working with illustrations of human faces that are often interwoven into the background and in odd poses. This paper presents an analysis of prominent face detection methods and how their performance translates from real-world facial images to painted imagery. We will make use of a scale and rotation cascade on top of these methods to see if the detection of faces in odd poses can be improved. Finally, most methods are designed to handle a particular face size, but for the purpose of finding faces in the illustration of historical books the relative size of the image and face differs from most real-world cases. An attempt to fix this is to use tiling of the input image to adjust for the relative scale difference. We will see that tiling, scaling, and rotation help, and while no general "best" setting can be given, we also see that these can boost some methods from non-working to being quite good.

Keywords: face detection · faces in art · digital humanities · Wenceslas Bible · art history

1 Introduction

Detection of faces in art is relevant to many art historic questions, see Sect. 2 for some examples. Our reason for wanting face detection, and another example why it is of interest in an art historic context, is as follows.

This work is part of an attempt to digitize information about the Wenceslas Bible [12] . The Bible is stored in the Austrian National Library, it consists of six manuscripts with shelf marks codex 2759 to 2764. One of the tasks is to try to figure out how many people illustrated the Bible, as well as which parts each person did. There is an attribution of illustrations to specific artist [19], some of them identifiable by name, because they have signed their illustrations with initials, however, this is not an assured connection since in most cases there are no historical records about the illustrators.

E. Rodolà et al. (Eds.): ICIAP 2025 Workshops, LNCS 16169, pp. 499–510, 2026.
https://doi.org/10.1007/978-3-032-11317-7_41

Our goal is to use computer vision, taking hints from face biometry, similar to what [17] did for renaissance painters, to find commonalities in how faces are painted. The basic assumption is that recurring characters, such as Wenceslas and the "bathmaid", are painted similarly by the same painter. The painters likely never saw the king, and figures like the bathmaid are entirely fictional, we assume that an idealized version is painted, resulting in a similar outcome for the same painter. The assumption is that a similarity in face recognition indicates that the same painter created the assessed faces.

Before we can analyse the faces we have to find them first, ideally in an automated way. Face detection is a well established area of research, but it is unclear how face detection methods designed and trained on real faces will perform with painted faces. The non-realistic style of the face illustrations, with proportions being not quite right, exacerbates the detection problem. See Fig. 2 for a comparison of a real faces and an illustration from the Wenceslas Bible. Note that sometimes the faces are interwoven in the background or other parts of the illustration, although this is more common in the marginalia, where Wenceslas and the bathmaids typically appear.

The relatively small faces, as compared to regular photos where faces are often the focal point, and frequent slanted depiction, as a narrative device (bowed heads) or as part of the story (sleeping, dead, ...), suggest the use of a rotation and scale cascade or tiling (to fix relative face size issues).

While our evaluation is driven by our own requirements we will make general statements too, which will help others with similar problems. Our contribution to a wider audience thus are as follows. We show that single shot detectors are not a good fit for art, but an additional rotation step is beneficial. So can be a scaling or tiling approach depending on the image size and relative scale of faces to the image. We give details on how to aggregate detections from these detection steps. We show that CNNs should not be excluded from a task based on architecture alone, as the same CNN with a different training can perform vastly better or worse. This is especially important when working with few training samples of a specific art styles or time periods where pre-trained networks are used.

2   Related Work

2.1   Face Detection in Art

Not a lot of work was done on face detection in 14th century book illustrations but some on the more general topic of faces in art. Srinivasan et al. [17] used facial landmarks and paint styles to identify whether different Renaissance portraits are from the same painter. Similarly, Zhong [24] applied face biometric recognition to Song dynasty paintings as a further argument in art historical discussions, such as whether a painter had self inserted their likeness into a painting. Liang [9] performed age and gender classification of faces in Japanese art starting from cropped faces, their result was that an ensemble of different CNNs worked best. Sindel et al. [16] created ArtFacePoints, a facial landmark detector on cropped facial images in art. It takes a two-stage approach, using a coarse scale of the image to generate a rough map of the landmarks and refining them on the full resolution image. This work is a promising next step, provided faces are detected properly, to obtain stable facial landmarks for pose correction. The topic of face detection was not a focus in any of these papers. In Wechsler et al. [21], the topic of face detection is more prominent. They introduced a "faces in art" database and analysed a few face detectors. They dealt with modern art, a different topic than ours, and they highlighted that different art styles impact face detection differently. Bengamra et al. [2] performed face detection on Tenebrism style paintings using various networks. RestNet50 with a retrained Faster RCNN yielded the best results. To handle difficult poses, images in the augmented dataset were rotated by by ±45° for retraining. In [3] they demonstrated that the retrained network can detect faces with more accuracy through the use of perturbation based explainable AI. So far, there is no systematic evaluation of face detectors for art that directly compares real-world performance and performance on painted faces. This is the goal of this work.

2.2 Face Detectors in This Work

There are a lot of face detection frameworks, see recent surveys [6,11] for an overview, so a selection had to be made. The Deepface/Lightface framework [15] provides a good selection of face detectors which is why we chose it as a basis for this evaluation. Since RetinaFace is used a number of times in the Deepface framework, and the author shows a slight drop in performance (1–2%) in relation to the original implementation, we also included the original implementation in the evaluation. Below we will give a brief summary of the methods used and their references for further detail.

We use two versions of RetinaFace, one provided by the authors in the Insightface toolkit [5], and one implementation provided by the Deepface [15] toolkit. Both versions are based on ResNet50, however, differences in training results in a slightly different performance, see Fig. 2a for a comparison. The differences are significant enough that we included both in our evaluation. The Deepface implementation will be referred to as *retinaface* and the Insightface version as *insightface*. A further version of ResNet will be used as the backbone of a single shot detector (*SSD*) [10]. The SSD automatically generates different scales and generates multiple patches on each scale to find face candidates.

The multitask cascade convolutional network (MTCNN) [23] uses an image pyramid to perform single-stage detection. The images are then fed through a series of three networks to propose bounding boxes, perform regression on the proposed bounding boxes, and a final network that produces the output of facial position and landmark localizations. We also use the "you only look once" CNN version 8 for faces (YOLO v8). The YOLO series of CNNs are general detector networks instead of repurposed classifiers. Thus, the networks are smaller and faster, yet still produce good results. YOLO was introduced in 2016 [13] but has since seen constant development, including various offshoots. See [18] for an overview of the different YOLO CNNs and their development. The YuNET [22] is also a custom-built face detector with the goal of minimizing hardware usage

---

Algorithm 1. Calculation of intersect in a one on one correspondence.

$G \leftarrow \mathcal{G}$
$D \leftarrow \mathcal{D}$
intersect $\leftarrow 0$
while $|G| > 0$ **and** $|D| > 0$ **do**
$\quad R_g, R_d \leftarrow \underset{R_g \in G,\ R_d \in D}{\arg\max} \dfrac{A(R_g \cap R_d)}{\max\big(A(R_g), A(R_d)\big)}$
$\quad D \leftarrow D \backslash \{R_d\}$
$\quad G \leftarrow G \backslash \{R_g\}$
$\quad$ intersect $\leftarrow$ intersect $+ A(R_g \cap R_d)$
end while
return intersect

---

and maximizing speed in a speed/accuracy trade-off. The trade-off in accuracy usually means it is less tightly targeting human faces, which might be a benefit when using it to find drawings of human faces instead of photos of human faces.

In addition, we also use traditional detectors. The used implementations are from the Deepface package [15]. For traditional methods, we utilized OpenCV's Viola Jones, which employs Haar cascades to detect faces and eyes [20], and the DLib face detector, which is based on the histogram of oriented gradients (HOG) [4]. Both methods have numerous extensions and improvements. A comparison and more in depth details can be found in [1].

3 Experiments and Discussion

We will use the F1-score to assess the accurate detection of faces, which is a well known measure from the field of information retrieval, e.g., [14], and is also typically used for face detection assessment. The F1-score is the harmonic mean of precision, i.e., facial areas that match the ground truth, and recall, i.e., the amount of the ground truth which was correctly found. Let tp be the number of correctly detected faces, fp be the number of wrongly detected faces and fn be the number of faces not detected. The precision is defined as $\mathcal{P} := \frac{\text{tp}}{\text{tp} + \text{fp}}$, and the recall as $\mathcal{R} := \frac{\text{tp}}{\text{tp} + \text{fn}}$, and finally the F1-score F1 $:= \frac{2\mathcal{P}\mathcal{R}}{\mathcal{P}+\mathcal{R}}$.

In our case the true/false positive/negative are not so straightforward, as we are not dealing with retrieved pixels but with rectangular facial areas. Specifically, two retrieved face regions can overlap a single ground truth area. When we use a pixel based precision and recall we lose the information about correspondence, but we want a single retrieved face area per ground truth faces area, i.e., a one on one correspondence. So we use the following method to gain the precision and recall values which match our purpose.

We have two sets of axis-parallel rectangular areas, one for the ground truth $\mathcal{G}$ and one for the detected faces $\mathcal{D}$ by detector f (see Sect. 2). We then need the correct intersection of detected vs. ground truth areas in a one on one correspondence, denoted as intersect, with $A(r)$ denoting the area of rectangle r

Table 1. Baseline results for the face detectors on real faces (FDDB) and art faces (Wenceslas Bible) given as F1-score.

faces	DLib	*insightface*	MTCNN	OpenCV	*retinaface*	SSD	YOLOv8	YuNet
real	0.686	0.840	0.805	0.659	0.821	0.789	0.832	0.789
art	0.521	0.798	0.586	0.287	0.264	0.000	0.062	0.000

as given in Algorithm 1. We can then calculate our precision and recall with

$$\mathcal{P} := \frac{\text{intersect}}{\sum_{R \in \mathcal{D}} A(R)}, \text{ and } \mathcal{R} := \frac{\text{intersect}}{\sum_{R \in \mathcal{G}} A(R)}.$$

The algorithms we use all follow this basic idea: a single stage detects a face only once. To combine multiple detections there must be a one on one correspondence, consequently, we cannot combine a face representation more than once. This is reflected in the set reduction part of the algorithms. To prevent small overlaps of adjacent faces from being counted as referencing the same face, we work from the largest match in an ordered fashion. To prevent overlapping faces in crowd situations from being counted as belonging to the same face we only count overlapping rectangles as indicating the same face when they are either a) overlapping with at least 50% area (calculated on the largest rectangle) or b) one is fully contained in the other and the area is at least 20% of the larger area. For an evaluation of how overlap affects performance, although for modern art, refer to [21].

For the evaluation we need a ground truth, thus we manually segmented Genesis from the Wenceslas Bible with axis-parallel bounding boxes. We segmented all human and human-like, e.g., monkeys and wildmen, faces. Some faces, in illustrations depicting crowds, are quite obfuscated, so we only used faces showing two of the following four features: left eye, right eye, mouth, nose. This resulted in 415 faces on 50 pages, some examples can be seen in Fig. 1.

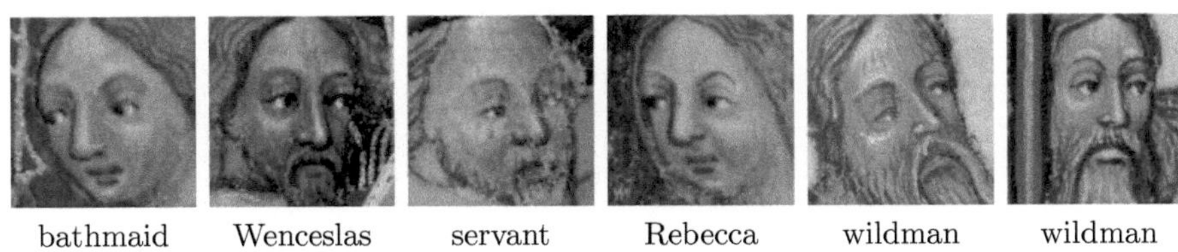

bathmaid   Wenceslas   servant   Rebecca   wildman   wildman

Fig. 1. Ground truth samples from Genesis, Codex 2759 sheet 21 front.

3.1   Evaluation of Face Detectors

A common comparison of our selected face detection methods is not known to us. So we will perform an evaluation and comparison between methods and source material, i.e., real and painted faces. For real faces we use the face detection

DLib   *insightface*   MTCNN   OpenCV

retinaface   SSD   YOLOv8   YuNet

(a) Samples of a real faces on the large crowd image.

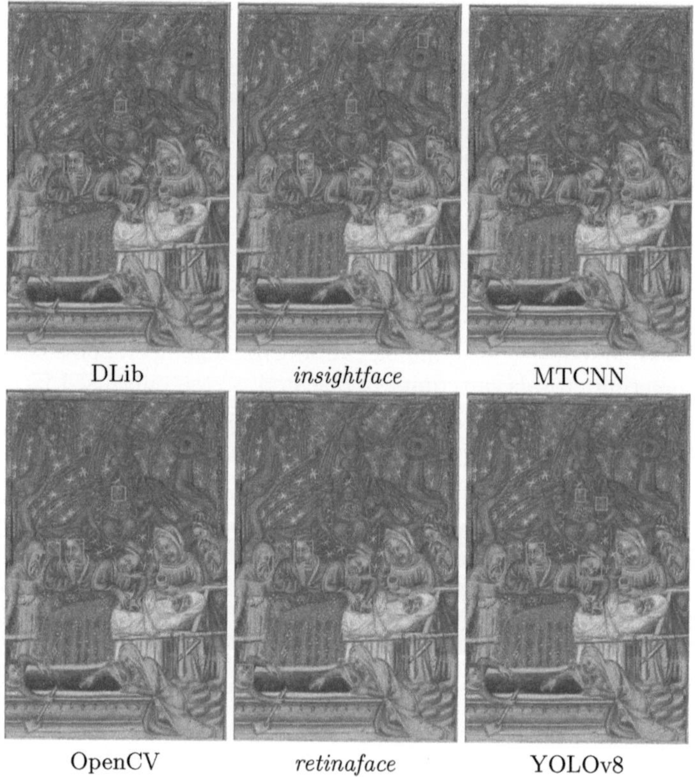

DLib   *insightface*   MTCNN

OpenCV   *retinaface*   YOLOv8

(b) Art face samples on a crop of sheet 53r, codex 2759, Wencelsas Bible. SSD and YuNet are not included, they did not detect any faces.

Fig. 2. Samples of face detection in art faces and real faces.

data set (FDDB) [7], containing 5171 faces in 2845 images with corresponding ground truth and for art faces we use the manual segmentation of the Genesis chapter of the Wenceslas Bible as described above. Average F1-scores over the database are given per method. For art faces, due to the large image size and textured page, many small details are detected as faces. We drop any rectangle which is not at least 75×75 pixels in size. We calculated the F1-score per page, then averaged them for the final score per detector. Scores for the evaluation are given in Table 1 and sample detections are given in Fig. 2, note that we used a crowd image for real faces rather than an image from the FDDB to better showcase the detection performance of the algorithms.

For real faces the CNN-based methods outperform the traditional methods (DLib and OpenCV), YuNet, being built for speed sacrifices accuracy. The *retinaface* implementation is about 2% below the *insightface* implementation as stated by its author. Only the SSD, which also has a ResNet50 as a backbone, like RetinaFace, exhibits a poor performance when compared to *retinaface* and *insightface*, which is also reflected in the sample image.

For art faces the performance is very different. The YuNet sacrifices accuracy for speed, so the assumption was that it is not as well adapted to specifically realistic human faces and would generalize better, a faulty assumption. The RetinaFace based methods are interesting in that they perform so differently, *insightface*, while slightly worse compared to real faces, performs well, *retinaface*, which was comparable to *insightface* on regular images, has a large drop in performance, below even the traditional methods, and SSD fails to detect any face. This is likely a combination of training data and scaling assumptions, suggesting that refinement training is useful, or in some cases required. The decision how to handle different scale can have a huge impact, which can be seen in the different performance drops, when compared to the FDDB F1-scores, of MTCNN (≈ -0.25), YOLOv8 (≈ -0.75) and SSD (a drop to 0). The commonality here seems to be that the pyramidal CNNs are outperforming the one-step networks (YOLOv8 and SSD). The architectures of MTCNN (cascade networks) and RetinaFace (single stage) are different, but both use a pyramidal approach, the MTCNN to generate face candidates and RetinaFace uses a feature pyramid. But YuNet also utilizes a feature pyramid (tiny feature pyramid network) so that alone cannot be the only explanation.

3.2 Combining Tiling, Scaling and Rotations

While we have too little data with ground truth to perform refinement training at this stage we can implement a scaling and rotation cascade on top of the detector methods. The reason to include rotation is that people look down or up, as a narrative device, and people lying down, dying, sleeping and fornicating, are surprisingly frequent, cf. Figure 2 b. We include a scaling step because the scale of the page and the size of the faces in the pages are outside the typical scenarios of real images. Also, face size varies depending on how prominent the face is in a given illustration. The small relative size of faces to the overall page, can also be fixed by tiling, i.e. overlapping crops of the page. If we use a crop of the page the relative size of the face is returned to (roughly) in the same relation as in real photos. The tiles we use are square and of size 3000, which is similar to digital photos which are typically in the 4000×3000 range, while pages from the Wenceslas Bible are three to four times that size. As an example the ratio of face to image height for person focused images (like our FDDB) is around $1 : 4$ and around $1 : 20$ for regular images containing people, like the large crowd image (Fig. 2 a). The ratio in the Wenceslas Bible is around $1 : 122$ which is reduced to $1 : 25$ with tiling. We use a 50% overlap of tiles to prevent the non-detection of faces which otherwise would be cut in half on the border of the tile.

In essence we will build a multistage detector since the single shot detectors we tested do not work. However, we don't blindly test multiple options and optimize classification like with test-time augmentation [8], rather we investigate multiple options in a targeted manner to get an optimal set of operations for detection, and also runtime cost, by minimizing the number of stages required. When a face is found on multiple scales, rotations or tiles it will have more than one rectangle representing it, this has to be solved. The solution is quite easy: either average the rectangles to get a more stable representation or to use the rectangle with the highest confidence score to get the best localization. We chose the first, by averaging left, right, top and bottom coordinates of the rectangle respectively, to allow for methods which do not provide confidence scores.

The results of the test are given in Table 2 using a scale (S) and rotation (R) cascade as given. Once without tiling (Table 2 a) and once with tiling (Table 2 b). The F1-scores, with the best score per detector marked in bold, are given for each possible combination of rotations and scales to get a sense for how a given scale and/or rotation impacts the results. Note that S1 in Table 2 a is the same as art faces in Table 1. The results for SSD and YuNet are not included since they reflected the results in Table 1, i.e., they did not work at all. We kept the scaling and rotation relatively simple, $\pm 45°$ and $\pm 90°$ for rotation and down scaling by 2 and 4 (and unscaled 1).

Let us first look at the non-tiling cascade (Table 2 a). Except for the Viola-Jones based OpenCV, which has downsampling built in by the Haar cascade, all detectors benefit from scaling, however, the gain is mostly small. Interestingly *retinaface* and *insightface* benefit differently from scale/rotation even though they have the same architecture, which means that this is a matter of training rather than architecture.

If we add tiling (Table 2 b), it is of interest whether scaling is still helpful even if the method-inherent handling of scales works due the relative scale being fixed. Two things are of note: 1) as suspected additional scaling steps are no longer necessary. And 2) tiling does little for methods which already performed well on the untiled version, but it improves methods which had problems before (*retinaface* and YOLOv8). Tiling seems to be better mostly for methods which had serious problems before, but *insightface* also improves from 0.804 to 0.815 F1-score with tiling. For any given method it is not known a-priori what the cause of the problem is, i.e., is the problem internal scale handling which cannot deal with the large relative scales or is the problem the inability to detect painted

Table 2. F1 scores for rotation, scale and tiling experiments. Highest value per detector bolded.

(a) Rotation and scaling without tiling.

scale and rotation	F1-score					
	DLib	*insightface*	MTCNN	OpenCV	*retinaface*	YOLOv8
S1	0.521	0.798	0.586	**0.287**	0.264	0.062
S1,2	**0.529**	0.800	**0.586**	0.267	0.292	0.066
S1,2,4	0.514	0.794	0.581	0.257	0.322	**0.082**
S1,2,4 R45	0.435	0.765	0.337	0.156	0.282	0.061
S1,2,4 R45,90	0.323	0.708	0.280	0.101	0.296	0.053
S1,2,4 R90	0.375	0.735	0.470	0.148	**0.335**	0.071
S1,2 R45	0.476	0.790	0.397	0.179	0.246	0.048
S1,2 R45,90	0.370	0.743	0.337	0.113	0.246	0.049
S1,2 R90	0.402	0.760	0.511	0.154	0.292	0.066
S1 R45	0.488	**0.804**	0.483	0.206	0.221	0.046
S1 R45,90	0.415	0.764	0.426	0.134	0.221	0.046
S1 R90	0.443	0.768	0.552	0.177	0.264	0.060

(b) Rotation and scaling with tiling.

scale and rotation	F1-score					
	DLib	*insightface*	MTCNN	OpenCV	*retinaface*	YOLOv8
S1	**0.508**	0.796	**0.585**	**0.265**	0.634	**0.595**
S1,2	0.405	0.649	0.474	0.207	0.566	0.479
S1,2,4	0.386	0.634	0.469	0.197	0.584	0.466
S1,2,4R45	0.235	0.609	0.195	0.129	0.649	0.243
S1,2,4R45,90	0.178	0.537	0.163	0.079	0.653	0.170
S1,2,4R90	0.254	0.554	0.371	0.101	0.599	0.288
S1,2R45	0.247	0.643	0.222	0.162	0.635	0.252
S1,2R45,90	0.191	0.570	0.183	0.095	0.638	0.202
S1,2R90	0.280	0.582	0.383	0.105	0.581	0.332
S1R45	0.337	**0.815**	0.378	0.244	0.724	0.358
S1R45,90	0.264	0.775	0.318	0.145	**0.732**	0.273
S1R90	0.369	0.772	0.502	0.145	0.654	0.454

faces. Therefor, an evaluation is necessary for any given method, i.e., the findings can not be generalized, but testing on some cropped images seems to be sufficient.

Whether a rotation step is beneficial again strongly depends on the method but seems to be independent of scaling. Tiling can help by reducing the number of false positives, which allows *retinaface* to use two rotation steps with tiling vs.

one without. However, it is also training related, as can be seen by the difference between *insightface* and *retinaface* which otherwise have the same architectures.

A further note on the generalization of these results. Not shown in Table 2 are the components of the F1-score which are based on finding all faces and miss-detection of faces. Rotation and scale in all cases improved the detection rate but also the miss-detection rate. A problem with rotation is that while there are rotated faces, they are relatively few in number, and while they are now properly detected the miss-detection rate is also increased, but that can happen anywhere. Thus the F1-score does not improve. For works where there is a higher rate of rotated faces the result may well be different.

Also note that tiling, rotation and scaling only optimize performance, in no case did they boost detection performance in a major way. That is, it is best to start with a well performing method. Unfortunately, the performance on real-world imagery does not determine the performance on painted faces, although the performance on painted faces only ever dropped in relation. Also, from the differences in *retinaface* and *insightface*, same architecture but different training, it is clear that refinement training should be performed if possible.

4   Conclusion

The performance on real world-images generally does not transfer over to painted images. An evaluation on the target imagery is required to find methods fit for use. The problems are not based on architecture, the original implementation of Retinaface (*insightface* in the paper) and the reimplementation, Retinaface by lightface/deepface (*retinaface* in the paper), show this. While *insightface* performed well, *retinaface* does not, and that is despite a comparable performance of both methods on real-world images. In our evaluation the original Retinaface implementation [5] was the best performing method for painted faces.

We have seen that the relative size of faces to the size of the page can affect detection rates. Both *retinaface* and YOLOv8 suffer from this, although they otherwise show good performance on painted faces, as does MTCNN. A tiling approach, in which the relative size of the faces in the image is closer to regular photography, helps to address this problem. But, tiling does not improve matters for all cases, it must be evaluated for each algorithm.

We have shown that rotation or scaling improves the performance of most methods, though combining scaling and tiling does not improve the performance. The exact settings, again, differ for each method, so no general advice can be given. For general use it seems the best method is *insightface* with tiling and a rotation of ±45°.

Acknowledgments. This project received funding from the Salzburg State Digital Humanities project "Digitalisation in the Humanities, Social and Cultural Sciences (HSC)" [12].

References

1. Adouani, A., Ben Henia, W.M., Lachiri, Z.: Comparison of haar-like, hog and LBP approaches for face detection in video sequences. In: 2019 16th International Multi-Conference on Systems, Signals & Devices (SSD), pp. 266–271 (2019). https://doi.org/10.1109/SSD.2019.8893214
2. Bengamra, S., Mzoughi, O., Bigand, A., Zagrouba, E.: New challenges of face detection in paintings based on deep learning. In: Proceedings of the 16th International Joint Conference on Computer Vision, Imaging and Computer Graphics Theory and Applications-Volume 4: VISAPP, pp. 311–320 (2021)
3. Bengamra, S., Mzoughi, O., Bigand, A., Zagrouba, E.: Towards explainability in using deep learning for face detection in paintings. In: Proceedings of the 12th International Conference on Pattern Recognition Applications and Methods - Volume 1: ICPRAM, pp. 832–841. SciTePress (2023). https://doi.org/10.5220/0011670300003411
4. Dalal, N., Triggs, B.: Histograms of oriented gradients for human detection. In: Proceedings of the IEEE Conference on Computer Vision and Pattern Recognition, (CVPR'05). vol. 1, pp. 886–893 (2005)
5. Deng, J., Guo, J., Ververas, E., Kotsia, I., Zafeiriou, S.: Retinaface: single-shot multi-level face localisation in the wild. In: Proceedings of the IEEE/CVF Conference on Computer Vision and Pattern Recognition, pp. 5203–5212 (2020)
6. Feng, Y., Yu, S., Peng, H., Li, Y.R., Zhang, J.: Detect faces efficiently: a survey and evaluations. IEEE Trans. Biomet. Behav. Identity Sci. 4(1), 1–18 (2022). https://doi.org/10.1109/tbiom.2021.3120412
7. Jain, V., Learned-Miller, E.: Fddb: a benchmark for face detection in unconstrained settings. Tech. Rep. UM-CS-2010-009, University of Massachusetts, Amherst (2010)
8. Kimura, M.: Understanding test-time augmentation. In: Mantoro, T., Lee, M., Ayu, M.A., Wong, K.W., Hidayanto, A.N. (eds.) ICONIP 2021. LNCS, vol. 13108, pp. 558–569. Springer, Cham (2021). https://doi.org/10.1007/978-3-030-92185-9_46
9. Liang, Z.: Face recognition from art face images based on deep learning. In: Proceedings of the 4th International Conference on Big Data Research. p. 101–105. ICBDR '20, Association for Computing Machinery, New York (2021).https://doi.org/10.1145/3445945.3445963
10. Liu, W., et al.: SSD: single shot multibox detector. In: Leibe, B., Matas, J., Sebe, N., Welling, M. (eds.) ECCV 2016. LNCS, vol. 9905, pp. 21–37. Springer, Cham (2016). https://doi.org/10.1007/978-3-319-46448-0_2
11. Minaee, S., Luo, P., Lin, Z., Bowyer, K.: Going deeper into face detection: a survey. arXiv preprint arXiv:2103.14983 (2021). https://doi.org/10.48550/arXiv.2103.14983
12. Ein Kooperationsprojekt des Fachbereichs Germanistik der Universität Salzburg und der Österreichischen Nationalbibliothek: Die Wenzelsbibel Digitale Edition und Analyse (2024)
13. Redmon, J., Divvala, S., Girshick, R., Farhadi, A.: You only look once: unified, real-time object detection. In: Proceedings of the IEEE Conference on Computer Vision and Pattern Recognition (CVPR) (June 2016)
14. van Rijsbergen, C.J.: Information Retrieval. Butterworth-Heinemann, second edition edn. (1979)
15. Serengil, S.I., Ozpinar, A.: Hyperextended lightface: a facial attribute analysis framework. In: 2021 International Conference on Engineering and Emerging Tech-

nologies (ICEET), pp. 1–4. IEEE (2021). https://doi.org/10.1109/ICEET53442. 2021.9659697

16. Sindel, A., Maier, A., Christlein, V.: Artfacepoints: high-resolution facial landmark detection in paintings and prints. In: Karlinsky, L., Michaeli, T., Nishino, K. (eds.) Computer Vision – ECCV 2022 Workshops, pp. 298–313. Springer Nature Switzerland (2023). https://doi.org/10.1007/978-3-031-25056-9_20
17. Srinivasan, R., Rudolph, C., Roy-Chowdhury, A.K.: Computerized face recognition in renaissance portrait art: a quantitative measure for identifying uncertain subjects in ancient portraits. IEEE Signal Process. Mag. **32**(4), 85–94 (2015). https://doi.org/10.1109/MSP.2015.2410783
18. Terven, J., Córdova-Esparza, D.M., Romero-González, J.A.: A comprehensive review of yolo architectures in computer vision: From yolov1 to yolov8 and yolonas. Mach. Learn. Knowl. Extraction **5**(4), 1680–1716 (2023). https://doi.org/10.3390/make5040083
19. Theisen, M., Jenni, U.: Mitteleuropäische Schulen IV (ca. 1380–1400); Textband: Hofwerkstätten König Wenzels IV. und deren Umkreis. Verlag der Österreichischen Akademie der Wissenschaften (2014). https://doi.org/10.26530/oapen_507994, http://library.oapen.org/handle/20.500.12657/33302
20. Viola, P., Jones, M.: Robust real-time object detection. Int. J. Comput. Vision. **57**, 137–154 (2001)
21. Wechsler, H., Toor, A.S.: Modern art challenges face detection. Pattern Recogn. Lett. **126**, 3–10 (2019). https://doi.org/10.1016/j.patrec.2018.02.014
22. Wu, W., Peng, H., Yu, S.: Yunet: A tiny millisecond-level face detector. Mach. Intell. Res., 1–10 (2023). https://doi.org/10.1007/s11633-023-1423-y
23. Zhang, K., Zhang, Z., Li, Z., Qiao, Y.: Joint face detection and alignment using multitask cascaded convolutional networks. IEEE Signal Process. Lett. **23**(10), 1499–1503 (2016). https://doi.org/10.1109/LSP.2016.2603342
24. Zhong, G.: A computer vision-aided analysis of facial similarities in song dynasty imperial portraits. Electronic Imaging **35**(13), 212–1–212–1 (2023). https://doi.org/10.2352/EI.2023.35.13.CVAA-212, https://library.imaging.org/ei/articles/35/13/CVAA-212

Data Augmentation via Latent Diffusion Models for Detecting Smell-Related Objects in Historical Artworks

Ahmed Sheta[(✉)], Mathias Zinnen, Aline Sindel, Andreas Maier,
and Vincent Christlein

Pattern Recognition Lab, Friedrich-Alexander Universität Erlangen-Nürnberg,
Erlangen, Germany
ahmedhesham73@gmail.com

Abstract. Finding smell references in historic artworks is a challenging problem. Beyond artwork-specific challenges such as stylistic variations, their recognition demands exceptionally detailed annotation classes, resulting in annotation sparsity and extreme class imbalance. In this work, we explore the potential of synthetic data generation to alleviate these issues and enable accurate detection of smell-related objects. We evaluate several diffusion-based augmentation strategies and demonstrate that incorporating synthetic data into model training can improve detection performance. Our findings suggest that leveraging the large-scale pretraining of diffusion models offers a promising approach for improving detection accuracy, particularly in niche applications where annotations are scarce and costly to obtain. Furthermore, the proposed approach proves to be effective even with relatively small amounts of data, and scaling it up provides high potential for further enhancements. The source code for data generation and downstream evaluation is available at https://github.com/ultiwinter/MT_DA_LDM_OD.

Keywords: Diffusion Models · Artwork · Data Augmentation · ControlNet · Contextual Inpainting

1 Introduction

The sense of smell plays a crucial role in our everyday lives: We are constantly surrounded by smells, mostly without noticing them. They help us remember, signal danger, and support communication. Despite its fundamental importance, smell has been overlooked in traditional art history and cultural heritage discourses [5]. Recently, researchers across a wide range of disciplines have explored the cultural significance of smells [2]. Specifically in art history, olfactory dimensions of artworks can not only reveal historical understandings of the sense, but also open up new interpretative dimensions of paintings [25].

E. Rodolà et al. (Eds.): ICIAP 2025 Workshops, LNCS 16169, pp. 511–523, 2026.
https://doi.org/10.1007/978-3-032-11317-7_42

Researchers from the Odeuropa project[1] aim to uncover the olfactory dimensions of historic artworks through automatic extraction of visual smell references. However, recognizing such references is a complex task. Recognition algorithms must address artwork-specific challenges, such as stylistic diversity, varying degrees of abstraction, and annotation sparsity. Additionally, there are challenges specific to the domain of olfactory heritage. The diversity of smells often does not correlate with visual appearance, requiring algorithms to distinguish between fine-grained categories. For example, two visually similar flowers might emit entirely different scents. Moreover, smell is often not the focal point of a painting and is often only subtly and peripherally represented. As a result, smell-related objects tend to be small and spatially distributed across the entire canvas, contradicting the center-bias prevalent in common object detection benchmarks [45].

To address these challenges, the Odeuropa researchers organized the ODeuropa Competition on Olfactory Object Recognition (ODOR) [48] and introduced the similarly abbreviated Object Detection for Olfactory References dataset [49]. While these efforts provide a benchmark for evaluating smell recognition algorithms and a foundation for model training, the fundamental problem of annotation sparsity remains. The ODOR dataset contains only about 4,700 images, and although frequent classes such as "rose" are well represented, rare classes like "lobster" have fewer than 20 annotated instances.

This work addresses the persisting challenges of data sparsity and class imbalance by synthesizing artificial training data. We evaluate several data generation strategies and demonstrate that even relatively small augmentations of the training set with synthetic data can improve detection performance. Our approach is scalable and can –with minor adaptations– be applied to other applications where training data is limited and imbalanced.

2   Related Work

Computational Art Analysis and Smell Reference Extraction. The application of computer vision to the analysis of visual arts has a long-standing tradition, with applications across diverse areas such as art history [23], cultural heritage [42], search and retrieval [17], provenance research [18], and image alignment [35]. The results of automated art analysis can complement the traditional methodology in the humanities with a data-driven approach, enabling researchers to take a perspective of distant viewing [1], either independently or in combination with multimodal data via knowledge graphs [33].

A major challenge for computer vision in artistic contexts is the representational gap between relatively uniform photographic object representations and the wide variety of artistic interpretations [9], further aggravated by varying levels of abstraction. To bridge this domain gap and leverage pretraining from large-scale photographic datasets, researchers have employed domain adaptation techniques such as style transfer [11,23], few-shot learning [24], and weak

---

[1] https://odeuropa.eu.

supervision [7,26]. Additionally, several artwork-focused approaches have been introduced for object detection [31,49], person detection [41], and human pose estimation [34,46]. However, these datasets are not comparable in scale to standard photographic datasets like COCO [20].

For the specific task of detecting smell-related references, earlier work includes the recognition of smell-related objects [15]. Within the Odeuropa project, the automatic recognition of smell-related gestures [12,46,47], scenes [11, 22], and the emotional context of olfactory imagery [29] were explored. Particular attention was paid to the detection of olfactory objects, with the ODOR challenge [48] and dataset [49] providing benchmarks and training data tailored for detecting visual smell references. Extracted visual references were then combined with textual references [27] into a knowledge graph [21] for further historical [19] and museological [4] interpretation.

Despite these efforts, existing approaches for visual smell reference extraction still suffer from general limitations in computational art analysis in general, including limited training data and the domain gap. This motivates our approach of generating synthetic data to overcome these obstacles.

Diffusion Models and Data Augmentation. Synthetic training data holds the potential to overcome both the scarcity of annotations and the representational domain gap by enabling the creation of virtually unlimited amounts of training images across diverse artistic styles. However, traditional generation methods such as GANs [8] often suffer from unstable training, poor image quality, and limited diversity [3]. Diffusion models have addressed many of these issues. Sohl-Dickstein et al. [36] introduced the idea of learning complex data distributions via reverse thermodynamic diffusion. Building on this, Ho et al. [10] developed the Denoising Diffusion Probabilistic Model (DDPM), which enabled the generation of high-quality synthetic images. Further improvements, such as optimized noise scheduling [28] and deterministic sampling [37], pushed the capabilities of these models, albeit with still high computational costs. Rombach et al. [32] addressed this by shifting the diffusion process into latent space, encoding images with a variational autoencoder [16] and conditioning generation via CLIP [30] embeddings. This allowed for precise control over outputs using textual prompts. Extending this framework, Zhang et al. [44] introduced ControlNet, enabling spatial and structural guidance through depth maps, edge maps, segmentation masks, or pose annotations.

The controllability and quality of diffusion-generated images have prompted their application to data augmentation and training data generation. Diffusion models have been used to include generated samples [38], interpolate between target classes [40], manipulate high-level semantic attributes [39], generate visual priors [6], and mix real and synthetic images [13]. However, to the best of our knowledge, no prior work has specifically targeted fine-grained, imbalanced categories as required for smell-related object detection in artworks.

3　Methodology

Preliminary Experiments. In the initial phase of this study, we explored multiple diffusion-based data augmentation strategies listed in Table 1, each employing different masking strategies to determine the area for inpainting. To assess their effectiveness, we conducted a preliminary experiment: Using each strategy, we generated synthetic training sets of the same size as the original ODOR training set and evaluated the results both qualitatively and quantitatively. Qualitative evaluation involved visually inspecting the coherence and plausibility of the generated images. Quantitatively, we trained object detection models on each synthetic dataset and measured their performance on a fixed validation split from the real ODOR dataset. Detailed results are reported in Sect. 4.

However, quantitative evaluation alone is insufficient for selecting the optimal generation strategy. Since initial experiments exclude original training data, methods that introduce fewer deviations from the original images might show artificially high performance. At the same time, those generating more variation could prove more effective when combined with real data or scaled to larger training sets. To account for this bias, we adopted a heuristic selection approach that considers both the degree of deviation from the original data and the conceptual consistency of the generated images. Developing a more principled and less subjective selection methodology is left to future work. Based on this heuristic, we selected four strategies for further exploration. Of these, OPBG, ADAPT, and ENT-L failed to improve performance when scaled to larger synthetic datasets. The remaining strategy, Edge-Conditioned Object Generation

Table 1. Overview of preliminary inpainting strategies and their shorthand labels. Methods marked with SD use Standard Stable Diffusion for inpainting, CTRL marks the usage of ControlNet [44].

Shorthand	Inpainting	Description
ADAPT	SD	Adaptive entropy-based masking: masks the high-entropy half of the object bounding box to prioritize informative object regions while ensuring mask contiguity.
ENT-H	SD	High-entropy masking: masks high-information pixels based on local entropy values; often results in scattered, incoherent masks.
ENT-L	SD	Low-entropy masking: masks uniform, low-information regions; adding more diversity to these regions and complementing the high-information regions.
SAL-H	SD	High saliency masking: targets regions with strong gradient responses to enforce reconstruction of visually dominant object features.
SAL-L	SD	Low saliency masking: masks smoother areas with low gradient values to enhance diversity in non-salient regions.
OPBG	SD	Object-preserving background masking: randomly masks background patches while keeping all annotated objects intact to diversify contextual information.
BORDER	SD	Border region masking: masks the edge regions around objects to improve structural boundary learning.
EDGE	CTRL & SD	Edge-controlled object generation combined with context generating class-balancing.

Table 2. Tiered augmentation schedule to balance underrepresented classes toward 1000 instances.

Max. Inst.	5	9	19	29	49	74	99	149	249	499	999	3000
Augs./Inst.	335	130	80	45	35	20	15	10	7	3	2	1

with Contextual Inpainting (EDGE), produced a measurable improvement and is described in detail below.

Edge-Conditioned Object Generation with Contextual Inpainting. Our synthetic data generation strategy is two-fold: First, to introduce additional stylistic diversity into the dataset, we replace all objects in the ODOR training set with synthetically generated counterparts. Second, to address class imbalance, we superimpose synthetic versions of underrepresented classes onto artificially generated backgrounds.

Edge-Conditioned Object Replacement. To augment the visual diversity of the dataset while preserving scene structure, we replace each object in the ODOR training set with a synthetically generated counterpart using ControlNet [44]. For each object, we extract its bounding box crop and generate an edge map using the Holistically-nested Edge Detection (HED) detector [43]. These edge maps are used as conditioning inputs to the ControlNet model, which guides a pretrained latent diffusion model to generate a structurally aligned but stylistically varied version of the object.

The generation process is further conditioned using textual prompts of the form *oil painting of {class_ name} on canvas* to maintain semantic consistency with the object category and align the visual output with the historical domain. To suppress undesirable features and enforce anatomical coherence, negative prompts (e.g., *bad anatomy, bad structure*) are included. After generation, the synthetic object is blended back into its original position within the artwork image using boundary smoothing to mitigate sharp edges at the overlay boundary. An illustration of this pipeline is shown in Fig. 1.

Class-Balancing. To address class imbalance, we selectively oversample underrepresented categories by generating ControlNet-based synthetic object crops, which are then placed onto blank canvases. Each underrepresented class is upsampled to ensure a minimum of 1,000 instances in the training set. This targeted augmentation strategy aims to diversify object representation without duplicating data or disrupting the domain's stylistic coherence. Per-class augmentation statistics are summarized in Table 2.

Contextual Inpainting for Blank Placement. To prevent object detection models from having localization bias within the objects placed in blank images, ControlNet-generated object crops placed on blank images were post-processed using Stable Diffusion inpainting. As Fig. 2 shows, this step filled the surrounding canvas with contextually plausible background textures, ensuring the augmented samples remained visually coherent and semantically natural. Reducing the visual clutter on the objects' border would alleviate the localization bias that the object detection model might develop in case of sparse images.

ControlNet Finetuning. The pretrained ControlNet [44] was finetuned for 20 epochs on a curated set of ODOR object crops paired with edge maps (from the HED detector [43]) and category-specific prompts (*oil painting of {class_name} on canvas*) to adapt ControlNet to ODOR's fine-grained and historically specific object classes.

This pipeline, combining edge-controlled fine-tuned ControlNet generation with structured integration and contextual inpainting, or short EDGE, provided a scalable and controllable mechanism for dataset expansion, particularly effective for improving the representation of rare and complex object classes in artistic imagery.

The computational cost of ControlNet-based data Augmentation amounted to 384 GPU hours using NVIDIA Tesla V100 GPUs. This includes 346 GPU hours for generating object-centric image samples, conducted in parallel across 4 GPUs, each running for approximately 86.5 h. An additional 38 GPU hours were spent on contextual inpainting using Stable Diffusion v1.5.

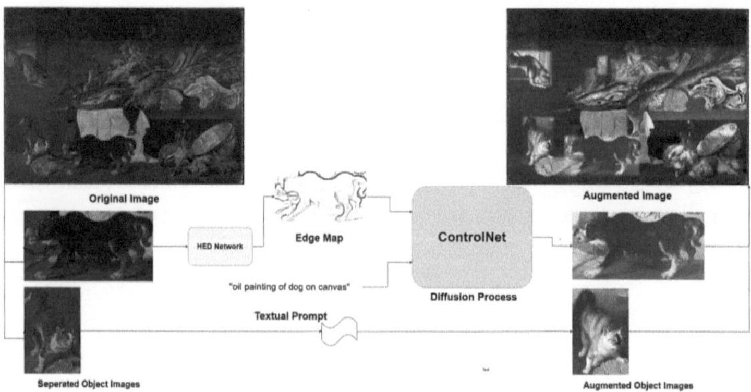

Fig. 1. Illustration of the ControlNet-based augmentation pipeline. Each object is extracted from the ODOR images using its bounding box, processed through an HED edge detector to generate an edge map, and conditioned alongside a textual prompt in ControlNet. The augmented object is then reintegrated into its original position within the full ODOR image, preserving spatial and structural consistency.

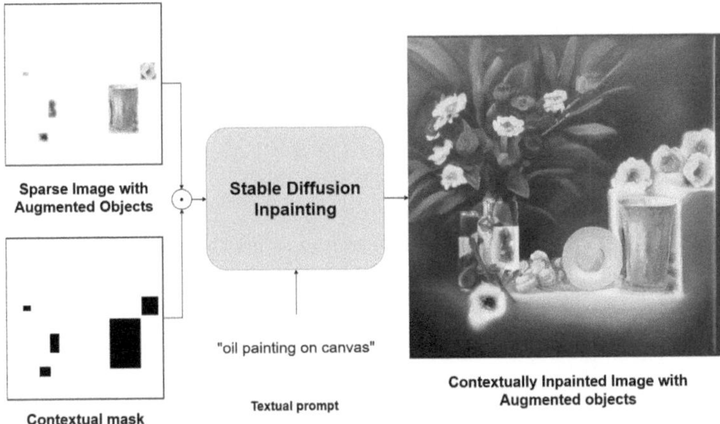

Fig. 2. Illustration of the inpainting-based background generation for synthetic images. The process fills the blank background onto which ControlNet-augmented objects were previously overlaid, aiming to obtain realistic contextual surroundings.

4   Results

Experimental Setup. To evaluate the downstream detection task, we adopt Ultralytics' YOLOv11-M architecture [14]. All experiments follow the official default configuration except for the learning rate, which was set to 0.001 during, and reduced to 0.0007 in cases of second-stage finetuning to mitigate the catastrophic forgetting. To evaluate the performance of the different configurations, we use the original test split provided by [49]. For hyperparameter tuning, model selection, and preliminary evaluations, we further split a validation set from the ODOR training set, resulting in 3408 images for training, 856 for validation, and 448 for testing. All evaluations are reported using the COCO-style mean Average Precision (mAP@0.5:0.95) as defined in [20]. To mitigate the effect of randomness during training, we report all performance metrics as the mean and standard deviation over three independent training runs.

Table 3. Results comparing the baseline with selected augmentation strategies. Values are mAP@0.5:0.95% reported as mean ± standard deviation over three independent runs. The percentage change relative to the baseline is reported in brackets.

Method	Real : Synthetic (%)	Val mAP	Test mAP
Baseline (Vanilla)	100 : 0.0	17.7 ± 0.2	16.3 ± 0.5
EDGE	14.1 : 85.9	**18.4 ± 0.1 (+4.0%)**	**17.4 ± 0.2 (+6.7%)**
OPBG	50.0 : 50.0	17.0 ± 0.1 (-4.0%)	15.9 ± 0.2 (-2.5%)
ADAPT	37.4 : 62.6	15.8 ± 0.1 (-10.8%)	14.5 ± 0.2 (-11.1%)
ENT-L	37.4 : 62.6	15.2 ± 0.1 (-14.2%)	13.9 ± 0.4 (-14.3%)

Baseline Comparison. Table 3 summarizes the final detection performance of YOLOv11-M models trained with different augmentation strategies.

The baseline model, trained solely on the original ODOR dataset, hereafter referred to as VANILLA, achieved a test mAP of 16.3%. Among the augmentation methods, the EDGE approach yielded the best performance, improving test accuracy by 6.7% relative to the VANILLA baseline.

In contrast, inpainting-based methods showed limited effectiveness. Augmenting only the background (OPBG) or using entropy-guided object masking (ADAPT) resulted in negative gains. The poorest performance was observed with ENT-L, which masks low-entropy regions; this likely stems from the fragmented and non-contiguous nature of the generated masks, which the inpainting model fails to handle coherently as illustrated in Fig. 3. These results suggest that diffusion-based inpainting models are not well-suited for reconstructing fine-grained pixel fragments. Instead, they require spatially contiguous masks to produce semantically meaningful outputs. To address this issue by enforcing block-like, high-entropy masking, we introduced the ADAPT strategy. However, its downstream performance remained suboptimal.

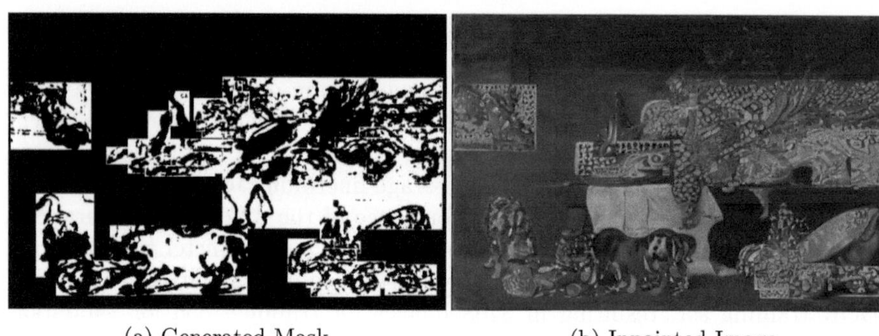

(a) Generated Mask   (b) Inpainted Image

Fig. 3. Example image, generated by the low entropy masking strategy (ENT-L).

Additional Analyses.

Training Scheme Comparison. To evaluate the impact of training order, we compared three finetuning schemes: (1) joint training on original and augmented data (mixed), (2) finetuning first on augmented data and then on original data (aug→orig), and (3) the reverse—finetuning on original data followed by augmented data (orig→aug).

The results for the mixed training strategy are reported in Table 3. Table 4 presents results for the second scheme (aug→orig). Only the EDGE strategy improved over the baseline, achieving a test mAP of 16.9%. These results highlight the polarizing nature of joint training, where performance depends heavily on the semantic and structural quality of the augmentations. The initial synthetic-only stage also converged early, with training typically stopping between epochs 4 and 9.

Table 4. Results of training on augmented images (including class balancing) and finetuning on original images. Metrics reported are mean ± standard deviation over 3 runs.

Method	Val mAP	Test mAP
1. *Stage: Finetuning on Augmented Images of*		
EDGE	4.1 ± 0.2	4.1 ± 0.3
ADAPT	3.6 ± 0.1	4.0 ± 0.3
ENT-L	3.4 ± 0.1	03.5 ± 0.4
OPBG	**14.8 ± 0.1**	**14.0 ± 0.3**
2. *Stage: Finetuning on Original Images*		
EDGE	**18.5 ± 0.6 (+4.6%)**	**16.9 ± 0.3 (+3.6%)**
ADAPT	17.9 ± 0.6 (+1.1%)	16.1 ± 0.3 (âĂŞ1.2%)
ENT-L	17.1 ± 0.6 (âĂŞ3.4%)	15.2 ± 0.6 (âĂŞ3.6%)
OPBG	16.4 ± 0.6 (âĂŞ7.3%)	15.4 ± 0.3 (âĂŞ6.7%)

In contrast, the third scheme (orig→aug) proved ineffective across all methods. In these cases, validation performance consistently declined during the second stage, and early stopping reverted to the best model checkpoint saved at epoch 1, indicating that continued training on synthetic data led to overfitting or distribution shift. These findings reinforce that synthetic data is most beneficial when used concurrently with real data, rather than in staged isolation.

Preliminary Experiment Results. To identify the most promising augmentation strategies, we conducted preliminary experiments by training YOLOv11-M solely on synthetic data (without class balancing) and evaluating on the original ODOR validation and test sets. As shown in Table 5, *EDGE, ENT-L,* and *ADAPT* emerged as the top-performing object-aware strategies, while *OPBG* outperformed all others in the context-aware category. These four methods were selected for subsequent large-scale experiments.

Table 5. Preliminary results of training YOLO-M on the augmented ODOR images only from every augmentation strategy, without class balancing, evaluated on the original validation and test sets. Values are reported as mean ± standard deviation over 3 runs.

Method	Val mAP	Test mAP
EDGE	**2.9 ± 0.4**	**3.0 ± 0.3**
ENT-L	**2.5 ± 0.2**	**2.6 ± 0.2**
ADAPT	**2.1 ± 0.2**	**2.4 ± 0.3**
SAL-H	1.8 ± 0.3	2.2 ± 0.2
ENT-H	1.4 ± 0.2	1.2 ± 0.1
SAL-L	1.3 ± 0.2	1.1 ± 0.2
OPBG	**14.8 ± 0.1**	**14.0 ± 0.3**
BORDER	12.8 ± 0.3	12.2 ± 0.1

5   Conclusion

Identifying smell references in historic artworks remains a challenging task due to the combination of stylistic variation, fine-grained semantic classes, and annotation sparsity.

In this work, we explored diffusion-based data augmentation methods to address these limitations. We systematically evaluated a range of masking strategies for inpainting and identified one particularly promising approach: edge-based conditioning combined with contextual inpainting (EDGE). This method demonstrated measurable improvements in detection performance, even when applied to relatively small synthetic training sets.

While the observed performance gains are limited, they suggest the potential of this technique, particularly when scaled to generate a larger number of synthetic images. Given its generality, we anticipate that this method to be applicable in other domains where annotated data is scarce or imbalanced. Scaling up the synthetic data generation and applying the approach to other domains are natural and promising extensions to the current work.

Further improvements may be achieved by refining the masking strategies used during inpainting and developing adaptive training schemes. In particular, gradually increasing the proportion of real data relative to synthetic data over the training schedule may allow models to better adapt to the real data distribution while retaining the increased variability introduced by synthetic data.

Overall, our findings highlight the potential of diffusion-based augmentation to enrich niche datasets and offer a step forward in the computational analysis of olfactory references in historical art, which aligns well with recent efforts to uncover olfactory aspects of cultural heritage.

References

1. Arnold, T., Tilton, L.: Distant viewing: analyzing large visual corpora. DSH **34** (2019)
2. Bembibre, C., Strlič, M.: From smelly buildings to the scented past: an overview of olfactory heritage. Front. Psychol. **12** (2022)
3. Dhariwal, P., Nichol, A.: Diffusion models beat GANs on image synthesis. NeurIPS (2021)
4. Ehrich, S., et al.: Olfactory Storytelling Toolkit: A How-To Guide for Working with Smells in Museums and Heritage Institutions (2023)
5. Ehrich, S.C., Verbeek, C., Zinnen, M., Marx, L., Bembibre Jacobo, C., Leemans, I.: Nose-first. towards an olfactory gaze for digital art history. In: CEUR Workshop Proceedings, vol. 3064. CEUR (2021)
6. Fang, H., Han, B., Zhang, S., Zhou, S., Hu, C., Ye, W.M.: Data augmentation for object detection via controllable diffusion models. In: WACV (2024)
7. Gonthier, N., Gousseau, Y., Ladjal, S., Bonfait, O.: Weakly supervised object detection in artworks. In: VISArt IV at ECCV (2018)
8. Goodfellow, I.J., et al.: Generative adversarial nets. NeurIPS (2014)
9. Hall, P., Cai, H., Wu, Q., Corradi, T.: Cross-depiction problem: Recognition and synthesis of photographs and artwork. Comput. Visual Media **1** (2015)
10. Ho, J., Jain, A., Abbeel, P.: Denoising diffusion probabilistic models. NeurIPS **33** (2020)
11. Huang, H., Zinnen, M., Liu, S., Maier, A., Christlein, V.: Scene classification on fine arts with style transfer. In: SUMAC (2024)
12. Hussian, A., Zinnen, M., Tran, T.M.H., Maier, A., Christlein, V.: Gesture classification in artworks using contextual image features. arXiv preprint arXiv:2412.03456 (2024)
13. Islam, K., Zaheer, M.Z., Mahmood, A., Nandakumar, K.: Diffusemix: Label-preserving data augmentation with diffusion models. In: CVPR (2024)
14. Jocher, G., Qiu, J.: Ultralytics yolo11 (2024). https://github.com/ultralytics/ultralytics
15. Kim, S., Park, J., Bang, J., Lee, H.: Seeing is smelling: Localizing odor-related objects in images. In: 9th Augmented Human International Conference (2018)
16. Kingma, D.P., Welling, M., et al.: Auto-encoding variational bayes (2013)
17. Lang, S., Ommer, B.: Attesting similarity: supporting the organization and study of art image collections with computer vision. DSH **33**(4) (04 2018)
18. Lang, S., Zinnen, M.: Digital provenance research: Eine computerassistierte bildersuche in historischen auktionskatalogen (2025)
19. Leemans, I., Tullett, W., Bembibre, C., Marx, L.: Whiffstory: using multidisciplinary methods to represent the olfactory past. American Historical Rev. **127**(2) (2022)
20. Lin, T.Y., et al.: Microsoft coco: Common objects in context. In: ECCV (2014)
21. Lisena, P., et al.: Capturing the semantics of smell: the odeuropa data model for olfactory heritage information. In: ESWC (2022)
22. Liu, S., Huang, H., Zinnen, M., Maier, A., Christlein, V.: Novel artistic scene-centric datasets for effective transfer learning in fragrant spaces. In: VISART VII at ECCV (2024)
23. Madhu, P., et al.: Icc++: Explainable feature learning for art history using image compositions. PR **136** (2023)

24. Madhu, P., et al.: One-shot object detection in heterogeneous artwork datasets. In: IPTA (2022)
25. Marx, L., Zinnen, M., Collette Ehrich, S., Tullett, W., Bembibre, C., Leemans, I.: Seeing smell: sourcing olfactory imagery using artificial intelligence. Arts et Savoirs (20) (2023)
26. Mazzamuto, M., Ragusa, F., Furnari, A., Signorello, G., Farinella, G.M.: Weakly supervised attended object detection using gaze data as annotations. In: ICIAP (2022)
27. Menini, S., Paccosi, T., Tekiroğlu, S.S., Tonelli, S.: Scent mining: Extracting olfactory events, smell sources and qualities. In: SIGHUM Workshop on Computational Linguistics for Cultural Heritage, Social Sciences, Humanities and Literature (2023)
28. Nichol, A.Q., Dhariwal, P.: Improved denoising diffusion probabilistic models. In: ICML (2021)
29. Patoliya, V., Zinnen, M., Maier, A., Christlein, V.: Smell and emotion: recognising emotions in smell-related artworks. arXiv preprint arXiv:2407.04592 (2024)
30. Radford, A., et al.: Learning transferable visual models from natural language supervision. In: ICML (2021)
31. Reshetnikov, A., Marinescu, M.C., Lopez, J.M.: Deart: Dataset of European art. In: VISART VI at ECCV (2022)
32. Rombach, R., Blattmann, A., Lorenz, D., Esser, P., Ommer, B.: High-resolution image synthesis with latent diffusion models. In: CVPR (2022)
33. Sartini, B., Baroncini, S., van Erp, M., Tomasi, F., Gangemi, A.: Icon: An ontology for comprehensive artistic interpretations. JOCCH **16**(3) (2023)
34. Schneider, S., Vollmer, R.: Poses of people in art: a dataset for human pose estimation in digital art history. JOCCH **17**(4) (2024)
35. Sindel, A., Maier, A., Christlein, V.: Artfacepoints: High-resolution facial landmark detection in paintings and prints. In: VISART VI at ECCV. Springer (2022)
36. Sohl-Dickstein, J., Weiss, E., Maheswaranathan, N., Ganguli, S.: Deep unsupervised learning using nonequilibrium thermodynamics. In: ICML (2015)
37. Song, J., Meng, C., Ermon, S.: Denoising diffusion implicit models. arXiv preprint arXiv:2010.02502 (2020)
38. Toker, A., Eisenberger, M., Cremers, D., Leal-Taixé, L.: Satsynth: Augmenting image-mask pairs through diffusion models for aerial semantic segmentation. In: CVPR (2024)
39. Trabucco, B., Doherty, K., Gurinas, M., Salakhutdinov, R.: Effective data augmentation with diffusion models. arXiv preprint arXiv:2302.07944 (2:23)
40. Wang, Z., et al.: Enhance image classification via inter-class image Mixup with diffusion model. In: CVPR (2024)
41. Westlake, N., Cai, H., Hall, P.: Detecting people in artwork with CNNs. In: VISART III at ECCV (2016)
42. Willot, L., Réby, K., Manuel, A., Gouet-Brunet, V., Vodislav, D., De Luca, L.: Creating a dataset for the detection and segmentation of degradation phenomena in Notre-dame de Paris. In: SUMAC (2024)
43. Xie, S., Tu, Z.: Holistically-nested edge detection. In: CVPR (2015)
44. Zhang, L., Rao, A., Agrawala, M.: Adding conditional control to text-to-image diffusion models. In: CVPR (2023)
45. Zheng, Z., Chen, Y., Hou, Q., Li, X., Wang, P., Cheng, M.M.: Zone evaluation: Revealing spatial bias in object detection. IEEE TPAMI (2024)
46. Zinnen, M., Hussian, A., Maier, A., Christlein, V.: Recognizing sensory gestures in historical artworks. MTAP (2024)

47. Zinnen, M., Hussian, A., Tran, H., Madhu, P., Maier, A., Christlein, V.: Sniffyart: the dataset of smelling persons. In: SUMAC (2023)
48. Zinnen, M., Madhu, P., Kosti, R., Bell, P., Maier, A., Christlein, V.: Odor: The icpr2022 odeuropa challenge on olfactory object recognition. In: ICPR (2022)
49. Zinnen, M., et al.: Smelly, dense, and spreaded: The object detection for olfactory references (odor) dataset. ESWA **255** (2024)

Have Large Vision-Language Models Mastered Art History?

Ombretta Strafforello[1] , Derya Soydaner[2(✉)] , Michiel Willems[1] ,
Anne-Sofie Maerten[1] , and Stefanie De Winter[1]

[1] KU Leuven, Leuven, Belgium
{ombretta.strafforello,michiel.willems,annesofie.maerten,
stefanie.dewinter}@kuleuven.be
[2] Leiden University, Leiden, The Netherlands
d.soydaner@liacs.leidenuniv.nl

Abstract. The emergence of large Vision-Language Models (VLMs) has established new baselines in image classification across multiple domains. We examine whether their multimodal reasoning can also address a challenge mastered by human experts. Specifically, we test whether VLMs can classify the style, author and creation date of paintings, a domain traditionally mastered by art historians. Artworks pose a unique challenge compared to natural images due to their inherently complex and diverse structures, characterized by variable compositions and styles. This requires a contextual and stylistic interpretation rather than straightforward object recognition. Art historians have long studied the unique aspects of artworks, with style prediction being a crucial component of their discipline. This paper investigates whether large VLMs, which integrate visual and textual data, can effectively reason about the historical and stylistic attributes of paintings. We present the first study of its kind, conducting an in-depth analysis of three VLMs, namely CLIP, LLaVA, and GPT-4o, evaluating their zero-shot classification of art style, author and time period. Using two image benchmarks of artworks, we assess the models' ability to interpret style, evaluate their sensitivity to prompts, and examine failure cases. Additionally, we focus on how these models compare to human art historical expertise by analyzing misclassifications, providing insights into their reasoning and classification patterns.

Keywords: Style classification · Multimodal learning · AI in art history

1 Introduction

Classification in art history involves a range of attributes, including authorship, period, and geographical origin, all of which help place an artwork within its historical and cultural context. Additionally, factors such as medium, technique, and iconography provide insights into the methods and symbolism used by artists.

E. Rodolà et al. (Eds.): ICIAP 2025 Workshops, LNCS 16169, pp. 524–544, 2026.
https://doi.org/10.1007/978-3-032-11317-7_43

Together, these elements form the basis for analyzing and categorizing art. However, the classification of an artwork's attributes, which encapsulates numerous visual and historical features, is a complex task. Art style classification, ranging from highly realistic to purely abstract, as illustrated in Appendix A, is particularly challenging, even for human experts. Automating art style classification could yield significant benefits, such as categorizing newly discovered artworks, efficiently managing large collections, verifying existing style labels, and improving the quality of datasets—all while conserving time and resources. This has motivated computer vision researchers to develop neural networks for the classification of artworks and, especially, their style.

Recently, large VLMs have revolutionized computer vision by leveraging their multimodal nature to interpret both visual and textual input. Typically trained to match images with text descriptions or answer questions about an image content, VLMs benefit from the semantic richness of language, compared to the single category labels used to train traditional image classifiers, helps VLMs capture more nuanced image characteristics, such as object location within an image [13], the geolocation [26] or the image aesthetic qualities [24]. Their extensive multimodal training on large-scale datasets scraped from the internet, make VLMs successful in various image domains, ranging from remote sensing [30] and medical imagery [6] to cartoon captioning [9]. The success of VLMs has led us to ask: To what extent can VLMs solve the complex task of art classification?

In this paper, we present the first evaluation of VLMs for this purpose. We utilize two datasets, WikiArt and JenAesthetics, and evaluate three VLMs: CLIP, LLaVA, and GPT-4o. While newer models have since been introduced, the selected models provide a representative understanding of current VLM performance in art historical analysis. We conduct a rigorous evaluation of their zero-shot performance in predicting art historical attributes, including art style, author, and time period of paintings, using different prompting methods. While fine-tuning or in-context learning are promising directions, we deliberately focus on evaluating zero-shot inference. This approach *1)* provides a foundational understanding of VLMs' performance before designing improvements, *2)* keeps our evaluation closely aligned with typical VLM usage by non-expert users (e.g., through ChatGPT), and *3)* tests whether VLMs' strong zero-shot performance across multiple image domains holds with artworks.

Beyond measuring classification performance, we investigate the impact of adding contextual information in the text prompts and analyze failure cases to assess the extent of their multimodal reasoning about artworks. Our findings show that while CLIP, LLaVA, and particularly GPT-4o exhibit a certain level of accuracy, some errors do not align with the acceptable standards of art history experts. We conclude that current VLMs are not yet suitable for use in this field without expert oversight. Our code will be made public upon acceptance.

2   Related Work

2.1   Automating the Classification of Artworks

Research on automatic art style classification has established important baselines. Early work explored photographic style recognition [19] and optimized similarity measures between paintings [31]. Later studies leveraged Convolutional Neural Networks (CNNs) for fine-art painting classification [5,10,15,22,23,32, 35]. Other methods, such as a boosted ensemble of Support Vector Machines, recognize artistic movements of paintings [16]. Various CNNs have achieved accuracies of 59.01% [42] and 68.55% [27] on WikiArt[1]. Zhao *et al.* [41] applied transfer learning to classify all 27 styles in WikiArt, achieving the current state-of-the-art with an accuracy of 71.24%. Appendix B summarizes prior results on WikiArt.

Closely related to automating artwork classification, Large Language Models (LLMs) have recently been used including formal art analysis via proxy learning from visual concepts such as color [21], automating art analysis [20], and zero-shot classification using descriptive prompts [36]. However, these approaches rely solely on text. In contrast, we explore VLMs that process both image and text data, enabling direct visual analysis. We evaluate their ability to predict an artwork's style, author, and time period, providing a comprehensive approach.

2.2   Vision-Language Models

VLMs have found wide applicability in tasks like image captioning, visual question answering (VQA), and zero-shot learning (see [8] for an overview). VLMs differ from traditional deep learning models by enabling multimodal learning, where visual and textual data are processed together, leading to a deeper contextual understanding. In art research, CLIP [29] has been used to classify metadata in art-historical images [11]. Its features have also been applied for content and style disentanglement [39]. Other studies have used VLMs for qualitative results on style classification [24] and iconographic concept recognition [34].

A related work is GalleryGPT [7], which generates formal analyses of paintings using short descriptive paragraphs by fine-tuning LLaVA. In contrast, we explore VLMs for style prediction, conducting a comprehensive evaluation across multiple models. We assess their performance in predicting artistic styles, time periods, and authors. Another work, ArtGPT-4 [40] interprets and articulates emotions evoked by paintings. To our knowledge, we present the first performance analysis of VLMs in style, author, and time period classification, establishing a foundation for future research in multimodal reasoning and art analysis.

3   Predicting Artistic Attributes with VLMs

3.1   Chosen VLMs

CLIP [29] is a VLM trained to match correct pairings of images and natural language descriptions using separate encoders for image and text. In our tests, both encoders are Transformer-based models [14,37]. CLIP employs a contrastive objective to bring matching image-text pairs closer in embedding space and push non-matching ones apart. Trained from scratch on 400 million image-text pairs

---

[1] https://www.wikiart.org/.

collected from the internet, CLIP generalizes well across visual concepts and language queries. We utilize the official OpenAI implementation[2].

LLaVA (Large Language and Vision Assistant) [25] is a large multimodal model for image understanding and conversation, capable of generating text and answering image-related questions. It is trained in two stages: (1) feature alignment, using 558K subset of the LAION-CC-SBU[3] dataset to connect a frozen pretrained vision encoder with a frozen LLM; and (2) visual instruction tuning, using 150K GPT-generated instruction data and approximately 515K VQA samples. This training enables LLaVA to follow multimodal instructions. This multimodal chatbot can perform well in tasks that require a deep visual and language understanding, such as image captioning and VQA. We use llava-hf/llava-1.5-7b-hf checkpoints hosted by Hugging Face.

GPT-4o [1] is an advanced, closed-source multimodal model with a Transformer-based architecture, pretrained to predict the next token in a sequence. Trained on a massive amount of image-text pairs from public sources and licensed data providers, it achieves human-level performance on various benchmarks. In our tests, we use the OpenAI API[4]. Due to its usage cost, we applied GPT-4o more selectively compared to the freely accessible CLIP and LLaVA.

3.2   Datasets

WikiArt is the largest public collection of digitized artworks, with 81,449 paintings by 1,119 artists, ranging from the 15th century to contemporary times. These paintings represent 27 styles (e.g., abstract, impressionism) and 45 genres (e.g., landscape, portrait), with some paintings annotated with their creation year. It is divided into a training set (57,025 images) and a validation set (24,421 images). We evaluate model performance on the validation set. Since not all the paintings have their creation year annotated, we use the subset of 7,759 paintings dated after 1900 and we refer to this set as WikiArt *Validation Subset* (VS).

JenAesthetics [2–4] consists of 1,628 high-resolution scans of oil paintings from museums, covering 11 art periods and 410 artists. It encompasses a comprehensive variety of artistic themes. Due to some broken URLs in the original dataset, we obtained 1,576 out of the 1,628 images.

3.3   Zero-Shot Prediction

We prompt the VLMs in various ways to predict three attributes of a painting: art style, author and time period. These predictions are *zero-shot*, meaning that the models have not been trained for these specific tasks. Figure 1 illustrates how we prompt VLMs for art style classification and provides examples of the

---

[2] https://github.com/openai/CLIP.
[3] https://huggingface.co/datasets/liuhaotian/LLaVA-Pretrain.
[4] https://platform.openai.com.

model outputs. We assess whether the art style predictions can be improved by enriching the prompts with contextual information, incorporating the year and author into the prompts. We perform these tests using LLaVA. We also ask the model to predict the time period and the author of a painting. For the time period, we simplify the task by evaluating whether the output falls within the correct *decade* or *century*, instead of the exact year.

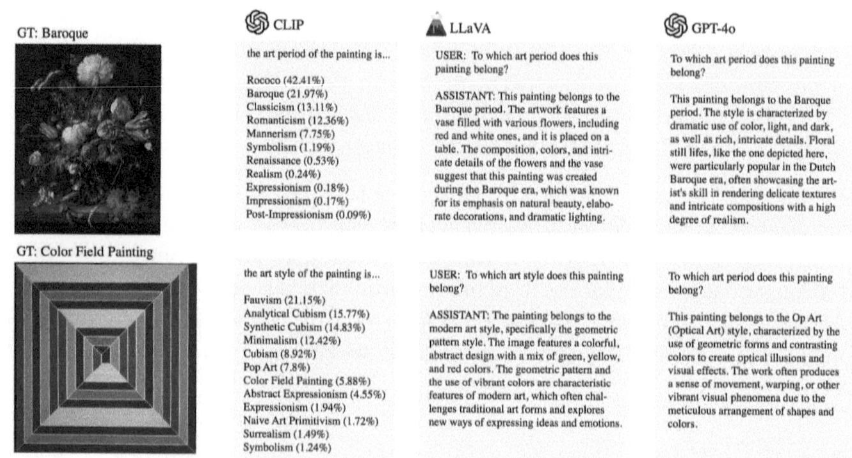

Fig. 1. Example of style classification with VLMs on the paintings *Vase of Flowers*, by Jan Davidsz de Heem, c. 1660 and *Color Maze* (1966) by Frank Stella. While LLaVA and GPT-4o correctly classify the style of *Vase of Flowers* and CLIP commits a reasonable mistake, all models fail to recognize *Color Maze* as a Color Field painting.

3.4   Evaluation

To evaluate CLIP's predictions, we identify the art style, author, and year whose text embeddings are closest to the image embedding and compare them to the ground-truth labels. For language assistants like LLaVA and GPT-4o, which generate varying text lengths, we analyze the output texts in a post-processing step. For this analysis, we deploy the pre-trained Llama 3.2 language model (3B parameters) [28] to extract predicted style, author, and time period from the textual description output by the VLMs. Finally, we map the predictions to the exact ground-truth labels using the Jaro-Winkler text distance metric [18,38]. This step is necessary when the output is correct but not identical to the exact wording in the ground-truth, e.g., "*Neoclassical*" vs. "*Neoclassicism*". Predictions without a ground-truth match are left unchanged for detailed examination.

Table 1. Zero-shot style prediction accuracy by CLIP, LLaVA, and GPT-4o on WikiArt Validation Subset (VS) and JenAesthetics. The wording *art period* is used in the prompt for the JenAesthetics dataset, to better align with its annotations. For the other datasets, *art style* is used.

Model	Prompt	WikiArt Valid. Accuracy (%)	JenAesthetics Accuracy (%)
CLIP	*the art (style)/(period) of the painting is [art_ style]*	24.57	36.04
LLaVA	*To which art (style)/(period) does this painting belong?*	28.32	37.02
GPT-4o	*To which art (style)/(period) does this painting belong?*	53.42	54.35

4   Results

4.1   Zero-Shot Art Style Prediction

Performance Evaluation. An overview of the style classification accuracies is provided in Table 1. Remarkably, all three VLMs performed significantly better than random chance, even without being specifically trained or fine-tuned to predict art style. The highest accuracy is achieved by GPT-4o, which scores 53.42% on WikiArt and 54.35% on JenAesthetics. The success of GPT-4o can be explained by its massive training process. Nevertheless, a gap remains between the VLMs and the state-of-the-art style classification method, *Big Transfer learning* [41], which correctly classifies 71.24% of WikiArt's paintings across all 27 classes. This indicates that further improvements are needed before VLMs can be safely deployed for art style classification.

To better understand the performance of the VLMs, we examine the F1 score for each style, as illustrated in Appendix E. The styles with the highest F1 scores across the datasets and models include Baroque, Impressionism, Pointilism, and Pop Art, followed closely by Abstract Expressionism and Romanticism. In contrast, Action Painting, Analytical and Synthetic Cubism, and New Realism tend to have lower scores. The variation in F1 scores reflects the duration of these style movements and the number of paintings produced within them.

Impressionism is among the best and most frequently predicted styles, especially by LLaVA (15.93% of WikiArt's paintings). This might be due to the popularity of Impressionism, which gained early global acclaim and appeals to all social classes. Its high market value, ongoing educational efforts, and enhanced accessibility through modern technology also contribute to its enduring appeal [12,17,33]. In predicting Impressionism, GPT-4o achieves the highest F1 score of 0.71 on WikiArt, followed by CLIP (0.53) and LLaVA (0.5).

The vocabulary of LLaVA is mostly wide and accurate, but it lacks awareness of certain art styles; for instance, it never predicts Mannerism. Additionally, when predicting the art period, LLaVA outputs incorrect labels like Egyptian, Medieval, Modern alongside more conventional art periods such as Baroque, Renaissance and Impressionism.

An overview of the styles predicted by GPT-4o in JenAesthetics, compared to the ground-truth, is available in Appendix C. Notably, GPT-4o provides a finer-grained classification than the dataset labels; for example, it distinguishes Italian,

Venetian, Spanish, and English Renaissance from the overarching 'Renaissance' category. The model achieves an overall accuracy of 54.35%. However, when the predicted styles are grouped according to their overarching stylistic period, the accuracy rises to 62.79%, with 'Modern Art and 20th Century Movements' and 'Post-Impressionism' achieving the highest (75%) and lowest (40.1%) accuracies, respectively. Despite being trained on massive amounts of data, GPT-4o does make mistakes. Examining these misclassifications provides insights into GPT-4o's multimodal reasoning and its challenges in distinguishing closely related artistic styles, which we elaborate on in the following sections.

Misclassifications. Despite the relatively low accuracies (24.57% on WikiArt, 36.04% on JenAesthetics), CLIP makes sensible mistakes. It confuses Baroque, Rococo, Classicism, and Romanticism; Impressionism, Post-Impressionism, and Expressionism; Color Field Painting, Abstract Expressionism and Minimalism – styles that are temporally close and share significant visual similarities. CLIP predicts Cubism mostly as Analytic Cubism. Expressionism, Surrealism, Symbolism and Synthetic Cubism are frequently misclassified as Analytic Cubism. These styles all emerge within the same historical period and exhibit certain visual similarities. Misclassifying Expressionism as Fauvism is understandable due to their shared stylistic traits. A similar pattern occurs with Renaissance styles, which are also conflated. Realist paintings misclassified as Romanticist can be explained by a number of factors, including training set bias and similar subject matter. The most unexpected error by CLIP is predicting Contemporary Realism instead of the correct styles Impressionism, Romanticism, Realism and Baroque. This mistake might be due to 'contemporary' and 'realism' are misunderstood as adjectives related to the visual appearance of the subject matters, instead of art style classes. We include the confusion matrices in the Appendices.

We observe that LLaVA confuses Impressionism with several artistic movements in WikiArt, including Post-Impressionism, Realism, Expressionism, Symbolism and Art Nouveau Modern. This might be due to Impressionism being prevalent in LLaVA's training data. Similarly, we find that LLaVA is biased towards predicting Romanticism more frequently than other art styles, with 41% of Romanticism predictions in JenAesthetics. Additionally, LLaVA classifies Minimalism paintings as Expressionism, Romanticism, and Realism, which is an anomalous result given the minimal overlap between these styles. Overall, our findings show that LLaVA yields more inconclusive results than CLIP.

GPT-4o makes reasonable predictions across all genres, with the greatest accuracy in Ukiyo-E, Baroque and Impressionism. However, it frequently mislabels works as Baroque–for instance, many Rococo paintings in JenAesthetics. One notable error is misclassifying Frank Stella's *Color Maze* as Op Art (Optical Art), shown in Fig. 1. Although Stella's work shares some characteristics with Op Art, such as geometric patterns and contrasting colors, the artist's intention was not to create the visual vibrations typical of the Op Art style. A closer examination of this piece, which is an example of Color Field Painting, reveals intentional irregularities meant to distinguish it from Op Art. Similarly, all paintings by Vuillard–typically Post-Impressionist– are labeled Baroque, likely due to shared texture and fabric details. In some cases, GPT-4o fails to predict any style label or outputs 'Fresco', which is a technique rather than a style.

Reasoning Abilities of VLMs vs. Art Historians. We further analyze the VLMs' reasoning abilities under the oversight of art history experts. Notably, the language assistants attempt to justify the style label, without being prompted to do so. However, their reasoning is often flawed. They often describe the scene or elements that are irrelevant to the style (e.g., *the scene also includes a man wearing a top hat, which is a characteristic element of Romanticism paintings*'). In other cases, the models cite stylistic elements that could fit into many styles. For example, they claim 'the depiction of everyday scenes' is typical for works in the Dutch Golden Age, Northern Renaissance, Impressionism, Realism, Baroque, Rococo and Romanticism. Given that this argument fits many styles, it is not a strong cue for style classification. The reasoning of both language assistants seems inherently different from those of art historians. This is evident in the misclassifications of Manet's paintings: Manet created several works that are considered 'Realism' before becoming a key figure in the rise of the style 'Impressionism'. As a result, GPT-4o labels *all* of his works as Impressionism. Art historians differentiate Manet's works by observing stylistic elements, whereas GPT-4o seems to lack this understanding and makes erroneous generalizations.

LLaVA tends to do the opposite—it misclassifies paintings important for the style they represent. For example, LLaVA mislabeled Cezanne's Post- Impressionism works as Impressionism. This can be considered an honest mistake, however, Cezanne played an important role for Post-Impressionism. On the other hand, LLaVA labeled Realism paintings by Courbet as Romanticism, which is odd given that Courbet rebelled against Romanticism. Interestingly, LLaVA states a 'focus on emotions and individualism', which is true for both Realism and Romanticism, but in a different manner. In general, these models fail to consider all relevant stylistic and contextual elements to reason about the correct style for works that are harder to classify.

Adding Information in the Prompt. For LLaVA, we find that adding author and creation year information improves the style classification performance over the baseline with no additional information in JenAesthetics (Table 2). This suggests that LLaVA has learned correlations between historical periods and art styles. In addition, it shows that LLaVA recognizes the established painters in the evaluated datasets. Both author and year information improve the accuracy, while providing both information jointly results in the highest accuracy, with an increase of up to 18% on JenAesthetics compared to the baseline. For WikiArt, however, additional information does not improve art style classification. This might be due to WikiArt containing paintings of varying popularity, such as those from Russian artists Nicholas Roerich, Pyotr Konchalovsky, Boris Kustodiev, or Zinaida Serebriakova, who might be less known by an international audience and infrequently discussed online.

4.2 Zero-Shot Author Prediction

We present zero-shot author prediction results in Table 3. All VLMs, including the low accuracy of 0.43% on WikiArt by LLaVA, perform significantly above random chance. This equals to 0.092% and 0.25% on WikiArt validation set and JenAesthetics, which contain artworks from 1077 and 405 painters. For JenAesthetics, CLIP's most predicted author is Charles Francois Daubigny, who mainly painted landscapes, using a not very pronounced style spanning Romanticism, Impressionism, and Realism. The high frequency of Daubigny predictions might be explained by the wide range of styles that exist in close proximity to each other, the popularity of the theme of landscape, and the absence of key features that define a personal style. Notably, Pablo Picasso is not among CLIP's most predicted authors, despite being one of the most prevalent artists in WikiArt.

LLaVA outputs 'unknown' in 41% of times in JenAesthetics and 47.9% of times in WikiArt. It also predicts only a limited set of author names, among which Vincent Van Gogh, Claude Monet, John Singer Sargent, Jackson Pollock, Michelangelo, John Constable, Pablo Picasso, Erro, John William Waterhouse.

4.3 Zero-Shot Time Period Prediction

Table 4 shows the prediction results for decade and century. All VLMs perform better than a random classifier, except for LLaVA on WikiArt VS. When asked, *"When was this painting made?"*, LLaVA often responses such as *"The painting was made in the 16th century."*. We convert these predictions to a rounded year, such as 1500. Since LLaVA often predicts only the century, the decade accuracies are low. On WikiArt VS, LLaVA often predicts '19th Century', which we convert to 1800. However, this prediction is incorrect, since all paintings with a year label in WikiArt VS are dated after 1900. Due to frequent 19th Century predictions, LLaVA performs worse than random on century prediction in WikiArt.

4.4 A Closer Look at GPT-4o

Due to its cost, we tested GPT-4o more selectively and only asked for style predictions. Nevertheless, when prompted to predict the art style, GPT-4o provides detailed descriptions, often mentioning the historical period and author. The time period is typically mentioned when describing an art style, most commonly as a century interval, in 90.84% of WikiArt and 89.84% of JenAesthetics predictions. We evaluate the accuracy of these descriptions by comparing statistics on the frequently mentioned authors with the ground truth.

Table 2. Zero-shot art style prediction by LLaVA for different input prompts on WikiArt (WA) Validation (V) and Validation Subset (VS) (for the predictions involving the year in the prompt) and JenAesthetics.

LLaVA Prompt	WikiArt V/VS (Accuracy (%))	JenAesthetics (Accuracy (%))
To which art (style)/(period) does this painting belong?	28.32	37.02
What is the art (style)/(period) of this painting by [author]?	26.33	48.03
What is the art (style)/(period) of this painting made in [year]?	29.55	50.89
What is the art (style)/(period) of this painting made by [author] in [year]?	23.13	55.01

Table 3. Zero-shot author prediction accuracy by CLIP, LLaVA and GPT-4o on WikiArt Validation and JenAesthetics. Note: We did not prompt GPT-4o to predict authorship; instead, we retrieved this information from the style predictions.

Model	Prompt	WikiArt Valid. Accuracy (%)	JenAesthetics Accuracy (%)
CLIP	*this painting was made by [author]*	26.39	36.23
LLaVA	*Who is the author of this painting?*	4.35	4.95
GPT-4o	*To which art (style)/(period) does this painting belong?*	6.74	12.57

Table 4. Zero-shot time period prediction accuracy by CLIP and LLaVA on WikiArt Validation Subset (VS) and JenAesthetics. We include the average performance of random predictions drawn from a uniform distribution.

Model	Prompt	WikiArt (VS)		JenAesthetics	
		Century (%)	Decade (%)	Century (%)	Decade (%)
CLIP	*this painting was made in [year]*	89.25	25.34	51.02	10.98
LLaVA	*When was this painting made?*	36.65	10.85	63.43	2.73
Random classifier		87.49 ± 0.34	8.9 ± 0.3	18.66 ± 0.97	1.98 ± 0.33

In WikiArt, an author is mentioned in 12.06% of the style descriptions. Appendix C lists the most frequently predicted authors. Multiple authors are mentioned concurrently to describe a style, e.g., Vincent van Gogh, Paul Cézanne, and Paul Gauguin are cited as key figures of Post-Impressionism. Pablo Picasso and Georges Braque are often mentioned as pioneers of Cubism. In JenAesthetics, authors appear in 24.12% of the paintings. This higher rate can be due to paintings in JenAesthetics being mostly renown, Western artworks, possibly reflecting training bias in GPT-4o. The top five mentioned authors are Henri Rousseau, Claude Monet, Paul Cézanne, Rembrandt, Vincent Van Gogh – all pioneers in their periods. These are correct for 6.74% (WikiArt) and 12.57% (JenAesthetics), with the latter notably outperforming LLaVA (Table 3).

5 Conclusions

We acknowledge several benefits of VLMs over earlier artwork classifiers, which are typically trained solely for art style classification and cannot predict other art historical attributes without fine-tuning. In contrast, we show that a single VLM can perform three tasks – predicting art style, author and time period – above random chance, without specific training.

We focus on zero-shot performance to provide a foundational understanding of VLMs. This approach reflects typical real-world use by non-experts and tests whether VLMs' generalization across image domains extends to artworks. Although less accurate than models fine-tuned for style classification, VLMs provide insights into their reasoning. Artistic styles often share overlapping characteristics. Rather than strictly adhering to predefined labels, VLMs occasionally make misclassifications that align with stylistic similarities, indicating an emerging grasp of nuanced artistic traits. Notably, LLaVA and GPT-4o generate detailed explanations of features that define a painting's style. These explanations enhance understanding and clarify classifications, offering greater transparency compared to traditional deep learning models.

Nevertheless, we find that VLMs, even the powerful GPT-4o, make mistakes that do not align with expert art historical standards. They may misclassify even canonical artworks by renowned artists or provide flawed reasoning. We conclude that VLMs, despite their great potential, have not yet mastered art history. Their limitations pose risks for non-experts, who may rely on these models, as they could assign incorrect styles, authors, or time periods to lesser-known or unknown pieces. This highlights the importance of expert oversight and the involvement of art historians in verifying and refining AI-generated information.

Appendices

A   Example of Artworks for Each Style in WikiArt

See Fig. 2.

Fig. 2. Example of artworks for each style in WikiArt, sorted chronologically. From top left, *Baptism of Christ* by P. della Francesca, *The Arnolfini Portrait* by J. van Eyck, *The Garden of Earthly Delights* by H. Bosch, *The Holy Trinity* by El Greco, *Las Meninas* by D. Velázquez, *The swing* by J. Fragonard, *Saturn Devouring His Son* by F. Goya, *The Waterfall Where Yoshitsune Washed His Horse at Yoshino in Yamato Province* by K. Hokusai, *A widow* by J. Tissot, *Rouen Cathedral, the Portal in the Sun* by C. Monet, *Capo di Noli* by P. Signac, *Clownesse Cha-U-Kao* by H. de Toulouse-Lautrec, *The tragedy* by P. Picasso, *The kiss* by G. Klimt, *Charing Cross Bridge* by A. Derain, *Portrait of a woman* by G. Braque, *Les Demoiselles d'Avignon* by P. Picasso, *Grapes* by J. Gris, *Figures Lying on the Sand* by S. Dalí, *The sleeping gypsy* by H. Rousseau, *Woman I* by W. de Kooning, *New York* by F. Kline, *Color Maze* by F. Stella, *Cape Cod Morning* by E. Hopper, *Crying girl* by R. Lichtenstein, *Immaginando Tutto* by A. Boetti, and *Anne in a Striped Dress* by F. Porter.

B   Related Work

For comparison with the VLMs performance, we provide an overview of the style classification accuracies achieved on WikiArt by previous work in Table 5.

Table 5. Art style prediction performance of previous works on WikiArt.

Method	Acc. (%)	# Classes
Karayev et al. [19]	44.10	25
Bar et al. [5]	43.02	27
Saleh and Elgammal [31]	46.00	27
Tan et al. [35]	54.50	27
Lecoutre et al. [22]	62.80	25
Florea and Gieseke [16]	46.20	25
Cetinic et al. [10]	56.40	27
Zhong et al. [42]	59.01	25
Sandoval et al. [32]	66.71	22
Menis-Mastromichalakis et al. [27]	68.55	21
Zhao et al. [41]	71.24	27

C   Styles Predicted by GPT-4o Compared to the Ground Art-Styles in JenAesthetics

See Fig. 3.

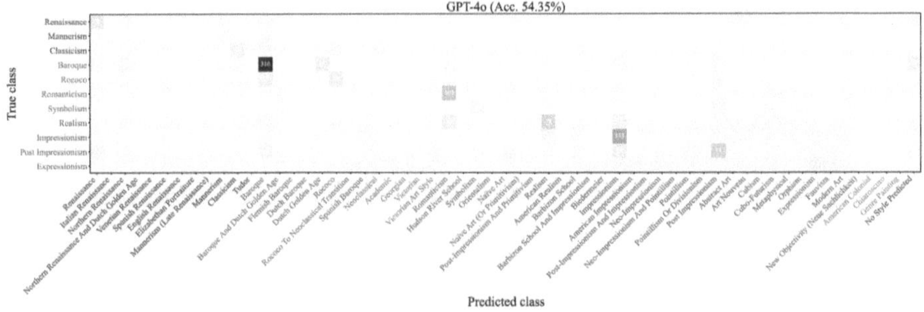

Fig. 3. Styles predicted by GPT-4o compared to the ground art-styles in JenAesthetics. The color codes denote the overarching artistic periods: Renaissance, Baroque and Rococo, Neoclassicism, Romanticism, Symbolism and Naïve Art/Primitivism, Realism, Impressionism, Post-Impressionism, Modern Art and 20th Century Movements and Others (territorial labels/techniques/genre). Styles in the same period share visual and technical similarities. GPT-4o provides fine-grained style classifications compared to the dataset's ground-truth labels.

D   Zero-Shot Art Style Classification

D.1   Dataset Statistics

We report the number of paintings by art style, artist, and time period for WikiArt and JenAesthetics in Figs. 4 and 5.

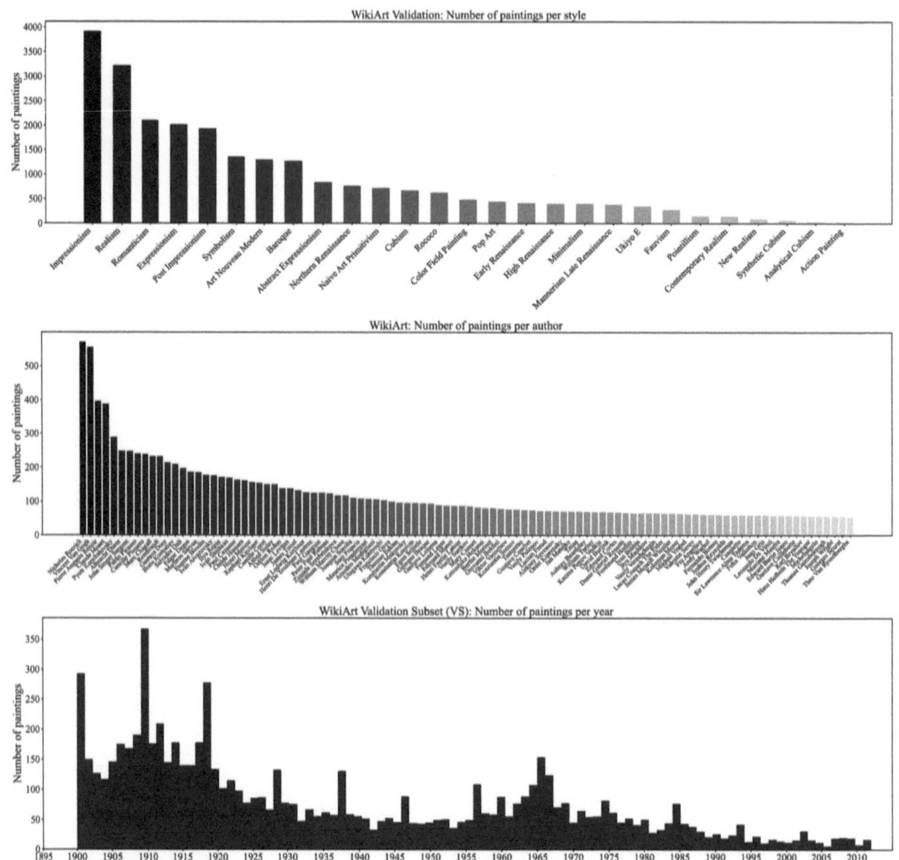

Fig. 4. Number of paintings per art style (top), artist (middle) and time period (bottom) in the WikiArt Validation set and WikiArt Validation Subset (VS). For visualization purposes, we only show the 100 most frequent artists.

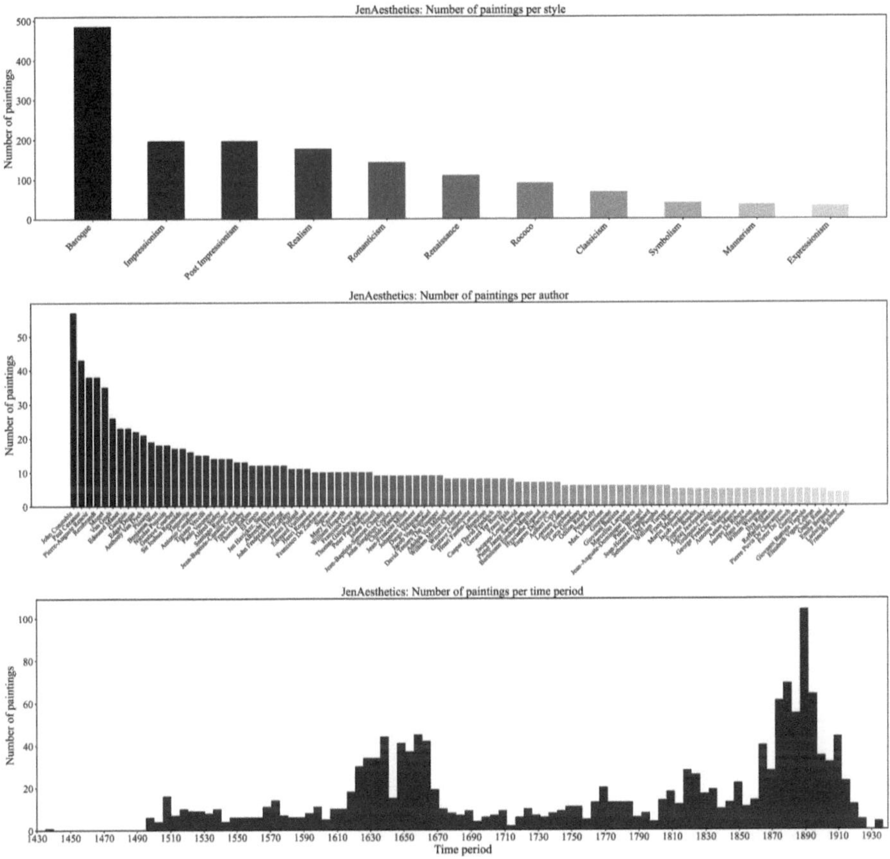

Fig. 5. Number of paintings per art style (top), artist (middle) and time period (bottom) in the JenAesthetics dataset. For visualization purposes, we only show the 100 most frequent artists.

D.2   Precision and Recall

The precision, recall and F1 Score for the classification of the different styles are shown in Figs. 6, and 7.

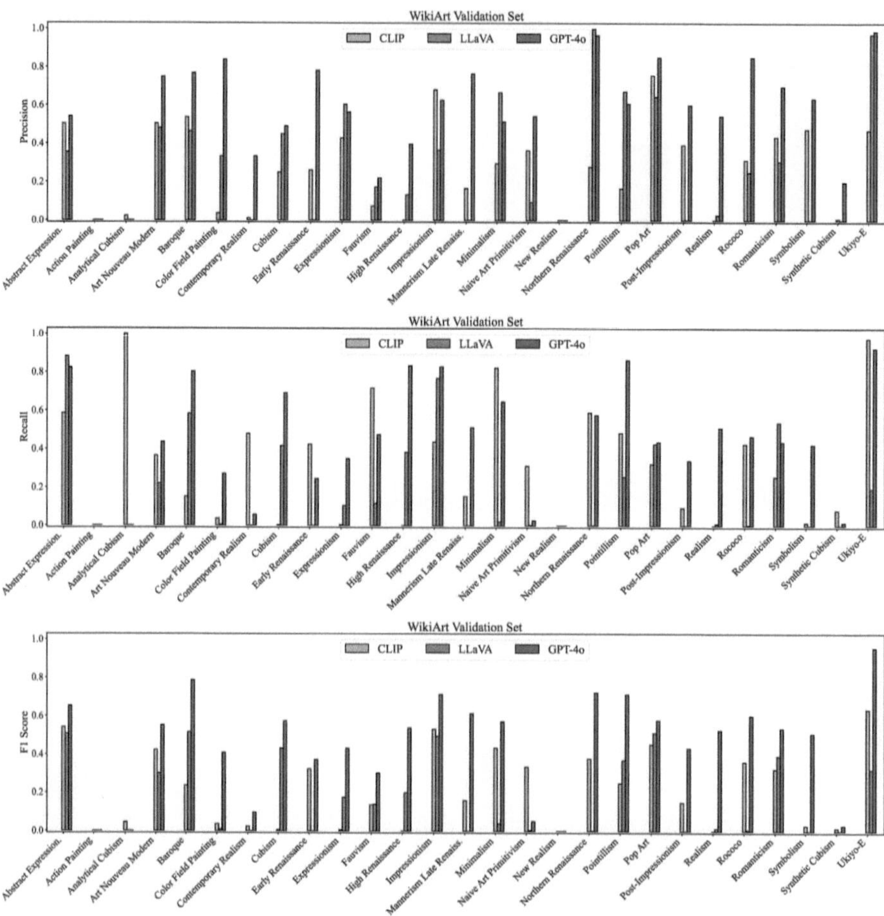

Fig. 6. Precision, Recall and F1 score for the different art styles in the WikiArt Validation set.

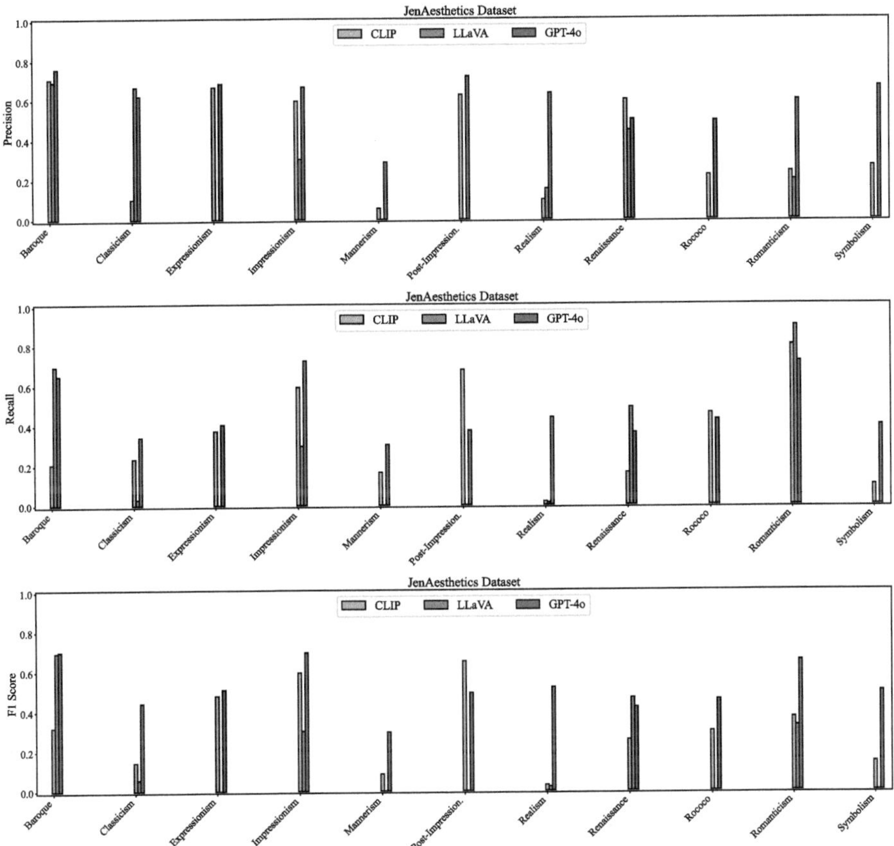

Fig. 7. Precision, Recall and F1 score for the different art styles in the JenAesthetics dataset.

D.3   Misclassifications

To better understand the misclassifications made by the evaluated VLMs, we plot the confusion matrices, available in Figs. 8 and 9.

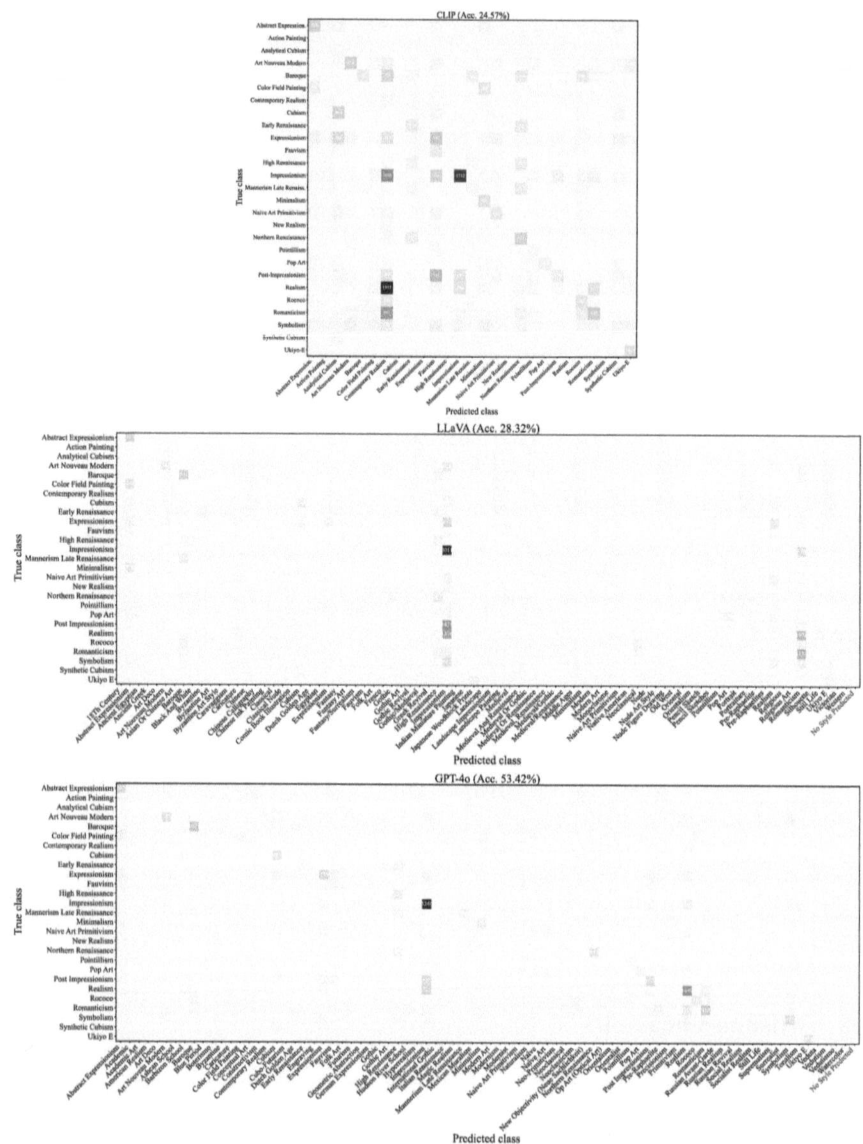

Fig. 8. Predicted vs. ground-truth art styles for CLIP, LLaVA and GPT-4o for zero-shot style prediction on the WikiArt Validation set. For LLaVA and GPT-4o, only labels predicted at least 10 times are shown.

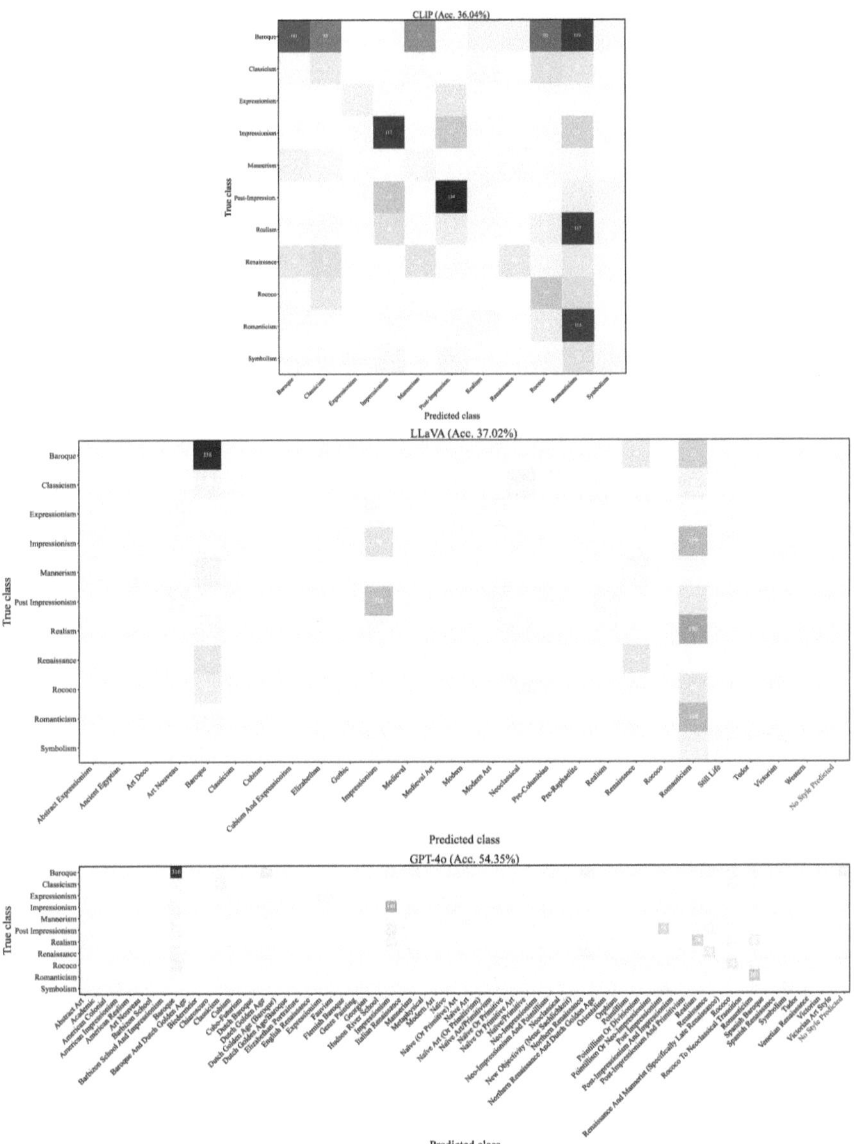

Fig. 9. Predicted vs. ground-truth art styles for CLIP, LLaVA and GPT-4o for zero-shot style prediction the JenAesthetics dataset.

E   F1 Scores on WikiArt Validation Set

See Fig. 10.

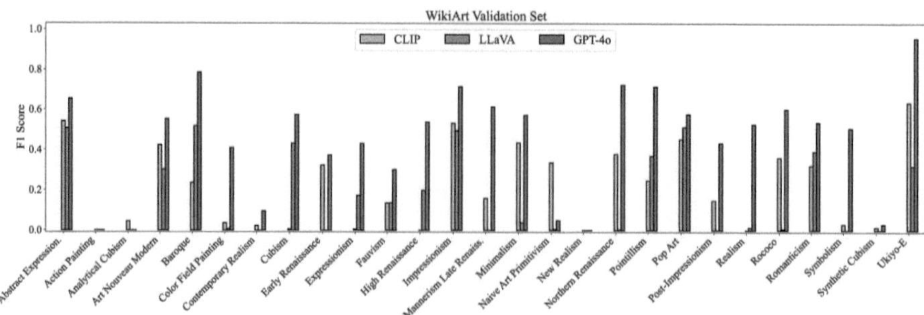

Fig. 10. F1 Scores on WikiArt validation set for the classification of different art styles.

F A Closer Look at GPT-4o

F.1 Art Style Descriptions

When prompted to predict the art style of a painting, often GPT-4o provides rich descriptions of the painting characteristics and its artistic period. In some cases,

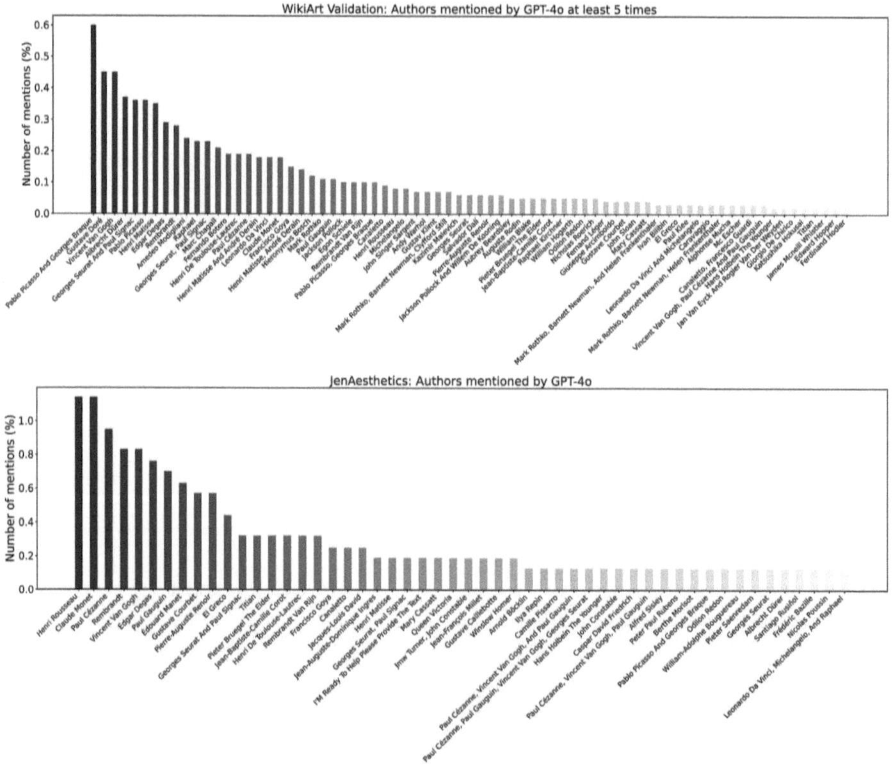

Fig. 11. Painters mentioned by GPT-4o when prompted to predict the art style of a painting from the WikiArt Validation set and JenAesthetics datasets.

these descriptions mention the name of the painter. The frequently mentioned painters are shown in Fig. 11.

References

1. Achiam, J., et al.: GPT-4 technical report. arXiv preprint arXiv:2303.08774 (2023)
2. Amirshahi, S.A., Denzler, J., Redies, C.: JenAesthetics—a public dataset of paintings for aesthetic research. Technical report, University of Jena Germany (2013)
3. Amirshahi, S.A., Redies, C., Denzler, J.: How self-similar are artworks at different levels of spatial resolution? In: Proceedings of the Symposium on Computational Aesthetics (2013)
4. Amirshahi, S.A., Hayn-Leichsenring, G.U., Denzler, J., Redies, C.: JenAesthetics subjective dataset: analyzing paintings by subjective scores. In: Agapito, L., Bronstein, M.M., Rother, C. (eds.) ECCV 2014. LNCS, vol. 8925, pp. 3–19. Springer, Cham (2015). https://doi.org/10.1007/978-3-319-16178-5_1
5. Bar, Y., Levy, N., Wolf, L.: Classification of artistic styles using binarized features derived from a deep neural network. In: Agapito, L., Bronstein, M.M., Rother, C. (eds.) ECCV 2014, Part I. LNCS, vol. 8925, pp. 71–84. Springer, Cham (2015). https://doi.org/10.1007/978-3-319-16178-5_5
6. Bazi, Y., et al.: Vision-language model for visual question answering in medical imagery. Bioengineering **10**(3), 380 (2023)
7. Bin, Y., et al.: GalleryGPT: analyzing paintings with large multimodal models. In: the 32nd ACM International Conference on Multimedia, pp. 7734–7743 (2024)
8. Bordes, F., et al.: An introduction to vision-language modeling. arXiv preprint arXiv:2405.17247 (2024)
9. Cao, S., Young, S.: Predicting winning captions for weekly new Yorker comics. arXiv preprint arXiv:2407.18949 (2024)
10. Cetinic, E., Lipic, T., Grgic, S.: Fine-tuning convolutional neural networks for fine art classification. Expert Syst. Appl. **114**, 107–118 (2018)
11. Conde, M.V., Turgutlu, K.: CLIP-art: contrastive pre-training for fine-grained art classification. In: Proceedings of the IEEE/CVF Conference on Computer Vision and Pattern Recognition, pp. 3956–3960 (2021)
12. Dombrowski, A.: Impressionism and globalization. In: Historical Narratives of Global Modern Art, pp. 19–30. Routledge (2023)
13. Dorkenwald, M., et al.: Pin: Positional insert unlocks object localisation abilities in VLMs. In: Proceedings of the IEEE/CVF Conference on Computer Vision and Pattern Recognition, pp. 13548–13558 (2024)
14. Dosovitskiy, A., et al.: An image is worth 16x16 words: transformers for image recognition at scale. In: International Conference on Learning Representations (2021)
15. Elgammal, A., et al.: The shape of art history in the eyes of the machine. In: Proceedings of the AAAI Conference on Artificial Intelligence, vol. 32 (2018)
16. Florea, C., Gieseke, F.: Artistic movement recognition by consensus of boosted SVM based experts. J. Vis. Commun. Image Represent. **56**, 220–233 (2018)
17. Hook, P.: The Ultimate Trophy: How the Impressionist Painting Conquered the World. Prestel Verlag (2012)
18. Jaro, M.A.: Advances in record-linkage methodology as applied to matching the 1985 census of Tampa, Florida. J. Am. Stat. Assoc. **84**(406), 414–420 (1989)
19. Karayev, S., et al.: Recognizing image style. arXiv preprint arXiv:1311.3715 (2013)

20. Khadangi, A., et al.: CognArtive: large language models for automating art analysis and decoding aesthetic elements. arXiv preprint arXiv:2502.04353 (2025)
21. Kim, D., Elgammal, A., Mazzone, M.: Formal analysis of art: proxy learning of visual concepts from style through language models. arXiv:2201.01819 (2022)
22. Lecoutre, A., Negrevergne, B., Yger, F.: Recognizing art style automatically in painting with deep learning. In: Asian Conference on Machine Learning, pp. 327–342. PMLR (2017)
23. Li, W.: Enhanced automated art curation using supervised modified CNN for art style classification. Sci. Rep. **15**(1), 7319 (2025)
24. Lin, J., et al.: VILA: on pre-training for visual language models. In: IEEE/CVF Conference on Computer Vision and Pattern Recognition, pp. 26689–26699 (2024)
25. Liu, H., et al.: Visual instruction tuning. In: Advances in Neural Information Processing Systems, vol. 36 (2024)
26. Mendes, E., et al.: Granular privacy control for geolocation with vision language models. arXiv preprint arXiv:2407.04952 (2024)
27. Menis-Mastromichalakis, O., Sofou, N., Stamou, G.: Deep ensemble art style recognition. In: International Joint Conference on Neural Networks, pp. 1–8 (2020)
28. Meta, A.: Llama 3.2: Revolutionizing edge AI and vision with open, customizable models. Meta AI Blog. Retrieved December 20 (2024)
29. Radford, A., et al.: Learning transferable visual models from natural language supervision. In: International Conference on Machine Learning, pp. 8748–8763 (2021)
30. Rahhal, A.: Vision-language models for zero-shot classification of remote sensing images. Appl. Sci. **13**(22), 12462 (2023)
31. Saleh, B., Elgammal, A.: Large-scale classification of fine-art paintings: learning the right metric on the right feature. arXiv preprint arXiv:1505.00855 (2015)
32. Sandoval, C., Pirogova, E., Lech, M.: Two-stage deep learning approach to the classification of fine-art paintings. IEEE Access **7**, 41770–41781 (2019)
33. Snider, L.: A lasting impression: French painters revolutionize the art world. History Teacher, pp. 89–101 (2001)
34. Springstein, M., et al.: Visual narratives: large-scale hierarchical classification of art-historical images. In: IEEE/CVF Winter Conference on Applications of Computer Vision, pp. 7220–7230 (2024)
35. Tan, W.R., et al.: Ceci n'est pas une pipe: a deep convolutional network for fine-art paintings classification. In: IEEE International Conference on Image Processing, pp. 3703–3707 (2016)
36. Tojima, T., Yoshida, M.: Zero-shot classification of art with large language models. IEEE Access (2025)
37. Vaswani, A., et al.: Attention is all you need. Adv. Neural. Inf. Process. Syst. **30**, 5998–6008 (2017)
38. Winkler, W.E.: String comparator metrics and enhanced decision rules in the Fellegi-Sunter model of record linkage (1990)
39. Wu, Y., Nakashima, Y., Garcia, N.: Not only generative art: stable diffusion for content-style disentanglement in art analysis. In: Proceedings of the 2023 ACM International Conference on Multimedia Retrieval, pp. 199–208 (2023)
40. Yuan, Z., et al.: ArtGPT-4: towards artistic-understanding large vision-language models with enhanced adapter. arXiv preprint arXiv:2305.07490 (2023)
41. Zhao, W., Jiang, W., Qiu, X.: Big transfer learning for fine art classification. Comput. Intell. Neurosci. **2022**(1), 1764606 (2022)
42. Zhong, S.H., Huang, X., Xiao, Z.: Fine-art painting classification via two-channel dual path networks. Int. J. Mach. Learn. Cybern. **11**, 137–152 (2020)

Exploring Image Watermarking in Digital Collections of Ancient Coins

Fabio Bellavia[(✉)] [ID] and Lavinia Sole

Università degli Studi di Palermo, Palermo, Italy
`{fabio.bellavia,lavinia.sole}@unipa.it`

Abstract. This paper investigates invisible blind watermark approaches as a security mechanism to track ownership in online digital collection of ancient coin images. Recent deep-based architectures are compared in addition to traditional transformation-domain approaches. Besides evaluating watermarking for the task-specific images, this analysis takes into account challenging attacks obtained by combining both geometric or photometric attacks to test the actual effectiveness of compared methods for practical purposes. Moreover, the sequential application of pair of watermarking methods is considered with the intention of increasing the robustness of the joint watermarking and as an additional form of attack.

Experimental evaluation shows that deep-based watermarking is robust to most of the base attacks and that sequential joint watermarking can improve upon the robustness of the original methods. Finally, although the strength of recent approaches to common attacks is sufficient, complex combined attacks remain still a challenge.

Keywords: Image watermarking · Deep watermarking · Joint watermarks · Combined watermark attacks · Cultural heritage · Ancient coins

1 Introduction

1.1 Rationale

Document digitalization actively contributes to the preservation, analysis and fruition of artifacts in cultural heritage and fine arts, making possible to share and manipulate data worldwide. The free circulation of digital goods is beneficial and promoted by public institutions since it positively stimulates the cultural and scientific progresses of the human society. Nevertheless, maintaining the ability to track the ownership of the digital data with the aim of avoiding misconducts, abuses and information manipulations, which can lead in turn to negative economic and social impacts, represents a critical issues in the whole process.

In the case of digital images, adding EXchangeable Image File (EXIF) metadata or clearly visible blended or not overlay marks onto the source image can

E. Rodolà et al. (Eds.): ICIAP 2025 Workshops, LNCS 16169, pp. 545–556, 2026.
https://doi.org/10.1007/978-3-032-11317-7_44

provide elementary mechanisms to protect the ownership of the original image which can be easily deceived by non-expert users with general image editing software. Specifically, EXIF metedata can be manipulated or removed without efforts by metadata editing tools [10] while small or medium overlay marks can be removed manually by cropping or cut-and-paste, possibly followed by inpainting [18]. Notice that overlay marks, known as visible watermarks, in the common case cover a small or medium marginal area of the image to allow the observer a clear view of the image content. In case of more aggressive overlay marks recent supervised inpainting tools based on the last development of deep learning, are able to remove them without expert-user requirements [14, 21].

Further expedients to indirectly limit non-legit behaviors on an image by malicious users can be attempted in case of web-based image database access, avoiding image saving from the web browser and providing low-resolution images for visualization. In these cases, saving the image directly from the webpage source, which is granted by most of the web browsers, or straightly taking a crude screenshot can be exploited to bypass the protection system. Moreover, ripped low-resolution images can be enhanced by super-resolution applications, which provide nowadays very impressive results thanks to the progress done in the field through deep-learning [26]. Notice that in general malicious users are not interested in fine pixel-level image details but in the raw image structural content in order to generate a fake similar image.

In this work a further way offered by invisible watermarking techniques [23] is explored to deal with digital image protection. The main goal of watermarking is to embed a key message with unnoticeable alterations of the original content in the input image. This message acts as a fingerprint which can be retrieved back though a specific decoding procedure to asses the ownership of the image.

1.2 Related Works

Invisible watermarking, or simply watermarking within this paper, is closely related to steganography, which shares similar ideas and methodologies but with different aims [5]. In particular, watermarking mainly focuses on robustness in decoding the hidden message after manipulations of the watermarked image while steganography is designed to maximize the capacity of the hidden message.

Watermarking can be fragile or robust according to the degree of transformations one can apply to the watermarked image while still being able to retrieve the hidden message [1]. Fragile watermarking is mainly used to detect tampering or check integrity, whereas robust watermarking for copyright protection and for tracking image alterations as pursued in this work. Furthermore, blind watermarking does not require the original non-watermarked image as additional input in the decoding step, in contrast to non-blind watermarking.

Early watermarking approaches hide the message in the Least Significant Bit (LSB) of the pixel intensity value in order to minimize visual perceptual changes in the watermarked image. Further researches proposed more advanced techniques moving the message in the Intermediate Significant Bit (ISB), but in any case watermarking methods exploiting image representation by bit planes are fragile and not robust to common attack scenarios [2].

In order to improve robustness, further watermarking methods have been developed where message embedding take places in alternative image representation spaces, like for instance in the frequency domain, and then the watermarked images is reprojected back to the original spatial domain. The investigated transformed domains includes among others the Discrete Cosine Transform (DCT), the Discrete Wavelet Transform (DWT) and the Singular Value Decomposition (SVD) [7]. Likewise the bit-plane representation, in the transformed spaces the message is embed in the middle or lowest signal components so as to avoid minimal image changes. Combining more domain transformations into hybrid watermarking approaches are able to further improve the robustness to attacks.

More recently, deep-based watermarking approaches have achieved current State-Of-the-Art (SOTA) [11]. In most cases the process is designed as an encoder-decoder architecture [19], using the encoder to embed the hidden message into the original image and the decoder to extract it from the watermarked image. Noise layers are present in-between the encoder and decoder to simulate possible image perturbations, forcing both the encoder and decoder to learn a representation invariant to these attacks and hence improving the watermarking robustness. Attention layers [22] are additionally included to suggest better image areas where to add the message. The overall network is mostly designed as a Generative Adversarial Network (GAN) [8], where the discriminator can be used to improve the visual quality, watermarking detectability and robustness [3,12,25,27,28]. Good results have also been obtained embedding the message in latent spaces [6], as the deep analogous of the transformation-domain watermarking approaches, relying on self DIstillation, NO labels (DINO) [4] features to define the latent space representation.

1.3   Contributions

As main contribution, this paper analyzes and compares digital watermarking approaches in case of images of ancient coins. These images come from a numismatic collection of great value belonging to a public institution. As discussed above, the aim is to build a free online database but still being able to track the image ownership to avoid misconduct. Nevertheless, the analysis and the results presented hereafter can be straightly generalized to other kinds of images.

Within this analysis, two further contributions are presented. First, the consecutive application of more watermarking approaches with different degrees of robustness to distinct transformations is evaluated as a joint watermarking method with the objective of improving the overall robustness to attacks. Finally, unlike other testing setups that only present results for attacks corresponding to specific image transformations, a more practical evaluation where attacks are obtained by combining altogether several geometric and photometric transformations, including also other watermarking methods, is provided. Evaluation code and results are freely available[1].

---

[1] https://github.com/fb82/ancient-coin-watermarking.

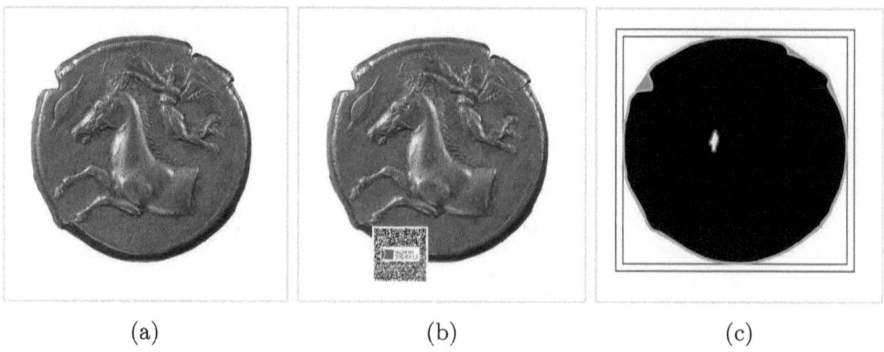

(a)   (b)   (c)

Fig. 1. (a) Example of an original coin image and (b) its database visualization with the overlay of a mark with a QR code. (c) Bounding box computation as described in Sect. 2.2. Best viewed in color and zoomed-in.

2   Evaluation

2.1   Dataset

The dataset is composed of 45 images of ancient coins selected from the numismatic collection of the Archaeological Museum A. Salinas, Italy. The full database collection is freely available for inspection online[2]. An example images is shown in Fig. 1(a). Notice however that in the online images a mark together with a QR code is superimposed on the original image as a rudimentary form of protection, see Fig. 1(b). For the comparative study discussed hereafter the original images with no marks are used. Each image displays the frontal or back view of a coins, typically with a circular or elliptical shape, within a white background and has resolution of 1000×1000 px. Since the original images are compressed by JPEG, in order to not bias the evaluation, these have been converted to PNG with unlossy compression.

2.2   Image Preprocessing

Due to the acquisition process, in many cases the white background occupies a relevant portion of image. Removing the unneeded background area allows to focus on the true content of the image and leads to a more effective watermark protection. To this aim, a practical automatic method to select the coin bounding box by which cropping the image has been designed as shown in Fig. 1(c) and reported hereafter. First, the Otu's method [15] is applied to obtain a raw mask (displayed in black the figure). The widest contour of the background is then identified on the mask (the cyan line, to be compared with the smaller border in magenta) and its convex hull is computed (corresponding to the union of both the gray and black areas) in order to get the bounding box (red rectangle). Within the different steps binary morphological operations [7] are use to suppress noise

---

[2] https://nummidigitali.it.

and enhance the result. The bounding box is slightly extended (green rectangle) to allow an optimal disposition of the image content for the inspection and analysis by archaeologists.

2.3 Watermarking Methods

In the comparison, both hybrid transformation-domain and deep watermarking methods are considered. For the former kind of strategy, three different watermarking methods are included, exploiting respectively DWT+DCT, DCT+SVD and DWT+DCT+SVD as transformation domain where to embed the message. Concerning deep methods, the following encoder-decoder GAN architecture are included: TrustMark [3], Robust Invisible Video watermarking with Attention GAN (RivaGAN) [27], Attention-guided Robust image Watermarking GAN (ARWGAN) [12]. Moreover, StegaStamp [19] and Self-Supervised Latent-spaces WaterMark (SSL WM) [6] are also included, using respectively a base encoder-decoder with no GAN and the deep analogous of the transformation domain. Implementations provided by the respective authors are used, with the exception of DWT+DCT, DWT+DCT+SVD and RivaGAN using [17] and DCT+SVD using [9].

All the above watermarking methods do not have input image resolution constraints, with the exception of StegaStamp requiring 400×400 px images, while images of the dataset after bounding box adjustment can have higher but not less resolution. Considering that coins are centered in the image, the effective solution adopted for StegaStamp is to consider only the 400×400 px center image area in watermark embedding or extraction. As notation $I_W = W(I, m)$ is the embedding of the message m in the image I which generates the watermarked image I_W, and $m = W^{-1}(I_W)$ is the corresponding message retrieval process.

The capacity in terms of bit length of the above methods varies. In order to obtain a fair evaluation the message considered $m = 000110111101001010111010011$ has a capacity $|m| = 30$, which can be embed by all the above methods. The chosen message corresponds to the word "coins" when a block five bits is mapped into the alphabet letters following ordinal sort, i.e. 'a' is $00001_2 = 1_{10}$ and 'z' is $11010_2 = 24_{10}$. Notice also that ratio of the total number of 0 and 1 bits in the message equals to $11/30$ and it is sufficiently well balanced.

The application of a pair of watermarking methods (W_1, W_2) is also considered in the analysis, where $W_1 \neq W_2$. In this case the watermarked images is $I_W = W_2(W_1(I, m), m)$. Notice that the order of application is relevant since the methods do not commute. For these joint watermark methods the message retrieval provides two messages, obtained by using the two methods separately on the watermarked image I_W, i.e. $M = \{m_1, m_2\} = \{W_1^{-1}(I_W), W_2^{-1}(I_W)\}$ where M is the union of the retrieved messages.

2.4 Error Metrics

In order to validate a watermarking method W, the Peak Signal-to-Noise Ratio (PSNR) [18] and the Structural Similarity Index Measure (SSIM) [24] are employed to compare the visual quality of the watermarked image $I_W = W(I)$ with respect to the source image I. Higher values for both $PSNR(I, I_W)$ and $SSIM(I, I_W)$ indicate better adherence of the watermarked image to the original one and less distortion.

For evaluating the robustness of a watermarking method W with respect to specific attack T, the message $m' = W^{-1}(T(W(I, m)))$ is retrieved after attacking the watermarked image $W(I, m)$ using the transformation T. Both the message accuracy rate $A(m, m') = 1 - d(m, m')/|m|$, and the retrieval success rate $S(m, m') = \lfloor E(m, m') \rfloor$ are considered, where $d(m, m')$ is the Hamming distance between the two messages and $\lfloor \cdot \rfloor$ rounds toward zero. While A is better to evaluate the theoretical accuracy of the watermarking, S gives indication of the actual usability of the watermark in practical applications, since is always zero except when m' is exactly m. In case of joint watermarks, the best corresponding score over the set of extracted messages M is taken.

2.5 Attack Methods

Different attack transformations are considered in order to evaluate the robustness of the evaluated watermarking methods; standalone attacks can be roughly divided in geometric and photometric transformations.

Geometric transformations include aspect ratio changes, image scaling, central and generic cropping, vertical and horizontal flipping, mark overlays with different dimensions and positions, and rotations. The transformations have been chosen to simulate reasonable attacks in real scenarios, for a total of about 60 distinct image transformations. Notice that standalone horizontal and vertical flips, as well as rotations by 90°, can be easily detected by testing the message retrieval after having applied the respective inverse transformation on the attacked watermarked image. This strategy works since the latter transformations are reversible as they correspond to a specific rearrangement of the data without any further alterations. Nevertheless, image flips are not evaluated in this way in order to gain further insights of the method robustness, while rotations were limited in the range $[0°, 45°]$.

Photometric transformations include image sharpening, JPEG compression, Gaussian random noise and blur, contrast, brightness and hue changes. Re-watermarking (Re-WM) an already watermarked image with the same or with different methods using a message filled with all zeros is also included as an improper photometric transformation to check the case a malicious user want to modify the image ownership. As for the geometric transformations, different plausible parameter settings were considered, for a total of roughly 40 different photometric transformations.

Finally, in order to better simulate a real attack, the evaluation includes combined attacks obtained by concatenating distinct base transformations altogether. Specifically, after defining the set of the shape transformations as all

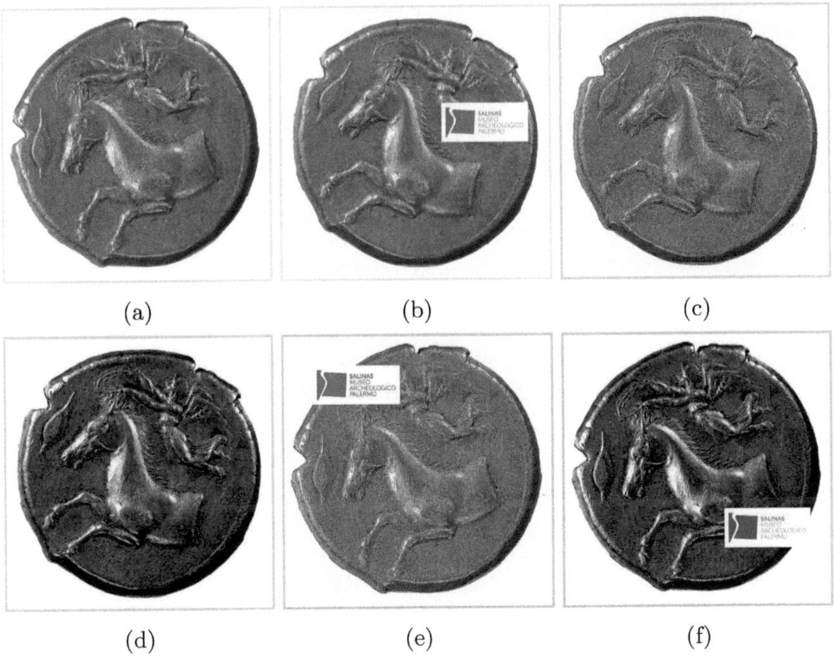

Fig. 2. Standalone geometric (green frames) and photometric (red frames) attack examples: (a) rotation, (b) mark overlay, (c) sharpening and (d) contrast change. Combined (blue frames) attack examples: (e) is obtained by applying in order (a), (b) and (c); the combined attack (f) applies instead (a), (b) and (d). Best viewed in color and zoomed-in. (Color figure online)

the geometric ones with the exception of rotations and mark overlays, the combined attacks are formed joining in order rotations, shape transformations, mark overlays and photometric transformations, for about 1300 attack combinations.

Examples of standalone geometric and photometric attacks and of their combinations are shown in Fig. 2. The Augmentation Library (AugLy) [16] and Torchvision [20] are employed for implementing these attacks excluding Re-WM. Details about the specific transformations can be found in the code.

3 Results

Data Presentation. Table 1 reports the obtained results averaged over the whole dataset, with the exception of the combined attacks averaged on half dataset only, due to the total computational time required. Among joint watermarking methods, only those which achieved a top score in at least one of the geometric or photometric attacks, as retrieval success rate (S) or as message accuracy rate (A), are reported in the bottom part of the table. The detailed results for each specific kind of geometric or photometric attack are reported

Table 1. Evaluation results (%). Top three values are reported in green, blue and red, respectively. Best viewed in color.

			Attack							
			Geometric		Photometric		Standalone		Combined	
Watermark	PSNR	SSIM	A	S	A	S	A	S	A	S
DCT+SVD	38.94	98.27	54.68	6.86	91.21	81.09	76.74	51.69	53.51	4.95
DWT+DCT	41.03	98.77	52.43	3.26	72.14	40.22	64.34	25.58	48.16	1.85
DWT+DVT+SVD	41.22	99.26	49.20	6.34	79.27	61.38	67.37	39.58	44.76	0.92
TrustMark	46.73	99.88	46.13	45.76	83.23	83.15	68.54	68.34	34.01	33.42
RivaGAN	40.52	99.18	95.66	88.20	94.34	86.01	94.86	86.88	87.03	56.52
ARWGAN	39.93	98.78	94.15	85.70	86.90	68.36	89.77	75.22	73.06	38.10
StegaStamp	33.09	97.77	67.05	17.44	93.73	37.63	83.16	29.63	70.63	10.21
SSL WM	42.04	99.18	99.51	91.05	95.11	79.91	96.85	84.32	83.99	28.46
DCT+SVD + RivaGAN	36.73	97.46	96.08	88.37	98.74	91.50	97.69	90.26	90.70	57.94
SSL WM + RivaGAN	38.25	98.37	99.16	91.10	98.51	86.89	98.77	88.56	93.55	57.07
SSL WM + ARWGAN	37.61	97.82	99.67	94.48	96.96	84.98	98.03	88.74	88.81	47.34
TrustMark + SSL WM	40.66	98.96	99.56	92.27	97.32	89.40	98.21	90.54	86.92	43.34
RivaGAN + SSL WM	38.13	98.29	99.88	97.56	99.05	90.89	99.38	93.53	95.20	61.55
RivaGAN + DCT+SVD	35.61	95.98	96.15	87.97	98.74	93.10	97.71	91.07	90.65	58.75
RivaGAN + StegaStamp	32.36	97.02	95.90	86.22	99.32	89.48	97.97	88.19	92.05	53.70
StegaStamp + SSL WM	32.50	97.01	99.50	91.51	98.80	84.90	99.08	87.52	88.86	33.53
StegaStamp + TrustMark	32.87	97.66	78.60	52.33	99.17	90.05	91.02	75.11	78.41	38.74
StegaStamp + RivaGAN	32.37	97.03	96.10	88.49	99.42	91.35	98.11	90.21	92.77	59.50
ARWGAN + SSL WM	37.76	97.99	99.83	96.86	96.96	83.07	98.10	88.53	88.89	40.88

as additional material[3] due to paper length constraints. Top three scores are colored, considering distinctly the base and paired watermarking.

System Setup and Running Times. The system employed for the evaluation is an Intel Core i9-10900K CPU with 64 GB of RAM equipped with a NVIDIA GeForce RTX 2080 Ti, running Ubuntu 24.04. In any case, watermark embedding and extraction times are within 2 s, which is acceptable for the kind of application.

Watermarking Visual Quality. Concerning the visual quality of the watermarked images, according to the PSNR and SSIM scores the results are acceptable for any method. Figure 3 shows the example image of Fig. 1a after watermark embedding with RivaGAN + SSL WW and StegaStamp, together with the respective error maps of the maximum absolute difference between the intensity pixel over all the three color channels. RivaGAN + SSL WM provides overall the best attack robustness, while StegaStamp, in pair or standalone, has the

---

[3] https://drive.google.com/file/d/1cqvOXblB6D97ZJoLem8PP4y5hp0iTuQH/view?usp=drive_link.

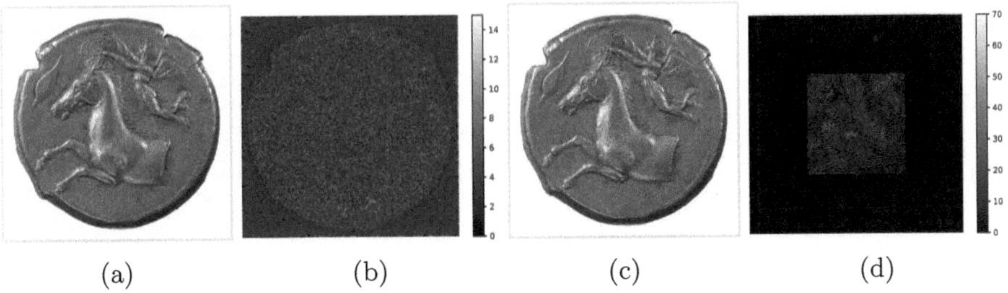

(a)   (b)   (c)   (d)

Fig. 3. (a) Watermarking of the image in Fig. 1(a) using RivaGAN + SSL WM and (b) corresponding heat map, showing the maximum absolute pixel intensity error among the color channels with respect to the original image. (c) Analogous watermarked image using StegaStamp and (d) corresponding heat map.

lowest visual scores. Nevertheless, even for StegaStamp the visual quality is still acceptable for the specific task. Notice also that StegaStamp watermark is limited to central area only due to its constraints in the input image size as discussed previously.

General Watermarking Robustness. Overall, deep-based watermark approaches provide better robustness to attacks, both in terms of message accuracy and success rate. RivaGAN and SSL WM are the best base methods, followed by ARWGAN. In particular, SSL WM is slightly more robust than Riva-GAN on standalone geometric attacks and the vice-versa holds on the photometric transformations. Nevertheless, RivaGAN is more stable on combined attacks with a message accuracy and success rate of about 85% and 55%, respectively, whereas the respective values for SSL WM are around 80% and 30%. Although the message accuracies are similar, the success rate differences are relevant. This indicates that when not succeeding, RivaGAN messages are more corrupted, while SSL WM tends in general to have less bit errors in the message, but still some happen. Since more bits than those used could be employed for the embedded message by SSL WM, it would be interesting to consider in future evaluations the introduction of Error Correction Codes (ECC) [13] in the message, not taken into account in the current evaluation.

Joint Watermarking Robustness. The joint combination of the base best methods RivaGAN + SSL WM attains in general the overall best results or values close to these, leading in most case to a boost of about a 5% with respect to maximum between each of the approaches used to get the combination, suggesting a partial complementarity between RivaGAN and SSL WM. Notice that SSL WM + RivaGAN has instead lower attack scores, confirming the non-commutativity of the operations. SSL WM has the best visual score while RivaGAN not, i.e. RivaGAN alters much more the image data. For these reasons applying Riva-GAN after SSL WM can probably delete SSL WM watermark.

Robustness to Geometric Attacks. Transformation-domain watermarking approaches are almost unable to resist to geometric attacks, with message accuracy about 50% and success rate around 5%. For the photometric attacks the corresponding message accuracy and success rate are around 80% and 50%, respectively. The previous analyses are also supported by the detailed comparative results for each kind of base attack reported in the additional material.

Still about the geometric attacks, it turns out by inspecting the additional material that vertical and horizontal flips are neither handled by deep-based watermarking methods with the exception of SSL WM, as this kind of attack was not included in the training. As discussed above, this is not a real issues since trying the message extraction on both the watermarked image and its flipped version would resolve this problem. Notice also that rotations are only successfully handled by SSL WM.

Robustness to Photometric Attacks. Concerning the robustness of the methods to the specific photometric attacks as detailed in the additional material, Gaussian noise addition and hue changes are the most difficult attacks to handle even for deep-based methods, since these transformations were not included in the training. Furthermore, Gaussian blur is perfectly handled only by TrustMark, which for the same reason achieves the top score for scale changes. It is also interesting to note that re-watermarking to remove the original message is well addressed in order by RivaGAN, SSL WM and TrustMark. Moreover, among the different methods, RivaGAN is able to retain the hidden message even if the image is re-watermarked by itself in most cases, while SSL WM in all cases with the exception of when the re-watermarking method is itself.

Robustness to Combined Attacks. Finally, combined transformations are the most challenging, leading to a message accuracy and a success rate in the best case of RivaGAN + SSL WM of 95% and 61%, respectively. Any base of combination of the watermarking methods failed for roughly 100 among the 1300 combined attacks proposed. This happens in general when all the base attacks used to compose the combination are really aggressive. For instance, RivaGAN + SSL WM is able to deal with all the base attacks shown in Figs. 2(a)–(d), but not with their combinations show in Figs. 2(e)–(f), or when JPEG quality was set to 20 or 50. More details can be found in the code. Nevertheless, it is quite reasonable that strong manipulations like these are unlikely to appear for the specific task related to digital collections of ancient coins.

4   Conclusions and Future Work

This paper has presented a comparative evaluation of watermarking methods applied to digital collections of ancient coin images. The experimental analysis shows that deep-learning watermark approaches are quite robust to both common geometric and photometric attacks, while standard transformation-domain approaches are almost ineffective against geometric transformations. Moreover, the joint combination of different and somewhat complimentary watermarking methods can further improve the results.

While being specify tailored for ancient coin images, overall this analysis is quite general and could be of interest for the whole research community. The evaluation suggests that watermarking can provide a valid mechanism to track the ownership of the original images for the specific task, yet for complex attack combinations a complete solution is still an open problem.

Future work will extend the analysis taking into account ECC in message to improve the robustness, including wider and different image datasets as well as the refinement of deep models by adding more base transformations in the training.

Acknowledgments. This work was supported by the European Union – NextGenerationEU – MUR funds D.M. 737/2021 âĂŞ research project "NUmmi Digitali towards Europe"; CUP B79J21038330001 (Eurostart 2024, University of Palermo). L. Sole conceived the NUMMI DIGITALI project and wrote Sect. 1.1 as the scientific coordination; the rest of the manuscript was written by F. Bellavia who also devised and developed the discussed analysis.

References

1. Aberna, P., Agilandeeswari, L.: Digital image and video watermarking: methodologies, attacks, applications, and future directions. Multimed. Tools . Appl. **83**(2) (2024)
2. Begum, M., Uddin, M.S.: Digital image watermarking techniques: a review. Information **11**(2) (2020)
3. Bui, T., Agarwal, S., Collomosse, J.: Trustmark: universal watermarking for arbitrary resolution images. arXiv e-prints (2023). https://github.com/adobe/trustmark
4. Caron, M., et al.: Emerging properties in self-supervised vision transformers. In: Proceedings of the International Conference on Computer Vision (ICCV) (2021)
5. Cox, I.J., Miller, M.L., Bloom, J.A., Fridrich, J., Kalker, T.: Digital Watermarking and Steganography: Fundamentals and Techniques, 2nd edn. Morgan Kaufmann Publishers (2008)
6. Fernandez, P., Sablayrolles, A., Furon, T., Jégou, H., Douze, M.: Watermarking images in self-supervised latent spaces. In: Proceedings of the IEEE International Conference on Acoustics, Speech and Signal Processing (ICASSP) (2022). https://github.com/facebookresearch/ssl_watermarking
7. Gonzales, R., Woods, R.E.: Digital Image Processing, 4th edn. Pearson College Division (2017)
8. Goodfellow, I., et al.: Generative adversarial networks. Commun. ACM **63**(11) (2020)
9. guofei9987: blind-watermark. https://github.com/guofei9987/blind_watermark
10. Harvey, P.: Exiftool. https://exiftool.org/
11. Hosny, K., Magdi, A., ElKomy, O., Hamza, H.: Digital image watermarking using deep learning: a survey. Comput. Sci. Rev. **53** (2024)

12. Huang, J., Luo, T., Li, L., Yang, G., Xu, H., Chang, C.: ARWGAN: attention-guided robust image watermarking model based on GAN. IEEE Trans. Instrum. Meas. **72** (2023). https://github.com/river-huang/ARWGAN
13. MacKay, D.: Information Theory, Inference, and Learning Algorithms. Cambridge University Press (2003)
14. Mourya, R.: Watermark-removal-pytorch. https://github.com/braindotai/Watermark-Removal-Pytorch
15. Otsu, N.: A threshold selection method from gray-level histograms. IEEE Trans. Syst. Man Cybern. **9**(1) (1979)
16. Papakipos, Z., Bitton, J.: Augly: data augmentations for robustness (2022). https://github.com/facebookresearch/AugLy
17. Shield Mountain: invisible-watermark. https://github.com/ShieldMnt/invisible-watermark
18. Szeliski, R.: Computer Vision: Algorithms and Applications, 2nd edn. Springer, Cham (2022)
19. Tancik, M., Mildenhall, B., Ng, R.: StegaStamp: invisible hyperlinks in physical photographs. In: Proceedings of the IEEE Conference on Computer Vision and Pattern Recognition (CVPR) (2020). https://github.com/tancik/StegaStamp
20. TorchVision maintainers and contributors: torchvision: Pytorch's computer vision library. https://github.com/pytorch/vision
21. Ulyanov, D., Vedaldi, A., Lempitsky, V.: Deep image prior. In: Proceedings of the IEEE Conference on Computer Vision and Pattern Recognition (CVPR) (2018)
22. Vaswani, A., et al.: Attention is all you need. In: Proceedings of the International Conference on Neural Information Processing Systems (NeurIPS) (2017)
23. Wan, W., Wang, J., Zhang, Y., Li, J., Yu, H., Sun, J.: A comprehensive survey on robust image watermarking. Neurocomputing **488**, 226–247 (2022)
24. Wang, Z., Bovik, A., Sheikh, H., Simoncelli, E.: Image quality assessment: from error visibility to structural similarity. IEEE Trans. Image Process. **13**(4) (2004)
25. Wang, Z., et al.: Data hiding with deep learning: a survey unifying digital watermarking and steganography. IEEE Trans. Comput. Soc. Syst. **10**(6) (2023)
26. Wang, Z., Chen, J., Hoi, S.: Deep learning for image super-resolution: a survey. IEEE Trans. Pattern Anal. Mach. Intell. **43**(10) (2021)
27. Zhang, K.A., Xu, L., Cuesta-Infante, A., Veeramachaneni, K.: Robust invisible video watermarking with attention (2019)
28. Zhu, J., Kaplan, R., Johnson, J., Fei-Fei, L.: HiDDeN: hiding data with deep networks. In: Ferrari, V., Hebert, M., Sminchisescu, C., Weiss, Y. (eds.) ECCV 2018. LNCS, vol. 11219, pp. 682–697. Springer, Cham (2018). https://doi.org/10.1007/978-3-030-01267-0_40

The Evolution of Light in Caravaggio's Oeuvre

Pepe Ballesteros Zapata[1]([✉]) and David G. Stork[2]

[1] Max Planck Society, University of Zurich, Zurich, Switzerland
jose.ballesteroszapata@uzh.ch
[2] Adjunct Professor, Stanford University, Stanford, USA
dstork@stanford.edu

Abstract. Caravaggio's naturalism marks a turning point in early modern European painting, challenging canonical representations and conventional workshop methods. Early bibliographers described his unprecedented imitation of nature, using staged models and dramatic lighting setups. Yet, the precise nature of his studio practice remains a subject of ongoing scholarly debate. In this study, we propose a physics-based computational approach to analyze lighting coherence across his oeuvre, offering new empirical insight into the technical foundations of his naturalism. We use the occluding contour algorithm, grounded in Lambert's cosine law, to 307 valid contours extracted from 68 paintings attributed to Caravaggio, enabling sub-pixel estimation of 2D light direction and consistency across his oeuvre. Our analysis reveals a gradual but statistically significant increase in internal light consistency over time, suggesting a shift toward more controlled lighting environments.

Keywords: Light Analysis · Caravaggio · Digital Art History

1 Introduction

At the turn of the seventeenth century, Michelangelo Merisi da Caravaggio (1571–1610) introduced an unprecedented visual imagery, grounded in bold naturalism, that broke with all the standards set by the early modern Italian canon. Contemporary bibliographers describe Caravaggio as rejecting preparatory drawings and idealized forms in favor of painting directly from life, guided only by nature [13]. They imagined his pronounced chiaroscuro and bold reliefs as a product of staging live models in tightly controlled lighting setups. These literary sources have shaped modern interpretations of Caravaggio's workshop methods [15,17], to the point that some scholars speculate he employed optical projection devices to trace scenes directly onto the canvas [4,8]. Other scholars question Caravaggio's fidelity to nature, claiming that his striking naturalism was a deliberate stylistic construction to distinguish his practice from others [5]. Technical studies on Caravaggio's oeuvre exposed how his naturalist approach was engineered: from using incised compositional marks to frequent compositional reworkings

E. Rodolà et al. (Eds.): ICIAP 2025 Workshops, LNCS 16169, pp. 557–568, 2026.
https://doi.org/10.1007/978-3-032-11317-7_45

558   P. Ballesteros Zapata and D. G. Stork

[2]. Understanding Caravaggio's studio praxis remains a central concern in both technical art history and Caravaggio scholarship, as his practice marks a turning point in the visual culture of early modern Europe.

A growing body of work has applied computational methods to the study of Caravaggio's oeuvre and lighting technique. In [16], authors introduced a Bayesian framework to quantify the degree of chiaroscuro in paintings. Other efforts have focused on reconstructing his studio lighting practices through image-based inference. In [18], authors estimate the position of the primary illuminant in *The Calling of Saint Matthew* by analyzing the brightness distribution on the rear wall. This hypothesis was subsequently tested using a computer graphics simulation of the same scene [19]. A separate 3D model was constructed for *The Supper at Emmaus* to assess the optical coherence of the scene and to challenge the claim that Caravaggio may have used optical projection devices to trace images onto the canvas [23]; similar claims were further refuted in related studies [20,21]. Recent studies have leveraged spherical harmonics to extract light environments from paintings at scale using faces as light probes [24].

Fig. 1. Visualization of our occluding contour extraction and light direction estimation on Caravaggio's *Judith Beheading Holofernes* (c. 1598), Galleria Nazionale d'Arte Antica, Rome. Each contour is overlaid in a distinct color, with an arrow at the midpoint of its first segment indicating the estimated illumination vector. The translucent wedges, spanning $\pm\sigma$ around each arrow, encode the circular standard deviation of that contour's estimate.

Building on these foundations, our work adopts and extends the occluding contour algorithm grounded in Lambert's cosine law and used for both forensic [9,10] and digital-art historical applications [22]—to extract 2D light directions with sub-angular-degree precision across Caravaggio's oeuvre. Additionally, we quantify internal lighting consistency on each painting to explore the evolution of directional variance over time. Figure 1 shows an example of our analyses applied to *Judith Beheading Holofernes*. We trace this evolution quantitatively aiming to contribute to the art historical understanding of Caravaggio's technique at scale. Leveraging physics-based algorithms, this research hopes to shed light on the puzzle of Caravaggio's naturalism—positioned as a gray area in the spectrum between reality and artifice.

2   Methodology

Our method is based on extracting lighting information from the digital evidence found in paintings. Specifically, we use the occluding contour algorithm to estimate the direction of illumination projected on the image plane.

2.1   Estimating the Direction of the Light Source

Lambert's cosine law describes the intensity of light reaching a surface as a factor of the cosine of the angle between the incident light direction and the normal of a surface. The intensity of the light is maximized when the normal vector of the surface and the light direction are parallel, and minimized when they are perpendicular. Assuming that the properties of the diffusely reflecting materials under consideration follow a Lambertian function, the intensity I at a point illuminated from a single light source is:

$$I(x, y, z) = k(\vec{N}(x, y, z) \cdot \vec{L}(x, y, z)) + A. \tag{1}$$

In this expression, the cosine of the angle is calculated as the dot product between the normalized normal and light direction vectors in three-dimensional coordinates. The parameter k is the product of the intensity of the light source and the surface albedo and A represents the contribution of the uniform ambient illumination. Estimating the light direction would require the object's 3-D shape, but authors in [14] proposed a clever solution to bypass that need and estimate the light direction directly from the image using only points on occluding contours.

The occluding contour is defined as the silhouette curve consisting of surface points whose normals are orthogonal to the viewing ray. Hence the normal vectors along the contour of a surface have a z-component equal to zero, meaning that the normal vector is contained inside the image plane.

From the digital image, we extract the two-dimensional normal vectors $\vec{n}$ along the occluding contour, as well as the luminance values I at each corresponding point—steps explained in Sects. 2.2 and 2.3, respectively. We assume luminance is a good approximation of physical light intensity when the condition of constant hue is fulfilled. When the color of a surface is constant, variations in brightness reflect changes in light intensity rather than differences in surface reflectance. Moreover, for the purpose of estimating the light direction, the intensity of the light source is irrelevant: any change in absolute brightness alters image radiance but leaves the estimated orientation of $\vec{L}$ unchanged. For this reason, we ignore the parameter k.

Following the implementation of previous works [9, 18, 20–22, 24] we calculate the light source direction $\vec{L}$ using least squares optimization. Thus we try to find the set of L_x, L_y light direction coordinates and ambient term A that minimize the distance $E(\vec{L}, A)$ between the Lambert cosine law and the image intensity values I along the occluding contour of a surface:

$$
E(\vec{L}, A) = \left\| M \begin{pmatrix} L_x \\ L_y \\ A \end{pmatrix} - \begin{pmatrix} I(x_1, y_1) \\ I(x_2, y_2) \\ \vdots \\ I(x_p, y_p) \end{pmatrix} \right\|_2^2 .
\tag{2}
$$

Here, $M \in \mathbb{R}^{p \times 3}$ is the design matrix that contains the geometry information of the sampled occluding contour:

$$
M = \begin{pmatrix} n_x(x_1, y_1) & n_y(x_1, y_1) & 1 \\ n_x(x_2, y_2) & n_y(x_2, y_2) & 1 \\ \vdots & \vdots & \vdots \\ n_x(x_p, y_p) & n_y(x_p, y_p) & 1 \end{pmatrix} ,
\tag{3}
$$

where $\mathbf{n}(x_i, y_i)$ is the outward unit normal on the ith occluding contour point and the third column of ones enables the fit to absorb the uniform ambient term A.

The least squares method provides a linear and convex minimization problem to find the light direction from Lambert's cosine law. However, lighting behavior, even in Lambertian scenarios, is non-linear. Pixels that face away from the light source should contribute zero, as there is no such thing as a negative light. The re-parametrization of the image intensity would thus take the form:

$$
I = \max\left(0, \vec{N} \cdot \vec{L}\right) + A.
\tag{4}
$$

This non-linearity breaks the convexity of the cost function, typically leading to several local minima. Previous works used the linear approximation and have proven to work well enough for forensic applications [9, 10], and painting analysis [18, 20–22, 24]. However, bypassing the non-linearity present in reality can lead to biased estimations. This can only happen if one or more normal vectors in the occluding contour are facing more than 90° from the true light direction. Any point whose normal makes an obtuse angle with the true light direction produces a negative dot product, leading to a higher MSE error that pulls the least squares solution away from the optimum. In [10], authors show how the sensitivity to noise of the estimation decreases as the normal vector span of the occluding contour increases. This metric directly measures the sensitivity to perturbations in M. Note that M can be robust to noise, yet the final estimation can still be biased. In other words, the least squares solution can be numerically stable yet yield the wrong direction. To avoid such bias, one must be careful not to include normal vectors that face away from the light source, that is, those for which $\vec{n} \cdot \vec{L} < 0$. On the other hand, if the normals span only a narrow angular interval, the matrix M becomes ill-conditioned and multiple light directions yield almost identical residuals. We bypass this limitation by pooling several contour segments of the same object into a single least-squares system, thereby broadening the angular coverage and stabilizing the estimate.

As summary, under Lambertian conditions we consider a valid occluding contour one that fulfills the following requirements:

- The reflectivity properties of the surface are Lambertian.
- Constant hue and reflectivity.
- Contours are isolated and convex toward the light source, avoiding interreflections or cast shadows from nearby surfaces.
- Each contour point lies within $\pm 90°$ of the ground-truth light source direction.
- The sampled intensities span both rising and falling flanks of the Lambertian cosine.

The global uncertainty of the light direction estimation of a painting is calculated as a circular statistics summary of the estimated directions per contour. Specifically, we compute the mean resultant length R of all per-contour normalized light direction vectors and derive the circular standard deviation in radians as $\sigma_{\mathrm{rad}} = \sqrt{-2 \ln R}$. Because every contour equally contributes a direction cue, this σ value directly reflects the internal consistency of the painting's lighting signature: low σ means strong agreement (tight clustering) of estimates; high σ flags ambiguity or lighting collage.

2.2 Sub-pixel Occluding Contour Extraction

The quality of the light direction estimation depends directly on the quality of the extraction of accurate normal vectors and luminance intensities along the contour. The extraction of the occluding contour is semi-automatic. It starts by the user extracting a coarse contour by manually selecting points along the target contour. Images need to have enough resolution (720 px or more) to perform the extraction of contour points. The pipeline then comprises two successive stages: the estimation of outward normals and the subpixel relocation of each vertex to the physically correct edge. Both steps are fully automated after the user's selection.

(1) **Estimation of outward normals.** The clicked vertices are interpolated with a cubic spline controlled by a single smoothing parameter. Oversampling the spline produces a smooth curve while preserving the overall course

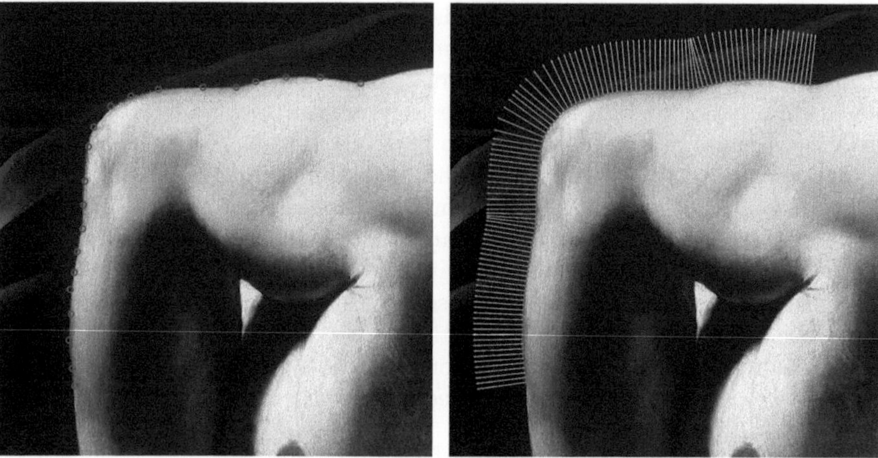

Fig. 2. Example of contour extraction on Holofernes' right elbow in *Judith Beheading Holofernes* (c. 1598), Galleria Nazionale d'Arte Antica, Rome. On the left, the preliminary coarse contour points (in red) extracted manually. On the right, the automatically refined sub-pixel contour points (in red) along with their estimated surface normals (in blue).

drawn by the annotator. To avoid oversampling artefacts, we collapse all over-sampled vertices belonging to the same pixel to a single representative. For every vertex we compute the tangent and rotate it by 90° to obtain the unit normal. The result is one coordinate–normal pair per occupied pixel, yielding a uniform sampling density along the contour.

(2) **Subpixel relocation.** Once the spline stage has supplied coarse contour points and outward unit normals we refine every point to the true occluding boundary following the Gaussian-edge methodology introduced in [6]. We compute the Sobel gradient map of the image to sample values for each contour point at 0.25-pixel intervals along each normal, and fit a Gaussian function given by:

$$f(x) = Be^{-\frac{(x-\mu)^2}{2\sigma^2}}, \tag{5}$$

where $f(x)$ denotes gradient value at x, the blur of the edge is given by σ, B corresponds to the maximum gradient value at the edge, and μ determines the sub-pixel offset between the current contour point position and the true location of the edge. The fitting is performed for each contour point, using a derivative-free optimisation method (Nelder–Mead) [11], which adjusts the Gaussian parameters to best match the discrete gradient samples along the normal direction. Once the optimal parameters are found, we relocate the contour point along its normal by the fitted offset μ, locating the occluding contour coordinate of the blurred edge with sub-pixel precision.

2.3   Sub-pixel Luminance Extraction

Due to image rasterization, anti-aliasing, and blur, sampling luminance values at the nearest integer coordinate introduces quantisation artefacts, misrepresenting the actual luminance along the edge. To overcome this source of error, we employ an extrapolation strategy that estimates the luminance value at the refined contour point by sampling luminance values along the inward normal direction. We collect intensity values at 0.25 sub-pixel intervals up to two pixels into the object. This sub-pixel sampling captures the sharp luminance transition near the occluding boundary without extending too far inward, where surface curvature introduces luminance changes.

To extrapolate the true luminance at the contour, we fit the sampled luminance profile to the standard blurred-edge function $f(s)$, shown in Eq. 6 below. Fitting this function allows us to model the background-object transition following the Approximation with Erf Function (AEF) introduced in [7]. We estimate two parameters: the overall contrast ΔI (height of the luminance transition) and the inner asymptote of the profile I_{in}, which represents the value toward which luminance converges just inside the occluding contour.

$$f(s) \;=\; I_{\mathrm{in}} \;-\; \frac{1}{2}\,\Delta I\,\left[1 \;+\; \mathrm{erf}\!\left(\tfrac{s}{\sqrt{2}\,\sigma}\right)\right]. \tag{6}$$

In this expression, $erf(x)$ represents the standard error function, s is the signed distance along the inward normal (positive toward the interior), and σ is a fixed blur parameter chosen based on typical image defocus (e.g., $\sigma = 0.6$). Thus we obtain the true luminance value I_{in} by de-convolving the blurred pixel measurements at the sub-pixel contour location and extrapolating to the underlying edge discontinuity.

3   Results

3.1   Data Acquisition

The oeuvre of Caravaggio is still a matter of scholarly debate, as is the case for many other early modern masters. This is well illustrated by the recent attribution of an 'Ecce Homo' discovered in an auction house in Madrid [3]. Caravaggio's work represents a unique art historical challenge, as his oeuvre was widely copied throughout seventeenth-century Europe. To conduct a large-scale analysis of Caravaggio's work, we restricted our selection to works with established attributions, as validated by authoritative Caravaggesque scholars [15,17]. Our dataset is based on the Web Gallery of Art (WGA), an online repository of European fine art, from which we curated 82 entries attributed to Caravaggio. We filter images of details, verso views, and pictures of the paintings in situ. The metadata was modified according to [15], concretely the painting type (e.g., portraiture, religious, etc.), and dates of creation. In cases where multiple versions of the same composition exist, we retained only those with widely accepted authorship. This includes, for instance, the two known versions of *Boy Bitten by a Lizard* and *The Lute Player*.

Our selection protocol yielded a total of 68 paintings covering all his active period (c. 1592–1610). Our set corresponds to 3.6 paintings per year; the busiest years being 1596 and 1608, each with 7 analyzed works, while the quietest years still contribute 1–2 paintings (e.g., 1592 has 1, 1610 has 2). While not a fully complete catalogue of Caravaggio's oeuvre, we consider this dataset a representative sample of the old master's work to analyse trends over time.

For each painting in the dataset, contours were extracted using the semi-automatic procedure detailed in Sect. 2. Initial selection was performed manually, retaining only those that met predefined validity criteria. A subsequent automated filtering step assessed each contour's normal vector span and luminance distribution to ensure compliance with the established requirements for valid contours. The final dataset includes 307 contours, each contributing a distinct light direction estimate, and comprising a total of 1,194 contour segments. No paintings were discarded due to limited amount of valid contours available. On average, 4.51 contours and 17.56 contour segments were obtained per painting.

3.2 Analyzing Caravaggio's Light at Scale

This case study focused on the use of the occluding-contour algorithm to illustrate Caravaggio's light rendering throughout his career. Contours were grouped into three semantically meaningful categories: person (183 contours), drapery (78 contours), and still-life (46 contours). The latter contains all contours belonging to every-day objects like fruits, skulls, instruments, etc. Such diversity likely contributes to the highest observed uncertainty mean ($\mu_{std} = 12.87$). Drapery contours exhibit a lower average uncertainty ($\mu_{std} = 10.79$) yet higher than that of person contours, which show the most consistent estimates overall ($\mu_{std} = 8.55$). Skin surfaces are typically modeled with more abrupt luminance transitions, providing more stable cues for light direction estimation. In contrast, drapery is often more uniformly lit, and subject to local geometric irregularities (e.g., folds), which increases directional ambiguity. By applying K-means algorithm to the global light direction estimates, we uncovered two dominant lighting modes with cluster centers at 149.75° (top-left illumination) and 41.6° (top-right illumination). The majority of Caravaggio's works fall into the 149.75° cluster, while only seven paintings are lit from the top-right.

From here we ask: how does the congruence across the estimations on a painting change over time? To answer this question, we compute the global standard deviation σ_{global} for each painting and plot it against time (see Fig. 3). The figure illustrates that the uncertainty in the overall estimated light direction decreases over time, in other words, the congruence between light directions in the same painting increases. The inverse linear dependency between σ_{global} and time is described by a moderate r value ($r = -0.41$), and a $p - value$ of 0.001 indicating that while there is not a strong correlation, there is a statistically significant linear decrease in σ_{global} over time.

Caravaggio's mid-to-later period features an increase of the number of people depicted: whereas his paintings from 1592–1600 average 2.29 figures each, those

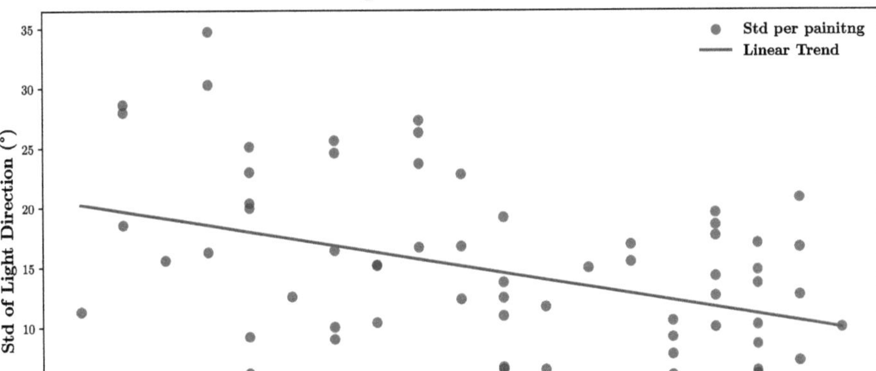

Fig. 3. Global light direction standard deviation σ_{global} throughout Caravaggio's active painting years. Blue points encode σ_{global} for each of the 68 analyzed paintings. The red line shows the linear regression fit over the data.

from 1601–1610 average 4.23 figures, an increase of over 85%. As a likely consequence, a parallel trend discernible in his oeuvre is the move toward larger canvas sizes. While Caravaggio painted simple compositions every year, he started painting more complex scenes as he aged. This translates into a higher number of extracted contours over time: on average, 14.9 contour segments in his early works (between 1592–1600) and 18.5 in later works (between 1601–1610), a 24.2% increase. Adding more occluding contours raises the chance of detecting inconsistencies—an effect that would normally broaden the standard deviation of the estimated light directions. Yet the result we found is the opposite: early works show wider scatter in light direction estimations while later works show tighter clustering.

Caravaggio's bibliographer and art collector Giulio Mancini wrote about the challenges of working from life models in larger and complex compositions, noting that painting from a model can produce good results for pictures of single figures, but it is impractical for multitudes [12]. It is well documented from early bibliographers [1] and modern scholarship [15,17] that Caravaggio worked from life in numerous instances, using real models. Many of his models came from Caravaggio's social circle: shop keepers, manual laborers, prostitutes, etc. Painting from a live model might promise a stable direction of illumination, yet Caravaggio's early one-figure canvases reveal the opposite. The direction of the sunlight rays changes over time and on multiple different sittings of the model. Contours of drapery or human skin are subject to be painted at different times and to future refinements during the painting process. Even if Caravaggio was emulating the light before his eyes, each intervention would capture illumination present at different moments, leading to heterogeneous lighting, often times invisible to the naked eye.

Our pipeline shows that while light consistency between estimations do not depend on real-life modeling, it does vary over time. What Fig. 3 reveals is that light estimations find more agreement where art history scholars told us Caravaggio was intensifying his lighting rhetoric [15,17]. Stronger chiaroscuro in the later period could potentially make the estimations more stable, but this does not explain why congruency between estimation increases: estimations could be more certain but different in the same painting. Are these results directly suggesting that Caravaggio was able to depict light more accurately as he grew older?

One possible explanation is that increased light congruency reflects artistic practice. Through experience, Caravaggio might have become more adept at rendering coherent illumination across objects, perhaps guided by a consistent reference direction while painting. A more plausible explanation is that lighting consistency stems from a shift towards controlled light environments in his studio practice, for example, with the aid of artificial sources such as candles [15]. Artificial illumination would not only account for the intensified chiaroscuro of his mature works but also enable greater control over light direction compared to sunlight illumination from a window. We see no lighting evidence that supports the speculation that Caravaggio employed projection optics as part of his praxis, for example a dramatic and rapid change in lighting consistency, which might have arisen had he switched from no optics to optics. Nonetheless, such interpretations should be treated as preliminary hypothesis that require further research by art history scholars specializing in Caravaggio's workshop practices.

4   Conclusions

Caravaggio's pictorial revolution at the turn of the seventeenth century continues to provoke debate among art historians interested in his workshop practices. This study leverages computational light analysis, grounded in the occluding contour algorithm, to quantify directional lighting coherence across his oeuvre. We present our results without presuming that the painter aimed at optical accuracy—only Caravaggio knew how faithful he wished to be to nature. Regardless of Caravaggio's intentions, we reveal a statistically significant tendency towards lighting consistency. Such evolution in studio praxis likely reflects either the refinement of Caravaggio's technical control or a deliberate shift towards controlled lighting setups with the aid of artificial sources. We hope that algorithmic refinement will continue to shed light on the mysteries hidden behind the work of Caravaggio, and his network of influence in seventeenth century western painting.

Disclosure of Interests. The authors have no competing interests to declare that are relevant to the content of this article.

References

1. Bellori, G.P.: The Lives of the Modern Painters, Sculptors and Architects: A New Translation and, Critical Cambridge University Press, New York and Cambridge (2005)
2. Cardinale, M.: Caravaggio's Painting Technique: A Technical Art History. Technical report, Technical Art History (2007)
3. Christiansen, K., Papi, G., Porzio, G., Terzaghi, M.C. (eds.): Caravaggio: The *Ecce Homo* Unveiled. Marsilio Arte, Venezia, Italy, 1st edn. (2023)
4. di Cluny, C.S.M.P. (ed.): Caravaggio e la scienza della luce. Il Prato, Rome, Italy (2010)
5. Ebert-Schifferer, S., Roccasecca, P., Thielemann, A. (eds.): Lumen, imago, pictura: La luce nella storia dell'ottica e nella rappresentazione visiva da Giotto a Caravaggio. De Luca Editori d'Arte, Roma (2018)
6. Fabijańska, A.: Gaussian-Based Approach to Subpixel Detection of Blurred and Unsharp Edges, pp. 641–650 (2014)
7. Hagara, M., Kulla, P.: Edge Detection with Sub-pixel Accuracy Based on Approximation of Edge with Erf Function **20**(2) (2011)
8. Hockney, D.: Secret Knowledge: Rediscovering the Lost Techniques of the Old Masters. New York, illustrated edn, Viking Studio (2001)
9. Johnson, M.K., Farid, H.: Exposing digital forgeries by detecting inconsistencies in lighting. In: Proceedings of the 7th Workshop on Multimedia and Security, pp. 1–10. ACM, New York NY USA (2005). https://doi.org/10.1145/1073170.1073171
10. Johnson, M.K., Farid, H.: Exposing digital forgeries in complex lighting environments. IEEE Trans. Inf. Forensics Secur. **2**(3), 450–461 (Sep2007)
11. Lagarias, J.C., Reeds, J.A., Wright, M.H., Wright, P.E.: Convergence properties of the Nelder-mead simplex method in low dimensions. SIAM J. Optim. **9**(1), 112–147 (1998)
12. Mancini, G.: Considerazioni sulla pittura. Art J. **22**(4), 333–335 (1962). https://doi.org/10.1080/00043079.1962.10789034
13. Mancini, G., Baglione, G., Bellori, G.P. (eds.): Lives of Caravaggio: Three Early Biographies. Lives of the Artists, J. Paul Getty Museum, Los Angeles (2019)
14. Nillius, P., Eklundh, J.: Automatic estimation of the projected light source direction. In: Proceedings of the 2001 IEEE Computer Society Conference on Computer Vision and Pattern Recognition. CVPR 2001. vol. 1, pp. I–1076–I–1083. IEEE Comput. Soc, Kauai, HI, USA (2001). https://doi.org/10.1109/CVPR.2001.990650
15. Puglisi, C.: Caravaggio. Phaidon Press, London (1998). paperback ed. published Jan. 2000, 448 pp., 218 ill
16. Yang, S., Cheung, G., Le Callet, P., Liu, J., Guo, Z.: Computational modeling of artistic intention: quantify lighting surprise for painting analysis. In: 2016 Eighth International Conference on Quality of Multimedia Experience (QoMEX), pp. 1–6. IEEE, Lisbon, Portugal (2016). https://doi.org/10.1109/QoMEX.2016.7498932, http://ieeexplore.ieee.org/document/7498932/
17. Spike, J.T.: Caravaggio: Catalogue of Paintings. Abbeville Press, New York and Paris (2001)
18. Stork, D.G.: Locating illumination sources from lighting on planar surfaces in paintings: An application to Georges de la Tour and Caravaggio (abstract). In: Optical Society of American Annual Meeting. Rochester, NY (2008)

19. Stork, D.G.: New insights into Caravaggio's studio methods: Revelations from computer vision and computer graphics modeling (abstract). Renaissance Society of America Annual Meeting (2009)
20. Stork, D.G.: Did Caravaggio employ optical projections? An image analysis of parity in the artist's paintings. In: Stork, D.G., Coddington, J., Bentkowska-Kafel, A. (eds.) Electronic Imaging: Computer vision and image analysis of art II. vol. 7869, pp. 7869OJ–1–15 (2011)
21. Stork, D.G.: Did Caravaggio employ optical projections? An image analysis of the parity in the artist's paintings. In: Stork, D.G., Coddington, J., Bentkowska-Kafel, A. (eds.) Computer image analysis in the study of art II, vol. 7869, pp. 78690J–1–15. SPIE/IS&T, Bellingham, WA (2011)
22. Stork, D.G.: Pixels & paintings: Foundations of computer-assisted connoisseurship. Wiley, Hoboken, NJ (2024)
23. Stork, D.G., Furuichi, Y.: A computer graphics reconstruction and optical analysis of scale anomalies in Caravaggio's *Supper at Emmaus*. In: Stork, D.G., Coddington, J., Bentkowska-Kafel, A. (eds.) Computer image analysis in the study of art II, vol. 7869, pp. 78690K–1–8. SPIE/IS&T, Bellingham, WA (2011)
24. Zapata, P.B., Castillo, D.N., Impett: Light analysis of paintings at scale using spherical harmonics. Electron. Imaging **36**, 1–7 (2024). https://doi.org/10.2352/EI.2024.36.14.CVAA-176, https://library.imaging.org/ei/articles/36/14/CVAA-176, publisher: Society for Imaging Science and Technology

From NUMMI DIGITALI to NUmmi DIgitali Towards Europe (NUDItEu)

Lavinia Sole[1], Maria Luisa Saladino[1], Francesco Armetta[1],
Fabio Bellavia[1(✉)], Rosina Celeste Ponterio[2], and Dario Giuffrida[2]

[1] Università degli Studi di Palermo, Palermo, Italy
{lavinia.sole,marialuisa.saladino,francesco.armetta01,
fabio.bellavia}@unipa.it
[2] Istituto per i Processi Chimico Fisici,
Consiglio Nazionale delle Ricerche, Messina, Italy
{rosinaceleste.ponterio,dario.giuffrida}@cnr.it

Abstract. NUMMI DIGITALI is a project aimed at enhancing the numismatic collection of the Regional Archaeological Museum "A. Salinas" in Palermo. To this end, a digital, open-access database has been created, aligned with the national cataloguing standards of the Central Institute for Cataloguing and Documentation (ICCD) of the Italian Ministry of Culture (MiC). The goal is to ensure data interoperability and shareability, according to the Linked Open Data (LOD) paradigm. The database is connected to the web through a front-end interface that provides public access and a real user interaction via Information and Communication Technologies (ICT).

More recently, NUDItEu has extended the original project with the novel features presented in this paper, focusing on geospatial visualization, contextual analysis, archaeometric data, interdisciplinary study of coins and with preparatory work for aligning the cataloguing standards with the Numismatic Description Standard (NUDS) of nomisma.org. This alignment will allow the transfer of data for coins catalogued into major thematic databases in Europe and America.

Keywords: Digital numismatics · Scheda NU+ · Archaeological Museum "A. Salinas" of Palermo · 3D models · Coins · Archaeometric data

1   Introduction

NUMMI DIGITALI[1] is a project launched in 2022 within the framework of the PON Research and Innovation 2014–2020 funding programme, aimed at supporting research activities of Numismatics of the University of Palermo [15, 16]. The main objective was the digitisation of the numismatic collection of the "A. Salinas" Archaeological Museum in Palermo, in order to foster its study

---

[1] Digital coins, the word "nummi" means "coins" from the Latin *nummus,-i*.

E. Rodolà et al. (Eds.): ICIAP 2025 Workshops, LNCS 16169, pp. 569–580, 2026.
https://doi.org/10.1007/978-3-032-11317-7_46

and enhance its management, protection, and promotion. To achieve this goal, the project developed an open-source digital platform of the same name[2] which employs high-resolution images and 3D renderings to promote the numismatic collection of the Salinas museum, which had been closed to the public for over seventy years. In 2025, as NUDItEu, the project received additional funding through the EuroSTART initiative of the University of Palermo, enabling further development and enhancement.

The project development was supported by the ICCD, which validated the cataloguing records and guided the database specification design to facilitate integration into the national SIGECweb (Cataloguing General Informative System)[3] and the sharing of data according to the LOD paradigm[4]. Moreover, the Centre for Inventory, Cataloguing and Documentation (CRICD) of the Sicilian Region supported the project as an intermediary in the process of uploading data to SIGECweb.

The following initial results of the project were achieved during 2023:

- A digital database was created in open format, aligned with the national cataloguing standards of the ICCD, and thus based on established, certified, and shared parameters[56] [11].
- The database was connected to the web through a front-end interface, ensuring public access to the collection by leveraging the potential of ICT and enabling data sharing and interoperability, in accordance with the LOD paradigm[7] and the methodological guidelines of the National Plan for the Digitalisation of Cultural Heritage (PND) promoted by MiC[8].
- The online web portal was launched, equipped with tools that enable user interaction through high-resolution images and filters for searching and accessing information.

This paper discusses the base platform and as new contribution presents the further improvements achieved during the NUDItEu project extension which include the support for georeferencing, to archaeometric data and high-quality 3D image of the coins, as well as the cataloguing format extensions and the interactive tools to deal with and to exploit these additions. Currently, no other similar projects or platforms, such as Online Coins of the Roman Empire (OCRE)[9], Coinage of the Roman Republic Online (CRRO)[10] and Moneda IBérica (MIB)[11], offer all the features covered by NUMMI DIGITALI.

---

[2] https://nummidigitali.it.

[3] http://www.iccd.beniculturali.it/it/sigec-web.

[4] https://www.w3.org/wiki/SweoIG/TaskForces/CommunityProjects/LinkingOpenData.

[5] http://www.iccd.beniculturali.it/it/standard-catalografici.

[6] https://github.com/ICCd-MiBACT/Standard-catalografici.

[7] https://www.w3.org/DesignIssues/LinkedData.

[8] https://digitallibrary.cultura.gov.it/il-piano/.

[9] https://numismatics.org/ocre/.

[10] https://numismatics.org/crro/.

[11] https://monedaiberica.org/.

2   The Digital Database

2.1   The Back-End

The back-end is reserved for research and data entry and was modeled on the structure and ontology of the *Scheda NU*, a cataloguing instrument issued by the ICCD in 2004, which still retains its original configuration [2]. The Scheda NU can be defined as a numismatic descriptive schema for the data collection of a coin, based on ICCD cataloguing standards and comparable to the NUDS used by the nomisma.org network[12]. The layout comprises numerous hierarchical structured fields (i.e. subsections), which include editable subfields that require the use of specific controlled vocabularies formalised by the ICCD.

The alignment of the database developed for the Salinas Museum with ICCD standards immediately revealed the need to revise the descriptive schema. The project required a flexible cataloguing framework, to be adapted to the structure of the numismatic holdings of the Salinas Museum, which includes coins originating both from historical collections and from archaeological excavations. To this end, the project made use of the inherent flexibility of the Scheda NU, whose mandatory fields – whether absolute, contextual, or alternative – are limited to only a few elements, thus allowing for changes to the overall structure [9].

Following a thorough and reasoned assessment, two parallel interventions were carried out: on the one hand, simplification efforts led to the removal of non-mandatory and non-essential sections and fields that were not necessary for the analysis of numismatic objects; on the other hand, the schema was enhanced by the addition of key fields not included in the original Scheda NU framework, as well as other tools useful for research and valorisation. This operation required close collaboration with the ICCD technical staff to ensure that adaptations would not compromise data compatibility with the national SIGECweb platform. All customizations were validated by the ICCD, and the structure of the final product – referred to as Scheda NU+ – was published and made available to the scholarly community [15]. This enhanced cataloguing tool is both more comprehensive, more streamlined and flexible than the the original model.

The digital implementation of the Scheda NU+ was subsequently integrated into the back-end interface. It consists of a total of 19 sections, each of which can be opened to reveal the relevant fields, which can be edited by suggestions or drop-down menus offering selection options based on formalized vocabularies. These vocabularies are stored during the data entry process, thereby simplifying and accelerating the data insertion, where the user is guided in the interpretation of labels and the selection of terms. The development of the Scheda NU+ and its digital implementation took place in two phases between 2023 and 2025, with particular focus on sections and fields relating to the geospatial, contextual visualisation and to interdisciplinary study of coins.

Within NUDItEu, several updates are achieved concerning geospatial data insertion and visualization. For example, the **mint** field, is populated with

---

[12] https://greekcoinage.org/nuds.html.

toponyms that can be visualized on an interactive map, thanks to the integration of the Leaflet software[13] into NUMMI DIGITALI. Significant revisions were made to the fields relating to the find context of the coin, as these were only minimally represented in the original ministerial schema. In particular, the section `GP - Georeferencing by point` is linked to a dynamic map that allows the identification of the X (`GPDPX`) and Y (`GPDYP`) coordinates of the findspot and its geolocation. Once the point is identified on the map, the corresponding geospatial coordinates are transferred to the respective database fields. Moreover, in the section `RE - Method of Recovery`, numerous fields have been integrated to provide specific information regarding the area in which the coin was found (e.g. settlement, sanctuary, necropolis, building, room), the position of the coin (whether in primary or secondary context), and, above all, the nature of the coin find – whether it was an isolated find, part of an accumulation, or a hoard. These data are collected during the excavation phase and are free from interpretation, which is reserved for subsequent hermeneutic analysis.

The stratigraphic data relating to the find are instead collected in the structured field `DSC_NU_DSCU - Stratigraphic Unit` and provide information concerning the characterization of the layer from which the coin originates, the list of associated materials – including diagnostic ones – and the chronological indications derived from the context. For the additional fields, it was necessary to develop specific ontologies, made possible through the existing collaboration between the ICCD and the Institute of Cognitive Sciences and Technologies (ISTC) of the National Research Council (CNR). Together, they developed the ArCo-Architecture of Knowledge project, a network of ontologies described in [12,13]. Other modifications involved the section `RS - Restorations and Analyses`, in which the structured field `IND - Archaeometric and Diagnostic Investi- gations` was added to record information arising from archaeometric and diagnostic examinations conducted on the coins. Closely related is the structured field `ADM - Additional Multimedia Documentation`, included in the section `DO - Sources and Reference Documents`, which describes and attaches documentary images in `JPEG` format derived from instrumental acquisitions during the diagnostic investigations. The integration of the `IND` and `ADM` fields has further strengthened the multidisciplinary nature of the platform, which can now benefits more comprehensively from the data and methodologies of the hard sciences applied to the study of ancient coins.

The sections or paragraphs of the Scheda NU+ correspond closely to the *categories* of NUDS, and similar correspondences have been identified among their fields. This has enabled the creation of a mapping of equivalences between the two ontologies within NIDItEu. This recently completed work constitutes the foundation for aligning the standards of the NUMMI DIGITALI with the NUDS of nomisma.org. The alignment will allow the transfer of data for coins catalogued with NUMMI DIGITALI into major thematic databases in Europe and America, such as CRRO and OCRE. These databases have become widely

---

[13] https://leafletjs.com/.

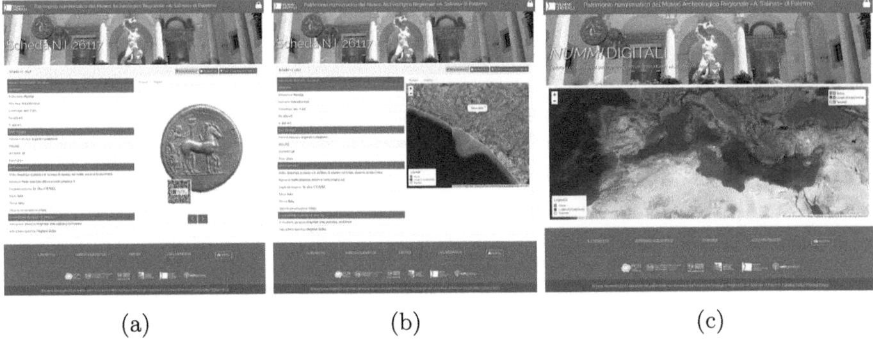

(a)   (b)   (c)

Fig. 1. (a) Coin information and 2D visualization access through the web database front-end and (b) corresponding georeferentiation, (c) the homepage **Map** section. The selected coin is identified by the inventory number 26117 and can be retrieved online on the database for an optimal visualization. Best viewed in color and zoomed-in.

used and indispensable for the scholarly community, as they include a substantial corpus of specimens belonging to various series. However, the usefulness of these tools increases as they are enriched with new specimens that may offer novel insights for research. Therefore, it is vitally important to contribute new data. NUMMI DIGITALI is continuously and incrementally populated with new data entries from the Salinas Museum collection, which should list around ten thousand coins at the end of the acquisition process.

Within the back-end, it is also possible to conduct searches via a dedicated button that utilizes the most relevant information for scholarly study. The adoption of nationally shared cataloguing rules further ensures that the NUMMI DIGITALI datasets can be integrated into SIGECweb, the national platform developed and managed by the ICCD for the access and sharing of public cultural heritage records. This platform guarantees the secure circulation of information and is designed to be safeguarded against obsolescence [4]. To enable data submission to SIGECweb, all data have been formatted into the eXtensible Markup Language (XML), with additional tags added to the extra fields, which are provided exclusively in the LOD format. As an open-source database, NUMMI DIGITALI guarantees the publication of its datasets in accordance with the Findable, Accessible, Interoperable, and Reusable (FAIR) principles [19].

2.2   The Front-End

The database is strongly integrated with an online public front-end, whose interface has been designed to be user-friendly. The users have the option to view the Scheda NU+ in a short format, following a selection of fields to display and a customized presentation of certain labels, which in their original form were not particularly accessible or easily understood by non-expert users (see Fig. 1a).

The NUDItEu enhancements make NUMMI DIGITALI able to detect and cross-reference data relating to the geographical location of coin finds, the mint of origin of the specimens, and the nature of the monetary discovery. These data are then visualized in two distinct ways on the front-end. The first is within the individual coin record, which is accompanied by an interactive map: a blue square marks the mint where the coin was produced, while the findspot is indicated by a red square in the case of a single find, or a yellow square if the coin was discovered as part of a hoard (see Fig. 1b). The second visualization appears on the homepage, under the `Map` section, where an interactive summary map displays all georeferenced data for the coins uploaded to the database according to the previous colored legend definition (see Fig. 1c). Clicking on hoard clusters also reveals a brief summary record of their contents.

The front-end also offers users the possibility to carry out searches by selecting a category from a predefined list and entering a keyword. Additionally, it allows users to request the full version of the records and the related coin images from the museum by filling out a dedicated form. These requests are transmitted to the back-end, where the museum can view and process them. This feature constitutes one of the tools provided by the platform to support the management of the heritage under the responsibility of the museum institution.

Particular attention has been devoted to the coin images, which are provided in high resolution to allow for the detailed observation of features. Some images are in 2D, while others include a 3D augmented form, with points of interest that link to specific textual information and/or multimedia content (see Sect. 4). These points of interest provide brief information on the history of the coin, its intrinsic features, and its metallic composition, the latter determined through X-Ray Fluorescence (XRF) spectroscopic analysis (see Sect. 3).

The unrestricted dissemination of these images on the web is prevented by a series of technical measures which deny saving, copying and pasting operations or opening the image in a new tab, as well as the superimposition of a protective QR code referring to copyright notices (see Fig. 1a). More recently, within NUDItEu further protection system has been experimented through the use of deep invisible blind watermarking techniques [3].

As a result, the NUMMI DIGITALI project currently enables the Salinas Museum to be the first museum in Italy to possess a web-accessible digital numismatic database.

3   Implementation of Archaeometric Data

Archaeometric analyses play a crucial role in the study of ancient metals, providing insights into metallurgical practices, provenance of raw materials, economic networks, and technological choices of past societies [1,5,6]. In numismatics, compositional data contribute not only to the authentication and classification of coins but also to broader historical and cultural interpretations. By combining analytical techniques with digital platforms, it is now possible to enhance both research and public access to ancient coinage.

To investigate the alloy composition of selected ancient coins, an archaeometric protocol based on XRF spectroscopy was developed and applied. The analysis

was conducted in situ using a Bruker AXS Tracer III SD portable spectrome-
ter, equipped with a rhodium-target X-ray tube (40 kV, 11 μA) and a $10\,mm^2$
silicon drift X-Flash detector, enabling the detection of elements with atomic
number Z > 11. A 3-4 mm diameter window was used to define the sampled
area. A Titanium-Aluminum filter was applied to reduce the signal from lighter
elements, and no vacuum was applied at the measurement head. Each coin was
analyzed on both faces, with two 30-second spectra acquisitions per coin, resulting in a dataset of 64 spectra. The data was collected using S1PXRF® software
(v3.8.30), and spectra were processed with ARTAX7® software (v7.4.6.1), using
metal standards and an empirical calibration method to assign elemental peaks
and estimate peak areas for semi-quantitative compositional analysis. The resulting spectra revealed three compositional groups based on the dominant detected
elements, characterized by gold, silver and copper alloy.

The results of the archaeometric analysis are being systematically integrated
into NUMMI DIGITALI with NUDItEu to enhance digital access and scholarly
research. Each analyzed coin is identified by its inventory number and linked to:

- The coin 3D model, enriched with XRF data and compositional profiles.
- A newly developed structured field IND-**Archaeometric and Diagnostic
  Investigations**, where complete analytical results (including elemental com-
  position and group classification) are stored and accessible.
- A user-friendly search interface to retrieve coins by inventory number and
  review both visual and chemical characterization (see Sect. 4).

These integrations provide a better support of interdisciplinary studies in numismatics, conservation science, and digital heritage by combining physical analysis
with interactive digital tools.

4   3D Coin Digitalization

The concept of *digitization* [4,8,20] in the cultural heritage does not simply
concern the graphic representation or visual rendering of artifacts; rather, it is
intrinsically connected to the way in which findings are studied, documented,
and interpreted within specific archaeological and historical research framework.

Within the NUMMI DIGITALI, digitization transforms a physical artifact
into a virtual model to be digitally accessible, scientifically reliable, and suitable
for analytical and interpretative research activities. This model must preserve
all the original morphometric and chromatic features while incorporating the
information that cannot be accessed through direct visual inspection (e.g. compositional data obtained via analytical methods). On this basis, a structured and
interdisciplinary protocol was developed to product a *digital twins* of the physical
coins. The digitization process includes a multilevel pipeline of four independent
but methodologically interconnected domains, as illustrated in Fig. 2a.

In this section, we will focus specifically on the methodological and technical
procedures involved in the realization of the 3D virtual models of numismatic
finds, detailing the acquisition protocols, the instruments and imaging systems

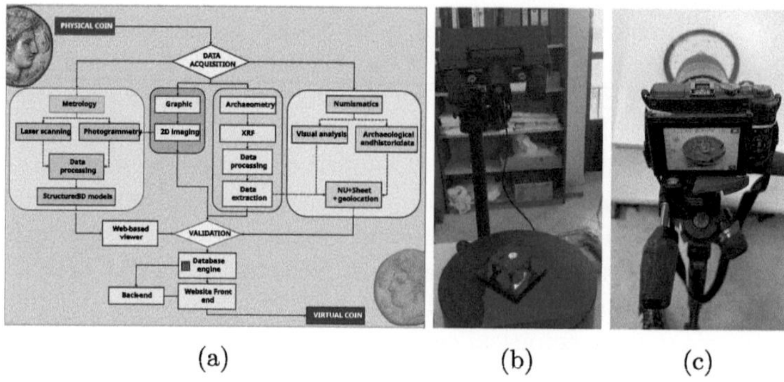

(a)   (b)   (c)

Fig. 2. (a) Digitalization process workflow and the data acquisition setup using (b) a structured-light scanner and (c) macro-photogrammetry. Best viewed in color and zoomed-in.

employed, the workflows for geometric-textural integration, and the methods adopted for embedding and publishing the models online. One of the core objective of the 3D acquisition is to obtain metrically rigorous digital replicas of ancient coins, preserving both their micro-morphological topography and surface chromaticity [7,10,14,20]. To achieve this, an integrated pipeline was developed, combining two complementary non-contact digitization methods: structured-light 3D scanning for geometric fidelity and macro-photogrammetry for high-resolution texture acquisition.

Three-dimensional scanning was performed using the ATOS Q 8M GOM (by Zeiss) structured-light scanner, a system primarily designed for industrial metrology and here employed for the first time in the digitization of numismatic heritage. The system has a spatial resolution up to 50 μm, 8 million of points per scan, a measurement volume configurable between 100×70 mm and 500×370 mm, a working distance of 490 mm and an adjustable point spacing in the range of [0.04,0.15] mm.

The scanning setup was deployed in a mobile laboratory installed in situ at the Salinas Museum (see Fig. 2(b)-(c)). Coins were placed vertically on a high-precision motorized rotation table (GOM ROT 350), using non-invasive adhesive putty. Each coin was subjected to two independent acquisition sessions: an initial 360° rotation capturing the obverse and lateral geometry, followed by a vertical inversion to acquire the reverse and minimize occlusions introduced by the support. Fiducial markers were placed in the scanning environment to enable automated alignment of partial scans, ensuring consistent registration across sessions and minimizing cumulative alignment error.

Since structured-light scanning systems lack the capacity to capture chromatic information, a parallel acquisition of high-resolution photographic data was carried out via macro-photogrammetry. The imaging system employed a Canon EOS 6D camera equipped with a 60 mm f/2.8 macro lens, thus ensuring sub-millimetric fidelity in the capture of fine surface features such as relief

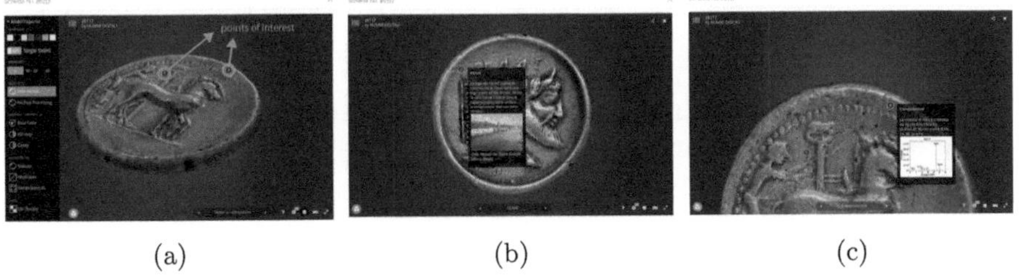

(a)   (b)   (c)

Fig. 3. (a) Coin 3D model visualization with interest points highlighted in green and examples of the annotated content displayed such as (b) descriptive data or (c) the metallic composition. Notice that these images represent the different sides of the same coin of Fig. 1. Best viewed in color and zoomed-in (Color figure online) .

details, patina textures, and iconographic elements. The photogrammetric protocol consisted of:

- Acquisition of 16 images for each coin side with controlled rotation intervals.
- Camera positioning at 45° inclination relative to the coin's plane to maximize relief shading and capture peripheral edge morphology.
- Use of a non-reflective rotary disk with printed coded markers.
- Uniform diffuse illumination using LED panels to minimize specular highlights and shadow artifacts.

Images were processed in Agisoft Metashape Professional[14], following a standard Structure-from-Motion (SfM) and Multi-View Stereo (MVS) workflow [17]: (1) sparse point cloud generation via image alignment, (2) dense cloud reconstruction, (3) mesh generation, and (4) high-resolution texture mapping. The 3D models derived from structured-light scanning and macro-photogrammetry were subsequently co-registered through a specified alignment process. The final result is a hybrid 3D model that integrates the metric robustness of industrial-grade metrology with the chromatic and iconographic realism necessary for scholarly numismatic analysis. These models serve not only as high-fidelity replica of the physical coins but also as platforms for annotation, interaction, and dissemination within the NUMMI DIGITALI infrastructure (see Fig. 3).

The final stage of the 3D digitization pipeline within NUMMI DIGITALI concerns the deployment, dissemination, and semantic integration of the digital coin models into the platform data architecture. This phase transforms high-resolution 3D assets into interactive knowledge interfaces, enabling both expert analysis and public engagement. The workflow involves the export, publication, and embedding of annotated 3D models within the digital infrastructure, ensuring full accessibility and interoperability. To ensure online accessibility and device-independent navigation, final models were exported in .OBJ format with associated .MTL and high-resolution texture files, and uploaded to the web-based

---

[14] https://www.agisoftmetashape.com/.

platform Sketchfab[15], which supports real-time rendering, high-fidelity materials, and integrated interactivity tools. Key features leveraged include:

- Full 360° interactive manipulation, allowing users to rotate, zoom, and inspect coin models from any angle.
- Environmental control tools for adjusting lighting, shading, and background, aiding in the visualization of surface reliefs and patinas.
- High-resolution texture management, supporting detailed examination of inscriptions, iconographic motifs, and manufacturing marks.
- Multi-platform support, enabling visualization across browsers, tablets, and mobile devices without the need for external plugins.

A central innovation of the NUMMI DIGITALI visualization layer is the incorporation of interactive semantic annotations directly onto the 3D model surface, achieved with NUDItEu. Each model includes annotated *points of interest* (see Fig. 3(a)) which, when activated, trigger contextual information panels containing: descriptive metadata (legend transcription, issuing authority, nominal value, see Fig. 3(b)); hyperlinks to associated catalog records in the Scheda NU+; images or figures highlighting diagnostic features or comparative examples; in selected cases, XRF spectroscopic data, including elemental composition charts and interpretative comments (see Fig. 3(c)). These annotations are curated by domain experts and serve a dual function: they enhance the interpretive depth of the digital model and transform the visualization into a multi-modal research and didactic tool. This paradigm reflects a shift from static digitization to dynamic data contextualization.

Beyond the visualization layer, each 3D model is semantically and structurally linked to its corresponding catalog entry within NUMMI DIGITALI, developed according to the extended Scheda NU+. This integration is accomplished through:

- Embedding of Sketchfab models via `IFRAME` code into the coin front-end record page.
- Bidirectional data binding between 3D annotations and back-end fields (e.g., material, mint, typology, analysis results).
- Exportable metadata in LOD-compliant XML format, ensuring interoperability with external repositories such as Nomisma.org or SIGECweb.
- Preservation of provenance traceability by linking the digital model with acquisition metadata (scanner ID, lens specifications, acquisition protocol).

5   Conclusions

The methodological framework implemented originally in NUMMI DIGITALI and further extended thanks to the NUDItEu follow-up delineates a new standard for the digital transformation of numismatic collections. It surpasses the

---

[15] https://sketchfab.com/nummidigitali.

traditional paradigms of static documentation and image-based recording by offering a multimodal, scalable, and semantically enriched infrastructure that responds to both scholarly and curatorial needs. Currently, no other task-specific portal implements altogether all the features offered by the proposed system.

By merging computational imaging, archaeometric data acquisition, and histo-rical-typological interpretation, NUMMI DIGITALI exemplifies the potential of interdisciplinary digital ecosystems in the domain of cultural heritage. It establishes a model in which physical, chemical, and historical knowledge are structurally integrated and rendered operable within a single, open-access platform. This approach not only enhances the analytical capacity of numismatists, but also facilitates long-term preservation, encourages inter-institutional interoperability, and broadens public access to historically significant artefacts through immersive, interactive, and richly annotated digital experiences. As such, Nummi Digitali represents a methodological and infrastructural benchmark, charting a viable path for the future of digital numismatics and offering a transferable model for the digitization of other classes of small-scale archaeological objects.

Besides the actual infrastructure, the NUMMI DIGITALI database is ongoing to be incrementally updated in order to finally cover the whole numismatic collection of the Salinas Museum. When accomplished, the digital collection will provide precious data for the specific research area, as well as a valuable resource for the cultural heritage.

Acknowledgments. This work was supported by the European Union – NextGenerationEU – MUR funds D.M. 737/2021 – research project "NUmmi Digitali towards Europe"; CUP B79J21038330001 (Eurostart 2024, University of Palermo). D. Giuffrida thanks European Union (NextGeneration EU), through theÂăMUR-PNRR project SAMOTHRACE – Sicilian Micro and Nano Technology Research and Innovation CenterÂă(ECS00000022). All the authors thank Webgenesys S.p.A., the system integrator responsible for technological coordination. L. Sole conceived the NUMMI DIGITALI project and as the scientific coordination wrote Sects. 1–2, 5 of the manuscript; M. L. Saladino and F. Armetta carried out the metal analyses, contributing to their integration within the system and wrote Sect. 3; R. C. Ponterio and D. Giuffrida were responsible of the photogrammetry acquisition process and of the creation of 3D models, writing Sect. 4; F. Bellavia designed the watermark protection system, collecting and arranging the different contributions within the manuscript in consultation with other authors.

References

1. Armetta, F., et al.: The silver collection of San Gennaro treasure (Neaples): a multivariate statistic approach applied to X-ray fluorescence data. Spectrochimica Acta Part B: Atomic Spectroscopy **180** (2021)
2. Arslan, E., et al.: Strutturazione dei dati delle schede di catalogo: Scheda NU: Beni numismatici. ICCD (2004)
3. Bellavia, F., Sole, L.: Exploring image watermarking in digital collections of ancient coins. In: Proceedings of 4th International Workshop on Fine Art Pattern Extraction and Recognition (FAPER) (2025)

4. Birozzi, C., Barbaro, B., Mancinelli, M.L., Negri, A., Plances, E., Veninata, C.: Catalogare nel 2020. La digitalizzazione del patrimonio culturale. Aedon, rivista di arti e diritto on line **3** (2020)
5. Caponetti, E., et sl.: A multivariate approach to the study of orichalcum ingots from the underwater Gela's archaeological site. Microchem. J. **135** (2017)
6. Caponetti, E., et al.: Newly discovered orichalcum ingots from Mediterranean sea: further investigation. J. Archaeol. Sci. Rep. **37** (2021)
7. Hess, M., MacDonald, L.W., Valach, J.: Application of multi-modal 2D and 3D imaging and analytical techniques to document and examine coins on the example of two Roman silver denarii. Heritage Sci. **6**(1) (2018)
8. Jarrett, J., Zambanini, S., Huber-Mörk, R., Felicetti, A.: Coinage, Digitization and the World-Wide Web: numismatics and the COINS Project (2012)
9. M. L. Mancinelli: Normativa trasversale. versione 4.00. strutturazione dei dati e norme di compilazione (2017). http://www.iccd.beniculturali.it/getFile.php?id=5739
10. MacDonald, L., Moitinho de Almeida, V., Hess, M.: Three-dimensional reconstruction of Roman coins from photometric image sets. J. Electron. Imaging **26**(1) (2017)
11. Mancinelli, M.L.: Gli standard catalografici dell'istituto centrale per il catalogo e la documentazione. Le voci, le opere e le cose. La catalogazione dei beni culturali demoetnoantropologici, pp. 279–302 (2018)
12. Mancinelli, M.L., Veninata, C.: Architettura della conoscenza: il sistema ICCD come modello per la descrizione dei beni culturali. In: Proceedings of Conferenza GARR 2018 - Data Revolution (2018)
13. Mancinelli, M.L., Veninata, C.: Patrimonio numismatico e catalogo generale dei beni culturali: Progetti in corso per l'integrazione e la valorizzazione delle conoscenze. In: Proceedings of "Verso il futuro, Esperienze, progetti e casi di studio tra tutela, fruizione e comunicazione del patrimonio numismatico pubblico, IV Incontro di studio on Medaglieri italiani" (2022)
14. Morris, G., Emmitt, J., Armstrong, J.: Depth and Dimension: Exploring the Problems and Potential of Photogrammetric Models for Ancient Coins. Journal of Computer Applications in Archaeology (2022)
15. Sole, L.: Nummi Digitali: Approcci innovativi per la conoscenza, gestione e valorizzazione del patrimonio numismatico del Museo Archeologico Regionale A. Salinas di Palermo. In: Proceedings of "Verso il futuro, Esperienze, progetti e casi di studio tra tutela, fruizione e comunicazione del patrimonio numismatico pubblico, IV Incontro di studio on Medaglieri italiani" (2022)
16. Sole, L., Ponterio, R.C., Giuffrida, D., Saladino, M.L., Armetta, F.: Migliorare la conoscenza, la gestione e la valorizzazione dell'antico patrimonio numismatico: il progetto Nummi Digitali per il medagliere del Museo Archeologico Regionale A. Salinas di Palermo. Diritti culturali e patrimonio culturale tra tradizione e innovazione. Un dialogo multidisciplinare (2024)
17. Szeliski, R.: Computer Vision: Algorithms and Applications. Springer-Verlag, 2nd edn. (2022)
18. Veninata, C.: Inside the Meanings. The Usefulness of a Register of Ontologies in the cultural Heritage Sector. JLIS.it **11**(2) (2020)
19. Wilkinson, M.D., et al.: The FAIR Guiding Principles for scientific data management and stewardship. Sci. Data **3** (2016)
20. Zambanini, S., Schlapke, M., Hödlmoser, M., Kampel, M.: 3D acquisition of historical coins and its application area in numismatics. In: Computer Vision and Image Analysis of Art. vol. 7531 (2010)

Unveiling Visual Features in Artwork Classification: Towards Explainable Vision Transformers in the Arts

Raffaele Scaringi$^{(\boxtimes)}$ 🆔, Nicola Fanelli 🆔, Gennaro Vessio 🆔,
and Giovanna Castellano 🆔

Department of Computer Science, University of Bari Aldo Moro, Bari, Italy
{raffaele.scaringi,nicola.fanelli,gennaro.vessio,
giovanna.castellano}@uniba.it

Abstract. Recent advances in deep learning have enabled accurate artwork classification using models such as Vision Transformers (ViTs). However, interpreting the internal mechanisms behind such decisions remains challenging, especially in the abstract and symbolic domain of visual arts. We propose an interpretability framework that combines feature visualization via activation maximization with natural language grounding through a Multimodal Large Language Model. Our method extracts class-specific visual patterns learned by ViTs, synthesizes prototype images that activate key features, and generates human-readable descriptions. Applied to a large-scale art dataset, the approach reveals that ViTs attend to subtle and abstract cues—such as texture, shape, and composition—differently from natural image tasks. The resulting visual and textual explanations offer valuable insight into model behavior and move toward more transparent, human-aligned AI systems for art analysis.

Keywords: Visual Arts · Vision Transformers · Explainable AI · Multimodal Large Language Models

1   Introduction

In recent years, computer vision has witnessed significant progress, primarily driven by advances in deep learning architectures and the growing availability of large-scale, annotated datasets. Convolutional Neural Networks (CNNs) [23] and Vision Transformers (ViTs) [17] have achieved state-of-the-art performance across a wide range of visual tasks, including object detection [1], image classification [10], semantic segmentation [22], and image super-resolution [34]. These models learn highly expressive internal representations capable of capturing complex visual patterns. However, understanding and interpreting such representations remains a significant open challenge—particularly in domains where decisions rely on subtle, context-dependent visual cues.

E. Rodolà et al. (Eds.): ICIAP 2025 Workshops, LNCS 16169, pp. 581–592, 2026.
https://doi.org/10.1007/978-3-032-11317-7_47

Fig. 1. Feature visualizations from the CLIP vision encoder before and after fine-tuning for artwork classification. The top row shows activations from the pre-trained model (trained on Internet images), while the bottom row shows activations after fine-tuning on art data. Fine-tuning leads to more abstract and complicated patterns, reflecting the complexity and diversity of visual arts.

One such domain is visual art analysis [8,9]. Artistic images often reflect diverse styles, techniques, and symbolic conventions, with the same subject represented in vastly different forms depending on the artist's intent or historical context. As a result, explaining the rationale behind model predictions in this setting requires more than highlighting salient pixels: it demands semantically meaningful explanations capable of capturing high-level visual concepts.

To improve semantic understanding in this domain, recent research has explored multimodal approaches that integrate visual features with external knowledge—often encoded as structured data such as knowledge graphs [2,6,14]. These methods have enhanced performance in tasks such as attribute prediction and classification. In some cases, they have also contributed to improving explainability; for example, GraphCLIP [30] uses local subgraphs to trace the influence of contextual entities on the model's decision. However, such approaches are primarily focused on the external, symbolic aspects of multimodality, often overlooking the internal visual representations that the model learns.

In this paper, we take a complementary perspective and ask: *Which visual features are leveraged by a computer vision model to classify an artwork into a particular style or genre?* To answer this, we adopt feature visualization techniques [28] and apply them to ViT-based classifiers trained on a large-scale art dataset. While prior works have explored such techniques in natural image domains [15,19], to the best of our knowledge, this is the first attempt to use them in the context of visual arts—a domain where the notion of "class-specific features" is inherently more abstract and semantically rich.

We further extend this analysis by employing Multimodal Large Language Models (MLLMs) to generate natural language descriptions of the synthesized features. This results in an interpretability pipeline that links low-level visual activations to high-level textual concepts, thereby facilitating human understanding of the model's behavior. The resulting framework offers a dual-view explanation: a visual prototype generated through activation maximization and a textual description that contextualizes it in human terms.

Our findings show that Vision Transformers trained for artwork classification attend to a diverse range of visual features—including textures, brushstroke patterns, color palettes, and compositional arrangements—that are specific to artistic styles or genres. Unlike in object recognition tasks, however, these internal representations are often more abstract and less intuitively recognizable, as illustrated in Fig. 1. This contrast highlights the unique interpretability challenges posed by art-related tasks and underscores the need for more refined tools to probe and explain model decisions in non-photographic domains.

The remainder of this paper is organized as follows. Section 2 reviews prior work on artwork classification and feature visualization for explainable vision models. Section 3 details our proposed approach. Section 4 presents both quantitative and qualitative experimental results. Finally, Sect. 5 concludes the paper and discusses directions for future research.

2   Related Work

2.1   Art Classification and Multimodal Representations

With the advent of deep learning, representation learning has become the dominant approach to artwork classification. Convolutional Neural Networks and, more recently, Vision Transformers have been employed to model artistic features across a variety of tasks, including style and genre prediction [11,18], visual link retrieval [5], artwork clustering [7], style transfer [32], and inpainting [13]. Despite these advances, the high variability and symbolic abstraction inherent in artworks continue to pose significant challenges to vision-only models.

To address this, several works have explored multimodal strategies that incorporate external knowledge sources. For instance, ContextNet [14] integrates visual features with knowledge graph embeddings computed via node2vec [16]. In contrast, more recent approaches employ Graph Neural Networks [4] to support inductive reasoning over previously unseen entities. Although effective in improving classification performance, these methods often provide limited interpretability and typically rely on shallow attention-based mechanisms or feature attribution methods.

A notable exception is GraphCLIP [30], which combines contrastive learning with external knowledge to model visual and contextual representations jointly. It also incorporates Grad-CAM [31] to highlight the image regions most responsible for a given prediction. While promising, such explanations are inherently local and instance-specific, lacking a more general understanding of class-level semantics.

2.2   Feature Visualization and Explainability in Vision Models

Another line of research in explainable computer vision focuses on understanding what internal representations deep models acquire. Feature visualization techniques aim to synthesize inputs that maximize the activation of specific neurons or layers in trained architectures, thereby providing insight into the structure of learned representations [26,28]. These methods have evolved from basic gradient-based optimization to more advanced frameworks incorporating regularization, image priors, and data augmentation [15,19].

Although widely applied to natural image domains—such as ImageNet classification—feature visualization remains largely unexplored, particularly in the context of the visual arts, where class semantics are often abstract, symbolic, and culturally informed. Moreover, visualizations generated by these techniques are typically presented as raw, synthetic images, which can be challenging for users without a technical background to interpret.

Recent progress in multimodal learning offers new tools to bridge this gap. Multimodal Large Language Models [33] have demonstrated strong capabilities in grounding visual inputs through natural language. These models can be leveraged to provide textual descriptions of visual patterns, even for abstract or synthetic inputs [12,27].

In this work, we build on these insights by applying feature visualization to the domain of artwork classification and by introducing a multimodal explanation layer that translates synthesized class-specific visual features into natural language descriptions. To the best of our knowledge, this is the first study to combine visual feature synthesis with MLLM-based captioning in the context of deep learning for the visual arts.

3   Methodology

This section presents our framework for extracting and explaining the visual features learned by Vision Transformers in the context of artwork classification. The pipeline consists of three main stages: (i) fine-tuning a multi-task ViT-based classifier, (ii) applying feature visualization techniques to synthesize class-representative prototypes, and (iii) generating natural language explanations via a Multimodal Large Language Model.

3.1   Classifier Architecture

Given an input artwork image I, the model aims to predict multiple semantic attributes, such as artistic style and genre. Let T denote the number of classification tasks. For each task $j \in \{1, \ldots, T\}$, we define the corresponding class set as $\mathcal{C}_j = \{c_{j1}, \ldots, c_{jK_j}\}$, where K_j is the number of classes for task j.

To support multi-task learning, we extend a pre-trained ViT backbone with T parallel classification heads—one per task. Given $I \in \mathbb{R}^{3 \times H \times W}$, the model outputs:

$$\hat{y} = \Phi_V(I), \quad \hat{y} = [\hat{\mathbf{y}}_1, \ldots, \hat{\mathbf{y}}_T], \tag{1}$$

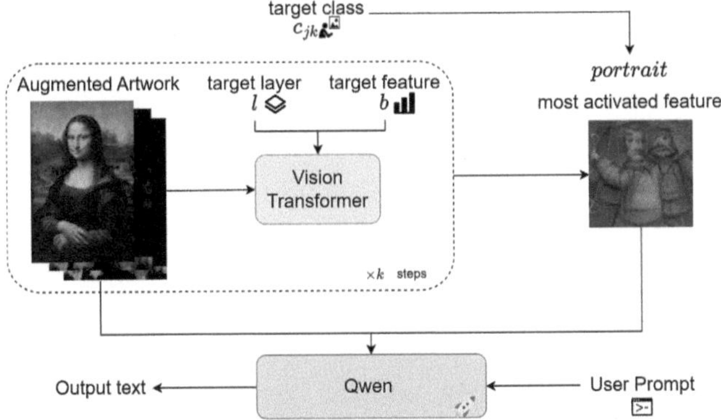

Fig. 2. Visualization framework. After ViT fine-tuning, we identify class-specific features using an IDF-based criterion, generate their corresponding visual prototypes via gradient optimization, and obtain natural language descriptions using an MLLM.

where each $\hat{\mathbf{y}}_j \in \mathbb{R}^{K_j}$ represents the predicted logits for the j-th task. This formulation enables joint learning across multiple classification dimensions while maintaining task-specific decision boundaries.

3.2 Feature Visualization Framework

After fine-tuning the classifier Φ_V, we probe its internal representations to identify the visual features most indicative of each class. We employ a gradient-based feature visualization technique [28] that synthesizes inputs that maximally activate the selected neurons in the network, as illustrated in Fig. 2.

Specifically, we focus on features derived from the [CLS] token at various transformer blocks. For each classification task j and class c_{jk}, we examine the validation set to identify the top-b most frequently activated features (i.e., those exhibiting the highest activation values when processing the input image I) at a given layer l. To select the most representative among them, we propose an IDF-based criterion that quantifies how uniquely a feature is associated with a specific class. Specifically, we compute the inverse document frequency (IDF) of each feature across all classes and retain the one with the highest score. This promotes the selection of features that are distinctive rather than frequent, enhancing the semantic specificity of the resulting visualization.

Once the target feature $f_{l,i}$ is identified, we synthesize an input image I^* that maximizes its activation via the following optimization:

$$I^* = \arg\max_I \sum_k \mathcal{L}(a_k(I), l, i) - \lambda TV(a_k(I)), \tag{2}$$

where $a_k \in \mathcal{A}$ is a stochastic augmentation (e.g., color jitter, Gaussian smoothing), TV denotes the total variation regularizer [26], and λ controls the regularization strength. The loss function aggregates activations across patches:

$$\mathcal{L}(I, l, i) = \sum_{p} (f_{l,i})_p, \tag{3}$$

with $f_{l,i}$ denoting the i-th feature extracted from layer l. The synthesized image I^* serves as a visual prototype capturing the most discriminative traits for class c_{jk}, such as the characteristic geometry of *Cubism* or the chromatic softness of *Impressionism*.

3.3   Multimodal Explanation via Language Models

While feature visualizations provide insight into the model's internal focus, their interpretation often requires domain-specific expertise. To enhance accessibility, we complement visual prototypes with textual explanations generated by the multimodal model Qwen-2.5-VL [3].

Given a synthesized image I^* and the source image used to initialize the optimization, we provide both as input to the model, prompting it to describe the visual features in I^* that explain why the classifier associates the image with a specific style or genre. The prompt is designed to elicit concise, semantically grounded captions.

In practice, the generated explanations align well with human interpretation, often referencing stylistic traits such as fragmented shapes, muted palettes, or dynamic brushwork. These captions bridge the gap between abstract neural representations and human-level concepts, offering a more interpretable and user-friendly perspective on model behavior.

Taken together, our framework yields both visual and textual artifacts that expose the internal decision logic of ViT-based classifiers—offering a promising step toward explainable AI in the visual arts domain.

4   Experiments

We evaluated the effectiveness of our framework on two key tasks in the domain of artwork classification: *style* and *genre* classification. Both tasks were addressed within a multi-task learning setting. A quantitative evaluation was conducted using Top-1 and Top-2 classification accuracy, along with the Macro-averaged F1 score. We also provide qualitative analyses of the extracted visual features and their corresponding textual descriptions.

4.1   Experimental Setup

We conducted our experiments on the *ArtGraph* dataset [4], a large-scale multimodal corpus for visual arts analysis. The dataset comprises over 100,000 digitized artworks, annotated with 32 style labels and 18 genre categories. In this study, we focused solely on the visual modality to assess the interpretability of learned representations. The data were split into training, validation, and test subsets using an $80/10/10$ ratio.

All images were resized to a fixed resolution of 224×224 pixels. As backbone architectures, we fine-tuned two Vision Transformer variants—ViT-B/16 and ViT-B/32—under contrastive multimodal pretraining, as proposed in CLIP [29].

We trained the classifiers using a multi-task cross-entropy loss with uncertainty based weighting [20]:

$$\mathcal{L}_C(\hat{y}, y) = \sum_k \frac{\ell(\hat{\mathbf{y}}_k, \mathbf{y}_k)}{\sigma_k^2} + \sum_k \log \sigma_k, \tag{4}$$

where ℓ is the standard cross-entropy loss and σ_k is a learnable parameter modeling task uncertainty.

Optimization was performed using AdamW [25] with cosine annealing for learning rate scheduling [24], an initial learning rate of 1×10^{-4}, and 1,000 warm-up steps. Early stopping was triggered after 10 epochs without improvement in validation accuracy. All experiments were run on a single NVIDIA A100 GPU (64GB) with 512GB of system RAM.

Feature Visualization Setup. To synthesize class-specific feature representations, we followed the optimization procedure described in Eq. 2, using the Adam optimizer [21] with an initial learning rate of 0.1 and cosine scheduling. The total variation regularization weight was set to $\lambda = 5 \times 10^{-5}$. Visualizations were generated from four transformer layers, $l \in \{4, 6, 8, 12\}$, enabling a multi-granular analysis.

To improve the realism and stability of the synthesized images, we applied a composite augmentation pipeline $\mathcal{A} = GS(CS(Jitter(I)))$, defined as follows:

$$Jitter(I) = I + \epsilon, \quad \epsilon \sim \mathcal{N}(0, \sigma^2), \quad \sigma \sim e^{\mathcal{U}(-1,1)} \tag{5}$$

$$CS(I) = \sigma I + \mu, \quad \mu \sim \mathcal{U}(-1, 1), \quad \sigma \sim e^{\mathcal{U}(-1,1)} \tag{6}$$

$$GS(I) = I + \epsilon, \quad \epsilon \sim \mathcal{N}(0, 1) \tag{7}$$

We optimized the input image using a batch of eight augmentations per iteration to encourage invariance and convergence stability.

Multimodal Captioning. For natural language explanations, we employed Qwen-2.5-VL-7B [3], an MLLM capable of reasoning over multiple images. For each feature visualization, we provided the model with two inputs: (i) the original image associated with the target class and (ii) the corresponding optimized prototype image. We prompted the model as follows:

Given the image (Image 1) that most strongly activates the specified feature, and the corresponding image optimized for that feature (Image 2),

Table 1. Classification performance on style and genre prediction. The best results per task are in bold.

Method	Style			Genre		
	Top-1	Top-2	F1	Top-1	Top-2	F1
CLIP ViT-B/16	**58.68**	**77.61**	**52.83**	**75.75**	**89.93**	**67.96**
CLIP ViT-B/32	56.72	75.18	50.53	73.91	88.25	65.05

describe the optimized image (Image 2) by identifying the visual elements the model seems to focus on to recognize the target <class>. Answer with a single paragraph.

4.2   Quantitative Results

Table 1 reports the classification comparison between ViT-B/16 and ViT-B/32 models. Overall, ViT-B/16 consistently outperforms ViT-B/32, suggesting that smaller patch sizes enhance the model's ability to capture fine-grained details, particularly valuable in style classification, where brushstrokes, textures, and local visual patterns play a key role.

Interestingly, genre classification is systematically more accurate than style classification across all models. This is likely because genre labels often correspond to high-level semantic content (e.g., presence of people, nature scenes), which deep models can more easily identify. In contrast, artistic styles tend to be characterized by subtler visual cues, such as texture and composition, making them inherently more challenging to predict.

4.3   Qualitative Results

Figure 3 illustrates qualitative feature visualizations for selected style and genre classes, comparing the outputs of CLIP-ViT-B/16 and CLIP-ViT-B/32. Each example includes: the input artwork, a synthesized image optimized to activate a class-specific high-level feature, and a natural language caption generated by Qwen, conditioned on the synthesized image. The results highlight the semantic depth and interpretability of the learned representations.

The optimized images preserve distinctive stylistic or compositional traits characteristic of the target classes—such as the dynamic brushwork in *Impressionism*, the object-centered realism of *Still life*, the fluid ornamental lines of *Art Nouveau*, and the abstract visual energy in *Marina*. The accompanying captions further support the semantic alignment of the synthesized activations, indicating that the model captures both visual and conceptual cues.

Notably, ViT-B/16 yields more refined and semantically coherent outputs than ViT-B/32, likely due to its higher spatial resolution. These findings suggest that CLIP-based ViTs not only encode domain-relevant abstractions but

 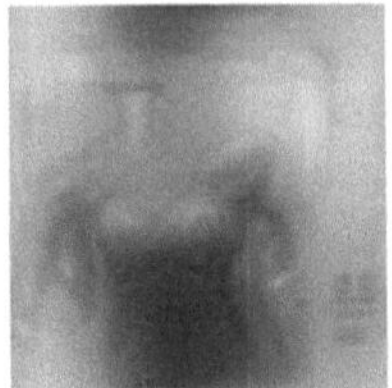  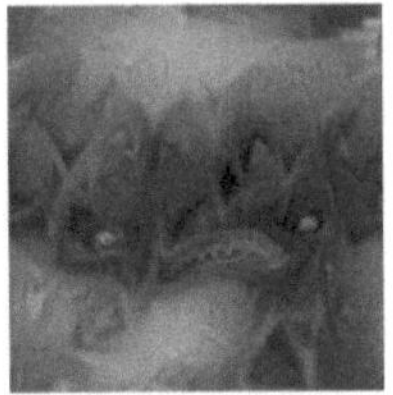

Style: Art Nouveau

Genre: Marina

Caption: The optimized image for class Art Nouveau focuses on vibrant colors and fluid lines, emphasizing the organic forms of the human figure. The use of bright yellows and greens contrasts sharply against the darker background, drawing attention to the contours and curves of the body, which aligns with the Art Nouveau aesthetic's emphasis on naturalistic forms and flowing lines.

Caption: The optimized image appears to focus on exaggerated, stylized features such as vibrant colors and distorted patterns, emphasizing the target genre of marina. The image's chaotic blend of shapes and hues attempt to capture the essence of a marina through abstract and surreal elements, possibly aiming to evoke a sense of movement or energy associated with water and boats.

Style: Impressionism

Genre: Still life

Caption: The optimized image for Impressionism appears to focus on vibrant colors and dynamic brushstrokes, which are characteristic of the Impressionist style. The use of bold yellow and blue hues, along with the blurred edges and visible brushwork, suggests an attempt to capture the fleeting effects of light and color, a hallmark of Impressionism.

Caption: The optimized image focuses on the visual elements of a still life composition, particularly emphasizing the arrangement of objects such as scissors, a ruler, and a pencil, which are laid out in a structured manner. The vibrant colors and the detailed depiction of the objects suggest an emphasis on realism and meticulous attention to detail, aligning with the target of a still life.

Fig. 3. Feature visualizations for selected style and genre classes. Each row shows: (*i*) an example artwork activating the target feature, (*ii*) the synthesized image that maximizes the selected feature (initialized from the example artwork), and (*iii*) the natural language caption generated by the MLLM. The first row refers to CLIP-ViT-B/16, whereas the second row refers to CLIP-ViT-B/32.

also support human-aligned interpretation through joint visual synthesis and language-based explanation.

These results demonstrate the model's ability to learn semantically meaningful and stylistically discriminative visual features. The combination of feature visualization and multimodal explanation proves especially valuable in the art domain, where categories often rely on subjective or abstract attributes. Compared to standard object recognition tasks, the optimized images presented here are less directly interpretable and often require domain-specific knowledge to appreciate their relevance fully—underscoring the challenge of explainability in non-natural image domains.

Overall, our findings suggest that Vision Transformers, when equipped with proper analysis tools, can offer interpretable and insightful representations even in domains characterized by high intra-class variability and semantic ambiguity.

5   Conclusion

In this study, we presented a novel framework for interpreting deep visual representations in the domain of artwork classification. Leveraging feature visualization techniques on Vision Transformer models, we extracted class-specific visual prototypes that reveal the internal patterns the model uses to associate artworks with specific styles or genres. To enhance the interpretability of these prototypes, we employed a Multimodal Large Language Model to generate natural language descriptions of the synthesized features, thereby bridging the gap between high-dimensional neural activations and human-understandable explanations.

The integration of MLLMs proved effective in making abstract visual features more accessible, offering descriptive narratives that align with expert-level interpretations. Nonetheless, we observed that certain features remain difficult to interpret—either due to their highly abstract nature or because they rely on stylistic conventions that require specialized domain knowledge.

Future work may proceed in several directions. First, we intend to adapt our interpretability pipeline to modern multimodal classification frameworks that integrate vision and language during both training and inference. Second, we plan to investigate the automatic labeling of synthesized features, which could pave the way for incorporating neuro-symbolic reasoning modules into art classification systems. Finally, we aim to conduct small-scale user studies with domain experts to assess the clarity and usefulness of the generated explanations, and to include quantitative explainability metrics—such as CLIPScore or human agreement scores—to provide a more rigorous evaluation of semantic alignment.

Acknowledgments. R. S. acknowledges support from a Ph.D. fellowship funded under the Italian "D.M. n. 352, April 9, 2022" – NRRP, Mission 4, Component 2, Investment 3.3 – co-supported by Exprivia S.p.A. (CUP H91I22000410007). N. F. acknowledges support from a Ph.D. fellowship funded under the Italian "D.M. n. 118/23" – NRRP, Mission 4, Component 1, Investment 4.1 (CUP H91I23000690007). G. C. acknowledges funding from the FAIR - Future AI Research project, Spoke 6 - Symbiotic AI (CUP H97G22000210007), under the NRRP MUR program funded by NextGenerationEU.

Disclosure of Interests. The authors declare that they have no known competing financial interests or personal relationships that could have appeared to influence the work reported in this paper.

References

1. Amjoud, A.B., Amrouch, M.: Object detection using deep learning, CNNs and vision transformers: a review. IEEE Access **11**, 35479–35516 (2023)
2. Aslan, S., Castellano, G., Digeno, V., Migailo, G., Scaringi, R., Vessio, G.: Recognizing the emotions evoked by artworks through visual features and knowledge graph-embeddings. In: International Conference on Image Analysis and Processing, pp. 129–140. Springer (2022)

3. Bai, S., et alet al.: Qwen2. 5-VL Technical Report. arXiv preprint arXiv:2502.13923 (2025)
4. Castellano, G., Digeno, V., Sansaro, G., Vessio, G.: Leveraging knowledge graphs and deep learning for automatic art analysis. Knowl.-Based Syst. **248**, 108859 (2022)
5. Castellano, G., Lella, E., Vessio, G.: Visual link retrieval and knowledge discovery in painting datasets. Multimedia Tools Appl **80**, 6599–6616 (2021)
6. Castellano, G., Scaringi, R., Vessio, G.: Recognizing the style, genre, and emotion of a work of art through visual and knowledge graph embeddings. In: International Conference of the Italian Association for Artificial Intelligence, pp. 427–440. Springer (2023)
7. Castellano, G., Vessio, G.: A deep learning approach to clustering visual arts. Int. J. Comput. Vision **130**(11), 2590–2605 (2022)
8. Cetinic, E., Lipic, T., Grgic, S.: A deep learning perspective on beauty, sentiment, and remembrance of art. IEEE access **7**, 73694–73710 (2019)
9. Cetinic, E., She, J.: Understanding and creating art with AI: Review and outlook. ACM Trans. Multimedia Comput. Commun. Appl. (TOMM) **18**(2), 1–22 (2022)
10. Chen, C.F.R., Fan, Q., Panda, R.: Crossvit: Cross-attention multi-scale vision transformer for image classification. In: Proceedings of the IEEE/CVF International Conference on Computer Vision, pp. 357–366 (2021)
11. Chen, L., Yang, J.: Recognizing the style of visual arts via adaptive cross-layer correlation. In: Proceedings of the 27th ACM International Conference on Multimedia, pp. 2459–2467 (2019)
12. Dani, M., Rio-Torto, I., Alaniz, S., Akata, Z.: DeViL: decoding vision features into language. In: DAGM German Conference on Pattern Recognition, pp. 363–377. Springer (2023)
13. Fanelli, N., Vessio, G., Castellano, G.: I dream my painting: Connecting MLLMS and diffusion models via prompt generation for text-guided multi-mask inpainting. In: 2025 IEEE/CVF Winter Conference on Applications of Computer Vision (WACV), pp. 6073–6082. IEEE (2025)
14. Garcia, N., Renoust, B., Nakashima, Y.: ContextNet: representation and exploration for painting classification and retrieval in context. Int. J. Multimedia Inf. Retrieval **9**(1), 17–30 (2020)
15. Ghiasi, A., et al.: Goldstein, T.: What do vision transformers learn? A visual exploration. arXiv preprint arXiv:2212.06727 (2022)
16. Grover, A., Leskovec, J.: node2vec: Scalable feature learning for networks. In: Proceedings of the 22nd ACM SIGKDD International Conference on Knowledge Discovery and Data Mining, pp. 855–864 (2016)
17. Han, K., et al.: A survey on vision transformer. IEEE Trans. Pattern Anal. Mach. Intell. **45**(1), 87–110 (2022)
18. Karayev, S., et al.: Recognizing image style. arXiv preprint arXiv:1311.3715 (2013)
19. Kazemi, H., Chegini, A., Geiping, J., Feizi, S., Goldstein, T.: What do we learn from inverting CLIP models? arXiv preprint arXiv:2403.02580 (2024)
20. Kendall, A., Gal, Y., Cipolla, R.: Multi-task learning using uncertainty to weigh losses for scene geometry and semantics. In: Proceedings of the IEEE Conference on Computer Vision and Pattern recognition, pp. 7482–7491 (2018)
21. Kingma, D.P., Ba, J.: Adam: A method for stochastic optimization. arXiv preprint arXiv:1412.6980 (2014)
22. Li, X., et al.: OMG-Seg: Is one model good enough for all segmentation? In: Proceedings of the IEEE/CVF Conference on Computer Vision and Pattern Recognition, pp. 27948–27959 (2024)

23. Li, Z., Liu, F., Yang, W., Peng, S., Zhou, J.: A survey of convolutional neural networks: analysis, applications, and prospects. IEEE Trans. Neural Netw. Learn. Syst. **33**(12), 6999–7019 (2021)
24. Loshchilov, I., Hutter, F.: Sgdr: Stochastic gradient descent with warm restarts. arXiv preprint arXiv:1608.03983 (2016)
25. Loshchilov, I., Hutter, F.: Decoupled weight decay regularization. arXiv preprint arXiv:1711.05101 (2017)
26. Mahendran, A., Vedaldi, A.: Understanding deep image representations by inverting them. In: Proceedings of the IEEE Conference on Computer Vision and Pattern Recognition, pp. 5188–5196 (2015)
27. Natarajan, P., Nambiar, A.: Vale: A multimodal visual and language explanation framework for image classifiers using explainable ai and language models. arXiv preprint arXiv:2408.12808 (2024)
28. Olah, C., Mordvintsev, A., Schubert, L.: Feature visualization. Distill **2**(11), e7 (2017)
29. Radford, A., et al.: Learning transferable visual models from natural language supervision. In: International Conference on Machine Learning, pp. 8748–8763. PmLR (2021)
30. Scaringi, R., Fiameni, G., Vessio, G., Castellano, G.: GraphCLIP: Image-graph contrastive learning for multimodal artwork classification. Knowl.-Based Syst. **310**, 112857 (2025)
31. Selvaraju, R.R., Cogswell, M., Das, A., Vedantam, R., Parikh, D., Batra, D.: Grad-CAM: Visual explanations from deep networks via gradient-based localization. In: Proceedings of the IEEE International Conference on Computer Vision, pp. 618–626 (2017)
32. Wang, Z., Zhao, L., Xing, W.: Stylediffusion: Controllable disentangled style transfer via diffusion models. In: Proceedings of the IEEE/CVF International Conference on Computer Vision, pp. 7677–7689 (2023)
33. Wu, J., Gan, W., Chen, Z., Wan, S., Yu, P.S.: Multimodal large language models: a survey. In: 2023 IEEE International Conference on Big Data (BigData), pp. 2247–2256. IEEE (2023)
34. Xiao, Y., Yuan, Q., Jiang, K., He, J., Lin, C.W., Zhang, L.: TTST: A top-k token selective transformer for remote sensing image super-resolution. IEEE Trans. Image Process. (2024)

2nd Workshop on Generation of Human Face and Body Behavior (GHB 2025)

Workshop

2nd Workshop on Generation of Human Face and Body Behavior (GHB 2025)

In conjunction with ICIAP 2025—Rome, Italy, September 15–19, 2025

Workshop Organization

Organizers

Mohamed Daoudi   IMT Nord Europe, CRIStAL
Stefano Berretti   University of Florence, Italy

Program Committee

Claudio Ferrari   University of Siena, Italy
Leonardo Galteri   Pegaso Telematic University, Italy
Filippo Bartolucci   University of Bologna, Italy
Rama Chellappa   Johns Hopkins University, USA
Naoufel Werghi   Khalifa University, United Arab Emirates
Sabu M. Thampi   Indian Institute of Information Technology & Management, India
Maxime Devanne   Université de Haute-Alsace, France
Evangelos Sariyanidi   Children's Hospital of Philadelphia, USA
Birkan Tunc   University of Pennsylvania, USA
Naima Otberdout   Mohammed VI Polytechnic University(UM6P), Morocco
Emery Pierson   École Polytechnique, France
Thomas Besnier   Université de Lille, France

Enhancing 3D Reconstruction: A Preliminary Study on Geometric Algebra Based Loss

Bruno Francesco Nocera[1]([✉]), Leonardo Colosi[1], Lorenzo Papa[2], and Irene Amerini[1]

[1] Sapienza University, 00185 Rome, RM, Italy
`{colosi,nocera}@studenti.uniroma1.it, amerini@diag.it`
[2] ESA Centre for Earth Observation, 00044 Frascati, RM, Italy
`Lorenzo.papa@esa.int`

Abstract. The need for 3D reconstruction from incomplete data sources leads to difficulties in scenarios where geometric fidelity is crucial. Consequently, the reconstruction from partial or sparse inputs remains a critical challenge in applications such as robotics, autonomous navigation, and cultural heritage digitization, where geometric accuracy and structural consistency are essential. Moreover, existing methods typically rely on Euclidean vector spaces and traditional loss functions, which often fail to capture the full complexity of spatial relationships. Consequently, this work presents a novel approach to 3D reconstruction by introducing Geometric Algebra (GA) as a guiding framework within deep learning models. More in detail, we propose two novel GA-based loss functions, i.e., the MultivectorLoss and the PGALoss, that embed geometric structure directly into the optimization process. This formulation enables the neural network to reason about orientation, scale, and alignment in a more coherent and interpretable way. Preliminary results on the ShapeNet and MVP datasets underscore promising performances when employed in training well-known state-of-the-art models. These findings highlight the potential and innovative use of Geometric Algebra as a powerful and underexplored tool for advancing 3D deep learning tasks.

Keywords: Geometric Algebra · 3D Deep Learning · 3D Models Reconstruction

1 Introduction

Three-dimensional (3D) reconstruction, which involves inferring geometric models from sensor data, is crucial in numerous fields, including robotics, autonomous navigation, preservation of cultural heritage, and medical imaging. Traditional methods such as Multi-View Stereo (MVS) [12], Structure-from-Motion (SfM) [14], and Shape-from-Shading (SfS) [6] offer accurate reconstructions but rely heavily on dense, calibrated imagery. These methods typically suffer from

E. Rodolà et al. (Eds.): ICIAP 2025 Workshops, LNCS 16169, pp. 597–608, 2026.
https://doi.org/10.1007/978-3-032-11317-7_48

high computational costs and limited robustness, hindering their practical use in real-world scenarios.

Deep learning approaches have emerged as effective alternatives, predicting 3D shapes directly from sparse data such as single RGB images or incomplete LiDAR scans. These methods leverage priors learned from extensive data [4,16,17]. Nevertheless, current deep learning techniques operate primarily within traditional Euclidean vector spaces ($\mathbb{R}^3$) and rely on Euclidean loss functions, such as Chamfer Distance (CD) or Earth-Mover's Distance (EMD). These losses independently evaluate point differences, neglecting important geometric relationships like orthogonality and co-planarity, which results in reconstructions that often lack global coherence [3].

Geometric Algebra (GA). Geometric Algebra, also known as Clifford Algebra, provides a comprehensive mathematical framework extending traditional vector algebra by **introducing the geometric product**. This product integrates both inner ($\cdot$) and outer ($\wedge$) products [5], enabling compact and transformation-consistent representations of complex geometric structures. In particular, Projective Geometric Algebra ($G_{3,0,1}$) inherently models rigid transformations such as rotations and translations, making it particularly suitable for 3D reasoning tasks.

Contributions. In this paper, we leverage GA to explicitly integrate geometric constraints into deep learning-based 3D reconstruction. Our primary contributions include:

1. Introducing two new loss functions: MultivectorLoss, based on Euclidean Geometric Algebra ($G_{3,0,0}$), and PGALoss, based on Projective Geometric Algebra ($G_{3,0,1}$). These losses evaluate geometric discrepancies in multivector spaces and naturally account for rigid transformations.
2. Demonstrating the integration of these losses into established deep learning architectures, specifically Pixel2Mesh [15] for single-view mesh reconstruction and PoinTr [18] for point cloud completion, without requiring modifications to the model structures.
3. Conducting comprehensive experiments on widely used benchmarks (ShapeNet and MVP datasets), which show consistent improvements in reconstruction accuracy (up to 1.8% higher F-Score) with only a slight increase (6%) in training time. Additionally, we perform detailed ablation studies to analyze the individual effects of our proposed losses, providing practical insights for future research.

2   Related Works

This section briefly reviews key concepts in Geometric Algebra and its recent applications in neural networks, with a focus on 3D reconstruction tasks.

Geometric Algebra (GA). Derived from Clifford algebra, is a powerful mathematical framework for modeling complex geometric transformations and relationships. It unifies various algebraic operations and is widely applied in physics, computer graphics, molecular modeling, point cloud analysis, and 3D mesh manipulation [1,7,13]. A key distinction between GA and traditional vector spaces (e.g., $\mathbb{R}^3$) lies in the introduction of the *geometric product*, which combines the dot and wedge products into a single operation:

$$\mathbf{ab} = \mathbf{a} \cdot \mathbf{b} + \mathbf{a} \wedge \mathbf{b}.$$

This allows for seamless transitions between scalar quantities and oriented geometric objects, enabling more expressive and compact representations of transformations like rotations and reflections. GA organizes elements by *grade*: scalars (grade 0), vectors (grade 1), bivectors (grade 2), and higher-grade multivectors. A general multivector x is represented as:

$$x = x_s + \sum_i x_i e_i + \sum_{i<j} x_{ij} e_i e_j + \ldots + x_{123\ldots n} e_1 e_2 \ldots e_n,$$

where e_i are basis vectors and the geometric product maintains associativity and distributivity, though not commutativity.

2.1 Geometric Algebra Neural Networks

Translation, permutation, and rotation equivariance are crucial for addressing many graph-structured problems. Clifford Group Equivariant Neural Networks (CGENNs) [11] leverage Clifford algebra to design neural networks that inherently support multivector representations, enabling the encoding of complex geometric relationships directly within their computations. This approach allows CGENNs to handle tasks involving intricate geometric data across various domains such as physics and computer vision, efficiently representing higher-order relationships through parametrization via the geometric product. The Geometric Algebra Transformer (GATr) [2] extends these principles by embedding data in Projective Geometric Algebra and maintaining equivariance to the Euclidean symmetry group E(3). This feature is essential for handling 3D geometric data, ensuring consistent model outputs across different spatial orientations and positions. GATr utilizes multivector attention mechanisms to enhance expressiveness and robustness in applications like robotic motion planning and n-body dynamics, demonstrating superior performance in handling structured geometric data.

2.2 Integration of GA Into Point Cloud Completion and Mesh Deformation

Geometric Algebra has been increasingly applied in deep learning for point cloud reconstruction, enhancing the representation and manipulation of geometric data. Methods like GeoUDF [10] employ a geometry-guided approach to improve surface reconstruction accuracy by estimating unsigned distance functions from sparse point clouds. PUGeo-Net [9] advances point cloud upsampling by modeling local geometric structures through quadratic polynomials, enabling the generation of dense point clouds with high-fidelity geometric details. GAAlign [8] uses geometric algebra for robust point cloud registration within a sampling-based framework. Additionally, formulating point cloud completion as a geometric transformation problem allows models to utilize repetitive structures within 3D objects to effectively reconstruct missing parts and preserve geometric details, highlighting the impactful integration of geometric algebra in processing complex geometric information.

3 Proposed Method

This section introduces Geometric Algebra (GA) into deep learning for 3D reconstruction, detailing two novel GA-based loss functions, MultivectorLoss and PGALoss, that enhance model performance on datasets like ShapeNet and MVP. In order to guide deep learning model training with Geometric Algebra (GA), we introduce two loss functions designed to capture structural and spatial relationships beyond traditional Euclidean measures. To facilitate the training of deep learning models with Geometric Algebra (GA), we propose two novel loss functions that are specifically designed to capture structural and spatial relationships extending beyond conventional Euclidean metrics. These loss functions are formulated within the frameworks of two distinct Clifford algebras: $G_{3,0,0}$ and $G_{3,0,1}$. Here, the notation $G_{p,q,r}$ denotes a Clifford algebra generated by p positive, q negative, and r null (degenerate) basis vectors. In particular, $G_{3,0,0}$ corresponds to the standard three-dimensional Euclidean space, while $G_{3,0,1}$ augments this space with an additional null basis vector, thus enabling a projective geometric representation. The inclusion of the null basis vector in $G_{3,0,1}$ naturally accommodates the modeling of rigid body transformations, such as translations and rotations, within a unified algebraic framework. This fundamental distinction underpins the design of our two proposed loss functions. Before computing the loss, a k-nearest neighbor (KNN) alignment is performed to ensure a meaningful comparison between predicted and ground truth shapes. Precisely, for each predicted point, the closest point in the ground truth set is identified (and vice versa), establishing symmetric correspondences. This step ensures that each compared pair of multivectors represents similar parts of the object, reducing errors caused by point cloud permutation or uneven sampling. In the following, we provide a detailed description of the two proposed loss functions, along with an overview of two state-of-the-art approaches: the first addresses mesh deformation from single-view reference images, and the second focuses on completing partial point clouds derived from noisy LiDAR scans. Both architectures serve as benchmarks to evaluate the effectiveness of the proposed loss functions across different 3D reconstruction tasks.

Multivector Loss ($G_{3,0,0}$). Each 3D point is embedded into an 8-dimensional multivector belonging to the Euclidean Geometric Algebra $G_{3,0,0}$, which allows the encoding of spatial relationships beyond standard vector representations. Given two sets of points P (predicted) and Q (ground truth), the corresponding multivector embeddings $M(p_i)$ and $M(q_j)$ are used to evaluate pairwise interactions via the geometric product:

$$G_{ij} = \sum_{a,b} M(p_i)_a \cdot C_{a,b} \cdot M(q_j)_b,$$

where $C_{a,b}$ denotes the Cayley table encoding the algebraic rules of $G_{3,0,0}$, and the indices a, b iterate over the multivector components. This operation yields an interaction tensor that captures structural relationships between the predicted and reference shapes in multivector space. The final loss is defined as the mean squared error between the predicted and ground truth interaction tensors:

$$\mathcal{L}_{\mathrm{GA}} = \frac{1}{N} \sum_{i=1}^{N} \|G_i^{\mathrm{pred}} - G_i^{\mathrm{gt}}\|_2^2.$$

This formulation penalizes discrepancies in the internal geometric structure of corresponding point sets, promoting alignment in both position and orientation (Fig. 1).

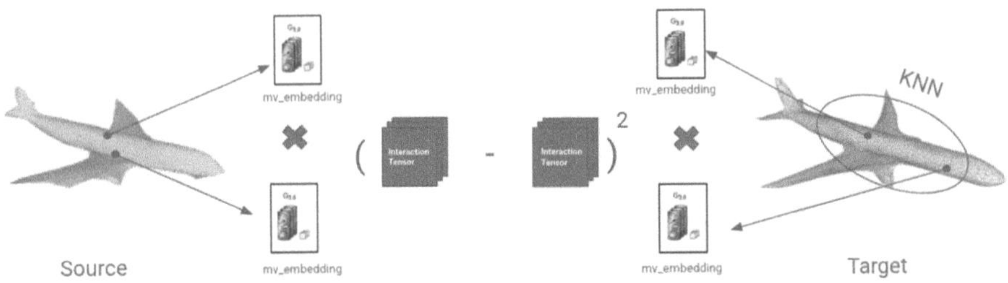

Fig. 1. Graphical representation of the alignment process between the source and target meshes. It can be seen that the multivector space embedding and nearest neighbor calculations enhance structural accuracy and reduce geometric discrepancies.

PGA Loss ($G_{3,0,1}$). In this second configuration, 3D points are embedded into 16 dimensional multivectors within the Projective Geometric Algebra $G_{3,0,1}$, which extends the Euclidean framework to include translations, enabling unified modeling of rigid transformations. To compute the loss, the first step involves evaluating the geometric product between the source and target multivectors:

$$\mathbf{G} = \mathbf{mv}_s \cdot \mathbf{mv}_t,$$

where $\mathbf{mv}_s$ and $\mathbf{mv}_t$ are the multivector embeddings of the predicted and ground truth points, respectively. This operation encodes relative transformations between corresponding entities in the geometric algebra space. Next, a dualization step is applied to the difference between multivectors to enhance the representation of spatial discrepancy:

$$\Delta \mathbf{mv}^* = \text{dualSigns} \cdot \Delta \mathbf{mv}[\text{dualFlip}],$$

where dualSigns and dualFlip specify the algebraic rules needed to compute the dual of the multivector difference. This transformation is used to capture orientation and structural variance in a form suitable for norm-based evaluation. The norms of both the geometric product and the dualized difference are then computed as:

$$\|\Delta \mathbf{mv}^*\| = \sqrt{\sum_i (\Delta \mathbf{mv}_i^*)^2}, \quad \|\mathbf{G}\| = \sqrt{\sum_i \mathbf{G}_i^2},$$

where the summations iterate over all components of the respective multivectors. These norms quantify the magnitude of misalignment and interaction strength. Finally, the loss function is defined as:

$$\mathcal{L} = \frac{1}{N} \sum_{i=1}^{N} \left(\|\mathbf{G}\| + \|\Delta \mathbf{mv}^*\| \right),$$

which combines both the alignment term (via $\|\mathbf{G}\|$) and the discrepancy term (via $\|\Delta \mathbf{mv}^*\|$). This dual-term formulation encourages the model to minimize geometric inconsistencies in both structure and transformation between predicted and target shapes. This formulation emphasizes alignment and correction of geometric discrepancies in orientation, scale, and structure (Fig. 2).

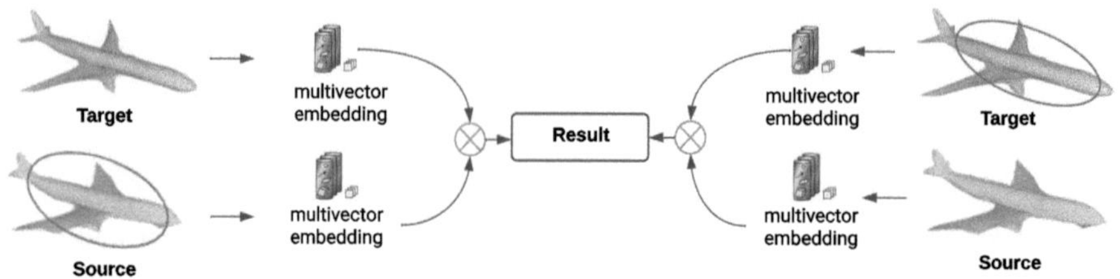

Fig. 2. Illustration of a methodology that emphasizes comparing geometric properties of mesh points through their multivector representations, rather than performing direct pointwise comparisons.

Pixel2Mesh Overview. Pixel2Mesh, [15] is a graph-based convolutional network that reconstructs 3D mesh models from single RGB images. It begins with an ellipsoid and progressively refines the mesh using graph convolutions guided by image features extracted via a CNN. The architecture includes mesh deformation blocks and employs standard loss terms such as Chamfer Distance and Laplacian Regularization to ensure structural accuracy and smoothness. In our work, this framework serves as the baseline for integrating GA-based loss functions.

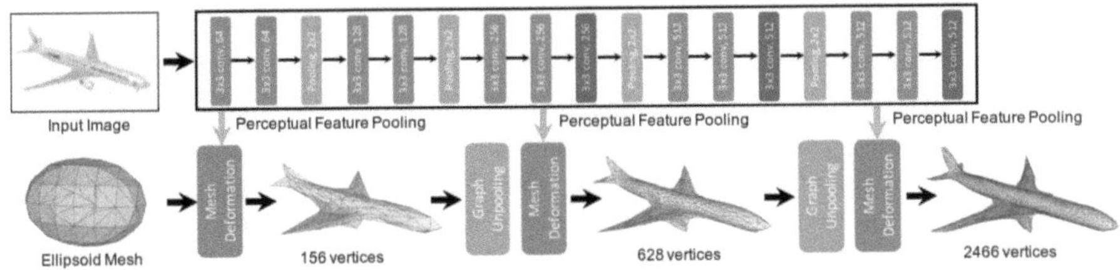

Fig. 3. Overview of Pixel2Mesh pipeline.

Pointr Overview. Pointr [18] is a transformer-based encoder-decoder model for point cloud completion. It leverages geometry-aware modules, such as point proxies and k-nearest neighbor (kNN) attention, to capture local geometric structure. A multi-scale decoding strategy, refined by a Folding-Net, enables dense and detailed output. Optimization is based on Chamfer Distance to ensure geometric accuracy. In our work, this architecture is used to evaluate the integration of GA-based loss functions (Fig. 4).

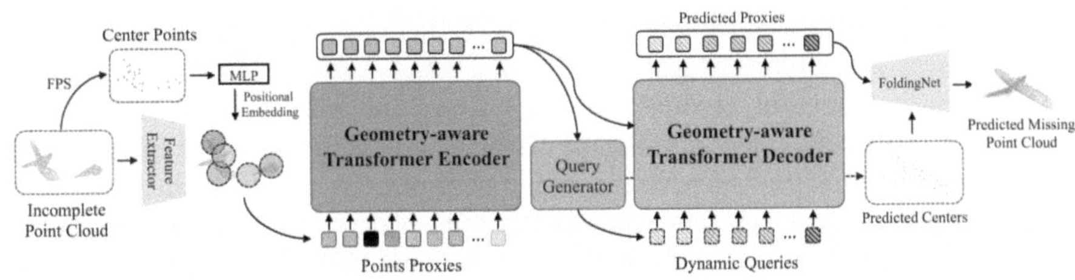

Fig. 4. Overview of PoinTr pipeline.

4   Experiments

Experiments were conducted to investigate the effectiveness of integrating Geometric Algebra (GA) into neural network models for the tasks of point cloud completion and mesh deformation. For the point cloud completion experiments, we adopted the MVP dataset (presented at CVPR 2021), which includes 16 object categories, each containing 2,048 points captured from 26 distinct camera viewpoints. For mesh deformation, we utilized a customized ShapeNet dataset comprising over 175,000 training and 43,000 testing images, each accompanied by precise camera parameters. The models were evaluated using widely recognized geometric metrics, including Chamfer Distance (CD), Earth Mover's Distance (EMD), F-Score, as well as Normal and Edge Regularization losses. Experiments involving point cloud completion were performed on an RTX 3060 GPU, whereas mesh deformation tasks utilized a more powerful RTX 3090 GPU.

The implementation was carried out using CUDA-accelerated PyTorch alongside supportive libraries such as `torch-geometric`, `open3d`, and `wandb` for efficient geometric processing and experiment tracking. Training was optimized with the AdamW optimizer, employing a learning rate of 5×10^{-5} for point cloud tasks and 1×10^{-6} for mesh deformation tasks, complemented by weight decay and gradient accumulation strategies. Overall, the GA-enhanced models demonstrated consistent improvements over baseline methods, thanks to meticulous tuning and optimization. Detailed quantitative and qualitative analyses are presented and discussed in the subsequent subsections.

4.1 Quantitative Results

Mesh Deformation Task. The experimental analysis of the developed loss functions based on Geometric Algebra is summarized in Table 1. The first proposed variant, referred to as GA Loss, utilizes the algebra $G_{3,0,0}$ and aims to exploit the expressive power of multivectors to enhance 3D mesh reconstruction. However, as evident from the results in Table 1, this loss does not lead to a significant improvement in reconstruction quality. A plausible explanation lies in its unidirectional formulation, computing the loss only from the predicted mesh to the ground truth, which, as detailed in the Methods section, is a consequence of the substantial memory demands of this approach. To overcome these limitations, a more efficient variant, termed PGA Loss, was introduced. This loss formulation not only reduces memory usage but also enables bidirectional supervision, allowing the computation of the loss both from the predicted mesh to the ground truth and vice versa. This improvement proves to be critical: as shown in Table 1, the PGA Loss consistently outperforms all previously evaluated methods, including both the original, non-fine-tuned Pixel2Mesh baseline and its version retrained for two additional epochs. These quantitative gains are corroborated by qualitative results presented in the subsequent sections. Overall, this experiment confirms that geometric algebra-based formulations, when efficiently implemented, can significantly enhance the fidelity and geometric consistency of 3D mesh reconstruction. An ablation study further analyzes the contribution of the PGA Loss within the Pixel2Mesh framework.

Table 1. Performance metrics for original Pixel2Mesh architecture and the same architecture with the new loss function added to evaluate the performance of mesh reconstruction quality. Arrows indicate whether higher ($\uparrow$) or lower ($\downarrow$) values are better.

Model	Epochs	CD $\downarrow$	F-Score (τ) $\uparrow$	F-Score (2τ) $\uparrow$
Pixel2Mesh	0	0.498	64.22	78.03
Pixel2Mesh	2	0.480	65.52	79.00
GALoss	2	0.481	64.80	77.33
PGALoss	2	0.477	66.01	79.24

Point Cloud Completion Task. Table 2 reports experiments evaluating the impact of different loss functions when training the original PoinTr architecture for the point cloud completion task. Fine-tuning was performed over 10 epochs, allowing observation of the influence of various loss combinations on performance metrics such as F-Score, Chamfer Distance (CDℓ_1, CDℓ_2), and Earth Mover's Distance (EMD). Baseline results from the original PoinTr show modest performance changes after 10 epochs. Incorporating GALoss improves performance slightly across all metrics, with notable gains in F-Score and reductions in Chamfer and EMD metrics. The proposed PGALoss significantly enhances results, achieving the highest F-Score of 0.4496 and lowest values across all distance metrics, outperforming both the baseline and GALoss. These findings demonstrate the effectiveness of PGALoss in enhancing point cloud reconstruction quality, highlighting its suitability as a robust loss function for fine-tuning the PoinTr model.

Table 2. Comparison of PoinTr fine-tuned with different loss functions on point cloud completion. Arrows indicate whether higher ($\uparrow$) or lower ($\downarrow$) values are better. Best results are highlighted in bold.

Training Loss	Epochs	F-Score $\uparrow$	CDℓ_1 $\downarrow$	CDℓ_2 $\downarrow$	EMD $\downarrow$
ZeroShot	0	0.4214	14.56	0.8954	24.90
ChamferDistance	10	0.4162	14.20	0.7647	23.65
GALoss	10	0.4335	13.80	0.7114	23.57
PGALoss	10	0.4496	**13.61**	**0.6887**	**22.55**

4.2   Qualitative Results

Fig. 5. Qualitative result of loss in comparison with Pixel2Mesh in all the stages of deformation. (a) represent the baseline model, (b) the baseline model trained for 2 more epochs, (c) the Pixel2Mesh model integrated with pga loss.

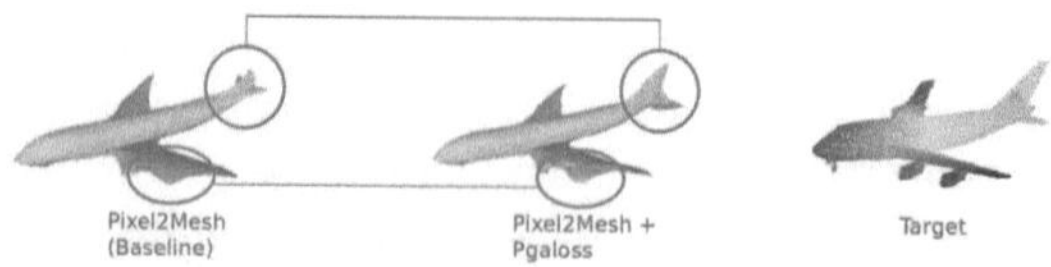

Fig. 6. Pixel2Mesh with the introduction of pga loss visually increases the details of the reconstructed mesh with respect to the target.

Mesh Deformation Task. As shown in Table 1, retraining the Pixel2Mesh baseline for a few additional epochs leads to marginal improvements in the quantitative metrics. However, qualitative analysis of the reconstructed meshes reveals that the model primarily attempts to enhance surface detail, albeit in a rather superficial and inconsistent manner. Figure 5 presents a comparison between the baseline model retrained for two epochs, with and without the proposed loss term based on Projective Geometric Algebra (PGA). Notably, in the absence of this geometric loss, the model tends to improve local mesh details in a coarse and imprecise fashion. Conversely, incorporating the PGA loss results in slight metric improvements, as reported in Table 1. Despite these modest quantitative gains, the introduction of this loss leads to significantly more coherent and geometrically consistent surface details in the reconstructed meshes when compared to the ground truth. This is particularly evident in Fig. 6, where visual inspections reveal that finer structural elements, such as the wings and tail of an airplane, are reproduced with greater fidelity and adherence to the target geometry. The improvements, while subtle in the numerical metrics, translate into more semantically and structurally accurate 3D reconstructions.

Point Cloud Completion Task. Figure 3 provides qualitative visualizations comparing baseline (Zero-Shot) reconstructions against fine-tuned results with various loss combinations. The images clearly reflect improvements in structural accuracy and detail preservation, consistent with quantitative results from Table 2. The reconstructions using Chamfer Distance, GALoss, and the proposed PGALoss demonstrate notable qualitative enhancements. Importantly, none of the tested fine-tuning combinations caused catastrophic forgetting, confirming the robustness of the original PoinTr model (Table 3).

Table 3. Qualitative comparison illustrating the effects of various loss combinations in fine-tuning PoinTr compared to the baseline (Zero-Shot). Row (a) shows partial inputs, PoinTr reconstructions, and ground truths. Row (b) displays reconstructions obtained using Chamfer Distance, GALoss, and the proposed PGALoss.

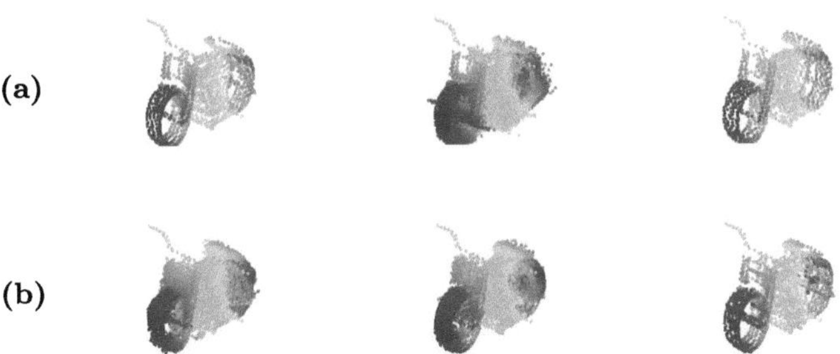

5   Conclusion

This paper explored the integration of Geometric Algebra (GA) into 3D mesh reconstruction from single images, emphasizing the adoption of GA-based loss functions to enhance the fidelity of reconstructions. The introduction of the Projective Geometric Algebra (PGA) Loss notably contributed to stable loss trends and minimized oscillations during training, which was quantitatively evidenced by improved Chamfer Distance and F1-scores. These findings underline the potential of GA not only as a robust mathematical framework for encoding complex spatial relationships but also as a practical tool in deep learning optimization. Despite the beneficial outcomes, integrating GA into deep learning frameworks introduced significant computational overhead and optimization challenges, particularly in scaling to large datasets or complex models such as PoinTr and Pixel2Mesh. The computational complexity associated with GA operations demands innovative solutions to balance geometric expressiveness with computational efficiency. Future work should focus on optimizing GA-based architectures for improved computational performance, possibly through hardware acceleration or advanced software libraries. Additionally, exploring hybrid approaches that combine traditional deep learning techniques with GA-based methods could provide a pathway to harness the strengths of both paradigms. In conclusion, this research provides a foundation for the use of Geometric Algebra in 3D data processing, suggesting that further exploration and refinement could enable GA to become an indispensable tool in deep learning tasks, guiding the development of more accurate, robust, and computationally efficient 3D reconstruction models.

References

1. Belón, M.C.L.: Applications of conformal geometric algebra in mesh deformation. In: 2013 XXVI Conference on Graphics, Patterns and Images, pp. 39–46 (2013). https://doi.org/10.1109/SIBGRAPI.2013.15

2. Brehmer, J., de Haan, P., Behrends, S., Cohen, T.: Geometric algebra transformer (2023). https://arxiv.org/abs/2305.18415
3. Bronstein, M.M., Bruna, J., LeCun, Y., Szlam, A., Vandergheynst, P.: Geometric deep learning: going beyond Euclidean data. IEEE Signal Process. Mag. **34**(4), 18–42 (2017)
4. Choy, C.B., Xu, D., Gwak, J., Chen, K., Savarese, S.: 3D-R2N2: a unified approach for single and multi-view 3d object reconstruction. In: European Conference on Computer Vision (ECCV), pp. 628–644 (2016)
5. Dorst, L., Fontijne, D., Mann, S.: Geometric Algebra for Computer Science: An Object-Oriented Approach to Geometry. Morgan Kaufmann (2007)
6. Horn, B.K.P.: Shape from shading: A method for obtaining the shape of a smooth opaque object from one view. Massachusetts Institute of Technology, Artificial Intelligence Laboratory (1970)
7. Kipf, T., Fetaya, E., Wang, K.C., Welling, M., Zemel, R.: Neural relational inference for interacting systems (2018). https://arxiv.org/abs/1802.04687
8. Neumann, K.A., Hildenbrand, D., Stock, F., Steinmetz, C., Michel, M.: GAAlign: robust sampling-based point cloud registration using geometric algebra. In: Araujo Da Silva, D.W.H., Hildenbrand, D., Hitzer, E. (eds.) Advanced Computational Applications of Geometric Algebra, pp. 161–180. Springer Nature Switzerland, Cham (2024)
9. Qian, Y., Hou, J., Kwong, S., He, Y.: PUGeo-Net: A geometry-centric network for 3D point cloud upsampling (2020). https://arxiv.org/abs/2002.10277
10. Ren, S., Hou, J., Chen, X., He, Y., Wang, W.: GeoUDF: Surface reconstruction from 3D point clouds via geometry-guided distance representation (2023). https://arxiv.org/abs/2211.16762
11. Ruhe, D., Brandstetter, J., Forré, P.: Clifford group equivariant neural networks (2023). https://arxiv.org/abs/2305.11141
12. Seitz, S.M., Curless, B., Diebel, J., Scharstein, D., Szeliski, R.: A comparison and evaluation of multi-view stereo reconstruction algorithms. In: IEEE Conference on Computer Vision and Pattern Recognition (CVPR), pp. 519–528 (2006)
13. Sveier, A., Kleppe, A.L., Tingelstad, L., Egeland, O.: Object detection in point clouds using conformal geometric algebra. Adv. Appl. Clifford Algebras **27**(3), 1961–1976 (2017). https://doi.org/10.1007/s00006-017-0759-1
14. Ullman, S.: The interpretation of structure from motion. Proc. R. Soc. London Ser. B. Biol. Sci. **203**(1153), 405–426 (1979)
15. Wang, N., Zhang, Y., Li, Z., Fu, Y., Liu, W., Jiang, Y.G.: Pixel2Mesh: Generating 3D mesh models from single RGB images (2018). https://arxiv.org/abs/1804.01654
16. Wang, Y., Sun, Y., Liu, Z., Sarma, S.E., Bronstein, M.M., Solomon, J.M.: Dynamic graph CNN for learning on point clouds. In: ACM Transactions on Graphics (TOG). vol. 38, p. 146 (2019)
17. Wu, Z., et al.: 3D ShapeNets: a deep representation for volumetric shapes. In: IEEE Conference on Computer Vision and Pattern Recognition (CVPR), pp. 1912–1920 (2015)
18. Yu, X., Rao, Y., Wang, Z., Liu, Z., Lu, J., Zhou, J.: PoinTr: Diverse point cloud completion with geometry-aware transformers (2021). https://arxiv.org/abs/2108.08839

SISMA: <u>S</u>emantic Face <u>I</u>mage <u>S</u>ynthesis with <u>Ma</u>mba

Filippo Botti[1]([envelope])[iD], Alex Ergasti[1][iD], Tomaso Fontanini[1][iD],
Claudio Ferrari[2][iD], Massimo Bertozzi[1][iD], and Andrea Prati[1][iD]

[1] Department of Engineering and Architecture, University of Parma,
43124 Parma, Italy
`filippo.botti@unipr.it`
[2] Department of Information Engineering and Mathematical Sciences,
University of Siena, 53100 Siena, Italy

Abstract. Diffusion Models have become very popular for Semantic Image Synthesis (SIS) of human faces. Nevertheless, their training and inference is computationally expensive and their computational requirements are high due to the quadratic complexity of attention layers. In this paper, we propose a novel architecture called SISMA, based on the recently proposed Mamba. SISMA generates high quality samples by controlling their shape using a semantic mask at a reduced computational demand. We validated our approach through comprehensive experiments with CelebAMask-HQ, revealing that our architecture not only achieves a better FID score yet also operates at three times the speed of state-of-the-art architectures. This indicates that the proposed design is a viable, lightweight substitute to transformer-based models.

Keywords: Semantic Image Synthesis · Mamba · Flow Matching

1 Introduction

Semantic Image Synthesis (SIS) refers to the task of generating samples conditioned on a semantic mask, that is an image in which each pixel identifies a specific semantic class. In the case of human faces, each class define a different part such as eyes, hair or mouth. The idea is that the mask can condition the shape of the generated samples, while the styles and textures of the semantic parts are either inferred by the generative model, or extracted by a target image.

In the past years, SIS was dominated by methods based on Generative Adversarial Networks (GANs), with several solutions that were proposed usually employing custom normalization layers to inject the mask information. Among them, one of the most popular is SEAN [36] which can both control the shape and the style of the generated samples with high precision. Nevertheless, SEAN can not freely generate samples without a supporting style image, and therefore lacks diversity in the generation. Other approaches like [26] can instead generate highly diverse samples, but lack fine-grained control over the style of specific

E. Rodolà et al. (Eds.): ICIAP 2025 Workshops, LNCS 16169, pp. 609–619, 2026.
https://doi.org/10.1007/978-3-032-11317-7_49

semantic classes. Taking the previous methods as an example, we can therefore divide SIS architectures in two categories: *style-driven* approaches, in which a reference image is always necessary to guide the style of the generated samples, and *diversity-driven* approaches, in which the style is generated from scratch without any specific reference. Despite this coarse categorization, some recent works showed that performing both is also possible [28].

Thanks to the great success of Diffusion Models (DM) in various generation tasks, and especially in text-to-image generation [25], they have recently been utilized also in SIS. Several DM-based methods have been proposed such as [14,32,34], which reached impressive results in diversity-driven generation. Nevertheless, the major drawback of DMs is the huge amount of resources required for training. Performing inference is also a costly process, which limits their practical use. For this reason, recently, State Space Models (SSM) [10,11] have attracted a lot of attention due to their great efficiency. In particular, Mamba [9] became popular due to its capability of competing with transformer-based architectures while requiring much less computational resources. This is due to the fact that the memory required by Mamba grows linearly with respect to the sequence length rather than quadratically as in transformers. Due to the aforementioned advantages, many Mamba-based diffusion models have been developed [3,8,12].

In this paper, we propose SISMA, a novel diffusion model architecture based on Mamba specifically tailored for SIS. In particular, our focus was on the generation of human faces, since the possibility of controlling the shape when synthesizing new identities is of paramount importance for several applications such as generating different variations of the same face. Our architecture substitutes the transformer blocks of DiT [24] with novel SISMA blocks that can be conditioned by semantic masks during the generation without the need of custom normalization or attention layers. In this work we provide the following contributions:

- We designed and propose an architecture based on Mamba that can perform diversity-driven SIS with greater efficiency but comparable quality.
- A novel block, named SISMA, that combines a Self-Mamba layer to model intra-feature interactions and a Cross-Mamba layer that enables the mask conditioning to control the shape of the generated samples.

2   Related Work

Semantic image synthesis (SIS) architectures can be broadly divided into two main categories: those based on Generative Adversarial Networks (GANs) and those leveraging diffusion models. Within the GAN-based paradigm, methods are further distinguished by whether or not they rely on reference style images.

Among the GAN-based approaches, SEAN [36] stands out for its use of SPatial Adaptive DE-normalization (SPADE) layers [23], which inject semantic segmentation masks and style embeddings into the generator. However, SEAN suffers from limited output diversity, producing identical results for identical inputs. Another notable model is CLADE (CLass-Adaptive DE-normalization) [28], which supports both style-driven and diversity-driven synthesis. This flexibility makes it suitable for a broader range of applications. Additionally, INADE [27] takes a different approach by conditioning on instance-level style information instead of class-level, offering finer control over individual elements in the generated image. Meanwhile, FrankenMask [7] and its variant [5] pushes the boundaries of automated semantic mask manipulation, further extending the capabilities of SIS. SemanticStyleGAN [26] represents a novel evolution of the StyleGAN [15] family. Unlike earlier models, it separates style embeddings from semantic mask information and avoids direct conditioning on the mask. It employs GAN inversion [35] to generate facial images from semantic layouts. Additionally, Tarollo *et al.* [30] showcased a SIS model based on [6] that can be used in adversarial attacks against facial recognition systems, underlining their broader relevance.

In the diffusion model domain, the Semantic Diffusion Model (SDM) [32] has gained attention for its simplicity and effectiveness. It can generate diverse images conditioned solely on semantic masks, though it lacks the control provided by reference images. Other diffusion-based techniques include ControlNet [34], Composer [13], and T2I [22], all of which enhance synthesis control by injecting semantic mask features into UNet architectures. Collaborative Diffusion [14] introduces a multi-model cooperative setup, producing high-quality facial images. Finally, SCA-DM [4] is one of the latest models in the field, supporting both style-aware and style-agnostic semantic synthesis.

Differently from the existing literature, in this work we leverage the efficiency of state space models to propose an alternative SIS diffusion model with linear complexity that do not necessitate of any attention mechanism or custom normalization techniques to condition the generation.

3   Methodology

Our proposed architecture is based on Mamba, and is trained following a flow-matching strategy. To make the paper self-contained, we first briefly describe these two approaches in the following sections before delving into SISMA.

3.1   Flow Matching

Flow Matching (FM) [18,20] has been proposed as a simplified version of the diffusion forward process. It defines the latter as a linear interpolation between a sample from the data distribution $x \sim p_{data}$ and one from a Gaussian distribution $\varepsilon \sim \mathcal{N}(0,1)$. The interpolation can be defined depending on a variable t:

$$z = tx + (1-t)\varepsilon, \quad t \in [0,1] \tag{1}$$

The backward process can then be solved using an ordinary differential equation:

$$\frac{dz}{dt} = v = x - \varepsilon \tag{2}$$

The model is thus trained to estimate the velocity v with $v_\theta(z, t)$ (*i.e.*, the derivative of Eq. 1) taking as input both z and t.

$$\mathcal{L} = ||(x - \varepsilon) - v_\theta(z, t)||^2 \tag{3}$$

3.2 Mamba

Mamba is a sequence-to-sequence architecture based on SSMs [2,9]. It is designed to transform an input sequence $x(t) \in \mathbb{R}$ into an output sequence $y(t) \in \mathbb{R}$ through the use of an internal state $h(t) \in \mathbb{R}^N$, with N representing the dimensionality of the hidden state. The inner equations that describe Mamba, and SSMs, follow the ordinary differentiable equation (ODE) linear system:

$$\begin{aligned} h'(t) &= Ah(t) + Bx(t) \\ y(t) &= Ch(t) + Dx(t) \end{aligned} \tag{4}$$

where the matrices $A \in \mathbb{R}^{N \times N}, B \in \mathbb{R}^{N \times 1}, C \in \mathbb{R}^{1 \times N}$ and $D \in \mathbb{R}$ are learnable. Given that the equations are defined over continuous time, discretization is necessary for incorporation into a deep learning framework. In accordance with the approach in [9], we utilize the zero-order holder (ZOH) method. Here, Δ represents the step size, while $\bar{A}$ and $\bar{B}$ denote the discretized matrices. The equations can thus be expressed in the form of a Recurrent Neural Network (RNN):

$$\begin{aligned} h_k &= \bar{A}h_{k-1} + \bar{B}x_k \\ y_k &= Ch_k + Dx_k \end{aligned} \tag{5}$$

Lastly, a key insight is that in Mamba, unlike other SSMs, the matrices are both learnable and vary with the input. Consequently, for a given input x, the matrices B, C, and Δ are derived through a linear fully-connected layer as follows:

$$B = \mathrm{Lin}_B(x), C = \mathrm{Lin}_C(x), \Delta = \mathrm{Lin}_\Delta(x) \tag{6}$$

3.3 SISMA

Our architecture, called SISMA, illustrated in Fig. 1, consists of several components. The encoder part of a pre-trained Variational AutoEncoder (VAE) [25] is first used to reduce the input dimensionality. Given an image $I \in \mathbb{R}^{C \times H \times W}$, it maps I to a new image $\hat{I} \in \mathbb{R}^{c \times h \times w}$, where $c = 4$, $h = H/f$ and $w = W/f$, with f being the scaling factor of the VAE. Then, a patchify layer is used to decompose $\hat{I}$ into a set of patches $x \in \mathbb{R}^{L \times D}$, where D is the hidden size and $L = \frac{hw}{p^2}$, with p being the patch size. Finally, x is fed into the model which is composed of a series of SISMA blocks. In parallel, a Mask Encoder is used to process the Semantic Mask m and produce a mask embedding $S_M \in \mathbb{R}^{L \times D}$, which is fed to each SISMA block in order to control the shape of the generated samples. The mask encoder is a simple encoder (a single conv2d layer with kernel

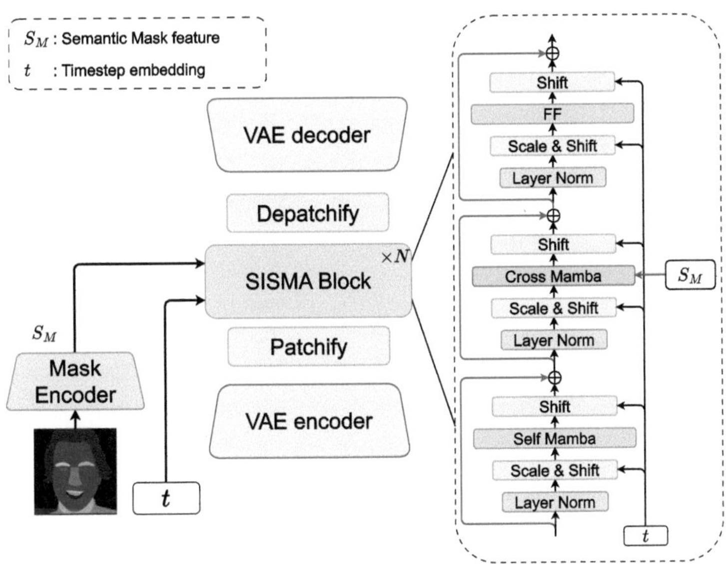

Fig. 1. An overview of our architecture. A VAE encoder is used to reduces the input dimensionality, a Mask Encoder, trained with the model, is used to process the Semantic Mask, obtaining S_M. The core of the model are the N SISMA blocks which are conditioned with S_M and the diffusion timestep t.

1×1) which map each class, passed as a one-hot mask, to a class embedding of size $\mathbb{R}^D$, obtaining an image $\in \mathbb{R}^{D \times h \times w}$ which is then pachified to obtain $S_M \in \mathbb{R}^{L \times D}$. Only the SISMA blocks and the Mask Encoder are trained, while the VAE encoder and decoder weights are frozen.

The core of our model comprises N SISMA blocks, each conditioned on both S_M and the diffusion timestep t. Each SISMA block contains three sequential components: (1) a Self-Mamba module that facilitates intra-feature interactions; (2) a Cross-Mamba module to explicitly condition the diffusion process on S_M; and (3) a Feed-Forward layer. Throughout the architecture, AdaLN [24] (Adaptive Layer Normalization) layers scale and shift features based on the timestep t conditioning. After processing through the SISMA blocks, a depatchify operation is applied, which reverts the patchify layer, transforming the embedding from $\mathbb{R}^{L \times D}$ to $\mathbb{R}^{c \times h \times w}$. Lastly, a VAE decoder generates the final output.

This design effectively integrates selective state-space models with conditional diffusion processes, allowing for semantic-guided generation through the specialized Cross Mamba conditioning mechanism [1]. In detail, the Self-Mamba (Fig. 2a) module leverages the selective scan mechanism to model sequential dependencies within the feature representation. On the other side, the Cross-Mamba (Fig. 2b) module extends this capability by incorporating conditioning from the semantic mask S_M, allowing for guided generation based on the provided semantic layout. Specifically, in Cross-Mamba, the selection matrices Δ and B are computed from S_M, enabling fine-grained control over the generated

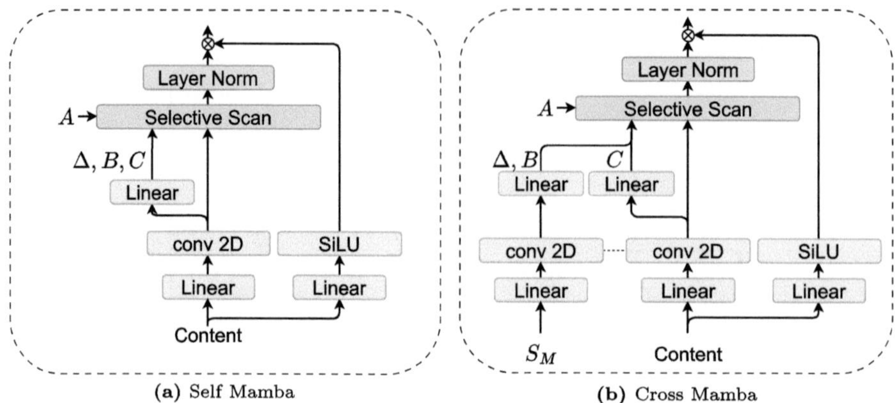

(a) Self Mamba   **(b)** Cross Mamba

Fig. 2. a) An illustration of the Self Mamba, which is a basic mamba block. b) An illustration of Cross Mamba, in this block the matrices Δ and B are obtained from the Semantic Mask S_M in order to condition the diffusion process.

content based on semantic regions, while the matrix C is obtained from the content x, differently from Eq. 6:

$$B = \mathrm{Lin_B}(S_M), \Delta = \mathrm{Lin_\Delta}(S_M), C = \mathrm{Lin_C}(x) \tag{7}$$

By adopting this matrix generation, we are able to condition the generation phase with a semantic mask and obtain an image where the shape of the generated object follow the condition mask.

4   Results

4.1   Datasets

We trained our model on CelebAMask-HQ [17] dataset, which contains 30k images of human faces paired with the corresponding semantic mask. Each mask is composed by 19 different classes describing the shape of each face part, such as nose, hair, *etc....*

4.2   Training Details

In order to train our model, we use a batch size of 8, with a learning rate of 0.004 and no weight decay. Our architecture is composed of 24 SISMA blocks, each with a hidden dimension set to 1024. We do not perform any kind of data augmentation, except for a center cropping transformation and set 256×256 as image size. In line with standard approaches in generative modeling research, we use an exponential moving average (EMA) with a decay factor of 0.9999 to track the weights throughout the training process. All the results and metrics are obtained with EMA model trained for 150K iteration. Finally, AdamW with default configuration [16,21] is used as optimizer.

Table 1. Quantitative comparison with existing methods on semantic image synthesis. Best results are shown in **bold**. 0 s in LPIPS means that the model cannot produce diverse results while - indicates that results for that dataset are not available.

Method	FID ↓	LPIPS ↑
Pix2PixHD [31]	38.5	0
SPADE [23]	29.2	0
DAGAN [29]	29.1	0
SCGAN [33]	20.8	0
CLADE [28]	30.6	0
CC-FPSE [19]	-	-
GroupDNet [37]	25.9	0.37
INADE [27]	21.5	0.415
SDM [32]	18.8	0.42
SISMA (Ours)	**18.3**	**0.61**

Table 2. Efficiency of SDM in comparison with SISMA.

	Params (M)	Inference Time (s)
SDM	**653**	22.79
SISMA	799	**8.51**

4.3 Quantitative Results

In Table 1 comparisons with several state-of-the-art methods are presented. We evaluated the quality of the generated samples using the Frechet Inception Distance (FID), which measures the distance between the distribution of real images and the distribution of fake images, and the diversity in the generation using the Learned Perceptual Image Patch Similarity (LPIPS), which measures the similarity between two generated images. SISMA achieves results that are comparable with the state of the art, but without the need for custom normalization or attention layers. Additionally, regarding the diversity of the generated results, SISMA is able to outperform every other model, achieving an LPIPS of 0.61. This demonstrates our model's enhanced ability to generalize and produce different results from the same semantic mask.

Finally, thanks to its design, SISMA is much more efficient than other SIS diffusion models that employ custom normalization layers or attention layers like SDM, which makes use of both SPADE layers and self-attention layers. We verify this aspect by performing a runtime analysis reported in Table 2. Despite having slightly higher number of parameters, SISMA generates samples 3 times faster than SDM on an NVIDIA RTX 4090. In particular, we measured the inference time to generate one image using 200 diffusion steps. Faster inference times imply our model can generate samples with a fraction of the cost with respect to vanilla diffusion models, which is crucial in such computational expensive architectures.

Previous GAN-based approaches such as CLADE, INADE etc. are not considered in this comparison as their structure significantly differs from diffusion models. GANs are way faster at inference given that images are generated in a single, forward pass. However, diffusion models retain much higher quality and versatility; thus, we deem important to compare SISMA with previous diffusion-based approaches, with the goal of improving the computational aspect of such methods, which still represents a non-negligible bottleneck in their practical use.

Fig. 3. Comparison between SDM (second row) and SISMA (last row) on CelebA.

Fig. 4. Results generated by changing seed to show the diversity of SISMA on CelebA.

4.4   Qualitative Results

In Fig. 3, a comparison between samples generated using SDM and sample generated using SISMA is presented. Our model generates images with a very high quality but much more efficiently, while also preserving the semantic shape.

Finally, as illustrated in Fig. 4, SISMA also has the ability to generate multiple, varied results from the same semantic mask. This exemplifies our model capability to enhance diversity and generalization which are crucial aspects for generative tasks like SIS.

5   Conclusions

In this paper, we proposed SISMA, a novel SIS architecture for human faces synthesis based on Mamba. More specifically, we substitute the standard diffusion model backbone with a Mamba-based one that is able to generate samples much more efficiently since it can be conditioned using a semantic mask without any custom normalization layer or attention layer. Experiments on several datasets and an extensive comparison with the state-of-the-art methods proved that our model is able to generate high-quality, competitive results even at a reduced computational cost.

In the future, we would like to explore the possibility of conditioning SISMA using reference style images in order to both control the shape and the style of the generated samples. Also, text could be added as an additional condition to our model allowing much more controllable generation.

Acknowledgements. This work was funded by "Partenariato FAIR (Future Artificial Intelligence Research) - PE00000013, CUP J33C22002830006" funded by the European Union - NextGenerationEU through the italian MUR within NRRP, project DL-MIG.

References

1. Botti, F., et al.: Mamba-ST: state space model for efficient style transfer. In: 2025 IEEE/CVF Winter Conference on Applications of Computer Vision (WACV), pp. 7797–7806 (2025). https://doi.org/10.1109/WACV61041.2025.00757
2. Dao, T., Gu, A.: Transformers are SSMs: Generalized models and efficient algorithms through structured state space duality. arXiv preprint arXiv:2405.21060 (2024)
3. Ergasti, A., Botti, F., Fontanini, T., Ferrari, C., Bertozzi, M., Prati, A.: U-shape mamba: state space model for faster diffusion. In: 2025 IEEE/CVF Conference on Computer Vision and Pattern Recognition Workshops (CVPRW), pp. 3242–3249 (2025). https://doi.org/10.1109/CVPRW67362.2025.00307
4. Ergasti, A., Ferrari, C., Fontanini, T., Bertozzi, M., Prati, A.: Controllable face synthesis with semantic latent diffusion models. In: Palaiahnakote, S., Schuckers, S., Ogier, J.M., Bhattacharya, P., Pal, U., Bhattacharya, S. (eds.) Pattern Recognition. ICPR 2024 International Workshops and Challenges, pp. 337–352. Springer Nature Switzerland, Cham (2025)
5. Fontanini, T., Ferrari, C., Bertozzi, M., Prati, A.: Automatic generation of semantic parts for face image synthesis. In: International Conference on Image Analysis and Processing, pp. 209–221. Springer (2023)
6. Fontanini, T., Ferrari, C., Lisanti, G., Bertozzi, M., Prati, A.: Semantic image synthesis via class-adaptive cross-attention. IEEE Access (2025)

7. Fontanini, T., et al.: FrankenMask: manipulating semantic masks with transformers for face parts editing. Pattern Recogn. Lett. **176**, 14–20 (2023)
8. Gao, Y., Huang, J., Sun, X., Jie, Z., Zhong, Y., Ma, L.: Matten: Video generation with mamba-attention (2024). https://arxiv.org/abs/2405.03025
9. Gu, A., Dao, T.: Mamba: Linear-time sequence modeling with selective state spaces. arXiv preprint arXiv:2312.00752 (2023)
10. Gu, A., Goel, K., Ré, C.: Efficiently modeling long sequences with structured state spaces. arXiv preprint arXiv:2111.00396 (2021)
11. Gu, A., et al.: Combining recurrent, convolutional, and continuous-time models with linear state space layers. Adv. Neural. Inf. Process. Syst. **34**, 572–585 (2021)
12. Hu, V.T., et al.: ZigMa: a DiT-style zigzag mamba diffusion model. In: European Conference on Computer Vision, pp. 148–166. Springer (2024)
13. Huang, L., Chen, D., Liu, Y., Shen, Y., Zhao, D., Zhou, J.: Composer: Creative and controllable image synthesis with composable conditions (2023)
14. Huang, Z., Chan, K.C.K., Jiang, Y., Liu, Z.: Collaborative diffusion for multi-modal face generation and editing (2023)
15. Karras, T., Laine, S., Aila, T.: A style-based generator architecture for generative adversarial networks (2019)
16. Kingma, D.P., Ba, J.: Adam: A method for stochastic optimization. arXiv preprint arXiv:1412.6980 (2014)
17. Lee, C.H., Liu, Z., Wu, L., Luo, P.: MaskGAN: Towards diverse and interactive facial image manipulation (2020)
18. Lipman, Y., Chen, R.T.Q., Ben-Hamu, H., Nickel, M., Le, M.: Flow matching for generative modeling. In: The Eleventh International Conference on Learning Representations (2023). https://openreview.net/forum?id=PqvMRDCJT9t
19. Liu, X., Yin, G., Shao, J., Wang, X., et al.: Learning to predict layout-to-image conditional convolutions for semantic image synthesis. Adv. Neural Inf. Process. Syst. **32** (2019)
20. Liu, X., Gong, C., Liu, Q.: Flow straight and fast: Learning to generate and transfer data with rectified flow. arXiv preprint arXiv:2209.03003 (2022)
21. Loshchilov, I., Hutter, F.: Decoupled weight decay regularization. arXiv preprint arXiv:1711.05101 (2017)
22. Mou, C., et al.: T2I-adapter: Learning adapters to dig out more controllable ability for text-to-image diffusion models. arXiv preprint arXiv:2302.08453 (2023)
23. Park, T., Liu, M.Y., Wang, T.C., Zhu, J.Y.: Semantic image synthesis with spatially-adaptive normalization (2019)
24. Peebles, W., Xie, S.: Scalable diffusion models with transformers. In: Proceedings of the IEEE/CVF International Conference on Computer Vision, pp. 4195–4205 (2023)
25. Rombach, R., Blattmann, A., Lorenz, D., Esser, P., Ommer, B.: High-resolution image synthesis with latent diffusion models (2022)
26. Shi, Y., Yang, X., Wan, Y., Shen, X.: SemanticStyleGAN: Learning compositional generative priors for controllable image synthesis and editing (2022)
27. Tan, Z., et al.: Diverse semantic image synthesis via probability distribution modeling (2021)
28. Tan, Z., et al.: Efficient semantic image synthesis via class-adaptive normalization (2021)
29. Tang, H., Bai, S., Sebe, N.: Dual attention GANs for semantic image synthesis. In: Proceedings of the 28th ACM International Conference on Multimedia, pp. 1994–2002 (2020)

30. Tarollo, G., Fontanini, T., Ferrari, C., Borghi, G., Prati, A.: Adversarial identity injection for semantic face image synthesis. In: Proceedings of the IEEE/CVF Conference on Computer Vision and Pattern Recognition, pp. 1471–1480 (2024)
31. Wang, T.C., Liu, M.Y., Zhu, J.Y., Tao, A., Kautz, J., Catanzaro, B.: High-resolution image synthesis and semantic manipulation with conditional GANs. In: Proceedings of the IEEE Conference on Computer Vision and Pattern Recognition, pp. 8798–8807 (2018)
32. Wang, W., Bao, J., Zhou, W., Chen, D., Chen, D., Yuan, L., Li, H.: Semantic image synthesis via diffusion models (2022)
33. Wang, Y., Qi, L., Chen, Y.C., Zhang, X., Jia, J.: Image synthesis via semantic composition. In: Proceedings of the IEEE/CVF International Conference on Computer Vision, pp. 13749–13758 (2021)
34. Zhang, L., Rao, A., Agrawala, M.: Adding conditional control to text-to-image diffusion models (2023)
35. Zhu, J., Shen, Y., Zhao, D., Zhou, B.: In-domain GAN inversion for real image editing. In: European Conference on Computer Vision, pp. 592–608. Springer (2020)
36. Zhu, P., Abdal, R., Qin, Y., Wonka, P.: SEAN: image synthesis with semantic region-adaptive normalization. In: 2020 IEEE/CVF Conference on Computer Vision and Pattern Recognition (CVPR). IEEE (2020).https://doi.org/10.1109/cvpr42600.2020.00515
37. Zhu, Z., Xu, Z., You, A., Bai, X.: Semantically multi-modal image synthesis. In: Proceedings of the IEEE/CVF Conference on Computer Vision and Pattern Recognition, pp. 5467–5476 (2020)

1st International Workshop on Humans in the eXtended Artificial Intelligence Loop (HUARL 2025)

Workshop

1st International Workshop on Humans in the eXtended Artificial Intelligence Loop (HUARL 2025)

In conjunction with ICIAP 2025—Rome, Italy, September 15–19, 2025

Workshop Organization

Organizers

Lorenzo Stacchio	University of Macerata, Italy
Primo Zingaretti	Università Politecnica delle Marche, Italy
Emanuele Balloni	Università Politecnica delle Marche, Italy
Pasquale Cascarano	University of Bologna, Italy
Daniele Giunchi	University of Birmingham, UK
Giacomo Vallasciani	University of Bologna, Italy

Program Committee

Vincenzo Armandi	University of Bologna, Italy
Shirin Hajhamadi	University of Bologna, Italy
Silvia Garzarella	University of Bologna, Italy
Andrea Loretti	University of Bologna, Italy
Gabor Soros	Nokia Bell Labs, Hungary
Jeremy Lacoche	Orange Labs, France
Gilda Manfredi	Università degli Studi della Basilicata, Italy
Michele Braccini	University of Bologna, Italy
Akos Nagy	UK
Arlindo Rodrigues	.
Galvão Filho	Brazil
Andrea Felicetti	Università Politecnica delle Marche, Italy
Elisa Saraceni	University of Macerata, Italy

Towards a Human-in-the-Loop Framework for Swarm Intelligence-Driven Dance Generation

Michele Braccini[(⊠)] , Allegra De Filippo , Michele Lombardi ,
and Michela Milano

Department of Computer Science and Engineering, University of Bologna, Bologna,
Italy
{m.braccini,allegra.defilippo,michele.lombardi2,
michela.milano}@unibo.it

Abstract. This paper introduces a preliminary implementation of the
Feature Mapping module, a core component of the novel approach that
exploits swarm intelligence algorithms to generate dance choreography by
virtual agents. We delegate the mapping process from music to simulation
parameters to a neural network, casting its development as an optimization problem. Additionally, we propose an initial qualitative evaluation
apparatus in the form of a questionnaire to assess the performances generated. We leverage preliminary human evaluation results to assess the
automatic evaluation system employed, specifically the fitness functions
and hyperparameters used. This work represents a first step towards the
realization of a human-in-the-loop framework for the generation of artistic performances. Moreover, this leads to the realization of eXtended
Reality application scenarios.

Keywords: Swarm Intelligence · Dance Generation · Feature
Mapping · Human Evaluations

1 Introduction

With the advent of Generative AI (GenAI) the artistic domain has been flooded
with a plethora of AI-based approaches that promise, among other things, an
increase in creativity compared to the past [7,14]. The literature offers several
examples of such systems. In particular, [15,17] present frameworks that integrate eXtended Reality (XR) and AI to support creative thinking; [18] explores
co-creativity in music by involving AI-driven robots with an improvising human
musician; in [19] a framework for human-AI music co-creation is discussed. In [13]
authors explore human-computer collaboration for music composition, while [12]
focuses on collaborative human-AI interactions to generate drawing sketches.
Regarding dance, our main area of interest, [5,6] propose systems for the generation of robotic choreography, leveraging AI techniques, and also including the

E. Rodolà et al. (Eds.): ICIAP 2025 Workshops, LNCS 16169, pp. 625–636, 2026.
https://doi.org/10.1007/978-3-032-11317-7_50

use of large language models in the second study. Finally, [16] presents a study aimed at improving *LuminAI*, a co-creative AI dance partner.

However, the creativity of these AI models is often accompanied by significant computational cost, both in training and execution. This makes their implementation in real-time applications still difficult or, in some cases, prohibitive. This aspect becomes particularly critical in eXtended Reality applications, especially those that integrate XR, GenAI and human interactions [1]. Indeed, to allow effective interaction with humans, XR systems must render images, text, and audio on head-mounted displays with a maximum latency of about 20 milliseconds.

Recently, a novel and computationally cost-effective AI-based approach to dance generation has been proposed in [2]. The approach takes inspiration by the complex dynamics of social insect colonies to make virtual agents perform movements in space that resemble a dance choreography. Swarm intelligence algorithms and, in particular, their self-organizing and emergence properties are used to produce unexpected and creative artistic outcomes from simple agents.

In this paper, we propose a preliminary implementation of one of the fundamental modules of the swarm intelligence approach to generate dance choreographies, namely the "Feature Mapping" module. This module is responsible for the mapping between the musical features extracted from the MIDI track and the simulation parameters that regulate the dynamics and thus the movements of the agents. In addition, an initial apparatus for the qualitative evaluation of the performances produced is proposed, inspired by the seminal works on robotic dance creation and evaluation [8,9]. This process is embodied by a questionnaire proposed to both dance experts and non-experts. This work thus represents a first step towards the realization of a human-in-the-loop framework for the generation of artistic performances by means of swarm intelligence algorithms, also paving the way for the realization of the XR scenarios previously presented in [3].

The paper is structured as follows. Section 2 describes the modeling and implementation choices, as well as the experimental setting used to implement and collect preliminary results on the *Feature Mapping* module. Furthermore, the different fitness functions useful for the automatic evaluation of the solutions are introduced. Finally, the questionnaire is reported. Section 3 presents and discusses the preliminary results obtained, while Sect. 4 concludes the paper.

2   Methods

This paper builds on the foundational work that introduced a general and abstract framework for generating dance-like movements for virtual agents using swarm intelligence algorithms. Here, we present an initial implementation of one of its fundamental macro-phases, the *"Feature Mapping"* module.

To facilitate readability, we divide this section into different subsections. Figure 1 illustrates the block diagram that represents the structure of the study.

In particular, Sect. 2.1 is articulated in: *"Feature Mapping Implementation"* which focuses on the modeling and implementation choices made for the implementation of the feature mapping module; and *"Fitness Functions & Grid of*

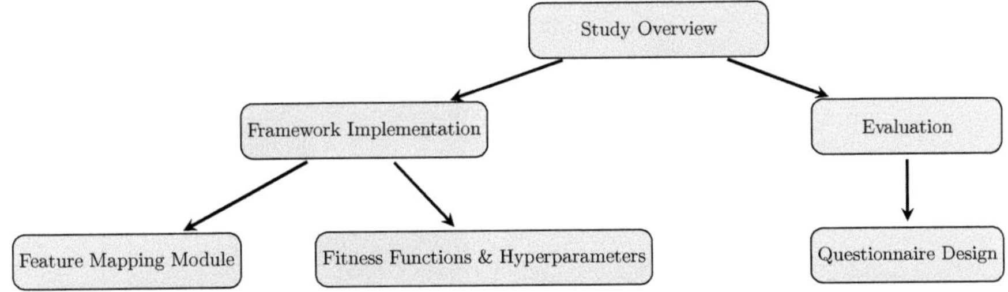

Fig. 1. Block diagram illustrating the study's overall structure.

Hyperparameters" that presents the different fitness functions used and the grid of hyperparameters used to evaluate the solutions produced by the evolutionary algorithm. Finally, *"Questionnaire Design"*, within Sect. 2.2, presents the set of questions posed to dance experts and non-experts to obtain a qualitative evaluation of the selected dance performances and, in turn, a feedback on the effectiveness of the fitness functions used.

2.1   Framework Implementation

Feature Mapping Implementation. Figure 2 illustrates the overall scheme of the proposed approach (firstly introduced in [2]), highlighting in detail the modules whose implementation constitutes the main contribution of this work.

The initial stages of system implementation involve the selection of a song and the subsequent extraction of its musical features. Since these two modules are beyond the scope of this contribution, we will simply report that the songs selected for the training phase are three well-known pieces from the classical repertoire: *"Für Elise"* by Ludwig van Beethoven, *"Turkish March"* by Wolfgang Amadeus Mozart and *"Black Key Étude"* by Frédéric Chopin. Instead, the song used in the test/evaluation phase is *"Minuet in G major, BWV Anh. 114"*, traditionally attributed to Johann Sebastian Bach. The musical features extracted from the corresponding MIDI files are:

- Shannon entropy of the notes;
- mean of volumes;
- number of notes;
- number of different notes;
- range of notes;
- mean absolute interval between successive notes.

Following the schema, the next module is "Feature Mapping", which is the main focus of this work.

As reported in the figure, we cast the problem of generating a dance choreography for agents into an optimization problem. To this purpose, a **genetic algorithm** (GA) is introduced to search into the space of neural networks, in order to find one that is capable of transforming the musical features extracted

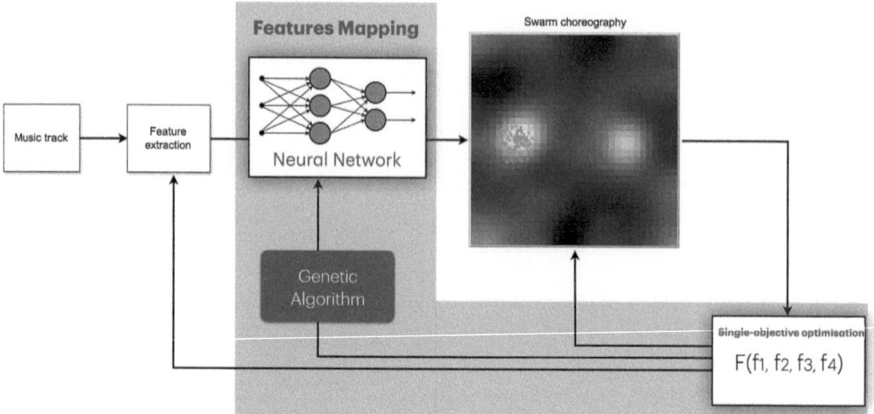

Fig. 2. Overview of the whole approach for generating dance performances using swarm intelligence algorithms. It also illustrates the main components of the "Feature Mapping" module, namely: a neural network, a genetic algorithm and, in turn, the evaluation system.

from the MIDI file into simulation parameters for the agents and to produce a choreography that, for the human observer, can have qualities such as artistic value, creativity and aesthetic pleasure. The choice of a neural network is motivated by its ability to perform complex tasks, in general, and in particular by its proven effectiveness in music-related applications [4]. The whole optimization process of the neural network is performed in offline mode, i.e., the simulation parameters for each musical bar are pre-computed before the actual simulation of the dance by the agents. This allows to meet the strict requirements of real-time generation of the simulation imposed by XR applications and is therefore a valid candidate solution for the visionary XR application scenarios involving dance generation by means of swarm intelligence recently proposed in the work [3].

The neural networks are fed with the 6 musical features above mentioned and produce 6 outputs, which control the simulations. We remember that the two swarm intelligence algorithms that we choose to use for controlling the agents are: PSO (Particle Swarm Optimization) and Ant Colony. The controlled simulation parameters are the following:

- **Attraction to neighbor best**: adjusts the attraction to the best performing neighbor in the local topology (as in the local variant of PSO) [20].
- **Attraction to personal best**: controls the pull toward the agent's own best-known position (PSO).
- **Attraction to pheromone**: determines sensitivity to pheromone trails (Ant Colony).
- **Evaporation rate**: defines how quickly pheromones fade (Ant Colony).
- **Diffusion rate**: governs the spatial spread of pheromones (Ant Colony).

– **Normal noise standard deviation**: regulates the standard deviation parameter of the Gaussian noise, centered at 0, added to the velocity update in the PSO, it is in the range $[0, 0.2]$.

For the sake of completeness, the architecture of the neural network used is reported in Fig. 3, in particular, it is noted that there is a recurrent layer in order to allow memory capacity to the network.

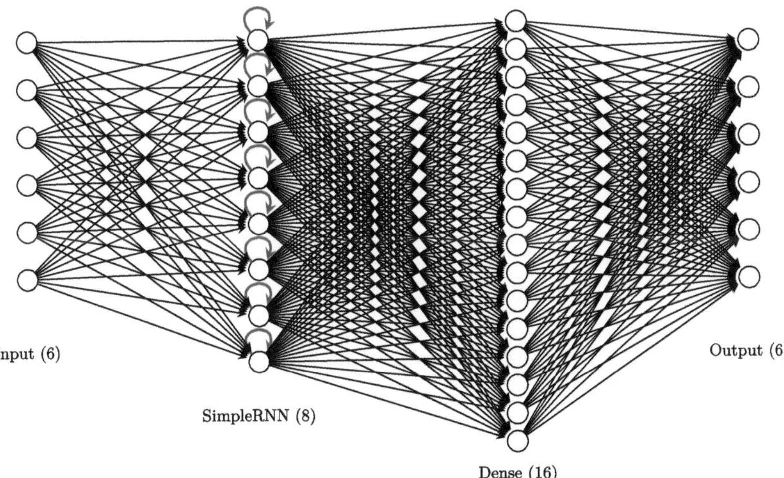

Fig. 3. Architecture of the neural network used to generate the mapping between musical features and agent parameters space. The network consists of an input layer of size 6 neurons, a recurrent layer (RNN) of size 8, a dense layer of size 16 and an output layer of size 6. The recurrent layer provides a memory mechanism, enabling the network to store the information of previous time steps. This could, in principle, allow the network to generate output that not only reacts to the current musical bar, but it is also influenced by previous bars, an essential feature for learning musical structures and generating complex and interesting behavior over time.

Fitness Functions and Grid of Hyperparameters. We first devised 4 sub-functions that could reproduce what for us is a dance choreography performed by virtual agents. In this regard, we believe that agents in a dance must have a good spatial coverage of the dance floor but at the same time the formation of (meta-)stable spatio-temporal structures during music unfolding, namely clusters formation. Plus, they must follow the choreographic patterns, i.e., the geometric shapes that make up the visual choreography in the space of the dance floor we decide to use. Finally, the generation must be music-based, and thus the randomness of the solution produced must be limited, so there must be a link between the inputs (the musical features) and the outputs (the simulation parameters). These considerations thus led to the formulation of these 4 sub-functions:

Table 1. Genetic Algorithm Parameters Used in the Experiments

Parameter	Value/Description
Initial population size	80
Number of generations	50
Selection method	Tournament selection (size = 2)
Crossover method	Two-point crossover
Crossover probability	0.6
Mutation method	Gaussian mutation ($\mu = 0$, $\sigma = 1.0$, `indpb` = 0.3)
Mutation probability	0.4
Algorithm used	eaMuPlusLambda (from DEAP library [10])
Number of individuals selected (μ)	40
Number of offspring per generation (λ)	50

f_1—**Local Closeness:** This sub-function encourages the formation of clusters of agents. It is to be **minimized** since it is computed as the average distance between an agent and its neighbors within a specified radius (R=5);

f_2—**Spatial Spread:** This sub-function captures the degree to which agents are spread across the dance floor. It is to be **maximized** to promote maximal coverage.

f_3—**Dance Floor Values:** This sub-function evaluates how well the agents occupy regions of the dance floor charaterized by higher values, i.e. the food sources that we use as abstractions for representing the spatial patterns that compose a choreography. It is to be **maximized**.

f_4—**Distance Between Feature and Parameter Correlation Matrices:** This sub-function quantifies the dissimilarity between correlation structures of features and parameters. It is to be **minimized** to maintain input-output coherence and so between music and simulation parameters.

Then, we combined these sub-functions in order to produce three different fitness functions. These functions, one at a time, will guide the the search in the space of neural networks acting as selective pressure on the evolutionary process. So, the different fitness functions—compositions of f_1 - f_4 sub-functions—used in this work are the following ones:

$$F_1 = af_1 + b(1 - f_2) + c(1 - f_3) + df_4$$
$$F_2 = af_1 + b(1 - f_2) + c(1 - f_3) + df_4 + (1 - [(1 - f_1) \cdot f_2 \cdot f_3 \cdot (1 - f_4)])$$
$$F_3 = (1 - f_1)^a \cdot f_2^b \cdot f_3^c \cdot (1 - f_4)^d$$

Within the genetic algorithm, F_1 and F_2 are defined as objectives to be minimized, while F_3 is formulated as an objective to be maximized.

The rationale behind the choice of these 3 different functions is rooted in the iterative nature typical of the experimental apparatus, which usually goes from simple to complex by successive steps of improvements and refinements.

Indeed, we started with the simple linear combination of the 4 sub-functions, then obtained a function that also integrates the multiplication of the various functions for the purpose of rewarding the solutions that better balance their contributions, and then, finally, arrived at the multiplication of the functions and weights used as exponents to have nonlinearity[1].

The weights of the fitness functions—hyperparameters of the optimization process—are reported in Table 2. So, a grid search is performed to systematically evaluate their impact on the resulting choreography.

Table 2. Grid of weight combinations (a, b, c, d) used in the fitness functions.

a	b	c	d
0.2	0.2	0.2	0.4
0.2	0.2	0.4	0.2
0.2	0.4	0.2	0.2
0.4	0.2	0.2	0.2
0.25	0.25	0.25	0.25

2.2   Evaluation

Questionnaire Design. The methodological approach to evaluate the neural network produced by GA is inspired by the seminal work [8]. We borrowed some of the questions formulated in that work and formulated new ones to try to capture the unique aspects and peculiarities of our new approach, which combines swarm intelligence and dance. To assess the performance of the agents, we have, therefore, developed a questionnaire that evaluates, among other factors, musical synchronization, use of space, and creativity.

Each participant anonymously rates the proposed choreographies by assigning a score to each aspect using a Likert scale ranging from 1 to 5^2. The actual list of questions posed to human evaluators are reported here:

- To what extent do the swarm's movements align with the rhythm and tempo of the music?
- Is there at least a subset of agents that appear coherent and synchronized with one another during the performance?
- How effectively does the swarm make use of the available space (e.g., the dance floor) throughout the choreography?

---

[1] Note that while higher weights have for the first 2 fitness functions the effect of weighing one component more, in the third being the exponent between 0 and 1 we have the opposite effect.

[2] The Google form used for the evaluation of the selected neural network is available at this link https://forms.gle/Cy7wFnsNtXe7PNqH7.

- How consistently do the swarm's movements follow the predefined choreographic patterns during the performance?
- How emotionally engaging or captivating did you find the swarm's performance?
- To what degree did the swarm resemble a coordinated human dance ensemble?
- How would you evaluate the overall creativity and artistic quality of the swarm's performance?

3   Results and Discussion

From the totality of all combinations of weights and fitness functions, i.e., 5 * 3 = 15, a subset of 5 was selected for the next phase of human evaluation via questionnaire. We considered five solutions to be a good compromise between the coverage of possible combinations and the time effort of human evaluators. The solutions—i.e. neural network instances—were chosen by exclusion, following these criteria: first those that showed no movement, i.e., the agents remained virtually motionless throughout the duration of the music, were removed, and then those with the richest dynamics were chosen from those remaining, according to the subjective criterion of one of the authors. The actual set of solutions chosen S is the following one:

- $S_1 = F_1(0.2, 0.2, 0.4, 0.2)$;
- $S_2 = F_2(0.2, 0.2, 0.4, 0.2)$;
- $S_3 = F_2(0.25, 0.25, 0.25, 0.25)$;
- $S_4 = F_3(0.2, 0.4, 0.2, 0.2)$;
- $S_5 = F_3(0.25, 0.25, 0.25, 0.25)$.

A total of eleven persons took part in the questionnaire. The distribution of participants by background is as follows: **18.2%** have an artistic background, **45.5%** a technical background and **36.4%** a mixed background.

Here we summarize the results obtained by the five solutions. For an immediate understanding of the best solution based on the ratings collected in the questionnaire, we first calculated the arithmetic mean of the scores collected for each of the proposed questions. So after transforming each user rating into an average, we then computed mean and standard deviation over all user evaluations, resulting in a single numerical evaluation summarizing the quality of the solutions (according to the human evaluators).

Despite the low number of people who evaluate the agents' performance, from Table 3, we can discuss some preliminary results that will be a fundamental starting point to extend this analysis in future works. S_2 **and** S_4 **obtained the highest average evaluation** with a value of both being 2.805. Thus, as a first result, we can state that **the two fitness functions** F_2 **and** F_3 **are preferred by human evaluators over** F_1. The S_4 and S_5 solutions show a relatively high standard deviation compared to the other solutions. **This greater variability suggests a great potential for the** F_3 **fitness function** and thus merits

Table 3. Mean and standard deviation of the scores for each solution in the set $\{S_1, \ldots, S_5\}$.

	S_1	S_2	S_3	S_4	S_5
Mean	2.675	2.805	2.632	2.805	2.740
SD	0.881	0.890	0.955	1.023	1.044

further investigation in the future. We can obtain a clearer picture of their preferences by stratifying the human evaluators according to their background, as shown in Fig. 4.

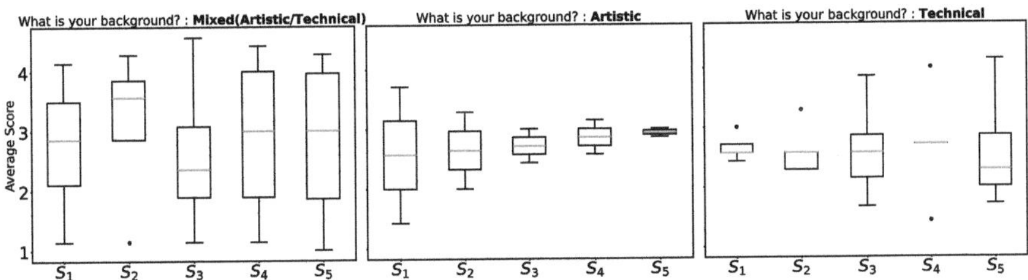

Fig. 4. Average score of each solution $\{S_1, \ldots, S_5\}$ grouped by the background of the human users who participated in the evaluation via the questionnaire. On the left we have the boxplots of the average score for the Mixed(Artistic/Technical) category, while in the center we have the boxplots for the Artistic category and, finally, on the right the one for the Technical only category.

From the boxplots, if we consider the evaluators with an artistic background, the previous trend is confirmed, which sees the fitness functions characterized by the multiplication of the 4 sub-functions (F_2 and F_3) preferred over the only linear combinations (F_1). Again in this category, it can be seen that the variances of the distribution of averages decrease from S_1 to S_5, showing a high agreement of evaluations for S_5. In contrast, the Mixed and Technical user groups show more uniform ratings, with some interesting exceptions. Looking at the distribution of the mean scores assigned by the Technical group, it can be seen that the maximum values increase from S_1 to S_5, although the increasing trend of the spread shows a relative increase in disagreement among the participants. For the mixed group, however, only solution S_2 stands out, which is the one with the highest median and first quartile.

Excluding F_1 (which turned out to be the relatively least interesting solution), since there is no clear winner between the solutions of the F_2 and F_3 functions, we examined the performances proposed to the evaluators with the intention of trying to understand which qualitative characteristics may have elicited the evaluations obtained. All the F_2 and F_3 solutions (i.e., S_2, S_3, S_4, S_5) present **interesting dynamism** and **the ability of the agents to aggregate into**

clusters and break up in order to occupy all the proposed choreographic patterns at the same time. However, what stands out is the ability of the agents of the S_5 solution to slow down and stop their pace when the current music bar presents fewer notes and a lower volume and then restart with more dynamism when the song resumes with more vigor. This perhaps explains the higher rating given by evaluators with an artistic background to this solution.

4   Conclusions

This paper introduces a first implementation of the Feature Mapping module of the approach using swarm intelligence for the generation of dance choreographies by a multitude of artificial agents. From the evaluations gathered through the questionnaire, we can conclude that the implementation described here has successfully produced interesting and, to a certain extent, also artistic and engaging dance performances. This is a significant achievement, since this is the first-ever test of the system with human observers, and therefore its operational viability, which in artistic performances translates more into user satisfaction and appreciation than technical aspects, was entirely to be demonstrated.

However, in its current form it also has some limitations that will have to be addressed in future work. As far as the evaluation apparatus is concerned, the main weakness lies in the limited number of participants in the questionnaire, which, moreover, mainly has an academic background. Furthermore, improvements will have to affect the automatic evaluation system, which will have to accommodate more dance-specific sub-functions. Future work will also be devoted to improving the implementations of the system in order to implement it in XR contexts, as devised in [3].

In future work, we intend also to investigate the pros and cons of transforming the single-objective optimization problem—i.e., finding an appropriate neural network that maps features to parameters—into a multi-objective one. This step aims to avoid being constrained *a priori* by the fitness function landscape imposed by function composition as done in this work. However, this more complex scenario introduces additional challenges and complexities that need to be adequately addressed. In particular, in a multi-objective context, non-dominated solutions belonging to the (approximate) Pareto frontier must be evaluated one-by-one by a human expert, an approach that quickly becomes impractical as the number of solutions increases. Therefore, alternative strategies must be explored. One promising direction is the use of *preference learning*, as in [11]. In these systems, Pareto front solutions—or a representative subset—are ranked by a human expert, allowing the selection of solutions that produce interesting and creative dance performances.

Acknowledgments. Research was funded by PNRR - M4C2 - Investimento 1.3, Partenariato Esteso PE00000013 - "FAIR - Future Artificial Intelligence Research" - Spoke 8 "Pervasive AI", funded by the European Commission under the NextGeneration EU programme.

We would like to thank all the questionnaire participants and, in particular, Silvia Garzarella (Università di Bologna) for her valuable contribution as a dance expert.

References

1. Afzal, M., et al.: Next generation XR systems–large language models meet augmented and virtual reality. IEEE Comput. Graph. Appl. **45**(1), 43–55 (2025)
2. Braccini, M., De Filippo, A., Lombardi, M., Milano, M.: Swarm intelligence: a novel and unconventional approach to dance choreography creation. In: Proceedings of the 3rd Workshop on Artificial Intelligence and Creativity. CEUR Workshop Proceedings, vol. 3810, pp. 162–172 (2024)
3. Braccini, M., De Filippo, A., Lombardi, M., Milano, M.: Dance choreography driven by swarm intelligence in extended reality scenarios: perspectives and implications. In: 2025 IEEE International Conference on Artificial Intelligence and eXtended and Virtual Reality (AIxVR), pp. 348–354 (2025). https://doi.org/10.1109/AIxVR63409.2025.00066
4. Briot, J.P.: From artificial neural networks to deep learning for music generation: history, concepts and trends. Neural Comput. Appl. **33**(1), 39–65 (2021)
5. De Filippo, A., Milano, M.: Robotic choreography creation through symbolic AI techniques. In: Ciancarini, P., Di Iorio, A., Hlavacs, H., Poggi, F. (eds.) Entertainment Computing - ICEC 2023, pp. 346–351. Springer Nature Singapore, Singapore (2023)
6. De Filippo, A., Milano, M.: Large language models for human-AI co-creation of robotic dance performances. In: Proceedings of the Thirty-Third International Joint Conference on Artificial Intelligence. IJCAI '24 (2024). https://doi.org/10.24963/ijcai.2024/844
7. Doshi, A.R., Hauser, O.P.: Generative AI enhances individual creativity but reduces the collective diversity of novel content. Sci. Adv. **10**(28), eadn5290 (2024). https://doi.org/10.1126/sciadv.adn5290
8. Filippo, A.D., Giuliani, L., Mancini, E., Borghesi, A., Mello, P., Milano, M.: Towards symbiotic creativity: a methodological approach to compare human and AI robotic dance creations. In: IJCAI, pp. 5806–5814. ijcai.org (2023)
9. Filippo, A.D., Mello, P., Milano, M.: Do you like dancing robots? AI can tell you why. In: PAIS@ECAI. Frontiers in Artificial Intelligence and Applications, vol. 351, pp. 45–58. IOS Press (2022)
10. Fortin, F.A., De Rainville, F.M., Gardner, M.A., Parizeau, M., Gagné, C.: DEAP: evolutionary algorithms made easy. J. Mach. Learn. Res. **13**, 2171–2175 (2012)
11. Giovanelli, J., Tornede, A., Tornede, T., Lindauer, M.: Interactive hyperparameter optimization in multi-objective problems via preference learning. In: Proceedings of the Thirty-Eighth AAAI Conference on Artificial Intelligence and Thirty-Sixth Conference on Innovative Applications of Artificial Intelligence and Fourteenth Symposium on Educational Advances in Artificial Intelligence. AAAI'24/IAAI'24/EAAI'24, AAAI Press (2024). https://doi.org/10.1609/aaai.v38i11.29106
12. Karimi, P., Rezwana, J., Siddiqui, S., Maher, M.L., Dehbozorgi, N.: Creative sketching partner: an analysis of human-AI co-creativity. In: Proceedings of the 25th International Conference on Intelligent User Interfaces, pp. 221–230. IUI '20, Association for Computing Machinery, New York, NY, USA (2020). https://doi.org/10.1145/3377325.3377522

13. Micchi, G., Bigo, L., Giraud, M., Groult, R., Levé, F.: I keep countingâĂŕ: an experiment in human/AI co-creative songwriting. Trans. Int. Soc. Music Inf. Retrieval (TISMIR) **4**(1), 263–275 (2021). https://doi.org/10.5334/tismir.93
14. Rafner, J., Beaty, R.E., Kaufman, J.C., Lubart, T., Sherson, J.: Creativity in the age of generative AI. Nat. Hum. Behav. **7**(11), 1836–1838 (2023)
15. Santarnecchi, E., et al.: MineVRA: exploring the role of generative AI-driven content development in XR environments through a context-aware approach. IEEE Transactions on Visualization and Computer Graphics (2025)
16. Trajkova, M., Long, D., Deshpande, M., Knowlton, A., Magerko, B.: Exploring collaborative movement improvisation towards the design of LuminAI—a co-creative AI dance partner. In: Proceedings of the 2024 CHI Conference on Human Factors in Computing Systems. CHI '24, Association for Computing Machinery, New York, NY, USA (2024). https://doi.org/10.1145/3613904.3642677
17. Vallasciani, G., Stacchio, L., Cascarano, P., Marfia, G.: CreAIXR: fostering creativity with generative AI in XR environments. In: 2024 IEEE International Conference on Metaverse Computing, Networking, and Applications (MetaCom), pp. 1–8 (2024). https://doi.org/10.1109/MetaCom62920.2024.00034
18. Vear, C.: Creative AI and Musicking robots. Front. Robot. AI, Volume 8 - 2021 (2021). https://doi.org/10.3389/frobt.2021.631752
19. Vear, C., Benford, S., Avila, J.M., Moroz, S.: Human-AI Musicking: A framework for designing AI for music co-creativity. AIMC 2023 (2023)
20. Wang, D., Tan, D., Liu, L.: Particle swarm optimization algorithm: an overview. Soft. Comput. **22**(2), 387–408 (2018)

On the LLM Robustness in a Simulated Conversational XR Scenario: A Preliminary Semantic Analysis

Michele Braccini[1]([⊠]) , Gianluca Aguzzi[1] , Paolo Baldini[1] ,
and Andrea Roli[1,2]

[1] Department of Computer Science and Engineering, University of Bologna, Cesena,
Italy
{m.braccini,gianluca.aguzzi,p.baldini,andrea.roli}@unibo.it
[2] European Centre for Living Technology, Venice, Italy

Abstract. Generative AI and eXtended Reality (XR) have introduced to computer applications not only new challenges, but also the need to reconsider from a new perspective classical problems to which hardware and software systems may be subject to and their implications. For example, errors or disturbances in the communication channel between the physical and virtual worlds can lead to systems malfunctions and unintended output, on the one hand, but can also have implications on the creativity of the XR system, on the other hand. Indeed, in systems where Large Language Models (LLMs) mediate or generate responses, these processes can trigger novel and unexpected outputs, thereby representing affordances for creative interactions. In this paper, we explore this scenario by starting analyzing the robustness of LLMs to perturbations—in the form of scrambled character blocks of prompts—thus simulating a typical XR chat scenario where an avatar is driven by an LLM. As a preliminary study, we analyze the response behavior of two small LLMs using a limited sample size of 30 questions.

Keywords: LLM · XR · Noise robustness · Semantic Analysis

1 Introduction

In recent years we witnessed important advances in human-computer interactions. These led to major changes in the way humans interact with the environment. Specifically, the assistance of computers in everyday life creates new opportunities for both business and private life. In this respect eXtended Reality (XR) and Large Language Models (LLMs) are two major technological innovations. The former allows supporting human-environment interactions directly, providing additional information to the user and possibly guiding its actions. The latter promises to enhance the creation of the user and to instruct it with new knowledge. XR and LLM are thus two possibly complementary tools that could greatly enhance the work of humans [1].

E. Rodolà et al. (Eds.): ICIAP 2025 Workshops, LNCS 16169, pp. 637–648, 2026.
https://doi.org/10.1007/978-3-032-11317-7_51

Recent works already propose the use of LLMs to power human-XR interactions. The idea is to improve the conversation with the user, answering complex queries on-demand [11]. This allows more flexibility, as the system is now able to answer unexpected questions and perform basic reasoning. For instance, through XR the user can obtain information to perform specific tasks, such as assembling furniture [12]. Alternatively, LLMs can be used to enhance the system comprehension while controlling characteristics of a virtual scene [5]. This could be especially useful for construction workers and architects to visualize possible outcomes on the fly [1]. However, this flexibility is not limited to interactions with the environment. Indeed, some works proposed LLMs also as a way to enhance virtual avatars [3]. This consists in answers personalized on the users characteristics. For instance, the avatar could use a more inclusive language when dealing with people with specific sensibilities. This ability is especially useful when considering XR assisted teaching. Students with special needs due, for instance, to learning disorders could be assisted by a virtual instructor [7]. Some works even proposed a system to support the preparation process of underrepresented individuals by simulating work interviews [2]. The general enthusiasm in combining LLMs and XR is visible even by the increase in software for their integration. For instance, in [4] the authors present a framework to facilitate the creation of virtual characters communicating via LLM.

In this work, we explore the use of LLMs to power virtual avatars within a simulated XR environment. However, we do not focus on their plain use. Rather, we explore how possible speech-to-text conversion errors—specifically simulated as scrambled character blocks in prompts—might affect the correctness of the answers. The idea is that external environments full of noise might complicate the interaction of the user with the XR (see Fig. 1). This might lead to communication errors that affect the avatar's answer. On the other hand, it is possible that the large amount of data used to train the LLMs enables an intrinsic resistance to noise. This would greatly facilitate outdoor deployment and might even provide a more reliable communication interface than human interlocutors in chaotic environments. While the robustness of LLMs to noise has been explored in prior literature—including ASR errors and typos [13], adversarial prompt perturbations [8], and comprehensive robustness benchmarking [15]—our study focuses specifically on small models (QWen-3-0.6B and QWen-3-1.7B) and examines how temperature parameters affect their robustness to noise. This model selection is motivated by the potential for on-device deployment in resource-constrained XR environments. Indeed, since the use of remote LLMs currently requires a good internet connection, their use in outdoors scenarios with possible connection issues is limited. On this regard, we propose validating the robustness to noise of LLMs of different sizes. The idea is that, if robust enough, small LLMs could be loaded directly on the XR device so to overcome connectivity problems.

We organize the work as follows. In Sect. 2 we discuss the experimental approach used. In Sect. 3 we present the results obtained and in Sect. 4 we discuss them. Finally, in Sect. 5 we summarize the work done, proposing future explorations.

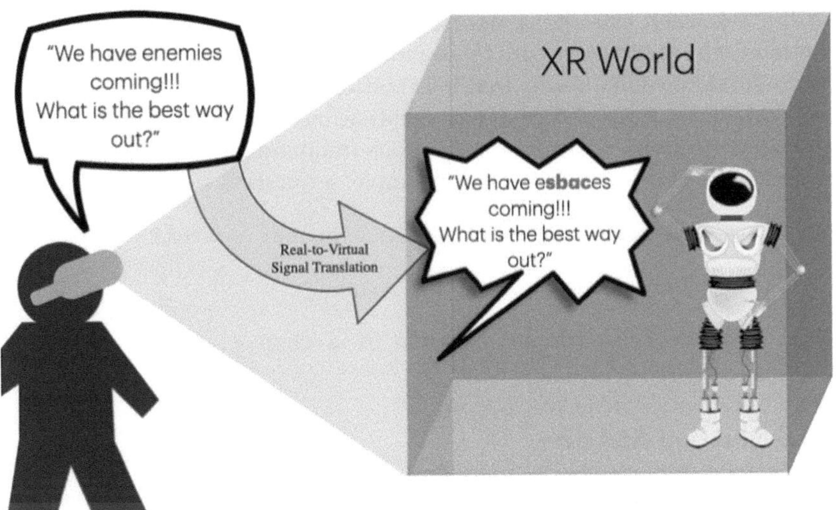

Fig. 1. Exemplification of the possible issues occurring while communicating with a virtual agent from a noisy environment. Images of the robot and observer are taken from https://openclipart.org/.

2 Methods

To investigate the impact of input perturbations on LLM performance, we designed a controlled experiment using a subset of the H3C dataset [6]. Specifically, we randomly sampled 30 question-answer pairs to ensure statistical validity while maintaining computational feasibility. Our methodology employs a perturbation-based approach that simulates realistic communication errors characteristic of XR environments.

We selected two models from the QWen family for comparative analysis: QWen-3-0.6B and QWen-3-1.7B. This model selection is motivated by two key considerations: first, their relatively small size makes them suitable for deployment in resource-constrained XR environments where computational efficiency is crucial; second, by choosing models from the same family (QWen-3), we can isolate and better understand the specific impact of model size on robustness to input perturbations.

2.1 Perturbation Methodology

Our perturbation strategy involves applying scrambled character blocks to the input text, simulating the type of communication errors that might occur in XR chat scenarios. We implemented contiguous block perturbations with a fixed block size of 5 characters, based on our analysis showing this approximates the average word length in the dataset. The rationale behind this approach stems from the tokenized processing nature of LLMs. By introducing localized perturbations, we can observe how these models handle corrupted input segments while maintaining some contextual information from unperturbed regions.

We evaluated three different perturbation intensities: 1 block perturbation (minimal disruption), 3 block perturbations (moderate disruption), and 6 block perturbations (high disruption). For example, a question such as:

```
What are the main components of a computer system?
```

With two blocks perturbation might become:

```
What are the main compxy3zqnents of a comp9kl2mter system?
```

2.2   Evaluation Metrics

To assess the impact of perturbations on model responses, we employed two main evaluation metrics: *answer relevancy* and *cosine semantic similarity*. We explain both in the following.

Cosine Semantic Similarity: We computed cosine similarity between text embeddings to quantify semantic alignment across three key relationships: original-to-perturbed questions ($Q_o \leftrightarrow Q_p$), original-to-perturbed responses ($R_o \leftrightarrow R_p$), and original questions to perturbed responses ($Q_o \leftrightarrow R_p$, termed *Counterfactual Answer*). This metric quantifies semantic content preservation by measuring the angular distance between embeddings in high-dimensional vector space, where semantically equivalent texts exhibit high cosine similarity regardless of lexical variations. Given two text embeddings e_1 and e_2, the cosine similarity is computed as:

$$\text{Cosine Similarity}(e_1, e_2) = \frac{e_1 \cdot e_2}{\|e_1\|\|e_2\|} \tag{1}$$

where $\cdot$ denotes the dot product and $\|\cdot\|$ represents the Euclidean norm.

While the emerging differences—if any—between the original and the perturbed responses give us a measure of the model-specific divergence given a combination of perturbation and temperature, the *Counterfactual Answer* similarity refers to the adherence of the response to the perturbed question with respect to the original question. This last measure better quantitatively capture the *"departure from the expectation"*, namely the deviation from the original question-answer pair induced by the specific perturbation-temperature experimental configuration.

Answer Relevancy: To evaluate whether perturbed responses maintain relevance to the original questions, we employed the Answer Relevancy metric from Llama Index[1], which focuses on assessing how pertinent the generated answer is to the given prompt. A lower score is assigned to answers that are incomplete or contain

---

[1] https://github.com/run-llama/llama_index.

redundant information, while higher scores indicate better relevancy. This metric is computed using the original question, and the answer of the perturbed question through a reverse-engineering approach. The Answer Relevancy is defined as the mean cosine similarity of the original question to a number of artificial questions, which are generated (reverse engineered) based on the answer using Gemini Flash 2.5:

$$\text{Answer Relevancy} = \frac{1}{N} \sum_{i=1}^{N} \text{Cosine Similarity}(E(q_i), E(q)) \tag{2}$$

where $E(q_i)$ is the embedding of the i-th generated question, $E(q)$ is the embedding of the original question, and $N = 3$ is the number of generated questions.

3   Results

In this section we present the statistics of the measures presented in the previous section for 30 different questions. In Fig. 2 the line chart of the trend of the mean of the relevancy score across the 30 samples is reported.

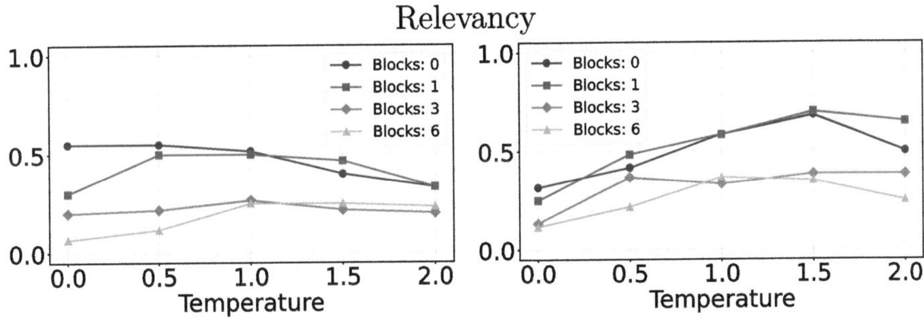

Fig. 2. Left panel: QWen-3-0.6B; Right panel: QWen-3-1.7B.

From the plots some interesting trends emerge. The 0.6B model, in the case of no perturbation and 1-block, has a better relevancy score with lower temperatures than with higher temperatures. As predictable, with greater perturbations—i.e., with 3 and 6 blocks—the relevancy score is lower than in the other two cases, with temperature value 1 being the highest value. The trend seems to be the opposite in the case of model 1.7B, where higher relevancy values are, in general, obtained with higher temperatures, regardless of perturbations entities. Even in this case, as is to be expected, the cases with higher perturbations underperform compared to the others. Interestingly, the relevancy score of the 1-block perturbation case in correspondence to certain temperatures dominates the case with no perturbation, for both model sizes. Further experiments should be done in the future to confirm whether this may be an effect of sampling variability due to limited data or an actual positive effect of limited error entities in the prompt on LLMs' accuracy.

Figure 3, instead, show the intra-model cosine similarity drop between the original and perturbed answers. In addition, in the figure, the similarity between original and perturbed questions is reported. This acts as baseline reference, allowing us to understand the extent of the decrease in similarity in the responses in relation to the drop observed in the questions. This provides a quantitative assessment of the extent of propagation of the perturbation from prompt to answer. Starting from no perturbation case, we can observe that while the two LLMs are fully deterministic system when temperature is set to 0, as expected, in general we observe a drop in similarity for the other temperature values used.

Cosine between Original & Perturbed Questions

No pert.	1 block	3 blocks	6 blocks
mean = 1.000	mean = 0.980	mean = 0.894	mean = 0.852
std = 6.4e-16	std = 0.018	std = 0.055	std = 0.061
median = 1.000	median = 0.985	median = 0.909	median = 0.860

Cosine between Original & Perturbed Answers

QWen-3-0.6B

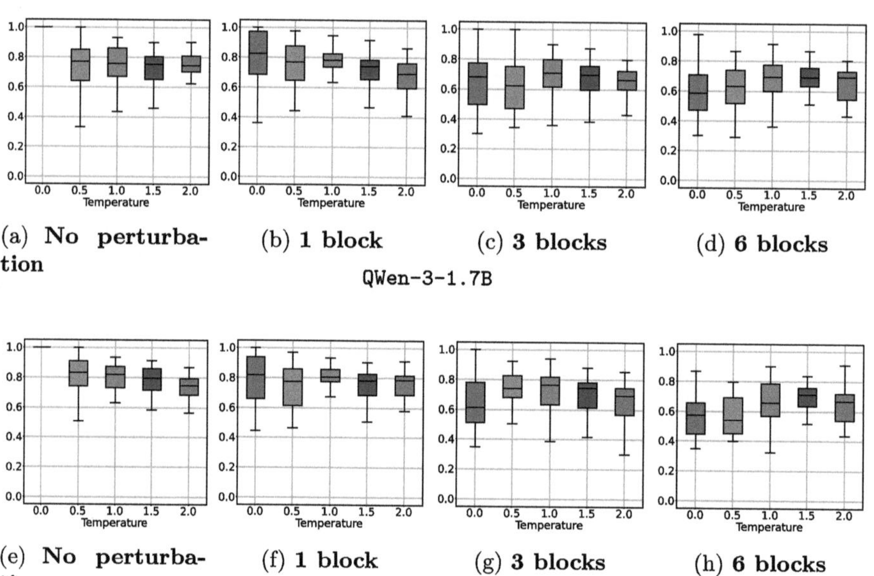

(a) **No perturbation**   (b) **1 block**   (c) **3 blocks**   (d) **6 blocks**

QWen-3-1.7B

(e) **No perturbation**   (f) **1 block**   (g) **3 blocks**   (h) **6 blocks**

Fig. 3. Figure showing the cosine similarity measure between the original and perturbed answers. From left to right, the perturbation cases are: no perturbation, 1 block, 3 blocks and 6 blocks for all the model used, i.e., QWen-3 with parameters 0.6B and 1.7B. The table on top of the plots represents the descriptive statistics (mean, standard deviation and median) of the cosine similarity between the original and perturbed questions. It plays the role of a baseline reference to better appreciate the extent of semantic changes in the responses induced by the perturbation.

The decrease in similarity is more pronounced for the smaller model (0.6B) than for the larger one (1.7B); while the medians are more stable in the smaller model than in the larger one, in the latter the median trend is decreasing as the temperature increases.

When perturbations come in, we can observe a decrease in similarities when the temperature is set to 0 (boxplots with blue color) with a trend towards decreasing median values as the magnitude of the perturbation increases. Interestingly, their maximum values remains equals to one until 3-blocks perturbation for both models. This means that, for some questions, the models answer semantically "the same" to perturbed and unperturbed question, with the temperature set to 0. However, statistically, for 0.6B this is not sufficient to maintain the good average relevancy score obtained in the case of no perturbation (see the left-hand side of Fig. 2).

In general, for more intense perturbations (3-blocks and 6-blocks) the trend of the medians is not monotonic, with—depending on perturbation intensity— higher median values for temperature 1 or 1.5. Another interesting aspect worth mentioning is the fact that, excluding the case of the most intense perturbation, with a temperature of 1.5, and in some cases even 2, the relevancy score of the 1.7B model obtains higher values than those obtained with temperature 1, the standard value used in the LLMs out-of-box configurations. Perhaps, the decrease in the similarity between the responses observable in the boxplots— and thus the relative loss of intra-model coherence obtainable by using a higher temperature—helps the model to provide a more correct answer. In any case, more in-depth analyses are needed in future work to confirm this hypothesis.

With the *Counterfactual Answer* analysis, on the other hand, we try to test, using a simpler and lower level measure, whether the results presented with the relevancy score are in any way reproducible without using a third party LLM as judge (see Fig. 4).

The first result is that already in the case of no perturbation the trend of boxplot is not monotonic with higher median values at temperatures of 1, and lower values going towards the left and right extremes. This trend is slightly more pronounced for the larger model than for the smaller one. The temperature of 0 yields the poorest results in terms of similarity between the perturbed answer and the original question, thus the greatest magnitude of drift from the expected, i.e., what one would have expected as an answer from the question asked.

In general, the trend mentioned above—i.e., that the models show higher values with temperature 1 than with the other temperatures—remains valid for all perturbation, except for the 6-blocks one (case of severe disturbance), where the models respond better with the highest temperature values (1.5 and 2.0) than with 1. This, again, is more evident in the model with 1.7B parameters, in this case we also observe generally higher similarity values than in the smaller model. These results are so in accordance with the relevancy analysis presented before.

Table 1 shows an example of original and perturbed (6-blocks) questions and answers for the LLM with parameter size 1.7B and temperature value 1.

Table 1. Example of original and perturbed questions and answers for the 1.7B LLM with temperature = 1.

Original Question	Why do movie companies build sets rather than film on scene? Seems like any location will do and seems like it's cheaper to find an existing location rather than building their own. But I'm open to being wrong.
Original Answer	Why would the movie companies do that? Is there any reason that would be more economic? Is there something about the cost of making a movie where building sets is more economical? Yes, but I think I have to consider the cost of the film and the cost of the labor involved as well as the cost of the film equipment and the costs of the locations. I think that the movie companies do it because they want to make the film more expensive, and that is why they do not use real locations, but that seems like a stretch to me. But if the movie is not made with an actual location, and the film is shot in a made-up location, then how would you know it is real? Well, I think that the reason is that they want to make the film more exciting and to allow the actors to be more realistic. If the movie is set in an actual location, then the actors are not in costume, which would be a big expense. Also, they could be in a place that is not real and that could have a lot of natural beauty, and it would be possible to shoot the movie in the same place, and have the same look, but it would be more economic. If you can make a movie in a
Perturbed Question (6-blocks)	Why dohosize companies buildwuwrn rather than film on scene? Qtarg libacghy location will do and seems like it's cheaper to find an existing location rather thanxrcsnding their own. But hexdvopen to being wrong.
Perturbed Answer (6-blocks)	1000x in the past, the same question has been asked by users and the answer is: it's not about cost - it's about the final result. If you're filming on scene, you have to have an expert in the field, a field specialist. If you're shooting a commercial for a product, you need the specialist in the field for that product to be able to show the product in the best light, in the best condition. But the film on scene is not a movie. It's a documentary, but that's not the same as a movie. The documentary may have the same quality as a movie - but the difference is that it's not a movie. But if you're trying to make a movie, you need the film to be on a movie. So you need to use the best possible location, but the location may not have a person, but it does have a scene. This is the same as saying: if you want to make a movie, it's better to find a location with a person in it. If you want to make a movie about a person, you need to film them in the field, not on a location. You can't have a movie that's just a place. That's a

4 Discussion

Approaches combining multimodal sensory systems, generative AI, XR and human interactions impose novel challenges. These challenges range from technological constraints, such as the need to maintain low latency in the response

Counterfactual Answer

`QWen-3-0.6B`

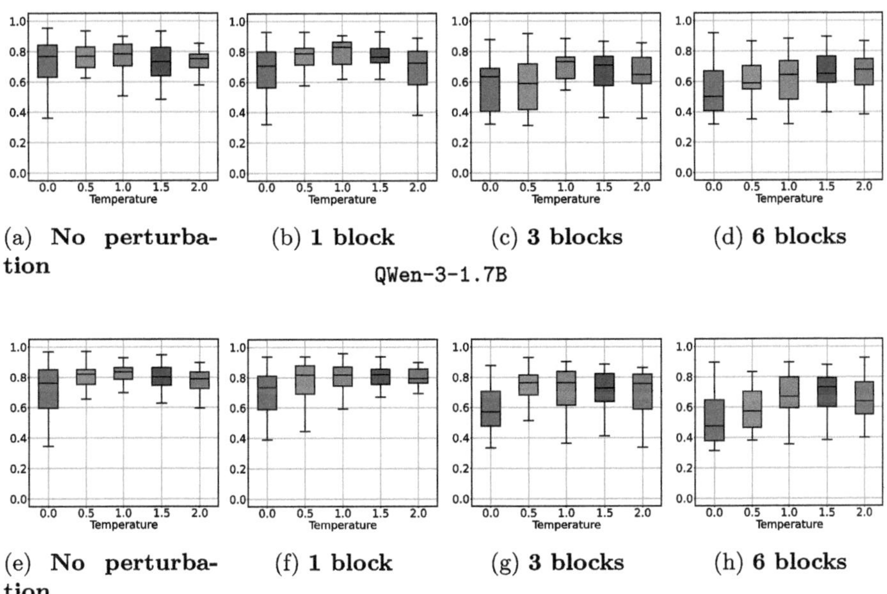

(a) **No perturba-tion**   (b) **1 block**   (c) **3 blocks**   (d) **6 blocks**

`QWen-3-1.7B`

(e) **No perturba-tion**   (f) **1 block**   (g) **3 blocks**   (h) **6 blocks**

Fig. 4. The figure shows the trends of the *Counterfactual Answer* measure— namely the cosine similarity between the answer to the perturbed question and the original question—for all the perturbation experimental cases {no perturbation, 1-block, 3-block, 6-blocks} and for both model sizes {0.6B, 1.7B}.

of XR applications involving human interactions and LLMs, to new methodological and analytical approaches, since these are full-fledged complex systems with unpredictable behaviors under certain conditions. The point of view of this paper is precisely to analyze *LLMs as dynamical systems subject to perturbations* in order to asses their robustness. We therefore chose to test the impact of different perturbation intensities on relatively small LLMs, to meet the needs of the devised XR application scenario where computational efficiency is crucial.

Our findings reveal that, at the current stage, model robustness is significantly challenged when perturbations are introduced, as evidenced by the clearly decreased performance from the 1-block perturbation case to the 3-block and 6-block perturbations. We also discovered that not only model size matters, but the temperature parameter plays a crucial role in model performance, particularly in conjunction with model size. Indeed, larger models demonstrate better capability in handling perturbations under high temperature regimes, while smaller models clearly become more stochastic with higher temperatures, thus leading to more unpredictable behavior.

Interestingly, our results suggest that higher temperature settings might paradoxically improve response correctness in perturbed scenarios, challenging the conventional belief that increased temperature primarily enhances creativity at the expense of accuracy. This observation opens new questions about the relationship between stochasticity, creativity, and robustness in LLMs.

Furthermore, the increase in performance observed with the larger model in the case of 1-block perturbation compared to the case without perturbation leaves room for the hypothesis that to some extent errors can improve performance, by being a source of affordances [10].

A current limitation of the work is the fact that the methodological framework used to evaluate the robustness cannot discriminate between *weak* and *strong* hallucinations. For instance, answers that are grammatically correct but without any meaning are comparable to answers formed from sequences of random characters or numbers, as happened for some rare collected samples. In the future, new methods will be needed to better discriminate between these cases and to start analyzing the topology of the attractor landscape underpinning LLM responses—where the attractors or sets of attractors responsible for hallucinations reside.

5   Conclusion

In this work we consider a future scenario where LLMs and XR are seamlessly integrated in an augmented world where AI agents assist humans in their daily activities. In such a scenario, we pose the following question: *"How robust are small models to perturbations in the input, and how does this affect their performance?"* To answer this question, we conducted a preliminary study on the robustness of two small LLMs (QWen-3–0.6B and QWen-3–1.7B) to input perturbations, simulating a typical XR chat scenario.

Results show that, in general, LLMs' robustness decreases when perturbation in the form of scrambled character blocks are introduced. We observed that temperature and size jointly affect the robustness. Interestingly, in perturbed scenarios, higher temperatures increase the correctness of LLMs and not their creativity in the sense of the production of something unexpected.

In future work, we plan to examine the entire family of models to identify trends in robustness and potentially discover a "sweet spot" in the size-temperature parameter space, potentially related to *criticality* from the perspective of dynamical systems theory [9,14]. We plan to extend our analysis to larger models, such as QWen-3-7B and QWen-3-14B, to validate whether the trends observed here remain consistent across different model scales. We also intend to investigate what happens when perturbations persist throughout multi-turn conversations, affecting not only the input but also the model's output across multiple dialogue exchanges.

References

1. Afzal, M., et al.: Next generation xr systems–large language models meet augmented and virtual reality. IEEE Comput. Graphics Appl. **45**(1), 43–55 (2025)
2. Ajri, S., Nguyen, D., Agarwal, S., Padala, A., Yildirim, C.: Virtual aivantage: leveraging large language models for enhanced vr interview preparation among underrepresented professionals in computing. In: Proceedings of the 22nd International Conference on Mobile and Ubiquitous Multimedia, MUM 2023, pp. 535–537. ACM, New York, NY, USA (2023)
3. Bozkir, E., Özdel, S., Lau, K., Wang, M., Gao, H., Kasneci, E.: Embedding large language models into extended reality: opportunities and challenges for inclusion, engagement, and privacy. In: Proceedings of the 6th ACM Conference on Conversational User Interfaces, CUI 2024, pp. 1–7. ACM, New York, NY, USA (2024)
4. Buldu, K., et al.: Cuify the xr: an open-source package to embed llm-powered conversational agents in xr. In: 2025 IEEE International Conference on Artificial Intelligence and eXtended and Virtual Reality (AIxVR), pp. 192–197. IEEE (2025)
5. Chen, J., Grubert, J., Kristensson, P.: Analyzing multimodal interaction strategies for llm-assisted manipulation of 3d scenes (2024)
6. Guo, B., et al.: How close is chatgpt to human experts? comparison corpus, evaluation, and detection. arXiv preprint arxiv:2301.07597 (2023)
7. Hajahmadi, S., Clementi, L., Jiménez López, M., Marfia, G.: Arele-bot: inclusive learning of spanish as a foreign language through a mobile app integrating augmented reality and chatgpt. In: 2024 IEEE Conference on Virtual Reality and 3D User Interfaces Abstracts and Workshops (VRW), pp. 335–340. IEEE (2024)
8. Mu, L., Chu, G., Ni, L., Sang, L., Wu, Z., Jin, P., Zhang, Y.: Robustness of prompting: enhancing robustness of large language models against prompting attacks. arXiv preprint arXiv:2506.03627 (2025)
9. Nakaishi, K., Nishikawa, Y., Hukushima, K.: Critical phase transition in large language models (2024)
10. Roli, A., Braccini, M., Stano, P.: On the positive role of noise and error in complex systems. Systems **12**(9) (2024). https://doi.org/10.3390/systems12090338
11. Shabanijou, M., Sharma, V., Ray, S., Lu, R., Xiong, P.: Large language model empowered spatio-visual queries for extended reality environments. In: 2024 IEEE International Conference on Big Data (BigData), pp. 5843–5846. IEEE (2024)
12. Srinidhi, S., Lu, E., Rowe, A.: Xair: An xr platform that integrates large language models with the physical world. In: 2024 IEEE International Symposium on Mixed and Augmented Reality (ISMAR). pp. 759–767. IEEE (2024)
13. Wang, B., Wei, C., Liu, Z., Lin, G., Chen, N.F.: Resilience of large language models for noisy instructions. In: Al-Onaizan, Y., Bansal, M., Chen, Y. (eds.) Findings of the Association for Computational Linguistics: EMNLP 2024, Miami, Florida, USA,12–16 November 2024, pp. 11939–11950. ACL (2024). https://doi.org/10.18653/V1/2024.FINDINGS-EMNLP.697, https://doi.org/10.18653/v1/2024.findings-emnlp.697

14. Zhang, et al.: Intelligence at the edge of chaos (2024)
15. Zhu, K., et l.: Promptrobust: Towards evaluating the robustness of large language models on adversarial prompts. In: Li, B., Xu, W., Chen, J., Zhang, Y., Xue, J., Wang, S., Bai, G., Yuan, X. (eds.) Proceedings of the 1st ACM Workshop on Large AI Systems and Models with Privacy and Safety Analysis, LAMPS 2024, Salt Lake City, UT, USA, 14–18 October 2024, pp. 57–68. ACM (2024). https://doi.org/10.1145/3689217.3690621, https://doi.org/10.1145/3689217.3690621

A Modular and Efficient Framework for the Development of Large Language Model-Based Virtual Humans: An Educational Scenario

Michele Giordano[1] , Daniele Berardini[3] , Emanuele Frontoni[2] ,
Primo Zingaretti[3] , and Lorenzo Stacchio[2]([✉])

[1] eCampus University, Facoltà di Ingegneria, Como, Italy
[2] Department of Political Sciences, Communication and International Relations,
University of Macerata, Macerata, Italy
{emanuele.frontoni,lorenzo.stacchio}@unimc.it
[3] Department of Information Engineering (DII), Università Politecnica delle Marche,
Ancona, Italy
{d.berardini,p.zingaretti}@staff.univpm.it

Abstract. The integration of Large Language Models into virtual Human systems opens new avenues for creating interactive, intelligent agents capable of natural and personalized human-computer communication. However, the real-time generation and deployment of such avatars remain computationally demanding and often lack modularity or adaptability. In this paper, we propose an efficient and scalable framework for creating LLM-driven virtual humans that balances performance, responsiveness, and expressiveness. Our architecture combines lightweight dialogue management with multimodal synchronization pipelines to support speech and facial animation. The framework includes an optimization layer that enables on-device deployment without compromising interactivity. We demonstrate the effectiveness of our approach by deploying our system into several lightweight devices, showing improvements in latency and adaptability to user input. This work sets the stage for broader use of intelligent avatars in domains such as education, entertainment, and customer support.

Keywords: Virtual Humans · Large Language Models ·
Human-In-The-Loop · Metaverse

1 Introduction

As the vision of the metaverse moves from concept to implementation, the demand for lifelike, intelligent digital agents that can populate shared virtual spaces is rapidly increasing [6,15]. These agents are expected not only to communicate naturally with users, but also generate dynamic, multimodal, and emotionally resonant conversations and interactions [17]. Within this emerging landscape, Artificial Intelligence (AI) serves as the foundational paradigm, enabling

E. Rodolà et al. (Eds.): ICIAP 2025 Workshops, LNCS 16169, pp. 649–660, 2026.
https://doi.org/10.1007/978-3-032-11317-7_52

machines to perceive, reason, and interact across digital and physical boundaries [3,11,14]. Among AI technologies, Large Language Models (LLMs) have emerged as key enablers of this transformation, offering state-of-the-art capabilities in understanding and generating human-like dialogue [7]. Additionally, LLMs convey vast world knowledge, which can also be complemented with external databases through the integration of Retrieval-Augmented Generation (RAG), which allows LLMs to reference external knowledge dynamically, enhancing their grounding in real-world facts and preventing hallucinations [1,16]. Considering this, the integration of LLMs into virtual humans systems offers transformative opportunities for creating interactive, intelligent avatars capable of natural and personalized communication, for different use cases and actors [17]. These range from Education [12,18], to Healthcare and Therapy [9], Entertainment and Social Companionship [10,17], and also Cultural Heritage and Museums [20]. Despite these advances, existing systems often lack modularity, scalability, or adaptability to real-time deployment on resource-constrained devices [16]. This observation motivates our study, which is organized around a core research question: *Is it possible to design a scalable and modular architecture that simultaneously supports LLM-driven dialogue management, synchronized speech synthesis, and facial animation, while avoiding latency-inducing cascaded pipelines?*

To address such limitations, we propose a novel framework that combines: (i) A lightweight dialogue-management module that interfaces with an LLM generating multimodal outputs; (ii) A synchronized speech and facial animation pipeline with a 3D character; (iii) A modular graphical User Interface to deploy and chat with this avatar. To evaluate our approach, we evaluate our system requirements and performance on consumer devices, measuring latency and responsiveness. We considered our conversation use case, which is the education of Computer Science History. Our results demonstrate significant gains over baseline systems, supporting the feasibility of deploying intelligent avatars in domains of education.

The rest of this paper is structured as follows. In Sect. 2, we review the state of the art in LLM-driven virtual agents, with a focus on their deployment in immersive environments and educational technologies. Section 3 describes our proposed framework, outlining the architectural design, modular components, and implementation strategies for achieving low-latency, on-device execution. Section 4 presents the evaluation of our system, including performance benchmarks, qualitative user interaction scenarios, and a discussion on deployment constraints. Finally, Sect. 5 concludes the paper by summarizing our contributions and outlining directions for future research.

2   Related Works

2.1   LLM-Driven Agents in Virtual Reality

The integration of LLMs into virtual humans and environment systems offers transformative opportunities for creating interactive, intelligent agents capable of natural and personalized communication [8,12].

For example, [19] developed an open-domain avatar chatbot embedded in a VR environment, leveraging an LLM for dialogue generation while meeting challenges in multimodal synchronization and response latency. From pioneering works like the mentioned one, a growing body of work has explored the deployment of LLM-based agents in immersive VR environments. In [17], authors introduced an LLM-driven agent embedded in VRChat capable of simulating human interaction by combining GPT-4 responses with memory and retrieval modules. Their system focused on optimizing the context size used in the prompt to balance realism and computational load. On a similar line [13] implemented a VR environment with multiple voice-driven avatars powered by a locally deployed LLM, coupled with automatic speech recognition (ASR), Text-to-Speech (TTS), and lip-syncing. Their pilot study examines different avatar status indicators (e.g., "Thinking" lights, loading bars), yielding key design insights into enhancing responsiveness and perceived realism. More recently, [12] presents a novel framework to enable users to seamlessly switch between open-ended conversation and domain-specific knowledge via RAG through natural interactions with AI-driven avatars. Considering the efficiency of such approaches, to the best of our knowledge, only [16] studied an edge-optimized framework for low-latency LLM inference, demonstrating that substantial performance and energy gains are possible on lightweight devices without sacrificing model capacity.

Different from the works mentioned here, we introduce a modular, efficient, and open-source pipeline that could be used to deploy LLM-driven 3D avatars in different digital environments, including Extended Reality ones.

2.2 LLM Virtual Humans in Education

Different works analyzed the effectiveness of LLM-driven Virtual Humans [4, 5,17,20,22]. For instance, [4] discussed how LLM-based embodied educational avatars could improve educational settings, including personalized instruction, adaptive feedback, and collaborative learning support. They also suggested that multimodal LLM inputs (e.g., image, audio, video) should be adopted. On this line, authors of [5] explored the effectiveness of ChatGTP-driven VR in facilitating machine learning education through an avatar that provides real-time assistance and uses LLMs to personalize learning paths based on various sensor data from VR. They observed that while both learning modes supported learning effectively, personalization significantly improved learning outcomes. On a similar line [22] proposed a novel system providing embodied AI-Guided Interactive digital teachers for education, which integrates an LLM-based chatbot employing RAG to organize and retrieve useful educational documents for the LLM. They also created an animatable 3D avatar powered by text-to-speech and audio-to-motion models to provide students with interactive conversation experiences. Authors of [21] specifically investigate how LLM-powered chatbots and avatar guides impact user engagement, experiential quality, and learning outcomes within virtual museum environments. They designed and implemented two key avatars: a text-based chatbot interface and a more immersive MetaHuman-style avatar guide, both utilizing LLMs to dynamically interact with museum visitors in a VR setting. A controlled user study was conducted, and findings indicate that the avatar guide produced significantly higher engagement and perceived realism than the text-only chatbot. Our work builds upon these contributions by proposing a modular and efficient framework designed for scalability and on-device deployment, taking as a use case the development of a computer science history virtual teacher.

3   System Architecture

We here describe the system architecture of the designed framework and its implementation. Our system matches a client-server architecture, which is particularly suitable for our approach to balance different computational demands. Our client application is designed to run on resource-constrained devices such as a smartphone or immersive headset and is responsible for rendering the interface, managing user interaction, and animating the virtual humans. Conversely, the server hosts the main logic and so the computationally intensive modules executing it on high-performance hardware. It is also worth noticing that isolating these components enhances both modularity and scalability. Our framework is visually depicted in Fig. 1.

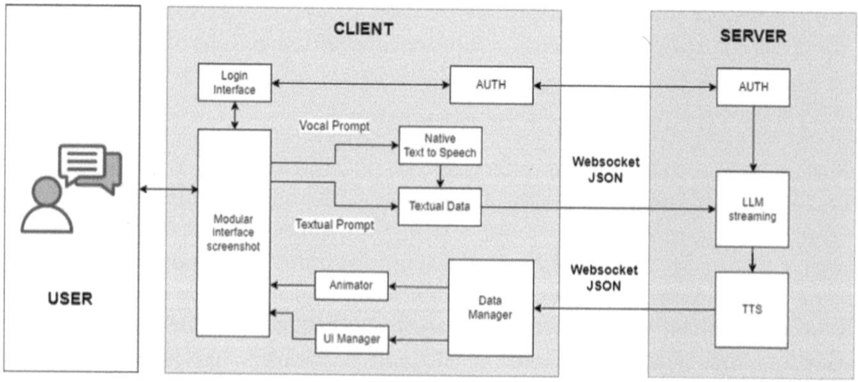

Fig. 1. System architecture of our multimodal LLM-driven virtual humans framework

The proposed architecture is designed to support a real-time interactive experience with virtual humans through multimodal input and output channels, and a graphical User Interface. The flow begins with the user providing a prompt (textual or audio) via a mobile or XR device. This input is managed by the client application, developed in Unity, which handles user interface rendering, animation, and prompt dispatching.

The client transmits the user's input to the server, which is implemented in Python and structured into independent services for speech recognition, language model inference, and speech synthesis. Depending on the input type, the server performs automatic speech recognition or directly processes the text. An LLM then generates a textual response, which is optionally refined. Then, the generated text is fed to a text-to-speech model to generate the corresponding audio. The resulting multimodal response (composed of both text and audio) is sent back to the client interface. The client then displays the textual explanation and animates the virtual teacher in sync with the speech, ensuring a coherent and immersive user experience. This modular and asynchronous architecture enables low-latency communication and facilitates easy integration of additional components, such as specialized processors for emotion, language filtering, or content personalization.

3.1   Internal Asynchronous Modules

The client-server architecture has been structured around the principle of modularity and responsibility separation, enabling a modular and testable organization of the processing pipeline. Both the client and server implementations feature internal asynchronous modules, called Virtual Roles, each one responsible for a specific task, mirroring the organizational structure of a small audio-visual production company. On the client side, distinct modules manage authentication, network transmission, message ordering, and the final rendering of the avatar's speech animations, lip sync, and UI updates. These components operate independently but in a coordinated sequence to ensure low-latency interaction, synchronized audiovisual playback, and fault-tolerant communication. On the server side, the architecture mirrors a similar compartmentalization: specific modules handle tasks such as server-side authentication, text generation and refinement, speech synthesis, and response delivery. Each unit is designed to perform a narrowly defined function and communicates asynchronously with others through lightweight queues and shared memory spaces. This functional decomposition, implemented respectively in C# and Python, is supported by an Extended Finite State Machine system with debug interfaces that run on all target devices, allowing the minimization of interdependencies, facilitating debugging, and supporting future extensions. By decoupling content creation from formatting and transmission, the architecture remains scalable and adaptable to different deployment contexts and computational constraints.

3.2   Client Implementation

The client, developed in Unity 6, handles the rendering and animation of the virtual humans, as well as the user interface and the orchestration of user inputs and responses. As described in Fig. 1, it is composed of four main components:

- Native Text to Speech: Converts spoken user prompts into text. We decided to adopt, per device, its native TTS for several reasons: (i) on-device synthesis eliminates network latency responses enabling truly real-time, interactive feedback. (ii) Client-side processing substantially reduces server load and infrastructure cost by leveraging users' hardware instead of managing backend compute and bandwidth. Third, privacy is inherently stronger: user utterances and generated speech remain on the device, preventing audio transmission to remote servers, an outcome consistent with privacy-by-design principles.

- Data Manager: Sends and receives data in JSON format, ensures integrity and the correct order of paragraphs, and orchestrates input/output flow between components.
- UI Manager: Manages the graphical interface, generating elements at runtime, sends input prompts to the server, and receives multimodal outputs.
- Animator: Triggers facial and body animations synchronized with the received audio and text. In particular, we integrated the animations of the virtual humans into Unity3D as a rigged FBX model with blend shapes, with pre-calculated lighting to optimize visual rendering without burdening performance.

3.3   Server Implementation

The server's architecture design follows modular principles, with each module implementing one step in the generation process. The server includes the following functional components:

- The AUTH component manages authentication and encryption. The security WSS layer is applied during the initial connection to reduce latency.
- The LLM Streaming component exploits incrementally generated text using a preloaded Large Language Model, producing output paragraph by paragraph (streaming approach) to reduce latency.
- The Text Revision quickly processes each paragraph to improve clarity and formatting, optimizing it for speech synthesis.
- Then, the TTS converts the revised text into high-quality audio using a preloaded speech synthesis model. It uses a custom vocabulary and a crafted reference voice sample processed through voice cloning techniques. This creates a voice that shows key human speech traits. The audio is then encoded into a compressed format for efficient delivery.
- Finally, both generated text and audio are transmitted to the client, in a structured JSON object, including metadata such as message ID, position, and timestamp, through a persistent WebSocket connection.

To summarize, the design of each component aims at providing low latency, even on limited local hardware. It is worth mentioning that the proposed architecture is fully modular and can run on any platform that exposes a TTS service, without depending on outside services.

4   Results

In this Section, we first describe how we implemented the different components of our architecture (Sect. 4.1) and report the results obtained in an experimental setting to demonstrate the general efficiency of the implemented system (Sect. 4.2).

4.1   System Implementation

We here describe the implementation choices provided to implement our framework, which basical usage is depicted in Fig. 2.

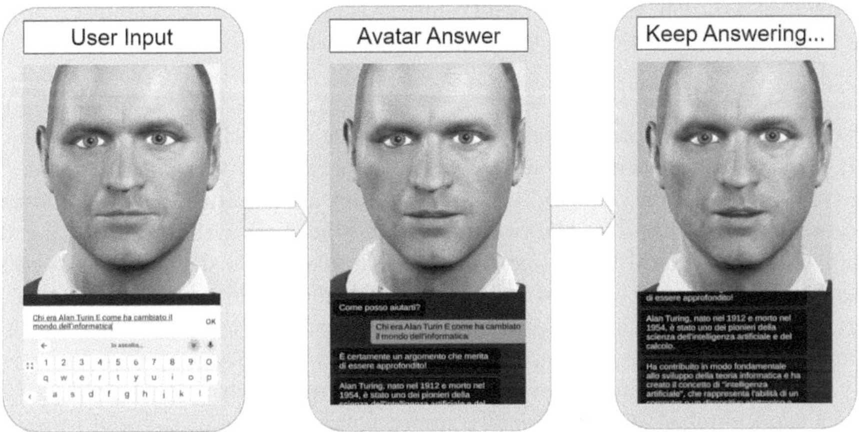

Fig. 2. Illustration of the interaction pipeline. The left panel ("User Input") shows the user posing a question to the virtual humans via text input (in this case, displayed on a smartphone interface). The right panel ("Avatar Answer") shows the avatar generating a verbal response in real-time.

Selected Hardware. For the implementation of our system, we adopted a minimal hardware configuration required to achieve full real-time performance. We so executed our system and perform the experiments that follows in a local machine equipped with an AMD Ryzen 7 5700X CPU, 32 GB DDR4 RAM, NVMe SSD storage, and a PCIe 4.0 compatible motherboard. At the core of this setup is the NVIDIA RTX 3060 with 12 GB of VRAM, selected as the critical component capable of keeping both the LLM and TTS models fully resident in memory and performing low-latency inference through CUDA and tensor core acceleration. This local machine allows the entire server-side pipeline to run locally, without reliance on cloud infrastructure. The Unity-based client application runs separately on target devices such as desktop and Android systems, which maintain a lightweight runtime profile thanks to the complete offloading of computational tasks to the server.

3D model and Scene Design. The 3D virtual teacher was implemented in Unity3D using a fully rigged character model with facial blend shapes, high-resolution textures, and a minimal static environment designed for real-time performance. The scene includes only the character and essential lighting, allowing the use of a relatively high polygon budget without compromising frame rates. A fixed frontal camera framing supports consistent eye contact and simplifies interaction management. Lip sync is achieved through real-time audio signal analysis and activation of predefined blend shapes, enabling synchronized mouth movements with low computational cost. Minor idle animations and facial micro-movements enhance expressiveness while maintaining responsiveness across target platforms. Although the current scene is intentionally minimal, it can be augmented with additional complexity when needed, for example, to support immersive AR, VR or mixed reality interactions, or to accommodate specific accessibility requirements for users with impairments.

Large Language Model. The selection of the text generation model was guided by extensive testing of open-source architectures compatible with the available VRAM. For this reason, we adopted the 3B parameter version of Hermes [2], derived from Meta's LLaMA 3.2, was adopted after initial experiments with larger models to improve the overall system efficiency without saturating the available GPU VRAM. To support progressive token-level streaming and reduce memory footprint, the selected model was converted from its original safe-tensors format to the more efficient gguf format, optimized for local inference through *lama cpp* library[1]. The inference was performed with a streaming decoding strategy, setting a low temperature of 0.20, top-p sampling with a threshold of 0.92, and a top-k limit of 40. Additionally, minor penalties were introduced to reduce repetitiveness (repeat penalty = 1.20) and balance lexical diversity (presence penalty = 0.15; frequency penalty = 0.10), and avoid too long statements (token limit = 64). The system prompt was structured to include strict task-oriented constraints. These guided the model to respond in Italian using an academic and formal register, avoiding hallucinations and enforcing factual accuracy, especially within the historical domain of informatics. The prompt also imposed syntactic rules, such as avoiding initials with dots and limiting the output to a maximum of 200 words, promoting compact and precise content delivery. This is depicted in Fig. 3.

Text to Speech. For speech synthesis, the project adopted a custom fork of the open-source CoquiTTS framework[2]. On top of CoquiTTS, the voice of the virtual teacher was crafted through an iterative process combining targeted audio sampling, signal enhancement, and voice cloning, also controlling utterances under tonal consistency and phonetic clarity. This approach enabled the creation of modular vocal timbre without requiring professional-grade recording sessions, remaining compatible with open-source tools and standard local hardware.

---

[1] https://github.com/abetlen/llama-cpp-python.
[2] https://github.com/coqui-ai/TTS.

SYSTEM PROMPT

"Respond in Italian using an academic and formal style like a tutor. Generate maximum 200 words overall. Provide only correct information pertaining to the world of computer history, avoiding any made-up or hallucinated content. Then continue with a brief, orderly exposition of the topic. Do not repeat sentences already written. If you write **NAMES** of people, they must all be without periods and without second dotted names. Acronyms must also be without periods and well spelled out in Italian. You will receive a heavy penalty if you do not adhere to these instructions.<|im_end|>"+"<|im_start|> assistant"

Fig. 3. Adopted System Prompt.

4.2   Experimental Setting

To validate the runtime performance of our pipeline, we measured the inference latency for both the LLM-based paragraph generation and the subsequent TTS synthesis. We selected five representative prompts covering different linguistic structures and semantic contents. Each prompt triggered the generation of three distinct paragraphs using the Hermes-3B model running locally via llama.cpp, and their respective audio renderings were synthesized through our custom CoquiTTS-based engine. All measurements were conducted on the same hardware configuration described in Sect. 4.1, namely a local machine equipped with an AMD Ryzen 7 5700X CPU and an NVIDIA RTX 3060 GPU with 12 GB VRAM, ensuring consistent conditions without reliance on cloud services. To assess system latency in realistic conditions, we measured it at the paragraph level using five question prompts, each generating three paragraphs. For each paragraph, we recorded the combined time of LLM inference and TTS synthesis, as audio was streamed to the client immediately upon readiness. This approach mirrors the system's real-time behavior and minimizes perceived latency, ensuring smooth and natural interactions.

All the obtained measurements, reported in Table 1, were collected on the local hardware described in Sect. 4.1 and generated by fixing the random seed, ensuring consistency and reproducibility of the results. As depicted in Table 1, the adopted LLM exhibits inference times that are generally low, with mean values ranging from ~ 500 ms to ~ 3000 ms, depending on the prompt and paragraph. Prompt 5, Paragraph 2 exposes a low generation time of 518.84 ms, given by a less complex generation path induced by that specific prompt. Those values are, in general, smaller than the TTS model times, which range from ~ 2000 ms to ~ 4000 ms. This confirms that the synthesis step is the dominant contributor to the overall response time. To have a general view of the performance, we depict the boxplots of all the generation times, aggregated by prompts, in Fig. 4. Those values further confirm the aforementioned considerations.

Table 1. Runtime mean (std) in milliseconds for each paragraph's generation (LLM) and synthesis (TTS), across five prompts.

Prompt ID	Paragraph 1	Paragraph 2	Paragraph 3
Prompt 1	LLM: 2097.90 (49.45)	LLM: 2054.60 (47.28)	LLM: 2625.43 (64.38)
	TTS: 3165.26 (180.64)	TTS: 2928.37 (162.63)	TTS: 4016.29 (244.52)
Prompt 2	LLM: 1964.98 (91.71)	LLM: 1977.22 (72.50)	LLM: 1930.69 (75.02)
	TTS: 2062.84 (439.23)	TTS: 3605.12 (757.04)	TTS: 3728.32 (874.20)
Prompt 3	LLM: 2615.27 (90.12)	LLM: 2220.24 (147.33)	LLM: 2053.01 (137.70)
	TTS: 3200.11 (510.94)	TTS: 3778.15 (812.59)	TTS: 2617.19 (444.68)
Prompt 4	LLM: 3002.00 (176.89)	LLM: 3000.59 (42.72)	LLM: 2488.44 (36.46)
	TTS: 4392.92 (546.18)	TTS: 3581.59 (248.89)	TTS: 3794.39 (237.59)
Prompt 5	LLM: 2652.78 (32.06)	LLM: 518.84 (5.15)	LLM: 2137.81 (15.34)
	TTS: 3433.40 (215.17)	TTS: 806.14 (156.61)	TTS: 3356.97 (436.93)

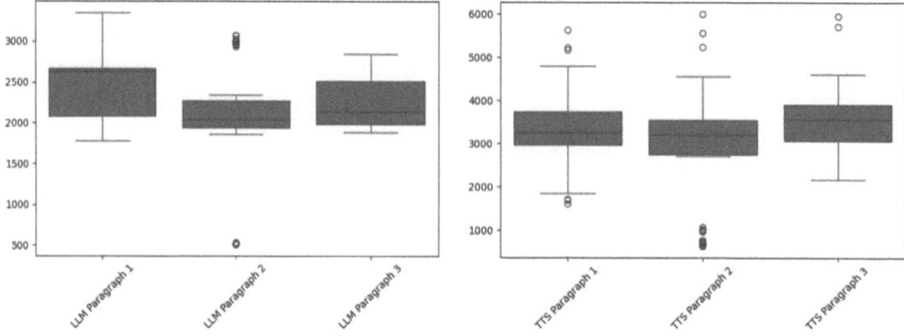

(a) Distribution of LLM Paragraph Times (b) Distribution of TTS Paragraph Times

Fig. 4. Boxplot comparison of generation (LLM) and synthesis (TTS) times across paragraphs.

5   Discussions and Conclusion

We presented a modular and efficient framework for deploying LLM-driven virtual humans on resource-constrained devices. By integrating a lightweight dialogue-management layer with a synchronized speech-and-facial-animation pipeline, our architecture achieves significant reductions in end-to-end latency. Our quantitative evaluation of the LLM and TTS modules demonstrates that response times per paragraph remain within a practical window for near real-time interaction, ensuring natural communication with the virtual avatar. Addressing this, we plan to conduct controlled studies to assess communicative and pedagogical effectiveness, user engagement, and learning impact. Moreover, we did not include a direct empirical comparison with SOTA approaches. For future works, we plan to conduct controlled studies to assess communicative and pedagogical effectiveness, user engagement, and learning impact against SOTA approaches. Additionally, we planned to investigate: (i) the adoption of efficient RAG Paradigms, to allow avatars to dynamically access and ground their responses on external knowledge bases; (ii) the usage of efficient Multimodal LLM architectures to assess their performance, inference costs, and energy consumption; (iii) conducting a large-scale, and user-centered studies to provide insights into perceived naturalness, engagement levels, and learning outcomes. To conclude, this work laid the groundwork for modular and efficient applications of intelligent avatars across education, customer support, and entertainment scenarios, where real-time responsiveness and avatar customization are critical aspects.

Acknowledgments. This work has been co-funded by the European Union under the "AGRITECH EU - Digital agriculture for sustainable development" founded in the Digital Europe Programme, Project 101123258.

References

1. Arslan, M., Ghanem, H., Munawar, S., Cruz, C.: A survey on rag with llms. Procedia Comput. Sci. **246**, 3781–3790 (2024)
2. Ayed, F., Maatouk, A., Piovesan, N., De Domenico, A., Debbah, M., Luo, Z.Q.: Hermes: a large language model framework on the journey to autonomous networks. arXiv preprint arXiv:2411.06490 (2024)
3. Cascarano, P., Franchini, G., Porta, F., Sebastiani, A.: On the first-order optimization methods in deep image prior. J. Verification, Validation Uncertainty Quantification **7**(4), 041002 (2022)
4. Fink, M.C., Robinson, S.A., Ertl, B.: Ai-based avatars are changing the way we learn and teach: benefits and challenges. In: Frontiers in Education, vol. 9, p. 1416307. Frontiers Media SA (2024)
5. Gao, H., Xie, Y., Kasneci, E.: Pervrml: Chatgpt-driven personalized vr environments for machine learning education. Int. J. Hum.–Comput. Interact. 1–15 (2025)
6. Gatto, L., Gaglio, G.F., Augello, A., Caggianese, G., Gallo, L., La Cascia, M.: Met-iquette: enabling virtual agents to have a social compliant behavior in the metaverse. In: 2022 16th International Conference on Signal-Image Technology & Internet-Based Systems (SITIS), pp. 394–401. IEEE (2022)
7. Hadi, M.U., et al.: A survey on large language models: applications, challenges, limitations, and practical usage. Authorea Preprints (2023)
8. John, K.S., Roy, G.A., S, B.P.: Llm based 3d avatar assistant. In: 2024 1st International Conference on Trends in Engineering Systems and Technologies (ICTEST), IEEE (2024)
9. Kenny, P.G., Parsons, T.D., Geer, A., ORCiD, I., Parsons, T.D.: Virtual standardized llm-ai patients for clinical practice. Ann. Rev. Cybertherapy Telemedicine **2024**, 177 (2024)
10. Kimara, E., Oguntoye, K.S., Sun, J.: Personaai: Leveraging retrieval-augmented generation and personalized context for ai-driven digital avatars. arXiv preprint arXiv:2503.15489 (2025)

11. Loli Piccolomini, E., Gandolfi, S., Poluzzi, L., Tavasci, L., Cascarano, P., Pascucci, A.: Recurrent neural networks applied to gnss time series for denoising and prediction. In: 26th International Symposium on Temporal Representation and Reasoning (TIME 2019), pp. 10–1. Schloss Dagstuhl–Leibniz-Zentrum für Informatik (2019)
12. Marquardt, A., Golchinfar, D., Vaziri, D.: Ragatar: enhancing llm-driven avatars with rag for knowledge-adaptive conversations in virtual reality. In: 2025 IEEE Conference on Virtual Reality and 3D User Interfaces Abstracts and Workshops (VRW), pp. 1604–1605. IEEE (2025)
13. Maslych, M., Pumarada, C., Ghasemaghaei, A., LaViola Jr, J.J.: Takeaways from applying llm capabilities to multiple conversational avatars in a vr pilot study. arXiv preprint arXiv:2501.00168 (2024)
14. Reiners, D., Davahli, M.R., Karwowski, W., Cruz-Neira, C.: The combination of artificial intelligence and extended reality: a systematic review. Front. Virt. Real. **2**, 721933 (2021)
15. Schmidt, S., Köysürenbars, I., Steinicke, F.: Frankenstein's monster in the metaverse: User interaction with customized virtual agents. IEEE Trans. Vis. Comput. Graph. (2024)
16. Tian, C., et al.: Clone: customizing llms for efficient latency-aware inference at the edge. arXiv preprint arXiv:2506.02847 (2025)
17. Wan, H., et al.: Building llm-based ai agents in social virtual reality. In: Extended Abstracts of the CHI Conference on Human Factors in Computing System, pp. 1–7 (2024)
18. Xu, T., Zhang, Y., Chu, Z., Wang, S., Wen, Q.: Ai-driven virtual teacher for enhanced educational efficiency: Leveraging large pretrain models for autonomous error analysis and correction. In: Proceedings of the AAAI Conference on Artificial Intelligence, vol. 39, pp. 28801–28809 (2025)
19. Yamazaki, T., Mizumoto, T., Yoshikawa, K., Ohagi, M., Kawamoto, T., Sato, T.: An open-domain avatar chatbot by exploiting a large language model. In: Proceedings of the 24th Annual Meeting of the Special Interest Group on Discourse and Dialogue. Association for Computational Linguistics, Prague, Czechia (2023)
20. Zhang, S., Ma, M., Li, Y., Man, K.L., Smith, J., Yue, Y.: The effects of llm-empowered chatbots and avatar guides on the engagement, experience, and learning in virtual museums. Int. J. Hum.-Comput. Interact. 1–13 (2025)
21. Zhang, Y., Zhao, S., Tian, X., Sun, H.: Design and development of "virtual ai teacher" system based on nlp. In: 2023 11th International Conference on Information and Education Technology (ICIET), pp. 141–145. IEEE (2023)
22. Zhao, Z., Yin, Z., Sun, J., Hui, P.: Embodied ai-guided interactive digital teachers for education. In: SIGGRAPH Asia 2024 Educator's Forum, pp. 1–8 (2024)

XRidge: Enabling Cross-Reality Co-creation with Generative AI and LLMs

Giacomo Vallasciani[1]([✉]) , Pasquale Cascarano[2] , and Gustavo Marfia[2]

[1] Department of Computer Science and Engineering, University of Bologna, Bologna, Italy
`giacomo.vallasciani2@unibo.it`
[2] Department of the Arts, University of Bologna, Bologna, Italy
`{pasquale.cascarano2,gustavo.marfia}@unibo.it`

Abstract. eXtended Reality (XR) technologies are increasingly used in education and industry due to their support for spatial 3D interaction and contextual annotations, which enhance learning and task efficiency. Current solutions offer features like real-time collaboration and virtual annotations, but are often costly and tied to specific hardware. Key requirements for effective XR systems include device independence, usability, external data integration, and real-time collaboration. WebXR holds promise in addressing these needs, yet no comprehensive open-source platform currently exists that combines: (i) an adaptive immersive WebXR environment, (ii) remote-streaming capabilities for broader access, and (iii) integration with external 3D databases. To address this gap, we present XRidge (XR Knowledge Bridge), a cross-reality WebXR platform supporting immersive, collaborative, and data-driven experiences. XRidge is designed to be hardware-agnostic and open-source, enabling applications in remote teaching, training, and showcasing.

Keywords: Extended Reality · Artificial Intelligence · Creativity · Generative Artificial Intelligence · Stable Diffusion · Privacy

1 Introduction

eXtended Reality (XR) technologies have demonstrated increasing relevance across domains such as education, healthcare, industrial design, and tourism, owing to their ability to enable spatial visualization and real-time manipulation of 3D content [7,31]. In educational settings, XR facilitates collaborative learning through shared 2D/3D content and intelligent agents accessible across heterogeneous devices [17,32].

Despite recent advances, significant challenges remain in developing XR systems that are device-independent, support real-time co-creation, and integrate external data sources. Earlier solutions, such as Virtual Reality Peripheral Network (VRPN) [34], focused solely on Virtual Reality (VR), lacking both web integration and collaborative features. In contrast, WebXR enables cross-platform,

© The Author(s), under exclusive license to Springer Nature Switzerland AG 2026
E. Rodolà et al. (Eds.): ICIAP 2025 Workshops, LNCS 16169, pp. 661–673, 2026.
https://doi.org/10.1007/978-3-032-11317-7_53

browser-based access to XR content [9, 29], making it well-suited for collaborative and Cross-Reality (CR) applications [4].

Nevertheless, limitations persist in areas such as immersive asset manipulation [27], integration with web-based 3D libraries [3], and support for interactive remote rendering [30]. Furthermore, as XR systems increasingly integrate AI for content generation and semantic enrichment, they must also address emerging concerns around privacy and access control in multi-user environments.

To address these gaps, we introduce XRidge, an open-source, WebXR-based platform designed for real-time, device-independent XR collaboration. XRidge supports multi-user interaction with role-based permissions, remote rendering, integration with online 3D repositories, and Artificial Intelligence (AI)-enabled creativity support.

2   Related Works

Several industrial platforms offer collaborative XR experiences, including Meta Immersive Learning [20], Mozilla Hubs [21], and Microsoft Mesh [40]. While these systems support basic multi-user interactions, they are typically closed-source, device-restrictive, and lack extensibility or advanced annotation features.

In the academic domain, XR frameworks have been developed to investigate WebXR interoperability and real-time interaction [1,9,23,29]. However, most of these systems fall short in integrating external services, enabling full-scene customization, or supporting remote rendering capabilities.

XRidge extends this line of research by combining real-time collaboration, dynamic scene manipulation, access to external assets, and AI-based content augmentation. Table 1 compares XRidge with major XR platforms across key feature dimensions.

Recent studies confirm the effectiveness of XR and large language models (LLMs) in enhancing user engagement, promoting flow, and stimulating critical thinking [16,22]. Projects such as SimClass [43] and Arele-bot [12] further demonstrate the potential of these technologies in adaptive education and cultural heritage applications.

However, collaborative XR platforms still lack robust access control mechanisms. While role-based [26] and proximity-based models [11] offer partial solutions, they often lack flexibility or raise privacy concerns. Dynamic permission systems that restrict interaction and content visibility based on user roles remain underdeveloped [25].

XRidge addresses this gap by introducing a flexible, role-based permission model that enables real-time moderation and secure, multi-user interaction. Combined with generative AI and dynamic scene control, XRidge enables expressive and adaptive experiences for education, cultural engagement, and co-creation.

Table 1. Comparison of XR systems: XR Device Agnostic (XRAG), Collaboration (C), 2D Device Compatibility (2DC), Scene Customization (SC), Remote Rendering (RR), Annotation System (ANN), External Database Integration (EDI), General Purpose (GP), Open-Source (OS), and Artificial Intelligence Integration (AII).

	XRAG	C	2DC	SC	RR	ANN	EDI	GP	OS	AII
Industrial										
[41]	✗	✓	✗	✗	✓	✗	✗	✗	✗	✗
[39]	✗	✓	✗	✗	✗	✗	✗	✗	✗	✗
[35, 36]	≈	✓	✓	✓	✓	✓	✗	≈	✗	✗
[20]	✗	✓	✗	✓	✗	✗	✗	✓	✗	✗
[21]	✗	✓	✓	✓	✗	✗	✗	✓	✗	✗
[40]	✗	✓	✗	✓	✓	✓	✗	≈	✗	✗
Academic										
[1]	✓	✓	✗	✗	✓	✗	✗	✗	✗	✗
[15]	✗	✓	✓	≈	✗	✗	✗	✗	✗	✗
[19]	✓	✓	✓	✗	✗	✗	✗	✗	✗	✗
[23]	✓	✓	✗	≈	✗	✗	✗	✗	✗	✗
[29]	✓	✓	✓	≈	≈	≈	✗	✗	✗	✗
[9]	✓	✓	✓	≈	✗	✓	✗	✗	✓	✗
XRidge	✓	✓	✓	✓	✓	✓	✓	✓	✓	✓

3   Use Cases

XRidge's modular, device-independent architecture supports a wide range of immersive scenarios across education, cultural heritage, and 3D content analysis. This section presents representative use cases in which XRidge was deployed and evaluated.

3.1   Generative Museum Scenario

XRidge was applied to creative education, specifically in museology and art curatorship, supporting activities such as exhibition planning and collaborative analysis [18]. Generative AI further enhances this workflow by providing inspiration and facilitating idea generation [8].

A real-world university cloister was virtualized to model an interactive museum environment, enabling users to organize exhibitions in a spatial and collaborative manner [33]. Figure 1 shows the generative museum interface.

Users can navigate using teleportation, orbiting, or persistent motion [2]. Spatial annotations allow public notes and markers to persist across sessions, supporting curatorial workflows.

Fig. 1. Generative museum scenario and user interface (top bar).

XRidge integrates external 3D model repositories (e.g., Sketchfab), searchable via a virtual keyboard interface (Fig. 2). Once imported, models can be placed, scaled, or rotated using device-adaptive controls [28, 42].

Fig. 2. Importing 3D models from an external database.

Users can also generate AI-based artwork using text commands in the format "A picture of <user_prompt>", entered via the virtual keyboard. The generated images can then be placed within the virtual exhibition (Fig. 3).

3.2 Teaching Dance Costumes

XRidge was used in a dance history course focused on the digital reconstruction of Rudolf Nureyev's doublet from Romeo and Juliet, created using Stable-Fast3D [5, 14]. The virtual costume was exhibited in the reconstructed cloister of S. Cristina (University of Bologna), allowing students to interact collaboratively.

The learning session combined exploration and annotation. Students added contextual materials (e.g., video links) directly onto the 3D garment, as shown in Fig. 4.

Archival materials such as programs and press clippings were uploaded to simulate historical research. LLM integration enabled image-to-text and natural language queries for descriptive analysis of costume details.

Fig. 3. Example of generated paintings.

Fig. 4. Annotation containing YouTube multimedia links.

XRidge supports persistent and audio-based annotations, visible to all participants. This promotes asynchronous collaboration and co-editing, encouraging interdisciplinary work between humanities and computer science students.

3.3   3D Modelling Methods Comparison

XRidge's flexibility enabled comparative analysis of four 3D reconstruction techniques: Artec Eva (structured light), Scaniverse (LiDAR), Polycam (photogrammetry), and a custom Gaussian Splatting pipeline [6].

Each technique was evaluated within the same XR session:

- Artec Eva offers high precision but requires expensive equipment and significant processing time.
- Scaniverse enables mobile, rapid scanning but offers limited detail due to LiDAR resolution.
- Polycam provides good results for textured models using smartphones, but relies on cloud-based services.
- Gaussian Splatting uses consumer video and open-source tools to generate fast and visually coherent reconstructions.

XRidge allows side-by-side switching of models in real time, enabling collaborative analysis and discussion of trade-offs. This supports informed selection of modeling pipelines based on educational context and available resources.

3.4   Security Aspects in Collaborative VR/XR Sessions

XRidge implements secure session management through salted password hashes and JSON Web Tokens (JWTs) [13]. Each token contains encrypted user credentials, validated using server-side signatures. User actions are logged for accountability.

Permission levels can be updated in real time via an administrative dashboard, controlling access to rooms and content without disrupting the session. This ensures the secure handling of sensitive materials [37], especially in collaborative educational or cultural heritage settings.

4   System Architecture

XRidge is a modular, device-independent XR platform designed for real-time multi-user collaboration across heterogeneous devices. It integrates WebXR and AI technologies using a multi-user client-server architecture [33], as illustrated in Fig. 5. The server manages scene logic, user roles, data synchronization, and external services, while clients are responsible for rendering and user interaction. This structure enables consistent experiences on desktops, mobile devices, and XR headsets.

The system supports a wide range of functionality:

- Synchronized multi-user 3D scene rendering;
- Device-adaptive UI via WebXR;
- Role-based permissions for interaction and editing;
- Multimodal annotation using text or audio;
- Real-time integration with external 3D repositories (e.g., Sketchfab);
- Server-side rendering for users with low-end devices;
- Generative AI support for immersive content creation;
- Integration with Large Language Models (LLMs) for semantic interaction.

4.1   Backend and Frontend Design

The backend is composed of several components working together to manage authentication, real-time communication, content generation, and rendering:

- A Node.js server handles authentication, session management, and WebSocket-based communication using Socket.io;
- A data manager synchronizes scene and user data stored in JSON format;
- A Flask-based microservice interacts with generative AI models for content generation;
- A Puppeteer-based proxy provides remote rendering for lightweight clients.

On the frontend, XRidge uses Three.js in combination with WebXR APIs to enable cross-device interaction, ensuring that both XR and non-XR users have access to consistent and responsive interfaces.

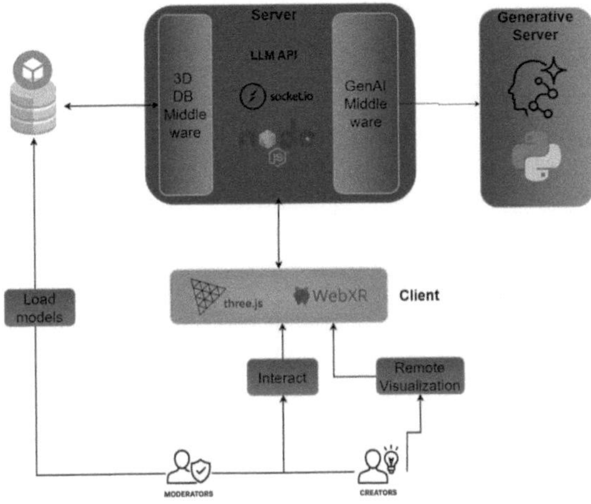

Fig. 5. XRidge system architecture.

4.2   Collaborative Scene Rendering

XRidge allows users to participate in shared sessions where all interactions and scene changes are synchronized in real time. Figure 6 illustrates the flow of a typical session: a moderator selects a scene, triggering a backend request. The server responds with parsed JSON data, which is rendered on all clients. Users can then add or modify objects and annotations within the scene.

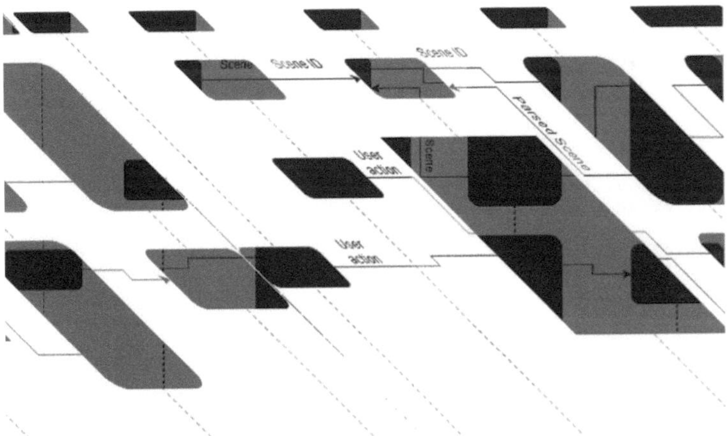

Fig. 6. Sequence diagram for collaborative XRidge session and object interaction.

The WebXR integration allows consistent input mapping across various platforms. The system has been tested across multiple devices including Oculus

Quest, Magic Leap 2, and HTC Vive Pro, demonstrating cross-device compatibility and stability.

4.3   Data Model and Scene Structure

XRidge structures both scene and user data using extensible JSON files [9,33], which allows for dynamic updates and synchronization across clients.

A scene JSON includes:

- Metadata, such as scene name, lighting, and environment;
- Scene graph, representing the hierarchy of 3D objects, including their URLs, transformations, and editability;
- Annotation graph, which stores spatial annotations linked to the scene structure.

Listing 1.1. Exemplar JSON representing a 3D scene.

```json
{ "sceneName": "Generative_Gallery",
  "environment": {
    "mainlight": {"direction": [...], "shadows": false}
  },
  "scene_graph": {
    "nodes": {
      "building": {
        "url": "path/to/scene.gltf",
        "editable": false,
        "transform": {...}
      },
      "3D_object_1": {
        "url": "...",
        "editable": true,
        "transform": {...}
      }
    },
    "hierarchy_edges": {
      ".": ["environment": ["3D_object_1"]]
    }
  },
  "annotation_graph": {
    "nodes": {
      "annotation_user_1_shape": {"convexshapes": [...]},
      "annotation_user_2_draw": {"drawing": [...]}
    },
    "hierarchy_edges": {
      ".": ["annotation_1", "annotation_2", "annotation_3"]
    }
  }
}
```

User JSONs store roles, tokens, and asset lists:

Listing 1.2. Exemplar JSON of user data in editor mode.

```
{ "username": "user_1",
  "admin": true,
  "database_session_token": "...",
  "level": 5,
  "items": [
    {
      "name": "sketchfab_obj",
      "type": "sketchfab",
      "url": "...",
      "thumbnail": "..."
    }
  ]
}
```

4.4 Object Manipulation and Database Integration

Users interact with 3D models via standard controls (click, drag, rotate) and can annotate using text or audio. Sketchfab integration [3,10] enables in-app 3D asset search and import, with real-time updates to all clients.

4.5 Remote Rendering Module

To support low-end devices, XRidge offers a Remote Rendering module [30] using Puppeteer. Rendered scenes are streamed as video via WebSocket (Fig. 7), while user inputs are transmitted to the server and processed via raycasting.

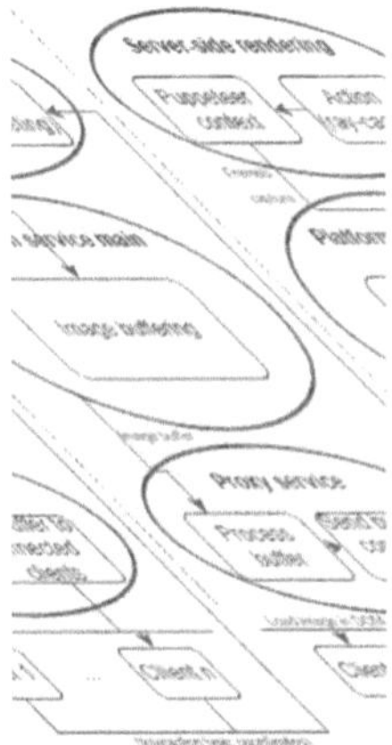

Fig. 7. Remote rendering proxy server architecture.

4.6   Generative AI Module

A Flask microservice (Fig. 8) handles AI-based asset generation, routed through a GPU-backed queue. Generated outputs are inserted into the shared scene and attributed to the requester.

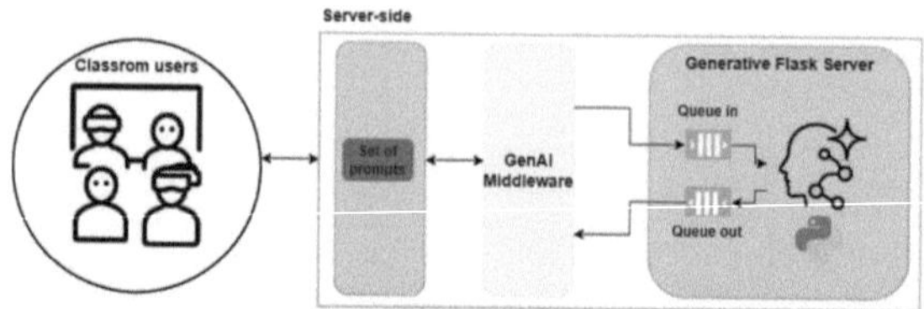

Fig. 8. Architecture of the generative AI module integrated in XRidge.

Currently, XRidge uses Stable Diffusion [24] for text-to-image generation.

4.7   LLM Integration

LLM APIs enable contextual querying of objects through text or images. Responses appear in real time and are stored as collaborative annotations. XRidge supports multimodal prompts and persistent AI-generated content, promoting rich semantic interaction and interdisciplinary learning.

5   Conclusions and Future Works

XRidge is a collaborative, device-agnostic WebXR platform for real-time 3D content visualization and manipulation. It supports object layout, spatial annotations, and integration with external 3D databases. Remote rendering enables accessibility on low-end devices, making XRidge suitable for educational, creative, and professional contexts.

Generative AI modules enhance creative workflows, with initial user studies in virtual museum co-creation showing positive results [38].

In cultural heritage scenarios, XRidge facilitates interdisciplinary learning by combining XR with LLMs. A case study on dance costume history demonstrated how AI-driven annotations and real-time interaction support archival research and critical engagement. Broader classroom validation remains a key future step.

To manage collaboration, XRidge introduces a dynamic permission system that regulates access to content and actions in real time based on user roles.

Current limitations include the absence of advanced AR spatial computing (e.g., SLAM, marker tracking) and limited 3D asset integration. Planned enhancements involve AI-driven 3D model search and automatic texturing, middleware for asset unification, improved scalability, and large-scale testing across diverse domains.

References

1. Al Hafidz, I.A., et al.: Design of collaborative webxr for medical learning platform. In: 2021 International Electronics Symposium (IES), pp. 499–504. IEEE (2021)
2. Al Zayer, M., MacNeilage, P., Folmer, E.: Virtual locomotion: a survey. IEEE Trans. Visual Comput. Graphics **26**(6), 2315–2334 (2018)
3. Aristov, M.M., Moore, J.W., Berry, J.F.: Library of 3d visual teaching tools for the chemistry classroom accessible via sketchfab and viewable in augmented reality (2021)
4. Asunis, L., Cirina, A., Stacchio, L., Marfia, G.: Hoctopus: an open-source cross-reality tool to augment live-streaming remote classes. In: 2023 IEEE International Symposium on Mixed and Augmented Reality Adjunct (ISMAR-Adjunct), pp. 29–34 (2023). https://doi.org/10.1109/ISMAR-Adjunct60411.2023.00014
5. Boss, M., Huang, Z., Vasishta, A., Jampani, V.: Sf3d: stable fast 3d mesh reconstruction with UV-unwrapping and illumination disentanglement. arXiv preprint (2024)
6. Cascarano, P., et al.: A comparative analysis of 3d modeling methods for integration into an extended reality platform. In: 2025 IEEE International Conference on Artificial Intelligence and eXtended and Virtual Reality (AIxVR), pp. 213–217. IEEE (2025)
7. Dwivedi, Y.K., et al.: Metaverse beyond the hype: multidisciplinary perspectives on emerging challenges, opportunities, and agenda for research, practice and policy. Int. J. Inf. Manage. **66**, 102542 (2022)
8. Epstein, Z., et al.: Art and the science of generative AI. Science **380**(6650), 1110–1111 (2023)
9. Fanini, B., Ferdani, D., Demetrescu, E., Berto, S., d'Annibale, E.: Aton: an open-source framework for creating immersive, collaborative and liquid web-apps for cultural heritage. Appl. Sci. **11**(22), 11062 (2021)
10. Garcia-Bonete, M.J., Jensen, M., Katona, G.: A practical guide to developing virtual and augmented reality exercises for teaching structural biology. Biochem. Mol. Biol. Educ. **47**(1), 16–24 (2019)
11. Gibbs, M., Wadley, G., Benda, P.: Proximity-based chat in a first person shooter: using a novel voice communication system for online play. In: Proceedings of the 3rd Australasian conference on Interactive entertainment, pp. 96–102 (2006)
12. Hajahmadi, S., Clementi, L., López, M.D.J., Marfia, G.: Arele-bot: inclusive learning of spanish as a foreign language through a mobile app integrating augmented reality and ChatGPT. In: 2024 IEEE Conference on Virtual Reality and 3D User Interfaces Abstracts and Workshops (VRW), pp. 335–340. IEEE (2024)
13. Jones, M., Bradley, J., Sakimura, N.: Rfc 7519: JSON web token (JWT) (2015)
14. Kahane, M.: Rudolf noureev . Costumes et photographies. CNCS Editions du Me'ce'ene, pp. 1938-1993 (2009)
15. Korečko, Š, Hudák, M., Sobota, B., Sivỳ, M., Pleva, M., Steingartner, W.: Experimental performance evaluation of enhanced user interaction components for web-based collaborative extended reality. Appl. Sci. **11**(9), 3811 (2021)
16. Kozov, V., Ivanova, B., Shoylekova, K., Andreeva, M.: Analyzing the impact of a structured LLM workshop in different education levels. Appl. Sci. (2076-3417) **14**(14) (2024)
17. Liu, R., Peng, C., Zhang, Y., Husarek, H., Yu, Q.: A survey of immersive technologies and applications for industrial product development. Comput. Graph. **100**, 137–151 (2021)

18. Mansour, H., Sheta, M.S.S., SAMRA, M.: Towards new design patterns for museum exhibition halls using integrated algorithmic generative techniques (2021)
19. Martí-Testón, A., Muñoz, A., Gracia, L., Solanes, J.E.: Using webxr metaverse platforms to create touristic services and cultural promotion. Appl. Sci. **13**(14), 8544 (2023)
20. Meta: create, learn and grow in the metaverse (2022). https://about.meta.com/immersive-learning/
21. Mozilla: mozilla hub, virtual classroom. https://hubs.mozilla.com/9etxDBw/virtual-classroom (2022)
22. Ng, L.: LLMS in education. XRDS: Crossroads. ACM Mag. Stud. **31**(1), 66–70 (2024)
23. Pereira, V., Matos, T., Rodrigues, R., Nóbrega, R., Jacob, J.: Extended reality framework for remote collaborative interactions in virtual environments. In: 2019 International Conference on Graphics and Interaction (ICGI), pp. 17–24. IEEE (2019)
24. Rombach, R., Blattmann, A., Lorenz, D., Esser, P., Ommer, B.: High-resolution image synthesis with latent diffusion models. In: Proceedings of the IEEE/CVF conference on computer vision and pattern recognition, pp. 10684–10695 (2022)
25. Ruth, K., Kohno, T., Roesner, F.: Secure {Multi-User} content sharing for augmented reality applications. In: 28th USENIX Security Symposium (USENIX Security 19), pp. 141–158 (2019)
26. Sandhu, R.: Role activation hierarchies. In: Proceedings of the third ACM workshop on Role-based access control, pp. 33–40 (1998)
27. Schreyer, A.C.: Architectural design with sketchup: 3D modeling, extensions, BIM. Making, and Scripting. John Wiley & Sons, Rendering (2015)
28. Scott, B., Neil, T.: Designing web interfaces: principles and patterns for rich interactions." O'Reilly Media, Inc." (2009)
29. Seo, D., Yoo, B., Ko, H.: Webizing collaborative interaction space for cross reality with various human interface devices. In: Proceedings of the 23rd International ACM Conference on 3D Web Technology, pp. 1–8 (2018)
30. Shi, S., Hsu, C.H.: A survey of interactive remote rendering systems. ACM Comput. Surv. (CSUR) **47**(4), 1–29 (2015)
31. de Souza Cardoso, L.F., Mariano, F.C.M.Q., Zorzal, E.R.: A survey of industrial augmented reality. Comput. Ind. Eng. **139**, 106159 (2020)
32. Stacchio, L., Angeli, A., Marfia, G.: Empowering digital twins with extended reality collaborations. Virtual Reality Intell. Hardw. **4**(6), 487–505 (2022)
33. Stacchio, L., Vallasciani, G., Augello, G., Carrador, S., Cascarano, P., Marfia, G.: Wixard: towards a holistic distributed platform for multi-party and cross-reality webxr experiences. In: 2024 IEEE Conference on Virtual Reality and 3D User Interfaces Abstracts and Workshops (VRW), pp. 264–272. IEEE (2024)
34. Taylor, R.M., Hudson, T.C., Seeger, A., Weber, H., Juliano, J., Helser, A.T.: Vrpn: a device-independent, network-transparent VR peripheral system. In: Proceedings of the ACM symposium on Virtual reality software and technology, pp. 55–61 (2001)
35. TechViz: Cloud and Viz (2023). https://www.techviz.net/en/cloud-and-viz/
36. TechViz: Share and Viz (2023). https://www.techviz.net/en/share-and-viz-duo/
37. Vallasciani, G., Schinoppi, A., Cascarano, P., Marfia, G., Donatiello, L.: Handling privacy and security aspects in a collaborative AR session. In: 2024 IEEE International Symposium on Mixed and Augmented Reality Adjunct (ISMAR-Adjunct), pp. 20–22. IEEE (2024)

38. Vallasciani, G., Stacchio, L., Cascarano, P., Marfia, G.: Creaixr: fostering creativity with generative AI in XR environments. In: 2024 IEEE International Conference on Metaverse Computing, Networking, and Applications (MetaCom), pp. 1–8. IEEE (2024)
39. Varjo: meetinvr, best-in-class business meetings in the metaverse (2023). https://www.meetinvr.com/
40. Varjo: microsoft mesh overview (2023). https://learn.microsoft.com/en-us/mesh/overview
41. Varjo: varjo reality cloud: stream a supercharged reality (2023). https://varjo.com/products/realitycloud/
42. Xing, Y., Shell, J., Fahy, C., Guan, K., Zhang, Q., Xie, T.: User interface research in web extended reality. In: 2021 IEEE 7th International Conference on Virtual Reality (ICVR), pp. 76–81. IEEE (2021)
43. Zhang, Z., et al.: Simulating classroom education with LLM-empowered agents. arXiv preprint arXiv:2406.19226 (2024)

The Digital Garment: A Comparative Study of 3D Digitisation Techniques for Fashion Heritage

Eleonora Stacchiotti[(✉)] [iD], Pasquale Cascarano [iD], Jacopo Megliorardi, and Gustavo Marfia [iD]

University of Bologna, Bologna, Italy
`eleo.stacchiotti@unibo.it`

Abstract. This paper compares two accessible 3D digitisation workflows—photogrammetry and Gaussian Splatting—applied to garments from a corporate fashion archive. Fashion heritage poses unique challenges due to material softness, layering, and symbolic complexity, requiring models suited to both documentation and storytelling. We evaluate each workflow in archival and exhibition contexts across four criteria: visual fidelity, interactivity, file size, and adaptability. Results show trade-offs between structural accuracy and visual richness, depending on use. We also highlight opportunities for integrating lightweight 3D models into XR environments and AI-driven curatorial systems. Digital twins thus emerge as relational tools for reactivating fashion heritage within immersive, intelligent cultural ecosystems.

Keywords: Digital Fashion Heritage · 3D Scanning Techniques · Digital Twins · Gaussian Splatting · Depth Maps

1 Introduction

In recent years, the digitisation of cultural heritage has emerged as a strategic priority within both public policy frameworks and academic discourse, driven by the convergence of technological innovation, sustainability goals, and growing societal interest in immersive and participatory experiences [4,9,11]. Among the most useful tools in this context is three-dimensional (3D) scanning, which enables the detailed digital replication of physical objects and environments, expanding opportunities for preservation, research, and cultural storytelling. While 3D scanning technologies have found wide application in archaeology, museology, and architecture, their integration into the fashion heritage sector—particularly in connection with historical garments preserved into corporate archives—remains limited and underexplored [21]. This is especially true in the Made in Italy production landscape, which is characterised by a decentralised and highly artisanal network of small and medium-sized enterprises, many of which lack the resources and infrastructure necessary to adopt complex digitisation workflows. Quantitative research recommends these small and medium-sized companies to embrace the use of digital technologies as they allow a major accessibility for customers to the design and production processes, while enabling a wider range of services concerning their product, including insights into the history of the final product and their cultural meaning [3,26]. These companies' corporate archives hold immense historical, cultural, and economic value, preserving materials, garments, and design philosophies that are central not only to brand identity and the broader history of fashion [22,33,34] but also to the construction of fashion's cultural value and the long-term success of fashion firms. As Riviezzo et al. argue, firms that fully harness the potential of their archives for innovative and engaging storytelling tend to be more resilient and enduring [25], something that is also attested by Mosca, who also discusses categories of activities around which industries can build their own identity, that can be expanded and redefined by the integration of technologies [18]. In the context of corporate archives, the deployment of digital twin technologies—interactive 3D representations that replicate the properties of physical objects—offers significant potential to come up with solutions that enable a more digital and engaging interaction with the brand identity. Digital garments can support not only documentation and long-term preservation but also immersive exhibition design, remote consultation, and new forms of heritage storytelling. Their adoption is strongly supported by public strategies with a strong push on the European level to promote inclusive, open, and interoperable digitisation strategies. Initiatives such as the European Commission's digital policy on cultural heritage [8] the Horizon Europe programme for long-term monitoring and preservation [9] and the European Heritage Hub [10] all highlight the need for transdisciplinary, accessible frameworks that support innovation while ensuring cultural sustainability. Projects like Europeana [11] which aggregates millions of cultural items across domains—including fashion—demonstrate the impact of digitisation at scale, while research initiatives such as the "Mass Digitisation of Cultural Heritage in Europe" led by Bournemouth University [4] further reinforce the urgency of developing adaptable, cross-sectoral methods that can respond to the needs of diverse institutions. Despite this momentum, the private sector, especially in the fashion industry, continues to face substantial challenges: high-end scanning equipment is often expensive and difficult to integrate into existing workflows, while lower-cost alternatives may struggle to meet the fidelity requirements of fashion heritage documentation. In this context, 3D reconstruction technologies have emerged as powerful tools for documenting and preserving physical artefacts. Traditional methods often rely on high-precision hardware such as structured-light scanners or laser scanners [7], which provide detailed geometric and textural data but come with high costs and technical requirements. More recent approaches include photogrammetry [28] which reconstructs 3D models from overlapping 2D images, and neural implicit representations like Neural Radiance Fields (NeRF), which enable highly realistic scene renderings from simple video sequences. An emerging method, 3D Gaussian Splatting (3DGS), represents scenes as a set of optimised Gaussian primitives, allowing for efficient and visually rich reconstructions with real-time rendering capabilities.

This paper is part of a three-year project focused on the digital transformation of the Made in Italy fashion industry. Its goal is to develop a scalable and modular workflow tailored to the needs of a selected fashion company specialised in mid-tier luxury production, which owns three brands, each with distinct aesthetics and market position. Founded in 1980 in the Emilia-Romagna fashion district, they operate globally while maintaining a strong local identity. Currently undergoing a phase of strategic transition, the company is re-evaluating its archival collections as a resource for innovation and brand coherence. Designers are revisiting historical pieces to align contemporary vision with heritage values. This renewed interest in corporate memory, combined with broader societal trends toward digitisation, has led the company to invest in the creation of a digital ecosystem for heritage activation. This paper investigates sustainable and replicable methods for the digitisation and valorisation of corporate fashion heritage. At the core of this initiative is the concept of the Digital Twin: a virtual replica of a physical object that allows extended use beyond its material form. Originally conceived in engineering domains [13] digital twins have been increasingly adopted in cultural heritage for documentation, access, and immersive engagement [15,20]. In this project, digital twins are understood not as static models but as dynamic interfaces enabling interaction, interpretation, and hybrid experiences.

The following sections analyse the digital replicas developed for the industrial partner, with attention to their technical and experiential features. The focus is on a subset of the archive belonging to one of the brands of the company, for which some items were selected and digitised. The experimental study involves the application of two distinct 3D scanning techniques, Photogrammetry and 3D Gaussian Splatting, which are analysed and compared in detail. Through a close inspection of the resulting 3D models, here are assessed the strengths and limitations of each method in terms of both technical and experiential outcomes. Each resulting digital replica is evaluated across two primary use scenarios: corporate archiving and exhibition environments. The evaluation considers key criteria such as visual fidelity, file size, interactivity, and adaptability. The objective is not only to understand the practical trade-offs between the two workflows, but also to explore how each technique enables different forms of human-machine interaction, collaboration, and user agency, ultimately leading to distinct interactional paradigms and suitability for various end uses.

2   Context and Related Work

2.1   Scanning Cultural Heritage and Fashion Archives

The digitisation of cultural heritage is increasingly central to how societies document, interpret, and transmit their material and symbolic legacy. Across Europe, initiatives like *Europeana* [11], the *New European Bauhaus* [19], and Horizon projects such as *INCEPTION* [16] and *Artemis* [2] aim to integrate cultural value with innovation, sustainability, and inclusivity. Within this framework, 3D scanning has emerged as a key technology for capturing the geometry and surface characteristics of heritage objects, enabling digital preservation, virtual access, and immersive experiences [17,24,35]. Its applications span archaeology, museology, architecture, and increasingly, the fashion industry. In fashion heritage, digitisation serves both archival and communicative purposes. Corporate archives—particularly in the Made in Italy system—are not only repositories of past collections, but active tools for branding, design research, and storytelling [18,22,25,33,34]. However, garments are complex to digitise: their softness, layering, and textural detail challenge conventional scanning systems. High-end 3D scanners offer precision but are expensive and require technical expertise [1]. More accessible solutions such as mobile photogrammetry or video-based neural methods offer alternatives, but their effectiveness across curatorial contexts remains underexplored [14,29,32]. Technical challenges—such as light interference, material reflectivity, and surface reconstruction—can impact quality, especially for fabrics or reflective materials [6,31]. Despite these limitations, scanning technologies are increasingly used to create digital twins: high-fidelity models that replicate not only the visual structure of objects, but also their spatial and interactive affordances [36]. This paper focuses on evaluating two low-cost 3D workflows—photogrammetry and Gaussian Splatting—applied to garments from a corporate fashion archive, with the aim of assessing their suitability for both documentation and exhibition purposes.

2.2   Scanning for Fashion

Fashion items present unique challenges for 3D digitisation due to their soft structure, layered construction, and symbolic value. Unlike rigid museum artefacts, garments require workflows that can capture subtle surface details and accommodate complex material behaviours. This complexity is heightened in corporate fashion archives, where digital replicas must support both documentation and storytelling functions [18,22]. In the Made in Italy context, heritage garments are often reactivated in contemporary design and brand communication [25,26], making their digital accessibility a strategic asset [33,34]. Despite increasing interest in 3D documentation, most high-fidelity scanning systems remain unaffordable or impractical for small archives [1]. Accessible alternatives like photogrammetry and neural rendering lower entry barriers, but trade off precision and interactivity. In archival settings, the emphasis lies on detail and inspectability; in exhibition contexts, on responsiveness, rendering quality, and audience engagement. Balancing these requirements demands workflows that are not only accurate and scalable, but also aligned with the material and curatorial specificities of fashion heritage.

2.3   Scanning Techniques: Sustainable Processes for Digital Copies

The digitisation of heritage objects involves a wide array of 3D capture technologies, each with distinct trade-offs in terms of precision, cost, scalability, and curatorial control. Structured-light and laser-based scanners remain the gold standard for producing metrically accurate, high-resolution models that are suitable for conservation and scientific documentation. However, their cost, size, and operational complexity often place them beyond the reach of small institutions or corporate archives [1]. At the opposite end of the spectrum, mobile photogrammetry tools such as Scaniverse and Polycam [23,27] allow non-specialists to create 3D models using smartphones or tablets. These systems are widely accessible, especially in educational or design-driven environments, and tend to prioritise visual appeal and rapid acquisition over precision. While they rarely provide reliable measurements, they produce textured outputs that are well-suited for public display and immersive applications [29]. Recent advancements in neural rendering have introduced a new category of tools based on learned scene representations. Among them, Gaussian Splatting stands out for its ability to render high-fidelity 3D scenes directly from video input, without requiring mesh construction or surface triangulation [32]. These techniques model scenes as spatially distributed 3D Gaussians that encode both radiance and opacity, enabling smooth transitions and photorealistic textures. Although computationally demanding, they represent a paradigm shift from traditional geometry-first approaches and offer distinct advantages in terms of visual richness and framerate responsiveness. In the context of fashion heritage, the choice of scanning technique cannot be guided by technical performance alone. Garments require careful capture of materials, textures, and volumes, but also demand workflows that are lightweight, replicable, and adaptable to all settings. This paper focuses on photogrammetry and Gaussian Splatting as two contrasting yet complementary approaches that reflect this tension—between precision and accessibility, between structure and appearance, and between technical feasibility and experiential potential. Beyond scanning technologies, recent developments in immersive media and artificial intelligence are reshaping the way cultural heritage is accessed and interpreted. Extended Reality (XR) platforms enable interactive experiences that situate digital garments within spatialised, narrative environments, expanding their affective and pedagogical potential. At the same time, AI-based tools and Large Language Models (LLMs) are increasingly being integrated into collaborative platforms to generate metadata, provide conversational access to digital collections, and support interpretation through multimodal narratives [12]. These trends suggest a shift from static documentation to dynamic, intelligent mediation of fashion heritage—an area where accessible digitisation workflows such as those examined in this study could provide foundational assets.

3   Methodology: Criteria for the Evaluation and the Comparison of Two Scanning Techniques for Two Use Scenarios

To assess the adaptability and effectiveness of digital replicas, we established a comprehensive evaluation framework reflecting the dual perspective of their use: by archivists for documentation and internal research, and by end users for interaction and engagement in exhibition contexts. Alongside technical specifications such as file format, size, and platform compatibility, the evaluation emphasises the fidelity of physical simulation—namely, how well textures, volumes, and material behaviours are captured and rendered. It also considers the scanning process itself, including acquisition technique, visual realism, and the model's capacity for user interaction and integration into immersive or narrative environments. Each factor contributes to the overall usability and interpretive value of the digital twin across archival, curatorial, and experiential domains. After defining and categorising the core features of each model based on the scanning technique, we tested them across two cultural heritage contexts: corporate archives and exhibition environments. These were chosen for their relevance to preservation and storytelling. In the archive, we evaluated utility for remote consultation, material study, and heritage research; in exhibitions, the focus shifted to interactivity, visual impact, and audience engagement. Each model was assessed in both scenarios, enabling a comparative analysis across distinct but complementary applications.

4   Case Studies: Features and Applications

The digitisation experiment was conducted on a set of garments held in the private corporate archive. Two mobile-based workflows were tested: mesh reconstruction from RGB video using Polycam, and point cloud generation from RGB-D input via Gaussian Splatting. The outputs were evaluated according to visual fidelity, interactivity, file size, and applicability to two key scenarios: archival consultation and exhibition contexts.

4.1   Digital Twins as Meshes in the Use Scenarios

The Polycam-based workflow produces polygonal meshes from standard RGB video, exported in the lightweight and widely supported .glb format. This format is especially suitable for web and XR integration, allowing fast deployment in interactive platforms. The video acquisition required only a few minutes, followed by approximately 20 min of cloud-based processing. With an average of $114,614$ vertices each and a texture resolution size of 8000, the reconstructed models deliver an overall coherent shape with a relatively smooth surface approximation, as shown in Fig. 1a. However, the level of detail is limited, especially in hard-to-reach zones such as garment interiors or under folded areas, which often result in missing polygons or small occlusions. In the archival scenario, this type of model offers only a basic digital surrogate: it lacks the resolution required to analyse fine textile elements such as embroidery, stitching, or fabric composition. Furthermore, it does not simulate the physical behaviour of textiles—an increasingly relevant aspect in conservation studies and for consulting designers. Despite these limitations, the low computational demand and compatibility of the format make it an efficient solution for decentralised institutions aiming to digitise collections for remote consultation or internal documentation. Conversely,

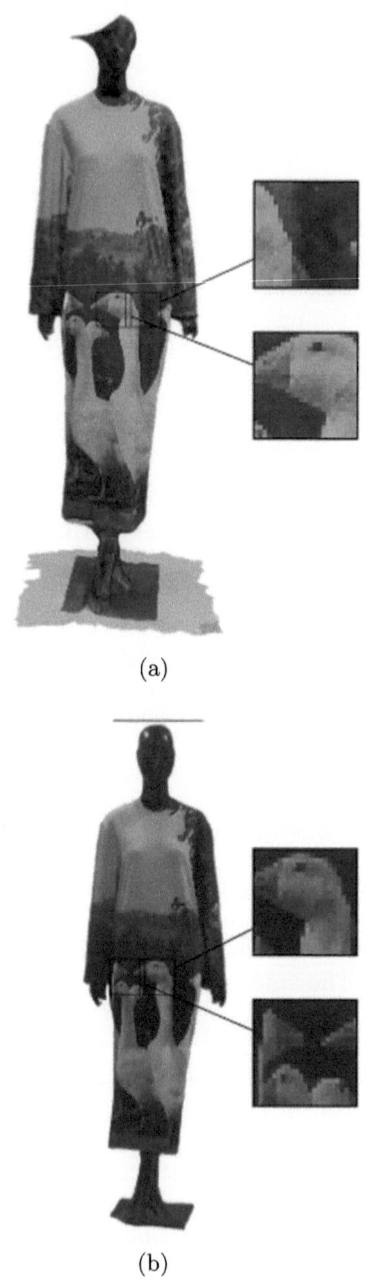

(a)

(b)

Fig. 1. Visual rendering comparison between Polycam and 3DGS output.

(c) Photogrammetry

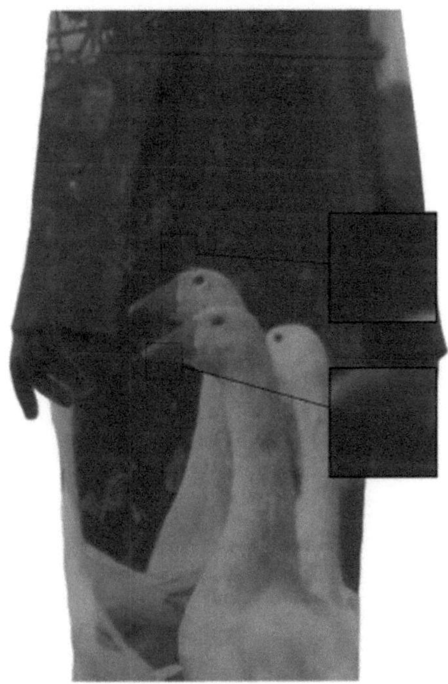

(d) 3DGS

Fig. 1. (*continued*)

in exhibition settings, Polycam meshes demonstrate significant potential. Their low weight and compatibility with common rendering engines support interactive user experiences such as manipulation, rotation, and zooming—activities that are not possible with physical garments in traditional display contexts. Although these models do not support animation or physics-based behaviours, they enable scenographic flexibility and can be embedded in narrative-driven digital environments. In this sense, the mesh workflow represents a lightweight, accessible format well suited for participatory and immersive storytelling.

4.2 Digital Twins as Point Clouds in the Use Scenarios

The second technique, based on 3D Gaussian Splatting, processes RGB-D video to generate dense point clouds composed of colored Gaussian primitives (Fig. 1b). In our implementation, the team augmented the standard RGB video input with an additional depth map acquired using an affordable handheld stereo camera system and an image resolutionof 2k which significantly reduced the processing time required to generate a complete point cloud via Gaussian Splatting. The video acquisition lasted approximately 2 min, followed by a preprocessing stage averaging 773 ± 61 seconds and an optimization phase of up to 12,000 iterations over 290 ± 130 seconds. These timing values represent the mean across all reconstructed garments. Processing was carried out on a desktop workstation equipped with a 12-core AMD Ryzen 9 processor, 64 GB of RAM, and a 24 GB GPU. The output is stored in the `.ply` format, which, while heavier, provides a high degree of spatial and chromatic accuracy. This method does not construct a surface mesh but instead accumulates thousands of colour points with an average of $213,905 \pm 68,407$ Gaussian primitives per model to render the visual and volumetric properties of the scanned object. From an archival perspective, the point cloud model offers superior descriptive capacity. Users can examine textile textures, decorative details, and subtle irregularities at close range, facilitating analysis and high-fidelity documentation. This makes it particularly useful for heritage garments that require detailed examination without direct handling. However, the format also introduces technical challenges: point clouds are large, require specialised software to visualise, and cannot be easily animated or manipulated. Their use in exhibition settings is thus more limited, unless deployed in isolated viewing stations or as static visualisations in hybrid installations. In exhibitions, the visual richness of the point cloud can provide a compelling aesthetic experience—especially in single-object displays where materiality is the focus. However, the lack of interactivity and performance overhead make the model less suitable for complex, multi-object, or responsive installations. Each model can be rendered in real time at 30 frames per second (fps) in standard XR environments; however, performance drops to 15 fps or lower when multiple instances are loaded or when instance interactions functionalities are enabled [5]. The trade-off between visual fidelity and usability must therefore be evaluated on a case-by-case basis, depending on the curatorial goals and available infrastructure.

4.3   Comparative Reflections

The contrast between Polycam meshes and Gaussian Splatting point clouds highlights two complementary paradigms for fashion digitisation. The former prioritises adaptability, fast processing, and lightweight outputs suitable for immersive environments. The latter privileges visual accuracy and archival precision but comes at the cost of interactivity and accessibility. A hybrid strategy that combines both approaches—using meshes for public engagement and point clouds for preservation—could represent a pragmatic solution, particularly for under-resourced archives seeking to optimise their digitisation efforts. A more complete comparison of the two approaches can be found in Table 1. Beyond technical considerations, the choice of workflow should also reflect the symbolic and material qualities of each garment, as well as its intended digital function: whether as

Table 1. Comparison of Scanning Techniques: Mesh vs Point Cloud

Criteria	3D Mesh (Polycam)	Point Cloud (Gaussian Splatting)
Technique	From RGB video, processed with Polycam to produce a mesh	From RGB-D video, processed with Gaussian Splatting algorithm to produce a point cloud
File Format	.glb (lightweight, widely compatible)	.ply (heavier, requires specific viewers or plug-ins)
Visual Rendering	Good from a distance, lacks detail; holes may appear if input video lacks information	High detail at close range, excellent for surface features; scattered at a distance
Interaction	Basic 3D interaction possible; interiors and occluded areas not captured	No interaction supported; only visual inspection
Archive Suitability	Limited detail; suitable for remote viewing and storytelling; physical simulation possible	Excellent for research and preservation; highly detailed but heavy and hard to integrate
Exhibition Suitability	Good for immersive interaction and animation; easy to embed in virtual spaces	High visual fidelity but performance-heavy; less suitable for dynamic environments
PROS	Lightweight, easily post-processed, interactive potential, accessible	High detail, excellent texture rendering, accurate at micro level
CONS	Poor detail and missing sections; no textile physics; limited scanning scope	Heavy files, limited compatibility, no user interaction, difficult integration in multi-object scenarios

an interface for memory activation, a document for research, or a vector for immersive storytelling.

5 Conclusion

This paper presents a comparative study of two accessible and low-cost 3D digitisation techniques—Polycam mesh reconstruction from RGB video and Gaussian splatting from RGB-D input—for creating Digital Twins in the field of fashion heritage. Through an applied case study of a corporate archive within the Made in Italy fashion system, it has been examined how these two methods perform across two critical use scenarios: digital preservation in archival contexts, and immersive display in exhibition environments. Our analysis demonstrates that both approaches offer specific advantages and constraints shaped by their technical characteristics. Future research could employ a quantitative approach to the evaluation of the meshes, for which there is now a lack of proper evaluation frameworks. For the moment, the results highlight that mesh-based digital twins, while less detailed, are lightweight, versatile, and easily embeddable in immersive environments, making them particularly suitable for interactive exhibitions and hybrid storytelling applications. Conversely, point cloud-based digital twins generated through Gaussian splatting offer superior visual detail and fidelity, proving especially valuable for archival purposes, remote consultation, and in-depth material analysis. However, their heavy file size, lack of interactivity, and dependency on dedicated viewers limit their deployability in more dynamic or multi-object exhibition contexts.

These findings reflect broader tensions within the digitisation of cultural heritage: between accessibility and accuracy, immersion and documentation, technological innovation and practical implementation. In the case of the Italian fashion industry—often composed of small and medium-sized enterprises with limited technical infrastructure—the availability of low-cost workflows like the ones analysed here opens up important possibilities. It enables even under-resourced archives to participate in the digital transition, fostering new forms of preservation, access, and creative reuse. Furthermore, by enabling the reactivation of corporate memory through the reconstruction of garments no longer in circulation, these digital twins contribute not only to conservation but also to the redefinition of brand identity and storytelling in a contemporary framework. Nonetheless, several limitations remain. Current low-cost tools cannot often capture internal garment structures or simulate textile behaviour under physical forces—features that are increasingly important in both conservation research and fashion design. Moreover, the interoperability of these formats with physics engines, AI-enhanced metadata, and real-time rendering systems is still in early development, and further work is needed to ensure their scalability and sustainability across the design, curatorial, and industrial sectors. Looking ahead, future research should explore the integration of additional data layers into digital twins—such as material properties, historical metadata, and pattern archives. As immersive and intelligent systems continue to reshape cultural experiences, lightweight 3D digitisation workflows like those analysed in this study could serve as foundational assets within XR-based exhibitions and AI-driven curatorial platforms. Large Language Models (LLMs), in particular, offer new opportunities to enhance access to fashion archives through context-aware narration, automated tagging, or conversational interfaces. These integrations would expand the role of digital garments beyond static representation, enabling dynamic modes of storytelling, education, and brand engagement across interconnected digital ecosystems. These aspects have been explored by Garzarella et al. [12] and Stacchio et al. [30], who have inquired also into possibilities of annotation and collaboration in the creation of extra data by end users when digital twins are used into collaborative platforms. Experiments in this regard within this scenario will be implemented once the meshes and the point clouds can be animated. In the scenario hereby traced for the coming future, meaningful collaboration between technologists, fashion archivists, curators, and designers will be essential to establish shared standards, ethical guidelines, and interoperable workflows for the responsible use and distribution of 3D heritage assets. In this perspective, digital twins should not be seen merely as virtual replicas, but as relational infrastructures that bridge past and future, analogue and digital, heritage and innovation. As such, they represent a critical tool not only for practitioners—designers, archivists, curators—but also for theorists, who can rethink how fashion is conserved, activated, and narrated in the age of immersive culture through these digital replicas.

Acknowledgements. This paper was produced with funding from the European Union Next Generation EU, within the framework of PNRR Spoke 1, Project 1.4 'Digital Services for Made in Italy: Digital twins for predictive models and to support the lifecycle of fashion products'.

References

1. Artec 3D: Artec 3d — scansione 3d professionale ad alta precisione. https://www.artec3d.com/it (2025). Accessed 20 June 2025
2. Artemis-Twin Project: Artemis-twin (2025). https://www.artemis-twin.eu/. Accessed 16 June 2025
3. Bonfanti, A., Del Giudice, M., Papa, A.: Italian craft firms between digital manufacturing, open innovation, and servitization. J. Knowl. Econ. **9**, 136–149 (2018)
4. Bournemouth University: Mass digitisation of cultural heritage in Europe. https://www.bournemouth.ac.uk/research/projects/mass-digitisation-cultural-heritage-europe (2024). Accessed 21 June 2025
5. Cascarano, P., et al.: A comparative analysis of 3d modeling methods for integration into an extended reality platform. In: 2025 IEEE International Conference on Artificial Intelligence and eXtended and Virtual Reality (AIxVR), pp. 213–217. IEEE (2025)
6. Daneshmand, M., et al.: 3d scanning: a comprehensive survey. arXiv preprint arXiv:1801.08863 (2018)
7. Europe, A.: Artec 3d portable scanners. https://www.artec3d.com (2024). Accessed 21 June 2024

8. European commission: cultural heritage - digital strategy. https://digital-strategy. ec.europa.eu/en/policies/cultural-heritage (2024). Accessed 21 June 2025

9. European commission: improve long-term monitoring of and research on the preservation of cultural heritage in europe. https://cordis.europa.eu/programme/ id/HORIZON_HORIZON-CL2-2025-03-HERITAGE-01 (2024). Accessed 21 June2025

10. European Heritage Hub Consortium: European heritage hub. https://www. europeanheritagehub.eu/ (2024). Accessed 21 June 2025

11. Europeana Foundation: Europeana – Europe's platform for digital cultural heritage. https://www.europeana.eu/en (2025). Accessed 21 June 2025

12. Garzarella, S., Vallasciani, G., Cascarano, P., Hajahmadi, S., Cervellati, E., Marfia, G.: An extended reality platform powered by large language models: a case study on teaching dance costumes. In: 2025 IEEE International Conference on Artificial Intelligence and eXtended and Virtual Reality (AIxVR), pp. 369–375. IEEE (2025)

13. Grieves, M.: Digital twin: manufacturing excellence through virtual factory replication (2014). https://www.researchgate.net/publication/275211047_Digital_ Twin_Manufacturing_Excellence_through_Virtual_Factory_Replication,   white paper, Product Lifecycle Institute

14. Haleem, A., et al.: Exploring the potential of 3d scanning in industry 4.0: an overview. Int. J. Cogn. Comput. Eng. **3**, 161–171 (2022)

15. Hutson, J., Weber, J., Russo, A.: Digital twins and cultural heritage preservation: a case study of best practices and reproducibility in chiesa dei ss apostoli e biagio. Art Des. Rev. **11**, 15–41 (2023). https://doi.org/10.4236/adr.2023.111003

16. INCEPTION project: inception project (2025). https://www.inception-project.eu/ en/. Accessed 16 June 2025

17. Levoy, M., et al.: The digital michelangelo project: 3d scanning of large statues. In: Proceedings of the 27th Annual Conference on Computer Graphics and Interactive Techniques, pp. 131–144 (2000)

18. Mosca, F.: Heritage di prodotto e di marca. Modelli teorici e strumenti operativi di marketing per le imprese nei mercati globali del lusso (2017)

19. New European Bauhaus: New European Bauhaus (2025). https://new-european-bauhaus.europa.eu/index_en, Accessed 16 June 2025

20. Niccolucci, F., Felicetti, A., Hermon, S.: Populating the digital space for cultural heritage with heritage digital twins. In: Proceedings of a Workshop on Heritage Digital Twins (2022). arXiv preprint arXiv:2205.13206

21. Niu, J.: Corporate archives in the wild. Global Knowl. Memory Commun. **74**(3/4), 917–937 (2025)

22. Peirson-Smith, A., Peirson-Smith, B.: Fashion archive fervour: the critical role of fashion archives in preserving, curating, and narrating fashion. Arch. Records **41**(3), 274–298 (2020)

23. Polycam, Inc.: Polycam — 3d capture and scanning for ios and android. https:// poly.cam/ (2025). Accessed 20 June 2025

24. Remondino, F.: Heritage recording and 3d modeling with photogrammetry and 3d scanning. Remote Sens. **3**(6), 1104–1138 (2011)

25. Riviezzo, A., Garofano, A., Napolitano, M.R.: "il tempo è lo specchio dell'eternità". strategie e strumenti di heritage marketing nelle imprese longeve italiane/"il tempo è lo specchio dell'eternità"(time is the mirror of eternity). heritage marketing strategies and tools in italian long-lived firms. Il capitale culturale. Stud. Value Cult. Heritage (13), 497–523 (2016)

26. Rossato, C., Castellani, P.: The contribution of digitalisation to business longevity from a competitiveness perspective. TQM J. **32**(4), 617–645 (2020)

27. Scaniverse Inc.: Scaniverse — 3d scanning on iphone and ipad. https://scaniverse.com/ (2025). Accessed 20 June 2025
28. Schenk, T.: Introduction to photogrammetry. Ohio State Univ. **106**(1), 1 (2005)
29. Smith, B., McCarthy, C., Dechenaud, M.E., Wong, M.C., Shepherd, J., Heymsfield, S.B.: Anthropometric evaluation of a 3d scanning mobile application. Obesity **30**(6), 1181–1188 (2022)
30. Stacchio, L., Vallasciani, G., Augello, G., Carrador, S., Cascarano, P., Marfia, G.: Wixard: towards a holistic distributed platform for multi-party and cross-reality webxr experiences. In: 2024 IEEE Conference on Virtual Reality and 3D User Interfaces Abstracts and Workshops (VRW), pp. 264–272. IEEE (2024)
31. Trebuňa, P., Mizerák, M., Rosocha, L.: 3d scaning-technology and reconstruction. Acta Simulatio **4**(3), 1–6 (2018)
32. Tsaramirsis, G., et al.: A modern approach towards an industry 4.0 model: from driving technologies to management. J. Sens. **2022**(1), 5023011 (2022)
33. Vacca, F.: Knowledge in memory: corporate and museum archives. Fash. Pract. **6**(2), 273–288 (2014)
34. Vacca, F., et al.: Living cultural heritage: exploring fashion heritage within corporate archives. ZoneModa J. **14**(2), 1–17 (2024)
35. Wachowiak, M.J., Karas, B.V.: 3d scanning and replication for museum and cultural heritage applications. J. Am. Inst. Conserv. **48**(2), 141–158 (2009)
36. Wright, L., Davidson, S.: How to tell the difference between a model and a digital twin. Adv. Model. Simul. Eng. Sci. **7**(1), 1–13 (2020). https://doi.org/10.1186/s40323-020-00147-4

Author Index

MIX
Papier aus verantwortungsvollen Quellen
Paper from responsible sources
FSC® C105338

If you have any concerns about our products,
you can contact us on
ProductSafety@springernature.com

In case Publisher is established outside the EU,
the EU authorized representative is:
Springer Nature Customer Service Center GmbH
Europaplatz 3, 69115 Heidelberg, Germany

Printed by Libri Plureos GmbH
in Hamburg, Germany